AF526340

Tectonic Events Responsible for Britain's Oil and Gas Reserves

This volume is dedicated to: **N. L. Falcon, F. R. S., Hon. F. G. S.,** Chief Geologist of British Petroleum from 1955 to 1965. During this time it was estimated that BP controlled about 22% of the free world's oil reserves. This achievement was a result of the work carried out by many explorationists in BP but particularly the work of G. M. Lees, J. V. Harrison and N. L. Falcon.

In 1960 N. L. Falcon, together with Peter Kent, was responsible for the Geological Society Memoir which heralded the start of serious exploration interest in the British Continental Shelf. The large Groningen gas discovery and the influence of N. L. Falcon on contemporary thought were major factors leading to North Sea exploration.

This volume contains a representative sample of the knowledge resulting from 25 years of active exploration of Britain and its continential shelf. We believe that we can be proud of the results of that activity and in contemplating the achievement, acknowledge with gratitude the leadership provided by Norman Falcon.

GEOLOGICAL SOCIETY SPECIAL PUBLICATION NO 55

Tectonic Events Responsible for Britain's Oil and Gas Reserves

EDITED BY

R. F. P. HARDMAN
Amerada Hess Ltd
London
UK

J. BROOKS
Brooks Associates
Glasgow
UK

1990

Published by

The Geological Society

London

THE GEOLOGICAL SOCIETY

The Geological Society of London was founded in 1807 for the purposes of 'investigating the mineral structures of the earth'. It received its Royal Charter in 1825. The Society promotes all aspects of geological science by means of meetings, special lectures and courses, discussions, specialist groups, publications and library services.

It is expected that candidates for Fellowship will be graduates in geology or another earth science, or have equivalent qualifications or experience. All Fellows are entitled to receive for their subscription one of the Society's three journals: *The Quarterly Journal of Engineering Geology*, the *Journal of the Geological Society* or *Marine and Petroleum Geology*. On payment of an additional sum on the annual subscription, members may obtain copies of another journal.

Membership of the specialist groups is open to all Fellows without additional charge. Enquiries concerning Fellowship of the Society and membership of the specialist groups should be directed to the Executive Secretary, The Geological Society, Burlington House, Piccadilly, London W1V 0JU.

Published by the Geological Society from:
The Geological Society Publishing House
Unit 7
Brassmill Enterprise Centre
Brassmill Lane
Bath
Avon BA1 3JN
UK
(*Orders:* Tel. 0225 445046)

Distributor
USA
AAPG Bookstore
PO Box 979
Tulsa
Oklahoma 74101-0979
USA

First published 1990

British Library Cataloguing in Publication Data
Tectonic events responsible for Britain's oil and gas reserves.
1. North Sea. Natural gas deposits & petroleum deposits
I. Hardman, R. F. P. (Richard Frederick Paynter)
II. Brooks, J. (James) III. Series
553.280916336

ISBN 0–903317–55–9

Printed in Great Britain

Contents

Acknowledgements

Any conference is undoubtedly a team effort and the Bath Conference was no exception. It could not have taken place without the enthusiasm and dedication of the entire committee. However in any committee there are always those that act as catalysts and in particular we would like to pay tribute to the role of J. R. V. Brooks who through his vision, drive and enthusiasm was materially responsible for the success of the conference. John Fuller, Bob Stoneley and the other referees are particularly thanked for editorial assistance and this volume would not have been possible without their help.

A major vote of thanks is extended to the Mayor of Bath, Mrs McDonagh and the City of Bath for their facilities particularly the Guildhall, where the conference was held, and for the reception in the Pump Room. Much of the credit for the smooth running of the event must go to Alison Higgs, of the Bath Conferences office and to her, particular thanks.

The Conference secretarial staff were composed mainly of volunteers and for their splendid efforts our thanks are hardly enough. Particularly to Janet Hughes for the weekends she gave up, and to Liz Atkinson and Christine Brooks many thanks for remaining cool during the flood of registrations for the Conference.

Finally special thanks to those who supported the conference with financial contributions or in kind; here we would like to thank the President of the Geological Society, Derek Blundell; the Chairman of the Petroleum Exploration Society, John Parker; and the former Chairman, Brian Light; the Director of the British Geological Survey, Geoff Larminie; and all those companies who provided financial assistance. The role of Amerada Hess in providing facilities to allow organization of this conference is gratefully acknowledged.

R. F. P. Hardman
J. Brooks

We are greatly indebted to the following companies who sponsored the conference:

Abstracts: British Gas
Conference proceedings: Amoco UK Exploration Limited, Arco British Limited, BP Exploration, Chevron Exploration North Sea Limited, Deminex UK Oil and Gas Limited, Elf UK plc, Enterprise Oil plc, Esso Exploration and Production UK Limited, Kelt UK Limited, Lasmo plc, Occidental International Oil Inc., Phillips Petroleum Co. UK Limited, Shell UK Exploration and Production, Sun Oil Britain Limited, Texaco Limited, Ultramar Exploration Limited, Unocal UK Limited
Folders: Amerada Hess Limited
Music: Fina Exploration
Wine: The Robertson Group plc

The Precambrian, Caledonian and Variscan framework to NW Europe

M. P. COWARD

Department of Geology, Imperial College, London SW7 2BP, UK

Abstract: During the Precambrian and the Palaeozoic, the tectonics of NW Europe were dominated by the sequential accretion of different terrains on to the North American Craton, e.g. old continental crust of the Scandinavian Craton, magmatic arcs of the Avalon–Brabant Massif, Pentevrian continental crust and the Brioverian magmatic arc. Terrains were locally bounded by thrust packages, by NW–SE trending strike-slip – transform faults parallel to the accretion direction, and by NE–SW trending strike-slip faults defining localized oblique collision or, more generally, boundaries to zones of lateral continental extrusion and escape. Close analogies can be made with the Tertiary Makran–Himalaya–Tibet collisional zones. The Laxfordian/Caledonian/Variscan thrusts and more importantly the large-scale strike-slip faults, imposed a complex heterogeneity to the crust which critically influenced the subsequent extension directions and the siting of basin bounding faults and tectonic inversion, from Devonian times to the present day.

The relatively recent availability of seismic data has led to enormous advances in our understanding of the deep geology of the British Isles. In particular the deep seismic surveys obtained by the BIRPS Group (British Institutions' Reflection Profiling Syndicate) have shown variations in crustal thickness and middle to deep crustal tectonic fabric in the northern and western offshore regions of Britain and throughout the North Sea (see for example Cheadle *et al.* 1987; Freeman *et al.* 1988; Klemperer 1988). The combinations of shallow commercial seismic data, deep level BIRPS seismic data and conventional structural and stratigraphic field data, allow new models to be derived for the tectonic development of Britain. They show us the relationship between deep level tectonic fabrics and surface structures, which combine to produce the overall tectonic framework to Britain and adjacent parts of NW Europe. This paper aims to describe this structural framework and (i) review the pre-Mesozoic basement kinematics in Britain and adjacent parts of NW Europe, based on the the available seismic data, previous published structural data and reviews, and new work by the author, and (ii) discuss how different fabric intensities and styles influence subsequent Upper Palaeozoic and Mesozoic–Cenozoic basin development.

The pre-Mesozoic basement rocks of the British Isles range from the Lewisian gneisses of NW Scotland, dated at *c.*2900 Ma (Moorbath *et al.* 1969), to the Devonian and Carboniferous sediments involved in the Variscan fold and thrust belts of southern Britain. Much of the tectonic framework of NW Europe was developed during three major compressive tectonic episodes, i.e. the Laxfordian (1800–1700 Ma), the Caledonian (500–400 Ma) and the Variscan (400–300 Ma). Other compressive and strike-slip orogenic events affected NW Europe but their effects were largely obliterated by the major compressive events listed above.

Figure 1 shows the distribution of these tectonic events in four principal domains in Britain. Domain 1 consists of Lewisian rocks of the NW Caledonian foreland, which with their upper Proterozoic cover were originally part of the N American–Laurentian craton. Their dominant crustal fabric is of Laxfordian age. Domain 2 comprises rocks which were deformed and metamorphosed during the Caledonian tectonic event and can be subdivided into several sub-domains based on the orientation and age of the Caledonian fabric. Domain 3 is the SE foreland of the Caledonian fold belt and consists of a Late Precambrian magmatic arc, known as the Acadian or Cadomian magmatic arc or Brabant Massif. Domain 4 consists of rocks deformed and weakly metamorphosed in the Variscan tectonic events during Devonian to Carboniferous times.

The Lewisian Gneisses of Domain 1, NW Scotland and their Proterozoic cover

The Lewisian rocks of NW Scotland are quartzo-feldspathic gneisses and metasediments that were affected by several episodes of late Archaean to early Proterozoic deformation and metamorphism. According to the now generally accepted chronology (e.g. Sutton & Watson

From Hardman, R. F. P. & Brooks, J. (eds), 1990, *Tectonic Events Responsible for Britain's Oil and Gas Reserves*, Geological Society Special Publication No 55, pp 1–34.

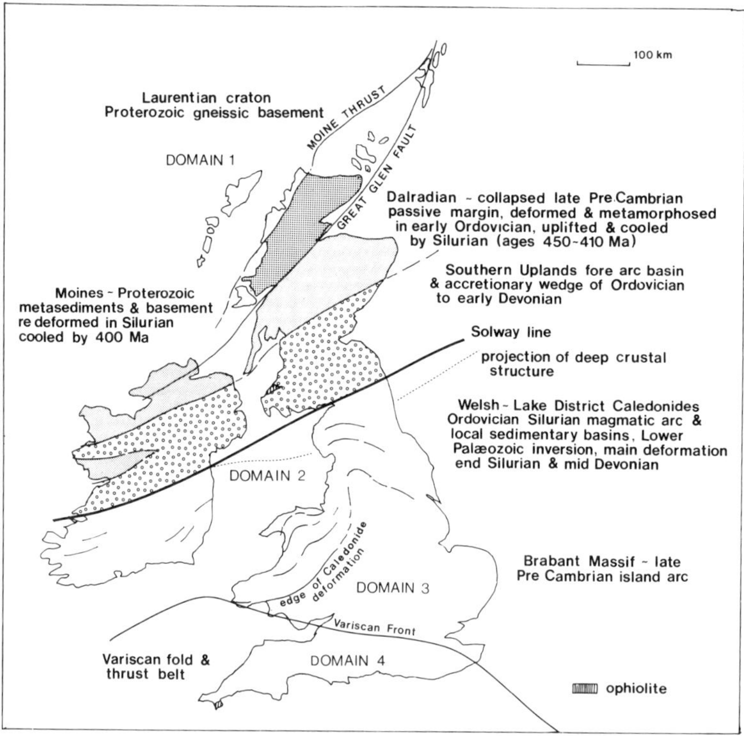

Fig. 1. Structural domains of Britain. See text for discussion.

1951; Park *c.*1970) the gneisses formed at *c.* 2900 Ma during high grade, locally granulite-facies metamorphism associated with imbrication and deep burial of sediments and granitic rocks. Slow cooling of the gneisses was followed by reactivation during the Inverian episode (2600–2400 Ma), associated with the development of upright folds and steeply dipping NW–SE shear zones (Coward & Park 1973). Scourian and Inverian age structures are cut by a swarm of dolerite dykes, the Scourie dyke suite of Sutton & Watson (1951). The subsequent Laxfordian deformation was heterogeneous and, on the mainland of Scotland, the Lewisian rocks can be divided into three zones, a central zone where the Scourian structures remain well preserved, and northern and southern zones, where Laxfordian reworking was more intense and dykes and Scourian fabrics have been reorientated into concordance (Fig. 2). On the Outer Hebrides the degree of reworking was even more intense, with only small pods remaining in which Scourian fabrics are preserved.

The Laxfordian deformation involved large-scale crustal thickening with an overthrust direction of SE to NW. The Loch Maree Group, which consists of over 3 km of late Proterozoic metasediments and metavolcanics (O'Nions *et al.* 1983), was thrust on to the Lewisian gneisses in the Gairloch region. Laxfordian granulite-facies metamorphic rocks were uplifted in South Harris on the Outer Hebrides, so that throughout the Lewisian outcrop large shear zones with relatively flat lying fabrics which have been subsequently deformed by upright folds can be observed (Coward & Park 1987). These upright folds probably detach on shear zones in the middle to lower crust (Coward 1990). Large scale moderately to steeply dipping shear zones at Laxford Bridge, Gairloch and South Harris (Fig. 2) are either the tilted boundaries to gently dipping shears or they represent oblique to lateral ramps where the shear climbed to different crustal levels perpendicular to thrust movement.

The early Laxfordian crustal thickening was followed by relaxation and extension, producing ductile shears with amphibolite-facies mineralogy (Coward 1990). This extension may have formed by collapse of the thickened Laxfordian orogenic zone, similar to the mode of crustal spreading in modern orogenic belts (e.g. Dewey 1988). Granitic crustal melts are prominent

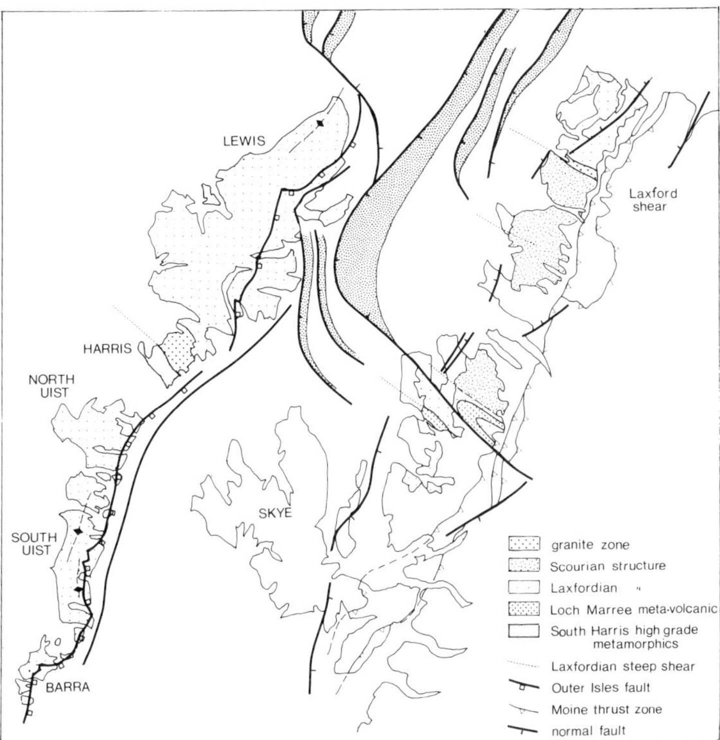

Fig. 2. Structures of the Lewisian Complex of NW Scotland and their influence on basin development in the Minches and Outer Isles Basins. Basement cut-out zones are shown by heavy stipple. From Coward *et al.* 1990.

along the major shear zones near Laxford and on South Harris (Fig. 2); they post-date the compressional shears and at Laxford they are synchronous with the extension. This extension was probably responsible for the uplift and emergence of Laxfordian high-grade metamorphic rocks and may also be responsible for the restoration of the Laxfordian crust to its present thickness, shown on the BIRPS surveys to be about 29 km (e.g. Brewer & Smythe 1984).

Cooling ages obtained from Laxfordian rocks range from *c.*1600–1400 Ma (Moorbath & Park 1971) and represent either the slow unroofing of the Lewisian complex following the main phase of extension or subsequent weak phases of uplift. Numerous faults with mylonite and pseudotachylite occur on the Outer Hebrides and may be of late Proterozoic age. In the Gairloch region of the Scottish mainland there are NW-trending pseudotachylite-bearing shear zones which pre-date the late Proterozoic Torridonian sediments. The Outer Isles fault zone was probably initiated as a Laxfordian (or later Proterozoic) structure and was subsequently reworked as a Caledonian thrust, as shown from the range in Proterozoic to early Palaeozoic K/Ar ages obtained from the phyllonites of the fault zone (D. Rex in Sibson 1977). However many of the mylonites show small-scale folds and shear bands indicating an extensional down-dip shear couple (Sibson 1977), and White & Glasser (1987) suggest that the phyllonites that occur close to the east coast of the southern Hebrides may be entirely the product of low-angle extensional faulting. This extension presumably post-dates the thrust movements on the Outer Isles fault; it may be Proterozoic, but is more likely to be of early Palaeozoic age.

The Stoer Group and the younger Torridonian Group, which overlie the Lewisian gneisses, are thick clastic sequences which give Rb/Sr ages of *c.*970 and 777 Ma respectively (Moorbath 1969). The upper Torridon Group of sediments forms a thick (>6 km) sedimentary wedge and Sr isotope data suggest derivation from a Laxfordian crust (Moorbath *et al.* 1969). Stewart (1982) suggested that the Torridonian basin developed from the erosion of uplited fault blocks formed as a result of regional extension. However on the mainland normal fault arrays of Torridonian age have not been detected.

Cheshire *et al.* (1983) and Kilenyi & Stanley (1985) extrapolate the Torridonian sequence from the mainland to the Minches and consider that the basal part of the half graben in the Minches is filled by Torridonian sediments. They suggest that extensional movements on the Outer Isles fault zone occurred during the late Proterozoic, as this fault system marks the western boundary of the Minches Basin. However the correlation between the Torridonian sediments seen onshore and the basal sediments of the Minches Basin is by no means certain. Other seismic interpretations (e.g. Enfield in Coward *et al.* 1989) place the Torridonian sediments as basement to the Minches half graben and suggest that the lowest sediments in the half graben are of Devonian age. The provenance of the thick Torridonian sediments seen onshore was to the NW, presumably from an area of uplift along the Outer Hebrides (Williams 1969) and the palaeoflow direction was from NW to SE. This is opposite to the expected palaeoflow direction if the Outer Isles fault had been a master bounding fault to the Minches Basin in Torridonian times. It is more likely that the Torridonian sediments were unrelated to regional crustal extension, but were deposited in a deep basin associated with late Proterozoic crustal thickening.

During Caledonian compression there was movement along the Outer Isles fault (Sibson 1977), generally reactivating the Laxfordian fabrics. During post-Caledonian times, there was further reactivation by extensional faulting to produce deep Devonian half grabens NE of the Outer Hebrides and at least some, if not most, of the extension in parts of the Minches Basin and rifting in basins N of the Scottish mainland (Fig. 2). The Outer Isles fault was subsequently reactivated several times during the late Palaeozoic and Mesozoic (Coward *et al.* 1989).

The Caledonides of Domain 2

The Caledonian orogenic belt extends from northern Norway and Greenland to the southern Appalachians and formed as a result of the closure of an early Atlantic (Iapetus) ocean during the early Palaeozoic. The main tectonic units in the British Isles are shown in Fig. 1. The orogenic belt formed by the accretion of magmatic arcs and continental fragments on to the North American continental craton.

The Caledonides NW of the Great Glen

The Caledonides of the NW Highlands of Scotland, NW of the Great Glen, consist of Proterozoic metasediments (the Moines) with minor basic and acid intrusives. They are intensely foliated and metamorphosed to upper greenschist and amphibolite facies and give syntectonic metamorphic ages of *c.*460 Ma and mineral ages which probably reflect later cooling and uplift at 430–400 Ma (Brewer *et al.* 1979; Johnson *et al.* 1985). Along the N coast of Scotland the Moines show a complex interlayering of intensely foliated metasediments, metabasite intrusions and earlier basement, due to isoclinal folding and closely spaced WNW-directed thrust imbrication (Butler & Coward 1984; Barr *et al.* 1986). The foliation dips *c.*150 to the E or SE in the western part of the section but has steeper dips, sometimes nearly vertical, in the east.

The Moines were thrust to the WNW over a foreland consisting of Lewisian basement, Torridonian arkosic sandstones and Cambro-Ordovician shelf sediments. In the northern part of the thrust zone in Sutherland, a minimum overthrust displacement of 54 km is shown by the restoration of imbricated Middle and Upper Cambrian sediments (Butler & Coward 1984) and the Moine Thrust cannot have cut up through basement or lower Cambrian sediments within 54 km of its present outcrop trace. However, seismic profiles from offshore northern Scotland (e.g. Fig. 3) show dipping reflectors in the middle crust which are probably related to the Moine Thrust and to shear zones in its hanging-wall. These reflectors are much farther W than would be predicted from onshore geology. Presumably this offset is the result of a NW trending lateral ramp, close to the N Scottish coast.

The Great Glen Fault defines the eastern limit of the Moines. Its age, sense and amount of offset are disputed. A pre-Devonian, sinistral displacement of *c.*2000 km was suggested by Morris (1974) and Van der Voo & Scotese (1981) based on palaeomagnetic interpretations, but the general consensus (Mykura 1976; Smith & Watson 1983) suggests a maximum of a few hundred kilometres sinistral displacement at the end of the Caledonian orogeny, with later Permo-Carboniferous dextral movements of a few tens of kilometres (e.g. Ziegler 1982; Watson 1985).

The British Caledonides SE of the Great Glen Fault

SE of the Great Glen Fault, the full history of the Caledonian orogen is preserved and involved (i) the development of a rift basin

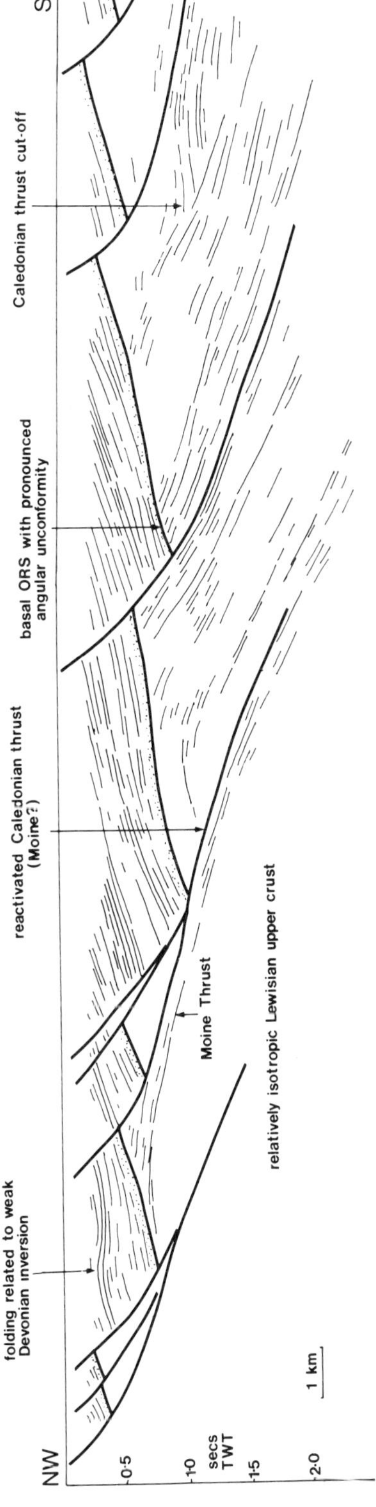

Fig. 3. Line drawing of a seismic section through the West Orkney Basin, to show the principle Caledonian and Devonian structures – part of line WOB 19, Western Geophysical Survey (see also Enfield & Coward 1987).

leading to a passive oceanic margin during the late Precambrian to Cambrian; (ii) the Early Ordovician collapse of this passive margin during the early stages of reversal in plate motion; (iii) the obduction of ophiolitic material during the Early Ordovician; (iv) the development of a magmatic arc; (v) the development of a fore-arc basin and (vi) the final docking of this fore-arc basin and its associated accretionary sediment wedge with a southern magmatic arc during the Late Silurian. This deformation involved both dip-slip and strike-slip displacements. Some authors emphasise the importance of strike-slip movements relative to dip-slip displacements. Thus Bluck (1985), Soper & Hutton (1984) and Hutton (1987) favour the strike-slip accretion of up to six allochthonous terranes, similar to the terrane accretion in the western USA (e.g. Coney 1989).

SE of the Great Glen, in the Grampian Highlands, the Grampian Moines comprise late Proterozoic metasandstones which pass up, without any major unconformity, into the Dalradian sequence. The Grampian Moines, which may be the time equivalents of the Torridon Group, rest on but are interleaved tectonically with slices of basement, dated at *c.*1100 Ma (Piasecki & van Breeman 1979). The lower Dalradian consists of syn-rift and post-rift sequences of sandstones, shales and limestones deposited on the NW side of a rift basin (Anderton 1982; 1985). Early Dalradian facies and thickness variations can be considered as due to deposition in different fault blocks, defined by SE-dipping listric or straight faults and associated NW trending transfer faults (Harte *et al.* 1984). One such transfer fault is observed as a major lineament on regional gravity and magnetic data (Fettes *et al.* 1986; Hall 1987). As extension accelerated, subsidence rates increased and the upper Dalradian sediments were deposited in a series of turbidite basins (Anderton 1985; Soper & Anderton 1984). Thinning of the lithosphere was associated with intense local igneous activity. The uppermost Dalradian sediments, which are of Arenig age and occur only in the southern Highlands, were deposited as a distal facies away from the basin margin, probably on the continental slope.

The extensional faults were reactivated in a major compressional episode of Ordovician age resulting in large-scale positive structural inversion. The distal sediments of the upper Dalradian were intensely folded into a large fanning fold complex, whose axial surface varies in dip across the Highlands. A simplified cross section through the Dalradian structures is shown in Figs 4b & 5. The wide range of structural orientations is typical of inversion tectonics (cf. Gillcrist *et al.* 1987), but unusual on such a large scale. No basement rocks were involved in these large-scale fold structures. The NW boundary of the Dalradian sequence is sometimes marked by a major shear zone, such as the Port Skerrols Thrust, which carries younger Dalradian rocks on to older Grampian Moines and probably represents the reactivated basin-bounding fault. The early large-scale fold fan is cut by NW verging ductile thrusts such as the Boundary Slide (e.g. Roberts & Treagus 1979). These folds and ductile thrusts are associated with the thickening of the sedimentary pile and were followed by local high-grade metamorphism associated with this tectonic burial (Wells & Richardson 1979).

Subduction continued to the SE of the Highlands, to generate arc-related magmatism and produce large bodies of gabbro and granite in the Highlands, above the NW dipping subduction zone (Fig. 4b). Flakes of ophiolitic material were obducted over the thickened Dalradian sedimentary pile (Dewey & Shackleton 1984), producing the Unst ophiolite complex on Shetland and the Highland Border Group of ophiolitic debris at the northern edge of the Midland Valley. S of the Midland Valley, the Ballantrae ophiolite was obducted over olistostrome material and formed the basement to a later fore-arc basin and its accretionary sediment prism in the Southern Uplands (McKerrow *et al.* 1977; McKerrow 1987). The accretionary prism finally docked with a magmatic arc on the southern side of the Iapetus margin during Late Silurian to Early Devonian times (McKerrow 1987).

All the structures SE of the Dalradian Highlands are dominated by SE-verging thrusts and folds. These range in age from the accretionary thrust structures of the Southern Upland fore-arc basin, to the late collisional structures of the Highland Boundary Fault, the Lake District and Wales. Their dip probably partly reflects the dip of the Iapetus subduction zone. Beamish & Smythe (1986) and Freeman *et al.* (1988) map reflectors in the middle to lower crust, which they associate with the suture. The Caledonian suture is often taken to lie along the Solway–Cheviot line (Fig. 1), although the most prominant crustal reflectors lie a few tens of kilometres to the south. The reflectors presumably represent mid-crustal shear zones associated with, if not on, the suture.

In the Southern Uplands and Lake District, deformation was associated with low-grade metamorphism and also the development of a locally strong penetrative cleavage. Assuming

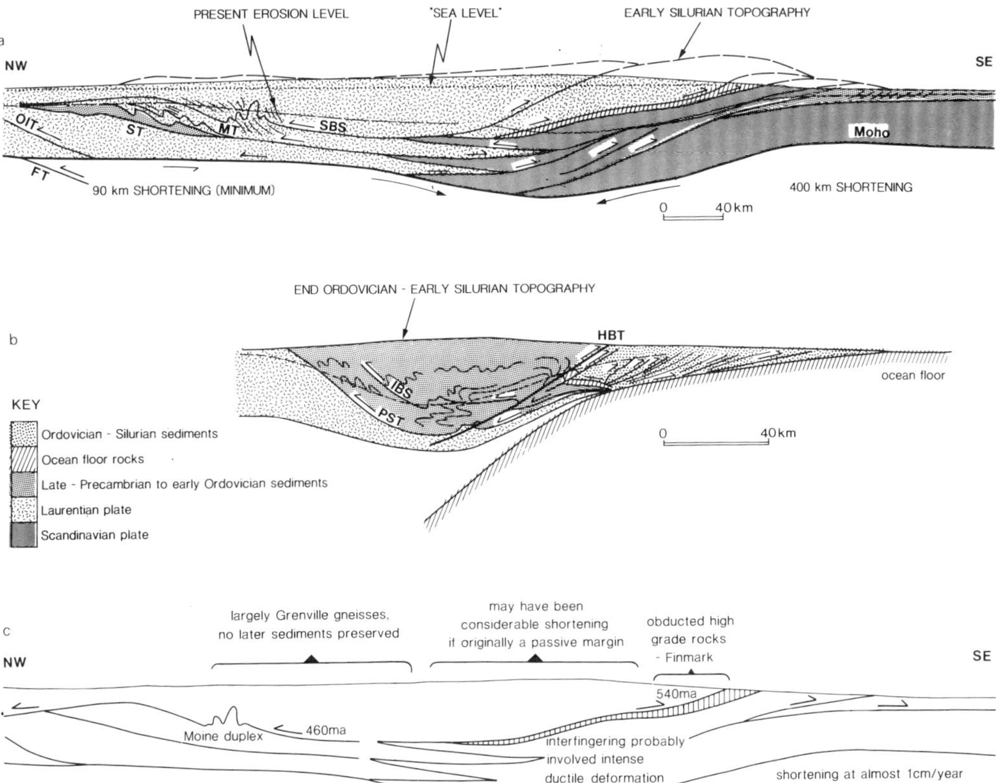

Fig. 4(a). Schematic cross section through the Scottish–Norwegian Caledonides during the late Silurian, assuming restoration of the Great Glen Fault system, based on cross sections by Barr *et al.* (1986), Butler & Coward (1984), Hossack & Cooper (1987), Coward (1983), McGeary & Warner (1985). FT, Flannan Thrust; OIT, Outer Isles Thrust; ST, Sole Thrust of the Moine thrust zone; MT, Moine Thrust; SBS, Sgurr Beag Slide. The shortening estimates are from Butler & Coward (1984) and Hossack & Cooper (1987). The line of section is approximately along line A–A' of Fig. 9. (b) Schematic section through the Caledonides SE of the Great Glen, during the Early Silurian. Dalradian structures are simplified from regional structural studies (e.g. Roberts & Treagus 1979), the southern structures are simplified from McKerrow *et al.* (1977). PST, Port Skerrols Thrust; IBS, Iltay Boundary Slide; HBT, Highland Boundary Fault. The line of section is approximately along line B–B' of Fig. 9. (c) Simplified line drawing of (a) above, discussing some of the principal tectonic features. Vertical line shading: obducted ophiolitic rock.

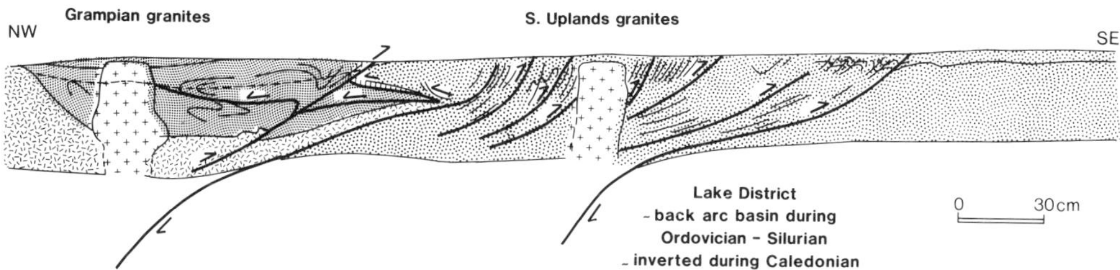

Fig. 5. Simplified section from the Great Glen Fault to the English Midlands, showing the present attitude of the structures – modified from Freeman *et al.* (1988). The line of section is approximately along B–B' of Fig. 9.

an allochthonous source for sediments in the accretionary prism, Bluck (1985) and McKerrow (1987) argue that important strike-slip movements occurred in the region of the Southern Uplands and along the trend of the Highland Boundary Fault during and following Caledonian collison. However, no steep strike-slip shear zones have been recognized from surface mapping or from deep seismic data (e.g. Freeman *et al.* 1988) and any major strike-slip movements must have been accommodated on NW-dipping faults or have been subsequently destroyed by these faults.

SE of the suture the basement consists of uppermost Precambrian intrusive and volcanic rocks and volcaniclastic sediments, overlain and intruded by lower Palaeozoic magmatic material. The Lower Palaeozoic sediments vary in thickness, associated with basin development and there were important phases of folding, local uplift and inversion, probably related to the growth of the magmatic arc (Fig. 6). The major deformation in Britain, south of the suture zone, occurred at end Silurian to early Devonian times and was associated with collisional thickening of the magmatic arc and its related sedimentary basins (Figs 5 & 6). The Caledonian foliations change strike across England and Wales, tracing out festoon-like patterns (Fig. 1) and these may be related to the indentation of different thicknesses of crust, as a result of early Palaeozoic basin development.

The Caledonides of southern Scandinavia

There is a marked difference in the character of the Caledonian structures on both sides of the North Sea. In Scandinavia, Caledonian tectonics involved collision between the N American craton and a large southern Scandinavia craton. The latter comprises Archaean to Proterozoic gneisses overlain by late Precambrian and Lower Palaeozoic sediments. Plate collision began during the Ordovician, but several hundred kilometres of crustal shortening and SE directed overthrusting continued until the late Silurian. The SW boundary of the Scandinavian craton lay approximately along the eastern margin of the Central Graben of the North Sea, along what is often considered to be the extension of the Tornquist Line (Fig. 7) (see Bergstrom 1984; Kumpas 1984; Pegrum 1984). This zone formed a fundamental crustal boundary in the Caledonides, between the continent–continent collision of the Scandinavian Caledonides to the NE and the continent–arc collision of the British and North American Caledonides to the SW.

Figure 7 shows a sketch map of the Tornquist

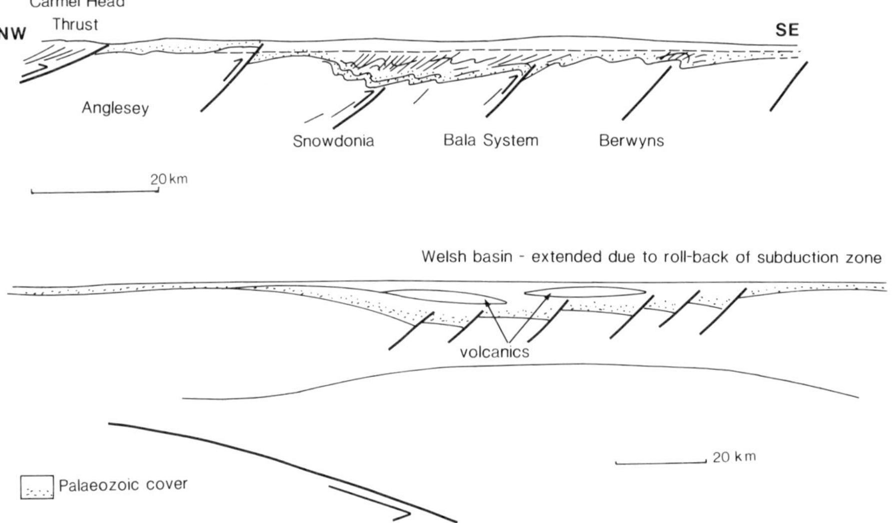

Fig. 6. Top: Simplified section through the Welsh Caledonides to show the SE verging folds and thrusts and Palaeozoic thickness variations. Bottom: Diagramatic section through the Welsh Caledonides to illustrate the model of back-arc extension above a SE-dipping subduction zone, antithetic to the NW-dipping subduction zone below the Southern Uplands and Grampian Highlands (Fig. 4b). This model explains the different early Palaeozoic phases of stretching and thickness — facies variations, the phases of tectonic inversion and the periods of volcanic activity (see Watson & Dunning 1979 and references therein).

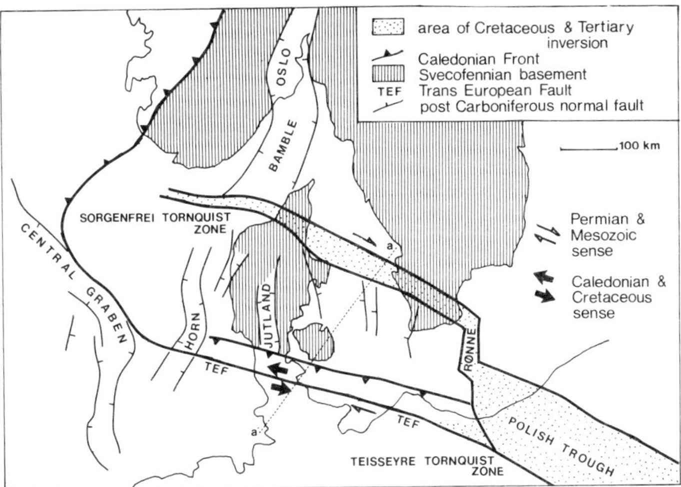

Fig. 7. Map of southern Scandinavia to show the position of the Tornquist zones and the Trans European Fault (from EUGENO Working Group 1988). See text for discussion.

zone in southern Scandinavia and northern Germany, based on the European Geotraverse data (EUGENO–S Working Group 1988). The western continuation of the Tornquist Line, the Sorgenfrei–Tornquist Line, appears as a tear fault of Palaeozoic to Tertiary age and in particular is associated with tectonic inversion in the late Cretaceous and early Tertiary. Scandinavian crust, dated by mineral ages from deep well cores at 800–900 Ma, similar to the ages of the Sveconorwegian orogenic belt of southern Sweden (Fig. 7), continues to an approximately WNW trending zone which cuts across southern Denmark and northern Germany. This zone, termed the Trans European Fracture Zone (EUGENO–S Working Group 1988), is a direct continuation of the eastern and main segment of the Tornquist Line in eastern Europe (Fig. 7). The Trans European Fracture Zone (TEF) is drawn just south of the limit of Caledonian deformation. North of the Caledonian Front, Lower Palaeozoic sediments sit in a small graben overlying Precambrian basement. South of the Front, the Lower Palaeozoic rocks carry a penetrative cleavage and low-grade metamorphism. They give Ar/Ar mineral ages of 530–400 Ma, with a peak of 450–440 Ma (Frost *et al.* 1981; Zeigler 1982; Liboriussen *et al.* 1987). As the Caledonian rocks are only observed in well cores, the tectonic transport directions cannot be determined from kinematic indicators, but only from the regional tectonic pattern. Elsewhere in the Scandinavian Caledonides, the overthrust direction was to the ESE, parallel to the strike of the TEF and the Caledonian Front across southern Denmark. Therefore the TEF is probably a zone of strike-slip movement, possibly with some local transpression or transtension. (Figs 9 & 10, modified by movement on the Great Glen Fault, Fig. 11.) A modern analogy would be the Chaman–Quetta fault zone in western Pakistan (Dewey *et al.* 1988), which marks the surface expression of the transform boundary of the indenting Indian plate (Fig. 12). East of this fault zone there was continent–continent collision between the Indian and Asian plates. West of the fault there was continent–arc collision in southern Afghanistan and in the Makran.

The Caledonide orogen

Figures 9 & 10 suggest the possible relationship between the Scandinavian Caledonides, the British Caledonides, SE of the Great Glen and the Caledonides to the NW of the Great Glen. The Moines and Moine Thrust underwent at least 90 km shortening (Butler & Coward 1984) between 460 and 400 Ma. This is long after the cessation of fold and thrust activity in Dalradian rocks to the SE of the Great Glen. Thrust tectonics in the Moines occurred at the same time as the growth of the accretionary prism in the Southern Uplands. Hence it is difficult to find the driving mechanism for the Caledonian deformation NW of the Great Glen, if the Moines were deformed in their present position relative to the rest of Britain (e.g. Coward 1983). A more favourable explanation is that the thrust structures of the Moines formed as

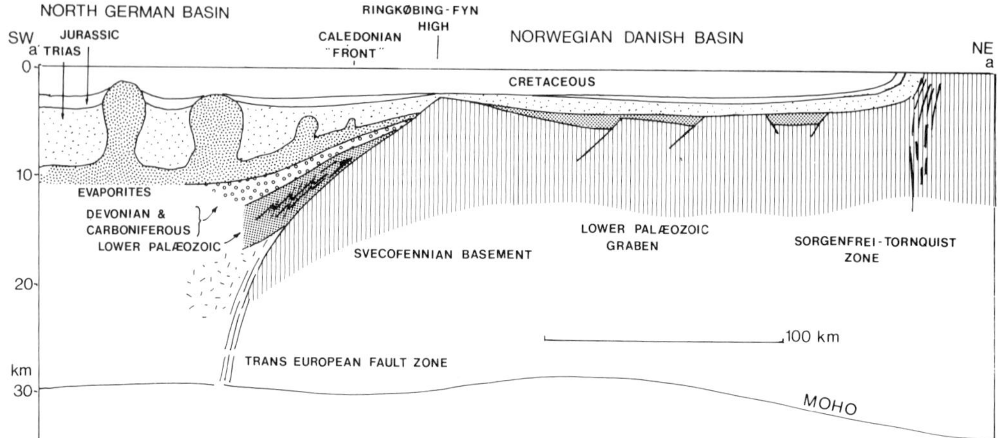

Fig. 8. Section across the Trans European Fault, the edge of the Caledonian deformation and the Sorgenfrei Tornquist Zone — from EUGENO Working Group (1988). The section line is shown in Fig. 7.

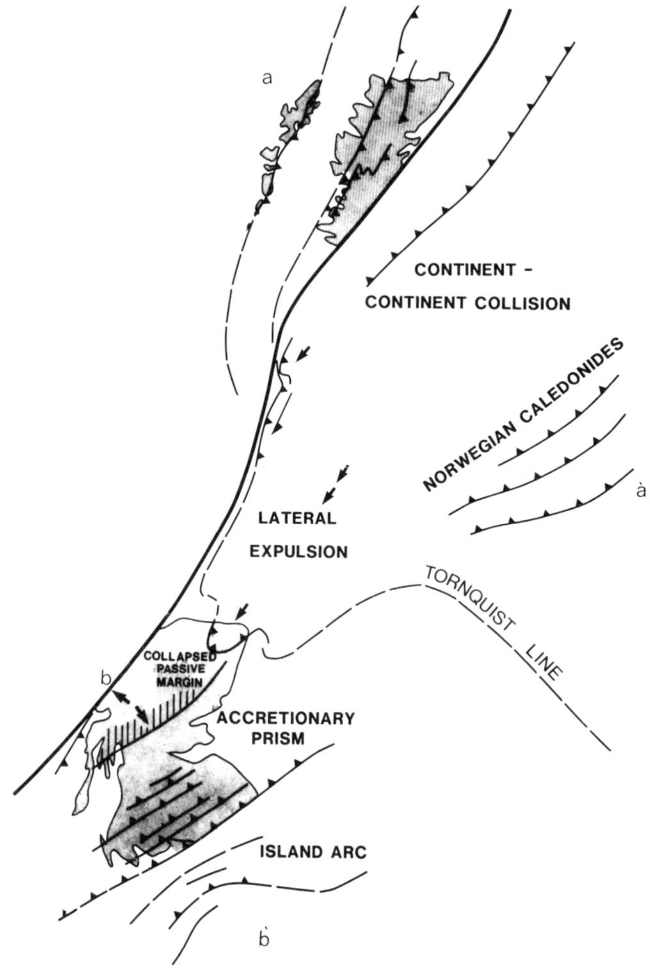

Fig. 9. Simplified map of the Caledonides after restoration of the Great Glen Fault.

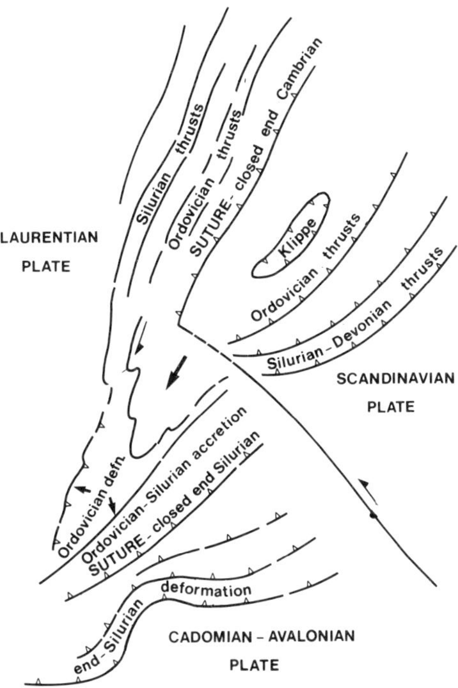

Fig. 10. Diagram to show the different events in the Caledonides, after restoration of the Great Glen system.

part of the continent–continent collision between the Scandinavian and North American cratons and were then displaced sinistrally during the late phases of Caledonian deformation. Figure 10 shows the principal subdivisions of this continental collision tectonic model. Figure 4a shows a schematic section across the Moine and Scandinavian collisional belt at end Silurian times and compares this with the contemporaneous section across the Dalradian–Southern Uplands (Fig. 4b). Modern analogies would be comparative sections across the Himalayas–Pamirs in Pakistan–western China, formed by continent–continent collision between India and Asia, and across the continent–arc collision zone in southern Afghanistan and eastern Iran (Fig. 12).

Figure 10 shows the style of deformation across the suggested Moine–Scandinavian orogenic belt. Ocean closure and obduction of high-grade rocks began during the early Cambrian and slices of ophiolitic material are preserved in the higher thrust sheets of western Norway (Nicholson 1979). There may have been a quiescence in plate convergence during the late Cambrian and early Ordovician, but the major shortening in the NW Highland Caledonides and in Norway took place during the later Ordovician and Silurian at a shortening rate of *c*. 1 cm/year (Hossack & Cooper 1987). The amount of shortening across this part of the Caledonides is not known, because of the uncertain relative positions of the NW Highlands and Scandinavia, the unknown amounts of strike-slip movement and the unknown amount of shortening that may have been taken up by deformation of the original passive margin. In northern Norway, kinematic movement indicators suggest considerable along strike ductile displacements (Hossack & Cooper 1987) and there are similar suggestions for SW-directed thrust movements in Shetland and NE Scotland (Fig. 9, with data from Flinn 1985; Coward 1983).

The post-Caledonian Devonian Basins

Caledonian compressional deformation was closely followed by extension, forming Devonian basins in, for example, the Minches, the Orkneys and Caithness, the Moray Firth, E and W Shetland, western Norway, the Midland Valley and Northumberland Basin and the Danish sector of the North Sea. Many of these basins formed by reactivation of earlier Caledonian fabrics, so that the Devonian faults have

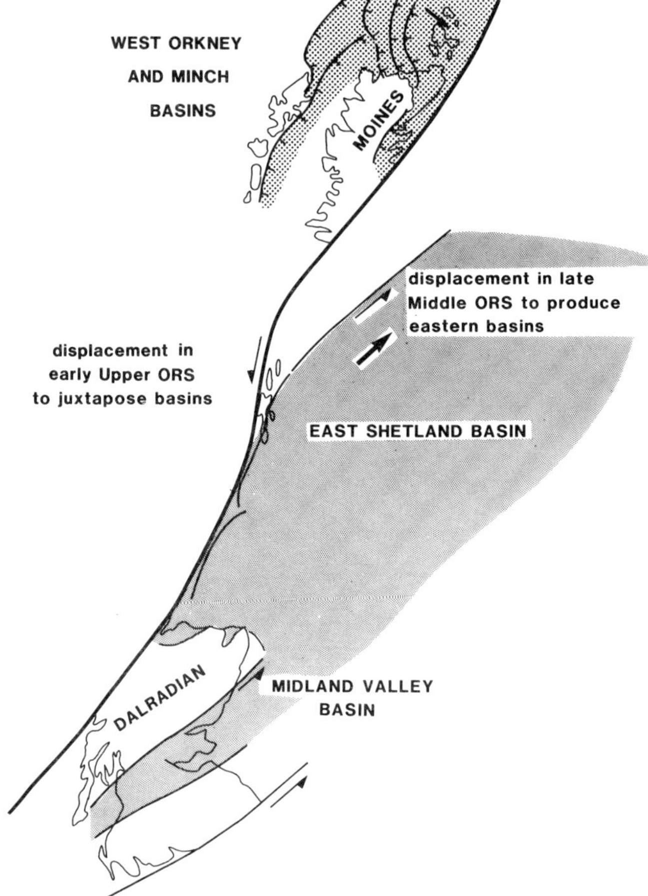

Fig. 11. Simplified map to show the generation of ORS basins east of the Great Glen Fault, associated with sinistral displacement on the fault.

similar strikes and dips to Caledonian thrusts. However not all Devonian faults were dip-slip; strike-slip movements occurred on both gently and steeply dipping faults. Another analogy can be made with the Tibetan–Himalayan region (e.g. Dewey 1988; Dewey *et al.* 1988), where much of the late Tertiary extension occurred by dip-slip reactivation of earlier major thrusts (Royden & Burchfiel 1987; Coward *et al.* 1987) but there was also considerable strike-slip fault movement and extension along the strike of the belt. The Devonian basins of Britain and NW Europe show the local development of intermontane basins which appear to have subsided gradually to become more lacustrine and sometimes marine. They show several unconformities and locally intense deformation related to phases of tectonic inversion.

The Orcadian and West Orkney Basins

The Devonian (or Old Red Sandstone–ORS) deposits of Caithness, the Orkney Isles and the Walls Peninsula of SW Shetland unconformably overlie an irregular landscape of Caledonian metamorphic rocks. They were deposited in a large intermontane basin known as the Orcadian Basin (Donovan *et al.* 1974). The Lower ORS (Siegenian–Emsian) has a restricted distribution and in western Caithness passes up into Middle ORS (Eifelian–Givetian) without any important stratigraphic break. The base of the Middle ORS oversteps the Lower ORS and onlaps basement rocks to the west. Unconformities are present at the base of the Middle ORS on the Orkney mainland and in southern Caithness, where the Lower ORS is considered

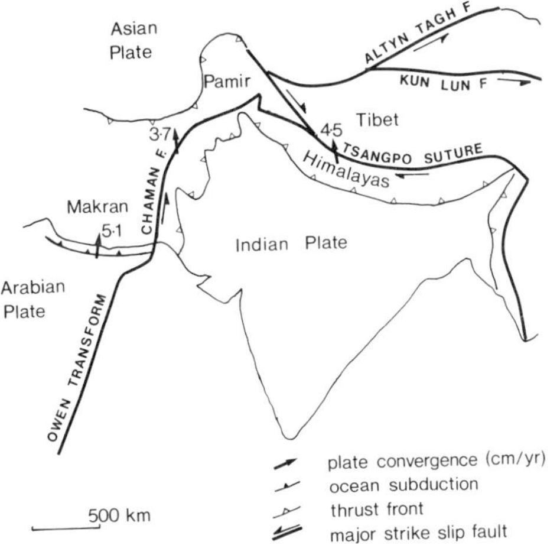

Fig. 12. Map of the Markran–Himalayas–Tibet collision zone, between the Indian and Asian plates, for comparison with Caledonides collision – see text for discussion – From Coward *et al.* (1987) and Dewey *et al.* (1988).

to have been uplifted, eroded and folded prior to deposition of the middle ORS.

The Lower ORS of the Orcadian Basin is characterized by coarse lenticular breccias and conglomerates, interfingered with finer bedded sandstones. These sediments were deposited within small isolated, mainly playa-filled intermontane basins (Mykura 1976). Local thick fanglomerates suggest deposition in basins bounded by active fault scarps. The Middle ORS in Caithness and Orkney began with the deposition of relatively quiescent lacustrine sandstones up to 4 km thick in Caithness. There was a gradual regression in later middle ORS times to dominantly alluvial sedimentation. Palaeoflow was from SW to NE, parallel to the general trend of the basin (Foster 1972), although locally the sediment supply was from the NW as shown by the presence of Torridonian and Cambrian clasts.

There is an important unconformity between the Middle and Upper ORS throughout the Orcadian Basin. On the Walls Peninsula of SW Shetland, a 9 km thick sequence of middle ORS thinly bedded sandstones was deformed by polyphase folding before the intrusion of the Sandwich plutonic complex, dated at 360 ± 11 Ma (Mykura 1976). On the Orkneys the onset of upper ORS sedimentation was marked by an unconformity followed by calc-alkaline lavas, dated at *c*.370 Ma (Halliday *et al.* 1977), followed by high-energy fluvial deposits, locally with aeolian dunes. These upper ORS deposits are considered to be of Frasnian age (Rogers *et al.* 1989).

The deep structure of the Orcadian Basin can be seen from seismic data in the West Orkney Basin (WOB), offshore northern Scotland (e.g. Fig. 3). Up to 4 km of sediments infill the larger of the half-graben and the average sediment thickness is in the order of 2 km. The basin faults and sediment fill can be traced close to the northern Scottish and Orkney coasts and correlations made with middle ORS deposits onshore (Coward & Enfield 1987). The WOB shows considerable extension, with maximum stretching factors of up to 1.6. However most of the sediments seen on the seismic data show evidence of faulting and no post-rift sediments are preserved. The latter may have been removed by subsequent late Devonian or Carboniferous tectonic inversion, or alternatively they were never deposited in this basin.

The NW edge of the WOB is formed by faults which are strongly curved in plan and possibly curved in section, while the east part of the basin is formed by faults which are approximately straight in plan and section. On the deep seismic data (e.g. the MOIST line, see Brewer & Smythe 1984; Cheadle *et al.* 1987 and the DRUM line, see McGeary & Warner 1985) all these faults dip to the ESE but cannot be traced to a depth greater than 18–20 km. They are parallel to reflectors in the basement, interpreted as Caledonian shear zones. On commercial speculative data (Coward & Enfield 1987; Coward *et al.* 1989), faults with listric geometries, which shallow towards a thin package of reflectors dipping 10°–15° to the ESE, can be mapped. These shallow reflectors are considered to represent structure within the Caledonian basement, possibly the offshore continuation of the Moine Thrust Zone (Coward *et al.* 1989). Thus much of the extension in the WOB developed on faults that flatten onto a gently dipping detachment, interpreted as the Moine Thrust, or onto major shear zones in the Moines. Half-graben observed offshore on seismic data are also observed onshore (Coward *et al.* 1989) and in each case coincide with the position of a major ductile shear zone (Coward *et al.* 1989), the extensional faults occurring a short distance into the hanging-wall of the ductile shears.

The dominant effects of the Middle ORS inversion appear to be concentrated around the faults in the Orkneys and Shetland Isles. This inversion may be due to the final effects of Caledonian compression, or to strike-slip tectonics associated with movements on the proto-Great Glen Fault. From the trends of the folds on West Shetland and structures within the fault zones on the Orkneys, the shear couple along the Devonian Walls Boundary Fault–Great Glen Fault would be sinistral (Enfield & Coward 1987; Coward *et al.* 1989). Sinistral strike-slip displacement would not only generate tectonic inversion, with NE trending folds, but also NE–SW extension and could be the cause of local anomalous NE extension directions observed in the Bressay–Sumburgh Basin, in eastern Shetland (Fig. 11 see also Coward *et al.* 1989). The amount of sinistral strike-slip displacement along the Great Glen Fault must remain speculative. From the offsets of the Caledonian Front on Shetland and the Scottish mainland, Flinn (1985) suggests an offset of 100–200 km. A large sinistral offset is required if the configuration of the Caledonian thrust belt is anything like that shown in Fig. 9. However from correlations of ORS facies across the Great Glen Fault in the Moray Firth, Rogers *et al.* (1989) suggest a small (<100 km) displacement. Strike-slip movements may be related to a late transpressive phase of Caledonian deformation (Soper & Hutton 1984; Hutton 1987) or alternatively to the effects of Acadian deformation in southern Britain and/or the Appalachians, which result in indentation tectonics and lateral expulsion of parts of the Caledonian belt (Fig. 13).

The Devonian Basins of western Norway

Two sets of Devonian basins occur W of Norway: (i) the Hitra basins, controlled by the Hitra Fault, which is probably a sinistral splay from the Great Glen system, and (ii) the Solund Basins which reactivate earlier Caledonian thrusts. The Solund to Hornelen Basins, shown in Fig. 14, comprise thick E-dipping Devonian sequences resting on Caledonian basement in the hanging-walls of listric low-angle faults

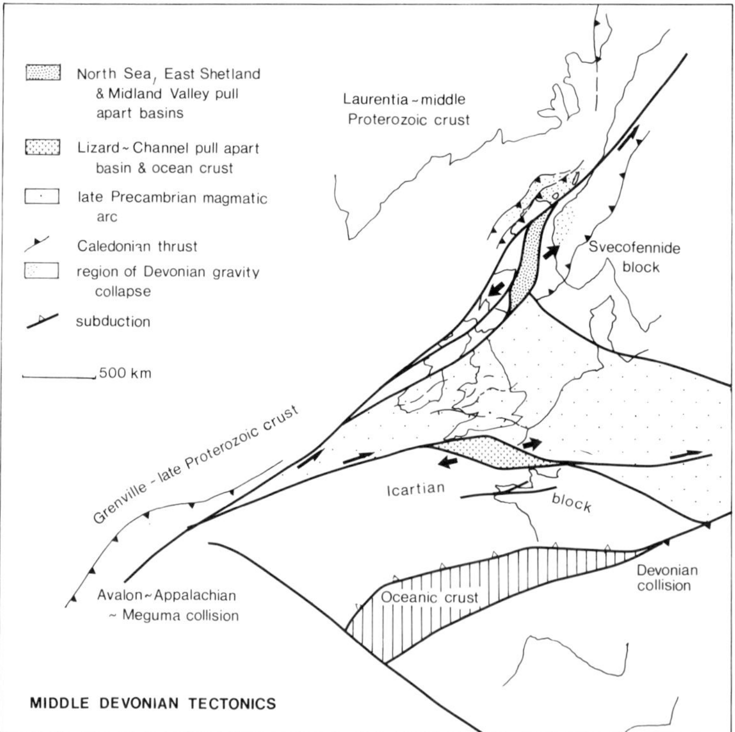

Fig. 13. Simplified map to show the Devonian tectonics of the North Sea and adjacent regions. During Devonian to early Carboniferous times, NW directed subduction occurred beneath NW France–southern Britain, generating back-arc extension during the early Carboniferous (Leeder 1982). However, in the Appalachians there was continued continent–continent collision (the Acadian phase), causing lateral extrusion of the British Caledonides, generating pull-apart basins along the major strike-slip zones. This model is similar to that proposed for Tertiary – Recent tectonics in Tibet and Anatolia (e.g. Dewey *et al.* 1987, 1989.).

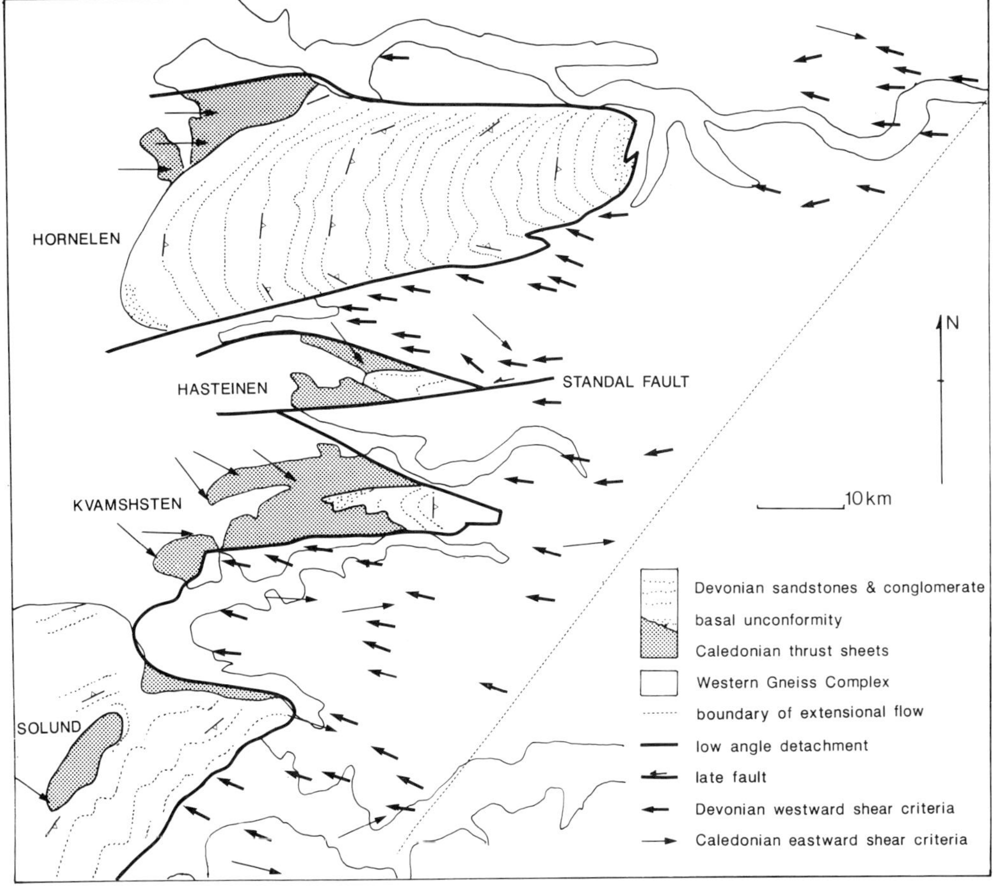

Fig. 14. Map of the Devonian basins of western Norway, from Seranne & Seguret (1987) and Seranne (1988).

(Hossack 1984; Norton 1986; Seranne & Seguret 1987; Seranne 1988). In the footwalls of these large faults the basement shows less intense Caledonian deformation, that is the Devonian extensional faults lie close to the boundary between Precambrian rocks, which were strongly affected by Caledonian deformation, and rocks which were far less affected. The faults are parallel or nearly parallel to ductile fabrics in the basement. Kinematic indicators in the basement rocks show that the Caledonian thrusts involved ductile shearing towards the ESE while, in the Devonian basin fill, the brittle extension was opposite, to the WNW (Seranne 1988; Seranne *et al.* 1989).

The Devonian displacements, in excess of 50 km, occurred along faults which, because of their low cut-off angles with bedding, initially must have had a gentle dip and therefore cannot be interpreted as rotated high-angle structures. Sediment facies vary from coarse sandstones and conglomerates close to the fault trace, to alluvial sands in the centres of the listric half-graben (Seranne 1988). Low-grade metamorphic minerals in the basin fill and deformation textures suggest that the basins were originally buried to a depth of *c.* 10 km (Seranne 1988), that is the basins were originally much larger. They presumably formed by collapse of the Norwegian Caledonides, reactivating the major basal Caledonian shear (Seranne 1988; Seranne *et al.* 1989). As the Caledonides were extended and unroofed, the basins were uplifted and subsequently eroded. The irregular listric form of the faults on the map may be partly original or due to cross folding associated with Devonian sinistral shear on the Hitra strike-slip system.

The Devonian Basins of the Midland Valley and the Scottish Borders

The late Silurian to early Devonian evolution of the Midland Valley is characterized by the continued subsidence of the Caledonian fore-arc basin (e.g. Leeder 1982). Subduction continued throughout the early Devonian as shown by the volcanics and granites north of the Highland Boundary Fault and in the Southern Uplands, and by the thick volcanic sequences in the Midland Valley. Uplift of these volcanic regions provided the source for thick fluvial red bed molasse deposits, locally up to 9 km thick in the Strathmore region on the northern flanks of the Midland Valley (Armstrong & Paterson 1970; Dewey 1982; Haughton 1988). Within the Midland Valley there are thick calc-alkaline, rhyolitic and basaltic volcanic sequences. In the Stonehaven region of the Midland Valley (Fig. 15), Upper Silurian (Downtonian) breccias, sandstones and sandy mudstones pass conformably upwards into the thick sequence of lower ORS conglomerates, sandstones and volcanics. The conglomerates, which interdigitate with the sandstones, thin towards the south. The source of the alluvial fans was from the NW.

Within the Lower ORS, there are large-scale folds which trend NE–SW and have steep south-facing limbs in the footwall of the Highland Boundary Fault. However the region NW of the Fault must have been an area of positive relief during Silurian and early Devonian times, as suggested by the source of some ORS pebbles and the pre-Devonian cooling history of the Dalradian rocks (Dempster 1985). The early to middle ORS movements on the Highland Boundary Fault and the related folds to the SE may be due to reworking of Caledonian fault structures in the basement. The fold axial traces are parallel to the trend of the Highland Boundary Fault and kinematic indicators in the ORS structures suggest dominantly dip-slip thrust movements towards the SE. NW–SE trending faults within the Midland Valley abut against the Highland Boundary Fault, suggesting that they either predate the thrust movements or, more plausibly, are strike-slip transfer faults of a similar age. In the hanging-wall to the Highland Boundary Fault, Dalradian recumbent structures, including isoclinal folds and

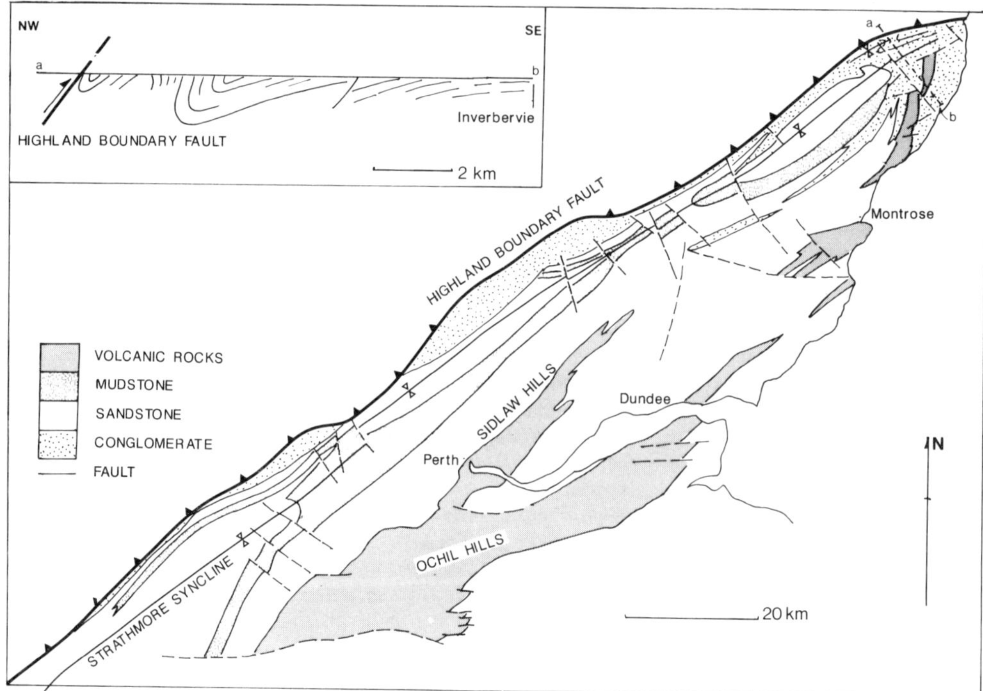

Fig. 15. Map of the NE part of the Midland Valley of Scotland, to show the distribution of the main folds and faults. Section a–b shows the large SE-verging folds in the ORS, formed possibly by reactivation of SE-verging basement structures.

cleavages, are deformed into a hanging-wall antiform, so that the early structures appear to be downward-facing (Shackleton 1958). The shape of the Highland Boundary Fault at depth is unknown but the reflectors, seen on the deep seismic line from offshore SE Scotland (e.g. Freeman *et al.* 1988) farther south, all dip at moderate angles to the north, suggesting that this might also be the dip of the ORS thrusts.

Strike-slip movements accompanied the final closure of the Iapetus Ocean and small-scale structures in the Southern Uplands and in Northern Ireland indicate that widespread sinistral wrench faulting continued both after the imbrication of the fore-arc basin sediments and after the intrusion of a younger dyke suite at 400 Ma. Clasts in the Lower Devonian sediments cannot be matched with the adjacent Caledonian basement (Bluck 1985; Haughton 1989) and some clasts have a provenance from the south and east, i.e. from within the Midland Valley. Lower ORS clasts have Sr isotope compositions which are more radiogenic than Southern Upland sediments and less radiogenic than Dalradian rocks (Haughton 1989), suggesting that all Dalradian, Southern Upland and Midland Valley sediments were derived from different basement sources. Haughton (1989) suggests that a basement similar to that of southern Greenland was a source for the Midland Valley sediments but a basement similar to the Grenville of the NE Appalachians acted as a source for the Southern Upland flysch. Thus some of the basin development in the Midland Valley and Northumberland Trough may have involved large scale wrench movements, with displacements of several hundred kilometres. The enormous thickness of the sediments may reflect their deposition in local pull-apart or transtensional basins.

During the middle Devonian, the Midland Valley was a region of uplift, followed by gradual slow subsidence throughout the late Devonian to Westphalian. In the Scottish Borders, the Lower Devonian is represented by volcanic sequences, followed by fluviatile Upper ORS, infilling a region of moderate relief. As in the WOB, this Middle Devonian uplift may be due to the final phases of Caledonian compression, or to the continued sinistral but now transpressional, strike-slip movements.

Fig. 13 summarizes the large-scale implications of the sinistral transtensional model. The NE–SW trending zones of strike-slip displacement of Devonian age have not been recognized in southern Scandinavia. Sinistral displacements in the Midland Valley have to link to similar displacement on the Hitra or more northerly shear systems of western Norway. Either there has been over 350 km of post-Devonian displacement on a continuation of the Tornquist Line, to offset these shear systems (Pegrum 1984), or the strike-slip faults must bend to produce a wide zone of transtension in the northern North Sea. The Great Glen Fault marks one edge of this diffuse transtensional zone, while the other boundary would lie along the Viking Graben. The extension direction in this proto-North Sea would be NE–SW. ORS deposits form the reservoir to the Buchan Field off NE Scotland and ORS rocks have been drilled in wells on the East Shetland Platform. Based on the interpretation of deep seismic lines. Beach (1986) suggested that there may be 10–12 km of Devonian sediments beneath parts of the North Viking Graben, with only a very thin remnant of the original Precambrian basement. If correct, this interpretation (Beach 1986) suggests very large localized extension of the Caledonian crust.

Devonian of the Anglo-Welsh Basins

SE of the Welsh Caledonides, the Lower ORS rests conformably on marine Upper Silurian sediments, except in SW Wales, where there is an angular unconformity. A Lower ORS regression, shown by the transition from slow shelf sedimentation through tidal/intertidal deposits up into river channel deposits, indicates a southerly migrating strandline of beaches and barriers (Allen 1974). During the middle ORS there was uplift and broad-scale folding, as represented by the middle ORS conglomerates of West Wales and the absence of sedimentation. In small intermontane basins, as in NE Anglesey, the ORS rocks were folded and cleaved. Throughout the ORS of the Anglo-Welsh Basin the palaeoflow direction was towards the south. The Lower ORS sediments include detrital metamorphic minerals but, in the Upper ORS, igneous and sedimentary rocks dominate (Allen 1974). This suggests uplift of the Anglo-Welsh Caledonides during the middle Devonian, cutting off the supply of metamorphic minerals from the Scottish Caledonides. In the upper ORS there was a transition to marine conditions by the late Devonian –early Carboniferous.

Devonian Environments

During the Devonian the sediments are dominated by molasse red bed facies, generally deposited in intermontane basins. No true foreland basin deposits occur in the British ORS;

there are no coarsening upward sequences and the basal ORS deposits often rest on deformed Palaeozoic rocks. Even in the Anglo-Welsh basins, SE of the Caledonian fold belt, the sediments suggest gradual infilling or uplift of the basin during early Devonian times, rather than the rapid deepening expected in a foreland basin. The deep sedimentary basins of the Orcadian, Shetland, West Norway and Midland Valley formed by large-scale extensional collapse of the thickened Caledonian crust, enhanced by localized pull-apart basin development in strike-slip zones. There is probably a large transtensional basin in the northern North Sea, of which the East Shetland Basin, parts of the Moray Firth and the ORS sediments in the Buchan Field may be part.

The strike-slip movements can be related to the late stages of Caledonian compression, possibly to lateral translation of crust away from zones of more intense late Caledonian compression in the Appalachian belt to the SW (Fig. 13). The strike-slip displacements are large, analagous to those of the strike-slip zones in Tibet which formed by the indentation of India into Asia (Fig. 12). The widespread Middle Devonian uplift and folding may record a late pulse of Caledonian compression, related to the final cessation of subduction. During the late Devonian there was continued gradual subsidence followed by marine incursions and renewed extensional faulting during the early Carboniferous.

Carboniferous Extension and Basin Development

The Carboniferous basins of NW Europe can be divided into two segments, separated by the NW–SE-trending Dowsing Fault Zone. To the west, the Caledonian basement was broken into a series of fault blocks, with pulses of NW–SE trending extension in the Tournaisian and Visean. To the NE, there was a morphologically smooth and almost horizontal depositional area throughout much of the central and southern North Sea.

The NW-trending Dowsing Fault is one of a series of faults which acted as transfer fault zones during early Carboniferous extension (Fig. 16). Dips of half-graben bounding faults

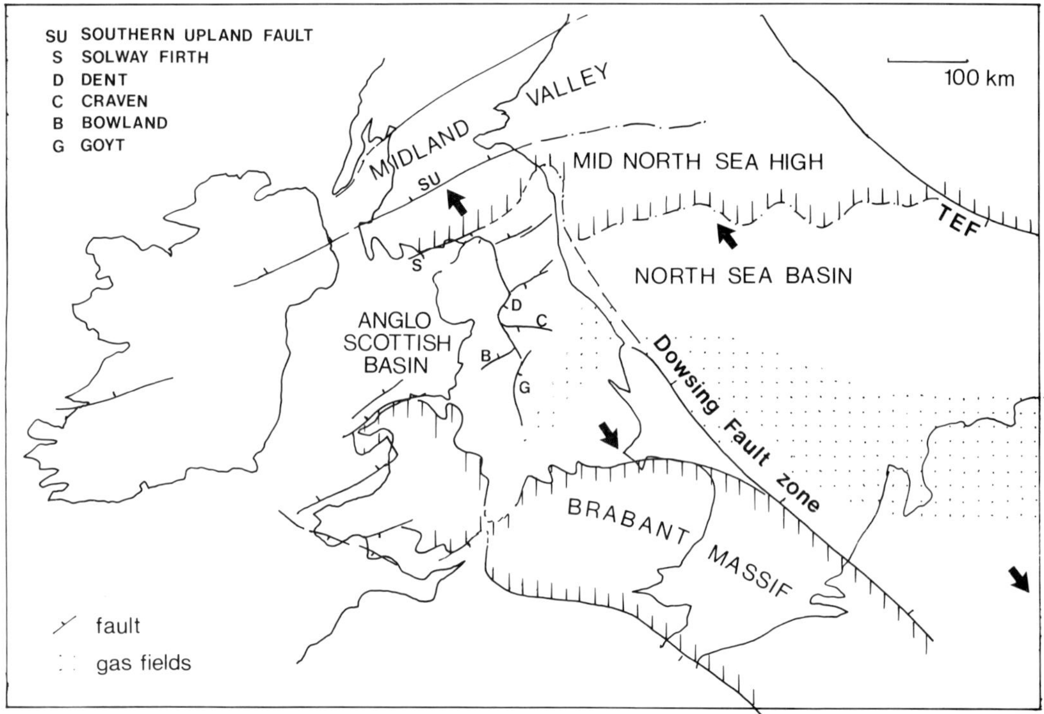

Fig. 16. Map to illustrate the principal early Carboniferous extensional and strike-slip faults. TEF, Trans European Fault. The distribution of the main gas fields (after Glennie 1986; Taylor 1986) shows the regions of greatest Permian subsidence.

change across these transfer zones. Thus in the western Pennines, the South Craven Fault was active during the Visean, transferring extension from a NW-dipping fault at the SE margin of the Bowland Basin, to a SE-dipping Middle Craven Fault (Gawthorpe 1987) (Fig. 16). There were also smaller transfer fault zones within individual basins transferring extension from one set of antithetic or synthetic faults to another set (Gawthorpe 1987).

West of the Dowsing Fault Zone, important basin bounding faults occur along the northern edge of the Solway Firth, the Dent Fault system and its continuation along the Lunedale Fault, the Craven–Bowland system and the NE edge of the Goyt Trough. Visean limestones vary in thickness from *c*.3 km in the Solway Basin to only 500 metres over the Alston fault block (Ord *et al.* 1988). In the Bowland Basin, basin-wide debris flows and slump deposits demonstrate two main episodes of extension during the late Chadian and the late Asbian (Gawthorpe 1987).

Most of the half-graben were infilled by the early Namurian, when steady subsidence and the influx of a fluvio-deltaic sequence of sandstones, coals and marine shales began. Cyclic alternations of marine and terrestial sediments from the Dinantian through to the Westphalian may reflect pulses of subsidence together with eustatic changes of sea-level associated with the earliest of the Permo-Carboniferous Gondwana glaciations (e.g. Ramsbottom 1974; Heckel, this volume). The main development of coals was in the Westphalian, when coastal plain sediments prograded southwards across Britain. An indication of the amount of subsidence is given by the 1 km thickness of Westphalian rocks in the Midland Valley and the 3 km thickness in the Lancashire–North Staffordshire coalfield (Ramsbottam *et al.* 1978). These syn-rift and post-rift sediment thicknesses in northern England suggest stretching factors in the order of 2, assuming a simple homogeneous lithospheric stretching model (Leeder 1982; Dewey 1982, after McKenzie 1978). In Scotland the stretching factor may be lower, although here estimates of stretch obtained from sediment thicknesses may be erroneous, due to the presence of Dinantian volcanics and hence a different sediment density and geothermal gradient. However the total stretch across the Anglo-Scottish Dinantian basins must have been in the order of 100 km.

In southern Ireland, Dinantian basins occur on the west coast in Clare and in Munster. In North Wales normal faults formed parallel to the Caledonian fabric along the Menai Straits and Anglesey. The wide distribution of Dinantian limestones and mudstones in Ireland suggests that the regional extension was similar to that of Britain.

The Brabant Massif remained as a tectonic high throughout the early Carboniferous and was onlapped by Namurian and Westphalian sediments, giving a steers head pattern of onlap similar to that of other thermal subsidence phases (e.g. Dewey 1982). South of this massif, thick Carboniferous sequences occur in South Wales, the Mendips, SE England and the Ardennes, along the northern edge of the Rheic Basin. Extension factors must have been similar to those of northern England, with over 2 km of Westphalian sediments preserved in South Wales. A minor but important phase of tectonic inversion occurred at the beginning of the Namurian, which resulted in folding the Lower Carboniferous sediments on the hanging-walls of some extensional faults. The South Wales coalfield is different from the Westphalian coalfields of northern England, in that rivers flowed into the basin from both north and south, indicating the presence of a Bristol Channel–N Devon Landmass (Leeder 1982). Some of the Westphalian sediments contain fragments of spilite and phyllite suggesting a source from the Variscan fold and thrust belt to the south. The uppermost Carboniferous sediments may be considered as a Variscan molasse.

The Dowsing Fault Zone marks the NE edge of the Brabant Massif and, in the southern North Sea and northern Germany, Carboniferous sediments were deposited in a broad basin, the NE boundary of which lies along the Tornquist Line and its continuation as the Trans-European Fault Zone (Fig. 16). In the North Sea the Westphalian reaches a thickness of 1.2 km in the Sole Pit Basin, close to the Dowsing Fault, while in N Germany the Namurian and Westphalian together reach over 2 km in thickness (Ziegler 1982). These figures suggest a lower stretching factor than that of the Anglo-Scottish basins, though as the basin is wide, the amount of stretching may be very similar. The difference in basin kinematics may reflect different Caledonian inheritance, in that the Anglo-Scottish and Irish basins overlay crust which had been deformed and thickened during the Silurian–Devonian times and the main stretch was concentrated in the region between the Iapetus Suture and the older Cadomian basement. The Carboniferous extension in the North Sea seems to have concentrated farther south, possibly along a continuation of the North German Caledonides.

The Variscan Tectonics of Domain 4

The Variscan tectonics of NW Europe have been interpreted as (i) a thin-skinned fold and thrust belt (Shackleton *et al.* 1982; BIRPS & ECORS 1986), overthrusting Lower to Upper Palaeozoic rocks to the NW, and (ii) the northern margin of a strike-slip orogen (Badham 1982; Sanderson 1984). The main belt of Variscan structures can be traced from South Wales to the Ardennes of northern France, Belgium and southern Germany. To the west, they continue into the southern Appalachians. In Britain there seems to be a spatial and temporal division between the Caledonide and Variscide orogenies. However in Germany and eastern Europe, Variscan deformation ranges in age from Devonian to late Carboniferous. In the southern Appalachians the Variscan and Caledonian deformations appear to be one semi-continuous process of arc accretion on to the American Craton.

Variscan tectonics of Germany and northern France

The internal, or Moldanubian, zone of the Variscides (Fig. 17) shows structures dominated by a gently dipping schistosity and gneissic banding, with overthrust sheets of eclogitic and granulitic material (Behr *et al.* 1980). The tectonic transport direction was towards the NW, as determined from linear fabrics and folds with strongly curvilinear axes (author's own unpublished observations). To the NW of this zone, the Rheno-Hercynian and Saxo-Thuringen zones show lower-grade metamorphism but similarly NW directed shears and thrusts. Together they indicate several hundred kilometres of crustal shortening. Within the Rheno-Hercynian zone upright to NW verging folds and thrusts probably decouple on a mid-crustal detatchment, as determined from deep seismic data (Meissner *et al.* 1981; Giese 1983). In Bavaria there are strike-slip faults and mylonite zones, which developed after the main thrust transport. These structures trend NW, parallel to the thrust transport direction in the Rhenish Massif and may be coeval with the thrusts in this Massif, as they do not cut across the external thrust zones.

In NW France, there may have been less crustal shortening during the Variscan, as the major thrusting is concentrated in southern Brittany and in the Massif Central. The rocks of northern Brittany and Normandy comprise

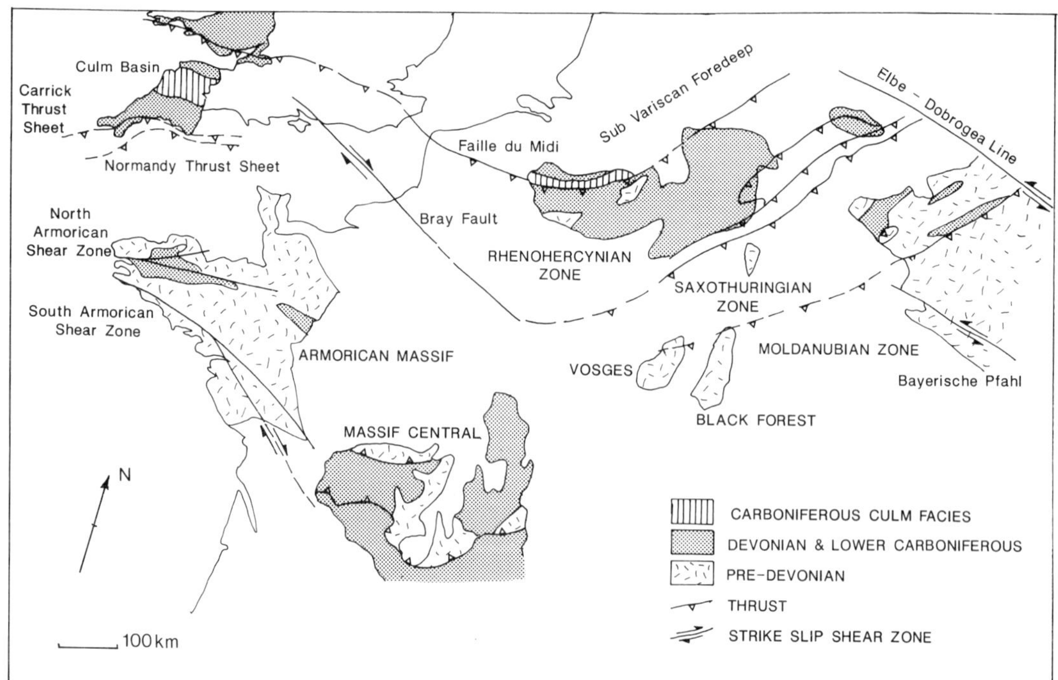

Fig. 17. Map to show the distribution of tectonic zones in the Variscides of NW Europe (after Coward & Smallwood 1984; Holder & Leveridge 1986a). See text for discussion.

Proterozoic basement (Autran *et al.* 1980), part of an Icartian basement block (see Fig. 13), which was deformed *c.*600 Ma ago by large scale shear zones and then unconformably overlain by relatively undeformed late Precambrian to Lower Palaeozoic cover sediments. There must have been a zone of strike-slip deformation along the eastern edge of the Armorican Massif, close to the present trend of the Bray Fault Zone, which separates the zone of thrust tectonics in the German Variscides, from the relatively weaker zone of deformation in northern Brittany (Coward & Smallwood 1984; Holder & Leveridge 1986a). Isotopic data from the German Variscides (e.g. Autran *et al.* 1980) show no evidence for Precambrian basement and the rocks themselves comprise late Precambrian to Palaeozoic sediments resting on the SE continuation of the Cadomian–Brabant magmatic arc. Presumably therefore, the Bray Fault Zone lies close to the edge of this old crustal block and acted as a major transform zone during Variscan tectonics, similar to the Tornquist Zone in northern Europe in Caledonian times.

The Variscides of Southern Britain

The Variscides of southern Britain form a continuation of the outer, or Rhenish, zone of the European Variscan belt (e.g. Holder & Leveridge 1986a). They are bounded in the east by a broad strike-slip zone trending from Kent to the Ardennes but as the thrust transport direction was dominantly towards the NW, across much of southern England and Wales, the Variscan Front is an oblique thrust structure (Figs 17 & 18).

The sediments involved in the British Variscides range in age from Devonian to late Carboniferous. Olistoliths in the Devonian flysch of South Cornwall contain shelf facies of Ordovician to Devonian sediments, comparable with similar rocks in France. However no older Lower Palaeozoic rocks occur in SW England and there are no overthrust sheets of older basement rock as in the southern Appalachians (e.g. Hatcher 1987). The Devonian sediments range from continental to shallow marine ORS in North Devon, to large thicknesses of reef and detrital limestones in South Devon. In Dyfed in South Wales, Powell (1989) described faults across which Devonian and Carboniferous sediments change markedly in thickness and facies, indicating extension during the Siegenian–Emsian and the Visean. In southern Cornwall, Devonian sediments are characterised by distal turbidites, probably deposited on thinned crust or the edge of a continental shelf. Dewey (1982) estimates a stretching factor of *c.*2 for N Devon, while the enormous thickness of Devonian sediments in South Cornwall, possibly up to 12 km (e.g. Holder & Leveridge 1986b), suggests that they were deposited on highly attenuated con-

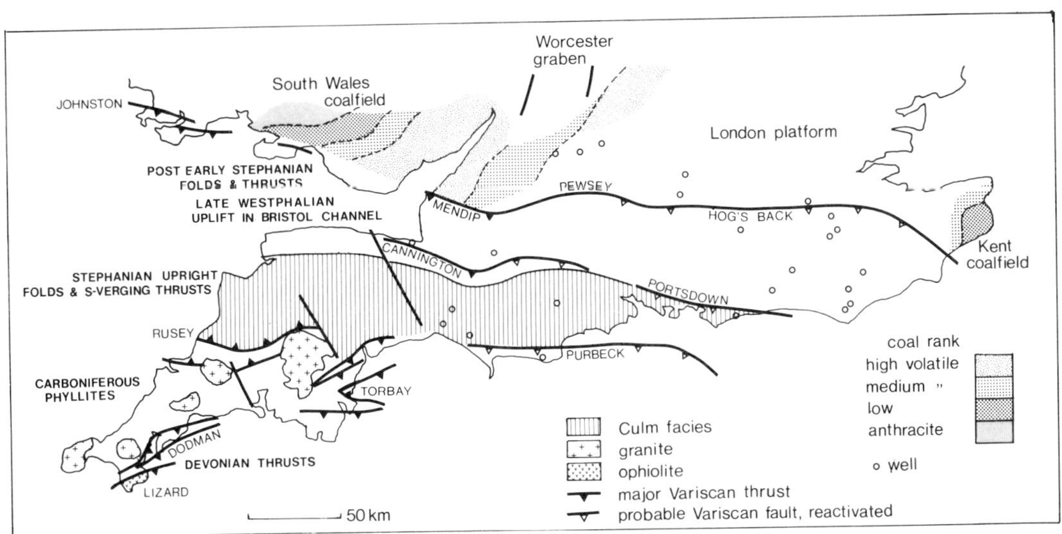

Fig. 18. Map of the principal structures in the Variscides of southern Britain (from Taylor 1986), showing the regions of thick-skinned thrust tectonics: the Mendip–Hog's Back and Cannington–Portsdown Faults, the regions underlain by Culm facies rocks and the rank of the Westphalian coal deposits. Note that the coal rank defines NE-trending zones, oblique to the Variscan Front, but approximately perpendicular to Caledonian and Variscan thrust directions and hence probably reflects tectonic inversion of deeper level structures.

tinental crust or on ocean floor. The Lizard ophiolite complex in South Cornwall, which is probably part of this ocean basin, gives a Sm/Nd isotope age of 375 ± 34 Ma, suggesting a Middle Devonian age of ocean floor generation (Davies 1984). Similarly Kennack Gneisses, which comprise mixed granite and basaltic magmas, give an Rb/Sr whole-rock isochron age of 369 ± 12 Ma (Styles & Rundle 1984) and are probably related to the ophiolite. Thus during the Devonian there must have been stretching of the southern part of the Brabant Massif in order to generate deep water flysch deposits and ocean-floor volcanics. Barnes & Andrews (1986) suggest that the Lizard represents new oceanic crust formed within the southern Cornwall basin. By rotating the cumulate layering in the Lizard gabbros to the horizontal and similarly rotating the dyke complex, so that their palaeomagnetic pole position coincides closely with the mean UK pole for the Silurian to Devonian, Hailwood *et al.* (1984) suggest an original NNW-spreading axis. Any other orientation requires rejection of the palaeomagnetic data (Barnes & Andrews 1986). If correct, this indicates that the Devonian extension direction was approximately NE–SW, oblique to the Variscan Front across southern Britain, but close to the trend of late Precambrian shear zones in northern Brittany (Brun & Balé 1989; Strachan *et al.* 1989). Possibly the Lizard ophiolite is part of a pull-part basin on a NE–SW trending shear system, similar to those operating in the Devonian basins of northern Britain. The sense of shear is unknown. However the indentation model for Devonian strike-slip basins, given in Fig. 14, suggests that there should have been dextral displacement in southern Britain during early to middle Devonian times.

The Variscan structure of SW Britain can be interpreted in terms of thin-skinned thrust tectonics (Shackleton *et al.* 1982). K/Ar mineral ages (Dodson & Rex 1971) suggest a diachroneity of deformation. Ages of 365–345 Ma have been obtained from the slates of South Cornwall, suggesting that the Lizard ophiolite was obducted and phyllites formed before the end of the Devonian. Holwill (1966) records pebbles of deformed Givetian limestone from the Upper Devonian sediments of Torbay (see also Coward & McClay 1983). Presumably these were derived from an uplifted thrust mass to the south. The overthrust direction, as obtained from lineations and sheath folds in the strongly deformed Lower Devonian sediments of South Cornwall, was towards the NW.

However, the majority of Devonian and Carboniferous slates give middle to late Carboniferous ages (340–320 Ma). In N Devon and Cornwall, the Namurian to Westphalian sediments are represented by the flysch-like Culm. These have been interpreted as syn-orogenic deposits derived from the Variscan thrusts (e.g. Isaac *et al.* 1982). However there are two contrasting sequences, to the north, in the Westward Ho region of North Devon, the rocks are dominantly deltaic and derived from a northerly source, while a more general basinal facies consists of mudstone and turbiditic sandstones with westward or eastward axial palaeocurrent directions. Much of the basinal debris was probably introduced from the northern delta slopes. Thus during or following Devonian thrust activity in South Devon, a large flysch basin developed in North Devon. This Culm facies can be traced across southern Britain from borehole data (Fig. 18 from Taylor 1986).

The Devonian and Carboniferous rocks of the Bristol Channel must have been uplifted at end-Westphalian times, as southerly derived debris flowed into the South Wales basin during late Westphalian times. There is also a pronounced angular unconformity between Westphalian and Stephanian rocks in the Bristol Channel region. Pre-Stephanian basin inversion and broad scale folding characterises much of the NE margin of the Brabant Massif and the southern North Sea basin. However the major phase of folding and thrusting in South Wales and the Mendips was post-Stephanian. Thin-skinned thrust zones characterise the deformation in the Westphalian sandstones, shales and coals of Pembrokeshire and parts of the South Wales coalfield (Trotter 1949; Coward & Smallwood 1984). Williams & Chapman (1986) suggest a thin-skinned interpretation for the Mendips thrust zone, with the thrusts detatching in ORS sediments. However Chadwick *et al.* (1983) propose a thick-skinned thrust model for the Mendips, based on the presence of moderately dipping seismic events in the upper to middle crust. In Pembrokeshire there is evidence for local thin-skinned thrusting in the Upper Carboniferous, breached by thick-skinned thrusts, such as the Johnston Thrust, which involve Precambrian and Palaeozoic basement (Coward & Smallwood 1984). There are facies and thickness changes in the Palaeozoic rocks across these thick-skinned thrusts, which are probably therefore reactivated earlier normal faults (Coward & Smallwood 1984; Powell 1989).

East of the Mendips there is evidence to show that the Variscan thrusts were reactivated several times in the Mesozoic to produce basin bounding faults, which were then tectonically inverted in the early Tertiary. Seismic data published by Kenolty *et al.* (1981) show that a concealed thick-skinned Variscan thrust controls the Vale of Pewsey anticline and probably the Ham and Kingslere anticlines. Similarly the Hog's Back monocline probably overlies a Variscan thrust (Smalley & Westbrook 1982; Taylor 1986). The trend of these moderately dipping thick-skinned thrusts, which form the Variscan Front in southern England, is shown on Fig. 18.

The Culm basin of North Devon shows a pattern of upright to north-verging folds in the north, passing to south-verging to recumbent folds and thrusts to the south. This major fold fan, which is about 50 km across, appears to be carried south on the gently dipping shears which outcrop in the Tintagel–Polzeath area. Large south-verging back-folds also occur in South Devon, steepening the earlier north-verging thrusts to vertical (Coward & McClay 1983; Coward & Smallwood 1984). Late Carboniferous mineral ages of 310–290 Ma have been obtained from the region of back-folding in S Devon and Carboniferous to Permian ages of 290–270 Ma have been obtained from the folded Culm rocks (Dodson & Rex 1971). This suggests that the back-folds and back-thrusts represent the late stages of Variscan crustal shortening.

Shackleton *et al.* (1982) make a rough estimate of over 150 km shortening across Devon and Cornwall. The thrust transport direction was almost generally towards the NW or NNW as determined from kinematic indicators on the fault planes and the trend of associated tear faults (Shackleton *et al.* 1982; Coward & McClay 1983). Fold hinges are generally normal to the thrust direction but are locally rotated into the NNW trend in more intensely deformed zones. In N Devon, South Wales and the Mendips, where thick-skinned thrust tectonics is evident, the structures trend approximately east–west, oblique to the general NW directed movements in the thin-skinned zones. This oblique trend may reflect the original trend to pre-Variscan basement structures which reactivated to give the thick-skinned thrusts. In Pembrokeshire there is evidence for *c.* 40° rotation associated with these oblique structures, based on palaeomagnetic data (McClelland-Brown 1983).

Variscan inversion of the Carboniferous basins

End-Carboniferous uplifts and folds characterize the basin-bounding faults of northern England and western Scotland (e.g. Gawthrope 1987). NE trending folds, which are often tight, occur in the Pendle–Skipton area of northern England associated with inversion of the Skipton–Clitheroe Basin (Gawthorpe 1987). Similarly, the folds in the Munster and Clare Basins of southern Ireland probably represent basin inversion rather than thin-skinned tectonics as suggested previously (e.g. Cooper *et al.* 1986).

Many of the NE-trending fault zones of Scotland show indications of dextral displacement during the Late Carboniferous and Permian. The E–W trend of normal faults and the NNE trend of folds in the Midland Valley suggests a dextral strike-slip shear couple to the region between the Highland Border and Southern Upland Faults. In the Shetlands the Walls Boundary Fault is a complex zone, with ambiguous shear criteria, although the dominant small scale shear indicators suggest that the main displacement is dextral. Flinn (1969; Flinn *et al.* 1979) suggests a dextral displacement of 65 km, based on a correlation of magnetic anomalies.

In East Shetland, tectonic inversion affects the Devonian sediments on the hanging-walls of the NE-trending faults. Kinks and thrusts affect the ORS on the Sumburgh Peninsula and large folds trend NNW across the region from Sumburgh to Lerwick, probably developed on the hanging-walls of reactivated normal faults. From their trend, these folds were formed by E–W compression, or more likely, were associated with dextral strike-slip movements on the Shetland fault system (Coward & Enfield 1977; Coward *et al.* 1989).

The Variscan Orogen

Throughout much of NW Europe and the North American Appalachians, the Variscan orogeny was but another episode in the history of progressive NW directed accretion of magmatic arcs and older continental fragments on to the stable North American Craton. In southern Britain however this history was punctuated by spasmodic crustal extension, particularly during the early Devonian and early Carboniferous. The early Devonian extension direction is uncertain although the faults had a WNW trend from South Wales to Kent and controlled the position of the margin of subsequent intense Variscan compression. From the trend of dykes in the Lizard complex, the extension direction can be shown to be NE–SW and this could be associated with a pull-apart basin in a major strike-slip shear as shown in Fig. 13. The Devonian normal faults along the Variscan Front could be oblique structures associated with this shear couple.

The early Carboniferous extension was more widespread across NW Europe, producing rift basins in northern Britain and western Ireland and a deep Culm facies basin from Cornwall to North Germany. Leeder (1982) suggests that this stretching was related to the development of a back arc basin, associated with NW-directed subduction of the Variscan Ocean beneath NW Europe.

Tectonic inversion occurred during the middle to late Devonian, obducting the Lizard ophiolite and uplifting Middle Devonian limestones, to provide a source for some Upper Devonian sediments. The Bristol Channel landmass was probably uplifted earlier, providing a sediment source for the Siegenian conglomerates of SW Wales. Minor inversion occurred during the early Namurian, folding the Lower Carboniferous rocks on the hanging-walls of the earlier normal faults. The main period of compression however occurred during the Westphalian–Stephanian, producing the thrust systems of Variscan orogenic belt.

Permo-Triassic subsidence

Deep Permo-Triassic basins formed in Cardigan Bay, the Celtic Sea and Western Approaches. They trend NE–SE perpendicular to the Caledonide and Variscan thrust directions and approximately parallel to the Caledonide fabrics, although oblique to the Variscan thrusts of South Wales. The basins are cut by NW–SE trending tear faults which link with strike-slip faults onshore in SW England. These tear faults cannot be traced onshore southern Ireland or northern France and hence must transfer displacement on to the basin-bounding faults offshore. The extension direction was NW–SE parallel to these tear faults and to small lateral ramps which offset the main bounding faults, especially along the SE margin of Cardigan Bay. The tear faults and lateral ramps are steep to vertical in dip and can be recognized on seismic data from the offset of prominant reflecting horizons and from associated flower-like secondary faults (Coward & Trudgill 1989).

Basement structures to the Celtic Sea graben can be examined onshore in SW England, where numerous extension faults dip NW. On the Devon and Cornish coasts there are both brittle and ductile extensional shears, including gently dipping low-angle detachments which rework the Variscan back-thrusts in the Rusey–Tintagel region. These extensional faults offset many of the earlier compressional structures, including the region where back-folds refold the earlier NW verging Variscan structures, leading to a complex confrontation of NW and SE verging folds.

Some of this extension may be associated with Carboniferous–Permian granite intrusion of the Cornubian Batholith. Granite sheets intrude several of the normal faults in the St Ives district of Cornwall and many of the Cornish tin lodes are related to the same NW–SE extension (Moore 1975). Seismic refraction studies by Brooks *et al.* (1984) and Doody & Brooks (1986) suggest that the Cornish granites have a flat base at about 10 km depth, that is they intruded as a horizontal sheet along the Variscan fabric, or were sliced across by a thrust or an extensional detachment (Shackleton *et al.* 1982). The extensional structures in SW England can be considered as Variscan analogues for the tectonics of the Basin and Range in the western USA; they form a zone of continental spreading above a crust weakened by tectonic thickening and by the addition of granitic rocks (e.g. Sonder *et al.* 1987).

South of Cornwall, the deep Plymouth Bay Basin contains a sequence of Permo-Triassic sediments over 10 km thick (BIRPS & ECORS 1986). The basin is not fault bounded, although some interpretations show small faults at the base of the Permian strata (Pinet *et al.* 1987). The basin overlies south-dipping reflectors, correlated with the Carrick, Lizard and Start Thrusts onshore (Day & Edwards 1983; BIRPS & ECORS 1986; Holder & Leveridge 1986b). Although the basin is deep, neither the top of the reflective lower crust nor the Moho are depressed or uplifted beneath the basin. A simple stretching model cannot be applied to the Plymouth Bay Basin; instead it is considered to have developed above a deep crustal ramp, probably where the flat detachments beneath the Cornish stretched upper crust ramp down to lower crustal levels. The onlap of sediments towards the NW, as seen on the Swat data (e.g. BIRPS & ECORS 1986), supports this model.

Volcanic rocks of Stephanian to Asselian age were extruded in graben and pull-apart structures in the North German-Polish plain (Zeigler 1982), probably related to NW–SE movements on the Tornquist Line. These movements which appear dominantly dextral are probably associated with Late Carboniferous inversion of basins in northern Britain; the Tornquist Line marks the NE boundary of these structures. Similar dextral movements can be postulated for the Dowsing Fault Zone, though here there are no volcanics.

In the North Sea there were two main areas of Permian subsidence (Glennie 1986), probably reflecting continued thermal subsidence following Carboniferous stretching. The Northern and Southern Permian Basins are separated by the Mid-North Sea–Rinkobing Fyn High, which probably represents the region of lowest Carboniferous stretching.

Discussion: Influence of basement structures on subsequent stretching

The Precambrian and Palaeozoic structures of NW Europe control subsequent basin development in several ways.

1. Control of subsidence patterns

(a) Generation of thermal subsidence. Crustal stretching during the early Carboniferous largely controlled the position and intensity of later Carboniferous and Permian subsidence. Much of the Permian subsidence in the North Sea is probably the result of thermal equilibration following Carboniferous extension. In North England a region of middle to late Carboniferous thermal subsidence overlies the zone of most pronounced Dinantian stretching, although here the subsidence pattern was modified by a late Carboniferous episode of locally intense tectonic basin inversion.

(b) Generation of crust and lithospheric mantle of anomalous thickness. In North Britain Devonian stretching affected crust previously thickened during the Caledonian Orogeny. At the peak of Caledonian compression and metamorphism, the crust of the NW Highlands was *c.*50 km thick (e.g. Coward *et al.* 1990) and hence considerable crustal stretching was necessary to reduce the topography to sea level. Early Devonian basins must have been cannibalised during subsequent Devonian stretching, so that only the later stages of the stretching history are preserved in the alluvial to lacustrine facies of the West Orkney Basin. As the lithosphere had probably not thermally equilibrated following Caledonian thickening, the thermal history of the WOB must have been very different from that of, for example, the Mesozoic basins of the North Sea, where Jurassic stretching affected lithosphere previously thinned during the Triassic (Barton & Wood 1984). The WOB must have had a lower geothermal gradient and hence been a much cooler basin than the Mesozoic basins of the North Sea.

2. Control of the positions of the later basins

(a) Modification of the strength of the lithosphere. Crustal thinning initially weakens the crust but, following thermal equilibration, the zone of thinned crust will become stronger than the adjacent zones of less thinned crust (Dewey 1982; Gillcrist *et al.* 1987). Similarly crustal thickening will initially strengthen the lithosphere, but as the lithosphere thermally equilibrates and the lower crust becomes hotter, it will weaken and hence lead to crustal spreading as in the Basin and Range (Sonder *et al.* 1987) or Tibet (Houseman & England 1986; Dewey 1988). Thus the zones of more intense Caledonian thickening in Scotland and Norway became the sites of intense Devonian extension (Norton 1986; Seranne & Seguret 1987; Coward *et al.* 1989).

(b) Generation of fundamental faults or zones of middle to upper crustal fabric which fail easily to produce basin bounding faults. Several late Palaeozoic and Mesozoic basins occur on the hanging-walls of earlier large scale thrusts. Examples are the Outer Isles and Minches Basins on the hanging-wall of the Outer Isles Fault (Brewer & Smythe 1984; Enfield & Coward 1987), basins on the hanging-walls of Caledonian thrusts along and to the south of the Solway Line (Hall *et al.* 1984), and the Celtic Sea and Bristol Channel Basins on the hanging-walls of Variscan thrusts offshore southern Ireland and South Wales (Cheadle *et al.* 1987; Brooks *et al.* 1988)

Large strike-slip fault systems, such as Tornquist Line, the Bray and the Great Glen Faults, appear to have acted as basement-bounding tear faults throughout several episodes of extension and inversion. Thus strike-slip faults in SW England developed during Variscan compression and were then reactivated during Permo-Triassic extension and Mesozoic–Cenozoic extension and inversion (Coward & Trudgill 1989).

As shown by the SHET deep seismic survey across the Walls Boundary Fault system north of Shetland (McGeary 1987), a major tear fault may produce a slight offset in the Moho. This and other major tear faults offset zones of different crustal fabric and strength, in that (i) the fabric may change strike and dip across the strike-slip fault and (ii) zones of weaker crust

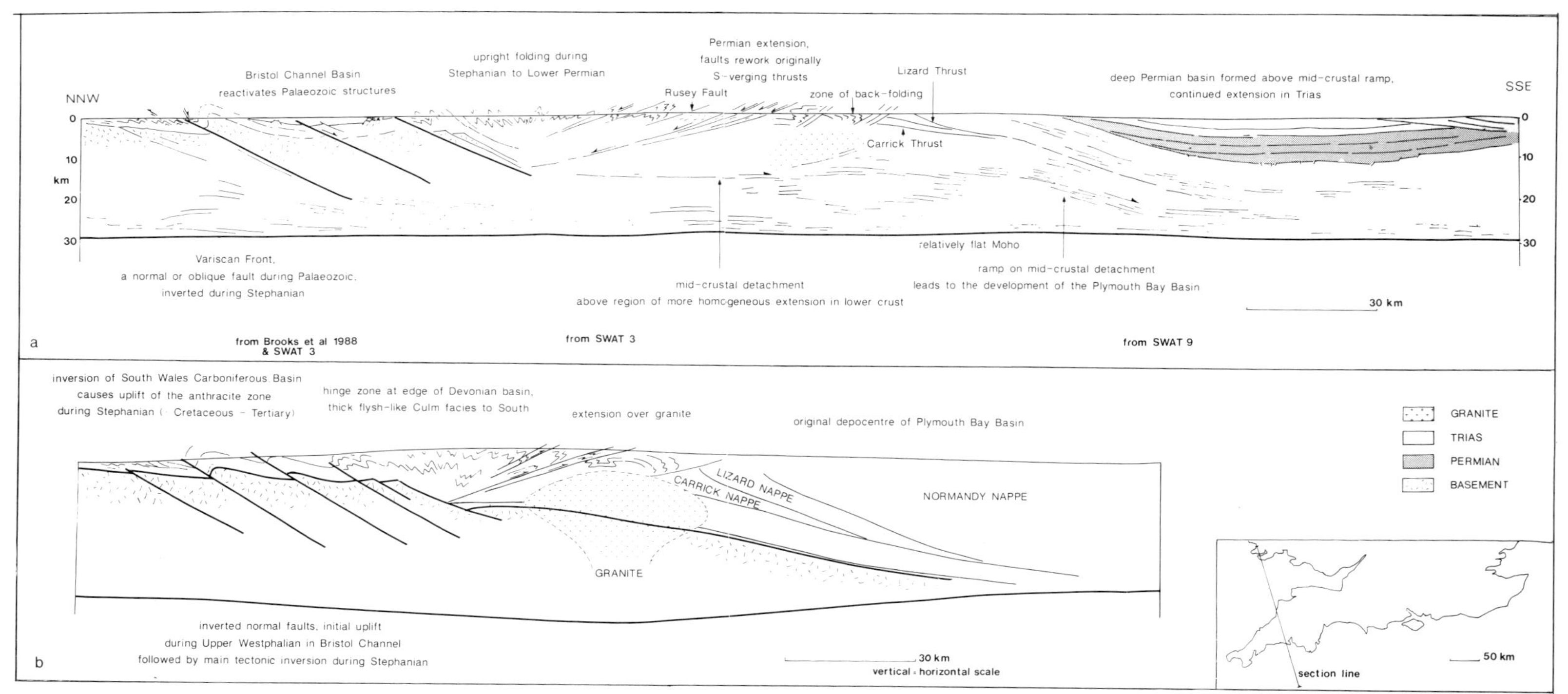

Fig. 19(a). Cross section through the Variscides of SW England from Pembrokeshire to the Plymouth Bay Basin, based on the SWAT survey (BIRPS & ECORS 1986), Pinet *et al.* (1987), Coward & Smallwood (1984) and unpublished work by the author. (b) Schematic partially restored cross section at the end of Variscan compression, before the main extension.

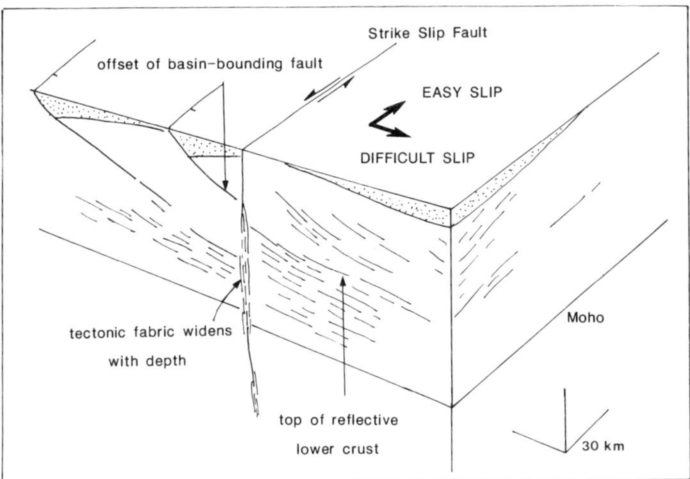

Fig. 20. Simplified block diagram to illustrate how major crustal-scale tear faults produce an anisotropy to the crust and tramline subsequent deformation (modified from Daly 1986).

formed, for example, of more quartz-rich material may abut against stronger crust formed of more feldspathic rock. The position and dip of middle to deep crustal reflectors differ across the Great Glen–Walls Boundary Fault on both the SHET survey, north of Scotland (McGeary 1987) and the WINCH survey, SW of Scotland (Hall *et al.* 1984). Furthermore a strike-slip fault may itself produce a steeply inclined ductile fabric at depth, as shown by field studies of ductile shear zones in high-grade metamorphic rocks (e.g. Coward & Park 1987). Hence major strike-slip faults produce irregular corrugations of the brittle–ductile layering in the lithosphere and also produce a 3D strength anisotropy (Fig. 20). This anisotropy produces an easy slip direction parallel to the large strike-slip fault and a more difficult slip direction perpendicular to the fault (Daly 1986). Hence subsequent crustal thickening or thinning is likely to be 'tramlined' (see Daly 1986; Gillcrist *et al.* 1987) by these fundamental faults.

It can be argued that the steeply dipping crustal anisotropies formed by major tear faults control the subsequent basin extension direction in the stretched crust and hence the trend of the transform structures as the basins grow. Thus earlier tear faults may essentially control the later plate divergence direction. It is only when true oceanic crust forms that this influence of crustal anisotropy ceases and the transform faults change trend with time. Thus during the Jurassic, the Atlantic Ocean developed along a wrench-rift system, from the Gulf of Mexico to Tethys. The continental break up and the early Atlantic transform faults generally followed Variscan structures and many of the Triassic–Jurassic graben of western Europe are bounded by Variscan tear faults.

The Great Glen Fault must offset the lower crustal decoupling zone of the West Orkney Basin. It therefore acts as a buttress to any further movement on this deep level decoupling zone and hence locks that part of the extensional fault system. This may explain the presence of a region of thick, relatively stable crust east of the Great Glen–Walls Boundary Fault system, i.e. east of the Orkney and Shetland Islands, where Mesozoic extension is slight. It may also explain why (i) the Great Glen Fault forms a lateral boundary to the Moray Firth Basin, probably controlling its opening direction and (ii) the northern continuation of the Great Glen system, along the More–Hitra Fault, forms the northern boundary to the Viking Graben.

(3) *Control of the dip of the master faults within the basin*

Following Navier-Coulomb failure criteria, a rock may fail along older fractures, rather than new fracture systems, if their cohesion, friction and orientations are correct. The ranges of orientations of reactivated fractures depend on the cohesion and friction and the 3D stress configuration. They have been calculated for a range of stress ellipsoid shapes and orientations (Jaeger & Cook 1979; Gillcrist *et al.* 1987).

Thus in a region which has undergone thrust tectonics, such as the Caledonides of Scotland

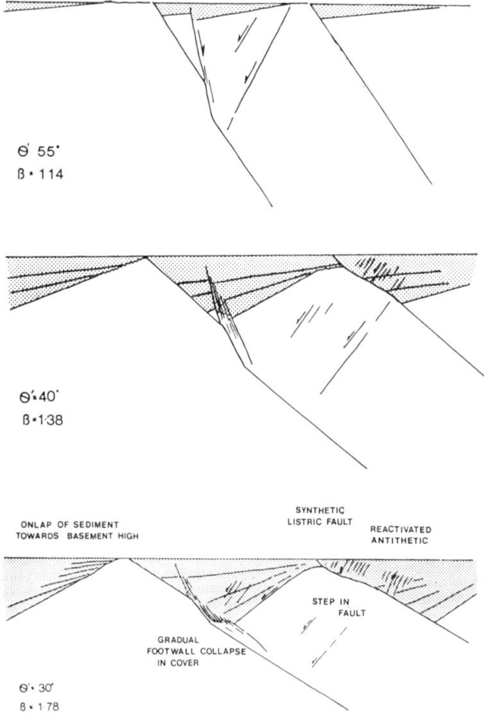

Fig. 21. Simplified sections to show how a step in a fault can generate extra strains in the hanging-wall or footwall. The bend may be produced from the reworking of an earlier anisotropy, e.g. an earlier thrust fabric with variable dips due to local back-steepening (see Butler & Coward 1984). In this particular example, the initial fault dip was taken to be 55°, and extension took place by block rotation. The hanging-wall is assumed to deform by antithetic fault slip. The uplift of the footwall and hence the amount of erosion, has been calculated using the subsidence model of McKenzie (1978) and Barr (1987), where

$$\text{subsidence} = 3.4\ (1 - 1/B)$$

where B is the stretching factor. An average sediment density is assumed; sediment is shown stippled.

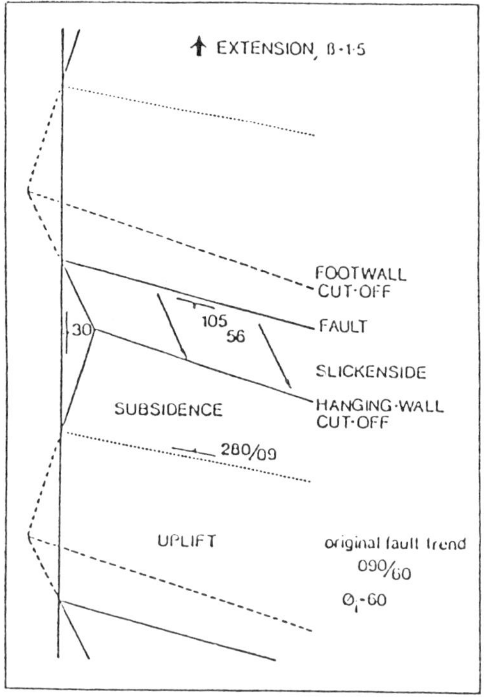

Fig. 22. Map of a region of block faulting, where the dip of the tear fault controls the axis of block rotation. In this model, the later N–S extension is parallel to an existing fault dipping 30° E. The later faults originally had an E–W strike and 60° dip but have been rotated into their present oblique orientation by the extension of $B = 1.5$. The resultant displacement direction on the fault is marked. The dotted line shows the tilt axis of the rotated blocks, separating relative subsidence from relative uplift, ignoring isostatic effects. Note this model assumes no boundary conditions, in that E–W extension is allowed during the later fault activity. If E–W extension were not allowed, the earlier fault would steepen, causing tectonic basin inversion.

and Norway, there may be a range of fabric orientations, from less than 30° dip, to over 75°, which may be favourable for subsequent reactivation. The Devonian low-angle faults of Norway (Seranne & Seguret 1987) and parts of the West Orkney Basin (Coward *et al.* 1989) originated from the reactivation of gently dipping Caledonian shear zones.

Where the basement fabrics are oblique to the subsequent extension direction, they may be in a favourable orientation for reactivation, depending on the shape of the stress ellipsoid (Gillcrist *et al.* 1987) and hence produce zones of oblique normal faults. Furthermore, where the earlier thrusts and ductile shears vary in orientation, as is to be expected in any major thrust zone, then the dip of the faults will also vary. Thus adjacent block-bounding faults may have different strikes and dips. Similarly a fracture may change strike and/or dip direction with depth, depending on the variation in orientation of the earlier basement fabric. Thus block-bounding faults may steepen or shallow with depth, similar to listric or stepped faults in growth fault environments (Fig. 21). Extension along a bent fault surface will lead to strains in the hanging-wall and/or footwall and therefore to a localized roll-over fold geometry and to

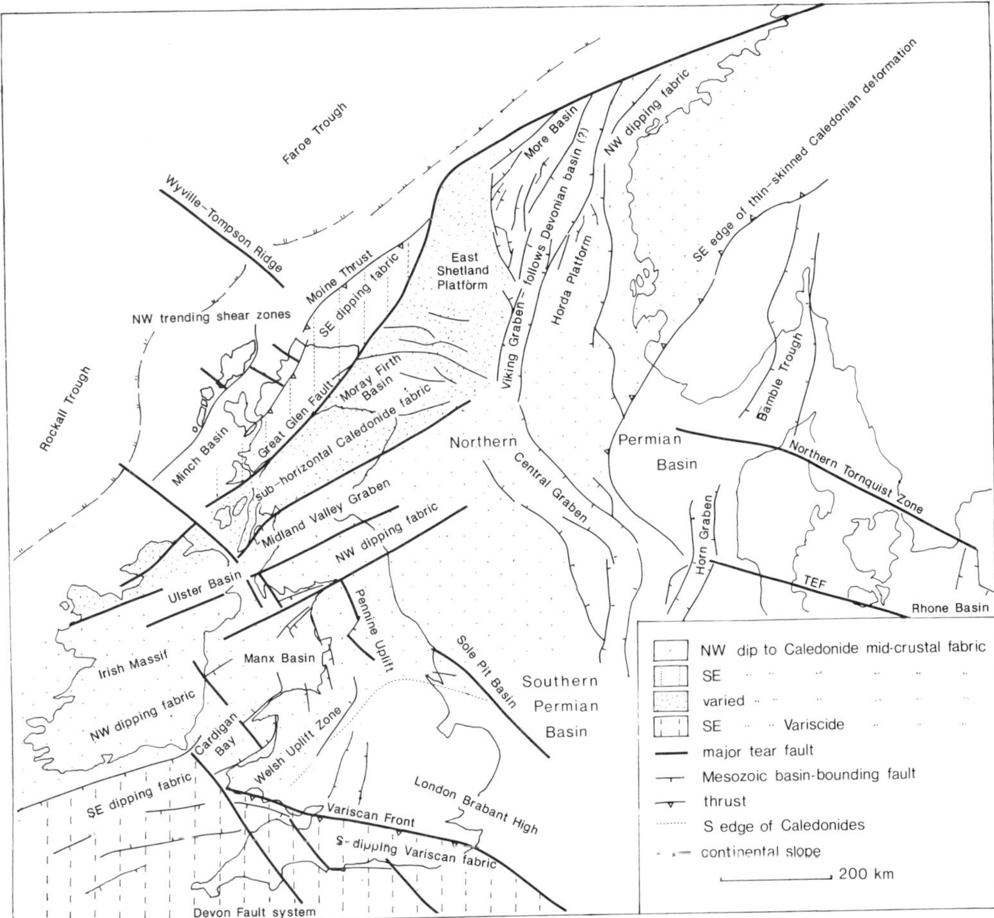

Fig. 23. Map showing the tectonic framework to NW Europe, including the dip of the Caledonian and Variscan mid-crustal fabrics and the trend of the major strike-slip faults. The important Mesozoic basin-bounding faults are shown (from Ziegler 1982).

systems of synthetic and antithetic faults (Fig. 21). Thus in regions of strong basement fabric, as in the northern North Sea or west of Britain, faults may change strike or dip with depth, leading to complex local strain patterns.

4. *Control of the dip of the main basin-bounding tear faults*

If the extension direction is parallel or closely parallel to an earlier fracture in the crust, this may reactivate to produce a tear fault, as discussed above. Where the early fracture is steep, and later extension takes place by block faulting (e.g. Morton & Black 1975; Jackson & White 1989), these blocks will rotate about the pole to the earlier fracture. However, where the early fracture has a moderate to gentle dip, the extensional fault blocks cannot rotate about a horizontal axis but will tend to rotate about an inclined axis parallel to the pole to the earlier fracture. This rotation will cause the extensional faults and the cut-off lines to trend oblique to the extension direction (Fig. 22). It may explain some of the oblique fault trends and tilts of fault blocks in parts of the Viking Graben. Here for example, fault blocks forming the Brent and Statfjord fields show simple tilting away from the main block bounding fault, while the blocks forming the Ninian and Gullfaks fields show more complex tilts, highly oblique to the main bounding faults. A knowledge of the basement structure beneath this part of the Viking Graben may help explain the different fault and rotational history of these fault blocks.

Conclusions

The crustal structure beneath NW Europe developed during three major orogenic episodes: the Laxfordian at *c.*1800 Ma, the Caledonian at *c.*600–400 Ma and the Variscan at *c.*400–300 Ma. These deformations formed the fundamental crustal architecture to NW Europe, producing:

(i) zones of local crustal anisotropy;

(ii) fundamental fault zones which offset layers of differing strength in the crust and hence tramline subsequent movement.

The basic fabric to NW Europe is as shown in Fig. 23. Plate accretion during the Laxfordian, Caledonian and Variscan was essentially NW–SE, producing fundamental tear faults with this trend. Oblique structures such as the Great Glen system probably developed as a result of intense crustal shortening, causing the crust to be squeezed out laterally, to produce a form of lateral continental escape. These oblique structures were also reworked extensively during subsequent basin development, generally buttressing the NW–SE extension, favoured by the more common Caledonian and Variscan faults.

References

Allen, J. R. 1974. The Devonian rocks of Wales and the Welsh Borderland. *In*: Owen, T. R., (ed.) *The Upper Palaeozoic and post-Palaeozoic rocks of Wales*. University of Wales Press, Cardiff, 47–84.

Anderton. R. 1982. Dalradian deposition and the late Precambrian–Cambrian history of the N Atlantic Region: a review of the early evolution of the Iapetus Ocean. *Journal of the Geological Society, London*, **139**, 421–31.

—— 1985. Sedimentation and tectonics in the Scottish Dalradian *Scottish Journal of Geology*, **21**, 407–436.

Armstrong, M. & Paterson, I. B. 1970. The Lower Old Red Sandstone of the Strathmore Region. *Report of the Institute of Geological Sciences* No. 70/12.

Autran, A., Breton, J.-P., Chantraine, J. & Chiron, J.-C. 1980. *Carte tectonique de la France*. 1:1M.

Badham, J. P. N. 1982. Strike slip orogens — an explanation for the Hercynides. *Journal of the Geological Society, London*, **139**, 493–504.

Barnes, R. P. & Andrews, J. R. 1986. Upper Palaeozoic ophiolite generation and obduction in south Cornwall. *Journal of the Geological Society, London,* **143**, 117–124.

Barr, D. 1987. Lithospheric stretching, detatched normal faulting and footwall uplift. *In:* Coward, M. P., Dewey, J. F. & Hancock, P. (eds.). *Continental Extensional Tectonics*. Geological Society, London, Special Publication, **28**, 75–94.

——, Holdsworth, R. E. & Roberts, A. M. 1986. Caledonian ductile thrusting in a Precambrian metamorphic complex: the Moine of northwest Scotland. *Bulletin of the Geological Society of America*, **97**, 754–764.

Barton, P. & Wood, R. J. 1984. Tectonic evolution of the North Sea basin: crustal stretching and subsidence. *Geophysical Journal of the Royal Astronomical Society*, **79**, 987–1022.

Beach, A. 1986. A deep seismic reflection profile across the northern North Sea, *Nature*, **323**, 53–55.

Beamish, D. & Smythe, D. K. 1986. Geophysical images of the deep crust: the Iapetus suture. *Journal of the Geological Society, London*, **143**, 489–497.

Behr, H., Engel, W. & Franke, W. 1980. *Guide to Excursion Munchberger Gneissemasse and Bayerischer-Wald*. Geology Institute and Museum, Gottingen, W. Germany.

Bergstrom, J. 1984. Lateral movements in the Tornquist Zone. *Geol. Foren. Stokh. Forh.*, **106**, 379–397.

Birps & Ecors 1986. Deep seismic reflection profiling between England, France and Ireland. *Journal of the Geological Society, London*. **143**, 45–52.

Bluck, B. J. 1985. The Scottish paratectonic Caledonides. *Scottish Journal of Geology*, **21**, 437–464.

Brewer, J. & Smythe, D. K. 1984. MOIST and the continuity of crustal reflector geometry along the Caledonide–Appalachian orogeny. *Journal of the Geological Society, London*, **141**, 105–120.

Brewer, M. S., Brook, M. & Powell, D. 1979. Dating of the tectono-metamorphic history of the southwestern Moine, Scotland. *In*: Harris, A. L., Holland, C. H. & Leake, B. E. (eds.), *The Caledonides of the British Isles – reviewed*, Geological Society, London, Special Publication, **8**, 129–138.

Brooks, M., Doody, J. J. & Al-Rawi, F. R. J. 1984. Major crustal reflectors beneath southwest Britain. *Journal of the Geological Society, London*, **141**, 97–103.

——, Trayner, P. M. & Trimble, T. J. 1988. Mesozoic reactivation of Variscan thrusting in the Bristol Channel area, UK. *Journal of the Geological Society, London*, **145**, 439–444.

Brun, J. P. & Balé, P. 1989. Cadomian tectonics in Northern Brittany. *In*: D'Lemos, R. S., Strachan, R. A. & Topley, C. G. (eds) *The Cadomian Orogeny*, Geological Society, London, Special Publication, **51**, 95–114.

Butler, R. W. H. & Coward, M. P. 1984. Geological constraints, structural evolution and the deep geology of the northwest Scottish Caledonides. *Tectonics*, **3**, 347–365.

Chadwick, R. A., Kenolty, N. & Whittaker, A. 1983. Crustal structure beneath southern England from deep seismic reflection profiles. *Journal of the Geological Society, London,* **140**, 893–911.

Cheadle, M. J., McGeary, S., Warner, M. R. &

MATTHEWS, D. H. 1987. Extensional structures on the western UK continental shelf: a review of evidence from deep seismic profiling. *In*: COWARD, M. P., DEWEY, J. F. & HANCOCK, P. (eds.). *Continental Extensional Tectonics*, Geological Society, London, Special Publication, **28**, 445–465.

CHESHIRE, J. A., SMYTHE, D. K. & BISHOP, P. 1983. *The geology of the Minches, Inner Sound and Sound of Raasay*. Report of the Institute of Geological Sciences, (83/6).

CONEY, P. J. 1989. Structural aspects of suspect terrains and accretionary tectonics in western North America. *Journal of Structural Geology*, **11**, 107–126.

COOPER, M. A., COLLINS, D. A., FORD, M., MURPHY, F. X., TRAYNER, P. M. & O'SULLIVAN, M. 1986. Structural evolution of the Irish Variscides, *Journal of the Geological Society, London*, **143**, 53–61.

COWARD, M. P. 1983. The thrust and shear zones of the Moine thrust zone and the Scottish Caledonides. *Journal of the Geological Society, London*, **140**, 795–812.

—— 1990. Shear zones at the Laxford Front, NW Scotland and their significance in the interpretation of lower crustal structures. *Journal of the Geological Society, London*, **147**, 279–286.

——, BUTLER, R. W. H., KHAN, M. A. & KNIPE, R. J. 1987. The tectonic history of Kohistan and its implications for Himalayan structure. *Journal of the Geological Society, London*, **144**, 377–391.

——, ——, CHAMBERS, A. F., GRAHAM, R. H., IZATT, C. N., KHAN, M. A., KNIPE, R. J., PRIOR, D. J., TRELOAR, P. J. & WILLIAMS, M. P. 1988. Folding and imbrication of the Indian sub-continent during Himalayan Collision. *Philosophical Transactions of the Royal Society of London*, A, **326**, 89–116.

—— & ENFIELD, M. A. 1987. The structure of the West Orkney and adjacent basins. *In*: BROOKS, J. & GLENNIE, K. W. (eds.). *Petroleum Geology of Northwest Europe*, Graham & Trotman, London, 687–696.

——, —— & FISCHER, M. W. 1989. Devonian basins of Northern Scotland: extension and inversion related to late Caledonian-Variscan tectonics. *In* COOPER, M. A. & WILLIAMS, G. D. (eds). *Inversion Tectonics*. Geological Society, London, Special Publication, **44**, 275–308.

—— & MCCLAY, K. R. 1983. Thrust tectonics of S. Devon. *Journal of the Geological Society, London*, **140**, 215–228.

—— & PARK, R. G. 1987. The role of mid-crustal shear zones in the early Proterozoic evolution of the Lewisian. *In*: PARK, R. G. & TARNEY, J. (eds.). *Evolution of the Lewisian and Comparable Precambrian High Grade Terrains*, Geological Society, London, Special Publication, **27**, 127–138.

—— & SMALLWOOD, S. 1984. An interpretation of the Variscan tectonics of SW Britain. *In*: HUTTON, D. H. W. & SANDERSON, D. J. (eds.). *Variscan tectonics of the North Atlantic region*, Geological Society, London, Special Publication, **14**, 89–102.

—— & TRUDGILL, B. 1989. Basin development and basin structure of the Celtic sea basins (SW Britain). *Bulletin of the Geological Society, France*, (8) t.v. 423–36.

DALY, M. 1986. Crustal shear zones and thrust belts: their geometry and continuity in Central Africa. *Philosophical Transactions of the Royal Society of London*, A, **317**, 111–125.

DAVIES, G. R. 1984. Isotopic evolution of the Lizard Complex. *Journal of the Geological Society, London*, **141**, 3–14.

DAY, G. A. & EDWARDS, J. W. F. 1983. Variscan thrusting in the basement of the English Channel and Western Approaches. *Proceedings of the Ussher Society*, **4**, 432–436.

DEMPSTER, T. J. 1985. Uplift patterns and orogenic evolution in the Scottish Dalradian. *Journal of the Geological Society, London*, **142**, 111–128.

DEWEY, J. F. 1982. Plate tectonics and the evolution of the British Isles. *Journal of the Geological Society, London*, **129**, 371–412.

—— 1988. Extensional collapse of orogens, *Tectonics*, **7**, 1123–1139.

—— & SHACKLETON, R. M. 1984. A model for the evolution of the Grampian tract in the early Caledonides and Appalachians. *Nature*, **312**, 115–121.

——, —— & CHANG CHENGFA 1988. The tectonic evolution of the Tibetan Plateau. *Philosophical Transactions of the Royal Society, London*, **327**, 379–413.

——, HEMPTON, M. R., KIDD, W. S. F., SAROGLU, F. & SENGOR, A. M. C. 1987. Shortening of continental lithosphere: The tectonics of eastern Anatolia – a young collison zone. *In*: COWARD, M. P. & RIES, A. C. (eds.). *Collision Tectonics*, Geological Society, London, Special Publication, **19**, 3–36.

DODSON, M. H. & REX, D. C. 1971. Potassium – argon ages of slates and phyllites from south-west England, *Quarterly Journal of the Geological Society*, **126**, 465–499.

DOODY, J. J. & BROOKS, M. 1986. Seismic refraction investigation of the structural setting of the Lizard and Start complexes, SW England. *Journal of the Geological Society, London*, **143**, 135–140.

DONATO, J. A. 1988. Possible Variscan thrusting beneath the Somerton Anticline, Somerset. *Journal of the Geological Society, London*, **145**, 431–438.

DONOVAN, R. N., FOSTER, R. J. & WESTOLL, T. S. 1974. A stratigraphical revision of the Old Red Sandstone of north-eastern Caithness. *Transactions of the Royal Society, Edinburgh*, **69**, 167–201.

ENFIELD, M. A. & COWARD, M. P. 1987. The West Orkney Basin, northern Scotland. *Journal of the Geological Society, London*, **144**, 871–884.

EUGENO-S Working Group 1988. Crustal structure and tectonic evolution of the transition between the Baltic Shield and the North German Caledonides (the EUGENO-S Project). *Tectonophysics*, **150**, 253–348.

FETTES, D. J., GRAHAM, C. M., HARTE, B. & PLANT, J. A. 1986. Lineaments and basement domains: an alternative view of Dalradian evolution. *Journal of the Geological Society, London*, **143**, 453–64.

FLINN, D. 1969. A geological interpretation of aeromagnetic maps of the continental shelf around Orkney and Shetland. *Geological Journal*, **6**, 279–292.

—— 1985. The Caledonides of Shetland. *In*: GEE, D. G. & STURT, B. A. (eds.), *The Caledonide Orogen–Scandinavia and Related Areas*, Wiley, Chichester, 1159–1172.

——, FRANK, P. L., BROOK, M. & PRINGLE, I. R. 1979. Basement – cover relations in Shetland. *In*: HARRIS, A. L., HOLLAND, C. H. & LEAKE, B. E. (eds.), *The British Caledonides Reviewed*, Special Publication of the Geological Society, London, **8**, 109–116.

FOSTER, R. J. 1972. *The solid geology of North-east Caithness*, PhD thesis, University of Newcastle.

FREEMAN, B., KLEMPERER, S. L. & HOBBS, R. W. 1988. The deep structure of northern England and the Iapetus Suture zone from BIRPS deep seismic reflection profiles. *Journal of the Geological Society, London*, **145**, 727–740.

FROST, R. T. C., FITCH, F. J. & MILLER, J. A. 1981. The age and nature of the crystalline basement in the North Sea Basin. *In*: ILLING, L. V. & HUDSON, G. C. (eds.), *Petroleum Geology of the Continental Shelf of North West Europe*. Heyden, London, 43–57.

GAWTHORPE, R. L. 1987. Tectono-sedimentary evolution of the Bowland Basin, N. England, during the Dinantian. *Journal of the Geological Society, London*, **144**, 59–71.

GIESE, P. 1983. The evolution of the Hercynian crust – some implications to the uplift problem of the Rhenish Massif. *In*: FUCHS, K., VON GEHLEN, K., MALZER, H., MURAWSKI, H. & SEMMEL, A. (eds.), *Plateau Uplift, the Rhenish Shield – a case history*, Springer, Berlin.

GLENNIE, K. W. 1986. Development of N.W. Europe's Southern permian Gas Basin. *In*: BROOKS, J., GOFF, J. & VAN HOORNE, B. (eds.), *Habitat of Palaeozoic Gas in N.W. Europe*, Geological Society, London, Special Publication, **23**, 3–22.

GILLCRIST, R., COWARD, M. P. & MUGNIER, J–L. 1987. Structural inversion and its controls: examples from the Alpine foreland and the French Alps. *Geodinamica Acta*, **1**, 5–34.

HAILWOOD, E. A., GASH, P. R. J., ANDERSON, P. C. & BADHAM, J. P. N. 1984. Palaeomagnetism of the Lizard Complex. *Journal of the Geological Society, London*, **141**, 27–35.

HALL, J. 1987. Geophysical lineaments and deep crustal structure. *Philosophical Transactions of the Royal Society, London*, **A317**, 33–44.

——, BREWER, J. A., MATTHEWS, D. H. & WARNER, M. R. 1984. Crustal structure across the Caledonides from the "WINCH" seismic reflection profile: Influences on the Midland Valley of Scotland. *Transactions of the Royal Society of Edinburgh, Earth Sciences*, **75**, 97–109.

HALLIDAY, A. N., MCALPINE, A. & MITCHELL, J. G. 1977. The age of the Hoy lavas, Orkney. *Scottish Journal of Geology*, **141**, 609–620.

HARTE, B., BOOTH, J. E. DEMPSTER, T. J., FETTES, D. J., MENDUM, J. R & WATTS, D. 1984. Aspects of the post-depositional evolution of the Dalradian and Highland Border Complex rocks in the Southern Highlands of Scotland. *Transactions of the Royal Society of Edinburgh, Earth Sciences*, **75**, 151–63.

HATCHER, R. D. Jr. 1987. The Moine thrust zone: a comparison with Appalachian faults and the structure of belts. *In*: FETTES, D. J. & HARRIS, A. L. (ed.) *Synthesis of the Caledonian Rocks of Britain*. NATO ASI Series, **175**, 247–257.

HAUGHTON, P. D. W. 1988. A cryptic Caledonian flysch terrane in Scotland. *Journal of the Geological Society, London*, **145**, 685–703.

—— 1989. Structure of some Lower Old Red Sandstone conglomerates, Kinkardineshire, Scotland: deposition from late-orogenic antecedent streams *Journal of the Geological Society, London*, **146**, 509–25.

HOLDER, M. T. & LEVERIDGE, B. E. 1986a. Correlation of the Rhenohercynian Variscides. *Journal of the Geological Society, London*, **143**, 141–147.

—— & —— 1986b. A model for the tectonic evolution of South Cornwall. *Journal of the Geological Society, London*, **143**, 125–134.

HOLWILL, F. J. W. 1966. Conglomerates, tuffs and concretionary beds in the Upper Devonian of Waterside cove, near Goodrington Sands, Torbay. *Proceedings of the Ussher Society*, **1**, 238–241.

HOSSACK, J. R. 1984. The geometry of listric growth faults in the Devonian basins of Sunnfjord, W. Norway. *Journal of the Geological Society, London*, **141**, 629–637.

—— & COOPER, M. A. 1986. Collision Tectonics in the Scandinavian Caledonides. *In*: COWARD, M. P. & RIES, A. C. (eds). *Collision Tectonics*. Geological Society, London, Special Publication, **19**, 287–303.

HOUSEMAN, G. A. & ENGLAND, P. C. 1986. Finite strain calculations of continental deformation. I Method and general results for convergent zones. *Journal of Geophysical Research*, **91**, 3651–3663.

HUTTON, D. H. W. 1987. Strike-slip terrain model for the evolution of the British and Irish Caledonides. *Geological Magazine*, **124**, 405–425.

ISAAC, K. P., TURNER, P. J. & STEWART, I. J. 1982. The evolution of the Hercynides of central SW England. *Journal of the Geological Society, London*, **139**, 521–531.

JACKSON, J. A. & WHITE, N. J. 1989. Normal faulting in the upper continental crust: observations from regions of active extension. *Journal of Structural Geology*, **11**, 15–36.

JAEGER, J. C. & COOK, N. G. W. 1979. *Fundamentals of Rock Mechanics*. Chapman & Hill Ltd., London.

JOHNSON, M. R. W. 1967. Mylonite zones and mylonite

banding. *Nature*, **213**, 246–247.

JOHNSON, M. R. W., KELLEY, S. P., OLIVER, G. J. H. & WINTER, D. A. 1985. Thermal effects and timing of thrusting in the Moine thrust zone. *Journal of the Geological Society, London*, **142**, 863–873.

KENOLTY, J., CHADWICK, R. A., BLUNDELL, D. J. & BACON, M. 1981. Deep seismic reflection survey across the Variscan Front of southern England. *Nature*, **293**, 451.

KILENYI, T. & STANLEY, R. 1985. Petroleum prospects in the north-western seaboard of Scotland. *Oil and Gas Journal*, 10–7, 100–108.

KLEMPERER, S. L. 1988. Crustal thinning of the northern North Sea quantified by deep seismic profiling. *Tectonics*, **7**, 803–822.

KUMPAS, M. G. 1984. Seismic interpretation of the Tornquist Zone in Denmark and Sweden. *Geol Foren Stokholm Forh*, **106**, 388–389.

LEEDER, M. R. 1982. Upper Palaeozoic Basins of the British Isles–Caledonian inheritance versus Hercynian plate margin processes. *Journal of the Geological Society, London*, **139**, 479–491.

LIBORIUSSEN, J., ASHTON, P. & THYGESEN, T. 1987. The evoution of the Fennoscandian border in Denmark. *Tectonophysics*, **132**, 21–29.

MCCLELLAND-BROWN, E. 1983. Palaeomagnetic studies of fold development and propagation in the Pembrokeshire Old Red Sandstone. *Tectonophysics*, **98**, 131–149.

MCGEARY, S. 1987. Non-typical BIRPS on the margin of the northern North sea: The SHET Survey. *Geophysical Journal of the Royal Astronomical Society*, **89**, 231–238.

— & WARNER, M. R. 1985. Seismic profiling the continental lithosphere. *Nature*, **317**, 795–797.

MCKENZIE, D. 1978. Some remarks on the development of sedimentary basins. *Earth and Planetary Science Letters*, **40**, 25–32.

MCKERROW, W. S. 1987. The Southern Uplands Controversy. *Journal of the Geological Society, London*, **144**, 735–736.

——, LEGGETT, J. K. & EALES, M. H. 1977. Imbricate thrust model of the Southern Uplands of Scotland. *Nature*, **267**, 237–239.

MEISSNER, R., BARELSEN, H. & MURAWSKI, H. 1981. Thin skinned tectonics in the northern Rhenish massif, Germany. *Nature*, **290**, 399–401.

MOORBATH, S. 1969. Evidence for the age of deposition of the Torridonian sediments of north-west Scotland. *Scottish Journal of Geology*, **5**, 154–70.

—— & PARK, R. G. 1971. The Lewisian chronology of the southern region of the Scottish mainland. *Scottish Journal of Geology*, **8**, 51–74.

——, WELKE, H. & GALE, N. H. 1969. The significance of lead isotope studies in ancient, high-grade, metamorphic basement complexes as exemplifited by the Lewisian rocks of north-west Scotland. *Earth and Planetary Science Letters*, **6**, 245–256.

MOORE, J. MCM. 1975. A mechanical interpretation of the vein and dyke systems of the SW England Orefield. *Mineralia Deposita*, **10**, 374–388.

MORRIS, W. A. 1976. Transcurrent motion determined palaeomagnetically in the northern Appalachians and the Caledonides and the Acadian Orogeny. *Canadian Journal of Earth Sciences*, **13**, 1236–1243.

MORTON, W. H. & BLACK, R. 1975. Crustal attenuation in Afar. *In*: PILGER, A. & ROSLER, A. (eds.), *Afar Depression in Ethiopia*, Schweizerbart'sche Verlagbuchhandlung Stuttgart, 55–65.

MYKURA, W. 1976. *Orkney and Shetland*. British Regional Geology, Institute of Geological Sciences, HMSO, Edinburgh.

NICHOLSON, R. 1979. Caledonian correlations: Britain and Scandinavia. *In*: HARRIS, A. L., HOLLAND, C. H. & LEAKE, B. E. (EDS.), *The Caledonides of the British Isles – Reviewed*, Geological Society, London, Special Publication, **8**, 3–18.

NORTON, M. G. 1986. Late Caledonian extension in western Norway: a response to extreme crustal thickening. *Tectonics*, **5**, 195–204.

O'NIONS, R. K., HAMILTON, P. J. & HOOKER, P. J. 1983. A Nd isotope study of sediments related to crustal development in the British Isles. *Earth and Planetary Science Letters*, **63**, 229–240.

ORD, D. M., CLEMMEY, H. & LEEDER, M. R. 1988. Interaction between faulting and sedimentation during Dinantian extension of the Solway basin, SW Scotland. *Journal of the Geological Society, London*, **145**, 249–259.

PARK, R. G. 1970. Observations on Lewisian chronology. *Scottish Journal of Geology*, **6**, 229–240.

PEGRUM, R. M. 1984. Structural development of the south-western margin of the Russian-Fennoscandian Platform. *In*: SPENCER, A. M. *et al.* (eds.), *Petroleum Geology of the North European Margin*, Norwegian Petroleum Society – Graham & Trotman, London, 350–369.

PINET, B., MONTADERT, L., MASCLE, A., CAZES, M. & BOIS, C. 1987. New insights on the structure and formation of sedimentary basins from deep seismic reflection profiling in Western Europe. *In*: BROOKS, J. & GLENNIE, K. (eds.), *Petroleum Geology of North West Europe*, Graham & Trotman, London, 11–31.

POWELL, C. M. 1989. Structural controls on Palaeozoic basin evolution and inversion in south-west Wales. *Journal of the Geological Society, London*, **146**, 439–46.

PIASECKI, M. A. J. & VAN BREEMEN, O. 1979. The "Central Highland Granulites": cover basement tectonics in the Moine. *In*: HARRIS, A. L., HOLLAND, C. H. & LEAKE, B. E. (eds.), *The Caledonides of the British Isles – Reviewed*, Geological Society, London, Special Publication, **8**, 139–144.

RAMSBOTTOM, W. H. C. 1974. Dinantian. *In:* RAYNER, D. H. & HEMMINGWAY, J. E. (eds.), *The Geology and Mineral Resources of Yorkshire*, Yorkshire Geological Society, 47–73.

ROBERTS, J. L. & TREAGUS, J. E. 1979. Stratigraphic and structural correlations between the Dalradian rocks of the SW and Central Highlands of Scotland. *In*: HARRIS, A. L., HOLLAND, C. H. & LEAKE, B. E. (eds.), *The Caledonides of the British Isles – Reviewed*, Geological Society,

London, Special Publication, **8**, 199–204.

ROGERS, D. A., MARSHALL, J. E. A. & ASTIN, T. R. 1989. Devonian and later movements on the Great Glen fault system, Scotland. *Journal of the Geological Society, London*, **146**, 369–372.

ROYDEN, L. H. & BURCHFIEL, B. C. 1987. Thin-skinned N–S extension within the convergent Himalayan region: gravitational collapse of a Miocene topographic front. *In*: COWARD, M. P., DEWEY, J. F. & HANCOCK, P. (eds.), *Continental Extension Tectonics*, Geological Society, London, Special Publication, **28**, 611–619.

SANDERSON, D. J. 1984. Structural variations across the northern margins of the Variscides in NW Europe. *In*: HUTTON, D. H. W. & SANDERSON, D. J. (eds.), *Variscan tectonics of the North Atlantic region*, Special Publication of the Geological Society, London, **14**, 149–165.

SERANNE, M. 1988. *Tectonique des Bassins Devoniens de Norvège: Mise en évidence de bassins sédimentaire en extension formes par amincissement d'une croûte orogénique épaisse*. Thèse de Doctorat, Université des Sciences et Techniques du Languedoc, Montpellier.

—— & SEGURET, M. 1987. The Devonian basins of western Norway, tectonics and kinematics of an extending crust. *In*: COWARD, M. P., DEWEY, J. F. & HANCOCK, P. (eds.), *Continental Extension Tectonics,* Geological Society, London, Special Publication, **28**, 537–548.

——, CHAUVET, A., SEGURET, M. & BRUNEL, M. 1989. Tectonics of the Devonian collapse basins of western Norway. Bulletin of the Geological Society, France, (8), t. v.

SHACKLETON, R. M. 1958. Downward facing structures of the Highland Border. *Quarterly Journal of the Geological Society, London*, **113**, 361–392.

——, RIES, A. C. & COWARD, M. P. 1982. An interpretation of the Variscan structures in SW England. *Journal of the Geological Society, London,* 139, 533–541.

SIBSON, R. H. 1977. Fault rocks and fault mechanisms. *Journal of the Geological Society, London*, **133**, 191–213.

SMALLEY, S. & WESTBROOK, G. K. 1982. Geophysical evidence concerning the southern boundary of the London Platform beneath the Hog's Back, Surrey, *Journal of the Geological Society, London*, **139**, 139–146.

SMITH, D. I. & WATSON, J. V. 1983. Scale and timing on the Great Glen Fault, Scotland, *Geology*, **11**, 523–526.

SONDER, L. J., ENGLAND, P. C., WERNICKE, B. P. & CHRISTIANSEN, R. L. A physical model for Cenozoic extension of western North America. *In*: COWARD, M. P., DEWEY, J. F. & HANCOCK, P. (eds.), *Continental Extension Tectonics*, Special Publication of the Geological Society, London, **28**, 187–201.

SOPER, N. J. & ANDERTON, 1984. Did the Dalradian slides originate as extensional faults? *Nature*, **307**, 357–359.

—— & HUTTON, D. H. W. 1984. Late Caledonian sinistral displacements in Britain: implications for a three-plate collisional model. *Tectonics*, **3**, 781–794.

STEWART, S. 1982. Late Proterozoic rifting in NW Scotland: the genesis of the "Torridonian". *Journal of the Geological Society, London*, **139**, 415–422.

STRACHAN, R. A., TRELOAR, P. J., BROWN, M. & D'LEMOS, R. S. 1989. Cadomian terrane tectonics and magmatism in the Armorican Massif. *Journal of the Geological Society, London*, **146**, 423–426.

STYLES, M. T. & RUNDLE, C. C. 1984. The Rb–Sr isochron age of the Kennack Gneiss and its bearing on the age of the Lizard Complex, Cornwall. *Journal of the Geological Society, London*, **141**, 15–19.

SUTTON, J. & WATSON, J. V. 1951. The pre-Torridonian history of the Loch Torridon and the Scourie areas in the North-west Highlands and its bearing on the chronological classification of the Lewisian. *Quarterly Journal of the Geological Society, London*, **106**, 241–296.

TAYLOR, J. C. M. 1986. Gas prospects in the Variscan Thrust Province of southern England. *In*: BROOKS, J., GOFF, J. & VAN HOORNE, B. (eds.), *Habitat of Palaeozoic Gas in N.W. Europe*, Special Publication of the Geological Society, London, **23**, 37–54.

TROTTER, F. M. 1949. The devolatilization of coal seams in South Wales. *Quarterly Journal of the Geological Society, London*, **104**, 387–419.

VAN DER VOO, R. & SCOTESE, C. 1982. Palaeomagnetic evidence for a large (~2000 km) sinistral offset along the Great Glen Fault during Carboniferous time. *Geology*, **9**, 583–589.

WATSON, J. V. 1985. Northern Scotland as an Atlantic–North Sea divide. *Journal of the Geological Society, London*, **142**, 221–243.

—— & DUNNING, F. W. 1979. Basement – cover relationships in the British Caledonides. *In*: HARRIS, A. L., HOLLAND, C. H. & LEAKE, B. E. (eds.), *The Caledonides of the British Isles – Reviewed*, Special Publication of the Geological Society, London, **8**, 67–91.

WELLS, P. R. A. & RICHARDSON, S. W. 1979. Thermal evolution of metamorphic rocks in the Central Highlands of Scotland. *In*: HARRIS, A. L., HOLLAND, C. H. & LEAKE, B. E. (eds.), *The Caledonides of the British Isles – Reviewed*, Geological Society, London, Special Publication, 8, 339–344.

WHITE, S. H. & GLASSER, J. 1987. The Outer Hebrides Fault Zone: evidence for normal movements. *In*: PARK, R. G. & TARNEY, J. (eds.), *Evolution of Lewisian and Comparable Pre-Cambrian High Grade Terrains*, **27**, 175–183.

WILLIAMS, G. D. & CHAPMAN, T. J. 1986. The Bristol – Mendip foreland thrust belt. *Journal of the Geological Society, London*, **143**, 63–73.

WILLIAMS, G. E. 1969. Characteristics and origin of a Pre-Cambrian pediment. *Journal of Geology, Chicago*, **88**, 1–14.

ZIEGLER, P. 1982. *Geological Atlas of Western and Central Europe*, Elsevier, Amsterdam.

Evidence for global (glacial-eustatic) control over upper Carboniferous (Pennsylvanian) cyclothems in midcontinent North America

PHILIP H. HECKEL

Department of Geology, University of Iowa, Iowa City, IA 52242, USA

Abstract: At least 20 Middle/Upper Pennsylvanian (Westphalian D/Stephanian A, B) major marine cyclothems, each consisting of a thin transgressive limestone, thin offshore dark phosphatic shale, and thick regressive limestone member, extend along outcrop for 600 km from the northern shelf region of Iowa, Missouri and Kansas, to the basinal region of central Oklahoma. Each is correlated lithostratigraphically throughout a gridwork of long cores in the northern region beneath Pleistocene cover, and in good exposures southward. This correlation is confirmed biostratigraphically, along the entire outcrop belt using a combination of conodonts, fusulinids and ammonoids. Most of the cyclothems are separated by thin terrestrial deposits with paleosols, but only rare deltas, for about the northern half of the outcrop distance, which rules out delta shifting as a general control for the vertical alternation of terrestrial and marine deposits across the broad shelf. On the northern shelf, all major cyclothems are traced across the cratonic Forest City basin and over the adjacent Nemaha uplift with little change, which rules out local differential tectonics as a general cause. Presence of Gondwanan glacial deposits at this time, in conjunction with the estimated frequencies of these Pennsylvanian cycles within the Milankovitch band of Earth's orbital parameters (which control variation in solar heating of the mid-latitudes), indicate that glacial eustasy, rather than distant orogenic movements, was the main control over the cyclothems.

The cyclic alternation of limestone and shale formations that dominates the Middle and Upper Pennsylvanian (Silesian) sequence (Fig. 1) along the Midcontinent outcrop belt (Fig. 2) has intrigued geologists ever since Weller (1930) and Moore (1931) first described it, and Wanless & Weller (1932) applied the term cyclothem to the component unit of repeating rock types. Weller (1930) invoked a model of periodic tectonism to explain both the overall alternation and the individual cyclothem. In contrast, Wanless & Shepard (1936) related both these features to eustatic changes in sea level brought about by waxing and waning of Gondwanan ice caps. More recently, autocyclic models of delta-shifting, first suggested by D. Moore (1959) in the Namurian of England, have been applied to cyclic sequences in the Appalachians (Ferm 1970) and Texas (Galloway & Brown 1973). In the meantime, Wanless (1964, 1967) suggested that the glacial eustatic model readily accommodates delta shifting as a mechanism to explain otherwise anomalous clastic wedges in many of the cyclothems. This view has been more fully developed by Heckel (1977, 1980, 1984a), who recognized the cyclothems as marine transgressive–regressive sequences, centred on the thin, non-sandy, black phosphatic ('core') shales, which represent maximum inundation of the shelf, and with most deltas forming during the succeeding regressive phases. The eustatic model has been applied to the sequence in Texas by Boardman & Malinky (1985) and Boardman & Heckel (1989).

Editors' note. The presence of a paper concerning North America in a volume dedicated specifically to tectonics in Britain may cause some raised eyebrows. The organizing committee specifically invited this paper in order to indicate the possibility that not all Britain's oil and gas reserves were due to tectonic movements. We felt that it would sound a cautionary note and allow us to approach the link of hydrocarbon reserves to tectonics with a more open mind.

Basic cyclothem

Although the term cyclothem has been applied to a number of different repeating lithic successions in the Pennsylvanian (Moore 1936, 1950; Weller 1958; see review in Heckel 1984b), current work on the Midcontinent sequence has established the basic transgressive–regressive cyclothem that characterises the sequence across the northern Midcontinent shelf from Kansas to Iowa (Fig. 2). In ascending order, this cyclothem (Fig. 3) consists of the following members, which resulted from a major

From Hardman, R. F. P. & Brooks, J. (eds), 1990, *Tectonic Events Responsible for Britain's Oil and Gas Reserves*, Geological Society Special Publication No 55, pp 35–47.

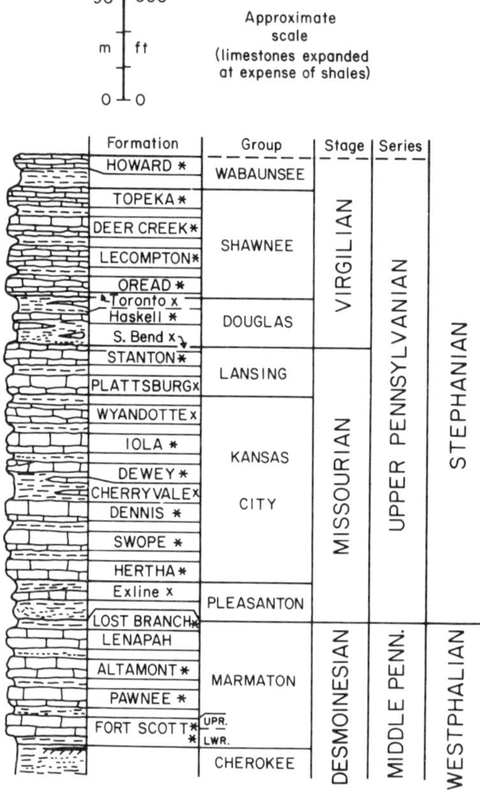

Fig. 1. Part of Middle-Upper Pennsylvanian (Westphalian D–Stephanian A, B) stratigraphic sequence along outcrop belt of Northern Midcontinent Shelf from Iowa to Kansas. Capital names in formation column denote marine, mostly limestone formations that include a major widespread cyclothem or intermediate transgressive–regressive cycle of deposition. Names in lower case denote members that represent mainly intermediate or minor cycles of deposition, not all of which are shown (see Fig. 7). * denotes presence of black phosphatic shale facies in offshore deposits across much of northern shelf, which helps to define that unit as a major cyclothem. x denotes presence of grey conodont-rich shale. Names of nearshore to terrestrial shale formations that separate limestone formations were omitted because of lack of space, but most are shown on Fig. 7.

rise and fall of sea level over the entire shelf, as interpreted on the basis of the following characteristics.

Transgressive limestone

The transgressive limestone, deposited in deepening water, is typically a thin (<1 m), marine skeletal calcilutite deposited below effective wave base during later transgression, but it locally includes calcarenites at the base deposited in shallower water during earlier transgression. The calcilutites typically are dark, dense and non-pelleted, with neomorphosed aragonite grains, and the calcarenites typically are overpacked and lack evidence of early marine cementation or meteoric leaching and cementation. This is because both facies remained in the marine phreatic environment of deposition until buried by higher marine strata of the cyclothem, which acted as a barrier to meteoric diagenesis. Thus, they underwent slow compaction before cementation, often by ferroan carbonates in a decreasingly oxygenated burial environment (Heckel 1983), in which much of the fine organic matter also became preserved in the rock.

Offshore ('core') shale

The offshore or 'core' shale, formed at maximum transgression, is typically a thin (<1 m), non-sandy, gray to black phosphatic shale deposited under conditions of near sediment starvation. In most cyclothems the water became deep enough for a thermocline to develop over much of the northern Midcontinent shelf. The thermocline reduced bottom-oxygen replenishment over shallower areas just enough to produce the gray dysoxic facies with only low-oxygen tolerant benthic invertebrates, such as certain crinoids and brachiopods (e.g. *Crurithyris*). It eliminated bottom oxygen over the deeper areas to produce the black anoxic facies with only pelagic fossil remains such as conodonts, fish debris, and under certain preservational conditions, radiolarians and ammonoids.

Sedimentation was so slow at this time in the northern Midcontinent that in the gray facies, aragonite fossils apparently were dissolved and calcite fossils were locally corroded (Malinky 1984). Evidence for this lies in the appearance of great numbers of originally aragonitic molluscs (snails, clams, ammonoids: see Boardman *et al.* 1984) now preserved as siderite, pyrite, or phosphorite in thicker developments of these gray shales in southern Kansas and Oklahoma. Ammonoids also are preserved locally in early diagenetic carbonate nodules ('bullion') in the black facies in Kansas and Missouri. In the first case, early rapid burial and mineralization, and in the second case early matrix mineralization, prevented sea-floor dissolution of these fossils in the colder, undersaturated waters below the top of the thermocline.

Quasi-estuarine circulation and upwelling associated with the thermocline in this deeper

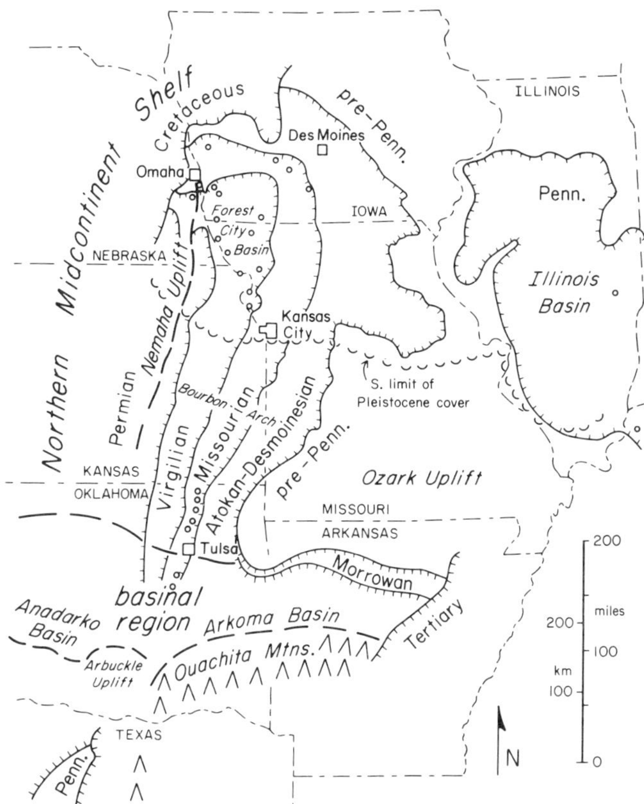

Fig. 2. Midcontinent Pennsylvanian outcrop belt with hachures in direction of dip, showing relation to generalized Pennsylvanian structural features. Nemaha Uplift, Bourbon Arch, and Forest City Basin were formed during Early Pennsylvanian time. Although Forest City Basin became largely filled by end of Middle Pennsylvanian (Desmoinesian) time, and entire region north of basinal region of Oklahoma acted as Northern Midcontinent Shelf, Nemaha Uplift and Bourbon Arch continued to exert subtle influence on depositional facies during later Pennsylvanian time. Shelf-basin boundary in Oklahoma shifted position through time. Circles denote locations of long cores.

phase of the shelf sea caused deposition in both the gray and black shale facies of non-skeletal phosphorite as peloids, laminae, and nodules. These are analogous to modern phosphorite nodules forming under similar conditions of periodic upwelling associated with a thermocline, in low-oxygen sediment on the offshore shelf along the coast of Peru (Kidder 1985).

Regressive limestone

The regressive limestone, deposited in shallowing water, is typically a thick (2–10 m), marine skeletal calcilutite deposited below wave base, grading upward into cross-bedded skeletal calcarenite, with algae and locally oolitic abraded grains, deposited above wave base in shallow water. In some cyclothems a distinctly different calcarenite appears at the base, associated with the offshore shale. This calcarenite contains only invertebrates (crinoids, brachiopods, bryozoans and encrusting foraminifers), often shows evidence of grain corrosion, and lacks any evidence of algae, grain abrasion, or cross bedding. It therefore must have formed below effective wave base and probably below effective photic base for the algae in this sea. It represents proliferation of invertebrates in deeper water as the thermocline weakened and sufficient oxygen returned to the bottom. Its overcompacted nature resulted from relatively deep burial before cementation, as in the transgressive limestone (Heckel 1983).

The tops of most regressive limestones, particularly northward, display sparsely fossiliferous, laminated to birdseye-bearing, lagoonal to

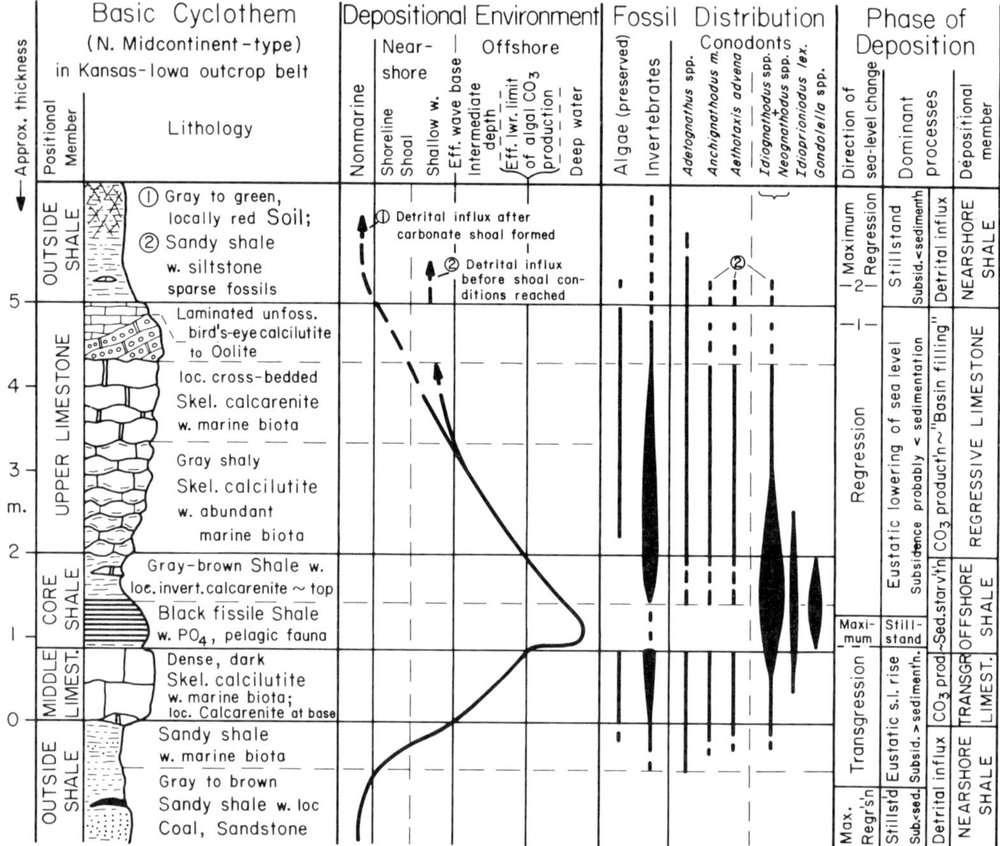

Fig. 3. Basic Midcontinent (Kansas) cyclothem characterizing with minor modification all major and many intermediate marine cycles of deposition across Northern Midcontinent Shelf, and representing one sufficiently slow and extensive inundation and withdrawal of the sea to produce both transgressive and regressive limestones and intervening black phosphatic shale facies across most of shelf. Phases of deposition reflect ranges of sea-level stand. Conodont information derived from Heckel & Baesemann (1975) and Swade (1985) supplemented by thesis and other unpublished data.

peritidal carbonates that represent passage of the strandline toward the basin during later regression. The tops of most regressive limestones also display subaerial exposure surfaces with such features as pitting, cracking, 'cryptokarst', and solution-tubes formed from plant rooting and infiltration of meteoric water, clay, and terrestrial organic matter. In local areas where meteoric water became saturated, laminar and pisolitic caliche formed on the surface of the limestone. Infiltration of oxygenating, undersaturated meteoric water farther down into the regressive limestone before much compaction took place also oxidized most of the original organic matter in the sediment, leached the aragonitic grains, and eventually became saturated enough to precipitate blocky calcite in both intergranular and moldic voids. This preserved original peloidal fabric, depositional packing of grains, and also porosity where cementation was incomplete. Thus the lighter-colored, more porous and more conspicuously sparry, upper regressive limestones stand in contrast with the darker, denser, overcompacted transgressive limestones and the lower, more offshore facies of the regressive limestones (Heckel 1983).

Nearshore ('outside') shale

The nearshore ('outside') shales (which lie outside the limestone formations comprising the three previously described marine members of the cyclothem) encompass a great variety of nearshore marine and terrestrial deposits on the shelf, all deposited at lower stands of sea level.

They include thick (up to 30 m) sparsely fossiliferous prodeltaic shales, which prograded out over regressive limestones during later stages of sea-level fall, particularly from Kansas City southward. In places these grade upward into delta-front and delta-plain sandstones and coals.

Outside shales also include thinner blocky mudstones, which range from grey to red in colour, from 0.2 to 2 m in thickness, and typically overlie exposure surfaces on regressive limestones. Some of those that have been studied in detail show upward decrease in crystallinity of illite and upward increase in mixed-layer and kaolinite proportions (Schutter & Heckel 1985), which are characteristics that suggest a weathered soil profile. Others display a number of micromorphological features of paleosols such as neostrians, argillans, skeletans, and various plasmic fabrics (Goebel *et al.* 1989). The blocky fabric of the palaeosol mudstone that is conspicuous on outcrop is a result of disturbance of the normally flat-lying clay minerals (characteristic of marine shales) by plant rooting, animal burrowing, and rain-water illuviation of clay minerals into holes formed by the organic agents and into cracks formed by periodic desiccation. In addition, these mudstones often contain irregular carbonate nodules, that show internal clotted, mottled, and cracked fabric characteristic of caliche (Prather 1985).

The blocky mudstones at several horizons in the sequence are overlain by the most widespread and thickest coals in the Midcontinent Pennsylvanian. These coals apparently formed in response to the early stages of sea-level rise of the succeeding transgression, which ponded fresh-water run-off to form broad swamps on the surface of low relief. These swamps then migrated shelfward ahead of the transgression. Coals at this position are particularly characteristic of the Middle Pennsylvanian (Desmoinesian) sequence, when the overall climate was wetter than during the Upper Pennsylvanian (Missourian) (Schutter & Heckel 1985).

Conodont information

The succession of conodont faunas reported at closely spaced intervals of all lithic units in outcrop and core sequences, by Heckel & Baesemann (1975) and Swade (1985), established a distinctive vertical succession of conodont genera that characterizes all the major Kansas cyclothems (Fig. 3) from the northern limit of outcrop to the basinal region of Oklahoma. Both grey and black facies of the slowly deposited offshore shales are characterized by high abundance of conodonts, hundreds to thousands per kilogram. These faunas are strongly dominated by *Idiognathodus* (includes *Streptognathodus*), with common *Neognathodus* (only in the Desmoinesian), *Idioprioniodus*, and, usually confined to the middle of the shale, *Gondolella*. In contrast, the rapidly deposited nearshore marine portions of the outside shales are characterized by low abundance of conodonts, from a few up to twenty or so per kilogram. These faunas are typically dominated by *Adetognathus*, sometimes subequally with *Idiognathodus* and *Anchignathodus*, but with *Idioprioniodus* and *Gondolella* conspicuously absent, and *Neognathodus* (Desmoinesian only) quite rare. The two limestone members contain faunas that are gradational and intermediate between those of the adjacent shale members, and these limestone faunas tend to be mirror images of one another, symmetrical about the 'core' shale that separates them (Fig. 3).

The distinctive differences in conodont faunas between offshore and nearshore parts of the cyclothems appear related to characteristics of the water masses that covered the shelf at different sea-level stands (Swade 1985). *Idiognathodus* (with *Neognathodus* in the Desmoinesian) apparently inhabited the normal, open marine, warm surface-water mass (Fig. 4), which covered most of the sea away from strong fresh-water influx, at all sea-level stands. *Idioprioniodus* probably occupied the lower, cooler water mass in the top of the thermocline, and therefore is found mainly in the offshore shale and adjacent deeper-water parts of the limestone members. *Gondolella* apparently lived in deeper, even cooler and possibly somewhat dysoxic water lower in the thermocline, and thus is more completely confined to the most offshore facies. These four genera probably were pelagic, as all are found in good numbers in the anoxic black shale facies, which lacks any definitely benthic fossils. At the other extreme, *Adetognathus* apparently inhabited the variable nearshore water mass, where it was tolerant of fluctuation in salinity and other conditions. *Anchignathodus* probably occupied a slightly more offshore environment associated with carbonate sediment, as it is most commonly found in the limestone members. Both these latter genera may have been benthic, as they are not found in the anoxic black shales.

Because these distinctive patterns in abundance and generic composition of conodont faunas are reasonably related to depositional environment and phase of deposition within the cyclothem, they can be used to identify the position of strata within the overall trans-

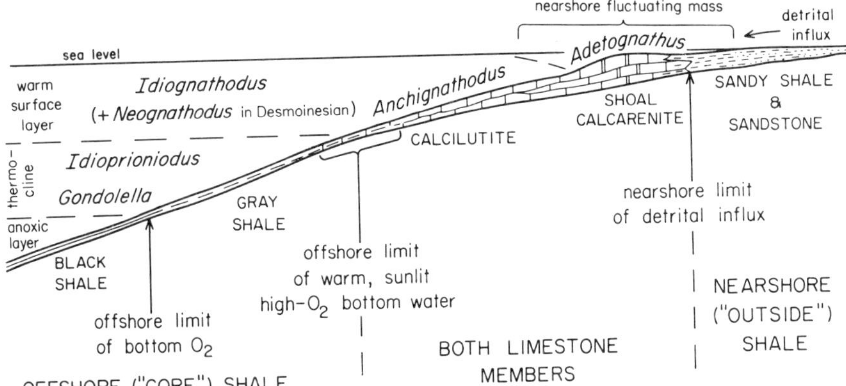

Fig. 4. Inferred living distribution of six major conodont genera in Midcontinent Pennsylvanian sea relative to major water masses developed at maximum transgression. Genus *Idiognathodus* here includes *Streptognathodus* of other authors. Transgression and regression of sea caused laterally equivalent rock types and various settled conodonts to become superposed to produce Kansas cyclothem (Fig. 3). Modified from Heckel & Baesemann (1975, p. 499) and Swade (1985, p. 50).

gressive–regressive sequence where the lithic sequence is ambiguous (e.g. entirely skeletal calcilutite or marine shale). In this way, thin grey offshore shales can be distinguished from thin grey nearshore shales that resulted from a single pulse of sediment influx; transgressive limestones can be distinguished from regressive limestones in areas where adjacent shales are unexposed; black offshore shales can be distinguished from black nearshore lagoonal shales; and black offshore shales can be traced laterally into grey offshore shales. The lateral tracing of the conodont-rich, sediment-starved, offshore 'core' horizons of the cyclothems has greatly helped to sort out the stratigraphy and environmental interpretation in areas of abrupt carbonate facies changes and thick enigmatic shale sequences, as is common in southern Kansas and Oklahoma and in most of the Texas sequence.

More recently, the species composition of the faunas of the offshore shales are being worked out (Swade 1985; Heckel 1990; Barrick & Boardman 1989). Discrimination of species among the most abundant genus *Idiognathodus*, supplemented at certain horizons by distinctive species of *Gondolella*, is establishing a biostratigraphic succession of conodonts within the offshore biofacies. Combining this with the succession of ammonoid faunas in the thicker offshore shales and of fusulinid faunas in the limestones is facilitating the correlation of each individual cyclothem across the Midcontinent (Figs 5 & 6) and with the Texas sequence (Fig. 7).

Other cycles

Because most interest has been focused on the classic cyclothem sequences, the portions of the sequence that do not fit readily into the basic pattern are only now receiving closer attention. In order for a cycle of transgression and regression of the sea to produce a classic Midcontinent (Kansas) cyclothem (Fig. 3), the inundation must have been slow enough to develop limestone over most of the shelf, and sufficiently extensive onto the shelf to become deep enough to establishe a thermocline and thus the black phosphatic facies over much of the shelf; furthermore, the withdrawal must have been slow enough and sufficiently free of detrital influx to develop limestone over most of the shelf, and it must have extended far enough basinward for either soils to form or for terrigenous detrital rocks eventually to cover much of the shelf. A marine horizon that lacks one or more characteristics of the basic Kansas cyclothem would have resulted if: (1) the inundation were too fast to allow carbonate formation; (2) the inundation were not far enough to become sufficiently deep to establish a thermocline for black shale formation; (3) the withdrawal were too fast to allow carbonate formation; (4) the withdrawal were not far enough to form a complete regressive sequence or (5) the withdrawal occurred during a time of overwhelming detrital influx.

Cycles have been classified into three categories for the purpose of further analysis, based largely on point (2) above, extent of trans-

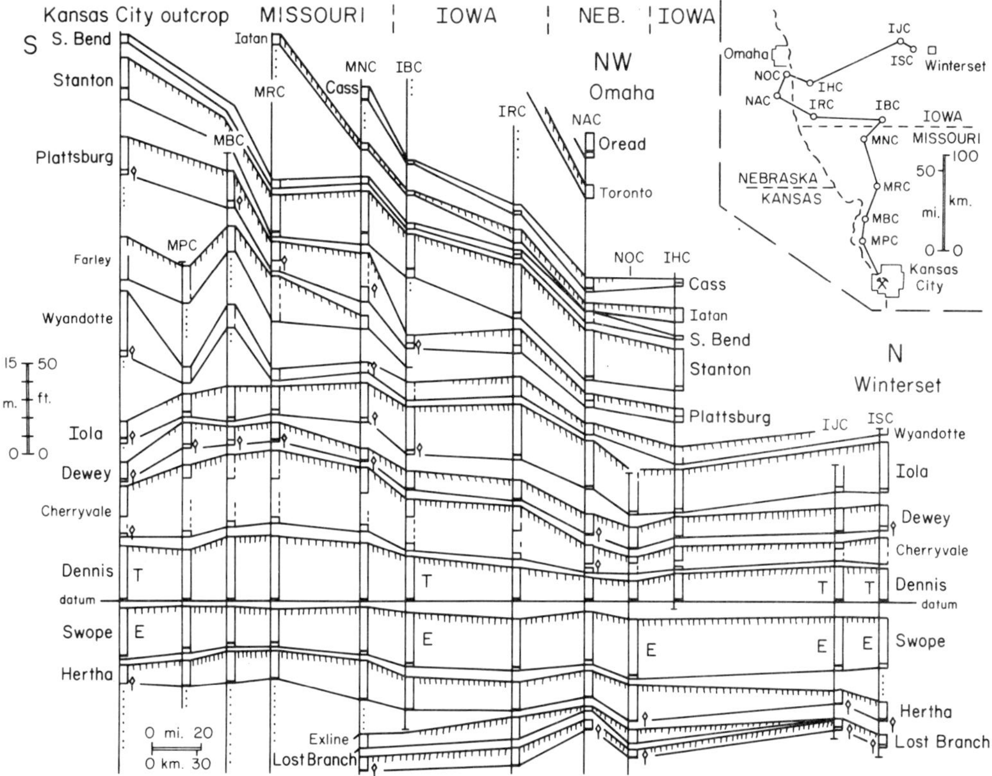

Fig. 5. Correlation cross section of lower part of Upper Pennsylvanian sequence (Missourian Stage = Exline–Iatan) on northern Midcontinent shelf from Kansas City to central Iowa, based on long cores held by respective state geological surveys. Named units are limestone formations; those labeled with largest letters are major marine cyclothems (with black lines near base representing black phosphatic shale); those with smaller letters are intermediate or minor marine cycles. Limestones are separated by shale formations (left unlabeled), which are mostly palaeosol mudstones where thin (shown by hachures on top of underlying regressive limestones). Shales contain rare deltas (dots show sandstones) only where thick (mainly below Hertha and above Iola cyclothems in northern Missouri). Tailed diamond symbols for conodont faunas and letters for fusulinids (E = *Eowaeringella ultimata*; T = lowest *Triticites*) show biostratigraphic control for correlation.

gression onto the shelf. Major cycles are those inundations far and deep enough onto the shelf to form a conodont-rich shale to the northern limit of outcrop in Nebraska and Iowa and generally develop enough of the other facies to be recognized as cyclothems over much of their extent. Intermediate cycles extend as marine horizons into Iowa and Nebraska, but carry conodont-rich horizons only on the lower shelf, and most have been recognized only controversially in some places as cyclothems or parts of them. Minor cycles typically extend as marine horizons only a short distance from the basinal region of Oklahoma into Kansas or Missouri, or represent a minor reversal within a more major cycle, and have not generally been recognized as cyclothems nor in some cases been named as separate units. The recognition of minor eustatic cycles requires correlation of the horizon over a reasonable geographic area, because in any one section or small geographic area, autocyclic processes such as delta shifting can produce minor cycles of deposition.

Possible controlling factors

The possible different factors that ultimately controlled formation of the Pennsylvanian cyclothems of Midcontinent North America have historically been recognized either as tectonic (Weller 1930, 1956), glacial eustatic (Wanless & Shepard 1936), or autocyclic, essentially delta shifting where marine deposits are involved (Ferm 1970; Galloway & Brown 1973). Delta

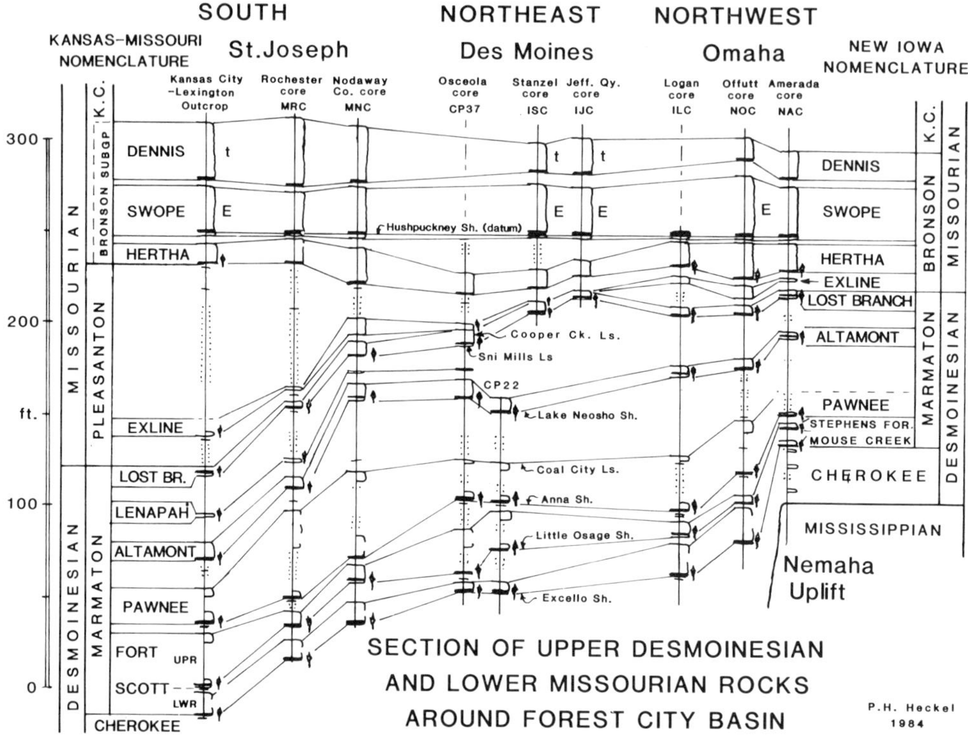

Fig. 6. Correlation cross section of Upper Desmoinesian and lower Missourian sequence on northern Midcontinent shelf from Kansas City through Iowa to Nebraska, based on long cores (see Fig. 2, 5) held by respective state geological surveys. Named units are limestone formations. Those with thick black lines at base (representing black phosphatic shales) are major marine cyclothems. Other units (e.g. top of upper Fort Scott; Coal City Ls.; Lenapah Ls.; Exline Ls.) are intermediate cycles. Fluvial and deltaic sandstones (shown by dots) are much more common in Desmoinesian and lowermost Missourian when climate was more humid than in rest of Missourian (Schutter & Heckel 1985), but marine horizons maintain lateral continuity between them. Moreover, all major cyclothems extend across Nemaha uplift (core NAC), an early Pennsylvanian structure that persisted as a topographic feature throughout Desmoinesian, as shown by thickening of Desmoinesian units away from it, and by presence of gray instead of black phosphatic conodont-rich core shales in Desmoinesian cyclothems on top of it. Tailed diamond symbols for condodont faunas and letters for fusulinids (same as on Fig. 5) show biostratigraphic control for correlation.

shifting is a local process that requires the presence of deltas throughout the vertical and lateral extent of the cyclic sequence and results in stratigraphic units with limited lateral extent. Glacial eustasy is a global process that requires the presence of large ice caps that wax and wane in the higher latitudes, and can result in distinct stratigraphic units of extremely widespread extent in stable cratonic areas and of potential correlatability on a global scale. Tectonic controls either can be local variations in uplift or subsidence, which would result in units of lateral extent limited by tectonic features, or they can be the more widespread effects of large-scale movements in major orogenic belts (as has been more recently developed in the Appalachians by Tankard (1986)), which could result in units of more widespread extent and potential correlatability.

Delta-shifting, as the basic control over the major Midcontinent cyclothems, can be readily ruled out on several counts. First, the extremely widespread extent of each of these laterally continuous major marine transgressive–regressive horizons (Fig. 5) covers a minimum area of remaining outcrop today of roughly half of the states of Iowa, Missouri, Nebraska, and Oklahoma, and all of Kansas, totalling perhaps 500 000 km^2, compared to a generous estimate of perhaps 5000 km^2 for the larger individual delta lobes (Gould 1970, p 9) in the Holocene deposits of the Mississippi River (one of the

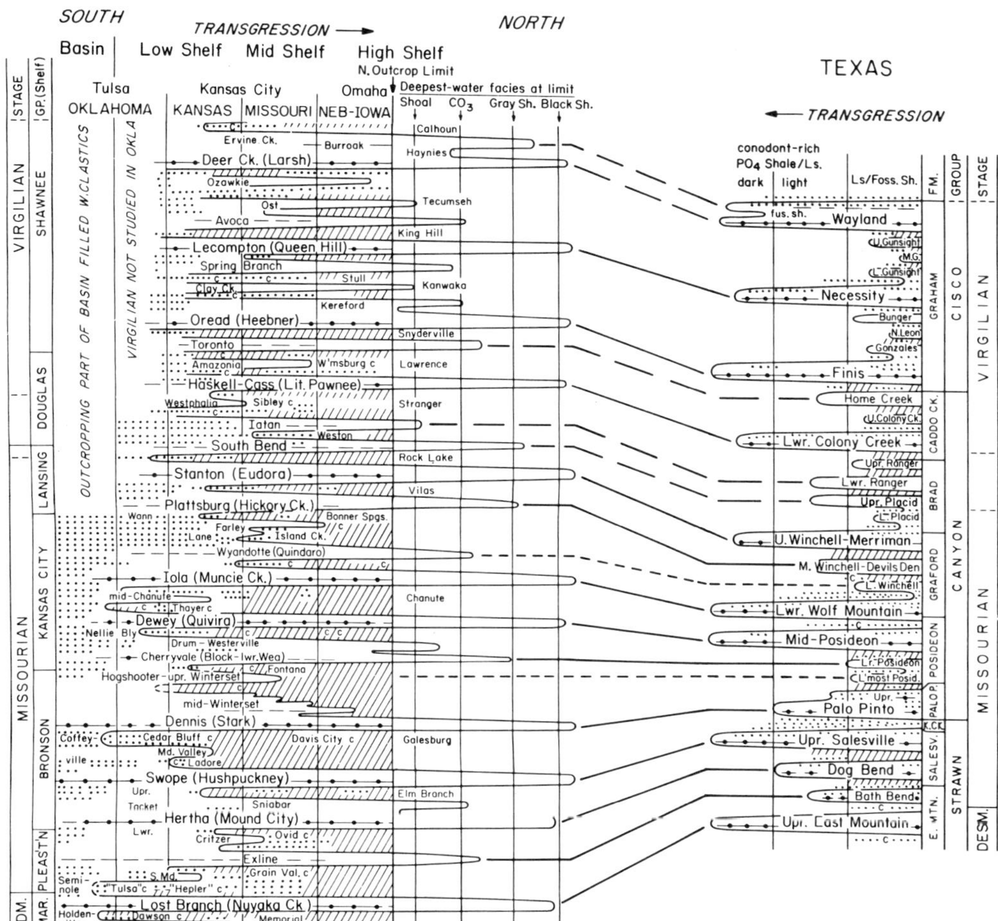

Fig. 7. Eustatic sea-level curve for part of Pennsylvanian sequence along Midcontinent outcrop (derived from Heckel 1986, as modified by Heckel 1990) and biostratigraphic correlation with curve for north-central Texas outcrop on other side of Arbuckle uplift (Fig. 2). Lowstand deposits (named on right of Midcontinent curve) include exposure surfaces and paleosols (///), fluvial-deltaic complexes (. . .) and coaly beds (c). Marine cycles (named on left side of Midcontinent curve and right side of Texas curve) contain highstand deposits, which include conodont-rich shales (named in parentheses), both black (solid line) and grey (dashed) with phosphate nodules (large dots). Size of letters in names of marine cycles reflects informal rank classification as major, intermediate, or minor. Correlations reflect (1) diagnostic faunas (solid lines) with first, last, sole, or acme occurrence of certain taxa, (2) compatible faunas (long dashes), and (3) positional matches (short dashes in selected examples) (from Boardman & Heckel 1989).

largest and most sediment-rich of modern delta systems). Second, as also illustrated by Fig. 5, deltas are simply lacking over much of the shelf area north of Kansas City, where palaeosols 1 to 2 m thick separate nearly all the Missourian cyclothems over perhaps 200 000 km^2. Third, the phosphatic black shales that mark the maximum inundative phase in the major cyclothems over most of the present outcrop area require thermoclines in minimum water depths on the order of 100 m (Heckel 1977), which is highly unlikely above shifting delta lobes at a fixed sea level stand. Fourth, in order for the well developed paleosols to form, long-term withdrawal of the sea is required; soils on modern active delta lobes are typically so immature as to be barely distinguishable from the original sediment, and their foundering beneath the sea critical to the delta shifting model would inhibit much further soil development after abandonment. Therefore, considering the widespread lateral continuity of both the maximum transgressive phosphatic shales and the maximum regressive palaeosols and exposure surfaces in close vertical alternation, the major Midcontinent cyclothems must have resulted from a significant rise of sea level, perhaps on the order of 200 m. At times the shoreline transgressed from the upper margin of the Anadarko–Arkoma basin of Oklahoma to well beyond the northern limit of outcrop in central Iowa, perhaps as much as 800 km, followed by a major drop in sea level with shoreline withdrawal a similar distance in the opposite direction. Even during late Desmoinesian time when the climate was more humid (Schutter & Heckel 1985) and terrigenous detrital deposits, including deltas, were more common in the cyclic sequence on the shelf north of Kansas City (Fig. 6), the major marine cyclothems with their conodont-rich black shales are still laterally continuous between the detrital units, and palaeosols are well developed on the tops of most of them.

Local differential tectonic movements in which variable rates of down-dropping of a basin adjacent to a faulted uplift could cause periodic transgression in the basin, followed by apparent regression as the basin filled with detritus, can also be ruled out as the control over the major late Desmoinesian and Missourian cyclothems. The main tectonic features in the northern Midcontinent were the faulted Nemaha uplift and adjacent Forest City basin (Fig. 2), which formed early in the Pennsylvanian and strongly affected local sedimentation in the basin through about mid-Desmoinesian time. Minor differential effects on late Desmoinesian sedimentation is shown by thinning of these units over the Nemaha uplift (Fig. 6, core NAC), but each individual major cyclothem (identified by its distinctive conodont fauna) is found within the thinned sequence upon the uplift, indicating that these marine transgressions inundated the uplift as well as the basin. Therefore, the marine inundation must have been controlled by forces other than those that differentially controlled the basin/uplift couplet. The presence of greater remaining topography during late Desmoinesian over the uplift is indicated by the lateral passage from black to grey phosphatic shales in each cyclothem at the locality upon the uplift (Fig. 6, core NAC). The younger Missourian cyclothems, however, show no detectable lateral change as they pass over the uplift, which must have been dormant by that time.

Thus we are left with a eustatic control. This classically has been considered glacial in origin (Wanless & Shepard 1936), which is strongly supported by the long-known presence of Gondwanan glacial deposits, and particularly by the more recent dating of the greatest amount of Gondwanan glaciation during Middle to Late Pennsylvanian time (Veevers & Powell 1987), when the Midcontinent cyclothems were best developed.

But eustatic changes in the Midcontinent could also be tectonic in origin, a result of periodic large movements in a more distant orogenic belt. The mechanism of episodic orogenic thrust loading causing repeated flexural subsidence in the foreland basin and resulting in transgression when the basin was down, followed by regression as it filled with sediment, has been applied to the Appalachian basin by Tankard (1986). It has been suggested to combine progressively with the dominant glacial eustatic control recognized in the Midcontinent to account for progressive differences in the cyclothems through the Illinois basin to the Appalachian basin, by Klein & Willard (1989). This type of mechanism could have had various different effects in distant areas. For example, it could cause transgression if the downwarping were transmitted that far, or conversely, it could cause regression if the foreland downwarp diverted enough marine water into the foreland basin from distant cratonic areas.

In order to discriminate between glacial and distant tectonic controls, the frequencies of the transgressive–regressive events provide important information. Analysis of the probable lengths of time during which each of the Midcontinent marine cycles of transgression and regression took place, based on sets of assumptions explained elsewhere (Heckel 1986), estimated a range of lengths of 235 000 to 400 000 years for the major cyclothems, 120 000 to 220 000 years for the intermediate cycles, and 44 000 to 120 000 years for the minor cycles. Even if the lower ends of the ranges are halved to accommodate the shorter duration of the analyzed sequence suggested by more recent radiometric dating (Leeder 1988), the estimated ranges for all types of cycles still fall within the range of periods of the earth's orbital cycles that constitute the Milankovitch insolation theory of control of the Pleistocene ice ages. These cyclic orbital parameters are: eccentricity, with two dominant periods, one about 413 000 years, and the other ranging from 95 000 to 136 000 years and averaging about 100 000 years; obliquity, with a dominant period near 41 000 years; and precession, with two dominant periods averaging 19 000 and 23 000 years

(Imbrie & Imbrie 1980). Their apparent control over the timing and duration of the Pleistocene ice ages suggests that they probably controlled that of the Pennsylvanian ice ages on Gondwanaland as well. Thus they were at least partly responsible for the frequent rise and fall of the North American Midcontinent Sea. Because the periods and amplitudes of particularly the 100 000-year eccentricity parameter are variable, they cause a somewhat irregular interference and amplification among all parameters throughout the succession of longer periods. Depending upon the exact nature of the linkage or mechanism of control over the waxing and waning of Gondwanan ice caps, this interference probably gave rise to the range of variation observed in the intermediate and the major Midcontinent cyclothems, which represent the strongest, most conspicuous effects of sea-level change during the Pennsylvanian.

It is apparent that only if the frequencies of episodic tectonic flexuring were coincidently within the same range as the Milankovitch cycles, would tectonic controls have likely played a major role in the genesis of Midcontinent cyclothems, considering the fairly well constrained frequencies of the cyclothems within that range. However, the range of periodicities suggested for the earlier Pennsylvanian (late Morrowan to late Atokan) transgressions thought to have resulted from foreland downwarp in the Appalachians is on the order of 3 to 5 million years (F. R. Ettensohn *in* Tankard 1986, p. 866). This is at least an order of magnitude less frequent than those of the Midcontinent cyclothems. Therefore distant orogenic movement of this sort is quite unlikely as an origin for the Midcontinent cyclothems, at this stage of our understanding.

Another line of evidence for glacial control of eustasy lies in the thinness of the algae-bearing transgressive limestone (Fig. 3) overlain by offshore deposits formed below the carbonate-producing warm sunlit surface water layer (Fig. 4). As developed by Heckel (1984a, p. 38), modern rates of shallow-water carbonate production of 0.5 to 3.0 mm/year would need to be outstripped by sea level rise in order to produce such a thin unit consisting of the material (algal blades and mud of presumably algal origin) that typically is involved in optimal carbonate production. Estimated rates of post-Wisconsinan (Flandrian) sea level rise at times between 10 000 and 7000 years B.P. range from 5 to 35 mm/year, which is sufficient to deepen the water fast enough to accomplish this. In contrast, modern basinal subsidence rates of 0.3 to 2.5 mm/year would essentially be accommodated by carbonate production to produce a limestone as thick as the amount of subsidence. Rates of sea level rise from distant tectonic movements are not known to be greater than rates of basinal subsidence, and would not likely achieve the potentially much higher rates of glacialeustatic rise.

The other imputed characteristic of glacial-eustatic control, that of ready correlatability of the cyclothems, has very recently received a strong positive test in an area that has implications for identifying tectonic control. Boardman & Heckel (1989) have biostratigraphically correlated all the major and intermediate cyclothems in a large segment of the Midcontinent sequence analyzed by Heckel (1986) with transgressive–regressive cycles of similar magnitude in north-central Texas, which is 400 km distant, across the Arbuckle uplift (Fig. 2) and close to the active Ouachita orogenic belt, a southwestward extension of the Appalachian trend. In spite of the conspicuous delta shifting in the regressive phases of the eustatic Texas cycles and in spite of the potential larger scale modifications from any tectonic flexuring originating from the nearby Ouachita belt, an extremely close correlation (Fig. 7) was achieved, in which all but one of even the minor cycles currently recognized in Texas have previously recognized counterparts in stratigraphic position in the Midcontinent. The correlation was based mainly on a combination of first, last, sole, or acme occurrences of conodonts, ammonoids, and fusulinids. It provides such a close match of both faunas and cycle magnitude in the two areas that it shows not only that the primarily glacial eustatic signal of the Midcontinent is strongly evident in the more detrital-rich Texas sequence, but also that any tectonic movements in the Ouachitas at that time had little effect on the Texas cyclic sequence other than possibly suppressing or enhancing cycle magnitude by one of the informal ranks in a few cases. Unpublished conodont data also indicate correlation of all the major cycles of roughly the same sequence with the major marine horizons in the Illinois basin.

The next logical step is to achieve correlation with the major marine horizons in the Appalachian basin. Preliminary discussions with biostratigraphers working there suggest that certain successions of one to three major Midcontinent marine cyclothems of this age have strong marine correlatives there, whereas others do not. This hints that with a firm biostratigraphic framework, we may be able to detect and measure a tectonic signal in the late Middle to Late Pennsylvanian of the Appalachian region, which would appear to be of significantly greater period length than the eustatic signal, thus closer to that suggested in Tankard (1986) for the transgressions he interpreted as tectonically controlled in the earlier Pennsylvanian of that region.

Conclusions

All the lines of evidence currently available strongly converge toward periodic glacial eustatic rise and fall of sea level as the principal cause for the distinctive marine cyclothems of Midcontinent North America. Their extreme lateral persistence, repeatedly superposing offshore marine sediment-starved phosphatic shales closely above paleosols and other exposure surfaces, for hundreds of kilometres across local tectonic features and largely in the absence of deltas, requires a periodic eustatic control. The presence of Gondwanan glacial deposits at this time, in conjunction with the estimates of the periodicity of the cyclothems within the range of those of the Earth's orbital cycles that are involved in Pleistocene glacial frequencies, and much shorter than those so far estimated for tectonic movement, strongly point to glacial rather than distant tectonic control of the eustasy.

Because of its global nature, glacial eustasy must have been an underlying control over Pennsylvanian stratigraphy in all areas. This would be true in western Europe, where involvement in tectonic movements and proximity to resulting local detrital sources allowed conspicuous structural complications and delta-shifting to mask the eustatic control by preventing the development of widespread limestone units. Both tectonic movement and glacial eustasy controlled the position of the shoreline in various places, and deltas dominated the areas near detrital sources when eustatic sea level was stable at low or high stand, or falling during regression. Glacial eustasy is evident in the older Westphalian and Namurian of Britain in the remarkable lateral persistence of the succession of transgressive dark marine shale horizons that have been correlated biostratigraphically among the complex tectonic elements (e.g. Ramsbottom 1979), and which also have a mean periodicity within the Milankovitch band according to recent British work (e.g. Leeder 1988).

Recognizing the broad control of eustatic events, we now stand at a threshold of biostratigraphic correlation of major cycles around and among basins so that 'event' correlation of the intermediate and minor cycles between them, can be tested as Boardman & Heckel (1989) have begun between the Midcontinent and Texas. From this we can document the relative extents of these eustatic events in other areas and evaluate the frequencies of other basic causes, such as tectonic movements, some of which at this early rudimentary stage of analysis appear to occur at significantly longer periods than glacial eustasy.

I thank the State Geological Surveys of Kansas, Oklahoma, Missouri, Illinois, Iowa, and Nebraska for various combinations of field and financial support and for access to long cores, the Allan and DeLeo Bennison Field Research Fund, many past and present students for long hours of work, M. R. Leeder for invitation to participate in this conference, and the Petroleum Exploration Society of Great Britain for making my attendance possible.

References

Barrick, J. E. & Boardman, D. R. 1989. Stratigraphic distribution of morphotypes of *Idiognathodus* and *Streptognathodus* in Missourian-lower Virgilian strata, north-central Texas. Texas Tech University Studies in Geology 2, II: Contributed Papers, 167–188.

Boardman, D. R. & Heckel, P. H. 1989. Glacial-eustatic sea-level curve for early Upper Pennsylvanian sequence in north-central Texas and biostratigraphic correlation with curve for Midcontinent North America. *Geology*, **17**, 802–805.

—— & Malinky, J. M. 1985. Glacial-eustatic control of Virgilian cyclothems in north-central Texas. *American Association of Petroleum Geologists, Southwest Section Meeting Transactions*, 13–23.

——, Mapes, R. H., Yancey, T. E. & Malinky, J. M. 1984. A new model for the depth-related allogenic community succession within North American Pennsylvanian cyclothems and implications on the black shale problem. *In*: Hyne, N. J. (ed.), *Limestones of the Mid-Continent*, Tulsa Geological Society Special Publication **2**, 141–182.

Ferm, J. C. 1970. *Allegheny deltaic deposits*. Society of Economic Paleontologists & Mineralogists Special Publication **15**, 246–255.

Galloway, W. E. & Brown, L. F. Jr. 1973. Depositional systems and shelf-slope relations on cratonic basin margin, uppermost Pennsylvanian of north-central Texas. *American Association of Petroleum Geologists Bulletin*, **57**, 1185–1218.

Goebel, K. A., Bettis, E. A. & Heckel, P. H. 1989. Upper Pennsylvanian paleosol in Stranger Shale and underlying Iatan Limestone, southwestern Iowa. *Journal of Sedimentary Petrology*, **59**, 224–232.

Gould, H. R. 1970. *The Mississippi delta complex*. Society of Economic Paleontologists & Mineralogists Special Publication **15**, 3–30.

Heckel, P. H. 1977. Origin of phosphatic black shale facies in Pennsylvanian cyclothems of Midcontinent North America. *American Association of Petroleum Geologists Bulletin*, **61**, 1045–1068.

—— 1980. Paleogeography of eustatic model for deposition of Midcontinent Upper Pennsylvanian cyclothems. *In*: Fouch, T. D. & Magathan, E. R. (eds) *Paleozoic paleogeography of west-central United States*, Society of Economic Paleontologists & Mineralogists, Rocky Mountain Section, Paleogeography Symposium I, 197–215.

—— 1983. Diagenetic model for carbonate rocks in Midcontinent Pennsylvanian eustatic cyclothems. *Journal of Sedimentary Petrology*, **53**, 733–759.

—— 1984a. Factors in Mid-Continent Pennsylvanian limestone deposition. *In*: Hyne, N. J. (ed.), *Limestones of the Mid-Continent*, Tulsa Geological Society Special Publication **2**, 25–50.

—— 1984b. Changing concepts of Midcontinent Pennsylvanian cyclothems. IX International Carboniferous Congress 1979, *Compte Rendu*, **3**, 535–553.

—— 1986. Sea-level curve for Pennsylvanian eustatic marine transgressive–regressive depositional cycles along Midcontinent outcrop belt, North America. *Geology*, **14**, 330–334.

—— 1990. Updated Middle-Upper Pennsylvanian eustatic sea-level curve for Midcontinent North America and preliminary biostratigraphic characterization. XI International Carboniferous Congress, 1987, *Compte Rendu*, Academica Sinica, Beijing, China.

—— & Baesemann, J. F. 1975. Environmental interpretation of conodont distribution in Upper Pennsylvanian (Missourian) megacyclothems in eastern Kansas. *American Association of Petroleum Geologists Bulletin*, **59**, 486–509.

Imbrie, J. & Imbrie, J. Z. 1980. Modeling the climatic response to orbital variations. *Science*, **207**, 943–953.

Kidder, D. L. 1985. Petrology and origin of phosphate nodules from the Midcontinent Pennsylvanian epicontinental sea. *Journal of Sedimentary Petrology*, **55**, 809–816.

Klein, G. deV. & Willard, D. A. 1989. Origin of the Pennsylvanian coal-bearing cyclothems of North America. *Geology*, **17**, 152–155.

Leeder, M. R. 1988. Recent developments in Carboniferous geology: a critical review with implications for the British Isles and N.W. Europe. *Proceedings of the Geologists' Association*, **99**, 73–100.

Malinky, J. M. 1984. *Paleontology and paleoenvironment of 'core' shales (Middle and Upper Pennsylvanian) Midcontinent North America*. PhD. Dissertation, University of Iowa.

Moore, D. 1959. Role of deltas in the formation of some British Lower Carboniferous cyclothems. *Journal of Geology*, **67**, 522–539.

Moore, R. C. 1931. Pennsylvanian cycles in the northern Mid-Continent region. *Illinois Geological Survey Bulletin*, **60**, 247–257.

—— 1936. *Stratigraphic classification of the Pennsylvanian rocks of Kansas*. Kansas Geological Survey Bulletin **22**.

—— 1950. Late Paleozoic cyclic sedimentation in central United States. 18th International Geological Congress, Great Britain 1948, Reports, **4**, 5–16.

Prather, B. E. 1985. An Upper Pennsylvanian desert paleosol in the D-zone of the Lansing-Kansas City Groups, Hitchcock County, Nebraska. *Journal of Sedimentary Petrology*, **55**, 213–221.

Ramsbottom, W. H. C. 1979. Rates of transgression and regression in the Carboniferous of NW Europe. *Journal of the Geological Society, London*, **136**, 147–153.

Schutter, S. R. & Heckel, P. H. 1985. Missourian (early Late Pennsylvanian) climate in Midcontinent North America. *International Journal of Coal Geology*, **5**, 111–140.

Swade, J. W. 1985. *Conodont distribution, paleoecology, and preliminary biostratigraphy of the upper Cherokee and Marmaton Groups (upper Desmoinesian, Middle Pennsylvanian) from two cores in south-central Iowa*. Iowa Geological Survey Technical Information Series **14**.

Tankard, A. J. 1986. Depositional response to foreland deformation in the Carboniferous of eastern Kentucky. *American Association of Petroleum Geologists Bulletin*, **70**, 853–868.

Veevers, J. J. & Powell, C. M. 1987. Late Paleozoic glacial episodes in Gondwanaland reflected in transgressive–regressive depositional sequences in Euramerica. *Geological Society of America Bulletin*, **98**, 475–487.

Wanless, H. R. 1964. Local and regional factors in Pennsylvanian cyclic sedimentation. *Kansas Geological Survey Bulletin*, **169**, 593–606.

—— 1967. *Eustatic shifts in sea level during the deposition of Late Paleozoic sediments in the central United States*. West Texas Geological Society Publication **69–56**, 41–54.

—— & Shepard, F. P. 1936. Sea level and climatic changes related to late Paleozoic cycles. *Geological Society of America Bulletin*, **47**, 1177–1206.

Wanless, H. R. & Weller, J. M. 1932. Correlation and extent of Pennsylvanian cyclothems. *Geological Society America Bulletin*, **43**, 1003–1016.

Weller, J. M. 1930. Cyclical sedimentation of the Pennsylvanian period and its significance. *Journal of Geology*, **38**, 97–135.

—— 1956. Argument for diastrophic control of late Paleozoic cyclothems. *American Association of Petroleum Geologists Bulletin*, **40**, 17–50.

—— 1958. Cyclothems and larger sedimentary cycles of the Pennsylvanian. *Journal of Geology*, **66**, 195–207.

Tectono-stratigraphic development and hydrocarbon habitat of the Carboniferous in northern England

A. J. FRASER[1] & R. L. GAWTHORPE[2]

[1]*BP Exploration, 301 St Vincent Street, Glasgow G2 5DD, UK*

[2]*Department of Geology, The University, Manchester M13 9PL, UK*

Abstract: Over 70 years of exploration in the Carboniferous of northern England has resulted in the discovery of rather modest recoverable reserves totalling 75 million barrels of oil and 27 billion cubic feet of gas. Nevertheless during this time the petroleum industry has amassed a substantial quantity of borehole and seismic information. Integrating this essentially subsurface database with information derived from outcrop studies has permitted a hitherto unachievable understanding of the Carboniferous in terms of its tectono-stratigraphic development and hydrocarbon habitat.

The strong NW–SE and NE–SE structural trends developed in the northern England Carboniferous were inherited from the late Palaeozoic Caledonian orogeny. These fault trends were consistently reactivated throughout the Carboniferous in both an extensional and compressional sense. The main influence on Carboniferous basin evolution in northern England was the Variscan collision-type orogeny. The Variscan plate cycle controlled the development of syn-rift, post-rift and inversion megasequences from late Devonian to early Permian times. Sequences developed within these Carboniferous megasequences are primarily controlled by episodic rifting and periodic fault reactivation with eustatic sea-level changes providing only minor control at the subsequence level.

The late Carboniferous–early Permian culmination of the Variscan orogeny is seen to be the main trap forming event. All hydrocarbon discoveries to date display some element of Variscan inversion in their geometry. Variscan tectonics have also exerted a subtle but important control on play fairway evolution. The main source rocks (pro-delta shales) are confined to isolated, syn-rift depocentres. Syn-rift siliciclastic reservoirs are also restricted to the rifted half graben. Carbonate grainstone reservoirs rim the margins of the deeper half graben where terrigenous input has been limited. Delta top channel and mouth bar reservoirs are best developed where they axially infill remnant syn-rift bathymetry.

Mesozoic burial, ensuring hydrocarbon generation post Variscan trap formation, is, however, the main control on the present day distribution of hydrocarbons in the Carboniferous of northern England. Several key areas where Carboniferous source rocks have generated significant hydrocarbons during the Mesozoic have been identified; the East Midlands being the most significant in terms of produced hydrocarbons and perceived future potential.

Recent developments in our understanding of the tectono-stratigraphic development of the Carboniferous in northern England (Gawthorpe 1987a, Gawthorpe *et al.* 1989; Fraser *et al.* 1990; Kimbell *et al.* 1989; Ebdon *et al.* 1990) have been made largely with the use of modern basin analysis techniques, and in particular, the application of sequence stratigraphy. Previous papers have concentrated on subsurface (seismic/wells) or surface (outcrop) data to construct a sequence stratigraphic scheme. Our approach integrates outcrop, well and seismic data to produce an integrated sequence stratigraphy for the Carboniferous of northern England. We will argue that the sequences described record tectonic events related to Variscan plate margin processes and that these events exert a fundamental control on the hydrocarbon habitat.

Exploration history

Over 70 years of petroleum exploration in the Carboniferous of northern England have resulted in the discovery of recoverable reserves totalling 75mmbbls of oil and 27bcf of gas. The first discovery was made at Hardstoft in Derbyshire in 1919 (Lees & Cox 1937) in fractured shelf carbonates of Dinantian age in a tight anticlinal structure. The first significant discovery was made by the D'Arcy Exploration Company in 1939 at Eakring in Namurian delta top channel and mouth bar sandstones with recoverable reserves of 7mmbbls. Important discoveries have since been made by BP at Beckingham/Gainsborough (13mmbbls oil, 6.5bcf gas) in 1959 and Welton (over 20mmbbls oil) in 1981, both in Namurian and early West-

From HARDMAN, R. F. P. & BROOKS, J. (eds), 1990, *Tectonic Events Responsible for Britain's Oil and Gas Reserves*, Geological Society Special Publication No 55, pp 49–86.

phalian A delta top sandstones. In general field sizes have been small with a mean of around 2mmbbls at a historical finding rate of one discovery for every four exploration wells. However, low development and production costs continue to make exploration of the Carboniferous onshore an attractive commercial proposition.

Regional structural framework

Caledonian inheritance

The structural development of the Carboniferous of northern England has its origins in the Caledonian orogeny (Leeder 1987; Bott 1976; Soper *et al.* 1987; Coward, this volume). The most recent models for the Caledonian orogeny involve a three plate configuration (Andre *et al.* 1986; Soper & Hutton 1984; Pharoah *et al.* 1987). Although the Iapetus suture remains the site of the main collision, the Midlands Microcraton has now been re-introduced as a northwards moving indentor which collided with Laurentia during the early Palaeozoic.

The Midlands microcraton represents a triangular-shaped terrane comprising relatively undeformed lower Palaeozoic platform sediments with the Welsh and East Midlands Caledonides wrapped around the apex (Fig. 1). The strong NE–SE trend of the Church Stretton, Bala and Pendle faults is linked to the underlying grain of the Welsh Caledonides (Fig. 2). The contrasting NW–SE trends, particularly of the Hoton,

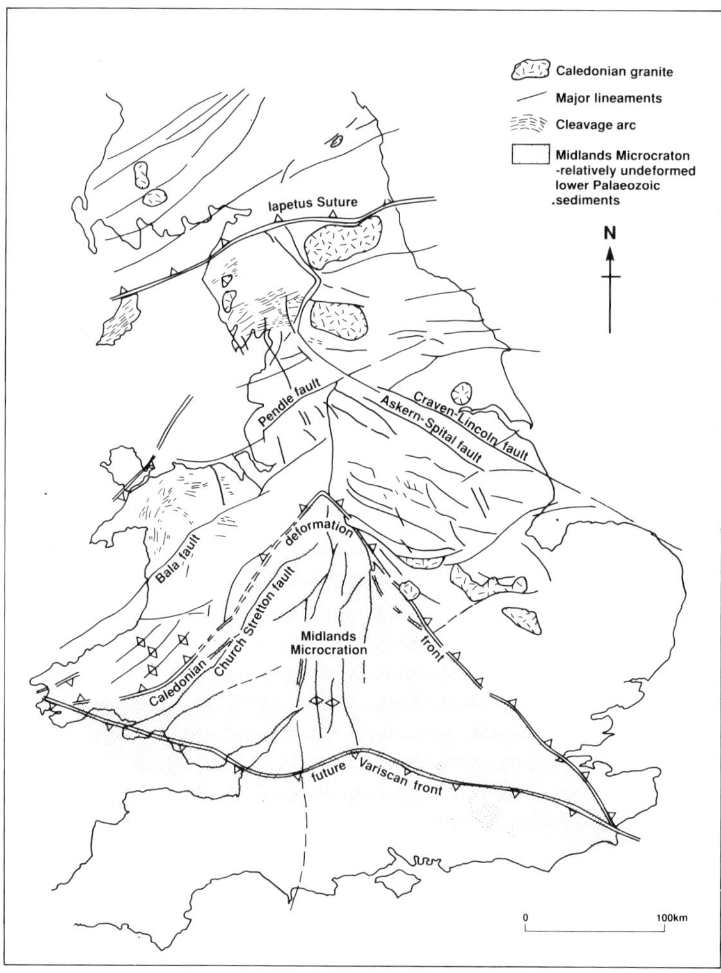

Fig. 1. Caledonian tectonic provinces of England and Wales (after Fraser *et al.* 1990; based on data from Turner 1949, Soper *et al.* 1987 and Pharoah *et al.* 1987).

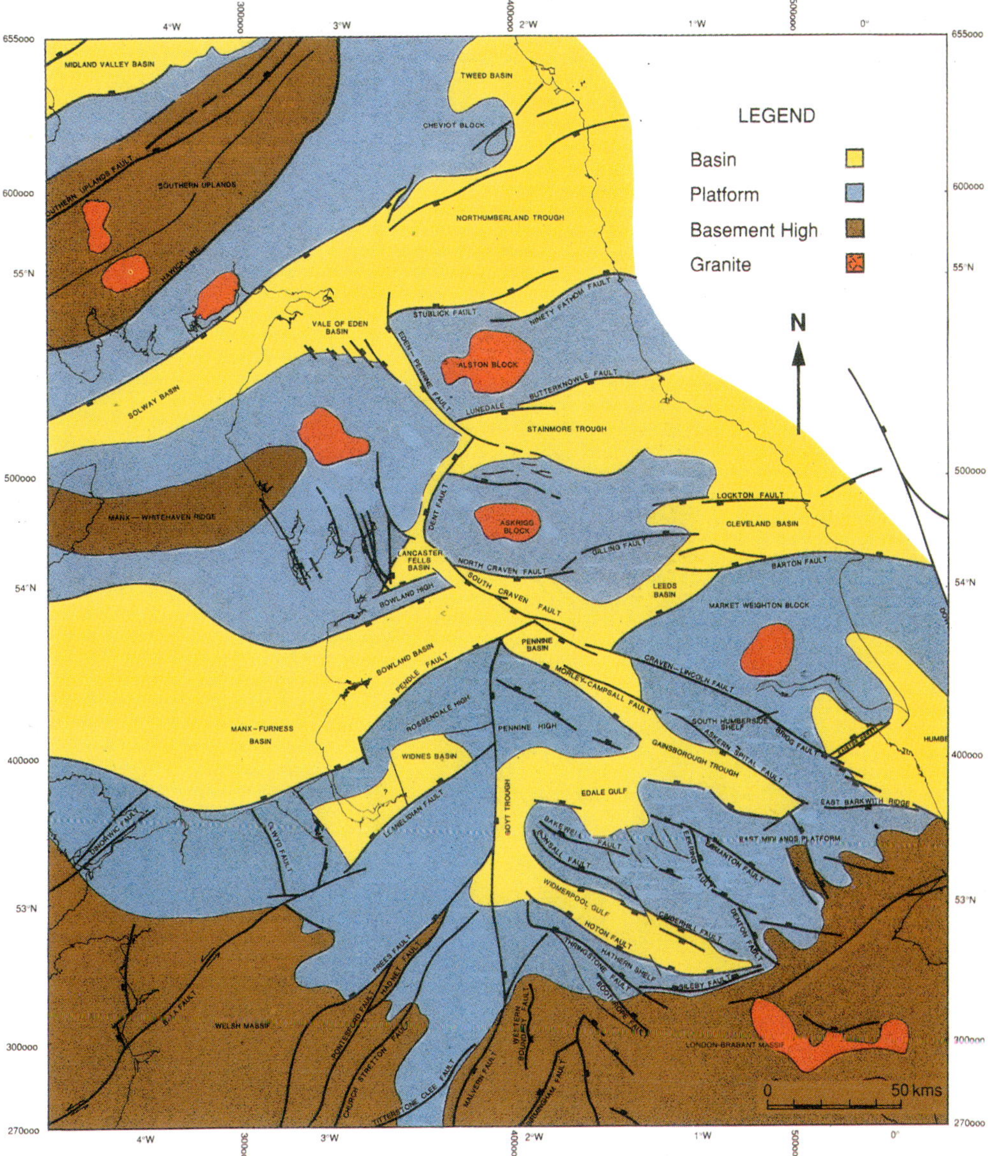

Fig. 2. Dinantian syn-rift structural elements in northern England.

Askern–Spital and Craven–Lincoln faults in the East Midlands, reflect an underlying deformed Caledonian (Charnian) thrust fold belt (Fig. 2). The final closure of oceans along the Iapetus and Tornquist sutures took place in the early Devonian (Soper *et al.* 1987). The subsequent late Palaeozoic and Mesozoic tectonic history of the region has been one of continued reactivation of these fundamental trends in extensional, compressional and strike-slip tectonic regimes.

Variscan Plate Cycle

The main plate margin process controlling the Carboniferous structural development of northern England was the formation of a collision-type orogenic belt in the Iberian–Armorican–Massif Central region of the Hercynides (Leeder 1987, 1988). Back-arc extension was established to the north of this orogenic belt by northwards directed subduction closing the Rheic Ocean. Within this back-arc system,

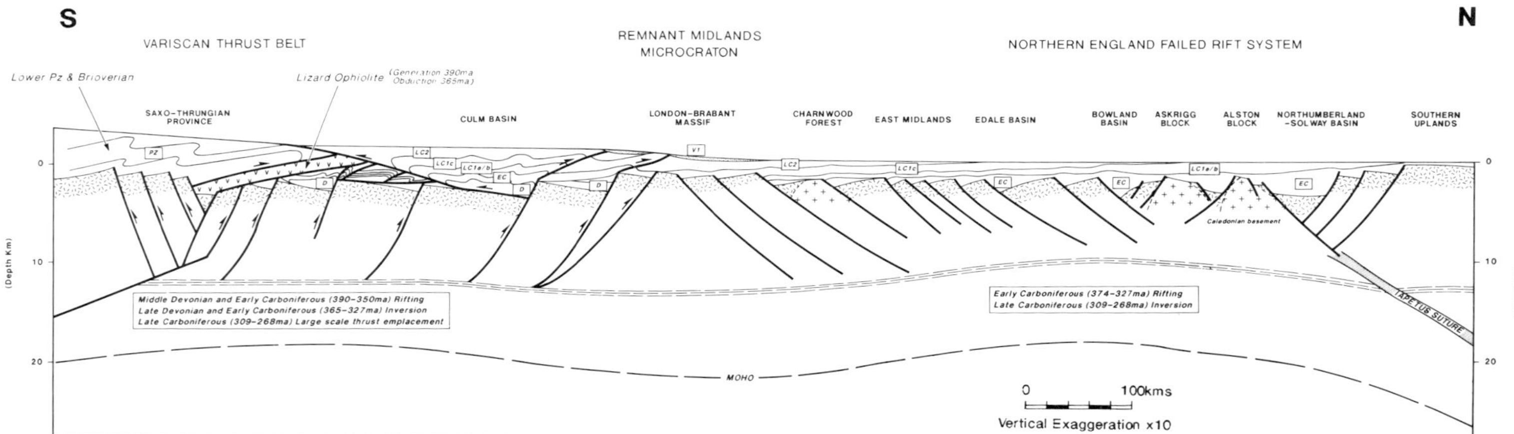

Fig. 3. Schematic crustal scale section showing the configuration of the Variscan thrust belt and the northern England rift system in late Westphalian C times. (Data from a number of sources including Leeder 1987, 1988; Sellwood & Thomas 1986, Wills 1973, 1978; Beamish & Smythe 1986, Whittaker *et al.* 1986). Legend: Pz, Lower Palaeozoic; D, Devonian; EC, Dinantian; LCla/b, Lower Namurian; LClc, Upper Namurian–Westphalian A; LC2, Westphalian A-B; Vl, Westphalian C.

extension was widespread throughout the late Palaeozoic (Fig. 3).

The northern England extensional province was situated some 500 km north of the main Rheno-Hercynian back-arc, separated by a horst comprising the remnants of the Midlands Microcraton (London–Brabant massif). This province is significantly younger (late Devonian–late Dinantian) than the back-arc basins lying to the south; extension having migrated northwards, possibly subsequent to the onset of ocean floor spreading within the back-arc in the late Devonian.

The late Devonian (Frasnian–Fammenian) onset of rifting in northern England initiated a series of linked, half grabens which were the precursors of the Northumberland-Solway, Stainmore, Bowland, Cleveland, Edale, Gainsborogh and Widmerpool basins (Fig. 2). These strongly asymmetric grabens were generated by N–S extension acting along a series of NW–SE and NE–SW trending faults which were themselves related to zones of earlier Caledonian structural weakness (cf. Figs 1 & 2). On depth converted seismic data the border faults appear as planar in section with detachments probably at mid-crustal levels and have similar geometries to major border faults in areas of active extention such as the Aegean and Basin and Range (e.g. Jackson 1987; Stein & Barientos 1985). Structural wavelength is greater in the NE England structural province where the presence of large Caledonian granitic bodies has maintained regional structural elevation of the Askrigg and Alston blocks (Fig. 2).

Pulsed rifting continued in northern England into the early Brigantian (Fraser *et al.* 1990) with reactivation of border fault zones of the graben during renewed phases of extension in the late Chadian–early Arundian and mid–late Asbian (Gawthorpe 1987a; Gawthorpe *et al.* 1989).

Subsidence during the Namurian and Westphalian times was broadly regional and essentially thermally driven. Some fault reactivation did occur during the early Namurian particularly on the Craven fault system (Bowland Basin) and the Askern–Spital fault (Gainsborough Trough).

The final closure of the Rheic ocean culminated in the Variscan orogeny from Westphalian C to early Permian times. The result of the orogeny was large-scale thrust and nappe emplacement in northern France, southern Belgium and southern England with corresponding crustal shortening on a regional scale. This shortening must have thickened the Carboniferous lithosphere and crust back up to the 'normal' magnitudes seen today. In northern England this was translated into inversion of pre-existing Dinantian extensional faults which resulted in the erosion of significant amounts of Carboniferous sediments.

Post-Carboniferous tectonics

Although not specifically within the remit of this paper, the influence of post-Carboniferous tectonics on the habitat of hydrocarbons within the Carboniferous needs to be addressed. Permian and Mesozoic extension associated with rifting in the Atlantic and Tethyan provinces modified trapping geometries and controlled the distribution of source rock maturation subsequent to Variscan trap formation. Tertiary uplift and erosion had the effect of freezing source rock generation and caused minor trap modification.

Rifting associated with early attempts to open the north Atlantic throughout the Permian and Triassic generated E–W extension on a series of N–S trending faults throughout western England. Existing NE–SW trending Caledonian lineaments were reactivated in a strike-slip sense as transfer faults linking the extensional basin system. Early Jurassic rifting in the Tethyan province resulted in renewed fault movement on NW–SE and E–W trending faults in northern England; for example the Hoton fault (Widmerpool Gulf) and Barton fault (Cleveland Basin) (Fig. 2). North Sea rifting in the late Jurassic and rifting in the Bay of Biscay/Rockall Trough areas in the early Cretaceous led to an important phase of extension throughout southern England which extended as far north as the Cleveland basin. Subsequent late Cretaceous thermal subsidence achieved maximum depths of burial of the Carboniferous over most of northern England.

Regional uplift of the onshore UK as a consequence of rifting between Greenland and Scotland from early Tertiary times induced a 1–2° easterly tilt on eastern England. An added complication during the Tertiary was the Oligo-Miocene culmination of the Alpine orogeny, closing the Tethyan ocean. In northern England E–W trending faults such as the Hoton and Barton faults (Fig. 2) were reactivated during this phase of inversion.

Sequence stratigraphic development

The sequence stratigraphic scheme for the Carboniferous of northern England has been based largely on the interpretation of modern, multifold, reflection seismic data. A threefold sub-

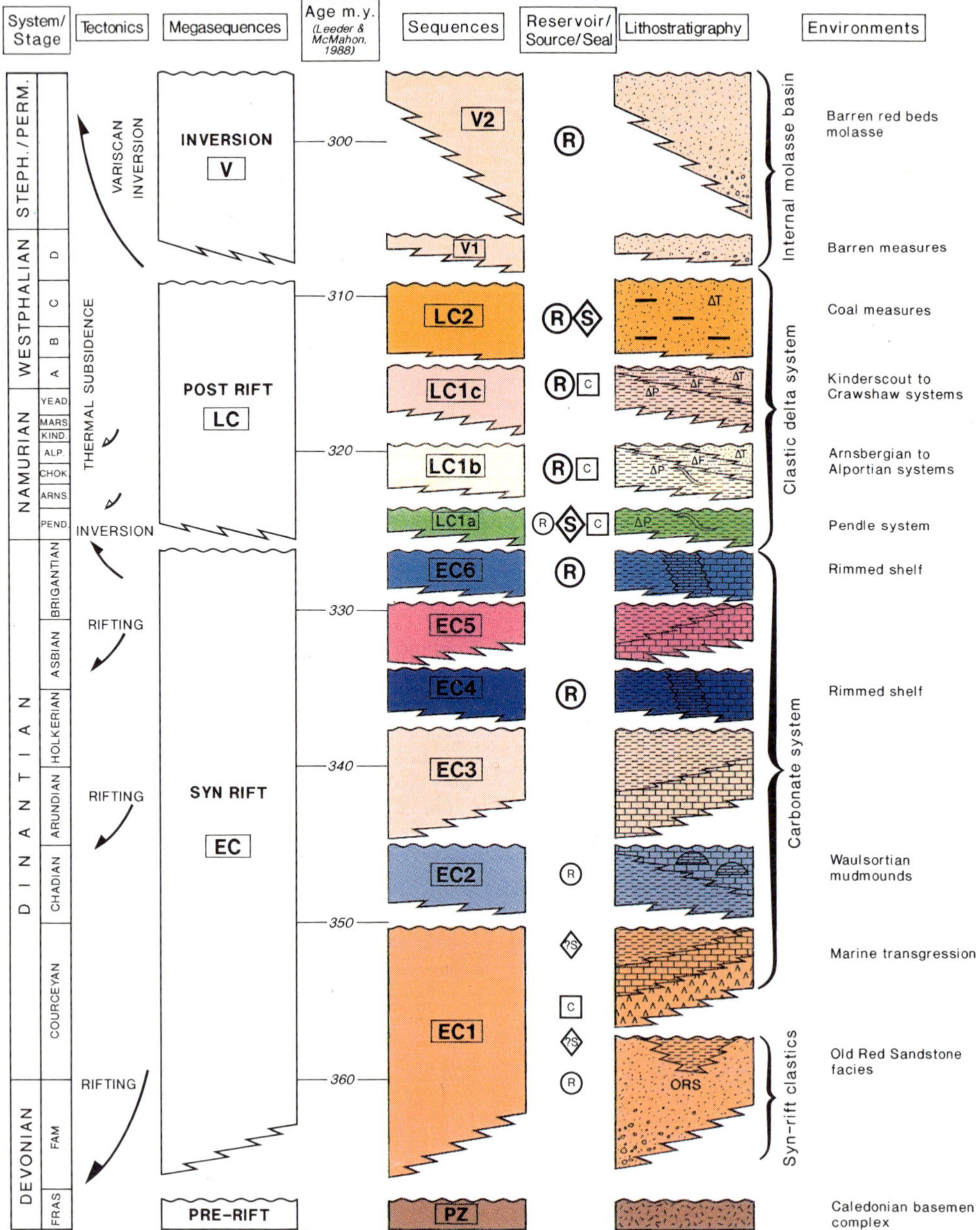

Fig. 4. Summarized stratigraphy of the Variscan plate cycle in the East Midlands showing megasequence and sequence development (modified from Fraser *et al.* 1990) △T = delta top, △F = delta front, △P = pro delta

division into plate cycles, megasequences and their constituent sequences is proposed, adapting the methodology of Hubbard *et al.* (1985a, b). Three megasequences have been identified, which can in turn be broken down into a series of depositional sequences (Fig. 4). These depositional sequences are individual stratigraphic units composed of a relatively conformable succession of genetically related strata and bounded at their top and base by unconformities or their correlative conformities (Mitchum *et al.* 1977).

There is currently considerable debate regarding the origin of sequence boundaries. Vail & Mitchum (1977) have emphasized the importance of eustatic sea-level changes as the major control on sequence development. More recently, Hubbard (1988) comparing the ages of sequence boundaries from a number of Jurassic–early Cretaceous passive margins has dismissed eustatic sea-level changes as a primary control and argued that tectonics and more specifically plate margin processes play the most important role. The cyclic nature of the Carboniferous in northern England has prompted equally active debate along similar lines (see for example Bott & Johnson 1967; Ramsbottom 1973, 1977, 1981; George 1978). In a sense the problem is one of scale. Boundaries that are likely to have been produced by eustatic sea-level changes (e.g. carbonate shelf cycles, Yoredale cycles and Upper Carboniferous goniatite bearing, marine bands are typically small scale and can only be delineated at the subsequence level. These events generally occur below seismic resolution for the northern England Carboniferous.

The Variscan plate cycle can be divided into *syn-rift, post-rift* and *inversion* megasequences (Fig. 4). These describe: a late Devonian-Dinantian rift controlled subsidence; a Namurian-Westphalian thermally driven subsidence; and a late Westphalian to early Permian inversion or foreland basin phase. Each megasequence represents an individual basin forming process, although all are ultimately controlled and driven by the evolving Variscan orogeny to the south. The development of the constituent sequences is discussed with particular reference to the Widmerpool Gulf regional seismic section (Fig. 5) and manual backstripping of this line (Fig. 6). A sequence stratigraphic correlation based on generalized sedimentary logs for the East Midlands, Bowland and Stainmore is presented in Fig. 7.

Syn-rift megasequence (late Devonian–early Brigantian)

The syn-rift megasequence exhibits a characteristic wedge shaped geometry (Figs 5 & 8) which is exemplified by a fourfold increase in sediment thickness across the northern bounding Askern-Spital fault of the Gainsborough Trough (Fig. 8). Isopachs for the upper part of this megasequence show the thickest sections to be confined to the individual fault bounded half graben (Fig. 9a).

Two main syn-rift depositional systems can be identified (Fig. 9b); (i) clastic fluvio-deltaic and (ii) carbonate platforms. During the syn-rift phase, tectonic subsidence generally exceeded the rate of sediment supply to the basins and the clastic deltas remained confined to the north of the region. As a consequence, the south of the area became sediment starved and carbonates accumulated on the platform areas with the development of rimmed shelf margins on both the footwall crests and hangingwall dipslopes of the deeper basinal areas. The half graben were themselves infilled by predominantly fine grained, hemi-pelagic, clastic deposits.

Sequence EC1 (late Devonian–early Chadian). Syn-rift I — fault controlled subsidence and initial development of the half graben. The sequence shows a characteristic wedge shaped geometry thickening into the border faults. The base shows a progressive onlap onto Lower Palaeozoic basement of the hangingwall dipslope. The top of the sequence is marked by a laterally continuous, high-amplitude reflector (Figs 5, 8 & 10). The internal character is of low-amplitude, laterally discontinuous events.

The initial syn-rift clastics ('Old Red Sandstone' facies) are poorly exposed at surface and are generally of local derivation, related to valley-fill processes. By analogy with other desert rift-basins (Leeder & Gawthorpe 1987) downlapping alluvian fan and fan-delta depositional systems are likely to be derived laterally from the footwall and hangingwall basin margins. The early syn-rift clastics proven by the Eakring-146 well (Falcon & Kent 1960) and the Whita and Annan Sandstones in the Northumberland Basin (Leeder 1974) represent examples of localized footwall-derived, and more sheet-like hangingwall-derived fan systems respectively.

The upper part of the sequence is characterized by onlap of the pre-rift sediments by carbonates and evaporites following the initial marine transgression into the rift system during the Courceyan. The sparse well penetration and poor seismic resolution of this sequence precludes any further breakdown of this interval at present.

Sequence EC2 (mid–late Chadian). Post-rift I — still stand or regressive phase characterized by carbonate ramp to rimmed shelf development. The sequence comprises high amplitude, laterally persistent reflectors which thicken up-dip along the hangingwall dipslope where hummocky downlapping clinoforms are identified (Fig. 5). In the Bowland Basin the sediment fill at this time is dominated by clean carbonate facies with shallow ramp grainstone shoals high on the hangingwall dipslope passing southward into deeper water wackestones/packstones

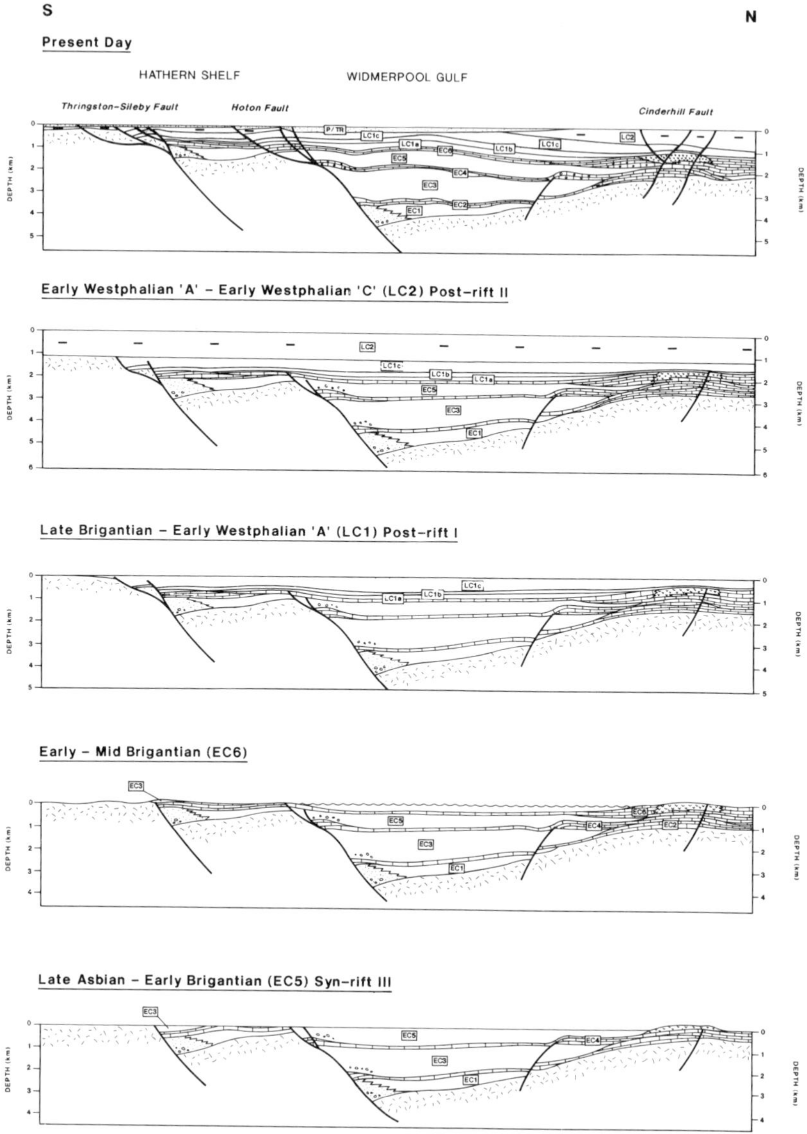

Fig. 6. Devono-Carboniferous basin development: Widmerpool Gulf and Hathern shelf, East Midlands (adapted from Ebdon *et al.* 1990).

towards the Pendle fault. Waulsortian buildups are well developed in these deeper ramp environments.

Sequence EC3 (Arundian–early Holkerian). Syn-rift II – reactivation of extensional faults causing fault block rotation and significant footwall erosion. The onset of the sequence is characterized by the development of boulder beds and slumps (event deposits; see Gawthorpe & Clemmey 1985, Gawthorpe *et al.* 1989) in the hangingwall and the rapid drowning of carbonate shelf margins in the East Midlands and Bowland as carbonate production was unable to keep pace with sea level rise. In the north of the province the Fell and Ashfell fluvio-deltaic systems were deposited in the Northumberland Basin and Stainmore Trough respectively during this time (Fig. 7). The progressive onlap of the pre-Arundian topography culminated in the

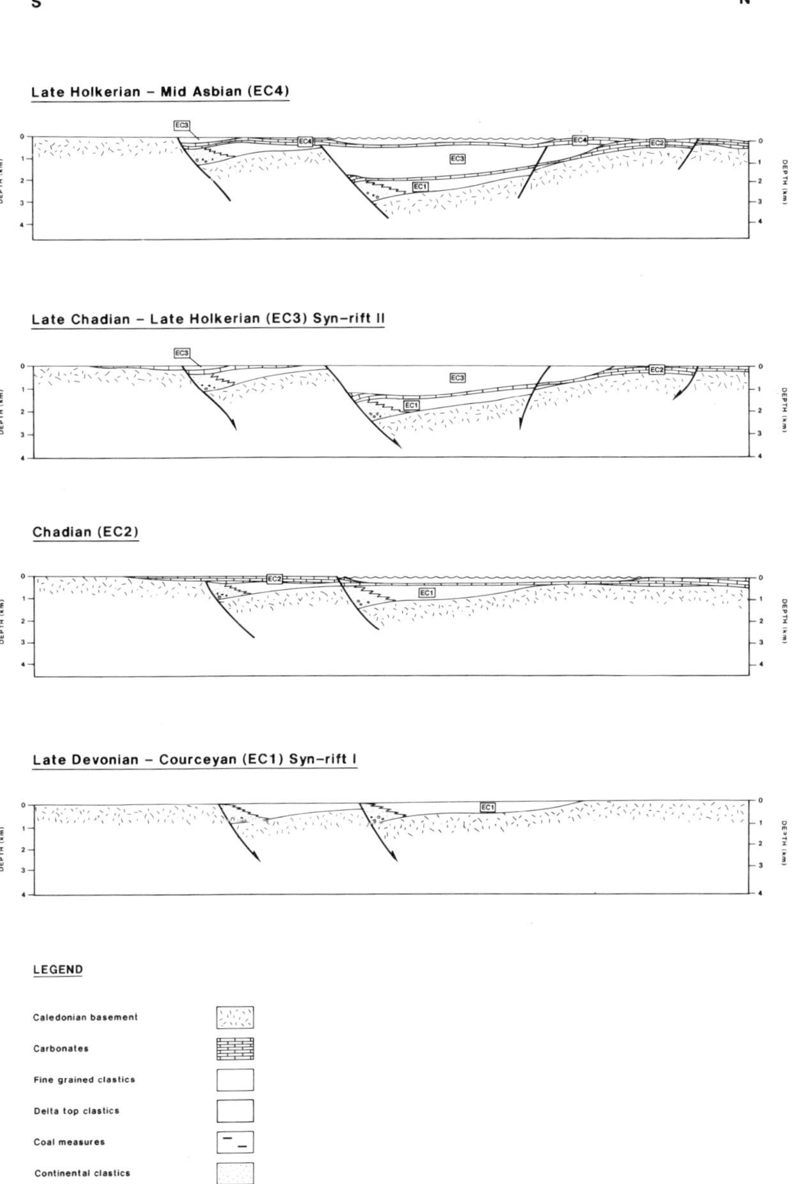

maximum transgression of carbonate platforms in northern England during the Holkerian (Strank 1987). The sequence shows a characteristic wedge shaped geometry and represents the most significant rift pulse, in terms of time and thickness, during the syn-rift megasequence. The base of the sequence shows a progressive onlap onto the underlying EC2 sequence (Figs 5, 8 & 10). Internally, the sequence consists of low-amplitude, high-frequency events which progressively onlap the hangingwall dipslope. Figure 11a shows boulders of proximal ramp and Waulsortian carbonates derived from footwall erosion of the South Craven fault system at the onset of EC3 rifting. These were deposited contemporaneously with and overlain by the Worston Shale Formation, a predominantly mud and silt grade clastic

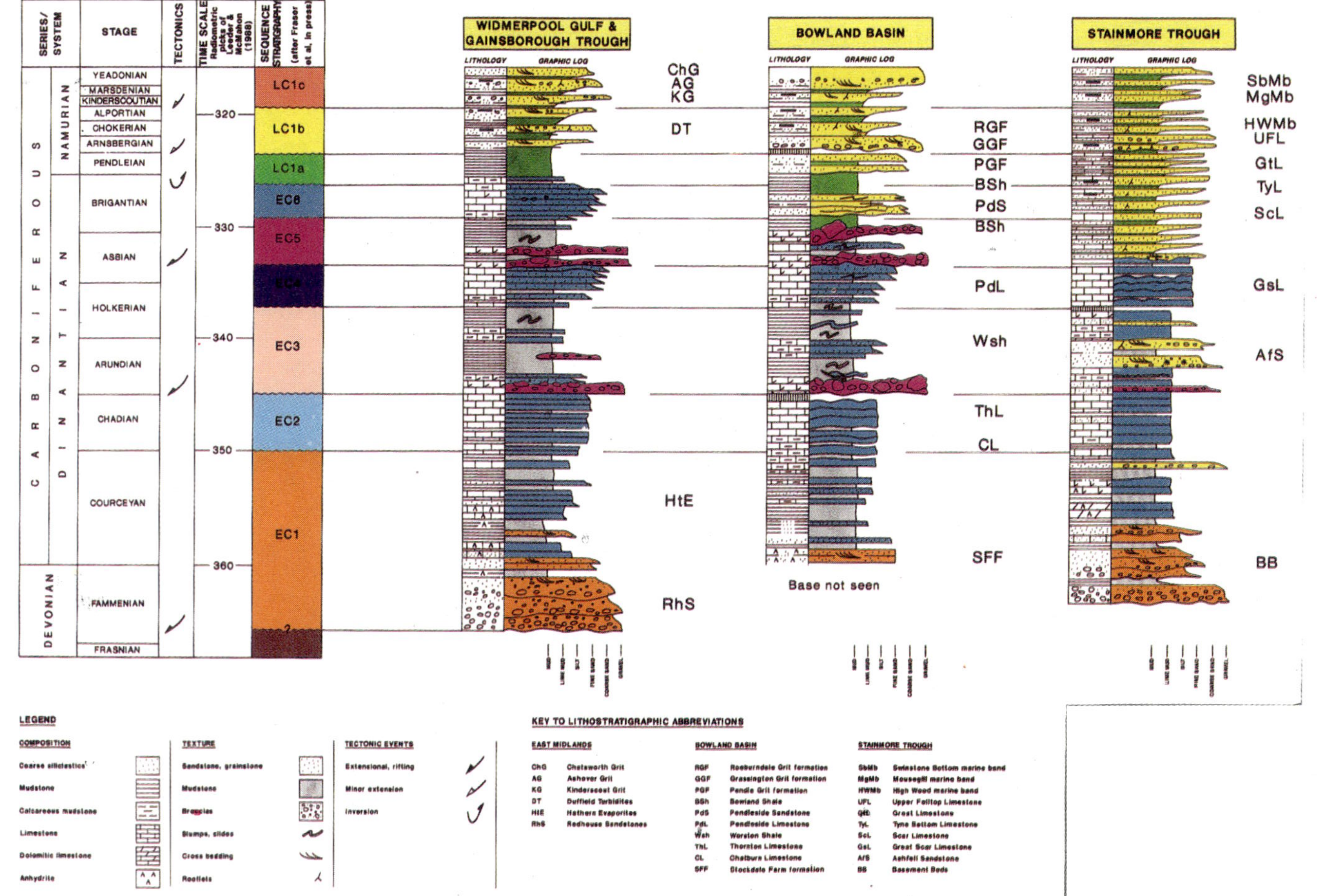

Fig. 7. Sequence stratigraphic correlation of generalized sedimentary logs from the Widmerpool Gulf/Gainsborough Trough, Bowland Basin and Stainmore Trough. Key to Lithostratigraphic abbreviations. *East Midlands*: ChG, Chatsworth Grit; AG, Ashover Grit; KG, Kinderscout Grit; DT, Duffield Turbidites; HtE, Hathern Evaporites, RhS, Redhouse Sandstones. *Bowland Basin*: RGF, Roeburndale Grit formation; GGF, Grassington Grit formation; PGF, Pendle Grit formation; BSh, Bowland Shale; Pds, Pendleside Sandstone; Pdl, Pendleside Limestone; Wsh, Worston Shale; ThL, Thornton Limestone; CL, Chatburn Limestone; SFF, Stockdale Farm formation. *Stainmore Trough*: SbMb, Swinstone Bottom marine band; MgMb, Mousegill marine band; HWMb, High Wood marine band; UFL, Upper Felltop Limestone; GHL, Great Limestone; TyL, Tyne Bottom Limestone; ScL, Scar Limestone; GsL, Great Scar Limestone; AfS, Ashfell Sandstone; BB, Basement Beds.

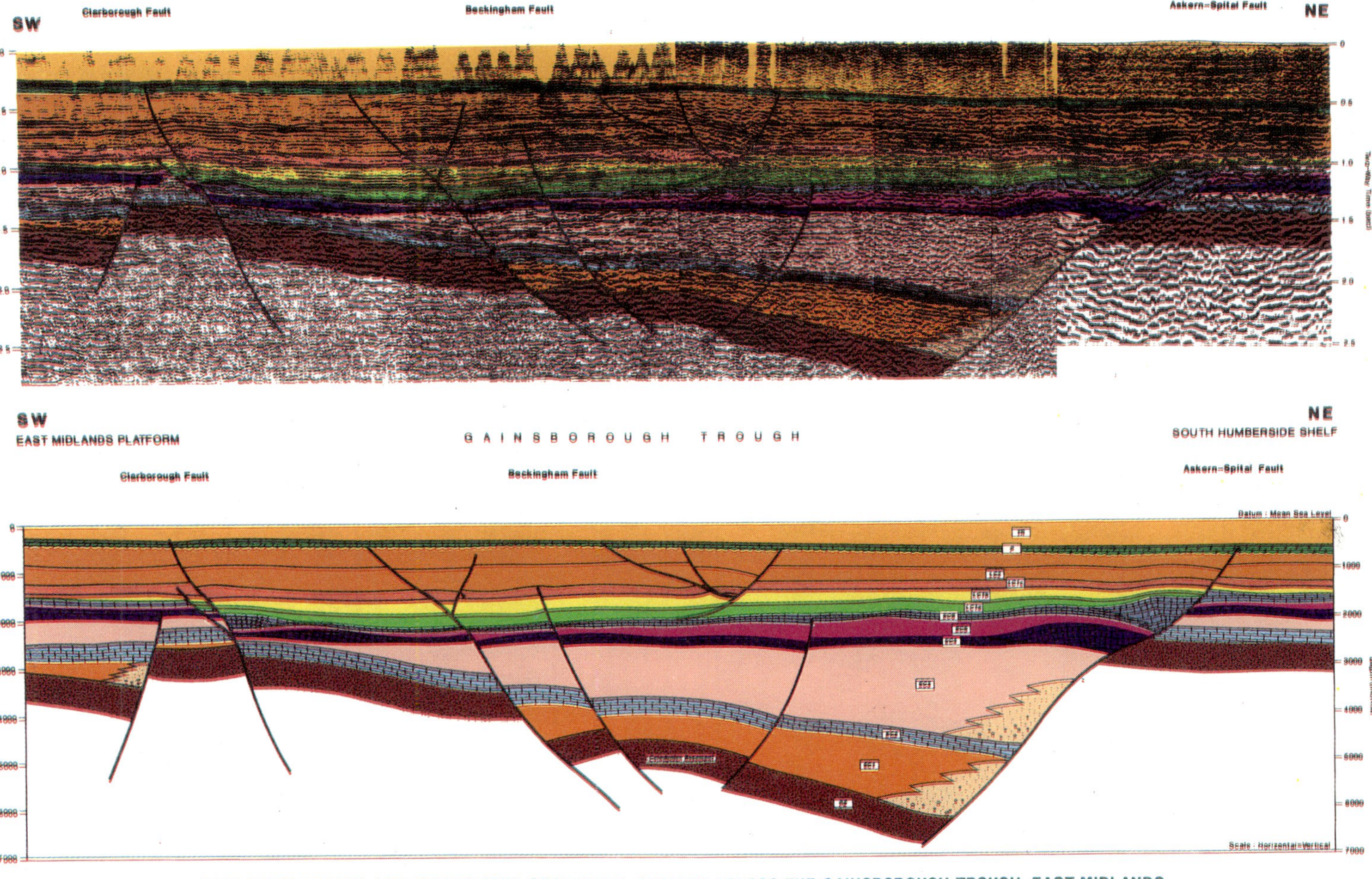

Fig. 8. Composite seismic and interpreted geological section across the Gainsborough Trough, East Midlands.

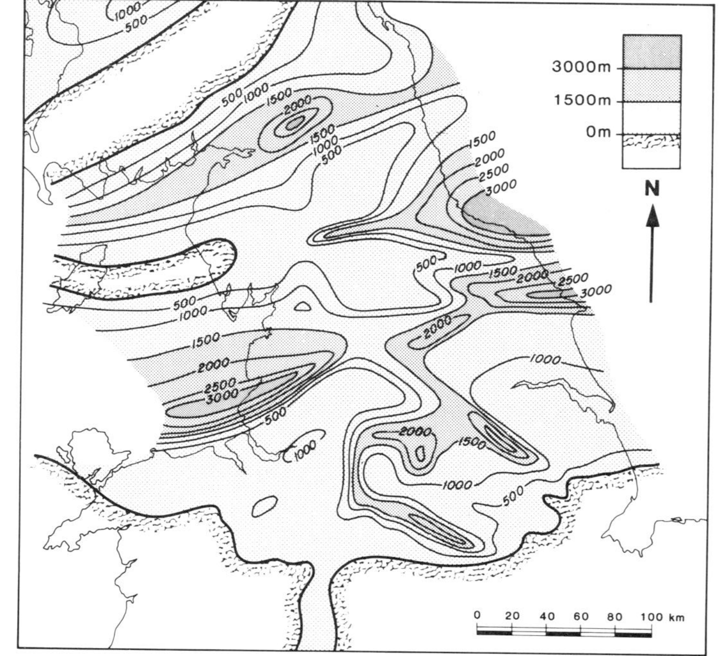

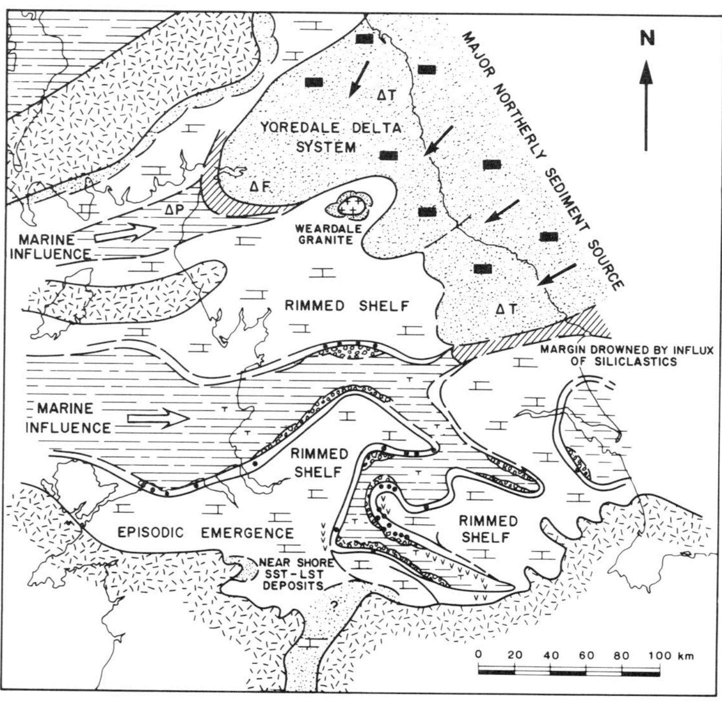

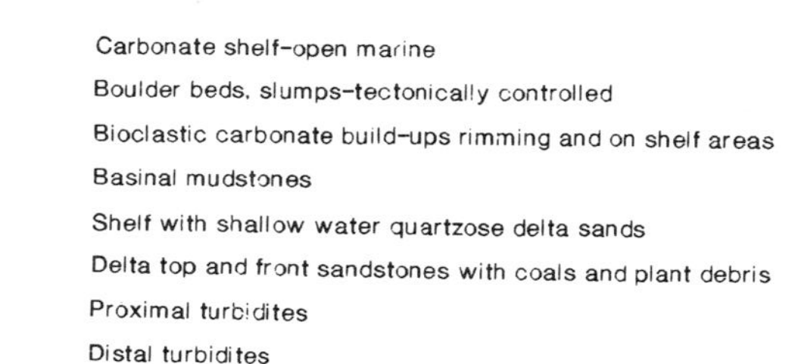

Fig. 9. Isopach and facies maps for the syn-rift megasequence. (a) Restored isopachs for syn-rift sequences EC2–EC6 (Chadian–early Brigantian). Data from George (1958), Leeder (1974). (b) Palaeofacies map for syn-rift sequence EC4 (late Holkerian-mid Asbian). $\triangle$T = delta top, $\triangle$P = pro-delta, v = volcanics. (Incorporates data from Stevenson & Gaunt 1971; Leeder 1974; Miller & Grayson 1982; Smith *et al.* 1985; Aikenhead *et al.* 1985; Gawthorpe 1987a, Grayson & Oldham 1987.)

deposit with thinly interbedded hemipelagics and distal calciturbidites, which forms EC3 in the Bowland Basin (Fig. 7). During the rifting phase periodic instability of the hangingwall dipslope led to the generation of syn sedimentary slumps and slides in sediments of EC3 age (Fig. 11b).

Sequence EC4 (late Holkerian–mid Asbian). Post-rift II – still stand or regressive phase characterised by carbonate ramp to rimmed shelf development in the East Midlands and Bowland. Here the sequence shows a marked thickening onto shelf areas (Fig. 5). In the basinal setting, EC4 comprises high amplitude, low frequency laterally continuous reflectors. Along fault scarps the sequence comprises high amplitude, low-frequency, high-angle reflectors which represent vertically and laterally accreting clinoforms. Along the hangingwall dipslope the sequence is largely characterless but in places high-amplitude, low-frequency downlapping clinoforms are identified prograding basinwards onto EC3 (Fig. 5). A sub-EC4 unconformity evidenced on seismic in Stainmore may be indicative of a late Holkerian basin inversion event perhaps coincident with the cessation of EC3 rifting (Fig. 10b). A similar angular unconformity can also be identified in the Widmerpool Gulf (Fig. 5).

In areas of shallow water, starved of siliciclastic sediment, shoaling upward cyclic shelf carbonates were deposited; for example the Great Scar Limestone and equivalents in western Stainmore and on the Alston and Askrigg blocks (Fig. 7). In the East Midlands and Bowland, these shelf carbonates pass basinwards through platform margin facies into hemipelagic lime mudstones and calciturbidites of the foreslope. This carbonate slope facies association shows a coarsening upward trend reflecting progradation of the rimmed shelf.

Sequence EC5 (late Asbian–early Brigantian). Syn-rift III – reactivation of the extensional fault regime with significant backstepping of the fault system in the Widmerpool Gulf where EC5 is well developed. By comparison, extension was milder during this phase in the Gainsborough Trough (cf. Figs 5 and 8). Elsewhere in the East Midlands and Bowland the extensive EC4 carbonate margins were drowned. Renewed footwall rotation resulted in strong erosional unconformities on the footwalls of the Hoton fault (Widmerpool Gulf; Fig. 12) and Craven fault (Bowland Basin) and the development of boulder beds and slumps in hangingwall settings. The sequence exhibits the characteristic wedge shaped geometry typical of syn-rift packages. Internally the sequence comprises low-amplitude, high-frequency reflectors which progressively onlap the hangingwall dipslope.

Volcanics related to the renewed extensional activity during EC5 (late Asbian to earliest Brigantian age) have been described from outcrop in Derbyshire; Lower and Upper Millers Dale lavas (Walkden 1977; Walters & Ineson 1981; MacDonald *et al.* 1984). These basalts are associated with a series of NW–SE trending faults (Cinderhill fault system) which bound the northern margin of the Widmerpool Gulf. Basaltic lavas and tuffaceous horizons of similar age have been encountered in exploration boreholes in the East Midlands along the same trend.

Regional stratigraphic correlation demonstrates considerable facies variations between the individual half grabens during EC5 (Fig. 7). The Widmerpool and Gainsborough Basins were receiving fine-grained clastic material and distal calciturbidites. The sequence in Bowland is dominated by the incoming of distal prodelta facies from the north (Lower Bowland Shale). In Stainmore, where shallower water conditions prevailed close to the clastic source, the onset of cyclic Yoredale sedimentation is apparent (Johnson 1960). The EC4/EC5 sequence boundary can be observed at outcrop in the northern part of the Bowland basin (Fig. 11c). As a means of demonstrating the link between the outcrop and seismic expression of sequence boundaries, the equivalent location on the Widmerpool Gulf regional seismic line is highlighted on Fig. 12. Strong erosional truncation on the footwall of the Hoton fault resulted in a marked angular unconformity on the Hathern shelf between EC1 and EC6 due to footwall uplift and erosion as a result of EC5 rifting. The erosional debris of this event (presumably shelf carbonates as in Fig. 11c) is represented by discontinuous high-amplitude, low-frequency events at the base of the EC5 sequence within the hangingwall (Fig. 12). These are overlain by hemipelagic lime mudstones comprising the remainder of the EC5 sequence, which has generated the observed seismic impedance contrast. These boulder beds may have some significance as potential reservoirs due to secondary porosity generation during burial diagenesis (Gawthorpe 1987b). This secondary porosity development can often be associated with hydrocarbons (Fig. 11d).

Sequence EC6 (early–mid Brigantian). Post rift III – basin inversion event resulting in a marked regressive phase over northern

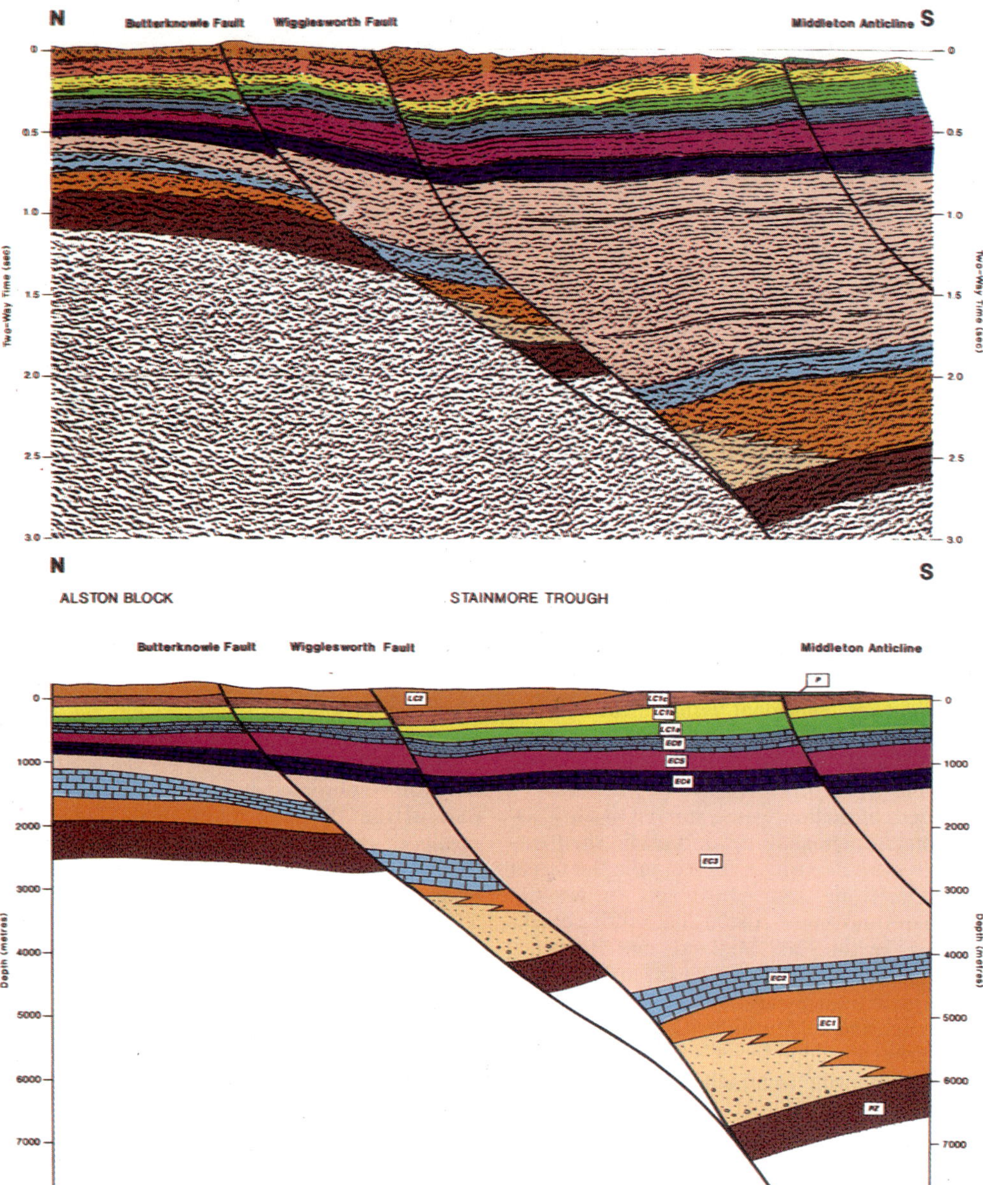

England. Seismic data from the Egmanton oilfield in the East Midlands (Fig. 13) shows an erosional unconformity in the hangingwall of the fault controlling the structure. This is interpreted as inversion of the original listric normal fault in the early–mid Brigantian and is apparently contemporantous with thrust and nappe emplacement in southwest England (Sellwood & Thomas 1986). The sequence is characterized by carbonate ramp to rimmed shelf development in the Widmerpool Gulf, Gainsborough Trough and Edale Gulf where water depth differences of up to 500 metres were established between shelf and basin. The sequence thickens markedly onto shelf areas. The base of the sequence exibits downlap onto EC5 along the hangingwall dipslope, becoming sub-parallel with EC5 into the basin (Fig. 5). Internally the EC6 shelf margins are comprised of complex oblique-sigmoid, high-amplitude, low-frequency clino-

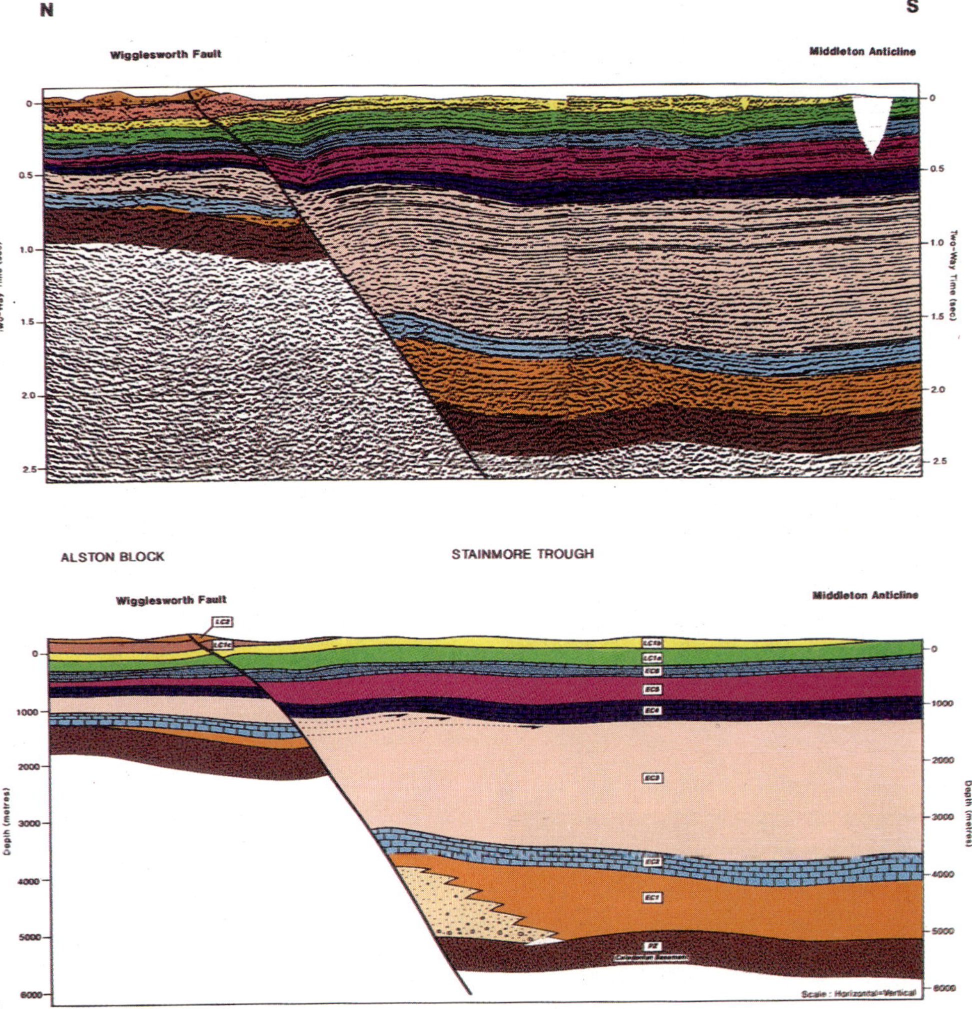

Fig. 10. Composite seismic and interpreted geological sections across western Stainmore. Surface geology based on Mills & Hull (1976), Burgess & Holliday (1979) and BGS solid geology sheets 31-Brough-under-Stainmore and 32-Barnard Castle.

forms (Fig. 5) which pass basinwards into high frequency parallel events. Mounded features up to 0.5 km across have been observed on the Hathern Shelf (Fig. 12). These are probably analogous to the Brigantian mudmounds of the Coalhills complex in Derbyshire (Walkden 1982).

During EC6 the Pendle delta system prograded southwards resulting in the deposition of siliciclastic turbidites (Pendleside Sandstone) in Bowland, effectively precluding further carbonate production in the basin (Fig. 7). Cyclic, shallow marine Yoredale sedimentation continued in Stainmore. The disconformity between the individual Yoredale cycles can be observed at outcrop but as they occur at a frequency below seismic resolution the regional significance of the individual events cannot be assessed.

Post-rift megasequence (late Brigantian–late Westphalian C)

Seismic data shows the post-rift megasequence to have been deposited regionally across northern England (Figs 5, 8 & 10). Isopachs for the post-rift section exhibit a classic bullseye pattern interpreted by Leeder (1982) as resulting from a phase of passive thermal subsidence (Fig. 14a).

1:10000
0
200m
400
600
800

WIDMERPOOL GULF

Cinderhill Fault

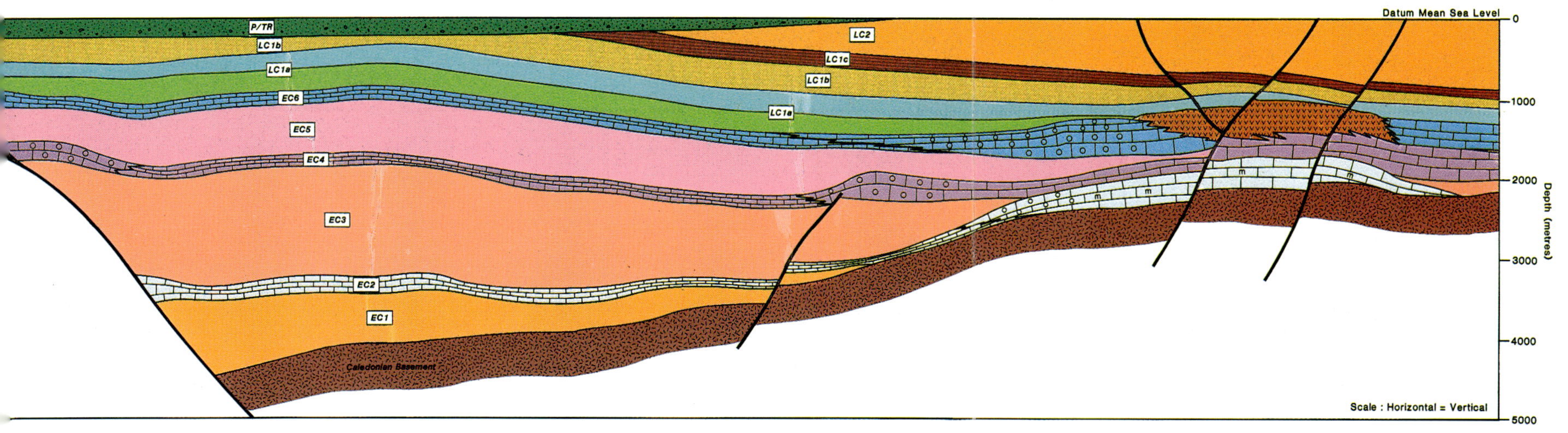

N

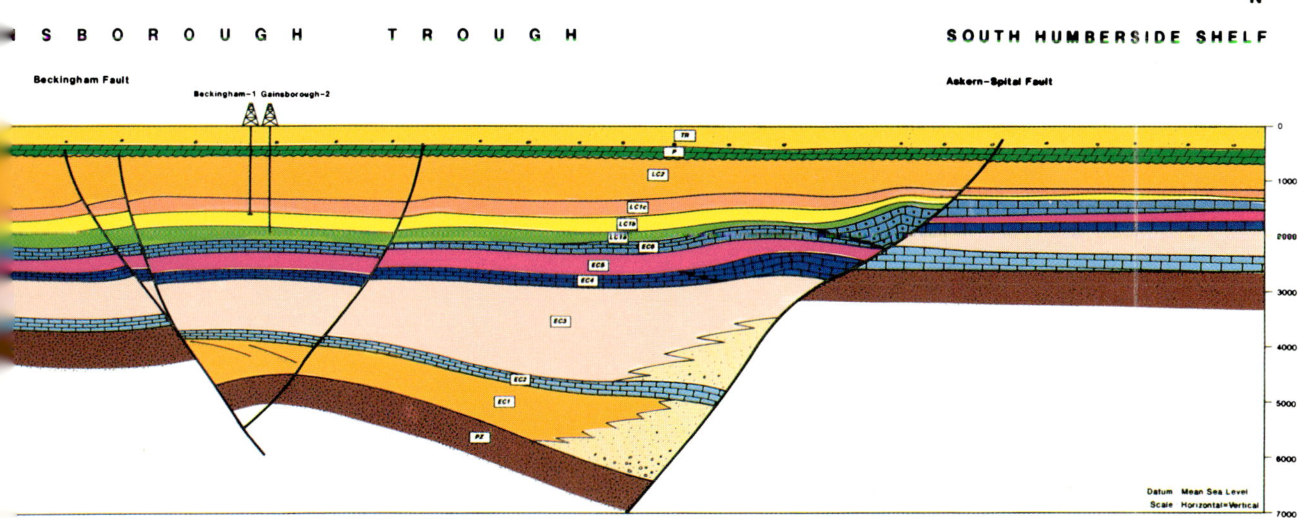

HATHERN SHELF

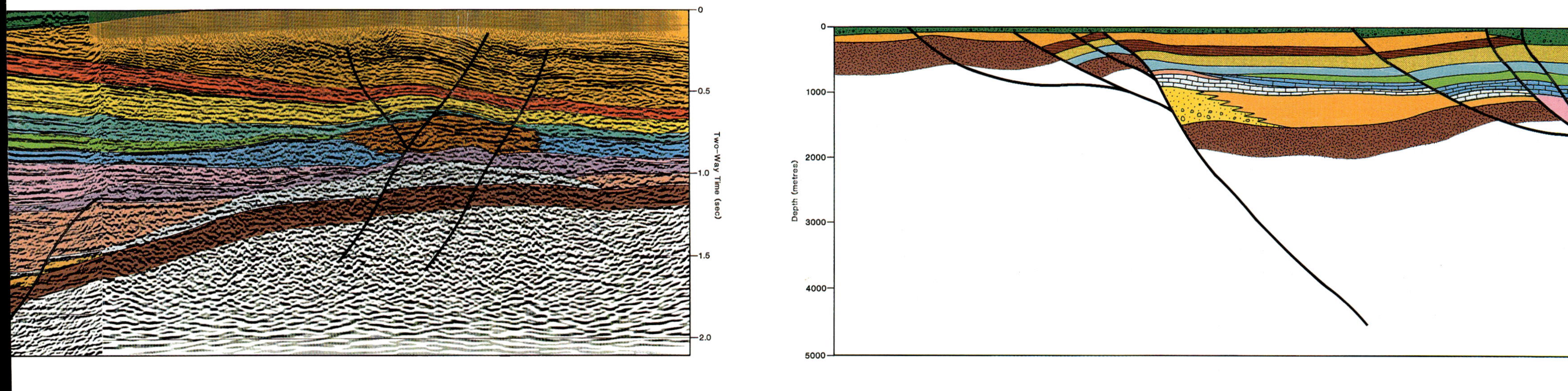

M I D L A N D S S H E L F

G A I

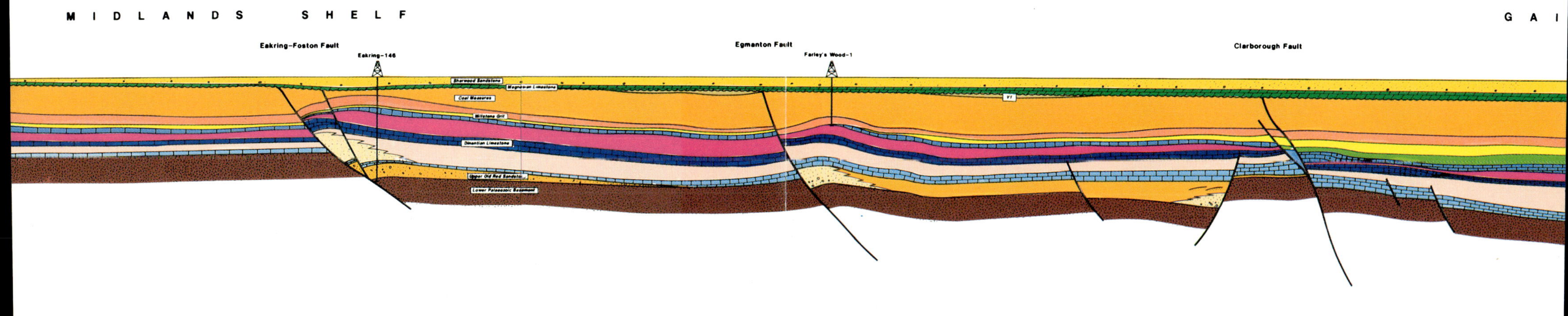

NORTH-SOUTH GEOLOGICAL SECTION ACROSS THE EAST MIDLANDS HYDROCARBON PROVINCE

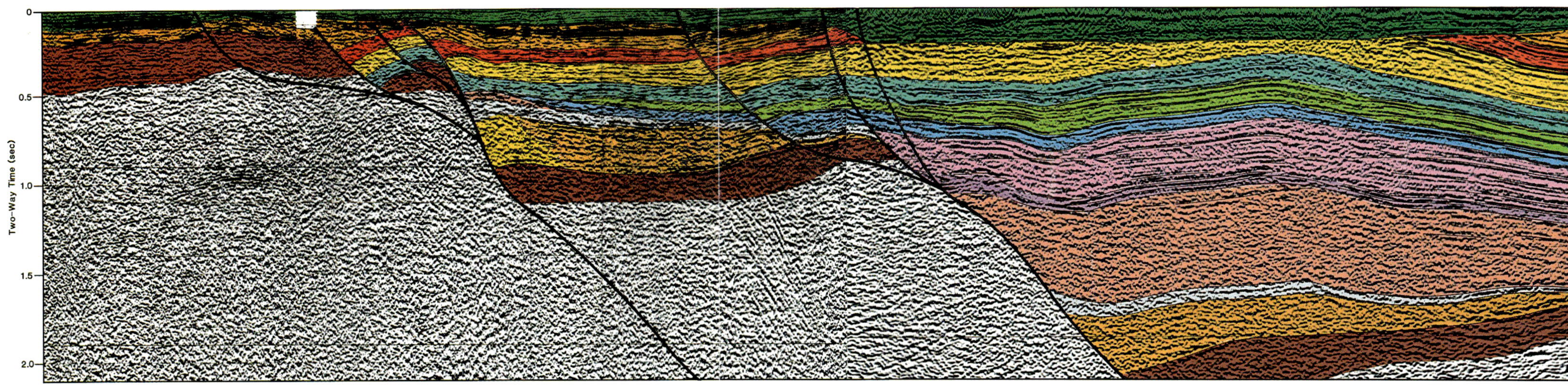

Fig. 5. Composite seismic and interpreted geological section across the Widmerpool Gulf and Hathern Shelf, East Midlands (after Fraser *et al.* 1990).

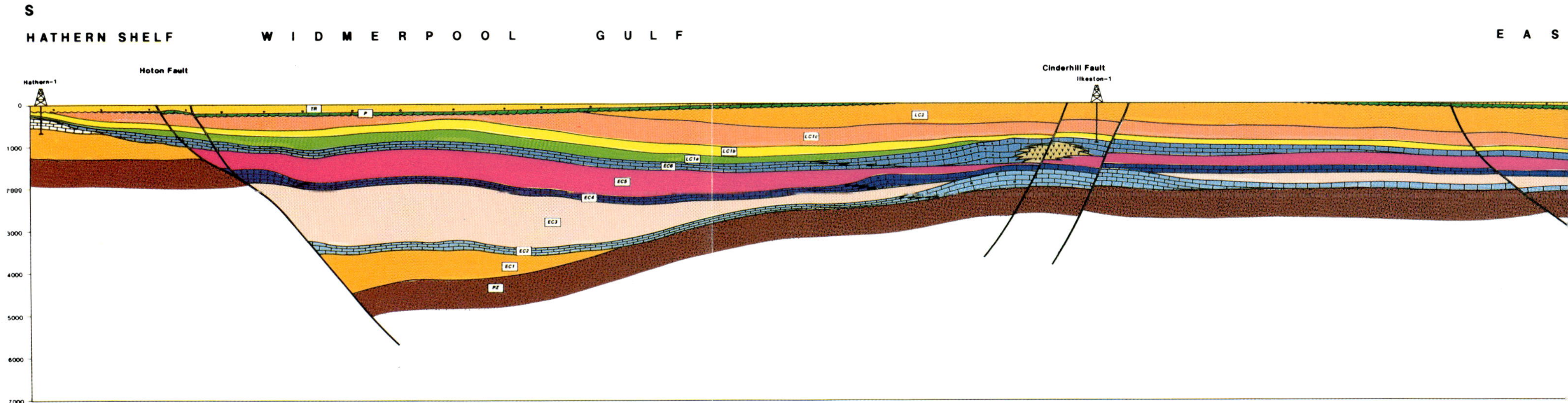

Fig. 18. North–South geological section across the East Midlands hydrocarbon province (based on depth converted regional seismic and well data). The section illustrates the fundamental difference in rift and inversion geometry between the major Widmerpool and Gainsborough half graben (NW–SE trending) and the smaller Eakring and Farley's Wood depocentres (NNW–SSE trending). Along the line of section oilfields are located in Variscan inversion anticlines at Rempstone, Eakring, Farley's Wood/Egmanton and Beckingham/Gainsborough.

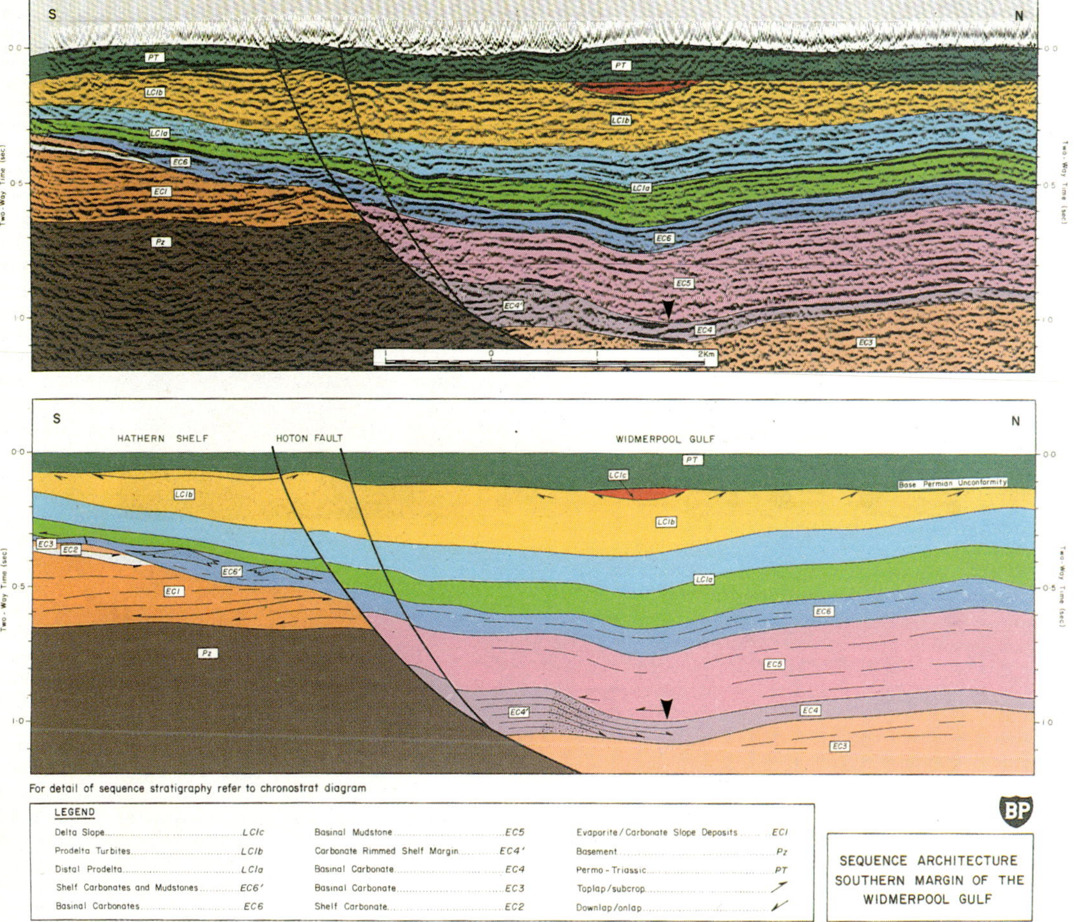

Fig. 12. Sequence architecture, southern margin of the Widmerpool Gulf, East Midlands, showing development of the EC4 footwall margin (EC4). Strong high amplitude reflectors at base of EC4 slope (indicated by arrow) are generated by boulder beds which represent footwall erosion of the Hathern shelf at the EC4/EC5 sequence boundary (cf. Fig. 11c) (from Ebdon *et al.* 1990).

Fig. 11. (a) Haw Crag (sequence EC3) northern Bowland: Boulders of proximal ramp and Waulsortian carbonates derived from footwall erosion along the South Craven fault system at the onset of EC3 rifting. (b) Worston Shale Formation. Croasdale Beck, northern Bowland basin (sequence EC3). Syn-sedimentary slumps and folds in interbedded hemipelagic mudstones and distal calciturbidites generated by slope instability during EC3 rifting. (c) Pendleside Limestone (EC4/EC5 sequence boundary), Green Pike, Bowland Basin. An excellent example of a major sequence boundary occurring in the middle of a lithostratigraphic unit. The bedded limestone units in the lower part of the picture are foreslope calciturbidites belonging to the EC4 sequence. These are unconformably overlain by shelf derived debris sourced from erosion of the footwall of the nearby Bowland High at the onset of EC5 rifting (cf. Fig. 12). (d) Pendleside Limestone (sequence EC5) mineral exploration borehole, south of Bowland High. Vuggy porosity in boulder beds associated with hydrocarbon shows.

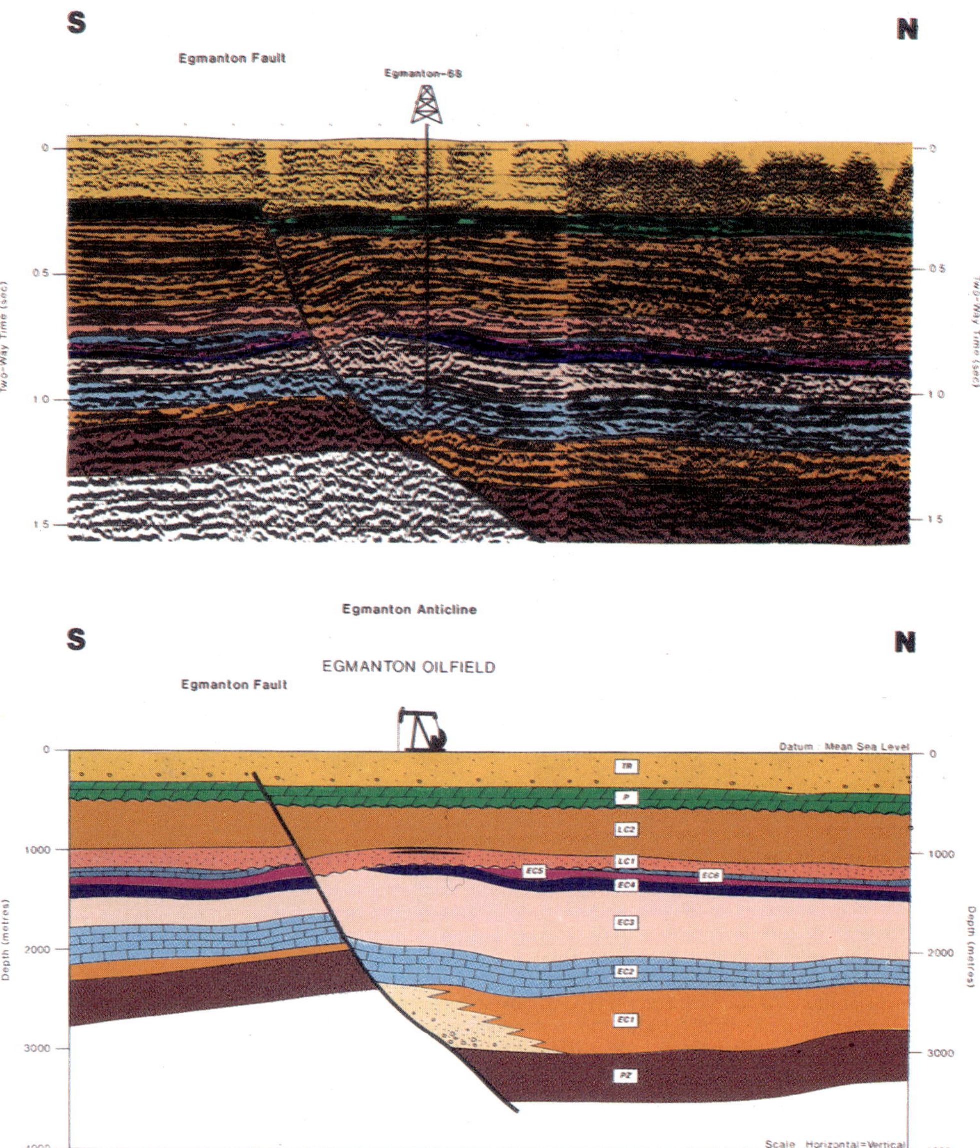

Fig. 13. Composite seismic and interpreted geological section across the Egmanton oilfield, East Midlands. Late Dinantian inversion is evidenced by the suggestion of subcropping EC3–EC6 sequences to the base of the post-rift (LC1).

A similar thermal sag has been identified further east in the Southern North Sea basin (Leeder & Hardman, this volume). The position of the post-rift depocentre in the south of the study area may result from a combination of factors: (i) the presence of extensive Caledonian granitoid bodies in the north of the area maintaining crustal buoyancy, (ii) the substantial water depths existing in the south of the area following the syn-rift phase and (iii) the presence of a thermal anomaly possibly related to an earlier pre-rift thermal bulge in the region. A representative palaeofacies map for the early post-rift (Fig. 14b) shows the southward progradation of the clastic fluvio-deltaic system (cf. Fig. 9b) as sediment supply overtook subsidence for the first time. This resulted in the burial of the last vestiges of the northern England car-

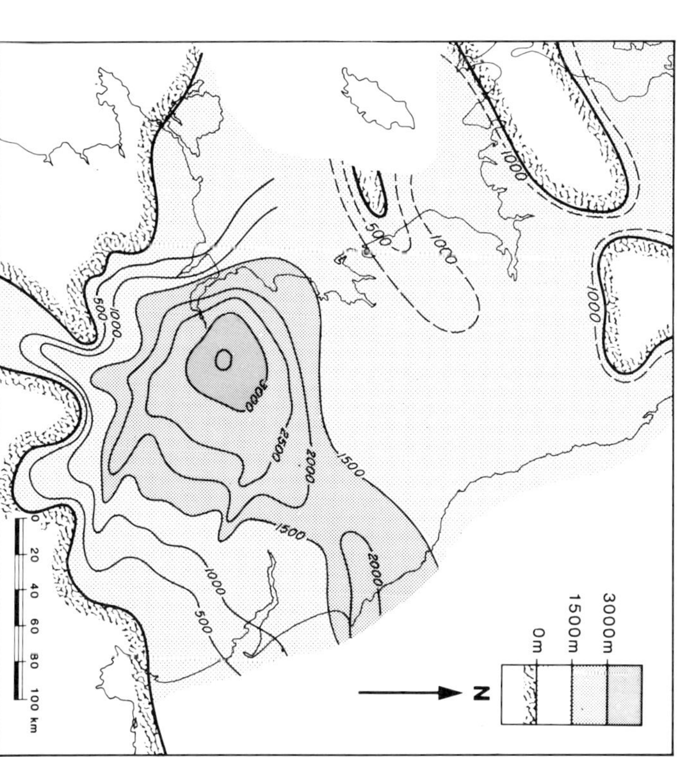

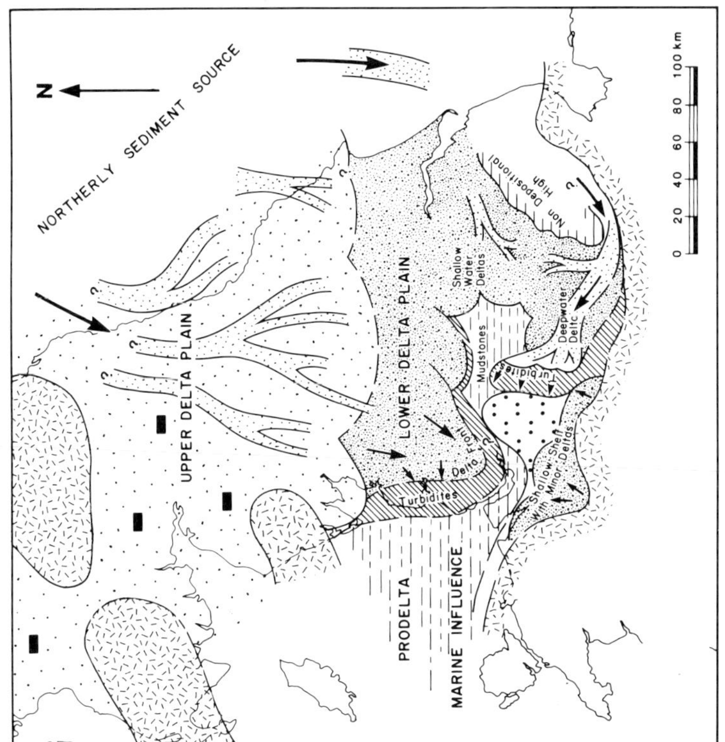

Fig. 14. Isopach and facies maps for the post-rift megasequence. (a) Restored isopachs for post-rift sequences LC1–LC2 (Namurian–Westphalian C). (Data from Wills 1951, Kent 1966, Calver 1968, Leeder 1982, Guion & Fielding 1988.) (b) Palaeofacies map for post-rift sequence LC1c (Kinderscoutian to late Westphalian A). (Data from Collinson *et al.* 1977, Reading 1964, Guion & Fielding 1988.)

bonate system in the late Brigantian-early Pendleian and the infill of the rifts with deltaic sediments. The backstripped seismic profile of the top EC6 (top syn-rift) reflector from Figs 5 & 8 indicates water depths of between 300 and 500 m for the two sediment starved half grabens between shelf and basin; the position of the shelf and basin being well constrained by borehole and seismic facies data. The observed thickening of the Namurian and Westphalian (post-rift) isopachs into the Widmerpool and Gainsborough half graben (Kent 1966) is thus partly explained in terms of progressive fill of the starved rifts by progressively shallower water deltaic systems.

Sequence LC1a (late Brigantian–late Pendleian). Onset of the post-rift thermal subsidence phase. The last vestiges of the carbonate system in the East Midlands were drowned and intra-shelf basins were developed following marine transgression over the post EC6 inversion topography (Gutteridge 1987, 1989). The Pendle delta system advanced from the northeast with relatively deep water (300–500 m), distal pro-delta, organic rich shales ponded in the starved Bowland, Edale, Gainsborough and Widmerpool basins (Upper Bowland and Edale shales). The sequence is thickest within the individual half graben, thinning progressively onto the hangingwall and footwall carbonate margins of the underlying sequence where it is extremely condensed (Figs 5 & 8). On seismic data, the sequence is characterized by high-amplitude, high-frequency, laterally continuous parallel reflectors which downlap onto EC6 within the basin and passively onlap the submerged margins (Fig. 7).

Sequence LC1b (late Pendleian–Alportian). Minor reactivation of extensional faults occurred on the Craven fault system (Bowland Basin) and the Askern–Spital fault (Gainsborough Trough). These fault movements are manifested in the pre-Grassington Grit unconformity on the Askrigg block (Arthurton *et al.* 1988; Dunham & Wilson 1985) and vertical offset of the Dinantian on the northern margin of the Gainsborough Trough (Fig. 10). Coarse siliciclastics progressively infilled the Bowland Basin with pro-delta organic rich shales deposited contemporaneously in the more distal Gainsborough Trough (Steele 1988) and Widmerpool Gulf (Gamma active shales, Fraser *et al.* 1990). LC1b is represented on seismic by high-amplitude, high-frequency discontinuous events. The base of the sequence shows occasional downlap onto LC1a in basinal areas and oversteps the remnant carbonate shelves where it onlaps earlier syn-rift (typically EC6) sediments.

Seismic data (presented herein) and confidential biostratigraphic studies (BP) show sequence LC1b to be condensed over the East Midlands shelf and probably throughout northern England. This may reflect large-scale switching of delta lobes into the southern North Sea area or more regional changes in climate/sediment supply in the hinterland to the north.

Sequence LC1c (Kinderscoutian–late Westphalian A). Mild reactivation of extensional faults occurred particularly in the eastern Gainsborough Trough where some 200 metres of relief developed on the NE–SE trending Scampton fault (Fig. 2) establishing a strong footwall unconformity in the area. The sequence is characterized by the continued and widespread progradation of the Kinderscout, Ashover and Chatsworth–Crawshaw delta systems infilling the Gainsborough Trough and Widmerpool Gulf. The Ashover delta filled the Widmerpool Gulf longitudinally in a SE to NW direction following the inherent Dinantian structural grain (Fig. 14b). As a consequence, the sequence is thickest in the previously sediment starved Widmerpool Gulf. The seismic character of LC1c consists of high-amplitude, low-frequency events which can be traced onto the shelf areas where the sequence is generally well developed. Lower delta plain conditions were widespread throughout northern England by the end of this interval (Guion & Fielding 1988).

Sequence LC2 (late Westphalian A–Westphalian C). The periodic tectonic activity evidenced on several of the major basin bounding faults during the early part of the post rift had almost ceased by the late Westphalian A. This also coincided approximately with the cessation of igneous activity in the East Midlands (Kirton 1984). During LC2, upper delta plain, coal swamp conditions were established over most of northern England, forming an internally drained basin north of the Wales-Brabant Massif (Guion & Fielding 1988). However erosion following the Variscan inversion, particularly in hangingwall settings, has resulted in a patchy present day preservation of the sequence. The sequence is characterised by laterally persistent, high-amplitude parallel reflectors which are generated by the many coal seams developed throughout LC2 (Figs 5 & 8).

Inversion megasequence (late Westphalian C–early Permian)

The effects of the Variscan orogeny are seen on

seismic sections by the generation of large-scale inversion anticlines in the hangingwalls of the existing basin bounding faults (e.g. Fig. 5). Early Permian peneplanation of these anticlines is evidenced by a pronounced subcrop to the base Permian throughout northern England. The degree of uplift and subsequent erosion of syn- and post-rift Carboniferous sediments has been estimated from regional stratigraphic (thickness) and vitrinite reflectance (palaeoburial) studies (Fig. 15a). The map highlights the extent to which basins such as Bowland and Northumberland, which are interpreted as lying normal to the NW–SE direction of maximum compressive stress, were inverted during the Variscan. The inversion megasequence (Fig. 15b) is characterized by the accumulation of continental red beds and molasse in intermontane troughs (essentially the footwalls to the syn-rift half graben) throughout northern England. These sediments are the first to display an important component of southerly derived material from the evolving Variscan mountain front in southern England (Besly 1988).

The nature and intensity of the inversion in northern England varies in a predictable sense with respect to fault trend and depth to detachment. NE–SW trending faults normal to the implied direction of maximum compressive stress such as the Pendle fault (Bowland basin) underwent strong inversion (Fig. 16). In contrast, NW–SE trending faults, such as the Askern–Spital fault (Gainsborough Trough), probably moved in a strike-slip sense with minimal vertical fault displacement (Fig. 8). The major basin bounding faults (NW–SE) display broad, long wavelength, inversion anticlines such as demonstrated on seismic in the Widmerpool Gulf (Fig. 5). These rarely show four-way dip closures and hence are seldom associated with hydrocarbon accumulations. In contrast the NNW–SSE faults with shallow detachments form tight, closed, inversion anticlines which constitute the major hydrocarbon-bearing traps in the East Midlands. Good examples of this style of inversion are seen at Eakring, Welton and Calow as discussed below.

Sequence V1 (Westphalian C–Westphalian D). This sequence records the initial basinwide Variscan inversion event involving the formation of a series of broad inversion anticlines and the widespread uplift and erosion of original syn-rift depocentres. The sequence records the first major input of southerly derived sediments with barren red beds and coarse pebbly sands and conglomerates accumulating in the distal part of the Variscan foreland (Besly 1988). Sequence V1 is poorly represented on seismic data in northern England as a result of widespread erosion below the base Permian unconformity. The sequence can be identified by basal onlap onto inversion anticlines in the hangingwalls of the major faults and is thickest in the synclinal areas between the inversion axes (Fig. 15b).

Sequence V2 (Westphalian D–early Permian). The onset of this the second and strongest pulse of the Variscan orogeny is dated as Westphalian D largely from subsurface evidence in the West Midlands. Its effect has been to strongly invert all major half grabens in northern England (e.g. Widmerpool Gulf and Bowland Basin, Figs 5 & 13). The event is probably a series of compressive pulses initiated in the Westphalian 'D'. However, lack of preserved section of this age precludes any further breakdown. The inversion can be demonstrated at outcrop by the tightly folded lower Carboniferous deposits in the Ribblesdale fold belt in the Bowland Basin (Gawthorpe 1987a). The sequence is characterized by continued uplift and erosion of Dinantian syn-rift basins and accumulation of molasse deposits in small intermontane troughs within an evolving foreland.

Tectono-sedimentary controls on the hydrocarbon habitat

The major tectonic controls on the hydrocarbon distribution in the Carboniferous of northern England are discussed in terms of trap formation and Mesozoic burial. The analysis of play fairway development considers the more subtle controls exerted by tectonic activity on the regional distribution of reservoir, source and seal facies.

Trap formation

All of the 30 or so fields discovered to date in the Carboniferous of northern England display some component of Variscan deformation in their geometry. Taking the East Midlands as an example, it can be seen that the oil and gas fields show a close relationship with the major fault trends (Fig. 17). Inspection of a depth converted regional seismic line across the province shows the major fields to lie in the hangingwalls of faults active during the Dinantian rift phase (Fig. 18). Sub-regional seismic lines with depth conversions are presented for several of the key East Midlands fields to illustrate the structural style associated with the

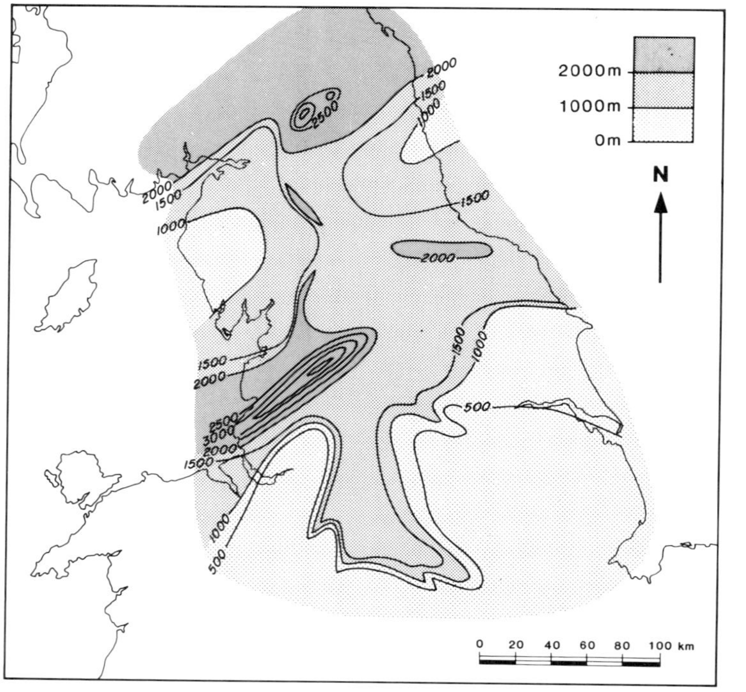

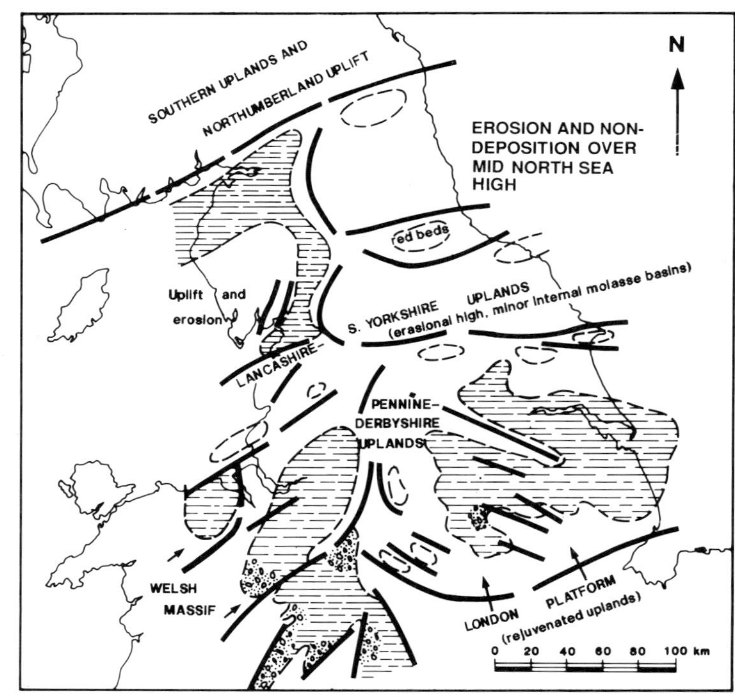

Fig. 15. (a) Estimated Variscan erosion (metres). Map compiled from a combination of regional seismic data and burial/uplift estimates from vitrinite reflectance and fission track data. (b) Palaeofacies map for the inversion megasequence V1–V2 (post Westphalian C). Data from well and seismic control, Wills (1951) and Besley (1988).

S N

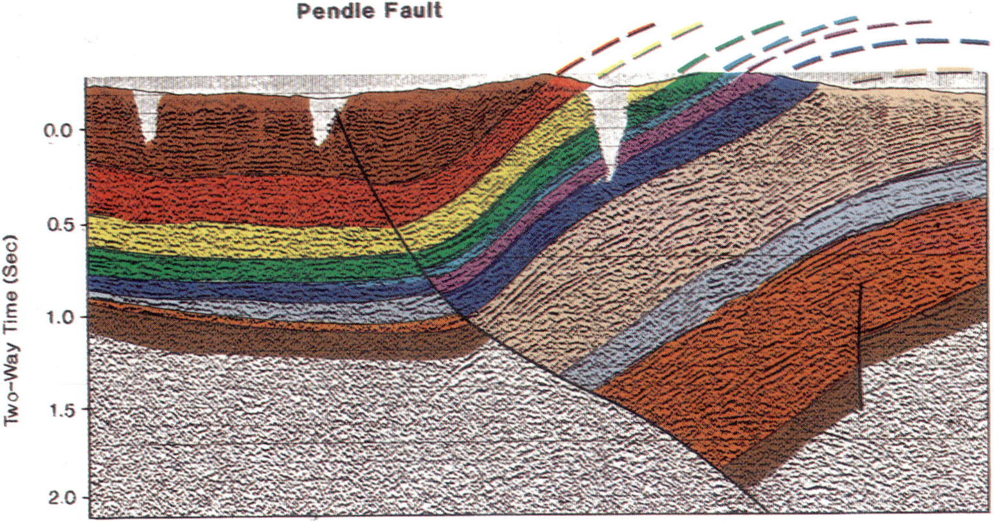

BOWLAND BASIN

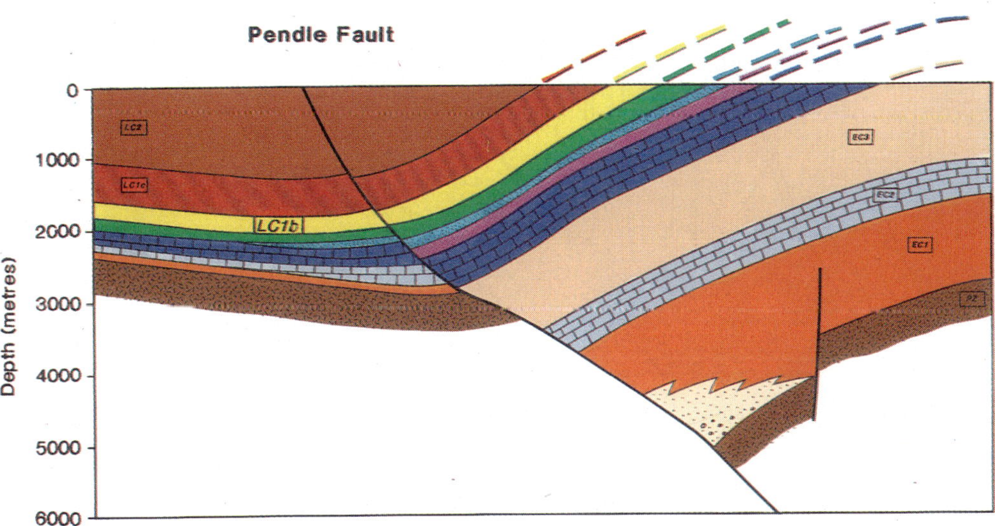

Fig. 16. Seismic and interpreted geological section across the Pendle monocline, southern Bowland basin showing strong Variscan inversion along existing southern border fault of the basin. Seismic data reproduced by kind permission of Amoco UK Ltd.

trapping geometry. For the location of the fields refer to Fig. 17.

Eakring–Dukes Wood (Fig. 19). The field occurs as a series of en echelon inversion anticlines adjacent to the NNW–SSE trending Eakring-Foston fault. Ultimate recoverable reserves are assessed at 7mmbbls of oil, mainly reservoired in a stacked series of delta top channel and mouth bar sands of Namurian–early Westphalian age (Sequence LC1c).

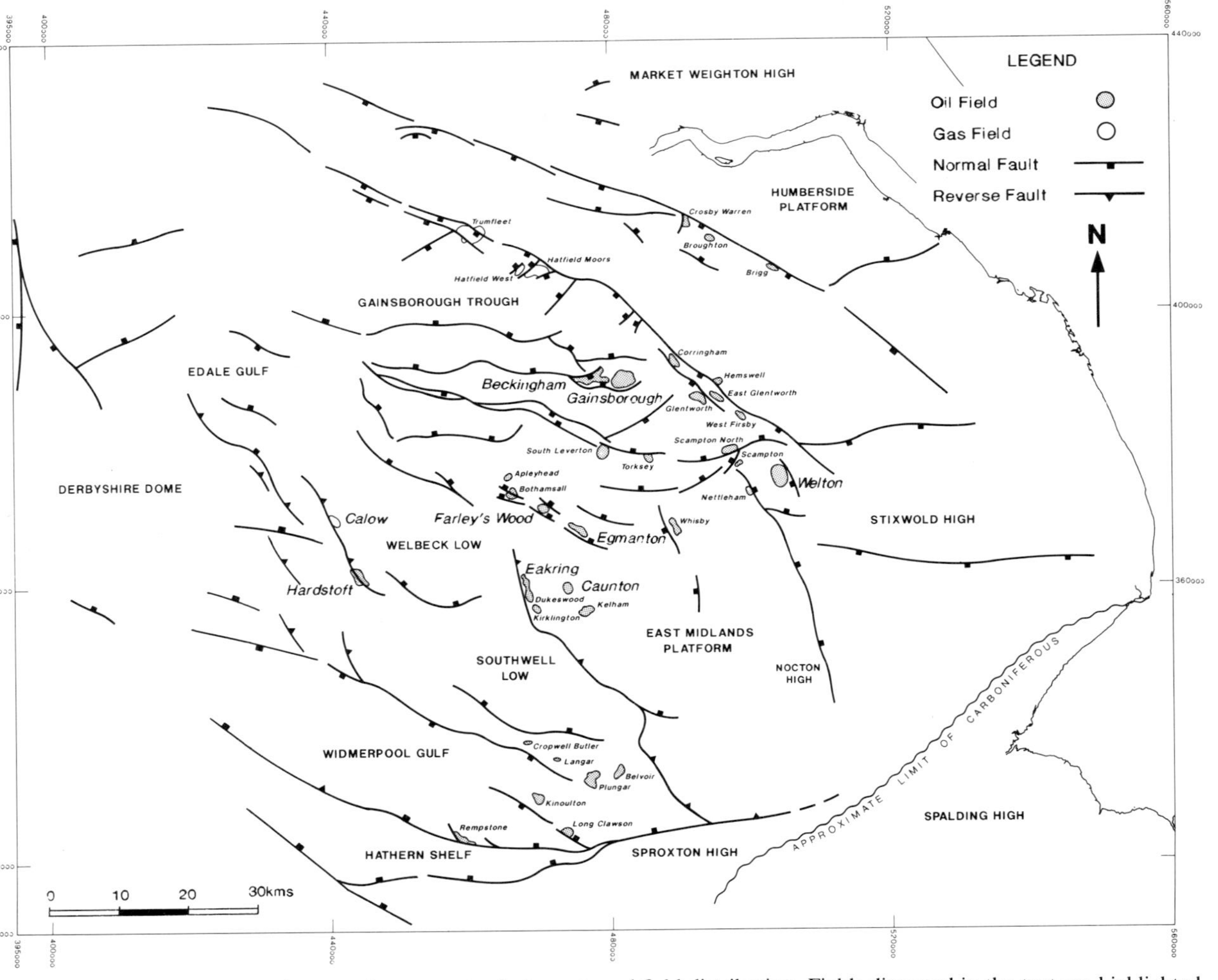

Fig. 17. East Midlands hydrocarbon province: structural elements and field distribution. Fields discussed in the text are highlighted.

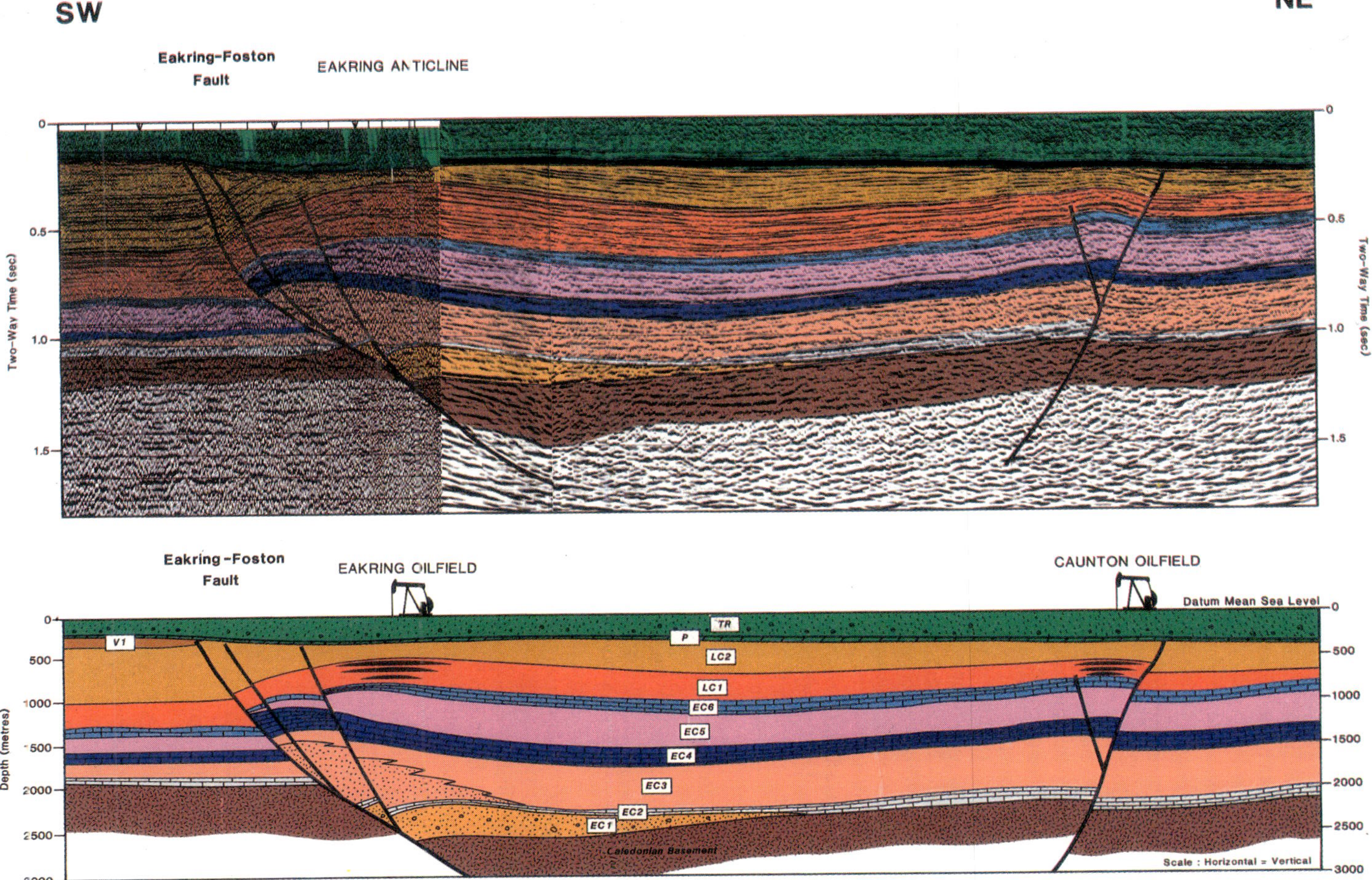

Fig. 19. Composite seismic and geological section, Eakring oilfield, East Midlands (after Fraser *et al.* 1990).

Calow (Fig. 20). The Brimington anticline is perhaps the best example of a tight inversion fold in the East Midlands. The structure is that of a ramp anticline formed on a shallow detachment within the Dinantian. The Calow field is a small gas accumulation (0.5bcf) residing in channel and mouth bar sands in sequence LC1c.

Welton (Fig. 21). Discovered in 1981, Welton is, to date, the largest hydrocarbon accumulation in northern England. Reserves are now assessed at well over 20mmbbls. Production is mainly from delta top channel sands within sequence Lc1c (Rothwell & Quinn 1988). Seismic shows the field to again result from Variscan compressional tectonics.

Egmanton (Fig. 13). The Egmanton field is an inversion anticline formed on a NW-SE trending fault to the south of the Gainsborough Trough (Fig. 3). The field was discovered in 1955 and recoverable reserves are assessed at 3.5mmbbls. The main component of trap formation was again the late Carboniferous, Variscan inversion, as evidenced by the strong subcrop to the base Permian unconformity (Fig. 13). However, the suggestion on seismic of subcropping Dinantian events at the base of the LC1 sequence may indicate a precursor Variscan inversion event.

Beckingham/Gainsborough (Fig. 8). The Beckingham/Gainsborough field was discovered by BP in 1959 in a complexly faulted series of stacked reservoirs of Namurian and Westphalian age (sequences LC1c/LC2). Recoverable reserves from the field are assessed at 13mmbbls of oil plus an additional 6.5bcf of associated gas. The complexly faulted anticlinal structure was apparently formed by inversion on a series of early syn-rift faults lying at depth within the Gainsborough Trough. Closure is complicated by the presence of shallow, listric faults which apparently sole out on Upper Dinantian (EC6) basinal carbonates.

Carboniferous play fairways

The sequence stratigraphic analysis of the Carboniferous described earlier represents a powerful tool with which to analyse the hydrocarbon system by constraining the regional distribution of reservoir, source and seal facies. This is used to develop a play fairway analysis for the Carboniferous.

Syn-rift clastics (Fig. 22). Late Devonian–Courceyan (EC1) syn-rift conglomerates and fluvial sandstones of 'Old Red Sandstone' facies form a potential play in the half graben throughout northern England. Two main areas of early syn-rift clastics can be identified (cf. Leeder & Gawthorpe 1987). Firstly, a narrow zone, a few kilometres wide, adjacent to the border fault zone and, secondly, a wider zone along the margins of the hangingwall dipslope. Well data, for example Caldon Low (Institute of Geological Sciences 1978), and Hathern-1 (Falcon & Kent 1960; Llewellyn & Stabbins 1968, 1970) suggest that sands of this age could be sealed by overlying Courceyan evaporites associated with an early marine transgression into the half graben. The early rifts may contain lacustrine (?marine) source rocks such as have been tentatively identified in the Gun Hill 1 well in the Goyt Trough (Lees & Tait 1945). In hangingwall settings, hydrocarbon prospectivity is reduced by the depth of burial of the reservoir facies and the consequent overmaturity of source rock facies.

The key to this play is the identification of the reservoir facies preserved in the footwall of EC3 border faults where burial depths are minimal (Fig. 23). The seismic example from the Widmerpool Gulf (Fig. 12) shows the case in point, where a wedge of EC1 early syn-rift clastics has been preserved on the footwall. Equivalent reservoir facies, the Redhouse sandstones, have been identified some 50 kms along strike in the BGS Caldon Low borehole (Aikenhead & Chisholm 1982).

Carbonate Shelf Margins (Fig. 23). Outcrop studies in Derbyshire and the Bowland Basin have suggested that Dinantian carbonate reservoirs are most likely to be found in the regressive Chadian (EC2), Asbian (EC4) and Brigantian (EC6) sequences when rimmed shelves were able to develop during periods of reduced tectonic activity. These grainstone margins are restricted to the central England rift system around half grabens such as the Widmerpool Gulf, Goyt Trough, Edale Gulf, Gainsborough Trough and Bowland Basin (Figs 2 & 9b).

The internal facies of the margins comprise carbonate shoals, build-ups, peri-platform talus, debris flows and turbidites: all potential reservoir facies. Boulder beds were developed on the shelf margins at the end Chadian (EC3) and late Asbian (EC5), associated with slope instability caused by renewed extensional tectonic activity at this time (Gawthorpe 1987a). These debris flows are seen at outcrop and in mineral exploration boreholes in the Bowland Basin, where they exhibit up to 30% secondary porosity where dolomitised (Gawthorpe 1987b) and show minor bitumen stain and mineralization (Fig. 11a).

The grainstone margins can also be identified in the subsurface on modern, high quality reflection seismic data (Fig. 5). The margins are characterized by a progradational sequence of laterally and vertically accreting clinoforms. These change character along the hangingwall dipslope into hummocky clinoforms and pass basinwards into a series of parallel reflections which are associated with distal calciturbidite sedimentary facies. The areal distribution of the carbonate margin play is thus limited to a 0.5 to 2 km wide fairway which rims the margins of the deeper half graben (Fig. 9b). The narrowness of the play fairway highlights the need to identify the margins on seismic, since present well control is of insufficient density, from an exploration point of view, to map the extent of the rimmed shelves in the subsurface.

In the East Midlands, the Brigantian (EC6) grainstone margin play is considered the most prospective given the presence of onlapping Namurian pro-delta mudstones which should provide both source and top seal (sequence LC1a). In the absence of fortuitous structural or fault dependent closure, success of the play requires lateral seal to be provided by the expected change from shelf margin facies into a tightly cemented cyclic shelf carbonate developed shelfwards of the margin. *Clastic delta systems (Fig. 24).* This play is by far the most important in the Carboniferous of northern England and has been described in some detail by Fraser *et al.* (1990).

Reservoirs are developed in both the delta top and pro-delta environments. Channel and mouth bar sands form the main producing reservoir facies on the delta top, with channel sands exhibiting the more favourable reservoir characteristics. Turbidite sands also form potential reservoirs in the pro-delta setting. Antecedent rift bathymetry exerted a significant control on the delta systems (Fig. 14b); delta top and mouth bar reservoirs being best developed where they axially infill remnant rift bathymetry.

Regional sealing facies on the delta top is provided by marine bands deposited during maximum flooding events and interdistributary bay fines. As discussed by Heckel (this volume) the marine bands are likely to have been glacio-eustatically controlled. The thin marine bands which overstep delta top environments have been shown to be adequate seals for oil accumulations, but are suspect for gas (Fraser *et al.* 1990). In the pro-delta, shale-rich environment, marine mudstones provide excellent seals for turbidite sandstone reservoirs.

The richest source rocks identified are interpreted as distal pro-delta mudstones which were deposited in advance of the southerly prograding delta systems (Fig. 25). In the East Midlands the pro-delta source facies (sequence LC1a) (Edale shale and equivalents) is well developed in the remnant half graben but is poorly represented on the East Midlands shelf.

The prodelta source facies are predominantly oil prone. Biomarker studies and carbon isotope analyses clearly indicate a mixed marine/terrestrial derived kerogen (Fraser *et al.* 1990). One explanation for the relationship between the deltas and the source facies involves the introduction of fresh water from the delta top to the isolated fault bound half graben causing stratification of the waters and resultant regional anoxia below the pycnocline.

Both the pro-delta and delta-top reservoir systems prograde across the distal pro-delta source rocks forming hydrocarbon migration pathways. Delta top coal swamps, developed regionally during the Westphalian (sequence LC2) over northern England, provide gas and occasional oil-prone source rocks which are likely to be in good communication with the delta top channel and mouth bar reservoirs.

Mesozoic burial history and hydrocarbon generation

As discussed earlier, the main trap forming event in northern England was the late Carboniferous–early Permian Variscan orogeny which created inversion anticlines in the hangingwalls of the major half graben border faults. The considerable uplift and erosion which resulted from this event led to the effective 'freezing' of source rock generation and migration. Therefore, Mesozoic burial leading to renewed generation from the source rocks, post-dating this trap forming event, is critical to the success of northern England as a hydrocarbon province. Several key areas of Mesozoic hydrocarbon generation have been identified (Fig. 26).

East Midlands. The dominant phase of generation in the East Midlands province occurred during Mesozoic times. This has been confirmed by recent fission track studies (Green 1989). Restored, pre-Tertiary uplift maps indicate that the basal Namurian pro-delta shales would have been generating liquid hydrocarbons over most of the area by late Cretaceous times. Gas gen-

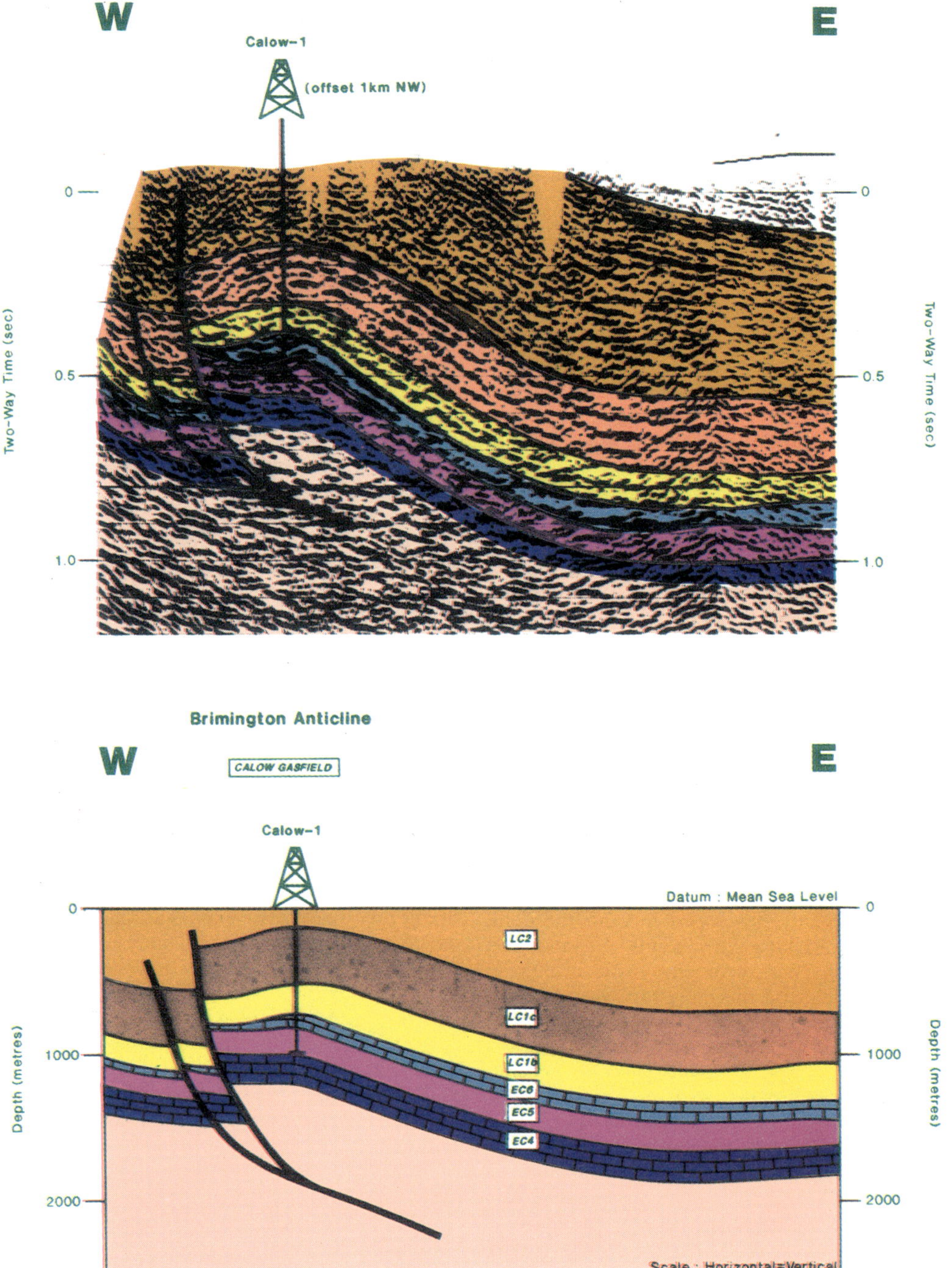

Fig. 20. Composite seismic and interpreted geological section across the Calow gasfield, East Midlands.

eration is likely to have been short lived and restricted to the deeper parts of the Gainsborough Trough and Widmerpool Gulf. All hydrocarbon generation was 'frozen' in the Tertiary by some 1000 metres of regional uplift. A degree of remigration of hydrocarbons took place in the late Tertiary as a result of a regional eastwards tilting of the basin by about

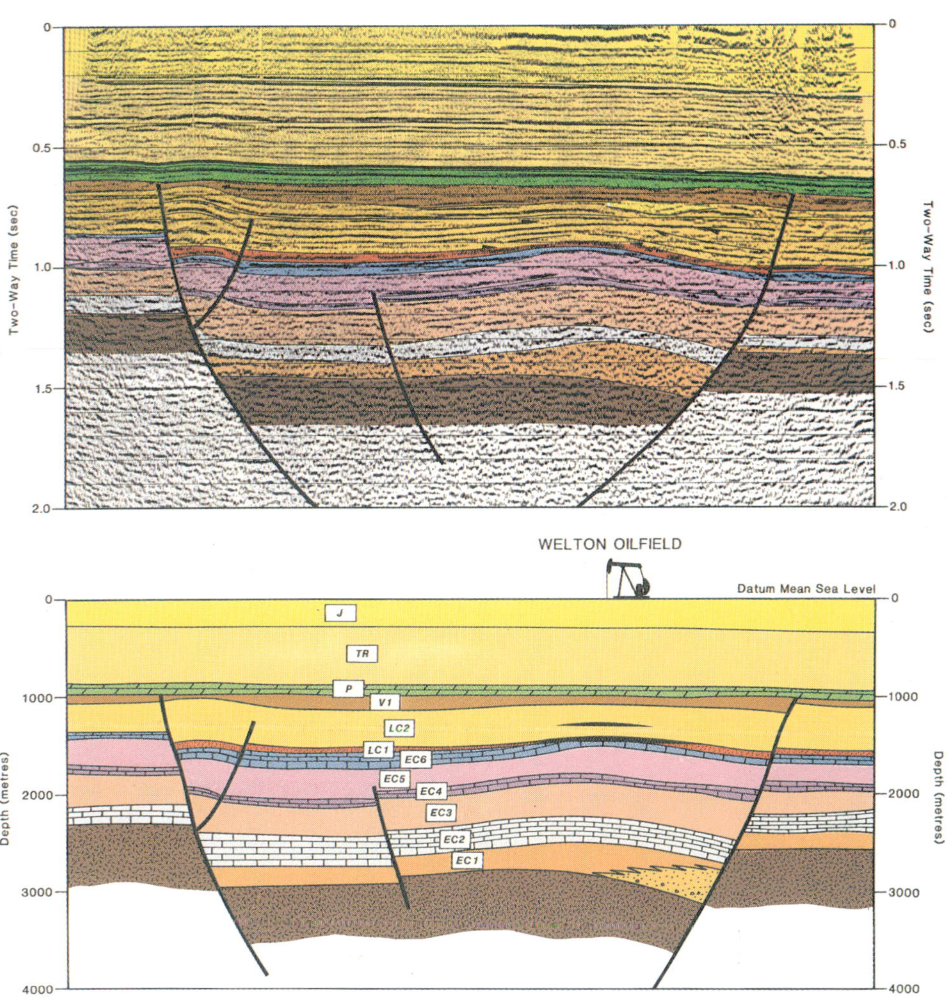

Fig. 21. Seismic and interpreted geological section across the Welton oilfield, East Midlands.

2° towards the Southern North Sea Basin (Fraser *et al.* 1990).

Cleveland Basin. The Cleveland Basin is characterized by superimposed Carboniferous and Cretaceous rift systems. The Carboniferous rift event formed the 'kitchens' for the deposition of the source facies, whereas late Jurassic–early Cretaceous extension matured the sediments through the oil window and into the gas zone by the mid Cretaceous. Generation peaked during the late Cretaceous and ceased during Oligocene basin inversion. Estimates of Tertiary uplift from sonic velocity and vitrinite data range from 1800 metres in the south of the basin (Malton) to over 3000 m in the north (Eskdale) in the hangingwall of the Lunedale fault.

SYN–RIFT SANDSTONES

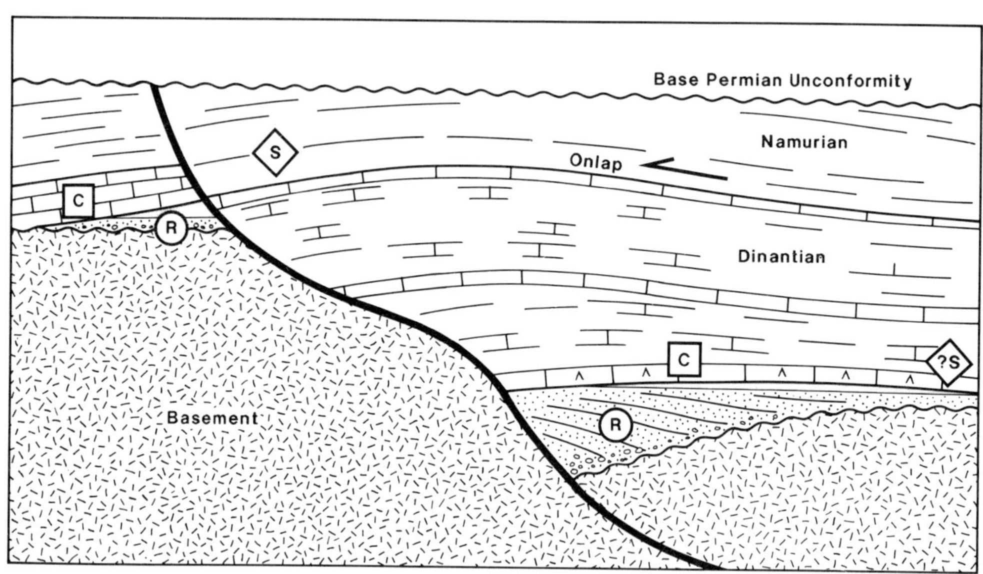

RESERVOIR : Upper Devonian - Courceyan syn-rift conglomerates and fluvial sandstones

SEAL : Courceyan evaporites and shales.

SOURCE : Early syn-rift (Courceyan) basinal shales - not proven.

Cross fault juxtaposition with pro-delta shales.

Carboniferous and Mesozoic charge.

CRITICAL FACTORS : (i) Excessive depths of reservoir burial in hanging wall.

(ii) Limited distribution.

(iii) Play relies heavily on unproven early syn-rift source.

Fig. 22. Early syn-rift clastic play fairway summary. Ⓢ = source, Ⓒ = seal, Ⓡ = reservoir

Manx–Furness. The important part of the burial in this area is related to the rapid subsidence of the basin during the Permo-Triassic. The onset of oil generation is taken as the mid-Triassic, with peak oil generation by the late Triassic and gas generation occurring from mid–late Jurassic through to the early Tertiary. This important phase of Mesozoic generation sourced the Morecambe gas field (5tcf reserves) in the offshore part of the basin (Ebbern 1981). This

CARBONATE SHELF MARGINS

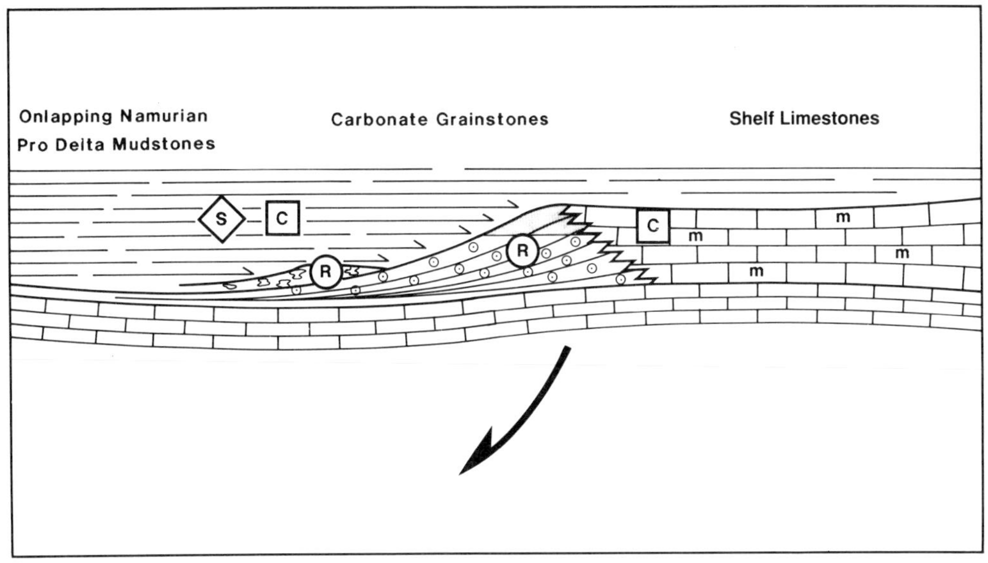

RESERVOIR : Chadian/late Asbian/early Brigantian grainstones

SEAL : Onlapping late Dinantian and Namurian pro-delta mudstones
Lateral facies change into tight shelfal limestones

SOURCE : Interdigitating and onlapping late Dinantian/early Namurian mudstones
(Bowland Shales and equivalents)

CRITICAL FACTORS : (i) Secondary porosity (dolomitisation) required.

(ii) Lateral seal.

(iii) Narrow belt, difficult to identify.

Fig. 23. Dinantian carbonate shelf margin play fairway summary.

phase of generation was frozen during the early Tertiary by uplift and erosion possibly related to rifting between Europe and Greenland.

Cheshire Basin. An important area of implied Mesozoic maturation and generation occurs in the Cheshire basin. Most of the subsidence in this area is of Permo-Triassic age with further passive infill of the area in the Jurassic and Cretaceous. The excessive Mesozoic burial suggests that gas is the only hydrocarbon phase likely to be encountered in the basin. This area was also affected by considerable uplift and erosion (*c.* 1000 m) in the Tertiary.

CLASTIC DELTA SYSTEMS

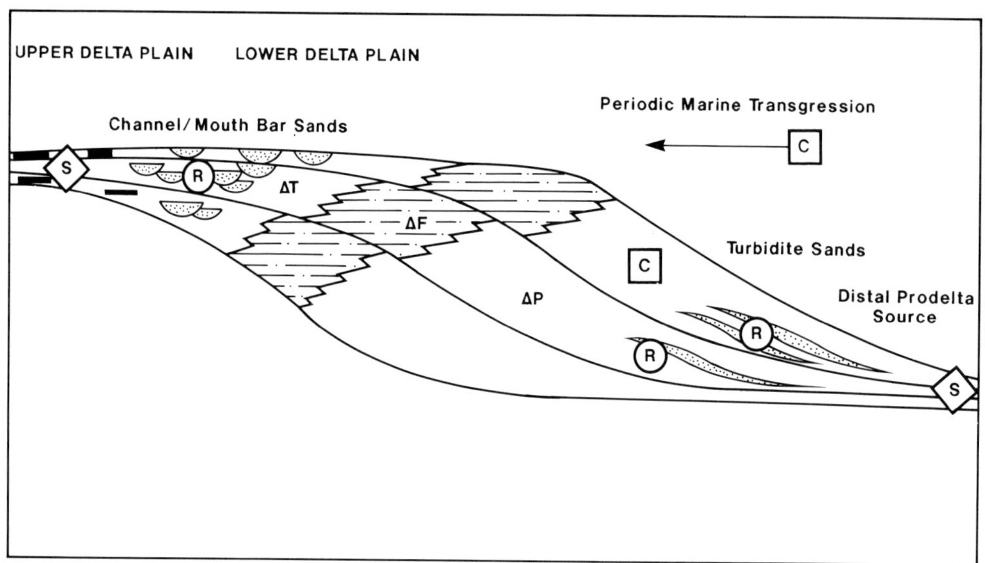

RESERVOIR : ΔT channel/mouth bar sands.

ΔP turbidite sands.

SEAL : ΔT marine bands – good; lacustrine/overbank muds – moderate/poor.

ΔP mudstones – good.

SOURCE : ΔT and ΔP reservoirs prograde over distal pro delta-source rocks.

– Delta top coals.

– Late Carboniferous/Mesozoic charge.

CRITICAL FACTORS : (i) Reservoir quality downgraded beyond 2500m burial depths.

(ii) Dinantian pro-delta source tends to be gas prone.

(iii) ΔT seals poor for trapping gas.

Fig. 24. Carboniferous clastic delta system play fairway summary. △T = delta top, △F = delta Front, △P = pro-delta

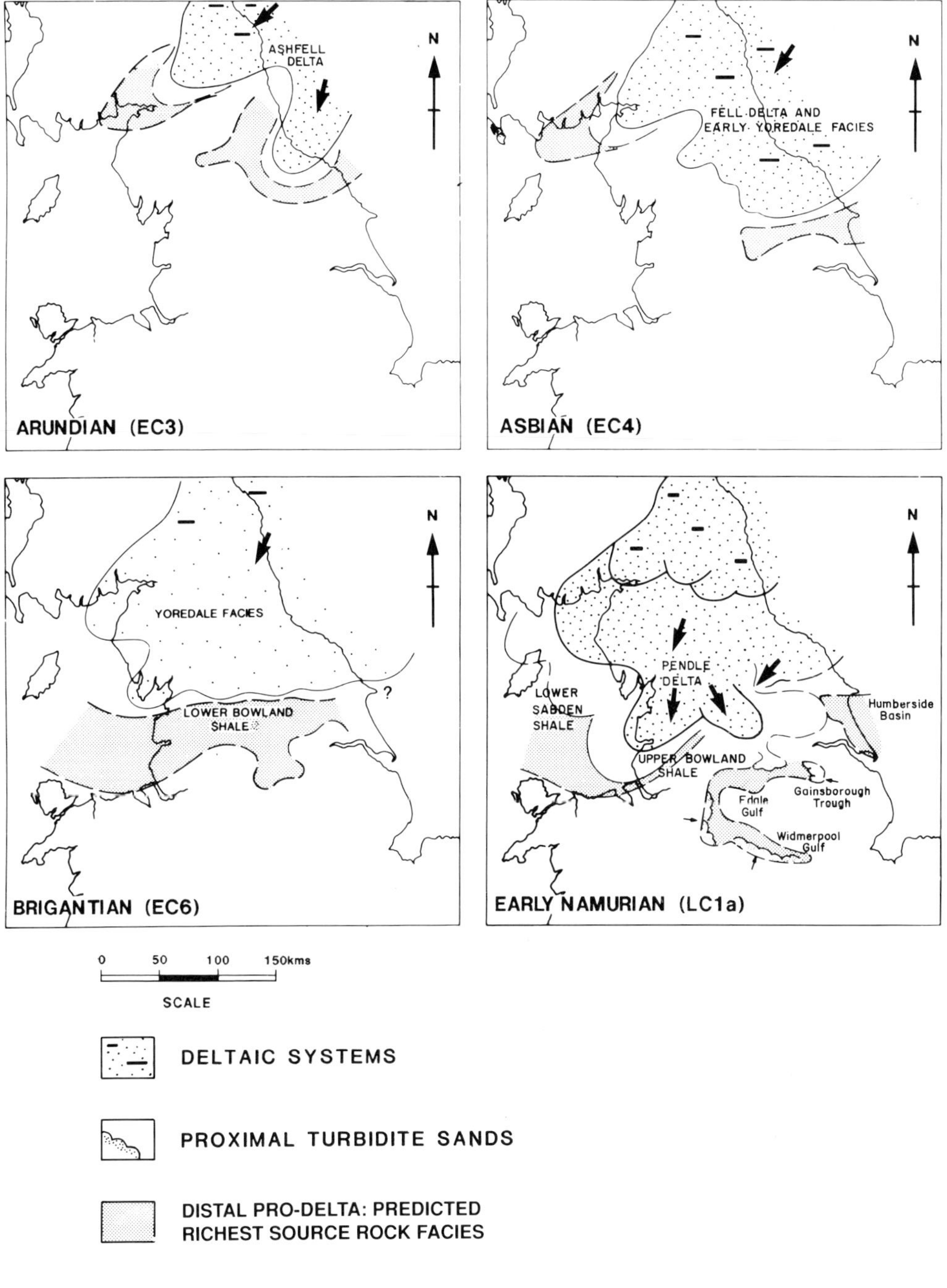

Fig. 25. Delta advance during the middle to late syn-rift and early post-rift and its relationship with source rock distribution and age in northern England.

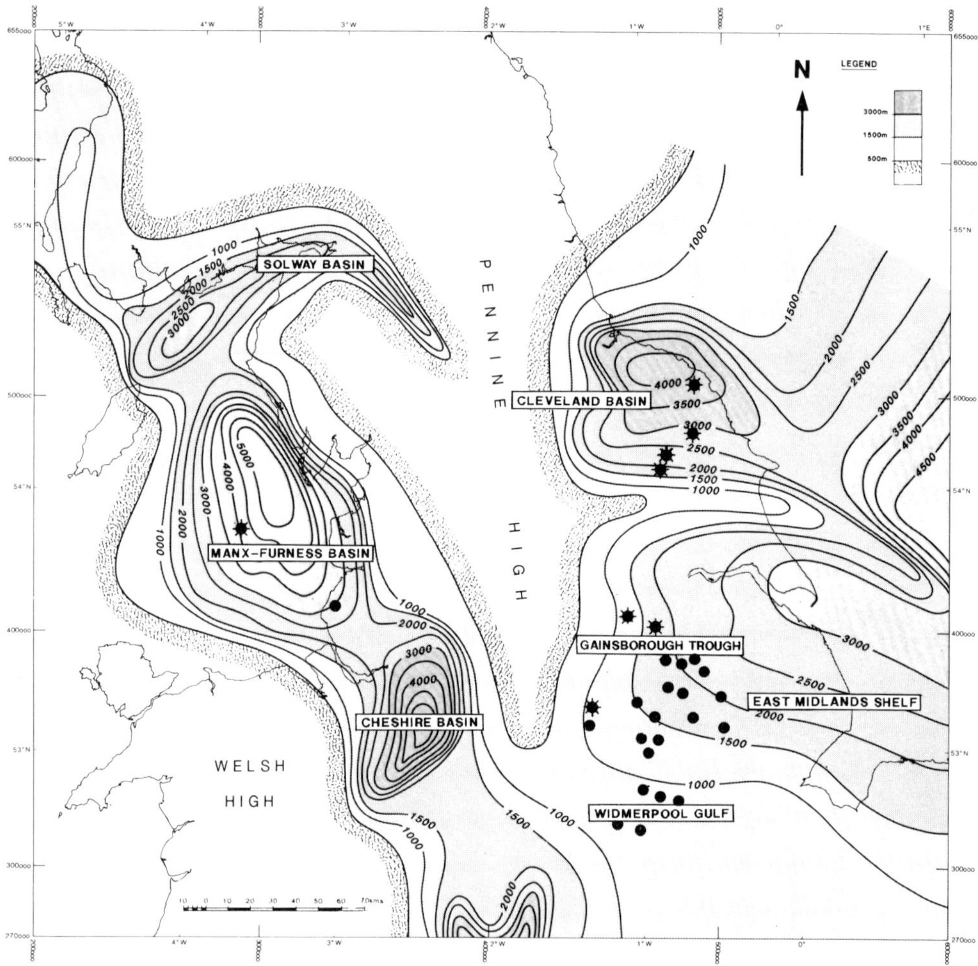

Fig. 26. Restored post-Carboniferous isopachs in metres and the location of the major hydrocarbon discoveries in northern England. Data from a number of sources including Whittaker (1985) and vitrinite/fission track studies. Note that gas is the main hydrocarbon phase in regions subject to over 3000 m of post Carboniferous burial and the East Midlands oilfields are concentrated in an area of moderate burial. There are no hydrocarbon discoveries along the axis of the Pennine high where Mesozoic burial has been negligible.

Solway Basin. Fission track studies around the Lake District (Green 1986) and isopach mapping suggest that this basin attained maximum generation during the Mesozoic. The greater part of the subsidence in this area is related to Permo-Triassic rifting. Onset of oil generation is thought to be in the late Triassic–early Jurassic with peak generation in the late Cretaceous. The basin has probably been subject to some 500 metres of late Tertiary uplift (Fraser *et al.* 1990).

Vitrinite reflectance and fission track data indicate that areas such as the central Pennines and the Northumberland and Stainmore Troughs reached maximum burial during the late Carboniferous. Significantly, this was prior to the trap forming Variscan inversion event; the only remnants of this earlier phase of generation being the widespread and well documented surface bitumen occurences in the region (Lees & Cox 1937).

Conclusions

The structures which controlled the Carboniferous tectono-stratigraphic development of northern England were inherited from the earlier Caledonian orogeny. These events imparted a strong NW–SE and NE–SW tectonic grain which is evidenced on surface and subsurface data throughout northern England. The subsequent Variscan plate cycle, which involved the closure of the Rheic and Rheno-Hercynian oceans, controlled the development of syn-rift, post-rift and inversion megasequences from late Devonian to early Permian times. Sequences developed within the Carboniferous are seen to be controlled by episodic rifting and periodic fault reactivation with eustatic sea level changes providing only minor control and only at the sub-sequence level.

In the south of the province during the Dinantian, carbonate environments were extensivly developed in syn-rift basins starved of terrigeneous clastics. The north of the region was dominated from early Dinantian times onward by a southward prograding terrigenous clastic delta system. The carbonate environments in the south were finally drowned in the early post-rift (Pendleian) when the supply of terrigenous clastic sediments outpaced subsidence for the first time, allowing rapid southward progradation of the delta systems.

Over 70 years of exploration in northern England has resulted in the discovery of 75mmbbl of recoverable reserves. Average field size is assessed at 2mmbbls at a historical finding rate of one discovery in four exploration wells. All hydrocarbon discoveries to date within the Carboniferous have shown some element of Variscan deformation in their geometry. Tectonics have also exerted a subtle, but important control on play fairway evolution. The main source rocks, developed in distal pro-delta environments, are restricted to the syn-rift depocentres. Syn-rift siliciclastic reservoirs, associated with the early rift phase are located within the isolated, rifted, half-graben. Carbonate grainstone reservoirs are controlled by the rift topography and rim the margins of the deeper half graben in the south of the province. Delta top channel and mouth bar reservoirs are best developed where they axially infill remnant rift bathymetry.

Mesozoic burial, ensuring hydrocarbon generation post-Variscan trap formation, is considered to be the main control on the present day distribution of hydrocarbons in northern England. Areas such as the central Pennines and Northumberland and Solway Basins which received limited Mesozoic burial are considered to have poor hydrocarbon potential. The Cleveland, Manx–Furness and Cheshire Basins where Mesozoic burial was excessive have gas as the major hydrocarbon phase. The East Midlands, where Mesozoic burial and post Variscan trap modification have been moderate, has therefore emerged as the most successful oil province in northern England.

Permission to publish this paper has been granted by the British Petroleum Co plc. We are also grateful to partners Gas Council (Exploration), Elf Oil and Gas Ltd and Blackland Oil for permission to publish seismic data from the East Midlands. Critical readings of the manuscript and many helpful suggestions were provided by R. J. Bailey and M. R. Leeder. We would also like to thank E. A. Kay, J. Hossack, S. D. Knott and B. C. Mitchener who made important contributions to the study. We also acknowledge the contributions of Dr G. A. L. Johnson and Dr R. E. Ll. Collier to our analysis of the Stainmore Trough. R. L. G. would like to thank the Durham University Research Foundation and the Society of Fellows for support whilst he was at the Department of Geological Sciences, University of Durham. Sincere thanks to Paul Batey for the drafting and Lisa Phelan who typed the manuscript.

References

Aikenhead, N. & Chisholm, J. I. 1982. *A standard nomenclature for the Dinantian formations of the Peak District of Derbyshire and Staffordshire.* Report of the Institute of Geological Sciences, 82/8.

——, —— & Stevenson, I. P 1985. *Geology of the country around Buxton, Leek and Bakewell.* Memoir of the British Geological Survey.

Andre, L., Hetrogen, J. & Deutsch, S. 1986. Ordovician–Silurian magmatic provinces in Belgium and the Caledonian orogeny in Middle Europe. *Geology*, **14**, 879–882.

Arthurton, R. S., Johnson, E. W. & Mundy, D. S. C 1988. *Geology of the country around Settle.* Memoir of the British Geological Survey.

Beamish, D. & Smythe, D. K. 1986. Geophysical images of the deep crust: The Iapetus suture. *Journal of the Geological Society, London*, **143**, 489–497.

Besly, B. M. 1988. Palaeogeographic implications of late Westphalian to early Permian red beds, central England. *In*: Besly, B. M. & Kelling, G. (eds). *Sedimentation in a Synorogenic Basin Complex: The Carboniferous of Northwest Europe*. Blackie, Glasgow, 200–221.

Bott, M. H. P. 1976. Formation of sedimentary basins of graben type by extension of the continental crust. *Tectonophysics*, **36**, 77–86.

—— & Johnson, G. A. L. 1967. The controlling mechanism of Carboniferous cyclic sedimentation. *Quarterly Journal of the Geological Society, London*, **122**, 421–441.

Burgess, I. C. & Holliday, D. W. 1979. Geology of the country around Brough-Under-Stainmore. *Memoir of the British Geological Survey.*

CALVER, M. A. 1968. The distribution of Westphalian marine faunas in northern England and adjoining area. *Proceedings of the Yorkshire Geological Society*, **44**, 479–496.

COLLINSON, J. D., JONES, C. M. & WILSON, A. A. 1977. The Marsdenian (Namurian R2) succession West of Blackburn: Implications for the evolution of Pennine delta systems. *Geological Journal*, **12**, 59–76.

DUNHAM, K. C. & WILSON, A. A. 1985. Geology of the Northern Pennine oilfield: Volume 2 Stainmore to Craven. *Economic Memoir of the British Geological Survey.*

EBDON, C. C., FRASER, A. J., HIGGINS, A. C., MITCHENER, B. C. & STRANK, A. R. E. 1990. The Dinantian Stratigraphy of the East Midlands: A seismostratigraphic approach. *Journal of the Geological Society, London*, **147**, 519–536.

EBBERN, J. 1981. The geology of the Morecambe gas field. *In*: ILLING, L. V. & HOBSON, G. D. (eds). *Petroleum Geology of the Continental Shelf of North-West Europe*, Institute of Petroleum, London, 485–493.

FRASER, A. J., NASH, D. F., STEELE, R. P. & EBDON, C. C. 1990. A regional assessment of the intra-Carboniferous play of northern England. *In*: BROOKS, J. (ed). *Classic Petroleum Provinces*. Geological Society, London, Special Publication **50**, 417–440.

FALCON, N. L. & KENT, P. E. 1960. Geological results of petroleum exploration in Britain 1945–1957. *Memoir of the Geological Society, London*. No. 2.

GAWTHORPE, R. L. 1987a. Tectono-sedimentary evolution of the Bowland Basin, N England, during the Dinantian. *Journal of the Geological Society, London*, **144**, 59–71.

—— 1987b. Burial dolomitisation and porosity development in a mixed carbonate – clastic sequence: An example from the Bowland Basin, Northern England. *Sedimentology*, **34**, 533–558.

—— & CLEMMEY, H. 1985. Geometry of submarine slides in the Bowland Basin (Dinantian) and their relation to debris flows. *Journal of the Geological Society, London*, **142**, 555–565.

—— GUTTERIDGE, P. & LEEDER, M. R. 1989. Late Devonian and Dinantian basin evolution in Northern England and North Wales. *In*: ARTHURTON, R. S., GUTTERIDGE, P. & NOLAN, S. C. (eds). *The role of tectonics in Devonian and Carboniferous sedimentation in the British Isles.* Yorkshire Geological Society (Occasional Publication No. 6), 1–23.

GEORGE, T. N. 1958. Lower Carboniferous palaeogeography of the British Isles. *Proceedings of the Yorkshire Geological Society*, **31**, 227–318.

—— 1978. Eustacy and tectonics: Sedimentary rhythms and stratigraphic units in British Dinantian correlation. *Proceedings of the Yorkshire Geological Society*, **42**, 229–262.

GRAYSON, R. F., & OLDHAM, L. 1987. A new structural framework for the northern British Dinantian as a basis for oil, gas and mineral exploration. *In*: MILLER, J., ADAMS, A. E. & WRIGHT, V. P. (eds). *European Dinantian Environments*. Wiley, Chichester, 33–59.

GREEN, P. F. 1986. On the thermo-tectonic evolution of northern England: Evidence from fission track analysis. *Geological Magazine*, **123**, 493–506.

—— 1989. Thermal and tectonic history of the East Midlands Shelf (onshore UK), and surrounding regions assessed by Apatite Fission Track Analysis. *Journal of the Geological Society, London*, **146**, 755–773.

GUION, P. D. & FIELDING, C. R. 1987. Westphalian A and B sedimentation in the Pennine basin, UK. *In*: BESLY, B. M. & KELLING, G. (eds). *Sedimentation in a synorogenic basin complex: The Carboniferous of northwest Europe*. Blackie, Glasgow, 153–177.

GUTTERIDGE, P. 1987. Diantian sedimentation and basement structure of the Derbyshire Dome. *Geological Journal*, **22**, 25–41.

—— 1989. Controls on carbonate sedimentation in a Brigantian intrashelf basin, Derbyshire. *In*: ARTHURTON, R. S. GUTTERIDGE, P. & NOLAN, S. C. (eds). *The role of tectonics in Devonian and Carboniferous sedimentation in the British Isles.* Yorkshire Geological Society (Occasional Publication No. 6), 171–187.

HUBBARD, R. J. 1988. Age and significance of sequence boundaries on Jurassic and early Cretaceous rifted continental margins. *American Association of Petroleum Geologists Bulletin*, **72**, 49–72.

——, PAPE, J. & ROBERTS, D. G. 1985a. Depositional sequence mapping as a technique to establish tectonic and stratigraphic framework and evaluate hydrocarbon potential on a passive continental margin. *In*: BERG, O. R. & WOOLVERTON, D. G. (eds). *Seismic Stratigraphy II*, American Association of Petroleum Geologists Memoir, **39**, 79–92.

——, —— & —— 1985b. Depositional sequence mapping to illustrate the evolution of a passive continental margin. *In*: BERG, O. R. & WOOLVERTON, D. G. (eds). *Seismic Stratigraphy II*, American Association of Petroleum Geologists Memoir, **39**, pp 93–115.

INSTITUTE OF GEOLOGICAL SCIENCES. 1978. IGS Boreholes 1977. *Report of the Institute of Geological Sciences*, No. 78/21, 1–24.

JACKSON, J. A. 1987. Active normal faulting and crustal extension. *In*: COWARD, M. P., DEWEY, J. F. & HANCOCK, P. L. (eds). *Continental Extensional Tectonics.* Geological Society, London Special Publication, **28**, 3–17.

JOHNSON, G. A. L. 1960. Palaeogeography of the northern Pennines and part of NE England during deposition of Carboniferous cyclothemic deposits. *International Geological Congress XXI Session Norden*, **12**, 118–128.

KENT, P. E. 1966. Structure of the concealed Carboniferous rocks of NE England. *Proceedings of the Yorkshire Geological Society*, **35**, 323–352.

KIMBELL, G. S., CHADWICK, R. A., HOLLIDAY, D. W. & WERNGREN, O. C. 1989. The structure and evolution of the Northumberland Trough from new seismic reflection data and its bearing on modes of continental extension. *Journal of the Geological Society, London*, **146**, 775–787.

KIRTON, S. R. 1984. Carboniferous volcanicity in England with special reference to the Westphalian of the E and W Midlands. *Journal of the Geological Society, London*, **141**, 161–170.

LEEDER, M. R. 1974. Lower Border Group (Tournasian) fluvio-deltaic sedimentation and the palaeogeography of the Northumberland Basin. *Proceedings of the Yorkshire Geological Society*, **40**, 129–180.

—— 1982. Upper Palaeozoic basins of the British Isles – Caledonide inheritance versus Hercynian plate margin processes. *Journal of the Geological Society, London*, **139**, 479–491.

—— 1987. Tectonic and palaeogeographic models for Lower Carboniferous Europe. *In*: MILLER, J., ADAMS, A. E. & WRIGHT, V. P. (eds). *European Dinantian Environments*, Wiley, Chichester, 1–19.

—— 1988. Recent developments in Carboniferous geology: A critical review with implications for the British Isles and NW Europe. *Proceedings of the Geologists' Association*, **99**, 73–100.

—— & GAWTHORPE, R. L. 1987. Sedimentary models for extensional tilt-block/half-graben basins. *In*: COWARD, M. P., DEWEY, J. F. & HANCOCK, P. L. (eds). *Continental Extensional Tectonics*. Geological Society, London, Special Publication, **28**, 139–152.

—— & MCMAHON, A. H. 1988. Upper Carboniferous (Silesian) basin subsidence in northern Britain. *In*: BESLY, B. M. & KELLING, G. (eds). *Sedimentation in a synorogenic basin complex: The Carboniferous of northwest Europe*. Blackie, Glasgow, 43–52.

LEES, G. M. & COX, P. T. 1937. The geological results of the recent search for oil in Great Britain by the D'Arcy Exploration Company Limited. *Quarterly Journal of the Geological Society, London*, **93**, 156–194.

—— & TAITT, A. H. 1945. The geological results of the search for oilfields in Great Britain. *Quarterly Journal of the Geological Society, London*, **101**, 255–317.

LLEWELLYN, P. G. & STABBINS, R. 1968. Demonstration: Core material from the Anhydrite series, Carboniferous Limestone, Hathern Borehole, Leicestershire. *Proceedings of the Geological Society, London*, **1650**, 171–186.

——, —— 1970. The Hathern Anhydrite Series, Lower Carboniferous, Leicestershire, England. *Transactions of the Institution of Mining and Metallurgy*, **79b**, B1–15.

MACDONALD, R., GASS, K. N., THORPE, R. S. & GASS, I. G. 1984. Geochemistry and petrogenesis of the Derbyshire Carboniferous basalts. *Journal of the Geological Society, London*, **141**, 147–159.

MILLER, J. & GRAYSON, R. F. 1982. The regional context of Waulsortian facies in northern England. *In*: BOLTON, K., LANE, H. R., & LEMONE, D. U. (eds). *Symposium on the palaeoenvironmental setting and distribution of the Waulsortian facies*. The El Paso Geological Society and University of Texas at El Paso, 17–30.

MILLS, D. A. C. & HULL, J. H. 1976. Geology of the country around Barnard Castle. *Memoir of the British Geological Survey*.

MITCHUM, R. U., VAIL, P. R. & THOMPSON, S. 1977. The depositional sequence as a base unit for stratigraphic analysis. *In*: PAYTON, C. E. (ed). *Seismic stratigraphy – applications to hydrocarbon exploration*. American Association of Petroleum Geologists Memoir, **26**, 53–62.

PHAROAH, T. C., MERRIMAN, R. J., WEBB, D. C. & BECKINSDALE, R. D. 1987. The concealed Caledonides of eastern England: preliminary results of a multidisciplinary study. *Proceedings of the Yorkshire Geological Society*, **46**, 335–369.

RAMSBOTTOM, W. H. C. 1973. Transgressions and regressions in the Dinantian: a new synthesis of British Dinantian stratigraphy. *Proceedings of the Yorkshire Geological Society*, **39**, 567–607.

—— 1977. Major cycles of transgression and regression (mesotherms) in the Namurian. *Proceedings of the Yorkshire Geological Society*, **41**, 261–291.

—— 1981. Eustasy, sea level and local tectonics, with examples from the British Carboniferous. *Proceedings of the Yorkshire Geological Society*, **43**, 473–482.

READING, H. G. 1964. A review of the factors affecting the sedimentation of the Millstone Grit (Namurian) in the Central Pennines. *In*: VAN STRAATEN, L. M. J. U., (ed.). *Deltaic and Shallow Marine Deposits*, Elsevier, Amsterdam.

ROTHWELL, N. R. & QUINN, P. 1987. The Welton Oilfield. *In*: BROOKS, J. & GLENNIE, K. (eds). *Petroleum Geology of North West Europe*. Graham & Trotman, London, 181–189.

SELLWOOD, E. B. & THOMAS, J. M. 1986. Variscan facies and structure in central S.W. England. *Journal of the Geological Society, London*, **143**, 199–207.

SMITH, K., SMITH, N. J. P. & HOLLIDAY, D. 1985. The deep structure of Derbyshire. *Geological Journal*, **20**, 215–225.

SOPER, N. J. & HUTTON, D. H. W. 1984. Late Caledonian sinistral displacements in Britain: Implications for a three plate collision model. *Tectonics*, **3**, 781–794.

——, WEBB, B. C. & WOODCOCK, N. H. 1987. Late Caledonian (Acadian) transpression in North West England: timing, geometry and geotectonic significance. *Proceedings of the Yorkshire Geological Society*, **46**, 175–192.

STEELE, R. P. 1988. The Namurian sedimentary history of the Gainsborough Trough. *In*: BESLY, B. M. & KELLING, G. (eds). *Sedimentation in a synorogenic basin complex: The Carboniferous of northwest Europe*, Blackie, Glasgow, 101–113.

STEIN, R. S. & BARIENTOS, S. E. 1985. Planer high angle faulting in the Basin and Range: Geodetic analysis of the 1983 Borah Peak, Idaho, earthquake. *Journal of Geophysical Research*, **90**, 11355–11366.

STEVENSON, I. P. & GAUNT, G. D. 1971. Geology of

the country around Chapel-en-le-Frith. *Memoir of the British Geological Survey.*

STRANK, A. R. E. 1987. The stratigraphy and structure of Dinantian strata in the East Midlands, UK. *In*: MILLER, J., ADAMS, A. E. & WRIGHT, V. P. (eds). *European Dinantian Environments*, Wiley, Chichester, 157–175.

TURNER, J. S. 1949. The deeper structure of Central and Northern England. *Proceedings of the Yorkshire Geological Society*, **27**, 280–297.

VAIL, P. R. & MITCHUM, R. M. 1977. Seismic stratigraphy and global changes of sea level, Part 1: overview. *In*: PAYTON, C. E. (ed). *Seismic stratigraphy – applications to hydrocarbon exploration*, American Association of Petroleum Geologists, Memoir **26**, 51–52.

WALKDEN, G. M. 1977. Volcanic and erosive events on an Upper Visean carbonate platform, north Derbyshire. *Proceedings of the Yorkshire Geological Society*, **41**, 347–367.

WALKDEN, G. M. 1982. Field guide to the lower Carboniferous rocks of the south east margin of the Derbyshire block: Wirksworth to Grangemill. *Publication of the Department of Geology and Mineralogy, University of Aberdeen.*

WALTERS, S. G. & INESON, P. R. 1981. A review of the distribution and correlation of igneous rocks in Derbyshire, England. *Mercian Geologist,* **8**, 81–132.

WHITTAKER, A. (ed). 1985. *Atlas of onshore sedimentary basins in England and Wales: Post Carboniferous tectonics and stratigraphy.* Blackie, Glasgow.

——, CHADWICK, B. A. & PENN, I. E. 1986. Deep crustal traverse across southern Britain from seismic reflection profiles. *Bulletin of the Geological Society, France*, **8**, 55–68.

WILLS, L. J. 1951. A palaeogeographical atlas of the British Isles and adjacent parts of Europe. Blackie, London.

—— 1973. A palaeogeographic map of the Palaeozoic floor beneath the Permian and Mesozoic formations in England and Wales. *Memoir of the Geological Society, London* **7**.

—— 1978. A palaeogeographic map of the Lower Palaeozoic floor below the cover of Upper Devonian, Carboniferous and Later formations. *Memoir of the Geological Society, London*, **8**.

Carboniferous geology of the Southern North Sea Basin and controls on hydrocarbon prospectivity

MICHAEL R. LEEDER[1] & MARTIN HARDMAN[2]

[1]*Department of Earth Sciences, University of Leeds, Leeds LS2 9JT, UK*
[2]*Amerada Hess Ltd, 2 Stephen Street, London W1P 1PL, UK*

Abstract: The thick sequences of Upper Palaeozoic strata present in the Southern North Sea Basin were probably deposited on a basement of Caledonian low-grade metamorphic rocks intruded by numerous late-Caledonian granitoids which form part of the subsurface extension of the Mid-European Caledonides linking the English Lake District and Northern Pennines to the Ardennes. This basement, along with the better known onshore areas of the British Isles and Ireland was intensely fragmented during late Devonian to mid-Carboniferous times by a major phase of extensional tectonics.

Active fault-bounded tilted blocks developed in the Dinantian with deposition of fluviatile redbeds, marine carbonate/deltaic clastic cycles and deltaic coastal plain facies. These are penetrated by a number of wells on the southern flanks of the Mid-North Sea High, but details of basin geometries and facies distributions remain unclear.

Crustal stretching appears to have given way to thermal subsidence by early to mid-Namurian times. This was coincident with a major climatic change, probably forced by glacial expansion in Gondwanaland. The major drainage system that resulted dominated sedimentation in the SNSCB, punctuated by numerous marine transgressions of undoubted glacio-eustatic origins. Within this broad depositional framework, Namurian facies were closely controlled by basement tilt block topography, with good evidence for a gradual southerly migration of the active fairways of fluviodeltaic sandbodies. Westphalian facies were dominated by the interplay between glacio-eustatic and sedimentary processes on a vast low-gradient alluvial plain with much less pronounced tectonic control.

By late-Westphalian B times a phase of basin inversion caused the gradual elevation of the area as a retro-arc foreland basin. Vadose diagenesis and pedogenesis occurred in the fluviatile facies of the so-called Barren Red Measures (BRM). With continued compressive deformation a series of over 50 NW–SE orientated folds, some associated with thrusts, developed. Significant erosion and truncation of these inversion anticlines was accompanied by the deposition of the BRM facies in characteristic growth fold basins, with marked onlap around the basin margins. The Carboniferous play in the SNSCB has been greatly influenced by the subsequent Mesozoic/Cenozoic geological history of the basin, including periods of subsidence and uplift. We demonstrate that variations in original detrital mineralogy and various diagenetic processes have exerted significant controls upon Carboniferous reservoir quality.

The coals in the Upper Carboniferous of the Southern North Sea Basin (SNSB) have long been known to source the huge Permian Rotliegendes gas accumulations (Eames 1975). Only in the last five years, with the availability of good seismic records below the Permian level and penetration by a number of deep wells have details of Carboniferous stratigraphic, sedimentological and structural development become clearer. Some of the wildcat wells into the Carboniferous were also deep stratigraphic obligation wells, whilst others, particularly in Quadrant 44, were gas discoveries within the Carboniferous sequence. These developments have lead to the recognition of the Carboniferous as an integral part of SNSB prospecting philosophy. In this paper we seek to outline the broad development of Carboniferous geological processes in the basin, which we henceforth call the Southern North Sea Carboniferous Basin (SNSCB).

During our study we have analysed 3000 km of regional seismic data, comprising the GECO SNS-83 and the NOPEC SNSRI-UK-87 grids (Fig. 1). The records of 212 wells that achieved significant penetration into the pre-Permian subcrop were utilized, 145 of which yielded significant stratigraphic information and 47 of which we designated as 'key wells' because of their deep penetration and/or significant cored intervals.

The pre-Upper Palaeozoic basement

Very little is known directly about the pre-Upper Palaeozoic basement in the SNSCB. Regional studies based upon the results of

From Hardman, R. F. P. & Brooks, J. (eds), 1990, *Tectonic Events Responsible for Britain's Oil and Gas Reserves*, Geological Society Special Publication No 55, pp 87–105.

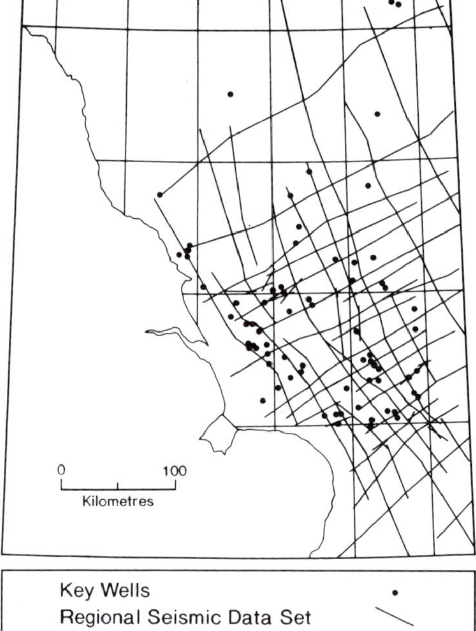

Fig. 1. Map to show the location of seismic lines (GECO/NOPEC) and key wells that make up part of the data base for the present study.

limited radiometric dating of short TD cores indicate that much of the Upper Palaeozoic in the basin is probably underlain by Caledonian (440–450 Ma) low-grade metamorphic rocks intruded by granitoids, such as that penetrated by A17-1 in the Netherlands sector (Frost *et al.* 1981, Fig. 2). This latter has an $^{39}Ar/^{40}Ar$ age of 350 Ma, probably indicating a late-Caledonian age of intrusion modified by a mild intra-Carboniferous thermal event analogous to that recorded in the onshore Weardale granite in the northern Pennines (Frost *et al.* 1981). In addition to this granitoid there is evidence from potential field (gravity) studies for at least three other such granitoid bodies (Donato *et al.* 1983). We name these bodies in this account (Fig. 7) as follows; north–central Quadrant 41 – *Teeside pluton*; east–central Quadrant 35/west–central Quadrant 36 – *Farne pluton*; southeast Quadrant 37/southwest Quadrant 38 – *Dogger pluton*.

It is evident from the above discussion that the basement to the SNSCB probably represents part of the branch of the Mid-European Caledonides that links the English Lake District and northern Pennines to the Ardennes (Pharaoh *et al.* 1987): Soper *et al.* 1987). There is good evidence from a number of wells in central

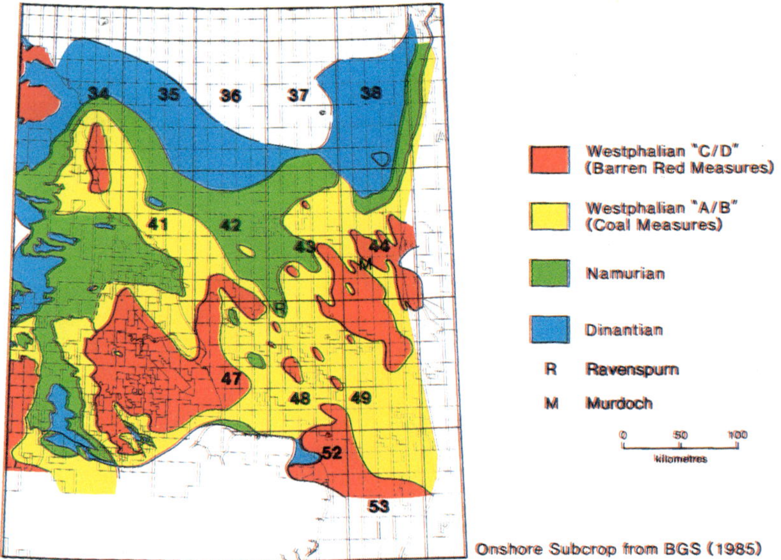

Fig. 2. Pre-Permian subcrop map for the Southern North Sea Carboniferous Basin (SNSCB) based upon the results of the present study. Major differences between this and previous published subcrop maps (Eames 1975; BGS 1985) arise from the obvious structural control upon the distribution of Namurian and Westphalian strata in NE–SW fold structures that formed during basin compression during late-Westphalian times (see Fig. 7). It is only recently that it has been possible to map these structures below the Permian unconformity from seismic data. R, Ravenspurn anticline; M, Murdoch anticline.

Quadrants 36, north and south 37, north and south 38 and south 30 that the Dinantian sequences that dip off the southern flank of the Mid North Sea High are underlain by a substantial thickness of Upper Devonian fluviatile clastics. In the central part of Quadrant 37 there is evidence from seismic that Permian strata may rest directly upon the Lower Paleozoic basement.

Stratigraphy and pre-Permian subcrop

Stratigraphic subdivision of SNSCB strata relies largely upon the palynomorph zonation proposed by Clayton *et al.* (1977). Making use of only the small amounts of sidewall or cuttings material usually available this approach can provide a reasonably accurate zonation, but only to approximately the stage level. The vagaries of preservation, depositional environment and burial diagenesis often conspire to reduce even this modest achievement.

In recent years the use of the natural gamma ray spectrometry (NGS) log tool has lead to increased precision in offshore Silesian subdivision and correlation (Leeder *et al.* 1990). This technique recognises the contribution of authigenic uranium, concentrated in marine anoxic black shales, to the total gamma ray spectra recorded on the conventional API log. Since there are some 80 Silesian marine bands recognised in onshore sequences the NGS log clearly has very considerable potential for refining offshore subdivision and correlation based in the first instance on palynological zonation. The use of NGS logs is not a simple matter however, since the uranium-rich black shale facies is not always developed. There are few such facies in certain Westphalian sequences penetrated in the eastern part of our study area (Quadrant 44). Under these circumstances, and as a vital additional check to the NGS log data, the carbon/sulphur (C/S) palaeosalinity method (Berner & Raiswell 1984) must be used on sidewall core material to identify marine horizons. A full discussion of the combined application of NGS logging and C/S analysis to SNSCB stratigraphy may be found in Leeder *et al.* (1989).

The results of our stratigraphic and seismostratigraphic mapping of the pre-Permian Silesian subcrop in the SNSCB are summarized on Fig. 2. A detailed account of the sedimentology of over 40 wells which penetrate the Carboniferous in the basin is given by McMahon (1990), whilst Besly (1990) provides a general review of basin development based mainly on interpretations of released well data. Cowans (1989) has recently summarised the general sedimentology and diagenetic development of Westphalian sequences in the SNSCB.

Dinantian

Dinantian sequences subcrop the Permian and dip southwards beneath the Silesian succession on the southern flank of the Mid North Sea High. They vary in character from thick fluviatile redbed sequences of Old Red Sandstone aspects in Quadrants 37 and 38, perhaps best developed as a >893 m sequence in 37/10-1, to alternating limestone/clastic sequences of fluvio-deltaic/marine cycles of Yoredale aspects in Quadrants 36, 42 and northern 44. The majority of wells in all of these areas are, however, rather poorly dated. Even so the facies present are all represented onshore in the Northumberland, Alston, Askrigg and Stainmore extensional syn-rift basins. Facies of Scremerston Coal Group (Upper Border Group) aspects and age occur in a >262 m sequence in Well 38/16-1. Elsewhere in the SNSCB, Dinantian sequences have not yet been penetrated, although there are recent indications from very deep wells that the earliest Namurian, and hence probably the late Dinantian is developed in a basinal mudrock facies in certain areas. This conclusion is borne out by the lack of strong seismic reflectors from the proposed Dinantian level in such areas, a feature probably due to the local absence of 'Carboniferous Limestone' facies in the deeper parts of tilt-block basins in the SNSCB south of the Mid North Sea High.

Namurian

Namurian strata subcrop the Permian as a broad belt flanking the Dinantian subcrop and also extend south as 'tongues' in the cores of at least five major SE-plunging anticlines (Cleveland, Sole Pit, Amethyst, Ravenspurn, 43/17 & 18 and 44/7 & 9). Namurian strata have also been penetrated by deep wells beneath the Westphalian subcrop in Quadrants 43, 44 and 48. Figs 3 & 4 show summary isopach and facies map for the Namurian which are based upon regional seismic interpretation and well correlation studies. As noted above refined correlations are now possible in the Namurian using a combination of NGS logging and carbon/sulphur analyses to detect the numerous marine bands of suspected glacio-eustatic origins (Leeder *et al.* 1990). Significant depocentres, possibly located over formerly active syn-rift lows are present in Quadrants 41, 43 and 48 where the maximum thickness exceeds 2000 m. Lithofacies evolution

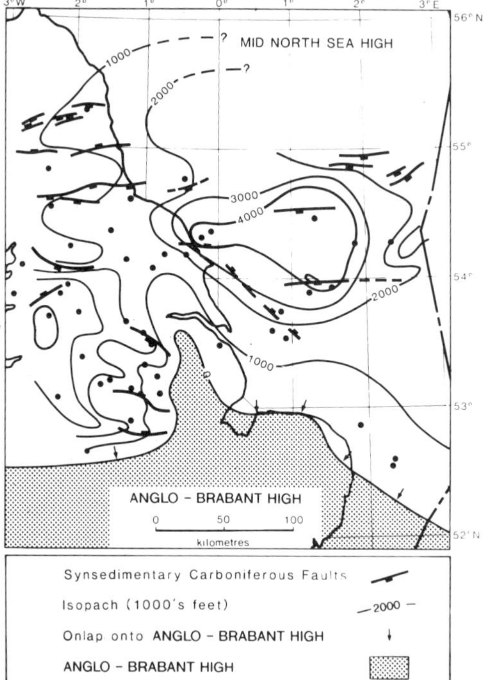

Fig. 3. Generalized Namurian isopach map to show the existence of a broad depocentre in the study area to the east of the East Midlands/Amethyst High. The syn-sedimentary Carboniferous faults were active from Dinantian to Mid-Namurian times and defined the margins of tiltblock structures that strongly influenced subsequent Namurian to Westphalian B sedimentation.

during Namurian times is best illustrated by well 48/3-3 which shows a large scale coarsening-upwards sequence some 1100 m thick (Leeder *et al.* 1985). This represents the progressive infill of a deep water, formerly active syn-rift basin. The oldest penetrated deposits, of Marsdenian age, are black shales deposited under anoxic deep water conditions. These pass upwards into delta front turbidites and delta slope deposits which are in turn overlain by coarse feldspathic fluvio-deltaic distributary channel sandbodies. These latter occur widely in Quadrants 43, 47 and 48 and represent the equivalents of well-known 'grits' (Kinderscout, Ashover, Chatsworth, Rough Rock etc) defined in the onshore Pennine Basin. The observed facies evolution is the product of repeated SW progradation of successive major fluvio-deltaic systems, frequently interrupted by rapid marine transgressions.

Regionally, a gradual southward migration of the axes of fluviatile deposition is seen with time (Fig. 4). Thus the area of northern 43 and 42 Quadrants was the first to have its deep water syn-rift basins infilled and traversed by river channels and northern Quadrant 48 was the last. Further south, in Qudrants 52 and 53, Namurian deposition was always dominated by fine-grained sediment.

Westphalian

Westphalian sequences subcrop the Permian extensively across the centre of the SNSCB. They are subdivided into Westphalian A/B and C/D units on Fig. 2. A Westphalian A/B depocentre with over 1000 m of section is located in the southern portion of Quadrant 43. This section thins southwards onto the flanks of the Anglo-Brabant High in Quadrant 53 (Tubb *et al.* 1986). Along much of the northern margin of this High, Westphalian rocks overstep the Namurian to lie on older strata in an onlapping geometry of 'steers head' type. Westphalian A and B sediments are mostly of fluvio-deltaic origin, but with thin marine mudrocks common

Fig. 4. Palaeogeographic maps for Namurian times based upon data from both onshore and offshore wells and from onshore outcrops. The latter leans heavily on the summaries of Jones (1980), Collinson (1988) and Steele (1988). Note the gradual southerly migration with time of basin fill sequences.
(a) *Early Pendleian.* Note the widespread development of syn-rift anoxic basins with their characteristic infill of black shale facies. To the north is the Yoredale cycle facies belt which was periodically transgressed by a carbonate shelf margin shown located over the Askrigg/Cleveland basin margin.
(**b**) *Late Pendleian.* Regional development in the northern part of the central Pennine and southern North Sea areas of the spectacular Grassington/Skipton Moor and Warley Wise/Pendle braidplains, river dominated deltas, bypassed slopes and underflow-dominated submarine fans.
(**c**) *Arnsbergian-Alportian.* Several former syn-rift lows are still clastic starved, with anoxic conditions periodically developed.
(**d**) *Kinderscoutian.* A major phase of fluvio-deltaic basin filling spreads southwards to the northern margins of Quadrants 47 & 48 and into the East Midlands tilt block terranes of Gainsborough and Edale.
(**e**) *Marsdenian.* The final phase of deep basin infill was accomplished as turbidite fans advanced over much of northern Quadrants 47 & 48 and westwards into the Widmerpool and North Staffordshire basins.
The scene is now set for the advance of the Rough Rock fluviodeltaic wedge across the infilled basins of much of the Southern NorthSea and northernEngland during Yeadonian times.

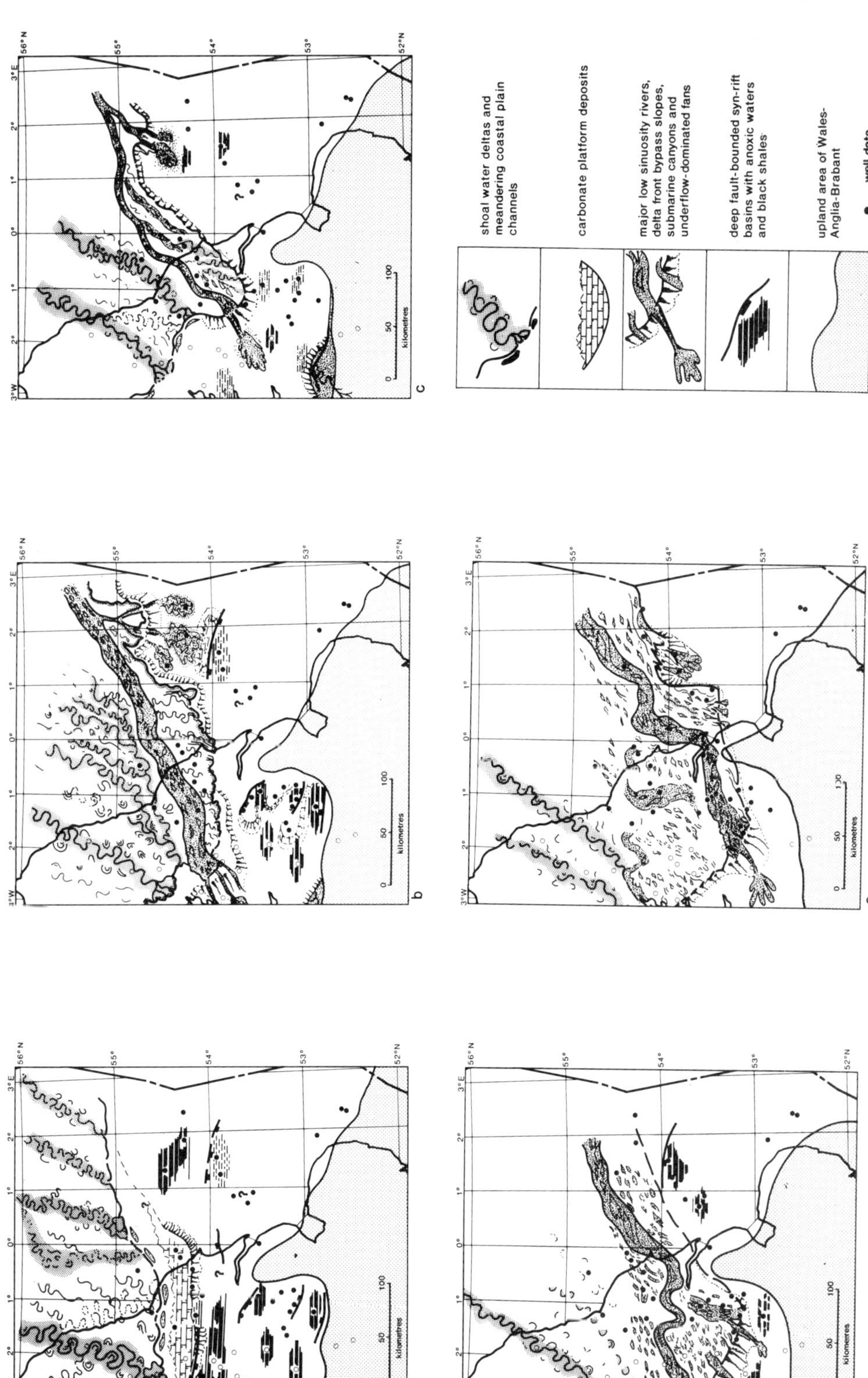
a
b
c
d
e
kilometres
0
50
100
3°W
2°
1°
0°
1°
2°
3°E
56°N
55°
54°
53°
52°N
shoal water deltas and meandering coastal plain channels
carbonate platform deposits
major low sinuosity rivers, delta front bypass slopes, submarine canyons and underflow-dominated fans
deep fault-bounded syn-rift basins with anoxic waters and black shales
upland area of Wales-Anglia-Brabant
well data

in Westphalian A successions in the western part of the Basin. Fluviatile sandstones throughout the Westphalian A/B show a marked eastwards increase in grain size. Stacked channel sandbodies characterise the Westphalian B in the southern part of Quadrant 44, a feature that has major implications for reservoir development and prospectivity.

Westphalian C/D sandstones are also predominantly fluvial in origin but these so-called 'Barren Red Measures' are generally characterised by red floodplain mudrocks with abundant ferralitic soil profiles and desiccation cracks. Coals and high gamma shales of marine origins are mostly absent whilst the mature sandstones are numerous but variable in thickness. These facies are thought to record the diachronous spread of well drained alluvium sourced from a drainage system that developed upon syn-depositional growth folds, such as the Murdoch and Ravenspurn Anticlines, within the basin during the late-Westphalian. Thus the large subcrops E and W of the Murdoch Anticline rest upon the earlier Westphalian with unconformity and marked onlap around the basin margins but with conformity in the basin centres (see later discussion and Figs 10 & 11). These sequences are thought to be derived as

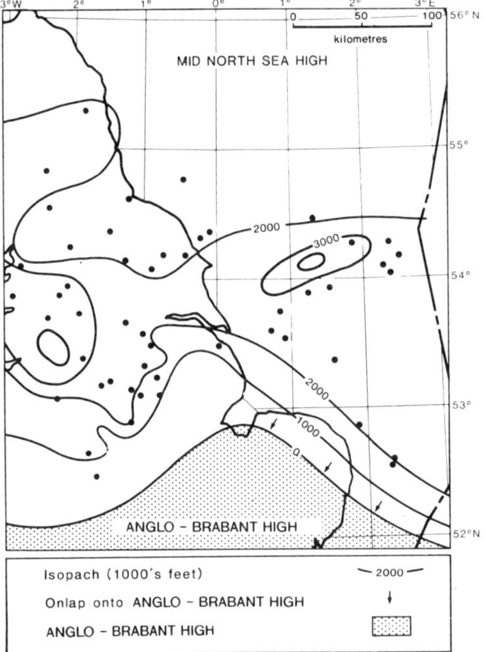

Fig. 5. Generalized Westphalian A/B isopach map to show the continued existence of a broad thermal 'sag' basinal form in the study area.

2nd cycle deposits from the Westphalian A/B strata within the anticline itself as the structure grew. Some small basins have been identified as bounded by normal faults, but we see little evidence for a renewed phase of regional extensional tectonics such as postulated by Besly (1988) from onshore outcrops in the English Midlands.

The significant stratigraphic break recognised at the base of the BRM in the area of Quadrant 53, adjacent to the Anglo-Brabant High, has been named the 'Symon Unconformity' by Tubb *et al.* (1986). This major seismic sequence boundary was attributed by these authors to uplift produced by lateral heat flow during the later phases of sag-type thermal subsidence. We prefer an interpretation linked to the regional effects of late Westphalian growth folding produced by regional compressional tectonics (see below).

Tectonic setting and major tectonic controls

Plate tectonic reconstructions suggest that the British Isles and offshore areas lay in an equatorial to subequatorial position north of the active Devonian/Carboniferous Hercynian orogenic belt. The basins created and inverted during this period are strongly influenced by this orogen and its gradual northward migration. The SNSCB is part of a large, retro-arc structure located on the Carboniferous 'foreland' immediately to the north of the older rigid terrane of Anglo-Brabant High and south of the granite-cored extension to the Southern Uplands. It is best viewed as an eastern continuation of the Pennine Carboniferous Basin which itself lies south of the granitoid-cored tilt blocks of the northern Pennines and north of the stable Welsh–Anglian–Brabant High. The two sub-basins are separated in the south by the East Midlands 'Shelf', an area of relatively slow subsidence intruded in the onshore by the Newark and Market Weighton (?Caledonian) granitoids. Recent gravity modelling (Donato & Megson 1990) over the offshore extension of this high (in the northern part of Quadrant 47) reveals the probable prescence of a further large area of subsurface granite which we name the Amethyst pluton.

The whole area suffered lithospheric extension in late Devonian to mid-Carboniferous times. Active fault-bounded half-grabens and tilted fault blocks developed, succeeded in Upper Carboniferous times by a post-rift phase of regional 'sag' subsidence caused by thermal re-equilibriation of the thinned lithosphere.

The magnitude of the original lithospheric extension cannot be directly determined since the crustal thinning was destroyed by basin inversion and shortening deformation in late Westphalian to Lower Permian times. Back-stripped subsidence curves for onshore Carboniferous basins (Leeder & McMahon 1983) indicate that substantial lithospheric thinning did occur, possibly by up to 50%. Such severe extension would have caused high post-rift geothermal gradients in Upper Carboniferous times and strong tectonic and topographic control on both facies and thickness distribution, particularly in the Lower and Middle Carboniferous.

The northward migration of the Hercynian thrust front in Westphalian C/D times caused renewed tectonic subsidence, in the foreland basins of South Wales–Kent–Calais, with the Anglo-Brabant High representing a peripheral bulge formed by flexural response to thrust loading (Besly 1988; Kelling 1988). In the SNSCB crustal shortening caused growth folding and the development of syn-tectonic unconformities along the margins of the growth folds and their surrounding aprons of fluviatile red beds which dominate late Westphalian sedimentation (see below).

Acceptance of the extensional model for Carboniferous tectonics leads to a predictable facies evolution as the early extensional basins, bounded by active fault systems, evolved into thermally subsiding basins. As shown in Fig. 6, the bathymetry caused by fault block rotation localized the infilling phase of Namurian submarine fan facies. As basin infill proceeded the earliest fluvio-distributary channels were localised in basinal lows over compacting deepwater mudstones. Westphalian sequences were much less influenced by remnant topography and the location of major fluvio-distributary sandbodies was determined by non-tectonic mechanisms such as variations in compaction, flood-induced river diversions and incision/aggradation cycles caused by glacio-eustatic sea level changes.

Following the general model put forward by Jackson (1979) we postulate that some of the syn-rift normal faults were reactivated as thrusts during shortening deformation in late-Westphalian times (see discussion below and Fig. 11). Good examples of such structures are seen onshore, for example in the Bowland Basin (Gawthorpe 1987) and the East Midlands (Fraser & Gawthorpe, this volume).

Structure

Seismic mapping has revealed a large number of structures (Figs 8–11) which affect the Carboniferous of the SNSB. Three main structural domains may be recognized.

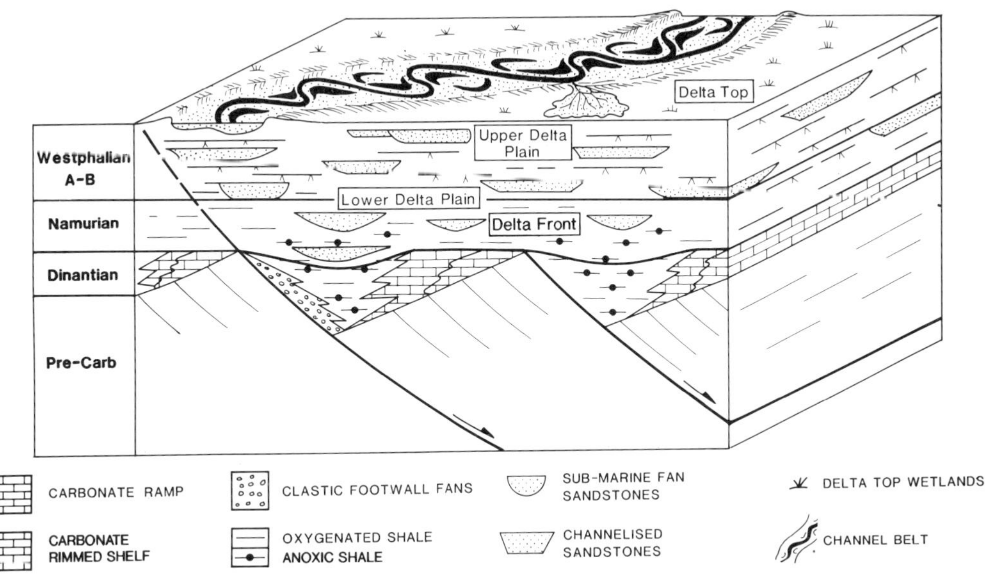

Fig. 6. Summary schematic block diagram to show the extensional tectono-stratigraphic model proposed for the Dinantian to Westphalian B evolution of the Southern North Sea Basin. Note that the diagram just represents one tilt-block array and that the position of the deeply buried Dinantian structures is difficult to locate, save around the Dogger Hinge Line area (see text for discussion).

Domain 1

This covers most of the basin from southern Quadrants 48 and 49 to central Quadrants 42, 43, and 44, and comprises an impressive 'en-echelon' foldbelt with many structures involving the Carboniferous sequence. They lie in a N–S 100 km wide zone between the Mid North Sea High and the Anglo-Brabant High (Fig. 7). The structures become increasingly difficult to recognize to the south where onlap occurs onto the Anglo-Brabant High (Fig. 7). The western margin of the domain is defined by the East Midlands High and its offshore extension into northern Quadrant 47 and is cored by large Caledonian granitoids, the Newark, Market Weighton and Amethyst plutons. Structural links with the South Pennine Basin occurred via the Cleveland Basin which had a growth-faulted northern margin defined by an extension of the Butterknowle Fault. It is interesting to note that the Teeside pluton, deduced from gravity studies in the area of 41/7 and 41/8, lies directly east of the onshore Weardale granite and north of this postulated growth fault.

Within the central fold belt a total of over 47 individual anticlines may be recognized (examples are provided by Figs 8–11). They trend consistently NW–SE but with a noticeable tendency for the most northerly folds to trend WNE–ESE as they approach the faulted margins of domain 3 adjacent to the Mid North Sea High. Fold wavelengths are typically in the range 5–12 km, with fold lengths of 6–25 km. The majority may be classified as non-cylindrical symmetrical buckle folds. They closely resemble those concentric buckle folds experimentally modelled by Dubey & Cobbold (1977). Some of the structures are associated with thrust faults (Fig. 9).

Three generations of folds may be recognized. (NB Folds of all three types were

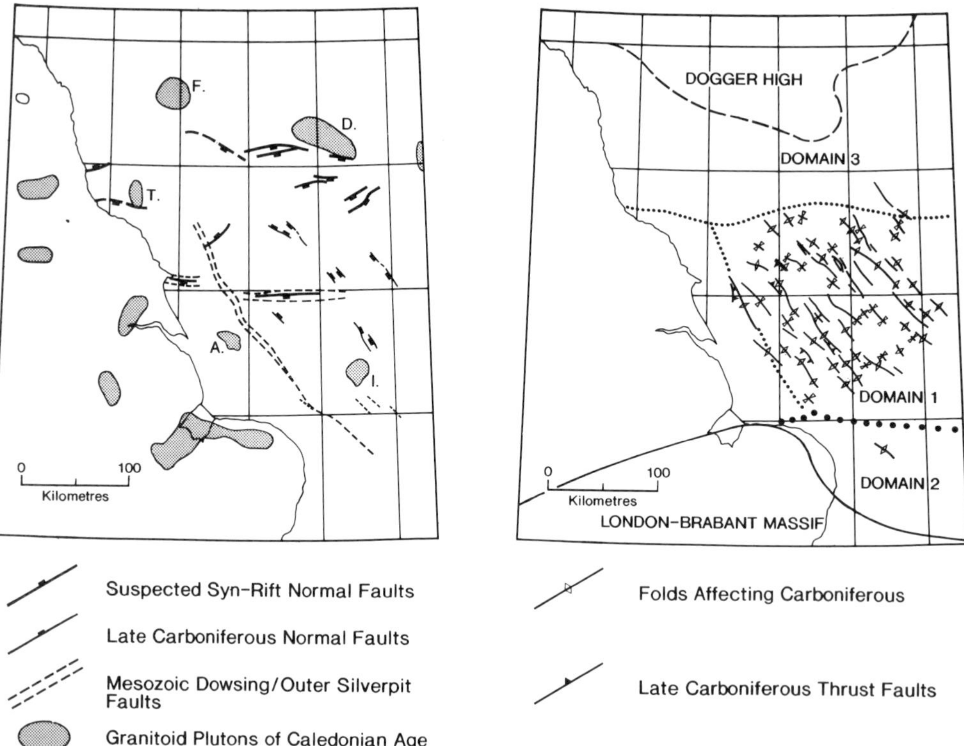

Fig. 7. Map to show the location of major folds and faults which affect the Carboniferous in the study area, together with the location of Caledonian granitoid plutons in the pre-Carboniferous basement. Note the location of the fold belt and of the position of the basin as a whole in an area of crust more-or-less free from the buoyant and buttressing effects of Caledonian plutons in the Lower Palaeozoic basement. Offshore plutons: A, Amethyst; D, Dogger; F, Farne; I, Indefatigable; T, Teeside.

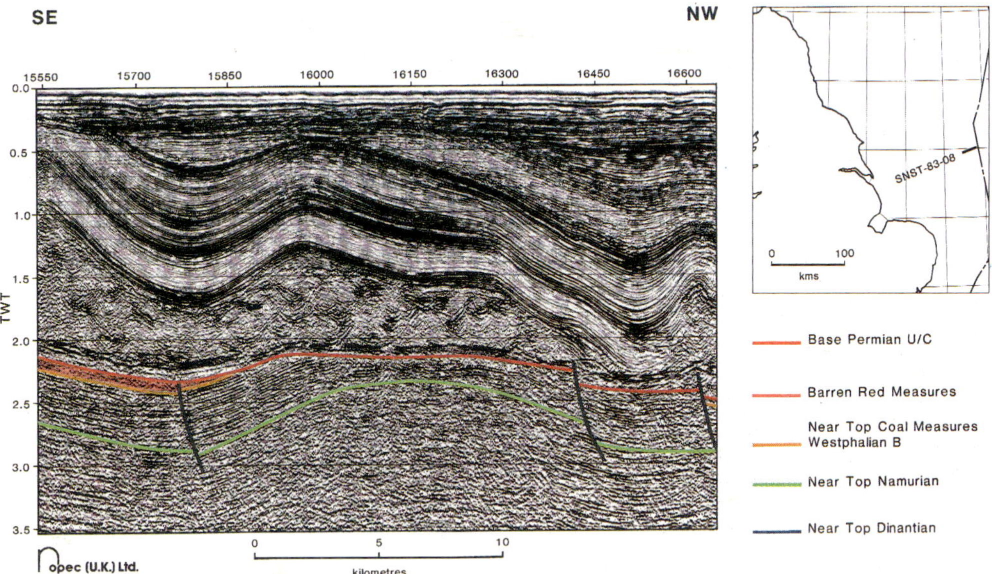

Fig. 8. Migrated seismic time section to show a typical pre-Permian fold structure in Block 49/4. This is part of the regional Silverpit anticlinal trend and shows truncated Westphalian A/B reflectors, probably coals, and a slight asymmetry to the SE. There is some evidence for the existence of 'Barren Red Measures' of Westphalian C/D age on the SE flank.

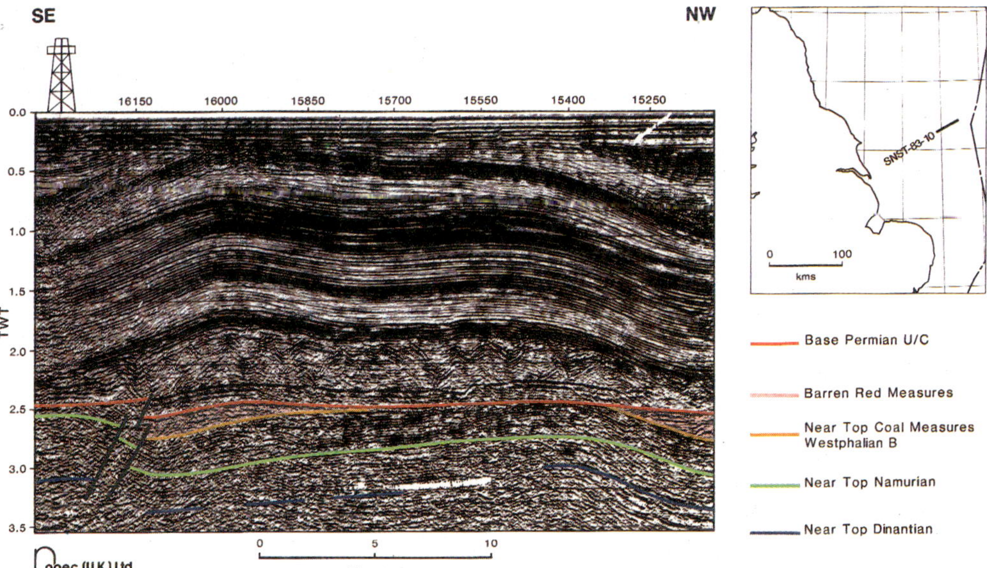

Fig. 9. Migrated seismic time section to show, in the SE, a fine thrust-related hangingwall anticline cored by early Westphalian A and Namurian strata and an adjacent footwall syncline which contains clearly onlapping Barren Red Measures facies. To the NE lies a broadly asymmetrical anticline with clear truncation of Westphalian A/B reflectors, probably a coal-rich interval, on its SE flank. On the NE side of this structure there are clearly onlapping BRM facies. Section across Blocks 43/17 and 44/21, 16 and 17.

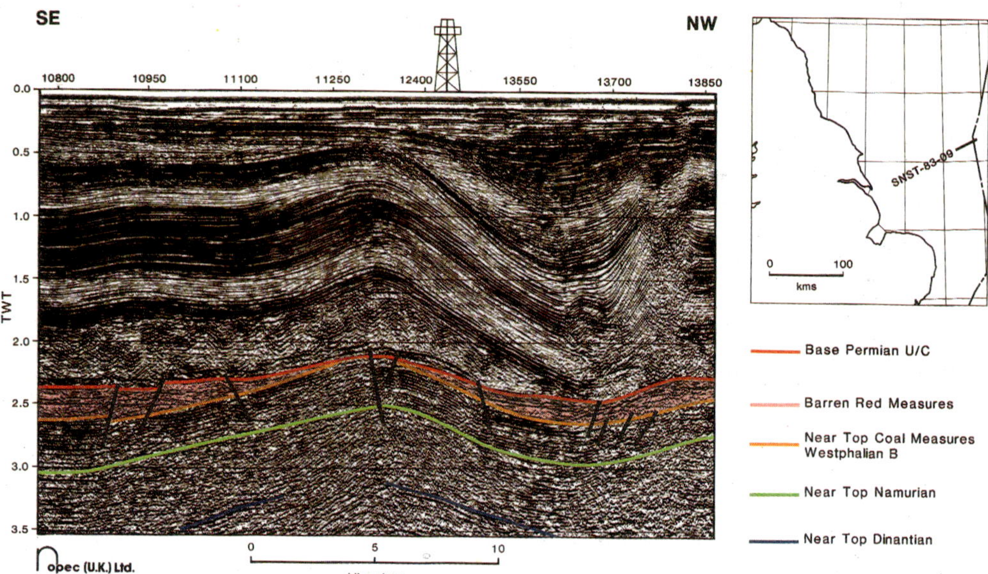

Fig. 10. Migrated seismic time section from Block 44/23 to show an asymmetrical anticline with no obvious intra-Carboniferous crest truncation. This is bounded by synclines in Barren Red Measures. The SE syncline shows unequivocal evidence for onlap onto tilted Westphalian A/B strata and thus has the status of a 'growth' syncline that was filled and onlapped as this and adjacent anticlinal structures grew.

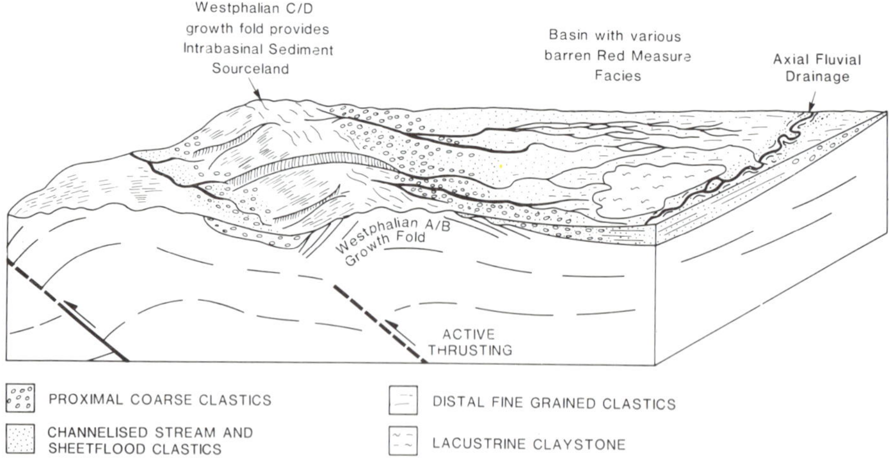

Fig. 11. Summary block diagram to show the growth fold/synclinal basin tectono-stratigraphic model for the Barren Red Measures in the study area. The model is inspired by the neotectonic behavior and stratigraphic relationships exhibited by the El-Asnam anticline in Algeria (King & Vita Finzi 1981).

carefully distinguished from artificial 'pull-ups' associated with thickened Zechstein evaporites in salt domes and pillars.

1. 47% of folds may be dated as pre-Permian since on the better seismic lines, intra-Carboniferous reflectors, particularly those from the

coal-bearing Westphalian and late-Namurian may be seen to be clearly truncated across the crests of structures (Figs 8 & 10). In many cases such geometries may be confirmed by subcrop mapping with the aid of biostratigraphically dated wells (Fig. 2). The majority of pre-Permian folds are symmetrical, with about equal numbers of the asymmetrical structures verging SW and NE. The NW–SE orientations of the folds parallels the trends seen in the Carboniferous subcrop of the onshore East Midlands (Smith 1987; Fraser *et al.* 1990). The trend may result from the 'fabric' of the Lower Palaeozoic basement (see Soper *et al.* 1987) influencing first the orientation of the syn-rift extensional structures which were then reactivated during shortening deformation as discussed briefly above. The folds are very difficult to recognize in the southern part of the study area, in Quadrant 53, where the Zechstein evaporites are highly irregular in thickness due to halokinesis and where Mesozoic inversion tectonics has often obscured pre-Permian structures.

We have clear evidence that Type-1 structures were true growth folds during the Westphalian C/D interval with conformable sequences of Red Measures occurring in adjacent synclines onlapping unconformably onto the partly eroded anticlinal flanks (Figs 9–11). By far the most significant such folds are the regional Murdoch Anticline, which trends NW–SE across southern Quadrant 44 and the Ravenspurn Anticline in southern Quadrant 43 (Fig. 2). Both of these structures are flanked by conformable sequences of Red Measures up to 750 m thick. Cores taken in the area around the Silverpit anticline in southern Quadrant 44 (such as the released well 44/21-1) reveal the presence of red beds with ferralitic soils and thick fluviatile sandstones containing very mature and well rounded pebble suites. These latter are presumed to have been reworked from the Westphalian A/B sequences that outcropped in adjacent anticlinal cores.

2. Many demonstrably pre-Permian folds also coincide with Mesozoic structures and it is inferred that these have been reactivated during Mesozoic inversion.

3. Folds that affect both the Carboniferous and the overlying Permian and Mesozoic sequences with no observed pre-Permian truncation or erosion are deemed to be post-Carboniferous, largely late Cretaceous, in age.

Domain 2

This includes the area southwards to the feather edge of the Carboniferous subcrop against the Anglian-Brabant High in southern Quadrants 52 and 53. As first discussed by Tubb *et al.* (1986) this domain is ramp-like in form, with steady southward thinning and progressive onlap of each of the Dinantian, Namurian and Westphalian sequences onto the High. Intra-Carboniferous structures and pre-Permian folds are impossible to identify because of a dense network of Mesozoic normal faults. To the south the SNSCB is bounded by the Anglo-Brabant High, onto which the Silesian sequence in the basin gradually onlaps (see Tubb *et al.* 1986). Recent gravity studies (Allsop 1987) show that this tectonic feature is also cored by an extensive (?Caledonian) granitoid which we name the North Norfolk Pluton.

Domain 3

The third domain occupies the southern flanks of the Mid North Sea High and is characterised by largely homoclinal strata separated by large normal faults of broadly east–west orientation which probably define intra-Dinantian tilt blocks. The domain represents the eastern extension of the northern English tilt block province. The High evidently represented an extension of the Southern Uplands Block during the Carboniferous. The density of seismic coverage over the High is really insufficient to define any offshore analogues with onshore structures but, as noted above, the recognition of two very large granitoid plutons from gravity studies (Farne, Dogger in Fig. 7) in Quadrants 35/36 and in 37/38 reinforce this proposal. We have defined major normal faults to the south of these areas which today throw down thick sequences of Silesian strata against the Dinantian which largely covers the Mid North Sea High. We propose the name Dogger Hinge Line for this zone. Some of these faults are proposed as synrift structures which show signs of growth. They are analogous to onshore structures such as the North Solway, Closehouse/Lunedale/Butterknowle and 90-Fathom/Stublick Faults which were active as syn-rift structures and which were frequently located at the margins of granitoid-cored tilt blocks and half grabens. To the north, the Dinantian sequences seem to thin gradually onto the core of the Mid North Sea High by onlap. A prominent reflector beneath the Dinantian is presumed to represent Lower Paleozoic basement. It must be assumed that during Silesian times the whole High was covered by a relatively thin onlapping Silesian sequence, subsequently removed by erosion in the pre-Permian inversion event.

Controls on reservoir quality

Introduction

As proposed above, Carboniferous syn-sedimentary tectonics has exerted a significant control upon the distribution of major reservoir objectives, particularly during lower to mid Namurian and late Westphalian times. The combination of this active tectonic influence, together with the erosion and redeposition associated with the end-Carboniferous inversion event, are considered to be the dominant factors in controlling Carboniferous prospectivity.

In detail however the Carboniferous play is further complicated by both the subsequent Mesozic/Cenozoic geological history of the Southern North Sea Basin and variations in original detrital mineralogy and diagenesis (see also the recent results of Cowans 1989). Together these effectively control the quality of reservoir rocks and the timing of hydrocarbon generation.

Diagenesis

Figures 12–14 show simplified paragenetic sequences for three unreleased Southern North Sea wells, referred to as X, Y and Z respectively. These wells have been choosen to illustrate stratigraphic and temporal variations in the diagenetic history of sandstones from the Namurian (X), Westphalian Coal Measures (Y) and Westphalian Barren Red Measures (Z). These samples are thought to be currently at or near their maximum palaeoburial depths. Furthermore the samples are all hydrocarbon-bearing and of similar grain size. Although there are distinct similarities in the overall diagenetic sequences observed in these wells, the relative abundance of individual mineral phases and the inferred effects of early burial compactional processes varies considerably from well to well.

Well X: Namurian. The original detrital mineralogy of these sandstones is characterised by mono- and poly-crystalline quartz, feldspar, mica and rock fragments of metamorphic and igneous origin. Mudrock intraclasts are also present in some samples. The sandstones are classified as arkoses. The main features of the paragenetic sequences are illustrated in Fig. 12. The initially high porosities have been reduced by very severe early mechanical compaction which occurred during very rapid early burial

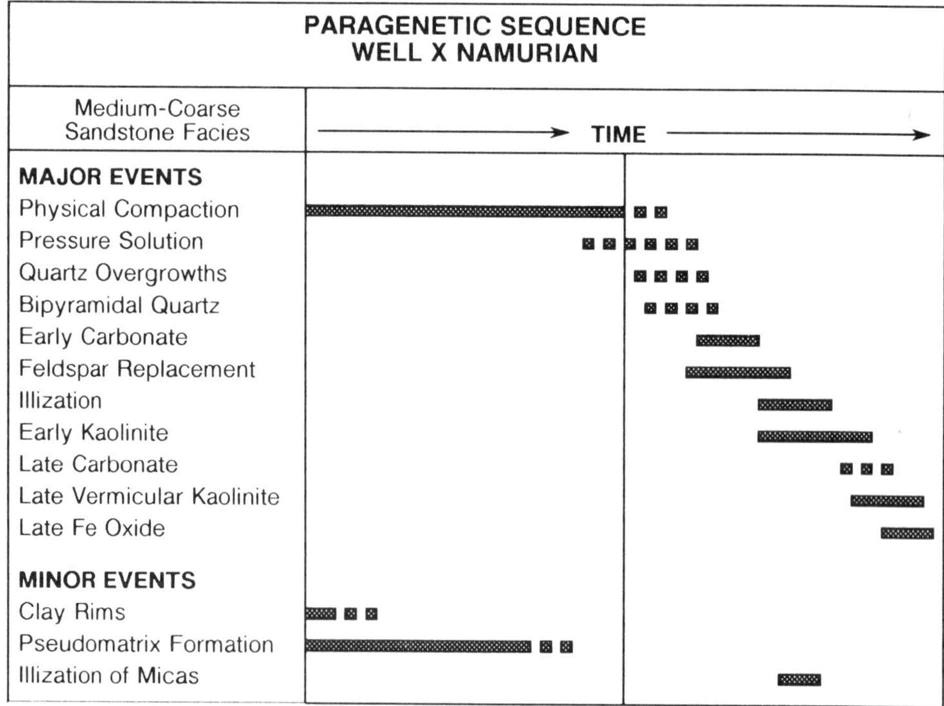

Fig. 12. Namurian paragenetic sequence for Well X (see text for discussion).

and appears to pre-date poorly developed quartz overgrowths.

Early diagenesis continued with limited and patchy growth of an ankerite/ferroan dolomite cement phase which occurs today as infills to isolated pore spaces and as remnants in more open pores. Illitization of mechanically infiltrated clay rims, micas and mud flakes appears to have taken place subsequent to dissolution of most of the early carbonate phase. This is thought to have occurred during uplift associated with end-Carboniferous inversion. Illitization is synchronous with the growth of a pervasive, very finely crystalline phase of kaolinite cementation which severely occludes much of the available pore space, particularly in finer grained sandstones.

The final major diagenetic event is a late phase of vermicular kaolinite overgrowth formation. This kaolinite replaces and etches both quartz and remnant carbonate phases and in coarser-grained samples its development governs the amounts of preserved secondary porosity.

Well Y: Westphalian Coal Measures. The original framework grains in sandstones sampled from this well consist of monocrystalline quartz with small amounts of polycrystalline quartz, mica, feldspar, rock fragments and mudflake intraclasts. The sandstone is classified as subarkosic with the numerous fresh-looking plagioclase grains suggesting a component of 1st cycle detritus.

The paragenetic sequence for these sediments begins with a phase of mechanical compaction. Early solution transfer and the associated precipitation of syntaxial overgrowths were instrumental in giving a moderate degree of grain framework stabilization during this phase (Fig. 13).

There is limited evidence for an early phase of halite cementation in some of the coarser grained sandstone samples. The timing of this phase is thought to be synchronous with end-Carboniferous inversion and subsequent deposition of hypersaline lacustrine mudrocks and evaporites of the overlying Rotliegend Silverpit Formation.

Renewed Mesozoic burial led to a phase of ferroan/magnesian-rich carbonate emplacement significantly reducing primary porosity, particularly in the finer grained facies. However, in coarser-grained sandstones the preservation

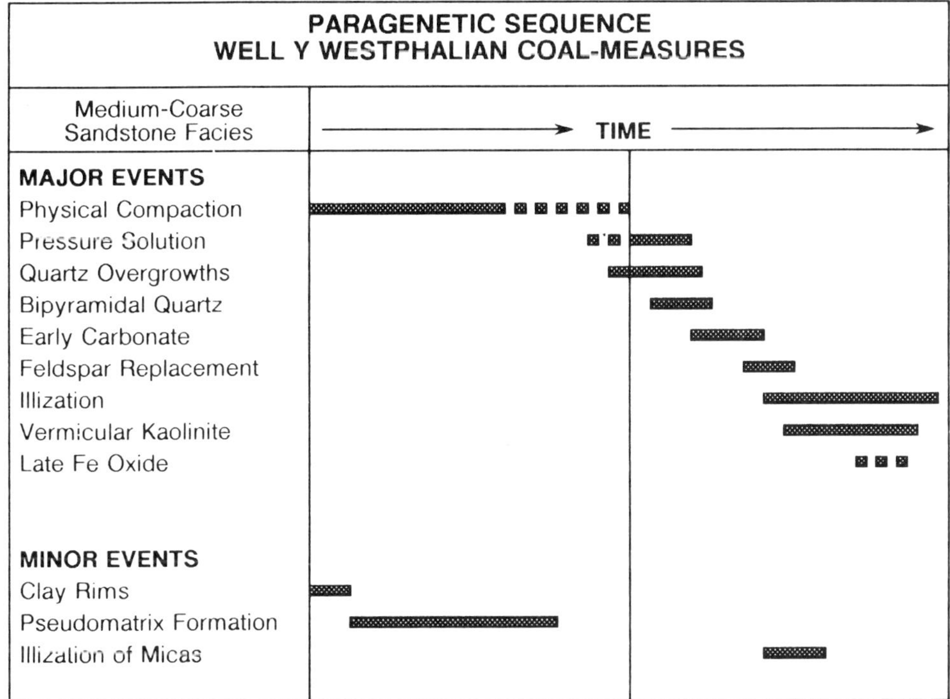

Fig. 13. Westphalian paragenetic sequence for Well Y (see text for discussion).

of this phase is much less pronounced. This is interpreted as a result of dissolution by migrating acid pore waters generated during decarboxylation reactions associated with hydrocarbon generation.

Final deep burial diagenetic products include illite and kaolinite which in places develop as complex intergrown morphologies and have a variable effect on the reservoir quality of the sandstones. In the finger-grained sandstones, these phases often completely infill the relatively small pore throats whereas those of the coarser samples are only partly occluded.

Well Z: Westphalian Barren Measures. BRM sandstones are mature quartz arenites indicative of either a long transport path, high degree of hinterland chemical weathering or a large component of multicycle weathering products. As discussed previously the latter interpretation is preferred.

As shown in Fig. 14 the paragenetic sequence commences with mechanical compaction and the infiltration of clays to form accretionary rims. With continued burial, precipitation of syntaxial overgrowths and early carbonate cement phases occurred. These combined to restrict the degree of porosity reduction by mechanical compaction during later burial. Silica cementation appears to have slightly post-dated the widespread precipitation of early carbonate. This latter phase, a microsparitic ferroan calcite cement, is now seen as pore-lining remnants in oversize and rhomboid-shaped pores. The inferred dissolution of this early carbonate phase appears to have occurred immediately prior to, or synchronous with, the dissolution of feldspar and illitization of mud-flakes, micas and pseudomatrix. These latter porosity/permeability destructive events are more highly developed in finer-grained samples.

Growth of pore-filling vermicular kaolinite during deep burial controls the degree to which the secondary porosity due to carbonate dissolution is preserved at the present day. The kaolinite is massive and blocky in form, includes substantial intracrystalline porosity and where well developed has a markedly detrimental effect upon effective porosity. The abundance of kaolinite in some samples suggests that precipitation occurred from externally-derived ionic species, since the original feldspar content is thought to have been very low.

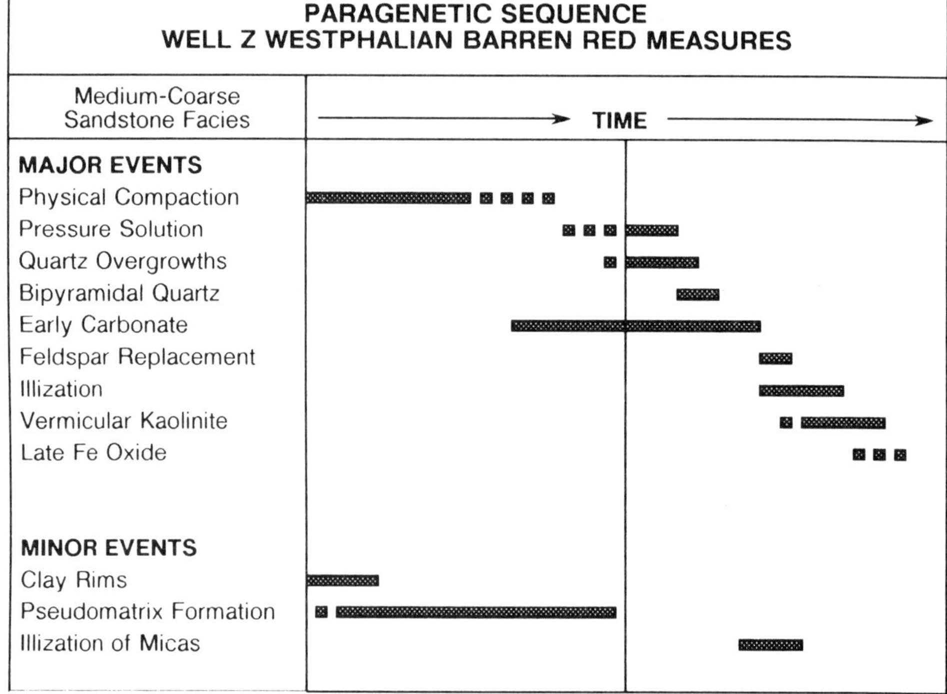

Fig. 14. Westphalian Barren Red Measures paragenetic sequence for Well Z (see text for discussion).

Discussion

The observations presented above suggest that diagenesis has exerted significant control upon Carboniferous reservoir quality. In general reservoir quality increases with increasing grain size, sorting and mineralogical maturity. As a consequence of this the thicker fluvio-deltaic sandtones are expected to form the only viable economic reservoirs (Fig. 15). Namurian sandstones however appear to an exception to this general trend and are separately discussed below. Finally, and perhaps most importantly, we have shown that the early precipitation of framework stabilising silica and carbonate cements (BRM reservoirs) both preserve original porosity in the face of mechanical compaction due to rapid early burial and give good secondary porosity production after dissolution has taken place.

Burial rate and maximum burial depth estimates for Namurian samples. The expectation that coarse-grained, moderately to well sorted Namurian sandstones would show reasonable to good reservoir characteristics does not seem generally to have been confirmed. This is thought to be due to a number of factors. Firstly it is important to recognize that there are fewer Namurian data points in the sample (Fig. 15) and a significant number of the wells are in relative close proximity to the axis of the Sole Pit inversion. Accordingly a greater maximum burial depth than conventionally deduced may have overprinted the facies control. Secondly we think that the very high rates of initial burial of Namurian sediments during the Namurian–Westphalian time interval would have caused severe mechanical compaction and thus an irredeemable porosity loss. It is likely that there are areas in the basin where grain framework stabilisation due to silica cementation was followed by early carbonate cementation and such areas will show more favourable reservoir qualities.

Figure 16 shows the relationship between maximum burial depth and present day measured porosity. Estimates of the maximum palaeoburial depth were made utilising a slightly

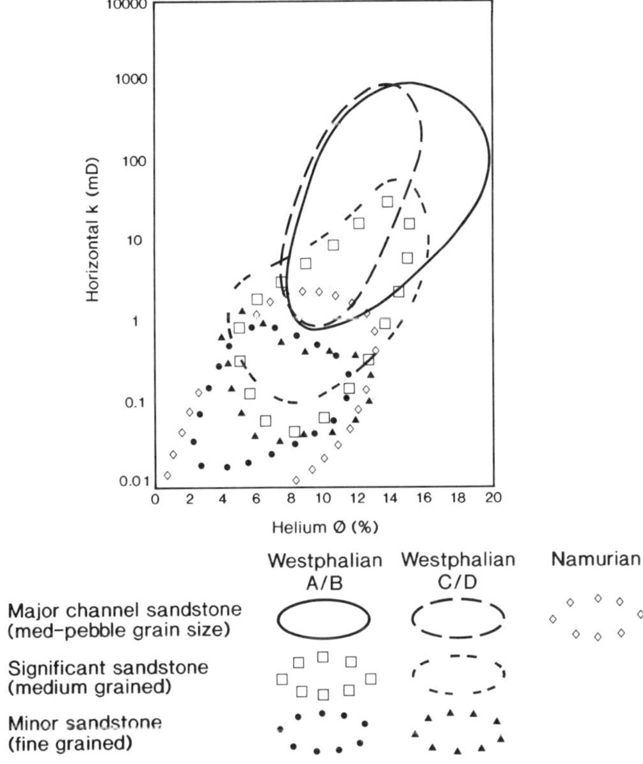

Fig. 15. Porosity-permeability crossplot for Carboniferous sandstones subdivided by lithofacies and sandbody thickness.

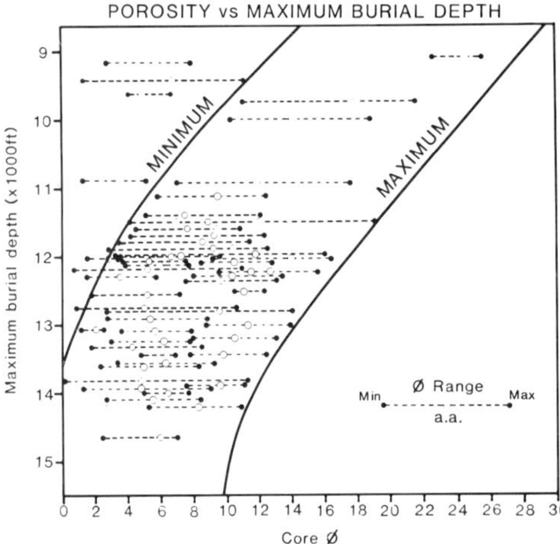

Fig. 16. Porosity versus maximum palaeo-burial depth for Carboniferous sandstones.

modified version of the vitrinite reflectance versus depth technique originally proposed by Cope (1986). Figure 16 shows that down to maximum palaeoburial depths of 14 000' (4266 m), porosity decreases linearly with depth, the gradient being approximately 3% porosity reduction per 1000' of burial. Below 14 000', maximum observed porosity appears to have reached a minimum value of 9.5% and very little further linear reduction with depth occurs.

It is significant that with the the exception of data from four wells, which plot outside the general trend, the range of porosities seen at any individual burial depth is readily explicable in terms of variations in facies and diagenesis, as discussed above. There is little evidence to support the presence of overpressuring. Close examination of well data which plots off trend shows that they are all located in the central and southern part of Quadrant 47. Recent gravity studies by Donato & Megson (1990) shows that there is likely to be a buried (?Caledonian) granitoid pluton situated under the Amethyst field in this part of Quadrant 47. The high heat flux expected from such a granite (cf such well known onshore examples such as Weardale and Wensleydale) may well be responsible for the relatively high maturity values seen in these wells. Thus the V_r values from the wells will underestimate palaeoburial since they are based on a regional average heat flux estimate. Indeed, recent results from fission track analysis (Duddy & Bray, pers. comm.) would lend support to the idea that the inversion magnitudes for this area of the SNSCB are underestimates.

Hydrocarbon generation history of Carboniferous source rocks

Carboniferous coals and shales are widely accepted as the source for the 24.5 × 10^{12} cu ft (693 × 10^5 m^3) of gas reservoired in Permian, Triassic and Carboniferous rocks in the Southern North Sea Basin (Cornford 1984). In general, geochemical analyses show Westphalian coals to be the most important gas prone source rocks. Interbedded shales within the Namurian to Westphalian sequence have more limited potential to generate gas and in places oil (Eames 1975; Lutz *et al.* 1975; Barnard & Cooper 1983).

Figure 17 shows a map of the present day vitrinite reflectance at topmost Carboniferous 'Grey Measures'. Large areas of the SNSCB currently lie within the gas generation window as defined by laboratory studies on the devolatilisation of Westphalian coals peformed by Higgs (1986). Within this area there are several distinct sub-basins which have different subsidence and inversion histories during the Mesozoic–Cenozoic period e.g. Solepit, Silverpit, Cleaver Bank, Amethyst Platform, Cleveland Basin

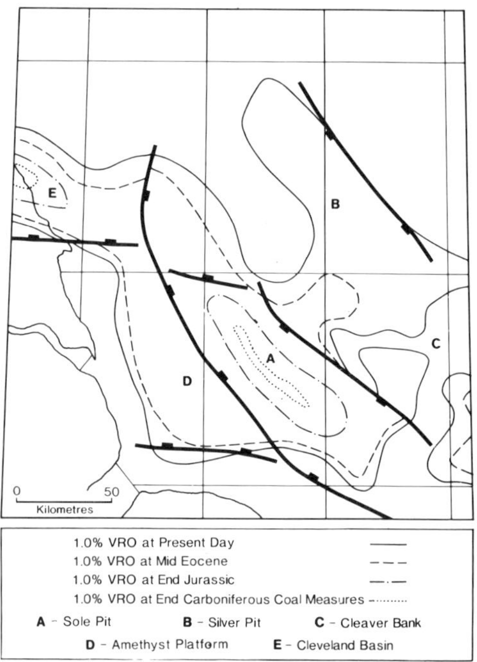

Fig. 17. Present day vitrinite reflectance at topmost Carboniferous 'Grey Measures' with superimposed maturity envelopes at topmost Carboniferous Grey Measures, for end-Carboniferous, end-Jurassic and mid-Eocene times.

(Glennie & Boegner 1981). The influence of these varying subsidence histories on the generation and migration of hydrocarbons has been addressed by generating a series of regional maturity maps for pre-Carboniferous inversion, end-Jurassic and mid-Eocene times. Simplified versions of the maturity envelopes taken from these maps are also shown in Fig. 17. Figure 17 was generated using a computer-based Lopatin style (Waples 1980) maturation modelling package, utilising data from over forty individual wells. This data included present day geothermal gradients derived from temperature readings made during downhole electric logging runs corrected for borehole cooling effects accompanied by data on the subsidence history of each well location. The latter were input as stratigraphic time-depth pairs using the time scales of Harland *et al.* (1982) for the post-Permian and Lippolt *et al.* (1984) for the Pre-Permian.

Data on the inversion episodes and past geothermal gradients were derived from a number of sources including Bunter Shale interval velocity studies, vitrinite reflectance-depth relationships and analystical geochemical data, particularly pyrolysis. The output from the modelling (time–temperature indices) was then converted to V_r equivalence using the Issler conversion and then compared to actual well data. 'Unknowns' in the model, such as inversion estimates and past geothermal gradient are then individually 'fine-tuned' such that a close fit is obtained between computed and observed maturity at both Mesozoic and Carboniferous levels. Figure 18 shows the subsidence profile and Fig. 19 shows the computed best-fit model for well 48/3-3 and the plot of observed versus calculated vitrinite reflectance. These models fit well with both the well data and previously published work of Cope (1986) and Glennie & Boegner (1981). However they are an interpretive best fit, particularly since many estimates for unknowns, such as past geothermal gradients, are uncertain. Further refinements to these models must await more widespread use of the results of palaeogeothermal indicators such as apatite fission track data.

As shown in Fig. 17 the results of the work discussed above suggests that gas generation may have commenced in limited areas in the central Solepit and Cleveland basin areas from within the Carboniferous section prior to end-Carboniferous inversion. By end-Jurassic times the area of mature source rocks had increased to encompass significant areas of the Solepit–Cleveland–Broad Fourteens trend of sub-basins. The location of these mature areas appears to be strongly controlled by the Dowsing Fault and Cleaver Bank uplift zone. By 50 Ma (Mid-Eocene) much of the currently mature areas of the Southern North Sea Basin had reached their maximum maturity, the major exception being the northern part of the Cleaver Bank and Quadrants 43 and 44, the Silverpit Basin, which only became mature following

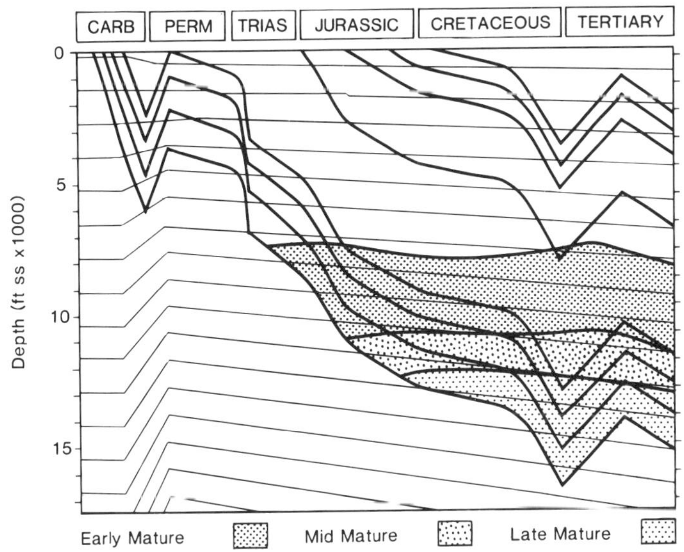

Fig. 18. Well 48/3-3 burial history.

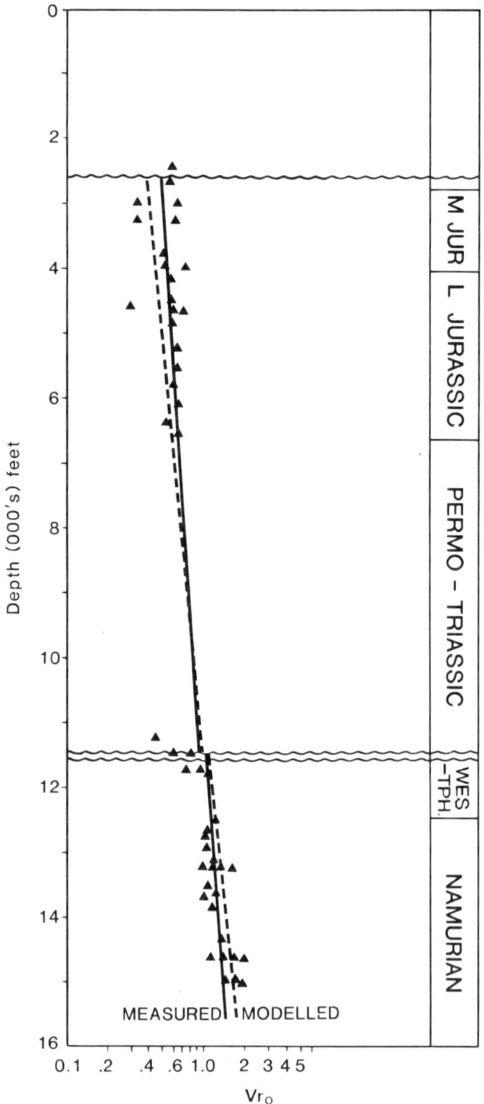

Fig. 19. Observed versus calculated vitrinite reflectances for 48/3-3.

significant subsidence and deposition in late Tertiary times. The implications of this for the prospectivity of the Carboniferous in Quadrants 43 and 44 are severe. Firstly reservoir rocks in these areas have remained open for the circulation of cement-precipitating pore waters for considerable times, many of the traps having been formed at the end-Carboniferous (see structural section above). Secondly these areas are currently at their maximum level of maturity and likely to be actively generating hydrocarbons today.

The authors are grateful to the management of Amerada Hess Limited for permission to publish this paper. We wish to thank R. F. P. Hardman for his enthusiastic support of our studies. We acknowledge the input given by many geoscientists, both at the University of Leeds and Amerada Hess, during the course of our work, particularly the contributions made by A. H. McMahon. Rob Gawthorpe is thanked for his helpful referees comments. The seismic data used to illustrate the paper is published courtesy of Nopec (UK) Limited.

References

Allsop, J. M. 1987. Patterns of late-Caledonian intrusive activity in eastern and northern England from geophysics, radiometric dating and basement geology. *Proceedings of the Yorkshire Geological Society*, **46**, 335–354.

Barnard, P. C. & Cooper, B. S. 1983. A review of geochemical data related to the NW European gas province. *In*: Brooks, J, Goff, J. & Van Hoorne, B. (eds) *Habitat of Palaeozoic Gas in NW Europe*. Geological Society, London, Special Publication, **23**, 19–33.

Berner, R. A. & Raiswell, R. 1984. C/S method for distinguishing freshwater from marine sedimentary rocks. *Geology*, **12**, 365–368.

Besly, B. M. 1988. Palaeogeographic implications of late-Westphalian to early Permian red-beds, central England. *In*: Besly, B. M. & Kelling, G. (eds) *Sedimentation in a synorogenic basin complex: the Upper Carboniferous of NW Europe*, Blackie, Glasgow, 200–221.

Clayton, G., Coquel, R., Doubinger, J., Gueinn, K. J., Lobaziak, S., Owens, B. & Streel, M. 1977. Carboniferous miospores of Western Europe: illustrations and zonation. *Mededelingen Rijks Geologische Dienst*, **29**.

Collinson, J. D. 1988. Controls on Namurian sedimentation in the Central Province basins of northern England. *In*: Besly, B. M. & Kelling, G. (eds) *Sedimentation in a synorogenic basin complex: the Upper Carboniferous of NW Europe*, Blackie, Glasgow, 85–101.

Cope, M. J. 1986. An interpretation of vitrinite reflectance data. *In*: Brooks, J., Goff, J. & Van Hoorne, B. (eds) *Habitat of Palaeozoic Gas in NW Europe*. Geological Society, London, Special Publication, **23**, 85–98.

Cornford, C. 1984. Source rocks and hydrocarbons of the North Sea. *In*: Glennie, K. (ed.) *Introduction to the Petroleum Geology of the North Sea*. Blackwells, Oxford, 171–205.

Cowan, G. 1989. Diagenesis of Upper Carboniferous sandstones: southern North Sea Basin. *In*: Whateley, M. K. G. & Pickering, K. T. (eds) *Deltas: sites and traps for fossil fuels*. Geological Society, London, Special Publication, **41**, 57–74.

Donato, J. A., Martindale, W. & Tully, M. C. 1983. Buried granites within the Mid North Sea High. *Journal of the Geological Society, London*, **140**, 825–837.

—— & Megson, J. B. 1990. A buried granite batholith

beneath the East Midland Shelf of the Southern North Sea Basin. *Journal of the Geological Society, London*, **147**, 133–140.

DUBEY, A. K. & COBBOLD, P. R. 1977. Noncylindrical flexural slip folds in nature and experiment. *Tectonophysics*, **38**, 223–239.

EAMES, T. D. 1975. Coal rank and gas source relationships – Rotliegendes reservoirs. *In*: WOODLAND, A. W. (ed.) *Petroleum and the continental shelf of NW Europe*, Institute of Petroleum, London, 191–203.

FRASER, A. J., NASH, D. F., STEELE, R. P. & EBDON, C. C. 1990. A regional assessment of the intra-Carboniferous play of northern England. *In*: BROOKS, J. (ed.) *Classic Petroleum Provinces*. Geological Society, London, Special Publication, **50**, 417–439.

FROST, R. T. C., FITCH, F. J. & MILLER, J. A. 1981. The age and nature of the crystalline basement of the North Sea Basin. *In*: ILLING, L. V. & HOBSON, G. (eds) *The Petroleum Geology of the Continental Shelf of NW Europe*, Heyden, London, 43–57.

GAWTHORPE, R. L. 1987. Tectono-sedimentary evolution of the Bowland Basin, N. England during the Dinantian. *Journal of the Geological Society, London*, **142**, 59–71.

GLENNIE, K. W. & BOEGNER, P. L. E. 1981. Sole Pit inversion tectonics. *In*: ILLING, L. V. & HOBSON, G. D. *The Petroleum Geology of the Continental Shelf of NW Europe*, Heyden, London, 430–57.

HARLAND, W. B., COX, A., LLEWELLYN, P. G., PICTON, C. A. G., SMITH, A. G. & WALTERS, R. 1982. *A geological time scale*. Cambridge University Press.

HIGGS, M. D. 1986. Laboratory studies into the generation of natural gas from coals. *In*: BROOKS, J., GOFF, J. & VAN HOORNE, B. *Habitat of Palaeozoic gas in NW Europe*. Geological Society, London, Special Publication, **23**, 113–120.

JACKSON, J. A. 1979. Reactivation of basement faults and crustal shortening in orogenic belts. *Nature*, **283**, 343–346.

JONES, C. M. 1980. Deltaic sedimentation in the Roaches Grit and associated sediments (Namurian R2B) in the southwest Pennines. *Proceedings Yorkshire Geological Society*, **43**, 39–67.

KING, G. C. P. & VITA-FINZI, C. 1981. Active folding in the Algerian earthquake of 10 October 1980. *Nature*, **292**, 22–26.

LEEDER, M. R. & MCMAHON, A. H. 1988. Upper Carboniferous (Silesian) basin subsidence in northern Britain. *In:* BESLY, B. M. & KELLING, G. (eds) *Sedimentation in a synorogenic basin complex: the Upper Carboniferous of NW Europe*. Blackie, Glasgow, 43–52.

——, RAISWELL, R., AL-BIATTY, H., MCMAHON, A. & HARDMAN, M. 1990. Carboniferous stratigraphy, sedimentation and correlation of Well 48/3-3 in the Southern North Sea Basin: integrated use of palynology, natural gamma/sonic logs and carbon/sulphur geochemistry. *Journal of the Geological Society, London*, **147**, 287–300.

LIPPOLT, H. J., HESS, J. C. & BURGER, K. 1984. Isotopische alter pyroklatischen Sanidinen aus Kaolin-Kohlentonsteinen als Korrealtion-smarken fur das miteleuropisch Oberkarbon. *Fortschritte in der Geologie von Rheinland und Westfalen*, **32**, 229–150.

LUTZ, M., KAASSHEITER, J. P. H. & VAN WIJKE, D. H. 1975. Geological factors controlling Rotliegend gas accumulations in the mid-European basin. *Proceedings of the 9th World Petrol Congress*, Applied Science, Barking, 93–103.

MCMAHON, A. H. 1990. *Silesian sedimentology and paleogeography of the UK Southern North Sea and adjacent onshore areas*. PhD Thesis, University of Leeds.

PHARAOH, T. C., MERRIMAN, R. J., WEBB, P. C. & BECKINSALE, R. D. 1987. The concealed Caledonides of eastern England: preliminary results of a multi-disciplinary study. *Proceedings Yorkshire Geological Society*, **46**, 355–370.

SMITH, N. J. P. 1987. The deep geology of central England: the prospectivity of the Palaeozoic rocks. *In*: BROOKS, J. & GLENNIE, K. W. (eds) *Petroleum Geology of NW Europe*, Graham & Trotman, London, 217–224.

SOPER, N. J., WEBB, B. C. & WOODCOCK, N. H. 1987. Late Caledonian (Acadian) transpression in NW England: timing, geometry and geotectonic significance, *Proceedings Yorkshire Geological Society*, **46**, 175–192.

STEELE, R. P. 1988. The Namurian sedimentary history of the Gainsborough Trough. *In*: BESLY, B. M. & KELLING, G. (eds) *Sedimentation in a synorogenic basin complex: the Upper Carboniferous of NW Europe*. Blackie, Glasgow, 102–113.

TUBB, S. R., SOULSBY, A. & LAWRENCE, S. R. 1986. Palaeozoic prospects on the northern flanks of the London-Brabant massif. *In*: BROOKS, J., GOFF, J. C. & VAN HOORN, B. (eds) *Habitat of Palaeozoic Gas in NW Europe*. Geological Society Special Publication, **23**,55–72.

WAPLES, D. W. 1980. Time and temperature in petroleum formation application of Lopatin's method to petroleum exploration. *Bulletin American Association Petroleum Geologists*, **64**, 916–926.

Timing, style and sedimentary evolution of Late Palaeozoic–Mesozoic extensional basins of East Greenland

FINN SURLYK

Geologisk Centralinstitut, Københavns Universitet, Øster Voldgade 10, 1350 København K, Denmark

Abstract: The late Palaeozoic–Mesozoic sedimentary basins of East Greenland record a complex series of events which eventually led to the successful opening of the Norwegian–Greenland Sea in the late Palaeocene. Major tectonic events and regional sea-level changes caused by plate movements and reorganizations, overprinted by the effects of local tectonics and associated sea-level changes are reflected in thick, unconformity bounded sequences. Caledonian crustal shortening and thickening culminated in the Silurian, and was succeeded by extensional collapse, vulcanism and intrusion of post-tectonic granites in the Devonian. Large, transtensional pull-apart basins were formed and became filled with continental red beds. Basin margins underwent repeated episodes of thrusting. A change to rifting took place in the latest Devonian. Rapid subsidence and continental deposition punctuated by episodes of igneous activity and contractional deformation continued from the Devonian until early Permian times for about 120 Ma leaving a record of alluvial fan and flood plain siliciclastics alternating with rhyolitic and alkaline basaltic intrusions and extrusions. Later basin subsidence reflected thermal cooling and contraction following the prolonged late Palaeozoic period of crustal thinning. Important phases of block faulting occurred at several intervals during Mesozoic times climaxing in the Volgian–Valanginian interval. Cenozoic basin inversions resulted in dramatic decoupled uplift and tilting of large blocks. The degree of Devonian extensional collapse, the position and orientation of Caledonian thrust planes, and the localization of mid- and late Palaeozoic transtensional strike-slip basins exerted a profound control on localization and style of the compartmentalized Mesozoic rift. The importance of pre-Mesozoic history, timing of Mesozoic tectonic events and the interplay of tectonism, and rates of subsidence, sea-level change and sediment influx have close parallels in the North Sea area. Examples from the exposed Mesozoic succession of East Greenland may thus serve as excellent predictive guides for subsurface geology around the British Isles.

The importance of tectonic style in controlling sedimentation is a currently fashionable research topic, and the possibility of predicting the nature and organization of sedimentary sequences in certain tectonic settings is of obvious importance in petroleum exploration.

The relative role of sea-level variations and tectonic activity in governing Jurassic basin evolution in East Greenland was discussed by Surlyk *et al.* (1981). The extreme difficulty in more precisely evaluating the importance of these factors was emphasized. Since then alleged eustatic sea-level curves have been considerably refined (Haq *et al.* 1987; Hallam 1988), and a regional sea-level curve has been constructed for East Greenland (Surlyk 1990). Several general models for the tectonic mechanisms of crustal stretching and subsidence have been presented (e.g. Badley *et al.* 1988; Lister *et al.* 1986; Rosendahl 1987; Rosendahl *et al.* 1986; Wernicke 1981; Wernicke & Burchfiel 1982). Thermal modelling of basin formation was initially based on the simple extension model of McKenzie (1978), followed by more complex models such as the dyke injection and the visco-elastic models (see reviews by e.g. Dewey 1982; Sclater & Célérier 1987). Recently it has been suggested that intra-plate stresses may be the causative factor of shorter-term sea-level changes (Cloetingh 1986; Cloetingh *et al.* 1985). These advances were essentially based on acquisition and public accessibility of high quality reflection seismic data, and an overwhelming number of basin simulations, and tectonic and sedimentation studies have appeared over the last few years.

In the case of well exposed, little deformed basins such as the late Permian–Mesozoic basins of East Greenland the situation is somewhat different. The nature and importance of the basin forming tectonics have been recognized for a long time (e.g. Vischer 1943; Maync 1947; Haller 1971; Surlyk 1977a), and detailed studies have been undertaken on syntectonic sedimentation on tilted blocks (Surlyk 1978a, b, 1984, 1987, 1989; Surlyk & Clemmensen 1983). Seismic data are only starting to appear from one area (Jameson Land). They will probably

From Hardman, R. F. P. & Brooks, J. (eds), 1990, *Tectonic Events Responsible for Britain's Oil and Gas Reserves*, Geological Society Special Publication No 55, pp 107–125.

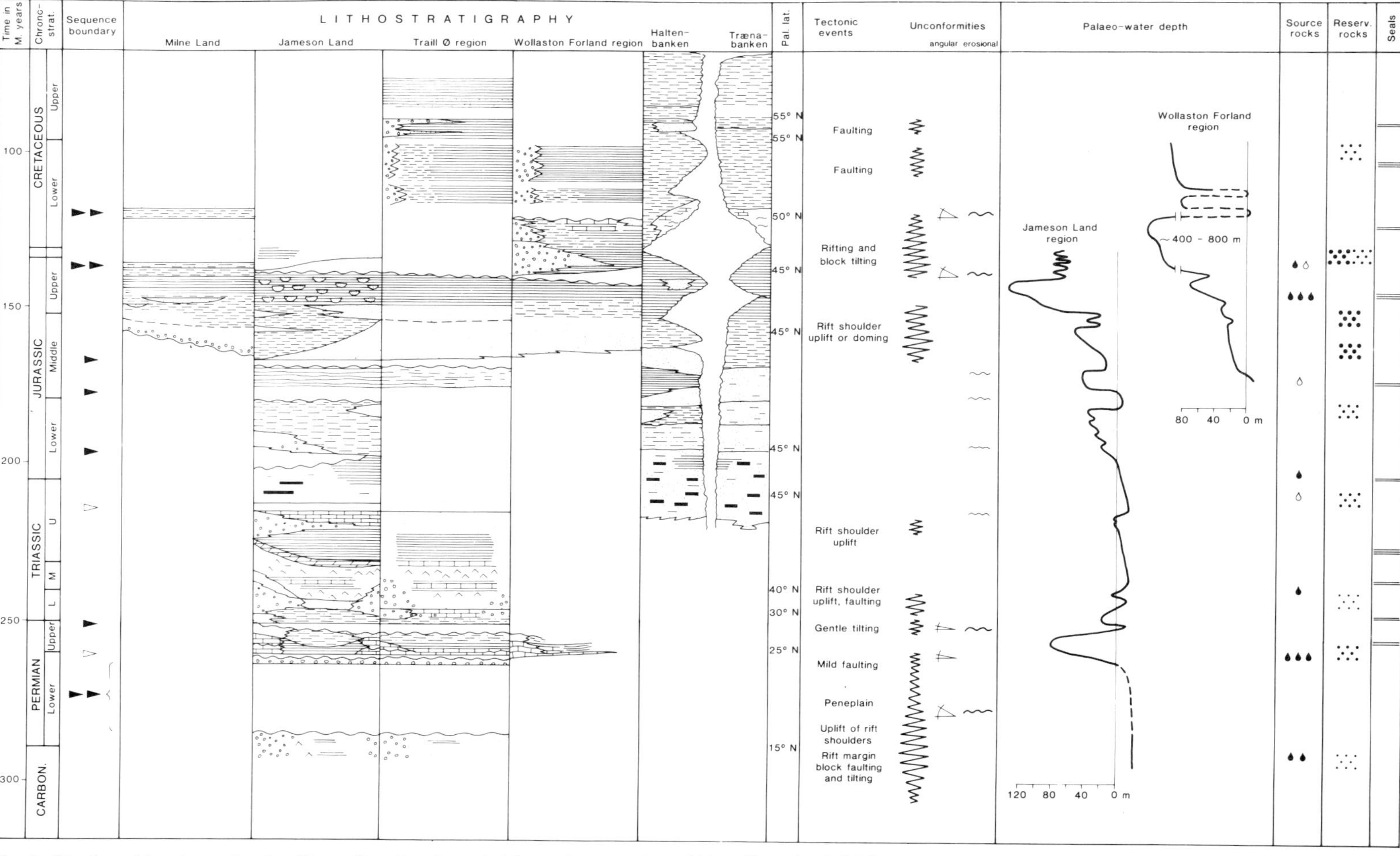

Fig. 1. Stratigraphic scheme for the Upper Permian through Mesozoic succession of East Greenland. Major tectonic events and sequence boundaries are indicated together with presence of potential hydrocarbon source rocks, reservoir rocks and seals. The estimated water depth curves represent the central basinal areas.

not add much to the understanding of the Mesozoic succession of this basin, while they will be of outstanding importance for the understanding of the deeply buried parts of the Devonian–Lower Permian succession.

The present paper focuses on the timing of tectonic events, and on the interplay between tectonism and rates of subsidence, regional sea-level changes and sediment influx in controlling and modifying the evolution of the East Greenland extensional basins (Fig. 1). Correct identification of the isolated or combined effects of these factors may result in development of predictive models which can aid interpretation of deeply buried sequences in and offshore East Greenland and on the conjugate Norwegian margin.

The text is organized following a sequence stratigraphic and depositional system format. Application of sequence stratigraphic — depositional systems concepts is possible due to the extensive 3D outcrops of the relatively undisturbed sequences. The marine sequences are bounded by major unconformities or their correlative conformities which commonly correspond to significant vertical facies changes. The Triassic continental succession cannot be easily interpreted within a sequence stratigraphic framework, and description and interpretation follow the main depositional systems.

Extensional collapse of the Caledonian mountain belt

Dewey (1988) suggested that many mountain belts follow a five-phase evolutionary pathway. Most of these phases can also be recognized in the Caledonides of East Greenland. Phases 1–3 involves lithospheric/crustal shortening and thickening, thrust and strike slip tectonics, slow thinning of the Thermal Boundary Convection Layer (TBCL) and slow uplift. This is followed by catastrophic convective erosion of the TBCL, resulting in rapid uplift and intrusion of post-tectonic granites. The first phases culminate in accelerating extensional collapse with lithospheric extension and thinning, and formation of deep non-marine basins accompanied by radial extensional spreading from elevated regions. Then follows a phase of post-extensional thermal recovery, thickening of the TBCL and progressive marine transgression.

In East Greenland the Caledonian orogeny culminated in the Silurian, and was followed by extensional collapse in the Devonian. The earliest sediments filling in the pre-extension topography are of latest early and middle Devonian age. The Devonian successions are 5–10 km thick and consist of conglomerates, sandstones and subordinate mudstones of alluvial fan, fluviatile and lacustrine origin. Earlier research is summarized by Haller (1971), Henriksen & Higgins (1976) and Friend *et al.* (1983). The tectonics and sedimentation facies of the Devonian basins are currently receiving detailed study (Larsen *et al.* 1989).

The Devonian basins are limited by major faults, and the basin fill is dominated by alluvial fan conglomerates along the margins, while fluviatile sandstones and lacustrine mudstones characterize the central parts. Synsedimentary folding, thrusting, uplift and emplacement of volcanic rocks took place intermittently during the middle and late Devonian. The acid volcanism was of local nature and includes flows and tuffs, while the basic intrusions and extrusions have a wider occurrence and are dominated by dykes and sills, tuffs and lavas being uncommon. Two relatively small granitic plutons have also been encountered.

The main tectonic events, which were termed the 'Hudson Land fold phases' by Bütler (1959) mainly affected the northern part of the Devonian basin.

The structural style of the Devonian basins indicates formation as deep volcanic pull-apart or oblique slip basins under dominant extension alternating with compressive events.

In the northern area the Devonian succession is well exposed and 8–9 km thick. It includes from below: the Middle Devonian Vilddal Supergroup (2.6 km) comprising conglomerates and sandstones deposited by eastwards flowing rivers; the Upper Devonian Kap Kolthoff Supergroup (3 km), which is composed of sandstones with subordinate conglomerates and acid volcanics. Drainage was from the eastern and western basin margins, and longitudinally towards the south in the central part; the Kap Graah Group (1.7 km) dominated by sandstones and conglomerates in the lower part. The basin was asymmetric and dominated by rivers flowing westwards from the eastern basin margin and longitudinally to the south close to the western margin. This drainage pattern was reversed in upper Kap Graah time when rivers flowed northwards; the Mount Celsius Supergroup (1.2 km) is dominated by lacustrine siltstones and sandstones deposited from northwards flowing rivers (Larsen *et al.* 1989).

In the Jameson Land area to the south, Devonian deposits are in contrast deeply buried under a thick Upper Palaeozoic–Mesozoic succession. Only along the northeastern and eastern basin margin are thick sequences exposed, including the Lower–Middle Devonian Kap Fletcher volcanics (1 km) overlain by Middle Devonian conglomerates, sandstones and siltstones (3 km). These occurrences are strongly suggestive of the presence of a deeply buried thickly developed Devonian succession in the Jameson Land Basin.

Devonian continental sedimentation continued uninterrupted into the Carboniferous, but the tectonic style changed to more uniform extension from latest Devonian times. In Jameson Land the area of sedimentation was extended outside the Devonian basin and at least 3–4 km of alluvial fan and floodplain sediments were deposited. Rotational block faulting took place in late Carboniferous – early Permian times along the basin margin over the shoulders of the underlying Devonian basin (Haller 1971; Surlyk *et al.* 1984, 1986; Larsen 1988).

Differential subsidence over the buried margin may provide some of the background for the nature of the faulting, but major basement-involving half-grabens were also formed outside the Jameson Land area towards the southwest, and to the north on western Clavering Ø (Fig. 2). This clearly shows that the late Carboniferous–early Permian was a period of intense extensional block-faulting heralding a long series of extensional events throughout late Permian–Mesozoic times. The border fault of the south western half-graben (Fig. 2) was formed by reactivation of a major Caledonian thrust.

The Carboniferous part of the successions is poorly dated, while the age of the Lower Permian is well documented by palynology (Piasecki 1984).

The marginal rift blocks were uplifted and peneplained before the late Permian transgression. The Upper Permian Foldvik Creek Group thus rests with a pronounced angular unconformity of up to 15° on the eroded and peneplained surface of the syn-rift sediments. Faulting continued into the late Permian, but with strongly decreasing intensity. The Upper Permian succession can thus be considered as representing the initial phase of subsidence by thermal contraction following the long period of extension by strike slip transtension and rifting (Surlyk *et al.* 1986).

Initiation of subsidence by thermal contraction

The Lower–Upper Permian boundary marks the most profound change in tectonic style and overall depositional environment in the post-Caledonian history of East Greenland (Fig. 2). The late early Devonian to early Permian succession was uplifted and deeply eroded along the basin margins during the final mid-Permian phase of rotational block-faulting, and a vast peneplain was developed. A widespread conglomerate-sandstone-mudstone sheet (Huledal Formation) was deposited on the peneplain with pronounced unconformable contact and marks the transition from a long period of crustal extension to a period of subsidence governed mainly by thermal relaxation of the stretched and thinned crust.

The thickness of the Huledal Formation is typically 20–30 m along the basin margins, with a maximum thickness of 160 m in Hold-with-Hope. It wedges out in the Wegener Halvø area. The thickness seems to increase towards the central parts of the Jameson Land Basin, perhaps reaching several hundred metres in the subsurface. Local thickness variations occur, especially where the lower part of the conglomerate was deposited along faults with active growth. Some of the syntectonic depressions appear to represent small grabens situated over rollovers adjacent to the main basin margin faults.

Along the margins, the formation consists of red and drab coloured scour and fill cross-bedded pebble and cobble conglomerates and sandstones, which pass into red mudstones towards the basin centre. Palaeocurrent patterns show that the drainage was away from the basin margins towards the centre.

Sedimentation took place in a bajada formed by coalescent, low gradient alluvial fans passing into a coarse-grained braidplain and extensive intermontane lakes. Well defined channel systems were not developed, and transport and deposition were mainly by flashfloods following torrential rain storms. Hot, arid conditions and strong evaporation led to the formation of gypsum impregnation of the distal bajada sediments.

A relative sea-level rise at the end of Huledal time is reflected by the occurrence of marine fossils in the uppermost parts of the conglomerates.

The late early Permian Huledal depositional system thus represents deposition in an intermontane basin in a period of decreasing extensional tectonics and beginning thermal subsidence, during the initial phases of a major late Permian rise in sea level. The base of the formation represents the most significant sequence boundary in the post-Caledonian predrift succession of East Greenland.

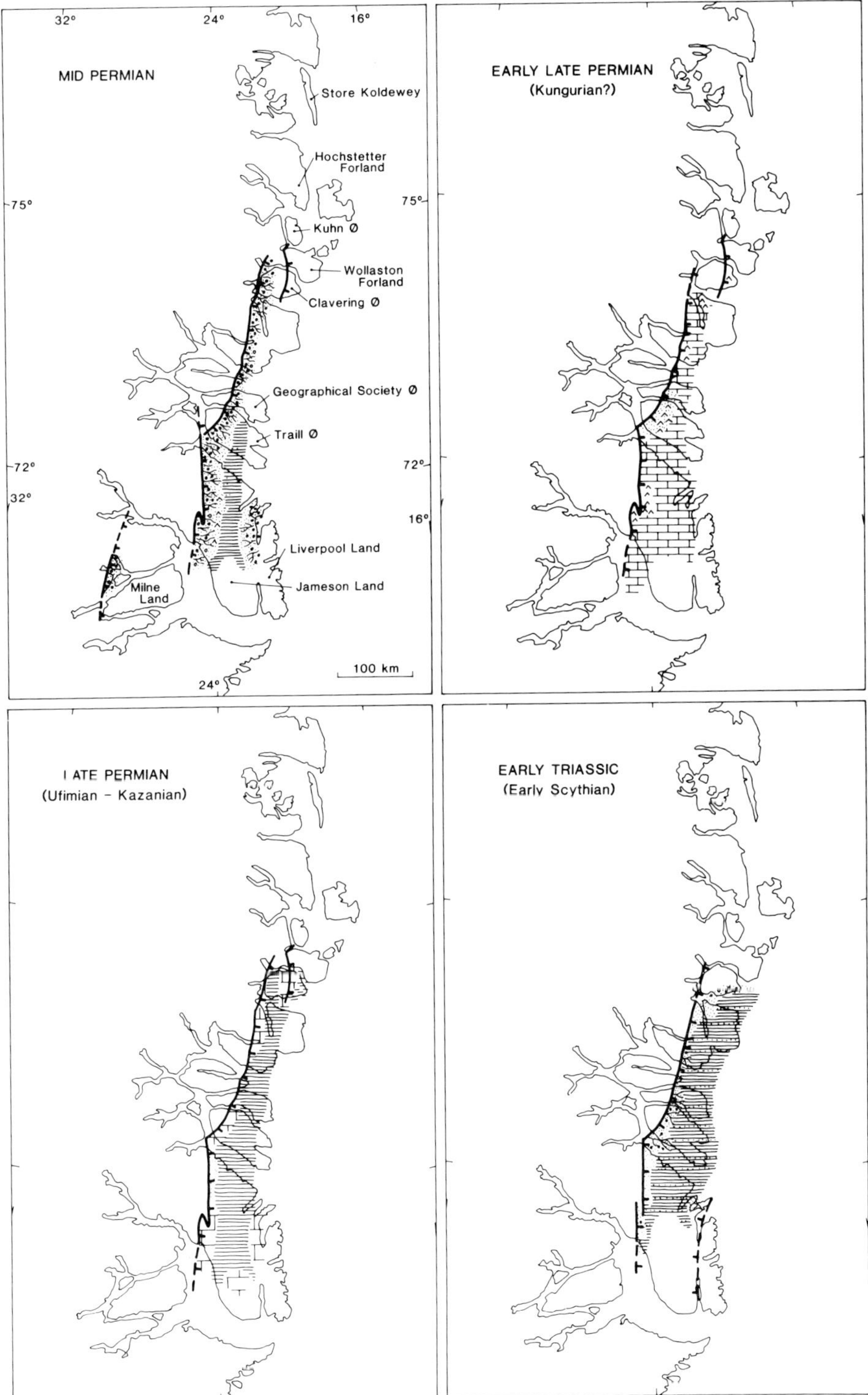

Fig. 2. Palaeogeographic maps with main fault zones indicated.

Late Permian hypersaline-to-normal marine bay: Foldvik Creek Group

The main part of the Upper Permian succession was formed under a regime of continued subsidence by thermal contraction with growth on some faults. Overall sea-level rise was punctuated by significant falls. The boundary between the mainly continental Huledal depositional system and the marine carbonate-evaporite-shale-sandstone succession of the Foldvik Creek Group is marked by an abrupt but conformable change in facies and represents an important sequence boundary (Figs 1 & 2).

The marine Foldvik Creek depositional system consists of two quite clearly demarcated temporarily distinct sub-systems, representing an early hypersaline phase, and late normal marine phase (Fig. 2). The initial phase of sea-level rise resulted in the formation of a shallow marine, hypersaline seaway with bays, islands and shoals. The hypersaline deposits are grouped in the thick and widespread Karstryggen Formation (Surlyk *et al.* 1986).

Along the western basin margin, the formation is dominated by shallow marine and supratidal limestones including algally laminated limestone, lime mudstone, oolitic grainstone, intraclast conglomerate, and penecontemporaneous evaporite. Thick successions of mainly laminated gypsum were deposited in marginal depressions localized over the hanging wall of mid-Permian tilted blocks.

The overall sea-level rise was punctuated by episodes of marked sea-level draw-down. This is reflected by the development of a number of karst horizons formed by subaerial dissolution and erosion. They show a high-amplitude topography and are associated with collapse breccias and reworked conglomerates.

Growth of carbonate mounds was initiated during the succeeding sea-level rise over topographic highs of the karsted surface, while black, bituminous shales were deposited between the mounds and in the deeper parts of the basin.

The carbonate build-ups belong to the Wegener Halvø Formation. They consist of limestones with a normal marine fauna of bryozoans, brachiopods, gastropods, and echinoderms. The initial stages are represented by algal stromatolite mounds and associated grainstone shoals. The major build-ups consist of core and flank facies. The cores have no rigid framework, and fenestrate bryozoans are the only potential sediment baffling organisms. However, massive amounts of early submarine cements provided sufficient rigidity to maintain the structural and topographic viability of the core. Sediment moved down the flanks by a variety of sediment gravity flows and the flank beds pass gradually into black basinal shales of the Ravnefjeld Formation. Flank deposits are compositionally similar to the core, but lacks the early submarine cement. The main formation build-up took place in a low-energy environment, but cross-bedded shoals developed at the crest of the build-up during the latest developmental stages. The larger build-up crests stood as much as 100 m above the surrounding seafloor.

The laminated, black shale facies of the Ravnefjeld Formation is well represented throughout the entire late Permian basin of East Greenland. The exposed thickness varies from a few metres along the western basin margin to about 60 m between the build-ups on Wegener Halvø. The thickness in the deeply buried central part of Jameson Land is not known, but is most likely of the order of 50–100 m. Significantly greater thicknesses are quite possible, but seismic data do not allow differentiation between the excellent source rock shales of the Ravnefjeld Formation and the overlying bioturbated non-source rock shales of the Oksedal Member.

The siliciclastics of the final basin fill phase are grouped in the Schuchert Dal Formation which was deposited in a period of relative sea-level fall where subsidence could not keep pace with sediment influx.

In the basinal areas this phase is marked by a change from the black, bituminous Ravnefjeld shales to grey or black fossiliferous mud and siltstones of the Oksedal Member. This unit has a low content of organic matter and is thoroughly bioturbated although laminated intervals still occur. The Oksedal shales coarsen upwards into the Bredehorn Member, which comprises fine- to medium-grained, micaceous sandstones with intercalated quartzite pebbles and limestone cobbles and boulders. The latter are extremely fossiliferous and are derived from only slightly older Upper Permian limestones exposed in synsedimentary uplifted and eroded fault scarps along the basin margin.

The Foldvik Creek Group thus represents a complex depositional system formed in a period of overall sea-level rise punctuated by important drops, followed by a final major fall in sea-level. Subsidence was relatively uniform but differential subsidence over buried fault-blocks and rejuvenation of some basin margin fault scarps attest to continued moderate tectonic activity in the transition from extension to thermal relaxation.

The lower sequence boundary is marked by a

facies change from the continental to marginal marine coarse siliciclastics of the Huledal Formation to the hypersaline carbonates and evaporites of the Karstryggen Formation. The upper sequence boundary is a major erosional, and in places angular, unconformity reflecting the maximum sea-level drawdown at end-Permian times. Formation boundaries and karst surfaces represent parasequence boundaries.

Early Triassic deltaic and shallow marine sandstones and shales: Wordie Creek Group

The major eustatic sea-level fall in the latest Permian led to emergence and erosion of the youngest Permian sediments throughout almost all of the Upper Permian basin. Deposition seems, however, to have continued uninterrupted across the Permian–Triassic boundary along the N–S basinal axes some of which were situated close to the western basin margins reflecting the relatively greater subsidence over the western, down-tilted part of the marginal fault blocks. In this area similar facies occur below and above the boundary, which can only be safely located on the basis of detailed biostratigraphy (Piasecki 1984). Elsewhere the Triassic sediments rest with marked unconformity on the eroded surface of the Upper Permian deposits (Trümpy 1969; Teichert & Kummel 1976; Vischer 1943). Major channels were cut down into the top of the Foldvik Creek Group and filled with coarse pebbly sandstone and conglomerate (Birkenmajer 1977). One of the channels on Wegener Halvø is 35 m deep, more than 200 m wide, and with a fill showing transport direction towards the southwest. Deep erosional channels filled with sandstone and conglomerate have also been observed at roughly the same stratigraphic level at Kap Stosch (Marcussen *et al.* 1988).

The area was, however, rapidly submerged in earliest Triassic time and a short-lived marine bay was formed, although the sea level still seems to have fluctuated widely (Fig. 2) (Clemmensen 1980a, b).

The marine Lower Triassic (Scythian) sequence is placed in the Wordie Creek Group. The group is up to 700 m thick and consists of grey and green shales with ammonites and fish fossils and subordinate grey or green glauconitic or arkosic sandstones. It shows onlap on the eastern basin margin. Small, high-gradient turbiditic deltas rapidly prograded southwards along the western basin margin to fill axial troughs or large erosive gullies (Surlyk *et al.* 1984). By the end of the early Scythian the marine basin was filled in, and during the remainder of the Triassic, deposition took place in continental environments, except for a brief quasi-marine interlude in the Middle Triassic.

The Wordie Creek seaway was closed towards the south and connected northwards with basins in North America and Siberia as indicated by ammonite faunas (Spath 1935; Tozer 1965; Trümpy 1969; Teichert & Kummel 1976).

The base of the Wordie Creek Group is a major sequence boundary. It reflects the marked fall in sea-level in the latest Permian which caused emergence and erosion of the Permian sequence. This was followed by rapid but fluctuating sea-level rise causing marine basin margin onlap. The upper sequence boundary is a major facies boundary caused by late Scythian basin margin uplift and relative sea-level fall.

Subsidence was mainly thermal with some basin margin faulting and fault-scarp rejuvenation.

Late Scythian–Anisian intermontane graben

In late Scythian time East Greenland was affected by a major phase of basin margin uplift and rapid fault-controlled basin subsidence leading to the formation of a N–S trending intermontane graben, about 300 km long and 75 km wide (Clemmensen 1978a, b, 1979, 1980a, b). Coarse-grained red alluvial fan sequences, up to 500 m thick, were formed along the marginal fault zones. They pass from both sides of the graben into sandy floodplain deposits with longitudinal drainage towards the north. A major barchanoid dune field was developed along the western margin (Clemmensen 1978a), while extensive inland sabkhas formed in the central parts of the graben.

Borderland uplift abated during the Anisian(?) and the deposits became increasingly finer grained. Aeolian dune sands continued to be deposited in the western part of the graben while variegated gypsiferous sandstone–mudstone cycles characterized the eastern part (Fig. 3).

Anisian(?) marine bay

In later Anisian(?) time the graben was transgressed and a warm, shallow marine embayment was formed. The deposits are up to 30 m thick, and drape a wide variety of continental deposits including alluvial fan, sabkha and desert dune facies. Grey calcarenites were deposited along the bay margins, while black mudstones were

Fig. 3. Palaeogeographic maps with main fault zones indicated.

laid down in the central parts. The mudstones are potential source rocks for oil.

The transgression came from the NE and the marine connection to a North Atlantic seaway was probably caused by fault movements along the Kong Oscar Fjord fault zone.

Anisian(?) — Ladinian(?) lake

The marine connection was closed in later Anisian(?) time and an extensive lacustrine system formed (Fig. 3). It is not known if this change is due to tectonic uplift of the eastern borderland, to sea-level fall, or both. Lack of biostratigraphically diagnostic fossils in most of the Middle Triassic also precludes comparison of facies changes with published sea-level curves.

The initial continental facies include up to 50 m of gypsiferous red siltstone and fine-grained sandstone with non-marine trace fossils. Sedimentation took place in floodplains and inland sabkhas.

In the Ladinian(?) 50–150 m of gypsiferous sandstone–mudstone cycles were laid down in sabkhas and desert lakes. Thin aeolian sandstones and halite pseudomorphs occur, while freshwater trace fossils are absent. The climate was semi-arid.

These facies give way upwards to 50–80 m of cyclical stromatolitic dolostone–mudstone–sandstone deposits, presumably of Ladinian age.

Towards the top of the succession relatively thick (4–5 m) medium-grained sandstones appear. They become thicker, coarser grained, and more frequent towards the north and may reflect tectonic uplift of a northern source area along a NW–SE cross fault in the Kejser Franz Joseph Fjord.

Carnian(?) — Rhaetian lacustrine and fluvial sedimentation

The intermontane lacustrine system continued into the late Triassic, but major facies changes can be noted. More than 200 m of red thin-bedded fine-grained sandstone–mudstone cycles were deposited. Gypsum and other evaporites are absent, while a diverse trace fossil assemblage occurs suggesting increasing humidity (Clemensen 1979; Bromley & Asgaard 1979).

The lacustrine red-beds are overlain by up to 250 m of Norian(?) – Rhaetian fluviatile red and grey sediments, forming a major coarsening-upwards unit possibly reflecting tectonic uplift of borderlands (Clemmensen 1980a). The main source area uplift took place along the western basin margin, as seen from palaeocurrents and facies trends. At the same time mild uplift in the region north of Jameson Land reduced the length of the depositional basin.

The fluviatile system changed through time from a relatively fine-grained meander belt and floodplain to a coarse-grained braidplain. A diverse trace fossil assemblage occurs in the fluviatile deposits (Bromley & Asgaard 1979).

A relative rise in sea-level and base-level is reflected in a change from the coarse braidplain sediments to a 60 m thick lacustrine or lagoonal unit of variegated limestones and mudstones passing upwards into grey dolomitic limestones, black mudstones and thin bone beds. The uppermost part of this succession contains freshwater and marine faunas indicating intermittent marine transgression of the basin from the northeast.

This unit also shows a gradual change from the red and variegated colours of the underlying Triassic deposits to yellow, grey and drab colours of the Rhaetian and younger, Jurassic deposits. This reflects a change to more humid, gradually cooler conditions corresponding to the slow northwards drift of Greenland.

The Triassic is 1000 to almost 2000 m thick in Jameson Land where it is best exposed and well studied. It is difficult to distinguish angular and erosional unconformities and thus well defined depositional sequences, and the lack of age-diagnostic fossils precludes identification of major hiati. Subsidence was apparently relatively uniform resulting from thermal contraction following the late Paleozoic extensional period overprinted by the effects of minor tectonism, notably in the early Triassic. Gradual onlap of the Liverpool Land block which formed the eastern graben margin attests to an increasing flexural rigidity which acted in concert with sediment load to induce downbending of the graben shoulders with time.

Northwards plate drift and the resulting decrease in temperature and increase in humidity, led to a shift from evaporitic red beds to grey and drab coal-bearing deposits.

Rhaetian-Hettangian humid floodplain and lake: Kap Stewart Formation

In mid-Rhaetian time the depositional basin became areally strongly reduced, only to cover the Jameson Land area where the 200–300 m thick Kap Stewart Formation was deposited (Fig. 3). The western basin margin was controlled by the old, post-Devonian Main fault. To the east the Liverpool Land block was onlapped and the precise location of the eastern boundary is not known, but it was probably located somewhere up the dip-slope of the down-flexed block. The northern boundary was formed by uplift of the area north of the Kong Oscar Fjord cross-fault zone. The southern basin margin is not exposed, but facies evidence suggests that it was not far south of the present day south coast of Jameson Land. It was possibly controlled by uplift along a cross-fault in the Scoresby Sundfjord.

The marginal southeastern part of the Jameson Land basin was characterized by distal alluvial fans passing into northwards flowing, low-sinuosity, braided bed-load rivers. Upwards the sandstones give way to plant-bearing, cyclical meander belt deposits with thin coal beds of swamp origin. Floral evidence shows that the overall climate was humid and still warm, with some differentiation between the seasons. The palaeolatitude was about 40°N in contrast to a mid-Triassic palaeolatitude of 20–30° N (Smith & Briden 1977).

In the central part of the basin the floodplain gave way to extensive black lacustrine shales, while wave-worked deltas prograded southwards from the northern borderland (Clemmensen 1976; G. Dam 1988, pers. comm. 1989).

Along the southeastern basin margin the top of the Kap Stewart Formation is marked by an important hiatus representing most of the Sinemurian Stage. A more gradual and complete transition to the overlying Neill Klinter Formation seems to occur in the central and northern parts of the basin and the basin margin unconformity apparently passes into the correlative conformity.

Pliensbachian–Toarcian tidal embayment: Neill Klinter Formation

The first fully marine inundation of the Jameson Land area since late Permian–earliest Triassic times took place in the Pliensbachian (Fig. 3). The 200–300 m thick Pliensbachian–late Toarcian Neill Klinter Formation was deposited in an extensive tidal embayment. The depositional basin was the same as for the preceding Kap Stewart system and was probably controlled by the same tectonic zones. It is likely that much of the Sinemurian Stage is represented by the transitional sequence in the central part of the basin, and that the hiatus between the Kap Stewart and the Neill Klinter depositional systems was absent or only minor here.

Up to 25 m of richly fossiliferous pebbly foreshore, high-energy sandstones of the Lower Pliensbachian Rævekløft Member were deposited unconformably on the Kap Stewart Formation along the eastern basin margin. The near-shore sandstones give way upwards to interbedded and strongly bioturbated tidal shelf sandstones and mudstones constituting the 110–220 m thick Gule Horn Member. It consists of four basinwide coarsening-upwards sequences which are possibly controlled by eustatic sea-level fluctuations. They are presently receiving detailed study by Dam (1988).

This depositional regime continued into late early Toarcian time, where richly fossiliferous bioturbated cross-bedded sandstones and muddy deep shelf sandstones of the 90 m thick Ostreaelv Member reflect continued overall sea-level rise. A total of at least six progradational parasequences forming an aggradational to retrogradational parasequence set can thus be recognized in the Neill Klinter Formation.

The Neill Klinter depositional system came to an end in the late Toarcian and the top of the formation marks a major sequence boundary. Fossil evidence shows that an open marine seaway connected East Greenland and NW Europe.

Deposition in this time interval represents the combined effects of steady subsidence and a high rate of sea-level rise punctuated by five phases of stillstand or minor falls. The formation can thus be interpreted as a somewhat irregular transgressive systems tract.

Latest Toarcian–Early Aalenian restricted marine embayment: Sortehat Formation

The Neill Klinter Formation is disconformably overlain by dark silty mudstones of the up to 100 m thick sheet-like Sortehat Formation. Thin sandstone beds become increasingly common upwards. The formation contains abundant restricted marine fossils and trace fossils, but ammonites are lacking, age diagnostic dinoflagellates are rare and the formation is accordingly poorly dated. Deposition took place in a relatively deep marine embayment probably with fluctuations in salinity and periodic poor oxygenation. A N–S sill over the eastern crestal area of the Liverpool Land block, where the formation has almost wedged out, may have exerted some hydrographic control. The age of the formation is indicated by the occurrence of late Toarcian ammonites in the top of the underlying Neill Klinter Formation and by ammonites of late Bajocian age in the basal part of the overlying Pelion Member. Scarce dinoflagellates suggestive of latest Toarcian-Aalenian age have been recovered.

The Sortehat depositional system was governed by continued, steady subsidence in a tectonically quiet period, under a high rate of sea-level rise leading to sea-level high-stand.

Bajocian–Late Callovian transgressive systems tract: wave worked deltas and sandy shelf of the Vardekløft Formation

Middle Jurassic deposition took place in two tectonically controlled en echelon arranged embayments, the southern Jameson Land embayment and the northern Wollaston Forland embayment (Fig. 4).

The base of the Middle Jurassic Vardekløft Formation corresponding to the base of the sandy Pelion Member is one of the most important Mesozoic sequence boundaries in East Greenland. It is isochronous in Jameson Land (Heinberg & Birkelund 1984) where it forms a marked, planar facies boundary with the Sortehat Formation. Unconformable relations are only recognized in a small fault-block on the south coast of Liverpool Land. Here, the Sortehat Formation is only 5–10 m thick, and Birkenmajer (1976) attributed the reduced thickness to the possible presence of an erosional unconformity at the base of the Pelion Member.

The incoming of the Pelion sandstones reflects the start of a period of borderland uplift, which lasted from the Bajocian to the middle Oxfordian. Major pulses of clastic influx took place in the Bajocian to middle Bathonian, the late Callovian and the middle Oxfordian.

The period was in addition characterized by the most important transgression since the late Permian (Surlyk 1977a, 1978a, 1990). Large areas north of Kejser Franz Joseph Fjord became covered by the sea and Middle Jurassic sandstones rest directly on peneplaned Caledonian basement rocks in the Wollaston Forland embayment.

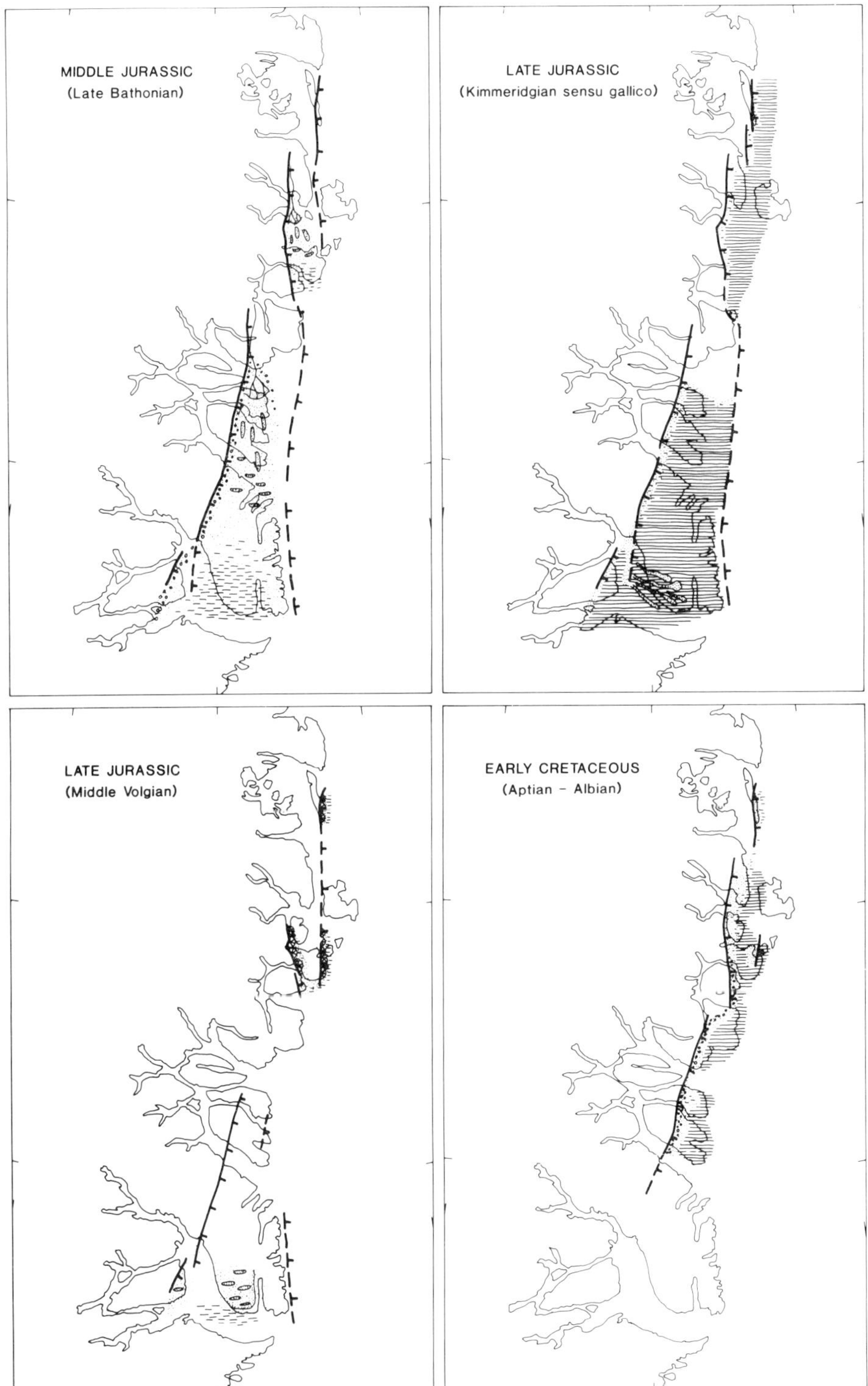

Fig. 4. Palaeogeographic maps with main fault zones indicated.

The base of the Pelion Member systematically youngs northwards in the Wollaston Forland embayment, and the Hochstetter Forland was first transgressed in late Callovian or early Oxfordian times (Surlyk 1978c). This was interpreted by Surlyk (1978a, b) and Surlyk & Clemmensen (1983) as reflecting northwards rift propagation overprinted by an overall rise in sea level.

The thickness of the Vardekløft Formation in the Jameson Land embayment increases from 160 m towards the south, where the Pelion Member has almost wedged out, to about 420 m in central Jameson Land, and at least 600 m in northern Jameson Land. The Pelion Member is composed of stacked coarsening-upwards units, but successions without any clear vertical organization also occur (Heinberg & Birkelund 1984). The coarsening-upwards cycles are here interpreted as reflecting large wave and tide dominated deltas prograding southwards in the elongate tectonic embayment. Fluvial or distributary channels and coal beds occur in the head of the bay on Traill ∅. Transport was mainly to the south, but bipolar currents recorded in the northern part of the embayment suggest tidal influence. Much of the sand was transported southwards in extensive sand dune systems. Giant-scale cross-beds with large-scale trough cross-bedded intrasets occur scattered and represent major outer shelf tidal or storm induced sand banks.

The Middle Jurassic sediments show a pronounced onlap along the western basin margin in Milne Land, where deeply weathered basement was transgressed and a steep rocky shore was developed. The lower part of the onlapping Charcot Bugt Member consist of late Bathonian conglomerates and cross-bedded sandstones. These variable deposits give way upwards to stacked upwards-coarsening sandy units. The upper part of the member consists of giant-scale cross-bedded sets of pebbly sandstone, up to 30 m thick, with E–SE dipping cross-strata. It is here suggested that these sets represent linear sandbanks of tidal and storm origin.

The influx of coarse clastics continued, but was overwhelmed by the Bathonian rise in sea-level, and the Pelion sandstones gave way to micaceous silty shales of the Fossilbjerget Member deposited in a quiet offshore sub-wave base environment.

The Pelion–Fossilbjerget boundary is relatively sharp, but markedly diachronous. It rises rapidly from basal Bathonian in southernmost Jameson Land over Middle Bathonian in central Jameson Land to Upper Bathonian or basal Callovian in north central Jameson Land. Only in the northern part of the Jameson Land embayment could the coarse clastic influx keep pace with the ongoing sea-level rise, and the Fossilbjerget Member wedges out here.

The offshore silty Fossilbjerget shales continued to be deposited in southern Jameson Land until the middle–late Callovian boundary. An early Callovian hiatus occurs in the most offshore sediment starved southern part of the embayment. The Pelion–Fossilbjerget couplet is interpreted as a major transgressive system tract composed of a complex retrogradational parasequence set. The sharp boundary between the Pelion sandstones and the Fossilbjerget shales was formed by shoreface retreat and erosion. It is thus not a sequence boundary. The mid-Fossilbjerget hiatus in the most distal, southern part reflects a high rate of sea-level rise and sediment starvation. It corresponds to the initial development of condensation phenomena approaching the maximum flooding surface.

The northern Wollaston Forland embayment had a rather similar history. The Bathonian–Callovian transgression led to gradual northwards onlap possibly accompanied by northwards basin propagation by down-faulting along cross-faults. Pelion Member sandstone deposition continued until the Callovian–Oxfordian transition but dating of this interval is rather poor. In the southern part of the embayment the unit consists of stacked upwards-coarsening units. Northwards the sequential organization disappears and the member consists of cross-bedded sandstones. The basal part of the succession reflects the northward shift of coastal, barrier-lagoon systems locally with thick coal-beds. The bulk of the member was deposited in a shallow marine embayment under strong tidal influence (Surlyk 1977a; Surlyk & Clemmensen 1983). The main transport was towards the south as in the Jameson Land embayment and several processes such as sand dune migration and construction of linear tidal banks can be recognized.

The Pelion Member is thus essentially developed as in the Jameson Land basin and forms a retrogradational parasequence set.

Late Callovian–Middle Oxfordian highstand systems tract: wave influenced deltas of the Olympen Formation

In the late Callovian a major clastic pulse occurred. In the Jameson Land embayment it is represented by the rapid progradation from the north of the dominantly sandy Olympen Formation, which downlaps onto the surface of the Fossilbjerget Member and almost wedges out to the south in central Jameson Land. The time interval is here in the offshore part of the basin characterized by marked condensation and hiati corresponding to the maximum flooding surface. The Olympen Formation is interpreted as a progradational parasequence set forming a high-stand systems tract.

In the Wollaston Forland embayment a stepwise vertical facies change occurred in the early Oxfordian or possibly latest Callovian. Stacked coarsening-upwards or non-organized parasequences of the Pelion Member gave way to finer grained, intensely bioturbated interlaminated silty mudstones and cross-laminated fine sandstones of the more than 100 m thick Jakobsstigen Formation. In Wollaston Forland this unit forms an overall coarsening-upwards succession which is composed of small-scale coarsening-upwards units, a few metres thick. The latter consist of a basal black laminated mudstone followed by bioturbated heterolithic mud and sandstone, and sandstone on the top (Surlyk 1977a). The sequential pattern shown by these mid and outer shoreface deposits was interpreted by Surlyk & Clemmensen (1983) as caused by transgression of extensive low gradient rather fine-grained tide and wave-worked deltas possibly caused by smallscale down-to-basin faulting along cross-faults. Northwards in Kuhn Ø they pass into cross-bedded tidal sandstones, and further north in Hochstetter Forland into coal-bearing coastal sandstones.

This depositional phase lasted until the late Oxfordian where also the northern region became deeply submerged, and the long, middle–early late Jurassic period of coarse clastic deposition came to a definitive end. The Jakobsstigen Formation and its proximal correlatives are interpreted as a progradational parasequence set representing a high-stand systems tract.

The Bathonian–middle Oxfordian thus represents a time interval of large-scale tectonic borderland uplift and a very high rate of coarse clastic sediment input. At the same time the most important regional transgression since the late Permian took place. Large areas became covered with marine sediments for the first time. The transgression and corresponding onlap proceeded from south to north apparently in a series of steps through the Bathonian and Callovian. The main coarse-clastic Bathonian–late Callovian sedimentary unit forms a major southwards prograding clastic wedge. The characteristic depositional motif is a coarsening-upwards parasequence tens of metres thick reflecting progradation of large tide and wave-worked deltas. The Bathonian–late Callovian interval is overall retrogradational, and represents a transgressive systems tract formed during maximum rate of sea-level rise.

The latest Callovian–middle Oxfordian interval is in contrast composed of a progradational parasequence set representing a highstand systems tract formed when rate of sea-level rise approached zero.

The rapid longitudinal fill of tectonic embayments with enormous amounts of coarse clastics resulted in a pronounced asymmetric N–S subsidence pattern. Up to 1200 m of coarse marginal marine sandstones were deposited in the northern head of the Jameson Land embayment during the latest Bajocian–middle Oxfordian time interval. This contrasts with about 160 m of sandstones and shales in the offshore part of the embayment in southern Jameson Land. The material entered the basin in the points of *en echelon* sidestepping of border faults (Surlyk *et al.* 1981). In these areas topographic depressions extended northwards from the basin axes. The elongate depressions controlled the localization of major river systems draining the updomed areas neighbouring the rift.

Late Oxfordian–Kimmeridgian deepening: Kap Leslie, Hareelv and Bernbjerg Formations

The overall rise in sea level initiated in the early Bathonian culminated in the late Oxfordian–Kimmeridgian where a relatively deep seaway was formed (Fig. 4). The sea level was now at its highest stand since the Caledonian orogeny. Only during mid and late Cretaceous time did the sea reach a comparable regional extent. The late Jurassic deepening and transgression is reflected by a vertical facies change to black or dark-grey shales. In the southernmost part of the Jameson Land Basin this facies transition was initiated already at the middle–late Callovian boundary. Only along the margin of the Wollaston Forland embayment did sandy shallow marine sedimentation still prevail. Even these areas received, however, muddy sediments from the start of the Kimmeridgian. This depositional phase first came to an end in the middle Volgian by the onset of a major tectonic phase.

The shale successions are placed in the Bernbjerg Formation in the areas north of Kong Oscar Fjord (Surlyk 1977a; Surlyk & Clemmensen 1983). The correlative unit in Jameson Land, the Hareelv Formation, is of more deep-water origin and contains thick turbiditic channel sandstones (Surlyk 1987).

The Bernbjerg Formation is up to 600–700 m thick, and is dominated by dark grey to black laminated mudstones and rhythmites. Graded sandstone beds, 10–20 cm thick, with erosional bases occur scattered or form amalgamated units up to 5 m thick. Ammonites and the bivalve *Buchia* occur in profusion. The fine details of the lamination show that deposition took place in a wide muddy shelf under considerable wave influence. The graded sandstones were probably emplaced during major storms. The formation is an excellent potential source for hydrocarbons. The lower part contains a relatively large proportion of terrigenous organic matter, but the Kimmeridgian interval is a rich, oil-prone source rock.

In the Milne Land area a N–S oriented sandy offshore bar was formed in the late Oxfordian landwards of a major intrabasinal fault zone. It was probably attached to the coast at its northern end where the coastline changed from a N–S to a NE–SW orientation. The sand was transported from the north by longshore currents. The sandstone bar is up to 90 m thick (Fürsich & Heinberg 1983). Grey, bioturbated sandy siltstones were deposited in the protected area behind the bar. Water depth increased rapidly seawards of the bar and the dark shales of the Hareelv Formation were deposited well below the reach of waves and storms. The formation is 200–500 m thick and contains thick, closely spaced sandstone bodies filling deep, steep walled gullies and elongate scours, or forming more regular, laterally extensive gully mouth or lobe deposits (Surlyk 1987). The sands were transported by erosive, high-density turbidity currents. A northeastern source area was represented by the short-lived rapidly prograding delta of the upper Olympen Formation, whereas the main, northwestern source area was the shallow, sandy offshore bar. The turbidity currents were probably triggered by earthquakes along the fault-controlled slope on the seawards side of the offshore bar. The turbiditic sandstone bodies are up to 50 m thick, hundreds of metres wide and may be more than 5000 m long in a downcurrent direction. They occur in a thick, oil-prone source rock, and may serve as a model for an unusual type of stratigraphic hydrocarbon reservoir.

The Hareelv Formation and its correlatives are interpreted as a new transgressive systems tract formed on top of the latest Callovian–middle Oxfordian high-stand systems tract. Highest rate of sea-level rise took place in the early Kimmeridgian and was reflected by a change in Milne Land from bioturbated silty shales and sandstones to 'oil shales', and elsewhere in East Greenland by a change to shales of high source rock quality.

Major rift phase: Middle Volgian–Valanginian

The most important phase of extensional tectonics since the early Permian was initiated in Middle Volgian times (Fig. 4). The area north of Kong Oscar Fjord was broken up into tilted blocks which became progressively fragmented and narrower with time. The Jameson Land area, south of the fundamental tectonic zone in Kong Oscar Fjord, underwent only mild tilting but enormous quantities of coarse sand were shed into the basin in the tectonically active period.

The depositional environments north and south of Kong Oscar Fjord were thus markedly different. The lithologies and volumes of syntectonic deposits were, however, relatively similar.

The tectonic style, stratigraphy and depositional environment in the northern region is well displayed in the Wollaston Forland region (Vischer 1943; Maync 1947, 1949; Surlyk 1978a, 1984, 1989).

Rotational block faulting and synrift deposition was initiated in the middle Volgian. The earliest deposits are coarse, poorly bedded, non-structured, marine gravity flow breccias composed of densely packed boulders and blocks derived from the basement rocks in the fault zone. The breccias have a steep, primary dip and are interpreted as representing a submarine talus.

The syntectonic deposits rapidly prograded eastwards and conglomerates, pebbly sandstones and interbedded mudstones progressively onlapped the tilted surface of the Bernbjerg Formation covering the dip slope of the hanging wall block. The first phase of talus cone-debris apron deposition rapidly filled the deepest part of the tectonic depression and a more organized, stable system was developed. It consists of a fringe along the fault scarp of gravelly fan deltas passing directly into coalescent submarine fans. Deposition was almost exclusively coarse grained and a wedge of gravity flow conglomerates, pebbly sandstones, and sandstones was formed, extending from the fault scarp to the basin axis about 15 km further east. The deposits have a complex internal organization. All beds wedge out within a few metres and mobilization of small, highly erosive gravity flows seems to have taken place from all parts of the fan surface.

This general type of deposition lasted from Middle Volgian to late Ryazanian time, corresponding to the main phase of block tilting and to the bulk of the syntectonic succession. At the Ryazanian–Valanginian transition a significant facies change took place caused by a regional rise in sea-level. The scarp area was transgressed and a shallow marine sandy shelf appears to have formed over the footwall block. The hangingwall still received conglomerates and sandy turbidites, but mudstones became increasingly dominant and touch the fault scarp in some places. Scouring was a less common phenomenon and the turbidites are parallel sided and better sorted. They are intercalated with calcareous, silty mudstones, but amalgamation is common. The submarine fans were thus better organized and more mud rich than the extremely coarse-grained Middle Volgian–Ryazanian deposits. Red condensed mudstones were deposited as a type of sea-mount facies over submerged structural highs, which were isolated from influx of coarse clastic material.

The tectonic phase faded out in the latest Valanginian. A short-lived uplift phase took place in the Hauterivian, and deposits of this age are thin or absent in most of East Greenland north of the Kong Oscars Fjord fault zone. Rift margin uplift following immediately upon rifting is a well known phenomenon.

The lack of rotational block faulting in the area south of the Kong Oscars Fjord fault zone prevented the development of steep slopes, and thus of sediment gravity flow. Instead a wide, level-bottomed seaway was formed which received enormous amounts of coarse pebbly sand which forms the Raukelv Formation.

The sands were deposited in sand dune and sandbank systems below wave base under the influence of storms and possibly also tidal currents (Surlyk & Noe-Nygaard 1989). Deposition was initiated with large-scale cross-bedded sandstones representing extensive fields of mainly southwards migrating sand dunes. They alternate with richly fossiliferous, bioturbated siltstones and fine-grained sandstones. The deepest offshore part of the embayment received mainly silty mudstones.

Above follow coarse-grained granule sandstones which overstep the lower units in eastern Jameson Land (Surlyk & Zakharov 1982). They consist of alternating sheets of large-scale cross-bedded units representing fields of southwards migrating sand dunes, and giant-scale eastward migrating cross-sets with set thicknesses varying from 15–50 m. The individual sets form extensive marker beds which can be followed for tens of kilometres. There are no signs of emergence or wave influence and deposition is interpreted to have taken place in slowly eastwards migrating linear sandbanks in a relatively deep shelf (Surlyk & Noe-Nygaard 1989). The material was probably transported southwards from the northern fault-scarp coastlines in extensive dune fields by coast-parallel currents.

The Raukelv Formation represents a rapidly prograding sandy shelf sequence. The relatively sudden transition from the Hareelv to the Raukelv Formation suggests that the change from relative sea-level rise to stillstand and fall was rapid. Omission surfaces on top of the sand dune and sandbank sheets are indicative of a number of short-term sea-level fluctuations in the middle Volgian to early Valanginian time interval. At least seven laterally very extensive omission surfaces have been documented by detailed mapping (Surlyk & Noe-Nygaard 1989).

An important late Volgian–Ryazanian basin margin hiatus occurs in Milne Land and suggest that sea level had reached its lowest position during that time interval.

Post-rift uplift phase: Hauterivian–Barremian

The rifting phase initiated in the middle Volgian faded out at the end of the Valanginian. Then followed a period of relative sea-level fall probably caused by a short pulse of immediate post-rift crustal uplift of the basin margin area.

In Milne Land the Volgian–Valanginian sandstones are overlain by up to 30 m of Hauterivian micaceous silty mudstones (Piasecki 1979). The overlying 100 m sandstones have not yet been dated. They were studied by Sykes & Brand (1976) who demonstrated the presence of a number of coarsening-upwards cycles interpreted as fan delta deposits. This unit is probably of late Hauterivian or Barremian age, and is the youngest succession preserved in the Jameson Land area.

In northern East Greenland, Lower Cretaceous shales rest with marked unconformity on the Volgian–Valanginian syn-rift deposits (Surlyk 1977a, b, 1978a). An Hauterivian interval, 10–20 m thick, has been recorded at the base of a monotonous Lower Cretaceous shale sequence which reaches a thickness of up to 1000 m. The thin Hauterivian interval is followed by some tens of metres of Barremian shales (H. N. Hansen, pers. comm. 1988).

Mid-Cretaceous transgression and thermal contraction

A drastic increase in subsidence rate possibly associated with a sea-level rise was initiated in the Aptian (Fig. 4). The greatest thicknesses are recorded from the central parts of the basin, where up to 500 m of Lower Aptian strata were deposited. Local thickness variations are to some extent controlled by the palaeotopography and differential subsidence over the Volgian–Valanginian tilted blocks, but the main pattern seems to reflect the formation of a broad thermal sag. Albian shales also occur in great thicknesses and about 400 m of Upper Albian shales have been recorded in central East Greenland. The precise age relations of these thick successions are, however, still poorly known.

Several events of basin margin faulting associated with the formation of fault scarps took place during the early Cretaceous. The deformations seem in all cases to be due to extensional faulting and there is no unequivocal evidence of strike-slip faulting or compressional tectonics.

The dominant Lower Cretaceous facies from the Aptian and onwards is dark, marine shale. Sandstones, pebbly sandstones, conglomerates and exotic blocks dominate along the western border fault and around local structural highs.

Thick conglomerates and pebbly mudstones (up to 80 m) are particularly common along the western border fault on Clavering Ø. The clasts comprise Caledonian crystalline rocks, Permian limestone and Jurassic sandstone. Most of the accumulations of coarse material were deposited by debris flows derived from the fault-scarp.

Syntectonic deposits also occur farther to the south along the main faults. The depositional conditions can thus be pictured as quiet sedimentation of mud in a relatively deep-water marine environment punctuated by density current deposition of very coarse material flowing eastwards down the slope away from fault-scarps. The basins were characterized by periods of poor circulation caused by eastern sills or barriers and eustatic sea-level high stand.

At the beginning of the Aptian transgression the sea formed narrow straits between the western mainland and eastern N–S islands over the crestal areas of the westwards dipping, hanging wall fault blocks. During the Albian, the sea onlapped the dip slopes of the blocks which were still not fully submerged by the end of the Albian. As the transgression proceeded, the islands shrank until they vanished in the Campanian (Donovan 1953, 1955).

The degree of submergence of the individual blocks during the Cretaceous varied depending on degree of tilt, topographic relief, block width and general subsidence.

The tectonic movements characteristic of the Aptian and Albian seem to have continued through the late Cretaceous, with culminations in the Cenomanian and the middle Turonian. The Upper Cretaceous reaches more than 700 m in thickness in the Traill Ø area and the facies are of the same type as in the Lower Cretaceous. Dark mudstones of relatively deep water origin dominate and along fault-scarp coastlines they interfinger with conglomerates and sandstones deposited from gravity flows.

Outliers of late Santonian–early Campanian and late Campanian

Marine mudstone with thin sandstones occur in the Traill Ø area and on Hold with Hope.

In southeast Greenland around Kangerdlugssuaq an isolated late Mesozoic–early Cenozoic succession underlies several kilometres of Paleogene plateau basalts (Soper *et al.* 1976). The lower boundary of the sediments is not known, but there are some indications that the Upper Cretaceous rests directly on Precambrian basement. The exposed sequence is only a few hundred metres thick and contains rocks of late Albian, middle early Cenomanian and Maastrichtian–Danian Age. The dominant facies is dark, marine mudstone with subordinate sandstones.

Conclusions

The sedimentary basins of East Greenland were initiated by extensional collapse of the Caledonian mountain belt, and the formation of deep, volcanic pull-apart basins in the middle Devonian–early Carboniferous.

The stress pattern shifted to more orthogonal rifting in the latest Devonian.

Rifting gradually died out in mid Permian times; rift shoulders were uplifted and marginal halfgrabens were eroded and peneplained.

A major regional transgression took place in the late Permian and a suite of hypersaline to normal marine deposits was laid down over the eroded surface of the syn-rift succession.

The Upper Permian sequence thus represents the initial phase of basin subsidence caused by thermal contraction following the long phase of crustal extension, but minor faulting still occurred.

The Permian–Triassic transition was marked by a major fall in sea-level and the top of the Permian succession was eroded along basin margins and over structural highs.

The sea rapidly encroached over the Permian basin in the earliest Triassic and a thick succession of shallow marine siliciclastics were deposited.

The sea rapidly retreated, however, and the remainder of the Triassic was characterized by deposition of continental red beds and related sediments, except for a brief marine interlude in the middle Triassic.

Faulting seems to have changed from listric to planar concomitant with the change from rift-controlled subsidence to thermal contraction.

During the early Triassic subsidence was accommodated along major, steep and probably planar faults and an intramontane graben was formed.

The Triassic–Jurassic transition was marked by a gradual change from red coloured continental deposits with evaporites to drab-coloured coal-bearing deposits reflecting climatic change accompanying the slow northwards drift of Greenland.

Jurassic–Cretaceous sedimentation was controlled by intermittent tectonic events and sea-level changes.

Major hiatuses reflecting sea-level fall occur in the Sinemurian, the late Toarcian and the Bajocian. Their duration decrease from the margins to the centre of the basin.

The most important transgression since the late Permian occurred at the Bajocian–Bathonian transition and large areas, especially in northern East Greenland became covered by the sea for the first time in post-Caledonian time.

At the same time, basin shoulder uplift and erosion resulted in a very high rate of clastic sediment influx.

Interpretation of the Middle and Upper Jurassic within a sequence stratigraphic framework allows a simple explanation for downlap surfaces and distal offshore hiatuses, and may provide the basis for prediction in offshore areas of East Greenland and Norway.

Sea-level rise was resumed in the Oxfordian and maximum sea-level high-stand was reached in the Kimmeridgian. This time interval was characterized by deposition of thick dark shale successions throughout East Greenland.

Fault activity started to increase in the late Oxfordian and a major rift event with tilting of fault-blocks was initiated in the middle Volgian and lasted until the early Valanginian.

Half-grabens in northern East Greenland were filled with thick wedge-shaped successions of submarine syn-rift conglomerates and sandstones. The Jameson Land basin subsided in a more regular way during this rift event and a thickly developed sandbank and dune sequence was deposited.

Rift margin uplift resulted in erosion of the top of the syn-rift successions and the Hauterivian and Barremian are absent or thinly developed in most of East Greenland.

Subsidence by thermal contraction ensued and a broad sag basin was formed in the early Cretaceous. Aptian–Albian shales are thus thickly developed in the central part of the basins and thin towards the margin.

Old basin margin faults were reactivated during several intervals during the Cretaceous, and coarse, marine fault-scarp deposits were formed in the Aptian, Albian, Cenomanian and Turonian.

Successful sea-floor spreading commenced in the early Paleocene with extrusion of thick, extensive plateau basalts.

The present paper is based on the work of numerous geologists. My own field work in the area over the last 22 years has been supported by the University of Copenhagen, the Carlsberg Foundation, the Danish Natural Science Research Council, several major oil companies, and the Geological Survey of Greenland (GGU). This paper is published with the approval of the GGU.

References

Badley, M. E., Price, J. D., Rambeck Dahl, C. & Agdestein, T. 1988. The structural evolution of the northern Viking Graben and its bearing upon extensional modes of basin formation. *Journal of the Geological Society, London*, **145**, 455–472.

Birkenmajer, K. 1976. Middle Jurassic near-shore sediments at Kap Hope, East Greenland. *Bulletin of the Geological Society of Denmark*, **25**, 107–116.

—— 1977. Erosional unconformity at the base of marine Lower Triassic at Wegener Ø, central East Greenland. *Rapport Grønlands geologiske Undersøgelse*, **85**, 103–107.

Bromley, R. G. & Asgaard, U. 1979. Triassic freshwater ichnocoenoses from Carlsberg Fjord, East Greenland. *Palaeogeography, Palaeoclimatology, Palaeoecology*, **28**, 39–80.

Bütler, H. 1959. Das Old Red-Gebiet am Moskusoksefjord. Attempt at a correlation of the series of various Devonian areas in central East Greenland. *Meddelelser om Grønland*, **160**, 5, 1–188.

Clemmensen, L. B. 1976. Tidally influenced deltaic sequences from the Kap Stewart Formation (Thaetic-Liassic), Scoresby Land, East Greenland. *Bulletin of the Geological Society of Denmark*, **25**, 1–13.

—— 1978a. Alternating aeolian, sabkha and shallow-lake deposits from the Middle Triassic Gipsdalen Formation, Scoresby Land, East Greenland. *Palaeogeography, Palaeoclimatology, Palaeoecology*, *24*, 111–135.

—— 1978b. Lacustrine facies and stromatolites from the Middle Triassic of East Greenland. *Journal of Sedimentary Petrology*, **48**, 1111–1128.

—— 1979. Triassic lacustrine red-beds and palaeoclimate: The "Buntsandstein" of Helgoland and the Malmros Klint Member of East Greenland. *Geologischer Rundschau*, **68**, 748–774.

—— 1980a. Triassic rift sedimentation and palaeogeography of central East Greenland. *Bulletin Grønlands Geologiske Undersøgelse*, **136**, 1–72.

—— 1980b. Triassic lithostratigraphy of East Greenland between Scoresby Sund and Kejser Franz Josephs Fjord. *Bulletin Grønlands Geologiske Undersøgelse*, **139**, 1–56.

CLOETINGH, S. A. P. L. 1986. Intraplate stresses: A new tectonic mechanism for fluctuations of relative sea level. *Geology*, **14**, 617–620.

——, MCQUEEN, H. & LAMBECK, K. 1985. On a tectonic mechanism for regional sealevel variations. *Earth and Planetary Science Letters*, **75**, 157–166.

DAM, G. 1988. Sedimentological studies of the fluviatile-shallow marine Upper Triassic to Lower Jurassic succession in Jameson Land, East Greenland. *Rapport Grønlands geologiske Undersøgelse*, **140**, 76–79.

DEWEY, J. F. 1982. Plate tectonics and the evolution of the British Isles. *Journal of the Geological Society, London*, **139**, 371–412.

—— 1988. Extensional collapse of orogens. *Tectonics*, **7**, 1123–1139.

DONOVAN, D. T. 1953. The Jurassic and Cretaceous stratigraphy and palaeontology of Traill Ø, East Greenland. *Meddelelser om Grønland*, **111**, 2, 1–150.

—— 1955. The stratigraphy of the Jurassic and Cretaceous rocks of Geographical Society Ø, East Greenland. *Meddelelser om Grønland*, **103**, 9, 1–60.

FRIEND, P. F., ALEXANDER-MARRACK, P. D., ALLEN, K. C., NICHOLSON, J. & YEATS, A. K. 1983. Devonian sediments of East Greenland VI: review of results. *Meddelelser om Grønland*, **206**, 6, 1–96.

FÜRSICH, F. T. & HEINBERG, C. 1983. Sedimentology, biostratinomy, and palaeoecology of an Upper Triassic offshore sand bar complex. *Bulletin of the Geological Society of Denmark*, **32**, 67–95.

HALLAM, A. 1988. A re-evaluation of Jurassic eustacy in the light of new data and the revised Exxon curve. *SEPM Special Publication*, **42**, 261–273.

HALLER, J. 1971. *Geology of the East Greenland Caledonides*. Interscience, London, 1–375.

HAQ, B. U., HARDENBOL, J. & VAIL, P. R. 1987. Chronology of fluctuating sea levels since the Triassic. *Science*, **235**, 1156–1167.

HEINBERG, C. & BIRKELUND, T. 1984. Trace-fossil assemblages and basin evolution of the Vardekløft Formation (Middle Jurassic, central East Greenland). *Journal of Paleontology*, **58**, 362–397.

HENRIKSEN, N. & HIGGINS, A. K. 1976. East Greenland Caledonian fold belt. *In*: ESCHER, A. & WATT, W. S. *Geology of Greenland*, Grønlands geologiske Undersøgelse, 183–246.

LARSEN, P.-H. 1988. Relay structures in a Lower Permian basement-involved extension system, East Greenland. *Journal of Structural Geology*, **10**, 3–8.

——, OLSEN, H., RASMUSSEN, F. O. & WILKEN, U. G. 1989. Sedimentological and structural investigations of the Devonian basin, East Greenland. *Rapport Grønlands geologiske Undersøgelse*, **145**, 108–113.

LISTER, G. S., ETHERIDGE, M. A. & SYMONDS, P. A. 1986. Detachment faulting and the evolution of passive continental margins. *Geology*, **14**, 246–250.

MARCUSSEN, C., LARSEN, P.-H., NØHR-HANSEN, H., OLSEN, H., PIASECKI, S. & STEMMERIK, L. 1988. Studies of the onshore hydrocarbon potential in East Greenland 1986–1987: field work from 73° to 76°N. *Rapport Grønlands geologiske Undersøgelse*, **140**, 89–95.

MAYNC, W. 1947. Stratigraphie der Jurabildungen Ostgrönlands zwischen Hochstetterbugten (75°N.) und dem Kejser Franz Joseph Fjord (73°N.). *Meddelelser om Grønland*, **132**, 2, 1–223.

—— 1949. The Cretaceous beds between Kuhn Island and Cape Franklin (Gauss Peninsula) Northern East Greenland. *Meddelelser om Grønland*, **133**, 3, 1–291.

MCKENZIE, D. 1978. Some remarks on the development of sedimentary basins. *Earth and Planetary Science Letters*, **40**, 25–32.

PIASECKI, S. 1979. Hauterivian dinoflagellate cysts from Milne Land, East Greenland. *Bulletin of the Geological Society of Denmark*, **28**, 31–37.

—— 1984. Preliminary palynostratigraphy of the Permian–Lower Triassic sediments in Jameson Land and Scoresby Land, East Greenland. *Bulletin of the Geological Society of Denmark*, **32**, 139–144.

ROSENDAHL, B. R. 1987. Architecture of continental rifts with special reference to East Africa. *Annual Review Earth and Planetary Science 1987*, **15**, 445–503.

——, REYNOLDS, D. J., LORBER, P. M., BURGESS, C. F., MCGILL, J., SCOTT, D., LAMBIASE, J. J. & DERKSEN, S. J. 1986. Structural expressions of rifting: lessons from Lake Tanganyika, Africa. *In*: FROSTICK, L. E. *et al.* (eds) Sedimentation in the African Rifts. *Geological Society Special Publication*, **25**, 29–43.

SCLATER, J. G. & CÉLÉRIER, B. 1987. Extensional models for the formation of sedimentary basins and continental margins. *Norsk Geologisk Tidsskrift*, **67**, 253–267.

SMITH, A. G. & BRIDEN, J. C. 1977. *Mesozoic and Cenozoic paleocontinental maps*. Cambridge University Press, Cambridge.

SOPER, N. J., HIGGINS, A. C., DOWNIE, C., MATTHEWS, D. W. & BROWN, P. E. 1976. Late Cretaceous–Early Tertiary stratigraphy of the Kangerdlugssuaq area, East Greenland, and the age of opening of the north-east Atlantic. *Journal of the Geological Society of London*, **132**, 85–104.

SPATH, L. F. 1935. Additions to the Eo-Triassic invertebrate fauna of East Greenland. *Meddelelser om Grønland*, **98**, 2, 1–115.

SURLYK, F. 1977a. Mesozoic faulting in East Greenland. *In*: FROST, R. T. C. & DIKKERS, A. J. (eds) *Fault tectonics in N.W. Europe*. Geologie en Mijnbouw, **56**, 311–327.

—— 1977b. Stratigraphy, tectonics and palaeogeography of the Jurassic sediments of the areas north of Kong Oscars Fjord, East Greenland. *Bulletin Grønlands geologiske Undersøgelse*, **123**, 1–56.

—— 1978a. Submarine fan sedimentation along fault scarps on tilted fault blocks (Jurassic-Cretaceous boundary, East Greenland). *Bulletin Grønlands geologiske Undersøgelse*, **128**, 1–108.

—— 1978b. Jurassic basin evolution of East Greenland. *Nature*, **274**, 130–133.

—— 1978c. Mesozoic geology and palaeogeography of Hochstetter Forland, East Greenland. *Bulletin of the Geological Society of Denmark*, **27**, 73–87.

—— 1984. Fan-delta to submarine fan conglomerates of the Volgian-Valanginian Wollaston Forland Group, East Greenland. *In*: KOSTER, E. H. & STEEL, R. J. (eds) Sedimentology of Gravels and Conglomerates. *Canadian Society of Petroleum Geologists, Memoir*, **10**, 359–382.

—— 1987. Slope and deep shelf gully sandstones, Upper Jurassic, East Greenland. *Bulletin of the American Association of Petroleum Geologists*, **71**, 464–475.

—— 1989. Mid-Mesozoic syn-rift turbidite systems – controls and predictions. *In*: COLLINSON, J. D. (ed.) *Correlation in hydrocarbon exploration.* Norwegian Petroleum Society and Graham & Trotman, London, 231–241.

—— 1990. A Jurassic sea-level curve for East Greenland. *Palaeogeography, Palaeoclimatology, Palaeoecology*.

—— & CLEMMENSEN, L. B. 1983. Rift propagation and eustacy as controlling factors during Jurassic inshore and shelf sedimentation in northern East Greenland. *Sedimentary Geology*, **34**, 119–143.

——, CLEMMENSEN, L. B. & LARSEN, H. C. 1981. Post-Palaeozoic evolution of the East Greenland continental margin. *In*: KERR, J. W. & FERGUSSON, A. J. (eds) Geology of the North Atlantic borderland. *Canadian Society of Petroleum Geologists, Memoir*, **7**, 611–645.

——, HURST, J. M., MARCUSSEN, C., PIASECKI, S., ROLLE, F., SCHOLLE, P. A., STEMMERIK, L. & THOMSEN, E. 1984. Oil geological studies in the Jameson Land basin, East Greenland. *Rapport Grønlands geologiske Undersøgelse*, **120**, 85–90.

——, ——, PIASECKI, S., ROLLE, F., SCHOLLE, P. A., STEMMERIK, L. & THOMSEN, E. 1986. The Permian of the western margin of the Greenland Sea — A future exploration target. *In*: HALBOUTY, M. T. (ed.) Future petroleum provinces of the world. *Memoir of the American Association of Petroleum Geologists*, **40**, 629–659.

—— & NOE-NYGAARD, N. 1989. Shelf sand banks and sand dune fields from the Volgian–Valanginian Raukelv Formation of Jameson Land, East Greenland. *Rapport Grønlands geologiske Undersøgelse*, **145**, 74–75.

—— & ZAKHAROV, V. A. 1982. Buchiid bivalves from the Upper Jurassic and Lower Cretaceous of East Greenland. *Palaeontology*, **25**, 727–753.

SYKES, R. M. & BRAND, R. P. 1976. Fan-delta sedimentation: an example from the Late Jurassic–Early Cretaceous of Milne Land, central East Greenland. *Geologie en Mijnbouw*, **55**, 195–203.

TEICHERT, C. & KUMMEL, B. 1976. Permian-Triassic boundary in the Kap Stosch area, East Greenland. *Meddelelser om Grønland*, **197**, 5, 1–54.

TOZER, E. T. 1965. *Lower Triassic stages and ammonoid zones of Arctic Canada.* Geological Survey of Canada Paper 65–12.

TRÜMPY, R. 1969. Notes on Triassic stratigraphy and paleontology of north-eastern Jameson Land (East Greenland). II Lower Triassic ammonites from Jameson Land (East Greenland). *Meddelelser om Grønland*, **168**, 2, 77–116.

VISCHER, A. 1943. Die postdevonische Tektonik von Ostgrönland zwischen 74° und 75°N. Br., Kuhn Ø, Wollaston Forland, Clavering Ø und angrenzende Gebiete. *Meddelelser om Grønland*, **133**, 1, 1–195.

WERNICKE, B. 1981. Low-angle normal faults in the Basin and Range Province: nappe tectonics in an extending orogen. *Nature*, **291**, 645–648.

—— & BURCHFIEL, B. C. 1982. Modes of extensional tectonics. *Journal of Structural Geology*, **4**, 105–115.

Rotliegend sediment distribution: a result of late Carboniferous movements

K. W. GLENNIE

Consultant, 4 Morven Way, Ballater, Grampian, AB3 5SF, UK

Abstract: The distribution of Rotliegend desert sandstones in Northwest Europe was mainly the outcome of four geological events that originated in the Carboniferous or earlier.

(1) Westphalian culmination of collision between north-moving Laurussia and the faster moving Gondwana resulted in the E–W trending Variscan Mountains, which were to be the main source of Rotliegend fluvial sediment over successor basins to the former back-arc foreland.

(2) During final convergence between Laurussia and Gondwana, the western Variscan orocline probably acted as a wedge that put Laurussia into transtension and, utilizing old lines of weakness, the Iapetus Suture between the Scottish-Greenland and Norwegian Caledonides and the Tornquist Line, initiated NW–SE and conjugate NE–SW systems of grabens.

(3) Late orogenic relative movement between Laurussia and Gondwana also caused strike-slip faulting, related rapid collapse of the Variscan Mountains and the extrusion of Lower Rotliegend volcanics through parts of the Variscan foreland. The end of volcanism was followed by thermal subsidence of the Permian basins, many of them at rates that exceeded those of sedimentation.

(4) The northward drift of Laurussia carried NW Europe from the equatorial zone of Carboniferous coal deposition to the latitude of a northern hemisphere trade-wind desert, where it lay in the rain shadow of the Variscan Mountains.

Major reserves of gas (and some oil) exist in the Lower Permian Rotliegend desert-sandstone reservoirs of Northwest Europe, and especially in what is known as the Southern Permian Basin; this gas-rich province extends eastward from the coast of England to Western Germany. Rotliegend reservoirs are also oil bearing at a few localities in the Northern Permian Basin and near the southern end of the Viking Graben. The economic necessity of fully understanding all aspects of the Rotliegend has stimulated extensive study of these rocks; the basins in which they were deposited, however, are the outcome of geological events that had their origins during the Carboniferous Hercynian Orogeny and even earlier.

In Europe, the Hercynian (Variscan) orogen extends from Spain and Brittany through Germany to Poland. Orogenesis began during the late Dinantian by continental collision between Laurussia and Gondwana (Fig. 1), and reached its culmination during the Westphalian. Relative movement between the two megacontinents was not brought to a complete halt by the collision, and their union to form the even bigger Pangaea seems to have been an uneasy one. Continuing slight differential movement between the Laurussian and Gondwanan components of Pangaea were sufficient to induce widespread tectonic activity that extended into the Permian, giving rise to a whole series of basins (e.g. Northern and Southern Permian Basins, Moray Firth and Western Approaches Basins), grabens (e.g. Oslo, Central and Worcester Grabens) and horsts across much of NW Europe, whose axes seem, collectively, to be aligned to most points of the compass (Fig. 2). The relationship between the orogenic movements inferred above and the resulting distribution of Lower Rotliegend volcanics and sediments and the volcanic-free Upper Rotliegend sediments forms the subject of this paper.

Outline of Carboniferous movements in Western Europe

In the early Carboniferous, Western Europe comprised three major units: firstly, a northern terrestrial area, inherited from the Devonian Oil Red Sandstone continent; secondly, an E–W trending shallow seaway extending across southern Britain to east Central Europe. This was flanked along its southern edge by (thirdly) a number of positive continental slivers (microcontinents; see Ziegler (1982, Encl. 9) for their present distribution) that now make up the crystalline core of the Variscan Orogen.

In the Namurian, much of that seaway became the site of equatorial coal-forming swamps. North of the Variscan Deformation Front, the

From Hardman, R. F. P. & Brooks, J. (eds), 1990, *Tectonic Events Responsible for Britain's Oil and Gas Reserves*, Geological Society Special Publication No 55, pp 127–138.

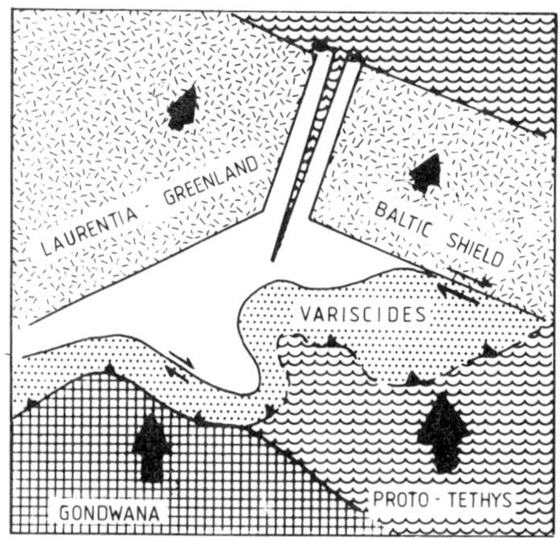

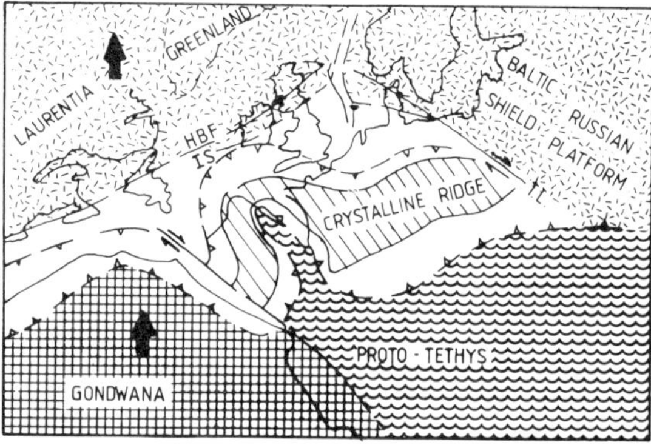

Fig. 1. (a) Cartoon depicting the megatectonic framework of the Variscan orogenic belt (Variscides). Gondwana collided with the southern edge of the slower, north-moving Laurussia (Laurentia plus Baltic Shield). The marked Variscan orocline probably acted as a wedge driven into the re-entrant between the Laurentian and Baltic shields, thereby initiating separation of the old northern Iapetus Suture and the creation of a proto-Atlantic — Viking and Central graben system. (b) The Variscan orogen with its crystalline core relative to some modern coastlines. HBF, Highland Boundary Fault; IS, Laurentian–Cadomian part of Iapetus Suture; TL, Tornquist–Teisseyre Line.

area of paralic coal deposition was to be replaced early in the Permian by Rotliegend desert sedimentation. The causes of these profound depositional, tectonic and climatic changes are discussed below.

During the Early Carboniferous, Laurussia and Gondwana were separated by the Proto-Tethys Ocean. To judge from the differing plate reconstructions of Morel & Irving (1978), Scotese *et al.* (1979) and Smith *et al.* (1981), both continents had been drifting northwards since at least the late Devonian, with Gondwana moving the faster of the two. Thus the Variscan Orogeny, which culminated in the late Westphalian, was not achieved by head-on collision but, rather, by Gondwana colliding with the southern edge of a slower moving Laurussia. Gondwana did not follow quite the same path as Laurussia, so effective convergence involved a relative south-easterly motion to Laurussia, which was to impart a degree of E–W motion between the two during later phases of the orogeny (Ziegler 1988). For such movements across the surface of the globe, old oceanic crust presumably had to be consumed in front of the moving continents and new crust created behind.

Lorenz & Nicholls (1984) suggest that the diverse slivers of older continental rock that now form the core of the Variscan Orogen should be considered collectively as Southern Europe. They believe that adjacent oceanic crust was subducted beneath both northern and southern flanks of Southern Europe, leading to considerable recrystallization of the older rocks forming the core of the growing orogen.

Carboniferous crust of oceanic origin has not

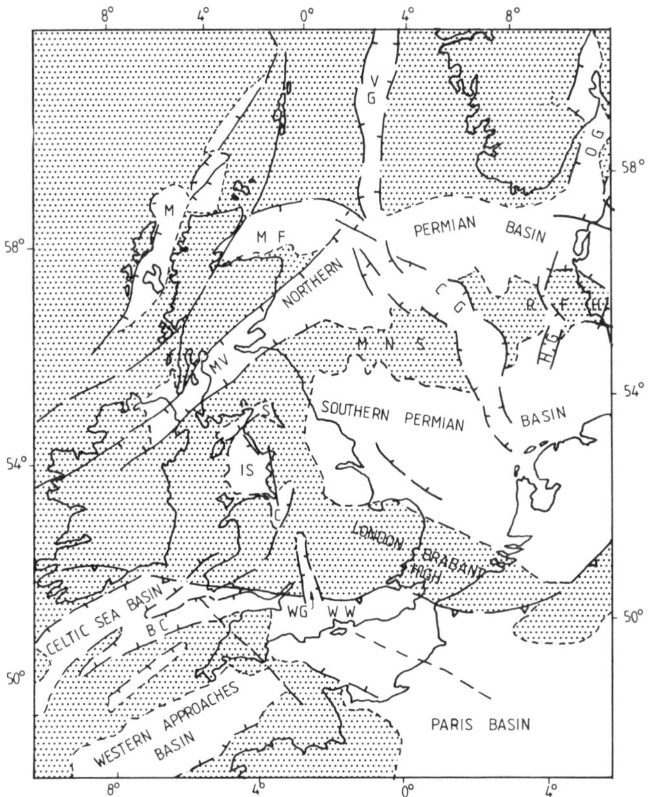

Fig. 2. Basins formed during the early Permian in the area surrounding the British Isles. BC, Bristol Channel Basin; C, Cheshire Basin; CG, Central Graben; IS, Irish Sea Basin; M, Minch Basin; MF, Moray Firth Basin; MV, Midland Valley of Scotland; OG, Oslo Graben; S, Solway Firth-Vale of Eden; VG, Viking Graben; WG, Worcester Graben; WW, Wessex–Weald Basin.

been positively identified within the Variscan Orogen. The serpentinites of the Cornish Lizard complex probably represent the obducted remnants of an older (Rheic) ocean that had already closed by the late Devonian in response to the northward movement of Armorica (Brittany), an interpretation that is supported by both radiometric ages and gravity surveys (Rollin 1986). The location of the early Permian Cornubian granites (280–270 Ma) rules out subduction of oceanic crust in the near vicinity, geometrical relationships limiting the location of Proto-Tethys to south of Armorica and the Massif Central rather than beneath the Culm Basin north of Cornubia and in Germany.

The interpretation of Lorenz & Nicholls (1984) to the effect that consumption of oceanic crust took place along the northern as well as the southern flanks of Southern Europe receives apparent support from the Harz Mountains, where there exists what Anderson (1975) interprets as an accretion wedge overlying a subduction zone. This area is associated with late Carboniferous basalts and has been intruded by granite of apparently similar age. As was the case in Cornwall, the proposed subduction zone north of the Hartz Mountains is too close to have been responsible for the granite intrusion, and a location for a single zone south of the mountains seems much more likely. Instead, the southward-dipping thrusts and fold axes in the Hartz Mountains probably resulted from the northward migration of intense orogenic compression, a process recognised elsewhere along the the northern margin of the orogen, for instance in Ireland (Cooper *et al.* 1986) and in the Mendips (Williams & Chapman 1986).

Leeder (1988) points out that although there is no palaeomagnetic evidence to support the existence of a wide Carboniferous ocean between the southern British Isles and the belt of metamorphic Variscides, the area could have developed as a back-arc seaway, possibly with a limited development of oceanic crust. Such a seaway would have originated in response to the northward dipping subduction zone along the Proto-Tethys margin of Southern Europe, and could thereby have created a temporary Southern Europe microcontinent. Indeed, Southern Europe itself is made up of several microcontinental slivers that were inherited from a complex set of Devonian and Carboniferous back-arc rifts (Ziegler 1988). Thus it is likely that northward-migrating crustal deformation during orogenesis resulted in the northern back-arc seaway being overridden by the thrusts and overturned folds of the northern Variscan Mountains. These thrusts depressed the successor Variscan foredeep basin, which also subsided isostatically under the load of sediment deposited over it during the Namurian and Westphalian.

Pre-Carboniferous background to orogenic framework

There are several aspects of late Carboniferous tectonism that indicate probable reactivation of lines of weakness inherited from earlier plate-tectonic events.

During the Cambrian, for instance, an arm of the Iapetus Ocean (Tornquist Sea) existed between Baltica and that part of modern Europe incorporating Central England and Denmark (Fig. 3). Evidence of the Ordovician oblique-slip closure of this seaway is now marked by the German-Polish Caledonides (Ziegler 1982) and the Tornquist–Teisseyre Line, a zone of strike-slip faults adjacent to the rigid edge of the Baltic Shield. During the final stages of the Variscan Orogeny, northerly directed compression was converted into widespread NW–SE oriented right-lateral strike-slip movement. This brought about the collapse of the newly formed mountain range, which in Germany and Poland was later covered by Rotliegend sediments. Similarly oriented strike-slip movements also transected the Variscan northern foreland; where these involved transtension, they gave rise to the widespread volcanism of the Lower Rotliegend.

The Variscan Orogeny was the outcome of collision between Gondwana and Laurussia. Laurussia itself formed during the late Silurian closure of the northern Iapetus Ocean, when Laurentia–Greenland collided with the Baltic Shield. The sedimentary sequences now incorporated into the Scottish Caledonides had been deposited along the western margin of the Iapetus Ocean whereas those of the Norwegian Caledonides were derived from the Baltic Shield; thrusting of the two sequences is away from each other Fig. 4). The suture marking the line of closure between the two fold belts will

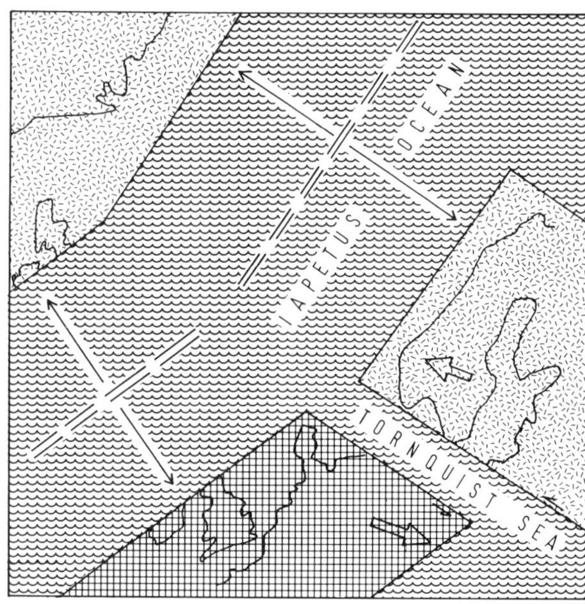

Fig. 3. Early Palaeozoic Iapetus Ocean separating Laurentia and Baltica-Southern Britain; the latter is itself separated by the Tornquist Sea. Late Ordovician closure of this sea left a suture (Tornquist Line) that was re-activated during the Variscan Orogeny.

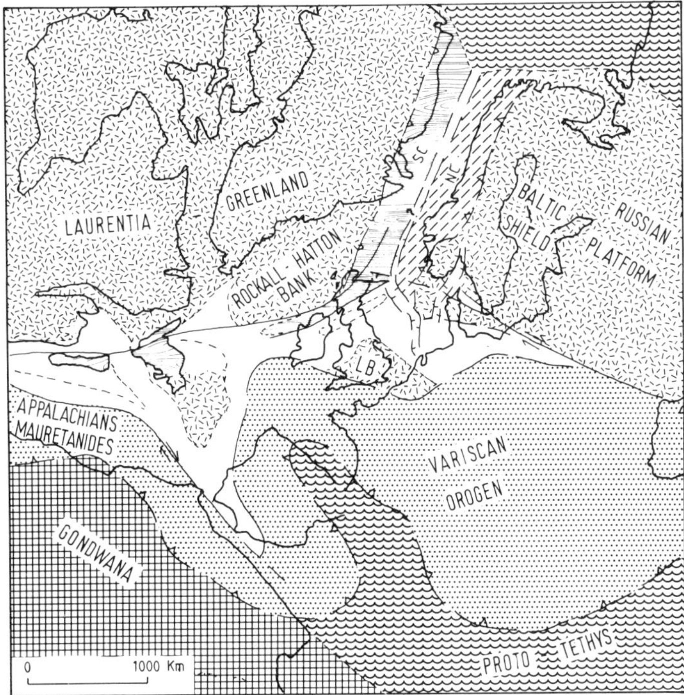

Fig. 4. The Scottish and Norwegian Caledonides flank, respectively, Laurentia-Greenland and the Baltic Shield at the time of the Variscan Orogeny. LB, London–Brabant Massif.

have formed a natural zone of weakness. Its reactivation during the latest Carboniferous or early Permian resulted in initiation of the N–S trending (today) Viking Graben of the northern North Sea and the Proto-Atlantic trough between Greenland and Norway.

The essentially non-metamorphic Lower Palaeozoic rocks of Southern Ireland, Wales, the Lake District, Germany and Poland (Fig. 5), on the other hand, took no part in the mid- to late Silurian (*c*. 410–420 Ma) final closure of the northern Iapetus Ocean. They were deposited in a roughly E–W trending (today's orientations) part of Iapetus flanking the northern margin of the Wales–London–Brabant Cadomian microcontinent (Ziegler 1982; Mason 1988). Thus this sequence was deposited on the south side of the Laurentian–Cadomian Iapetus and bore no direct relationship to the Lower Palaeozoic sediments that were accreted to Laurentia throughout much of the late Ordovician to early Devonian time span by sinistral oblique-slip movements (see e.g. Bluck & Leake 1986; Anderson 1987). The two sides of the Laurentian–Cadomian ocean are now separated by a north-dipping Iapetus suture (Beamish & Smythe 1986) that trends across Ireland and through the Solway–Northumberland Trough to the central North Sea, where it presumably transects the Mid-North Sea High.

Although the Variscan Orogeny resulted from N–S compression, its northern foreland was transected by rifts that seem to have formed under the influence of E–W tension. A southern prolongation of the Proto-Atlantic and, to a lesser extent, the Viking Graben, is directed towards a major oroclinal bend in the Variscan Mountains where they are draped around the Aquitaine–Cantabria microcraton (Figs 1 & 4). It is suggested that this arc acted as a northward-driven wedge, which, taking advantage of the old line of weakness between Scandinavia and Laurentia, played an important role in generating an environment of east–west tension across not only the Proto-Atlantic but, by southerly propagation, within the Variscan foreland area as well.

In Ziegler's (1988) reconstruction, the southerly prolongation of the Proto-Atlantic trends to the west of the British Isles along what is now the Rockall Trough and to the west of the Variscan orocline (Figs 1 & 4). Thus some late Carboniferous sinistral strike-slip movement could also be expected along the axis of

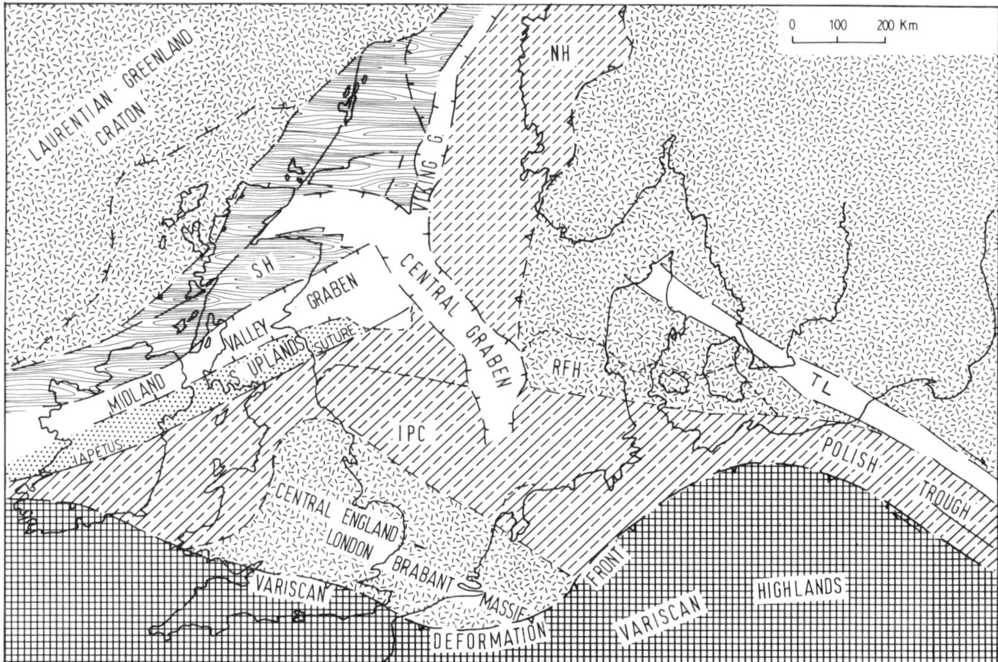

Fig. 5. Several distinct Caledonoid fold belts of NW Europe, which all underwent entirely different early Palaeozoic histories and had different effects on the Variscan Orogeny: (1) Scottish Highlands (SH) separated from (2) Norwegian Highlands (NH) by Northern Iapetus Suture (present Viking Graben); (3) Non-metamorphic Caledonides extending from Ireland to Poland (IPC), which is separated from Southern Uplands by SW Iapetus Suture and from Baltica by combination of Tornquist Line (TL) and sliver of Precambrian basement below Ringkobing–Fyn High (RFH).

the Proto-Atlantic, an interpretation that has some support from Leeder (1988). Whether or not oceanic crust of Permo-Carboniferous age was created in the Proto-Atlantic at this time is in dispute (see e.g. Haszeldine 1984; Haszeldine & Russell 1987; Leeder 1988; Ziegler 1988).

The evidence indicates that a combination of both sinistral and dextral transtension associated with the Proto-Atlantic and Tornquist–Teisseyre line, together with the relatively rigid NE flank of the London–Brabant Massif, all played a role in generating the system of N–S, NW–SE and conjugate NE–SW grabens that transect the Variscan northern foreland. Faults parallel to the Tornquist Line are undoubtedly responsible for the zones of transverse faults in the Central Graben described by Cartwright (1987), for example, and in North Germany, north-trending grabens now contain Lower Rotliegend gas-bearing aeolian sandstones (Gast 1988).

All was not transtension, however. By the end of the Carboniferous, the old Variscan northern foreland formed a more or less triangular area hemmed in to the NW and NE by erosional relics of the Laurentian and Scandinavian Caledonian mountains, and by a newly formed but, in many areas, highly sheared Variscan mountain range to the south. Rotliegend sediments were to be deposited over that triangle later during the Permian. Before that could happen, however, the area was also subjected to late Westphalian dextral transpression, which caused inversion and erosion of Carboniferous strata that had been deposited in areas adjacent to rigid basement. Examples include the Cleveland Hills Basin (Kent 1980), Sole Pit and Broad Fourteens Basins (Glennie & Boegner 1981; Van Wijhe 1987) and the Polish Trough, all of which trend NW–SE. At about the same time, the Pennine Trough was inverted into a N–S trending anticline, which was to become the effective western limit of the Southern Permian Basin.

The line of crustal weakness associated with the Sole Pit-Broad Fourteens axis of inversion was to be utilized again in the early Permian; the Sole-Pit, Broad Fourteens and West Netherlands sub-basins became local depocentres for Rotliegend and succeeding sequences, only to be inverted in the late Cretaceous and mid-

Tertiary in response to Alpine compression (Van Hoorn 1987; Van Wijhe 1987).

The Rotliegend

The Rotliegend has been mentioned several times as the ultimate recipient of events that had their origins prior to the Permian.

The early Permian Rotliegend sedimentary sequences of NW Europe can be divided into two, the Lower and Upper Rotliegend, on the presence or absence of associated volcanic rocks. Both sequences were laid down in an arid environment. The Upper Rotliegend, however, is more widespread and is probably best known as the economically important lower Permian sedimentary fill of the Northern and Southern Permian Basins (Glennie 1986).

Existing evidence indicates that the sediments associated with the Lower Rotliegend are best developed between roughly the North Sea Central Graben and the Baltic Shield (Fig. 6; see also Dickson *et al.* 1981). The Lower Rotliegend is composed essentially of volcanics with only relatively minor amounts of interbedded sediment. Lower Rotliegend igneous activity seems to have extended from around 295 Ma to perhaps 270 Ma. Recent radiometric dating of late Westphalian tuffs in Germany (Lippolt *et al.* 1984; see also review by Leeder 1988) has carried the Carboniferous–Permian boundary back to a probable age of 300 Ma. If correct, this means that the Lower Rotliegend volcanics no longer have a component within the Carboniferous (Stephanian) as widely believed for many years (e.g. Glennie 1986) but is entirely of early Permian age. Lower Rotliegend volcanics are most evident in North Germany and Poland and within the major Horn–Bamble–Oslo system of grabens (Fig. 6; Ziegler 1982; Oftedahl 1976). Their distribution is closely associated with the system of NW–SE horsts and grabens that parallel the Tornquist–Teisseyre Line, and with conjugate structures trending NE–SW (e.g. Oslo Graben, Bamble Trough). The influence of both Caledonian and Variscan movements is clear.

The thickest Rotliegend sequences are preserved in northern Germany in the area of most widespread Lower Rotliegend volcanic activity. Thermal cooling associated with this igneous activity will certainly have played a role in basin

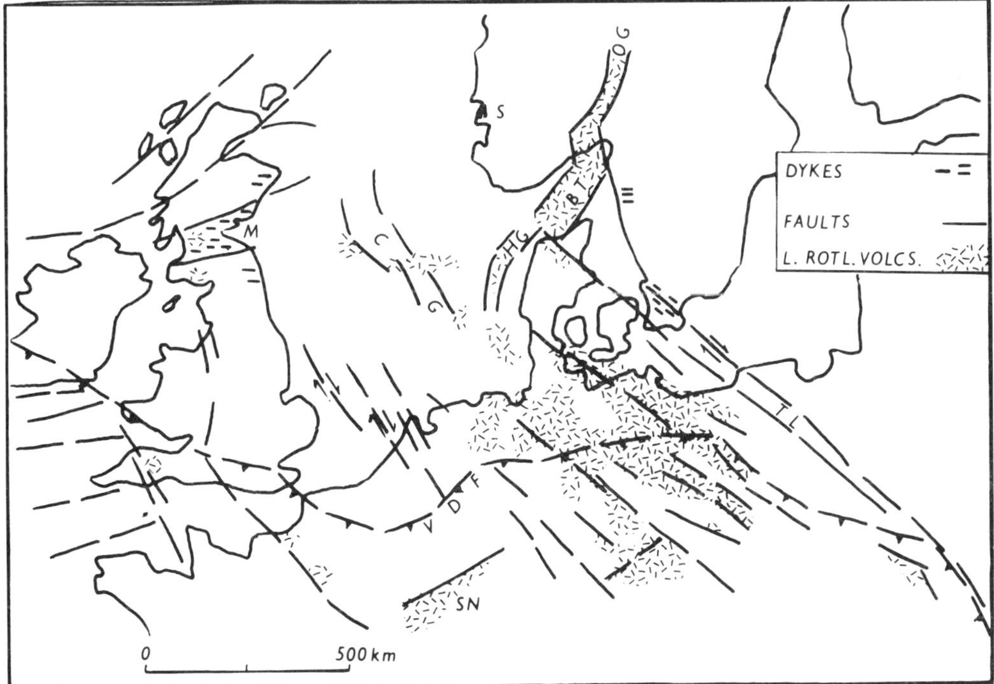

Fig. 6. Distribution of Lower Rotliegend volcanics, including time-equivalent dyke swarms. BT, Bamble Trough; CG, Central Graben; HG, Horn Graben; M, Midland Valley; OG, Oslo Graben; S, Sunnhordland; SN, Saar-Nahe Basin; TL, Tornquist Line; VDF, Variscan Deformation Front. Modified from Glennie (1986).

subsidence in that area, but cannot be invoked for other areas like the UK part of the Southern Permian Basin where Lower Rotliegend volcanic activity is unknown.

The Lower Rotliegend must now also include such well-known British igneous features as the Whin Sill and associated dykes in NE England (295 ± 6 Ma, Fitch & Miller 1967; see also Randall 1980), and the EW trending swarms of tholeitic dykes, also dated at around 295 Ma, that transect the Midland Valley of Scotland (Francis 1983) and probably extend across the North Sea to southern Sweden (Ziegler 1982, Encl. 12).

As Francis (1983) points out, the EW trend of the Scottish dyke swarms brought an abrupt end to the long history of NE–SW Caledonoid structural control on the area. Apart from the Mauchline Volcanics in Ayrshire, the surface expression this dyke system is not seen in Scotland, probably because of the effect of a long period of Permian and Mesozoic erosion. Of significance for the succeeding pattern of Upper Rotliegend sedimentation is that these dykes were developed on trend with the axes of both the Northern Permian Basin and the Mid North Sea–Ringkøbing-Fyn high.

Lower Rotliegend volcanics have been preserved on the flanks of and cutting the Mid North Sea–Ringkøbing-Fyn High, especially where the Central and Horn Grabens cut that high (Fig. 6). The Mid-North Sea High seems to be cored by early Devonian granites (Donato & Tully 1981), which would give it a degree of rigid buoyancy not found in adjacent basinal areas with thinner crusts. This might explain why the Northern and Southern Permian Basins are separated by a ridge. On the Danish side of the Central Graben, the Ringkøbing–Fyn High is underlain by Precambrian basement (Fig. 7), which was probably separated from the Baltic Shield by a splay of the Tornquist–Teisseyre fault system. Besley (1990) suggests that the Mid-North Sea–Ringkøbing–Fyn High, as a relatively positive area, may have been covered by a relatively condensed Carboniferous sequence.

In most parts of the North Sea area there was a fairly extensive period of erosion between the end of volcanic activity and the onset of Upper Rotliegend sedimentation. In the east, much of that erosion was probably caused by thermal uplift of the areas of Lower Rotliegend volcanic activity, the hiatus between Upper and Lower Rotliegend strata being known as the Saalian Unconformity.

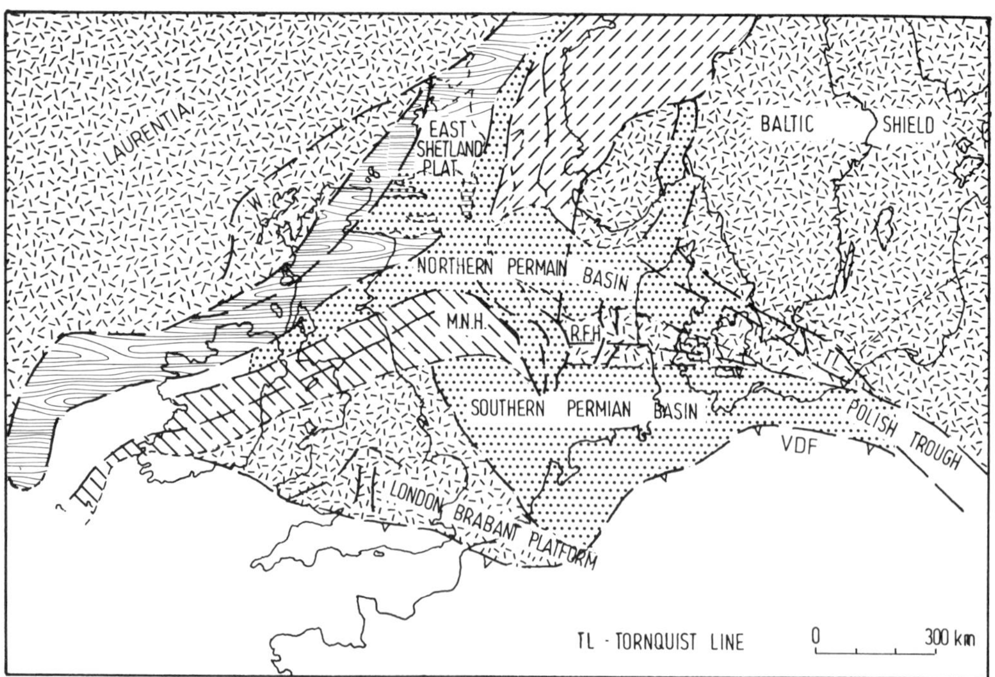

Fig. 7. The framework of Precambrian and Caledonide rocks that surround the Northern and Southern Permian Basins north of the Variscan Deformation Front.

Irrespective of whether there was Lower Rotliegend volcanic activity or not, the rate of basin subsidence in both the Northern and Southern Permian Basins was greater than that of Rotliegend sedimentation (Glennie 1986), so that sediment load played only a minor role in basin subsidence.

The orientation of both Northern and Southern Permian Basins is essentially parallel to the axis of the Variscan Mountains. The Southern Permian Basin is a successor basin to the late Carboniferous Variscan foredeep, and probably owes its origin to a degree of late orogenic back-arc extension. Crustal thinning, thermal uplift and erosion during back-arc extension, followed by crustal cooling, was possibly the order of events controlling basin formation, the hotter areas associated with most volcanic activity showing the greater degree of subsidence. The location of the Northern Permian Basin is harder to account for, but could also be associated with a degree of crustal extension and thermal cooling.

The sedimentary sequences of both the Lower and the Upper Rotliegend are terrestrial, and were deposited under an arid climate. The lack of fossils other than a few reptiles renders impossible their accurate dating by palaeontological means. Because of differential subsidence related to the distribution pattern of Lower Rotliegend volcanism, the age of the earliest Upper Rotliegend sediments in the Southern Permian Basin will increase towards northern Germany, a point confirmed by the greater number of halite beds in the desert-lake sequence of Germany (Fig. 8; see also, e.g. Gralla 1988) when compared with British waters. A similar pattern may occur in the Northern Permian Basin with the Bamble Trough acting as the low point.

Carboniferous–Permian climatic changes

During the Devonian–Permian timespan, Laurussia migrated passively northward. In the Devonian, much of the Old Red Sandstone continent lay within the trade-wind desert belt of the Southern Hemisphere. In the late Carboniferous, the Variscan northern foreland was the site of coal-forming equatorial forests, the source of the gas that locally fills Rotliegend sandstone reservoirs. Red beds of Westphalian D, however, already heralded the approach of Rotliegend desert conditions, partly in response to the area moving into the latitude of a Northern Hemisphere desert, but also because the Rotliegend basins lay in the rainshadow of the newly formed Variscan mountain range. Since that time, NW Europe seems to have been subjected to some 10° of clockwise rotation (see e.g. Glennie, 1986, Fig. 3.9).

The Variscan Mountains were the source of most of the sediment found in the Southern Permian Basin. To judge from Lower Permian coals that are preserved within the Saar and Paris basins, these mountains were within the equatorial rain belt, so that products of erosion were carried fluvially towards the Rotliegend basin centre. Those sediments that failed to reach the basin-centre Rotliegend desert lake (Fig. 8) were subjected to deflation under the influence of the prevailing trade winds, and were redeposited farther to the west as dune sands.

The prevailing wind in the Northern Permian Basin seems to have been from the NW, so that a semi-permanent area of high barometric pressure must have been centred over the Mid North Sea High (Glennie 1986). It is suspected that the surrounding relief inherited from the late Carboniferous Variscan Orogeny, together with the area's new-found location in the heart of a mega-continent played an important role in creating such a sharp change in wind direction over a relatively short distance. A similar sharp reversal in wind direction is found in the orientation of Pleistocene dunes in Western Australia.

Conclusions

It is clear from the foregoing that many of the geological events that took place in Western Europe during the earlier Palaeozoic helped to shape the late Carboniferous orogeny. The more important events may be summarized as follows.

1. During the early Carboniferous, both Gondwana and Laurussia were migrating northwards. The Variscan Orogeny began when Gondwana collided with the slower moving Laurussia.

2. North-dipping subduction along the south side of Laurussia had already caused back-arc spreading south of the Variscan foreland. In the early stages of the Variscan Orogeny, northward directed thrusting overrode the back-arc basin causing down-warping of the northern foreland, which became the site of thick sequences of Carboniferous Coal Measures. Much of this foreland was to become the depositional area of the Rotliegend.

3. The rigid pre-Devonian margins of the Laurentian and Baltic components of Laurussia formed an angle into which the Variscan Orogen was being squeezed. This resulted in: (a) right-lateral transpressional inversion and erosion of

Fig. 8. Distribution of main Upper Rotliegend facies in NW Europe. 1, Pre-Permian Basement; 2, Desert Lake; 3, Lake-margin sabkha; 4, Mixed dune & wadi; 5, Dune sand; 6, Faults; 7, Rotliegend depositional/ erosional edge; 8, Permian wind direction; 9, Fluvial transport direction. CG, Central Graben; DB, Dutch Bank Basin; HBF, Highland Boundary fault; HG, Horn Graben; IB, Irish Sea Basin; MNSH, Mid North Sea High; NPB, Northern Permian Basin; OG, Oslo Graben; RFH, Ringkobing-Fyn High; SPB, Southern Permian Basin. Modified from Glennie (1986); incorporates some facies data for the Northern Permian Basin from Sørensen & Martinsen (1987).

the Coal Measures along NW–SE lines of weakness during the latest Carboniferous; (b) earliest Permian dextral transtensional faulting parallel to the Tornquist Line, and the extrusion of Lower Rotliegend lavas across mainly the eastern foreland and adjacent northern Variscan Mountains; (c) possible sinistral faulting along the line of the Proto-Atlantic Ocean between Greenland and Norway, and (d) the creation of widespread half-grabens along mainly NW–SE and conjugate NE–SW trends.

4. Uplift of the Variscan foreland in response to crustal thinning and heating resulted in widespread erosion (Saalian Unconformity). The end of orogenesis and Lower Rotliegend volcanism was followed by crustal cooling and subsidence of the Rotliegend basins at a rate which exceeded that of subsidence.

5. The northward drift of Laurussia carried NW Europe from the zone of equatorial rain forest in the late Carboniferous to the latitude of a Northern Hemisphere desert by the earliest Permian. Sand-size components of the fluvial sediment carried north from the Variscan Mountains were deflated by the prevailing NE Trade Winds and redeposited as Rotliegend

dune sands; the finest sediment was deposited in a permanent terminal lake occupying the axis of the basin.

References

ANDERSON, T. A. 1975. Carboniferous subduction complex in the Hartz Mountains, Germany. *Geological Society of America Bulletin*, **86**, 77–82.

ANDERSON, T. B. 1987. The onset and timing of Caledonian sinistral shear in County Down. *Journal of the Geological Society, London*, **144**, 817–825.

BEAMISH, D. & SMYTHE, D. K. 1986. Geophysical images in the deep crust: the Iapetus suture. *Journal of the Geological Society, London*, **143**, 489–507.

BESLY, B. 1990. Carboniferous. *In*: GLENNIE, K. W. (ed.) *Introduction to the Petroleum Geology of the North Sea*. Blackwell, Oxford. 3rd Edition.

BLUCK, B. J. & LEAKE, B. E. 1986. Late Ordovician to early Silurian amalgamation of the Dalradian and adjacent Ordovician rocks in the British Isles. *Geology*, **14**, 917–919.

CARTWRIGHT, J. 1987. Transverse structural zones in continental rifts – an example from the Danish sector of the North Sea. *In*: BROOKS, J. & GLENNIE, K. W. (eds) *Petroleum Geology of North West Europe*, Graham & Trotman, London, 441–452.

COOPER, M. A., COLLINS, D. A., FORD, M., MURPHY, F. X., TRAYNER, P. M. & O'SULLIVAN, M. 1986. Structural evolution of the Irish Variscides. *Journal of the Geological Society, London*, **143**, 53–61.

DICKSON, J. E., FITTON, J. G. & FROST, R. T. C. 1981. The tectonic significance of post-Carboniferous igneous activity in the North Sea Basin. *In*: ILLING, L. V. & HOBSON, G. D. (eds) *Petroleum Geology of the Continental Shelf of North-West Europe*, Heyden, London, 121–137.

DONATO, J. A. & TULLY, M. C. 1981. A regional interpretation of North Sea gravity data. *In*: ILLING, L. V. & HOBSON, G. D. (eds) *Petroleum Geology of the Continental Shelf of North-West Europe*. Heyden, London, 65–76.

FITCH, F. J. & MILLER, J. 1967. The age of the Whin Sill, *Geological Journal*, **5**, 233.

FRANCIS, E. H. 1983. Carboniferous-Permian igneous rocks. *In*: CRAIG, G. Y. (ed.) *Geology of Scotland*, 297–324.

GAST, R. E. 1988. Rifting im Rotliegenden Niedersachsens. *Die Geowissenschaften*, **6**, 115–122.

GLENNIE, K. W. 1986. Early Permian Rotliegend. *In*: GLENNIE, K. W. (ed.) *Introduction to the Petroleum Geology of the North Sea*, Blackwell, Oxford. 63–85.

—— & BOEGNER, P. 1981. Sole Pit inversion tectonics. *In*: ILLING, L. V. & HOBSON, G. D. (eds) *Petroleum Geology of the Continental Shelf of North-West Europe*. Heyden, London. 110–120.

GRALLA, P. 1988. Das Oberrotliegende in NW-Deutschland – Lithostratigraphie und Faziesanalyse. *Geologisches Jahrbuch*, A **106**, 3–59.

HASZELDINE, R. S. 1984. Carboniferous North Atlantic palaeogeography: stratigraphic evidence for rifting, not megashear or subduction. *Geological Magazine*, **121**, 443–463.

—— & RUSSELL, M. J. 1987. The late Carboniferous northern Atlantic Ocean: implications for hydrocarbon exploration from Britain to the Arctic. *In*: BROOKS, J. & GLENNIE, K. (eds) *Petroleum Geology of North West Europe*, Graham & Trotman, London, 1163–1175.

KENT, P. E. 1980. Subsidence and uplift in East Yorkshire and Lincolnshire: a double inversion. *Proceedings Yorkshire Geological Society*, **42**, 505–524.

LIPPOLT, H. J., HESS, J. C. & BURGER, K. 1984. Isotopische alter von pyroklastischen sanidinen aus koalin-kohlensteinene als korrelationsmarken fur das Mitteleuropaische Oberkarbon. Fortschr Geol. *Rheinid u. Westf.* **32**, 119–150.

LORENZ, V. & NICHOLLS, I. A. 1984. Plate and intraplate processes of Hercynian Europe during the late Paleozoic. *Tectonophysics*, **107**, 25–56.

LEEDER, M. R. 1988. Recent developments in Carboniferous geology: a critical review with implications for the British Isles and N.W. Europe. *Proceedings Geologists Association*, 73–100.

MASON, R. 1988. Did the Iapetus Ocean really exist? *Geology*, **16**, 823–826.

MOREL, P. & IRVING, E. 1978. Tentative paleocontinental maps for the early Phanerozoic and Proterozoic. *Journal of Geology*, **86(5)**, 535–561.

OFTEDAHL, C. 1976. Northern end of European continental Permian – the Oslo region. *In*: FALKE, H. (Ed.) *The continental Permian in Central, West, and South Europe*. NATO Advanced Study Institute Series C. Reidel, Dordrecht, 2–13.

RANDALL, B. A. O. 1980. The Great Whin Sill and its associated dyke suite. *In*: ROBSON, D. A. (ed.). *The Geology of North East England*. The Natural History Society of Northumbria Special Publication, 67–75.

ROLLIN, K. E. 1986. Geophysical surveys on the Lizard Complex, Cornwall. *Journal of the Geological Society, London*, **143**, 437–446.

SCOTESE, C. R., BAMBACH, R. K., BARTON, C., VAN DER VOO, R. and ZIEGLER, A. M. 1979. Paleozeoc base maps. *Journal of Geology*, **87**, 217–277.

SMITH, A. G., HURLEY, A. M. & BRIDEN, J. C. 1981. *Phanerozoic Palaeocontinental World Maps*. Cambridge University Press.

SØRENSEN, S. & MARTINSEN, B. B. 1987. A palaeogeographical reconstruction of the Rotliegendes deposits in the Northeastern Permian Basin. *In*: BROOKS, J. & GLENNIE, K. W. (eds) *Petroleum Geology of North-West Europe*. Graham & Trotman, London, 497–508.

VAN HOORN, B. 1987. Structural evolution, timing and tectonic style of the Sole Pit inversion. *Tectonophysics*, **137**, 239–284.

VAN WIJHE, D. H. 1987. Structural evolution of inverted basins in the Dutch offshore. *Tectonophysics*, **137**, 171–219.

WILLIAMS, G. D. & CHAPMAN, T. J. 1986. The Bristol-

Mendip foreland thrust belt. *Journal of the Geological Society, London,* **143**, 63–73.

ZIEGLER, P. A. 1982. *Geological Atlas of Western and Central Europe*. Shell Int. Pet. Mij./Elsevier, Amsterdam.

—— 1988. *Evolution of the Arctic-North Atlantic and the Western Tethys.* American Association of Petroleum Geologists Memoir **43**.

The Triassic – early Jurassic succession in the northern North Sea: megasequence stratigraphy and intra-Triassic tectonics

R. STEEL & A. RYSETH

[1]*Norsk Hydro, Oslo, Norway*

[2]*University of Bergen, Norway*

Abstract: The occurrence of Triassic to early Jurassic mudstone and sandstone sequences in the northern North Sea basin, and their possible significance in terms of basinal subsidence pattern is reviewed. The latest phase (Scythian) in a period of crustal stretching is illustrated from the Horda Platform where block rotation has been accompanied by infilling (>2000 m) of alluvial deposits in a series of N–S oriented, asymmetric sub-basins.

The 'post-rift' succession is subdivided into three megasequences which, in their broad areal development over most of the northern North Sea basin, contrast with the rift-infill sequences. The late Scythian–Ladinian megasequence (PR1) documents an early phase of widespread, fine-grained, floodbasin deposition which was interrupted by prograding, sandy sheetflood lobes (Teist Fm). The latter were derived from marginal alluvial fan bajadas along the Norwegian hinterland and from certain segments of the East Shetland uplands. This phase of infilling culminated with basinwide, sand deposition (Lomvi Fm) from amalgamated fans and braided river systems. Although classified as 'post-rift', megasequence PR1 does show some 'growth' along the older rift lineaments.

The Carnian–early Rhaetian megasequence (PR2) records retrogradation of the earlier sandy systems towards the basin margins (lower Lunde Fm) followed by renewed establishment of extensive lacustrine and coastal floodbasin conditions (middle Lunde Fm.). Low-lying tracts within the northerly floodbasins allowed influx of brackish lagoonal waters from a sea to the north. The sequence culminates with repeated progradation of extensive sandy alluvial plains (lower and middle parts of upper Lunde Fm).

The early Rhaetian–Sinemurian megasequence (PR3) also records retrogradation of the earlier sandy fluvial systems (middle to upper parts of upper Lunde Formation), followed by the development of extensive, fine-grained floodbasins and occasional brackish-marine lagoons (uppermost Lunde Fm.) into which isolated sandy streams and fan deltas prograded from both East Shetland and Norwegian margins. Subseqeunt progradation of streams and fan systems eventually allowed the development of northward, through-flowing, axial river systems (Statfjord Fm). There are clear signs of more active intra-basinal tectonics during this third stage in the basin's post-rift history, with latest Triassic–earliest Jurassic depocentres tending broadly to occupy the sites of the future Viking and Sogn Grabens.

In the central North Sea the three-fold megasequence development cannot easily be recognized, though broadly time-equivalent units can be identified. The earliest phase was dominated by inland floodbasin deposition (Smith Bank Fm.) but this was replaced towards the Norwegian hinterland by, at first, lacustrine/lagoonal and inland sabkha deposition, and later, by sandy alluvial fan development (Skagerrak Fm.). The second phase was dominated by sandy alluvial deposition in the northern and eastern reaches but became muddy floodbasin-dominated farther west. The latest phase is only fragmentarily preserved and therefore poorly recognized because of major mid-Jurassic erosion over much of the central North Sea. Where preserved it tends to be sandy, especially in the Norwegian sector.

The pattern of initial retreat of sandy fluvial systems followed by extensive, fine-grained floodbasin development, subsequent progradation of sandy alluvium and eventual development of axial, through-flowing river systems appears to have been characteristic of each of the 3 post-rift intervals (particularly of the first and third, but also, albeit more crudely, of the second) and is interpreted in terms of an increasing then decreasing rate of relative base-level rise. This, in turn, is argued to have been caused largely by increasing then decreasing basinal subsidence rates.

Intra-Triassic tectonics are reflected in (a) early Triassic rifting, (b) three distinct phases of post-rift subsidence and (c) clear signs of differential fault movement and sediment growth along certain intra-basinal lineaments, particularly in the third post-rift interval.

From Hardman, R. F. P. & Brooks, J. (eds), 1990, *Tectonic Events Responsible for Britain's Oil and Gas Reserves*, Geological Society Special Publication No 55, pp 139–168.

The Triassic–earliest Jurassic succession of the northern North Sea has received relatively little attention largely because of limited commercial interest, the relatively poor resolution of seismic data at this level, the apparently monotonous nature of its non-marine sequences, and poor biostratigraphic control. A re-classification of the succession in the northernmost region into Teist, Lomvi, Lunde and Statfjord Formations (Vollset & Doré 1984) has improved our understanding of the stratigraphy. The present study, aided greatly by recent summaries and improvement in our understanding of Triassic biostratigraphy (Lervik *et al.* 1990; Nystuen *et al.* 1990; Eide 1990) re-examines the succession, highlighting sandstone and shale-rich intervals, their likely evolutionary relationship to each other and their probable significance in terms of a varying rate of base level change within the Triassic–early Jurassic basin. In addition we provide a new series of schematic palaeogeographic maps for the northern North Sea, and a discussion of the Permian-earliest Triassic 'rift' phase. There has been growing interest in the latter on account of the Triassic 'contribution' to total basinal subsidence in the region (Giltner 1987). The present study is focused primarily on the region north of 60°N, but we also attempt to correlate and show schematic palaeogeography for the central North Sea (Norwegian sector). Figure 1 shows the well data base, the reference wells, the main Triassic tectonic elements and the lines of correlation.

The Triassic succession (Teist, Lomvi and Lunde Formations) consists of alternations of mainly reddish, buff-coloured or green sandstones with siltstones and mudstones, and with frequent nodular carbonates. Evaporites and dolomites are locally developed in some regions. Arkosic sandstones alternating with red or grey mudstones characterize the early Jurassic interval (Statfjord Formation). The clastic deposits are mainly of fluvial, floodbasin and lacustrine origin (Clemmensen *et al.* 1980; Fisher 1984, Nystuen *et al.* 1990) whereas the carbonates are pedogenic (Steel 1974). Marine deposits occur sparsely at two levels in the Triassic, but dominate in the upper parts of the Statfjord Formation, where shallow marine sandstones separate the underlying fluvial units from thick marine mudstones of the Dunlin Group (Vollset & Doré 1984). The drilled succession in the Tampen Spur area (Fig. 2) and on the Horda Platform (Fig. 3), on either flank of the northern Triassic–early Jurassic basin, is about 2000 m and 3000 m thick, respectively. In the central North Sea thicknesses are usually much less (less than 1000 m, Fig. 4), but this is partly due to a variable but undetermined amount of Jurassic erosion (Lervik *et al.* 1990). Towards the Norwegian mainland, however, the succession appears to expand in thickness (1800 m thick in well 17/12–1) and becomes coarser (Fig. 5). Figures 2–5 illustrate details of the succession from chosen reference wells in these areas, which are discussed further below.

In the northern region the succession can be divided into a lower part which accumulated in a series of sub-basins with intervening tilted basement blocks and an upper part which appears to have filled the entire basin from the East Shetland Platform to the Horda and Øygarden fault zones. This is the basis for the 'rift' and 'post-rift' subdivision of the Triassic succession.

Early Triassic rift sequence

Northern North Sea

An early episode of tectonic activity in the northern and central regions of the North Sea, at least partly of Triassic age, has long been alluded to (Eynon 1981; Ziegler 1982; Wood & Barton 1983). Badley *et al.* (1984, 1988) have been more specific in their suggestions of an early Triassic crustal stretching in the form of rotated fault-blocks and syn-tectonic growth of a sedimentary sequence within the various down-tilted compartments. The significance of this early phase of extension has also been further highlighted by Giltner (1987) who has suggested that underestimation of thermal subsidence associated with Triassic extension has caused overestimates of late Jurassic extension in both Central and Viking Grabens. Total extension in the axial areas of the basin (β = 1.8), calculated by means of subsidence modelling, has been shown to consist of Triassic stretching by a factor of about 1.5 and late Jurassic extension of about 1.2 (Giltner 1987).

Figure 6, with new evidence on intra-Triassic tectonics, shows a seismic section crossing the Horda Platform (data from Nopec a.s.). Comparison of fault displacements at early Jurassic and older levels shows clearly the development of three Triassic sub-basins. The succession in sub-basin (b) between basement (drilled in well 31/6–1) and earliest Jurassic can be subdivided into two parts. There is an *upper package* (PR1–PR3) which shows reflectors sub-parallel to the earliest Jurassic strata, extends down to a level somewhat below the crests of the uptilted basement blocks, and is dated in well 31/6–1 as

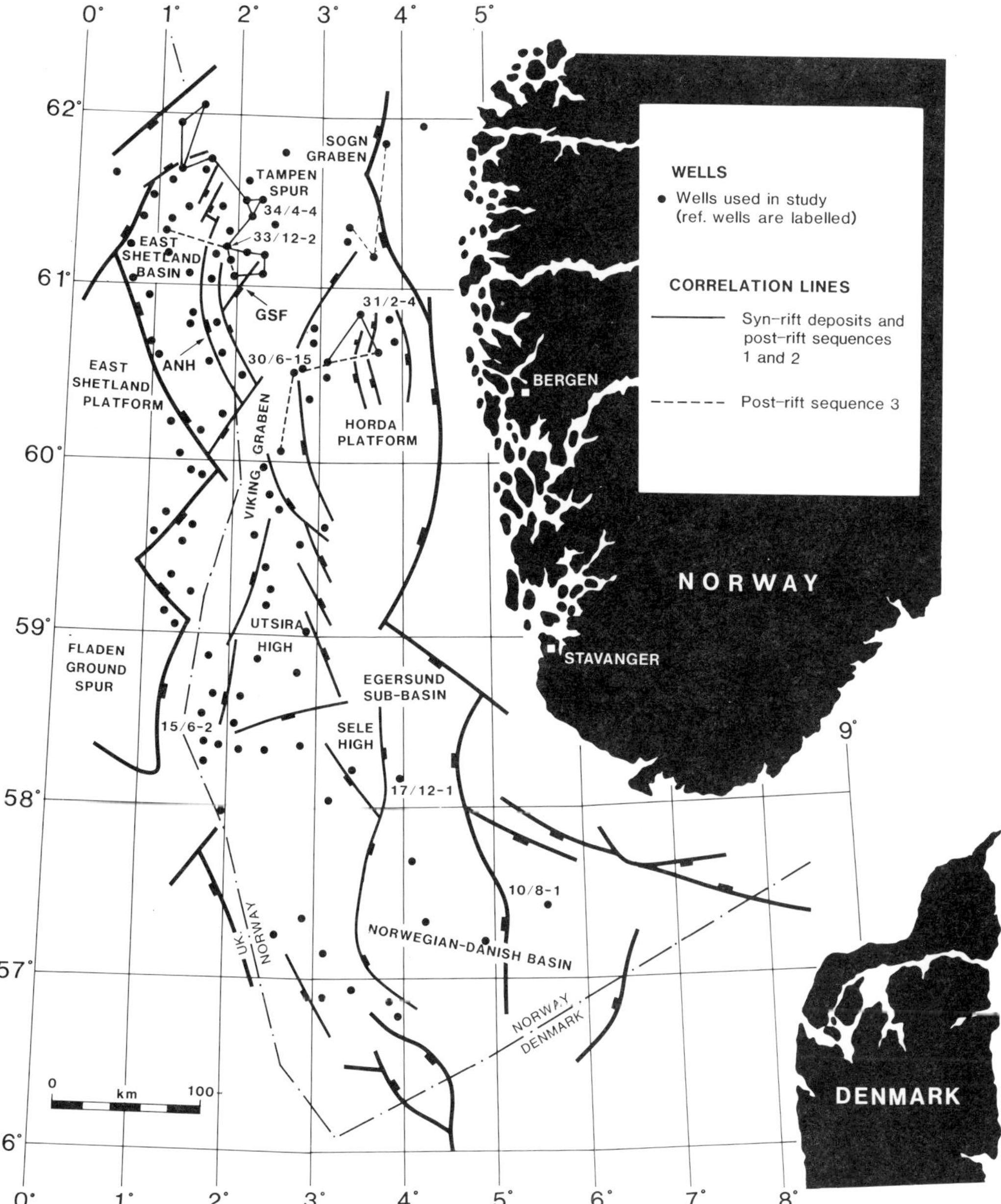

Fig. 1. Location map showing main wells, lines of section and Triassic lineaments. GSF: Gullfaks South fault; ANH: Alwyn–Ninian–Hutton fault.

being earliest Jurassic to early Triassic. The basis for the latter is a Scythian–Anisian age (from palynology) for shales within the Teist Formation, at a level some 400 metres above basement (Lervik *et al.* 1990). The *lower package* of strata, which appears to infill the basement topography is more than 2000 metres thick in sub-basin (b) and on the basis of the age assignment referred to above, is presumed to be Scythian and older. Sub-basin (c) also shows a lower, syn-rift package of strata but in this case the internal reflectors appear to converge rather

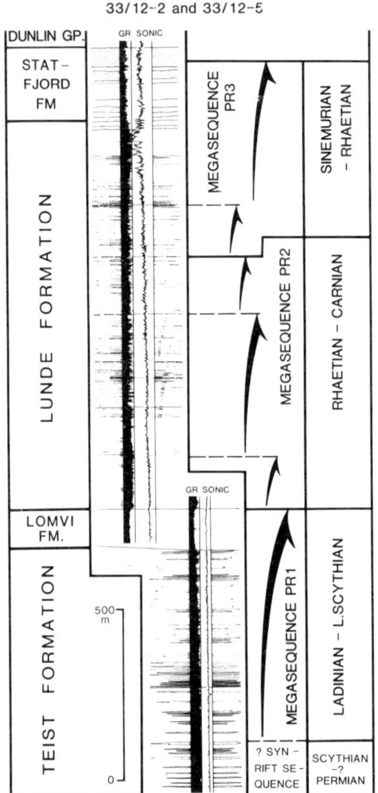

Fig. 2. Triassic reference wells for the Tampen Spur area: 33/12–2 and 33/12–5.

than onlap towards the crest of the tilted block. Sub-basin (a) probably lacks a syn-rift package and is therefore younger than the other sub-basins. The Triassic succession here is thinner and shows evidence of sedimentary growth against the Horda/Øygarden fault zone, all the way through the inferred Triassic–early Jurassic succession. Figure 6 suggests that megasequence PR1 is transitional between the syn-rift and post-rift strata, in as much as this megasequence shows clear signs of growth towards the master-fault in each sub-basin.

The evidence from Fig. 6 strongly suggests that rifting in the northern North Sea continued into Scythian times. The age of *initiation* of extension may by Permian or older, but is as yet undetermined. A schematic and highly speculative (because of scarcity of well data) palaeogeographic map for late Scythian times is shown in Fig. 10A.

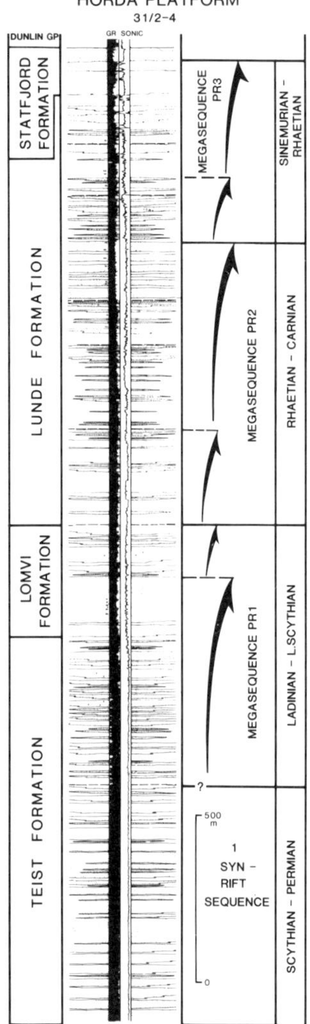

Fig. 3. Triassic reference well for the Horda Platform: 31/2–4.

Adjacent areas

An intra–Triassic discordance has been claimed to be present on the Tampen Spur area (Frost 1987). Here it was interpreted in terms of early extension (along reactivated Caledonian thrust planes) and infilling of a series of half-grabens, followed by later Triassic deposition flatly across the older tilted deposits.

Similar ?Permo-Triassic extension and half graben infill has been documented from the Moray Firth area (Frostick *et al.* 1988), the Minch and Inner Hebrides basins off north-west

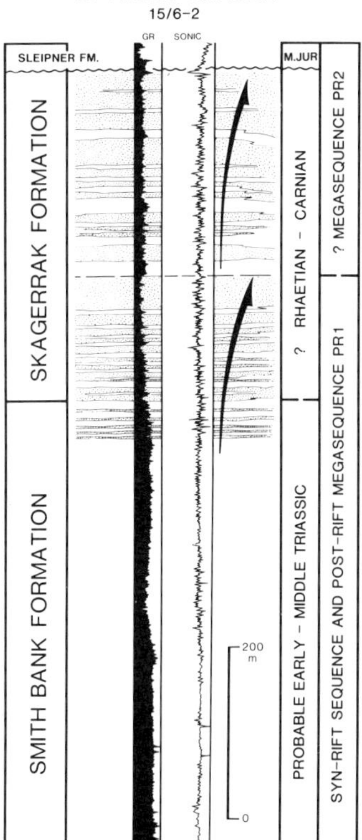

Fig. 4. Triassic reference well for the southernmost Viking Graben area: 15/6–2.

Scotland (Steel 1977) and from the MOIST-line off northern Scotland (Brewer & Smythe 1984). It is again likely that extension occurred along Caledonian or older thrust planes (Smythe *et al.* 1982) but there is no evidence in this region that block rotation terminated or was most severe in earliest Triassic times.

In the central North Sea region tectonic activity is often difficult to interpret because of salt movement, but an intra-Triassic unconformity and marked Triassic fault activity has been reported by Skjerven *et al.* (1983). On the other hand, Gabrielsen *et al.* (1986) reported N–S trending fault activity of 'pre-Zechstein' age in this region.

It is tempting, in the southern North Sea region, to relate the well documented Hardegsen discordance and tectonic phase with what has now been reported from farther north. The Hardegsen event caused uplift and erosion along the flanks of the Mid-North Sea–

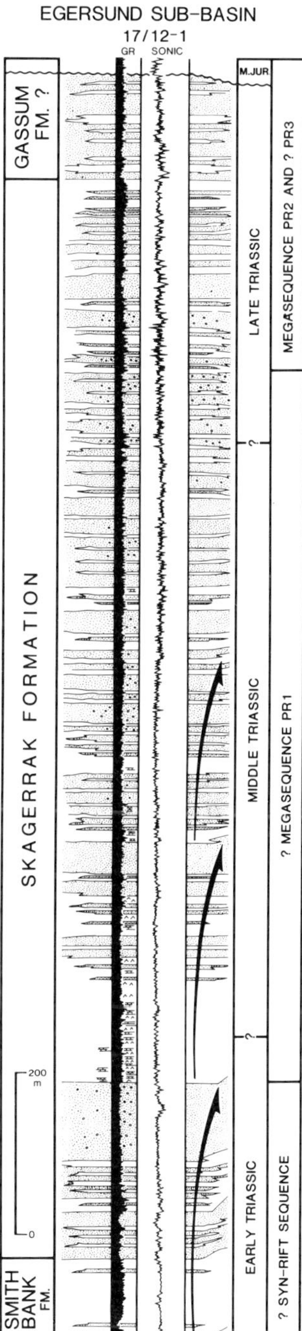

Fig. 5. Triassic reference well for the region south of Stord Basin: 17/12–1.

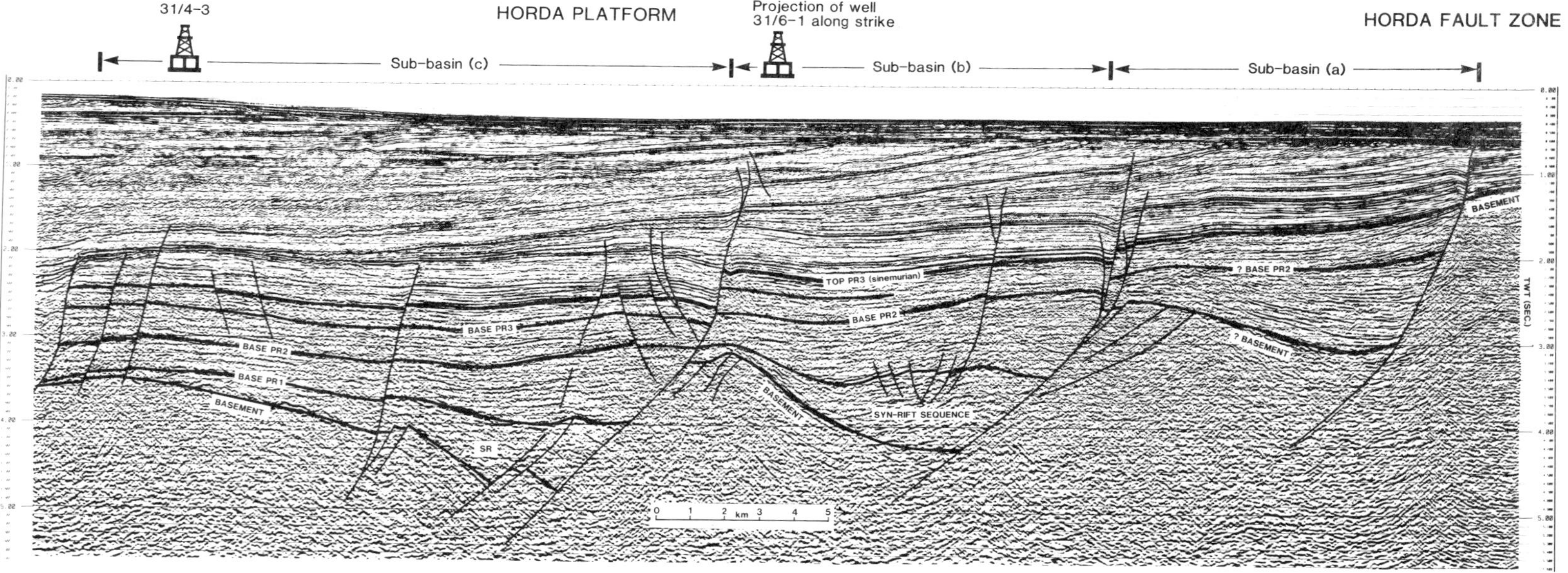

Fig. 6. Interpreted seismic section from the Horda Platform illustrating the ?Permian–Scythian 'rift' sequence and the middle Triassic to Sinemurian post-rift megasequences (PR1–PR3).

Ringkøbing Fyn High, the Pennine Massif and the London–Brabant Massif (Fisher 1984), thus creating minor basins and highs and infill of the topographic features. Early Triassic tectonic activity is well illustrated in the Horn Graben (Olsen 1983). In some regions of the southern North Sea, however, the Hardegsen unconformity appears to truncate large-scale domal swells, for example onshore the Netherlands (Nederlands Aardolie Maatschappij B. V. and Rijks Geologische Dienst 1980).

Post-rift megasequences

The Middle Triassic to Lower Jurassic succession, from near the level at which the underlying rotated fault blocks are 'drowned', up to the base of the Dunlin Group, is the main subject of this paper. This part of the succession is generally believed to represent a phase of regional subsidence due to thermal cooling (Badley *et al.* 1988). Our subdivision of this 'post-rift' succession into three megasequences (PR1–PR3) highlights what we suggest to be significant variation in the *rates* of relative base level change during this interval. It is suggested below that the overall upwards transition from sandstones up to mudstones and siltstones and back to sandstones, within any megasequence, probably reflects an initial increase then a decrease in the subsidence rate within that interval.

Definition of post-rift megasequences

Our subdivision of the late Scythian to Sinemurian succession (PR1-PR3, Figs 2 & 3) emphasizes the significance of the main sand-rich and siltstone/mudstone-rich intervals. The former correspond broadly with Lomvi, the lower part of upper Lunde and Statfjord Formations whereas the latter correspond with the middle to upper parts of Teist, middle Lunde and the uppermost part of Lunde Formations. Only the base of any megasequence is defined, and this is chosen at the stratigraphic level where a main siltstone/mudstone unit overlies the level of maximum progradation (maximum alluvial sandstone density) in the underlying megasequence. Reference to Fig. 7 shows that megasequence boundaries, thus defined, are more easily identified in the intermediate and basinal areas than towards the basin margins. Except where retrogradation of the alluvial sandy system towards the basin margin (and the corresponding spread of the fine-grained floodbasin facies towards the margin) has been rapid, the megasequence boundary (time-line) will occur *within* the sandstone facies at the margin, and there will be developed a sandy retrogradational wedge between the sequence boundary and the overlying siltstone/mudstone unit. This retrogradational wedge can consist of either non-marine, marine or mixed facies, and simply represents a relative rise of base level subsequent to maximum progradation of the fluvial systems.

The alternative level for defining the megasequence boundary would be where there was maximum retreat of the sandy alluvial system towards the basin margin area, referred to as the level of maximum retrogradation in Fig. 7. This level, however, suffers from the disadvantage that it is not easily recognized in the basinal area, and can occur some considerable distance above the level of maximum progradation (Fig. 7). In addition, towards the basin margins there is often a tendency for rapid aggradation of sand and for the lithostratigraphic boundary between sandstone (marginal) and mudstone (basinal) to rise steeply and in a diachronous manner. In this situation both the level of maximum retrogradation and the megasequence boundary may be difficult to recognize.

According to the boundary definition outlined above, our post-rift megasequences tend to have the following general characteristics.

1. They are up to 1 km in thickness and show either an overall asymmetric upward fining to upward coarsening (FUCU) trend or an overall crude upward coarsening (CU) trend.
2. Both within the lower, retrogradational (FU) and within the main progradational (CU) parts of the megasequence there is a development of smaller scale (100–400 m) sequences of CU character. These sequences also appear to be of basinwide extent.

The stratigraphic megasequences are not yet defined south of 60°N. Nevertheless, the broadly age-equivalent parts of the succession in the central North Sea (Figs 4 & 5) are briefly discussed below with reference to the paleogeographic maps.

Age of megasequences

The dating of the succession is not yet well constrained, and has been based on operator well reports. In addition, the dating has been checked against and supplemented from recently available Triassic biostratigraphic summaries (Lervik *et al.* 1990; Eide 1990) which use many of the same wells. The uncertainty in the dating is reflected in the overlap in age between successive megasequences (Figs 2, 3, 9 & 11).

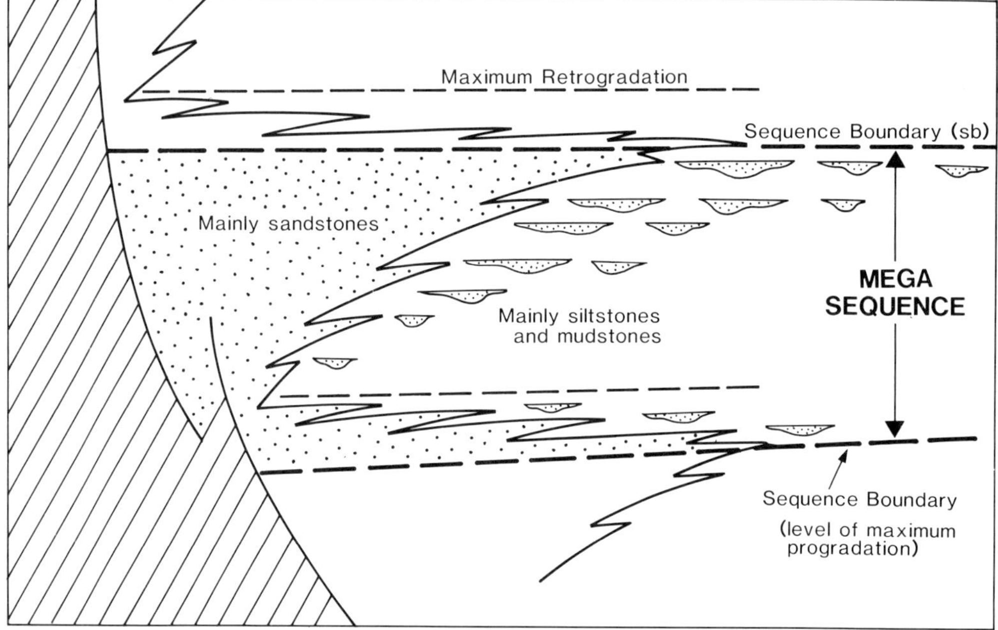

Fig. 7. Conceptual diagram illustrating some of the principles surrounding sequence stratigraphy in continental successions and how megasequence boundaries are defined in the present study.

Post-rift megasequence 1 (PR1)

Stratigraphy and facies

Post-rift megasequence 1 is of probable late Scythian to Ladinian age and consists in the northern North Sea of Teist and Lomvi Formations. The lowest part of the megasequence is rich in siltstones and mudstones, both on the Horda Platform and Tampen Spur areas, and the base can be identified in only a few deep wells (e.g. 31/2–4, Fig. 9) where the mudstones pass abruptly downwards into a more sand-rich sequence. Seismic data on the Horda Platform show the base of the sequence to correspond with a level just above the crest of the underlying tilted basement block (Fig. 6).

Correlation from the northern to the central North Sea region is speculative. However, using the biostratigraphy and correlation guidance suggested by Lervik *et al.* (1990), megasequence PR1 is probably represented by Smith Bank Formation mudstones in the southern Viking Graben area, and further south (Fig. 4). Further east a significant amount of Skagerrak Formation sandstones would be included, as well as intervals of dolomites and evaporites.

Teist Formation (Figs 2 & 3) consists of alternating sandstones and mudstones which originated from sandy sheet-flood and stream-flood events (Nystuen *et al.* 1990) generated mainly from alluvial fans along the margins of the basin and dispersed into mud- and silt-dominated terminal floodbasins and lakes. Detailed interpretation is hampered by a lack of cores through this formation.

The Teist Formation becomes sandier upwards in many areas and passes either gradually or more abruptly into the *Lomvi Formation* (Figs 2 & 3). The latter is a prominent sandy unit some 100–250 m thick which appears to blanket most of the basinal area (Fig. 10B), at least from 60°30′N to 62°N. Although an aeolian origin has been occasionally proposed for this formation, there is a distinct lack of diagnostic aeolian textures and lamination types (the formation is cored in well 34/7–6, Nystuen *et al.* 1990). In addition, its occasional pebbly nature and the abundance of small-scale (less than few metres in thickness) upward-fining motifs, of probable waning-floodwater origin, strongly indicate a fluvial origin, as also suggested by Vollset & Doré (1984). Where Lomvi Formation occupies a more basinal position it can attain an evaporitic character (e.g. well 33/12–5) (Lervik *et al.* 1990). The lithologies of *Smith Bank* and *Skagerrak Formations* are shown in Figs 4 & 5.

Basin development and paleogeography

Across the northwestern margin of the basin (Tampen Spur), megasequence PR1 is generally less than 900 m thick. Figure 8, an interpretative correlation across this margin, from about 62°N 1°E to the edge of the later Viking Graben, shows the overall CU character of the unit and the marked lateral facies change from a sandstone-dominated succession in the northwest to a siltstone/mudstone-dominated succession in the southeast. This diagram emphasizes that the megasequence is part of an asymmetric clastic wedge, cored by the sand-rich Lomvi Formation. Note that the megasequence boundaries (approximating time-lines) *cross* the lithostratigraphic boundaries within and around this clastic wedge.

Megasequence PR1 exceeds 750 m in thickness on the Horda Platform (Fig. 9), and thins to less than 600 m across the eastern reaches. Well control here is insufficient to document clear lithofacies trends laterally, but a crude upwards-coarsening (CU) can be seen through most of the unit. The seismic data in Fig. 6 show that PR1 thickens locally against each of the sub-basin masterfaults, illustrating a continued differential movement along the older rift lineaments during the middle Triassic.

Figure 10B shows a speculative and schematic paleogeographic map of a time interval (Anisian–Ladinian) late in the development of megasequence PR1. North of 60°N, bajada and alluvial plain outbuilding from both margins of the basin probably succeeded in producing a sand blanket over all but the central reaches of the basin, by a mergence of clastic wedges. The northerly arm of fine-grained deposits suggested in Fig. 10B reflects the splitting and fining (partly evaporitic) tendency (discussed above) within the basinal development of Lomvi Formation. The region south of 60°N has not been discussed in detail in this work, but a recent palynological review (Lervik *et al.* 1990) of the region south of the Utsira High has suggested possible correlation with the northern areas. Norwegian quadrants 2, 9 (west), 15, 16 and 17 as well as UK quadrants 21, 22, 29 and 30 contain shale-dominated sequences (Smith Bank Formation) of probable Anisian to Ladinian ages. The above-mentioned areas are therefore shown in Fig. 10B as dominated by floodbasin and lacustrine deposits. Further east towards the Norwegian mainland probable time-equivalent facies become sandy in quadrants 17 and 10 (Fig. 5). Figure 10B therefore portrays a Middle Triassic scene where alluvial clastic outbuilding from along most of the Norwegian hinterland provided the bulk of the sand to the central and northern North Sea basin, in addition to the sand provided from the most northwesterly reaches. The area south, south west and south east of the southern Viking Graben was dominated by mud deposition, with the exception of sand dispersed from a Scottish hinterland in the Moray Firth area, as well as possible local sand derivation from the Utsira High and Fladen Ground Spur areas.

We suggest that parts of the Sele High, the Utsira High, the Fladen Ground Spur and the region of East Shetland Basin west of the Ninian–Hutton–Alwyn alignment (Fig. 10) were upland areas during much of the middle Triassic, though there is no definitive proof of this.

Post-rift megasequence 2 (PR2)

Stratigraphy and facies

Megasequence PR2 is of Carnian to Rhaetian age and consists, in the northern North Sea, of the lower, the middle and much of the upper members of Lunde Formation. The lower part of the megasequence shows alternating sandstones and mudstones but becomes rich in mudstones and siltstones upwards. The lower boundary against the sandier Lomvi Formation is well marked in most wells, both in the Tampen Spur and Horda Platform areas. The megasequence is up to 750 m and 900 m thick respectively in these two areas.

In the central North Sea, the eastern regions as well as the area around the southern Viking Graben are dominated by Skagerrak Formation sandstones. Farther south and southwest, Smith Bank Formation mudstones become more prominent in megasequence PR2.

The lower member of Lunde Formation. Some 100 m (Tampen Spur) to 200 m (Horda Platform) thick, this consists of alternating sandstones and mudstones and represents a retreat or retrogradation of the earlier clastic wedge.

The middle member of Lunde Formation. This is 150–200 m thick, consists of thinly interbedded mudstones, siltstones and fine sandstones and stands out within the red-bed Triassic succession because of the grey and black colours of its fine-grained components, its bioturbation and marine algae, and the presence of wave-generated ripple-lamination. The unit is interpreted in terms of ephemeral fluvial deposits interbedded with lacustrine/lagoon-infill se-

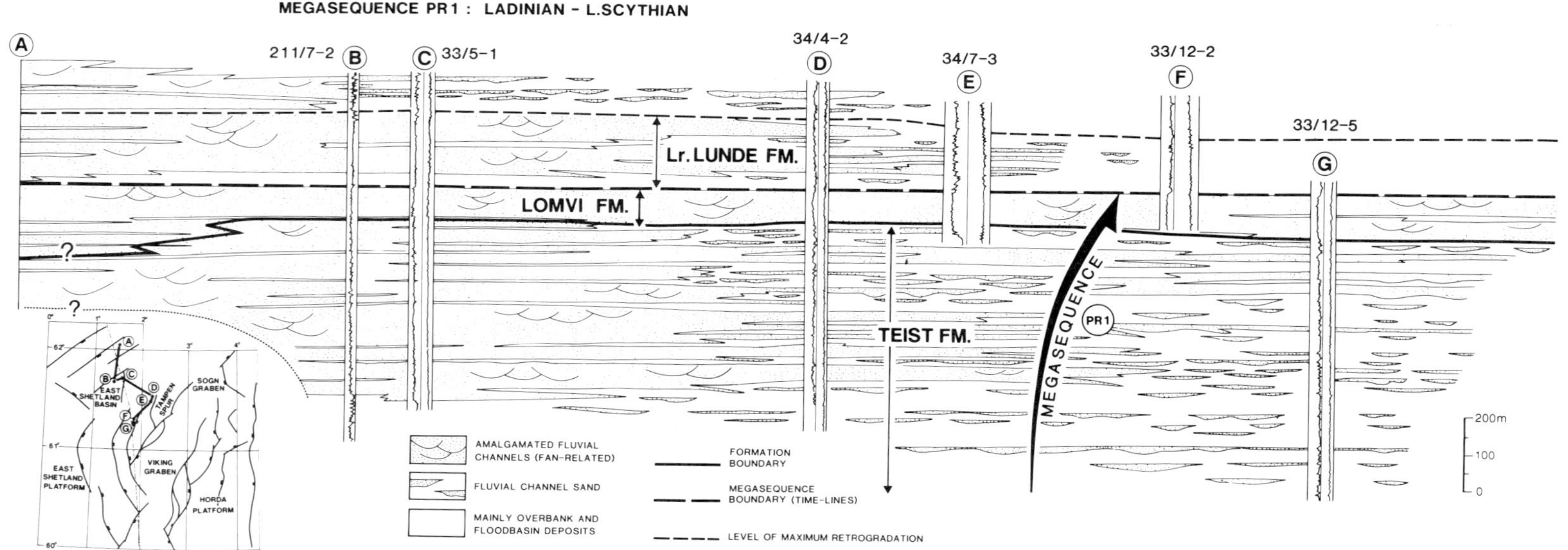

Fig. 8. Triassic megasequence PR1 in the Tampen Spur area.

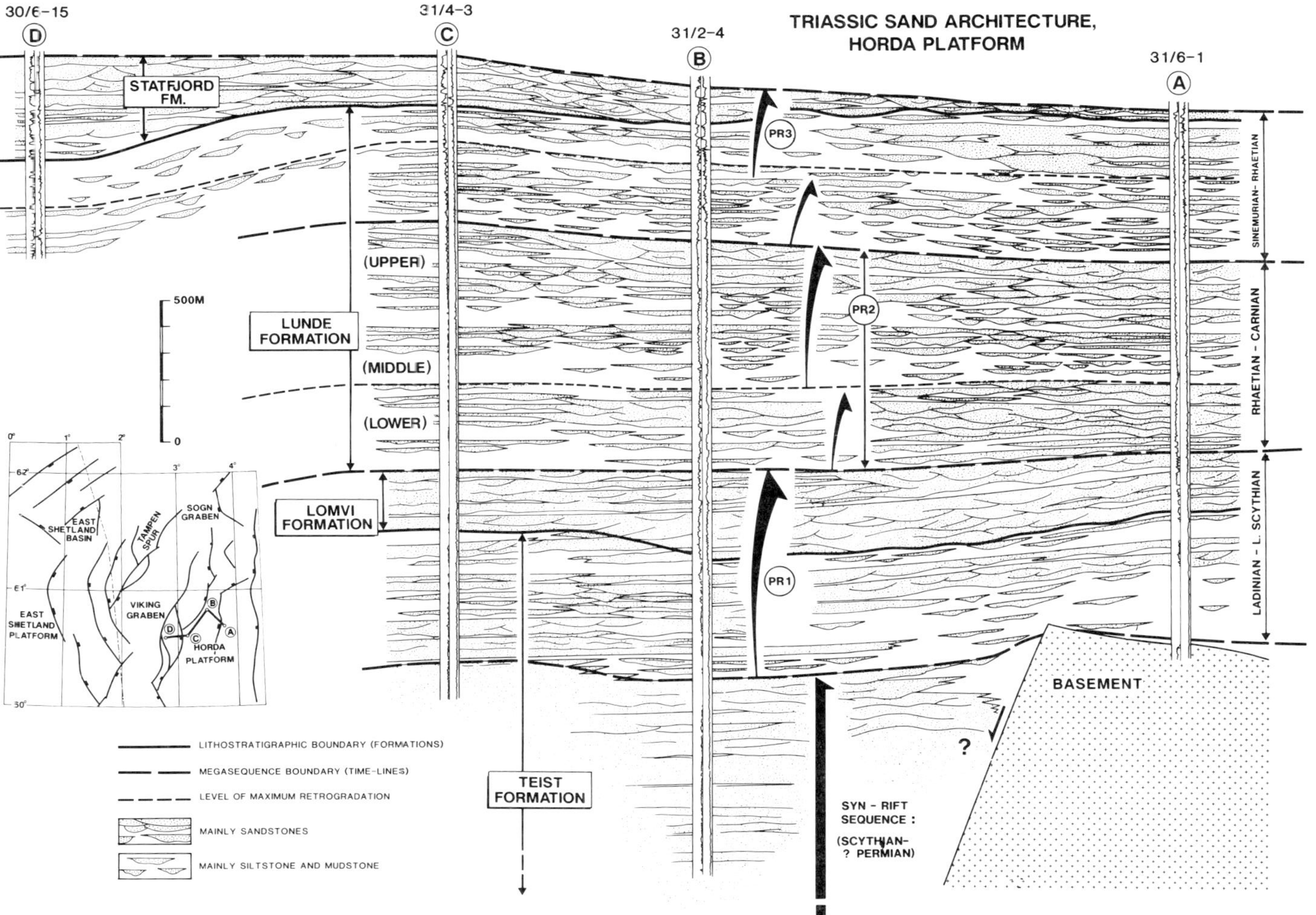

Fig. 9. Syn-rift and post-rift megasequences on the Horda Platform.

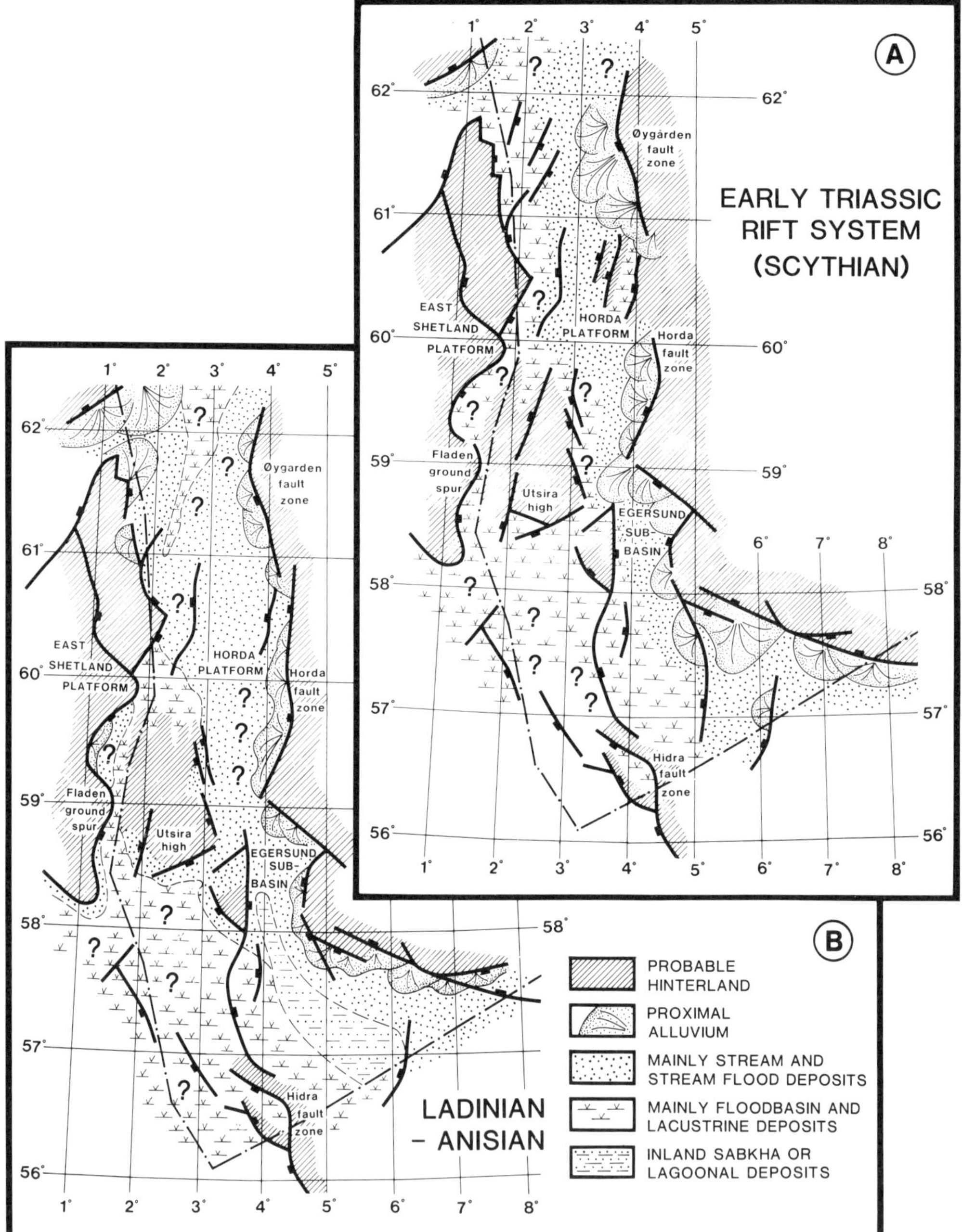

Fig. 10. Tentative and schematic Triassic palaeogeographic maps for (A) Scythian and (B) Ladinian/Anisian intervals in the northern and central North Sea areas.

quences. The latter are especially prominent, showing 1–2 m thick coarsening upward (CU) motifs which contrast strongly with the associated fluvial motifs which are similar in scale but have an upward-fining character.

The upper member of Lunde Formation. Excluding the mudstone – dominated upper half of the member which is placed within the overlying megasequence, this makes up the remainder of megasequence PR2, and consists of alternations of lensoid sandstone bodies, dominantly low-angle or flat-laminated, and red siltstones with root structures and pedogenic carbonate nodules. These sequences are interpreted in terms of floodbasin deposits, sheetfloods, and ephemeral stream channels of varying morphology and sinuosity (Nystuen *et al.* 1990). Small-scale (few metres or less) fining upward (FU) motifs betray the ephemeral and flood-dominated nature of deposition on the alluvial plains, whereas CU or FU-motifs on larger scales probably reflect lateral shifting and abandonment of flood or stream-tracts, or the progradation of alluvial lobes into terminal floodbasins. CU-motifs on 100 m scale or the FUCU character of the entire megasequence, features which appear to have a basinwide extent, probably reflect the outbuilding of major clastic wedges into the axial floodbasins, and were triggered by regional base-level change, as discussed below.

Within both reference wells in the central North Sea region (Figs 4 & 5) the *Skagerrak Formation* consists of alluvial sandstones alternating with thinner units of overbank and flood-basin siltstones.

Basin development and palaeogeography

Figs 9 & 11 illustrate the lateral development of megasequence PR2 over parts of Horda Platform and Tampen Spur, respectively. Despite the general tabularity of the sequence over great distances, an increasing sand content can be seen towards the basin margins, both in well 31/6–1 (Fig. 9) and in the wells in the north-westerly reaches of the basin (Fig. 11). The megasequence thickens locally against some of the active, intrabasinal faults, for example the Gullfaks South Fault (Fig. 1), and probably also across the western edge of the Horda Platform.

Figure 12 shows a speculative palaeogeography for an early and late phase within the interval represented by post-rift megasequence 2. During late Carnian–early Norian the landscape in both the northern and central North Sea was dominated by muddy floodbasins which contained many large lakes. In the north there was a marine connection, bringing lagoonal waters into the coastal floodbasins, as evidenced by the middle Lunde Formation deposits. The eventual persistence of this marine connection farther south into the primitive Viking Graben is unknown. Fluvial deposition had withdrawn some way towards the margins of the basin in many areas, though there were numerous narrow channel tracts leading out into the basinal areas (Fig. 12A). It is likely that floodbasin deposition also dominated the region south of the primitive Viking Graben, though the sandy Skagerrak fluvial systems were prograding out from the Norwegian hinterlands, as well as locally from the Fladen Ground Spur and the Utsira High.

By late Norian–early Rhaetian times the region had been transformed by the outbuilding of alluvial fans and rivers from the whole length of the Norwegian highlands and from the north-western hinterlands (Fig. 12B). By this stage, (lower part of upper Lunde Fm.) sandy alluvial systems dominated the entire basin, with the near-mergence of clastic wedges even in the axial regions. Tracts of floodbasin deposition possibly dominated only in the north where the marine connection had previously been, and along certain reaches of the primitive Viking Graben. The region south and southwest of the southernmost Viking Graben remained as a depocentre for mainly fine-grained deposits, except in the neighbourhood of high-standing blocks.

The palaeogeographical transition from Fig. 12A to 12B, showing the change from mud-dominated to sand-dominated landscape, is believed to have resulted from a general decrease in the rate of relative base level rise through the time period represented by megasequence PR2.

As for the Middle Triassic interval, it is likely that parts of the Utsira High and Fladen Ground Spur areas were uplands during Carnian–Rhaetian. The East Shetland Basin west of the Ninian–Hutton–Alwyn alignment is, however, likely to have been part of the Triassic basin. This region accumulated a thin Triassic succession which, although poorly dated, is possibly time-equivalent to PR2 (see below).

Post-rift megasequence 3 (PR3)

Stratigraphy and facies

The Rhaetian–Sinemurian interval (megasequence PR3) encompasses parts of the upper Lunde Formation, and the alluvial parts of the Statfjord Formation (Deegan & Scull 1977; Vollset & Doré 1984). In vertical sections the megasequence is recognized as a crudely FUCU interval above the sandstone-dominated upper parts of megasequence PR2. The upper sequence boundary is picked where the first marine influence can be recorded on a basinal scale, usually slightly below the top of the Statfjord Formation, at the base of a transgressive sandstone unit. In the reference wells on the

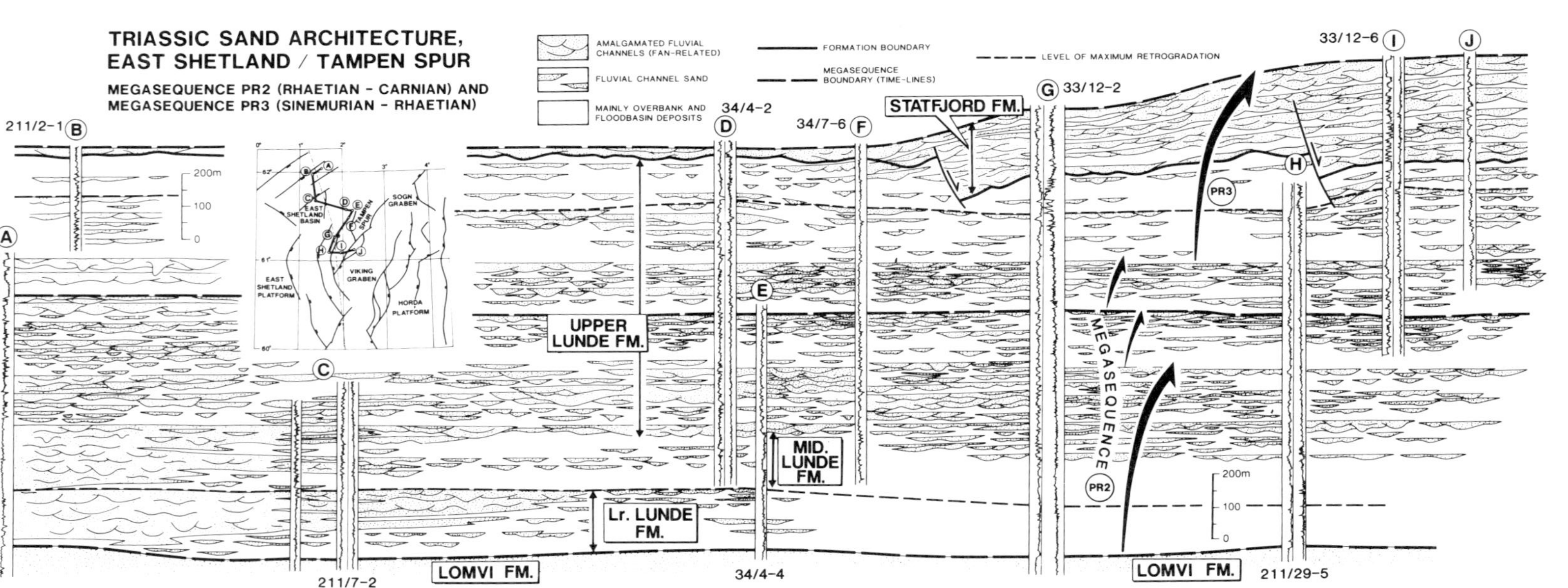

Fig. 11. Triassic megasequences PR2 and PR3 in the Tampen Spur area.

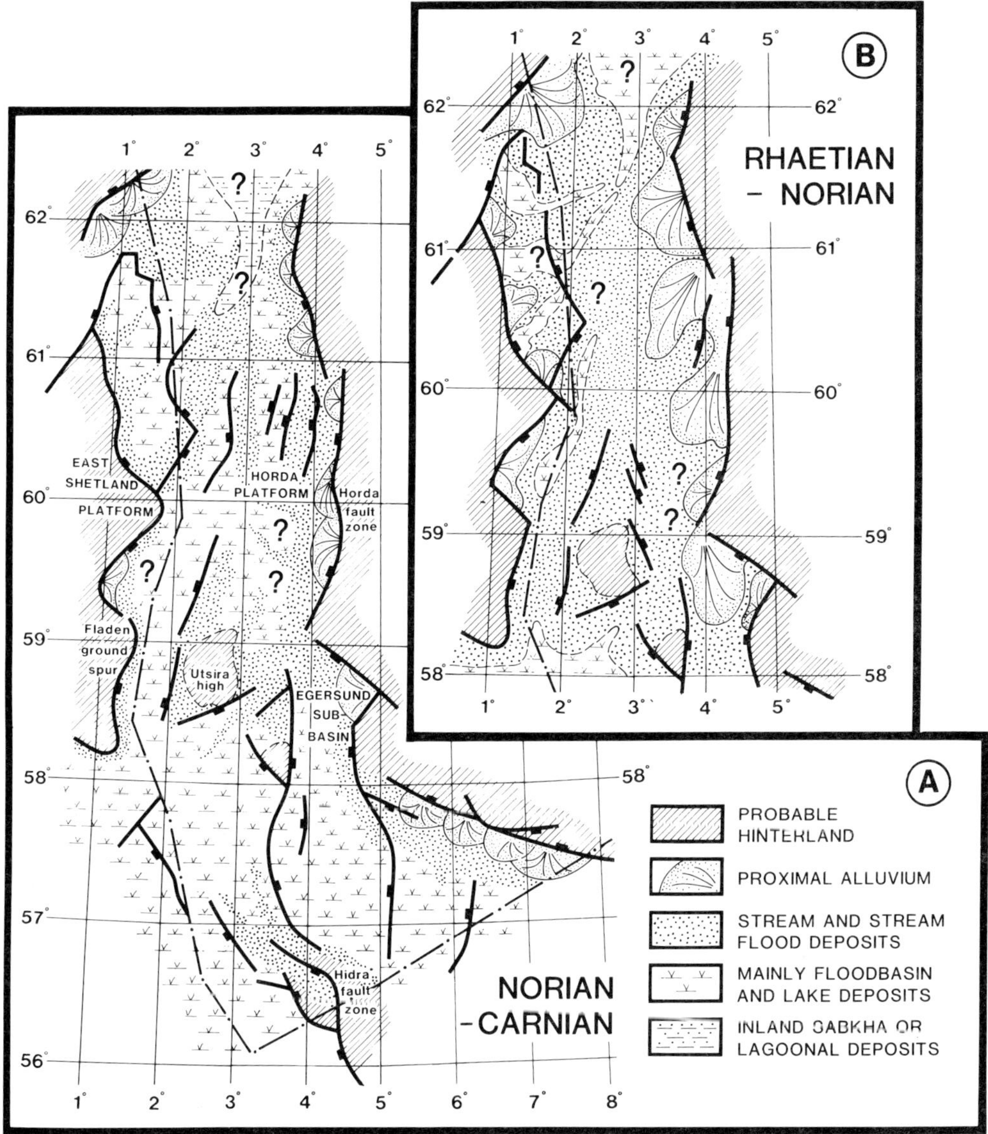

Fig. 12. Tentative and schematic palaeogeographic maps for megasequence PR2: (A) Norian-Carnian and (B) Rhaetian-Norian intervals in the northern North Sea.

Tampen Spur and the Horda Platform (Figs 2 & 3), the megasequence has a thickness of approximately 620 m and 550 m, respectively.

Figures 9 & 11 show that the megasequence comprises three main sub-units, including a lower sandstone–mudstone interval (middle to upper parts of upper Lunde Fm.), a middle mudrock-dominated interval (uppermost Lunde Fm.), and an uppermost sandstone–dominated interval (Statfjord Fm.). The boundary between the lower and middle sub-units is interpreted in terms of maximum retrogradation (Fig. 7), and it is therefore interesting to notice that Nystuen *et al.* (1990) recorded micropalaeontological evidence of some marine influence at this stratigraphic level from one locality on the Tampen Spur.

Definition of the base of the Statfjord Formation was originally related to the recognition of a basal, coarsening upward Lunde–Statfjord transition zone, as defined by Deegàn & Scull (1977) in the type well 33/12–2. Regional studies now show that this pattern is atypical, and we therefore choose to pick the base of the Statfjord Formation where there is a significant vertical increase in the sandstone/shale ratio above the shaly interval (Fig. 13) irrespective of age. This approach accords with Vollset and Doré's (1984) definition of the Statfjord Formation in well 30/6–1.

Biostratigraphic zonation of the megasequence requires detailed work, and systematic studies are available only from a few wells.

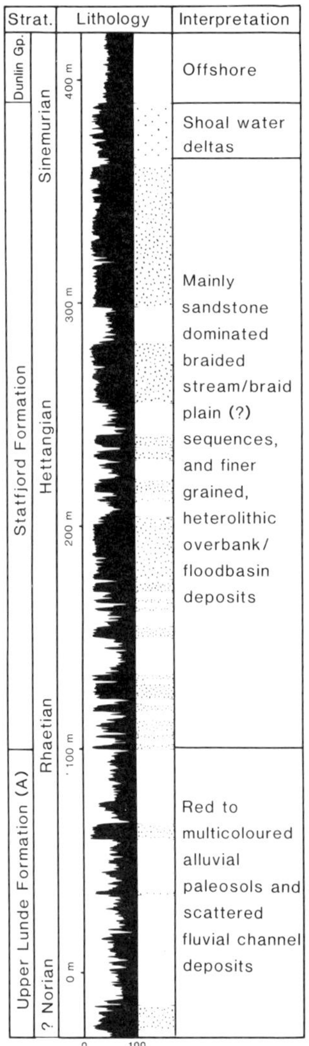

Fig. 13. Reference well for Statfjord Formation in the Oseberg area (30/6–15).

Generally however, the Triassic–Jurassic boundary is picked at the first downhole occurrence of the terrestrial palynomorphs *Riccisporites tuberculatus* and *Ovalipollis pseudoalatus* of Rhaetian affinity (L. Lømo, pers. comm.; Lervik *et al.* 1990). Our data indicate that this time-line falls *within* the Statfjord Formation. Biostratigraphic data also indicate a hiatus at the Triassic–Jurassic junction, with Rhaetian rocks apparently being succeeded by Sinemurian strata without any intervening Hettangian. This is particularly prominent along basin margins and locally on structural highs, though it is difficult to demonstrate any significant time-breaks from core data and wireline logs.

Sedimentary facies within the lower sandstone–mudstone interval are similar to those in the uppermost part of megasequence PR2, and a floodbasin and fluvial channel interpretation is invoked, despite the higher content of fine-grained deposits. A more pronouced fining-upward tendency of the main sandbodies may, however, indicate that streams had a more sinuous channel morphology at this stage than during deposition of the underlying beds (see also Nystuen *et al.* 1990).

Vertical lithology/facies changes within the middle and upper sub-units record a transition from semi-arid to more humid conditions, as well as the establishment of marine conditions at the on-set of deposition of the next megasequence (Sinemurian). Cored sections from the middle, fine-grained interval are composed of red mudrocks with zones of multicoloured grey, green, yellow and purple mottling, thin whitish and leached sandstones and scattered calcrete units, which probably accumulated within generally dry, well drained alluvial plains with strong pedogenic activity (see also Nystuen *et al.* 1990). Interbedded fine- to medium-grained sandstones with intra-clast lags and well developed vertical fining probably represent stream channel deposits. Prominent planar lamination in many sandstone bodies points towards deposition from highly turbulent and sand-choked flood-waters, and indicates that fluvial systems at this stage were of the ephemeral type (e.g. Tunbridge 1981).

The Statfjord Formation is composed of mainly coarse- to fine-grained sandstone beds (> 100 cm thick) with common lag conglomerates, cross-stratified, planar laminated, massive and ripple-laminated beds, intercalated with rooted and non-rooted mudrocks, thin (0–100 cm) and fine-grained sandstones and scattered coals. These deposits probably record the presence of streams of a more perennial character (braided, meandering), alternating with moderately to poorly drained overbank areas. Notably, the amount of red pigment decreases vertically and coal-bearing grey beds become more common.

Separating this alluvial facies association from the overlying marine Dunlin Shale is a relatively thin (2–30 m), commonly coarsening upwards unit, of bioturbated mudrocks and fine-grained sandstones intercalated with beds and sequences of coarser sandstones. This upper unit probably records a phase of marine incursion and progradation of shoal-water deltas prior to the main transgression of the Dunlin Sea. Notably, the oldest marine sporomorphs encountered in the Statfjord Formation are of Sinemurian age, whereas the establishment of fully marine conditions (Dunlin Group) seems to date from late Sinemurian/early Pleinsbachian. The boundary between fluvial and deltaic deposits marks a turnover from a progradational to a retrogradational stage, and is taken as the upper sequence boundary, whereas the top of the Statfjord formation probably represents a maximum flooding surface (*sensu* Galloway 1989) within the overlying megasequence.

Basin configuration and palaeogeography

The correlated section from the East Shetland Basin to the Viking Graben (Fig. 14) shows an apparent termination of the middle and upper sub-units against the Alwyn–Ninian–Hutton alignment (lineament 'X'; Fig. 14). The relationship between the upper Lunde Formation and the 'undifferentiated Triassic' interval in the East Shetland Basin (referred to as the Cormorant Formation; Deegan & Scull 1977) is uncertain, since we have not been able to identify the shaly Lunde interval in this area. For the Cormorant Formation, an upper Triassic (Norian/Rhaetian) age has been assigned (Brennand 1975), hence it should correspond to parts of the Lunde Formation, and it probably correlates with our post-rift megasequence PR2, or possibly also with the lower sub-unit of PR3.

Thin (< 30 m), transgressive sandstones of Sinemurian age (not included in PR3) characterize the Statfjord Formation to the west of the Alwyn–Ninian–Hutton alignment (Røe & Steel 1985). The regional importance of this complex fault system as a boundary separating a western area (East Shetland Basin) with a much attenuated, incomplete Statfjord Formation resting unconformably upon Triassic strata, from an easterly basin with a thick (100–200 m) Statfjord Formation with a more conformable relationship to the underlying units, has been documented along its entire N–S-trending trace (Johnson & Stewart 1985; Røe & Steel 1985; Johnson & Eyssautier 1987). Consequently, the stratigraphic relationships indicate that the East Shetland Basin to the west of this fault zone underwent a period of erosion or non-deposition during the latest Triassic/earliest Jurassic (see also Hay 1978).

In the Tampen Spur area, the lower sub-unit has a fairly uniform thickness of approximately 150 m to 170 m, and shows no major lateral variation in the fluvial sandstone density (*c.* 40%). On the Horda Platform, about 190 m is present in well 31/2–4, and both the thickness and sandstone density appear to be relatively constant (Fig. 9). This pattern suggests that the earliest phase of deposition during the Rhaetian–Sinemurian interval took place across a broad alluvial plain which experienced only a limited amount of intra-basinal differential subsidence.

The middle sub-unit varies in thickness from about 100 m to approximately 240 m (Fig. 14) across the Tampen Spur, and generally contains less than 15% fluvial sandstone. East of the Tampen Spur, on the flank of an assumed early Viking Graben (well E, Fig. 14), the unit reaches a thickness of about 300 m and contains nearly 50% fluvial sandstone. A high content of fluvial sandstone along the basin axis during a retrogradational stage is clearly a complication of the simple pattern outlined in Fig. 7. One possible explanation for this anomaly may be that the retrogradational stage was associated with fault-controlled (?) differential subsidence that caused major streams to concentrate in major topographic depressions, so that sandstone densities would increase both along basin margins and axes. This also accords with the situation across the Horda Platform, where the sandstone density of the middle sub-unit increases towards the basin margin.

On the Horda Platform (Fig. 15), the thickness (middle sub-unit) ranges from about 200 m (31/2–4) to *c.* 270 m (31/6–1), accompanied by lateral changes in the distribution of fluvial sandstones. To the east, 50–70% fluvial sandstone is observed (31/2–4, 31/6–1), whereas the unit contains approximately 25% fluvial sandstone on the western margin of the Horda Platform (30/6–15, 31/4–3), suggesting that this area was situated away from the main fluvial axis. No data on thickness and sandstone content are available from the Viking Graben, but very sandy deposits of presumed similar age have been drilled in the Sogn Graben area to the north (Fig. 16). This distribution of fluvial sandstones may indicate that major streams draining the eastern, proximal parts of the

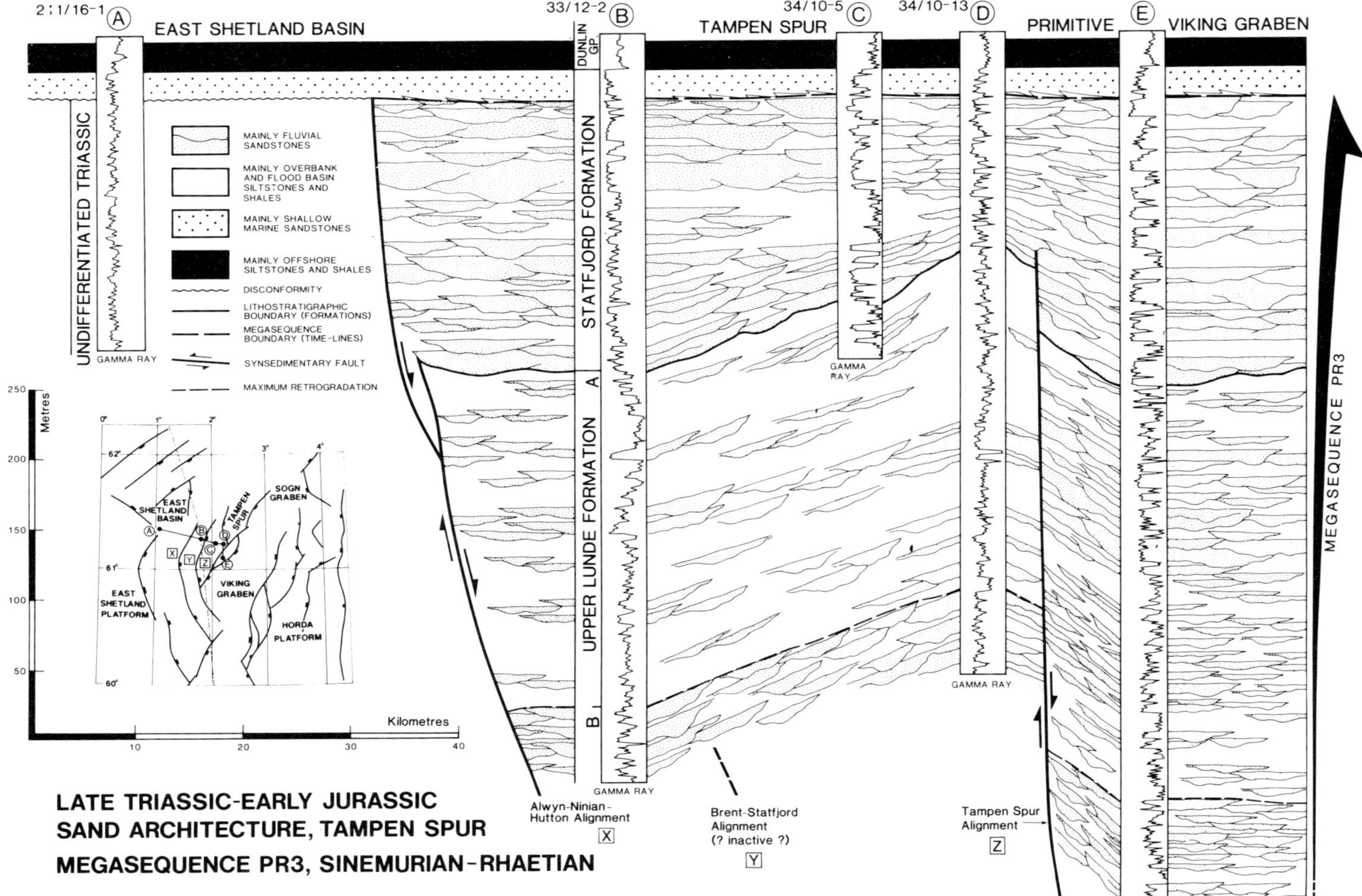

Fig. 14. Correlation panel for megasequence PR3 between the Alwyn–Ninian–Hutton fault zone and the primitive Viking Graben.

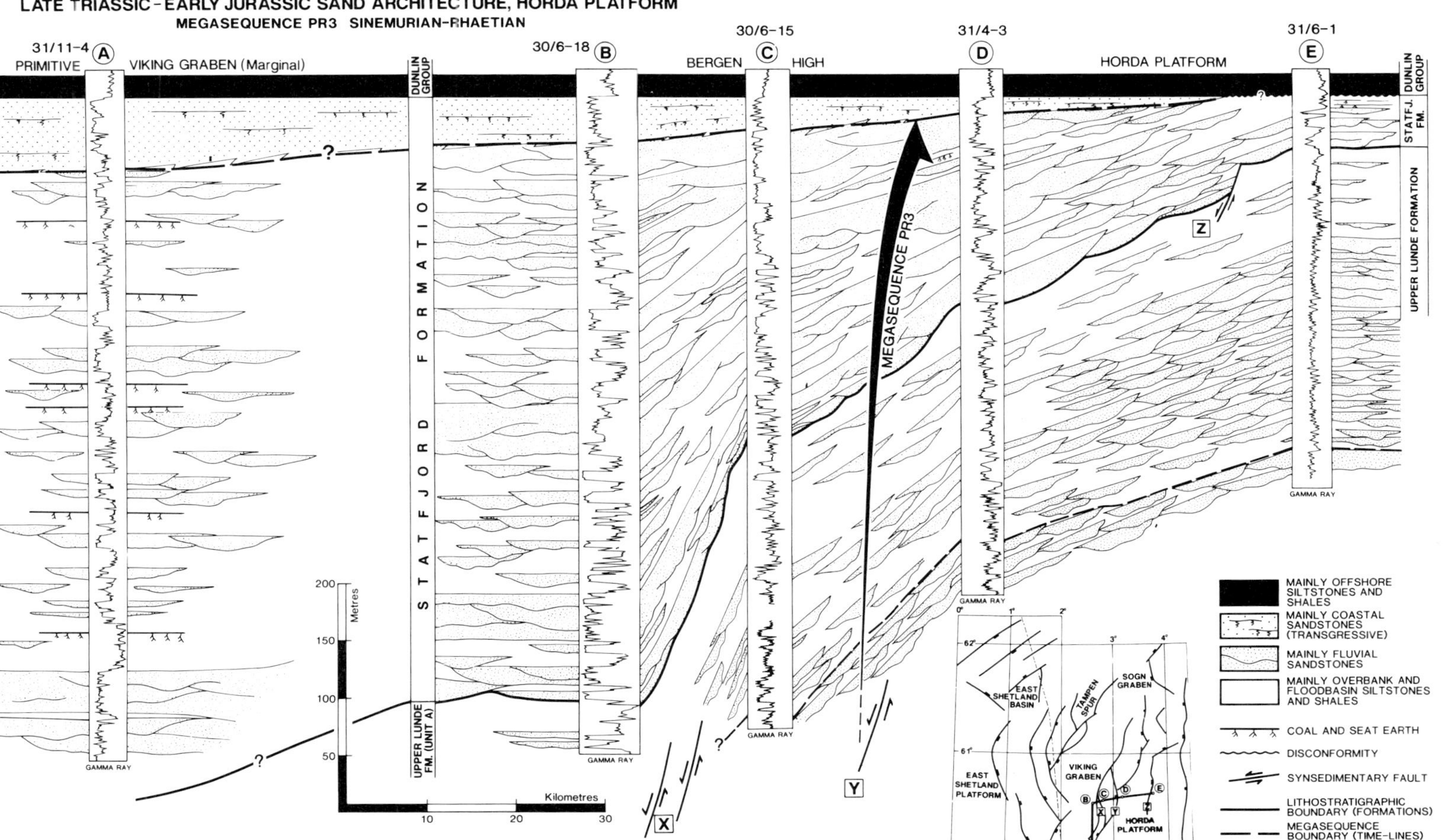

Fig. 15. Correlation panel for megasequence PR3 over the Horda Platform and into the Viking Graben area.

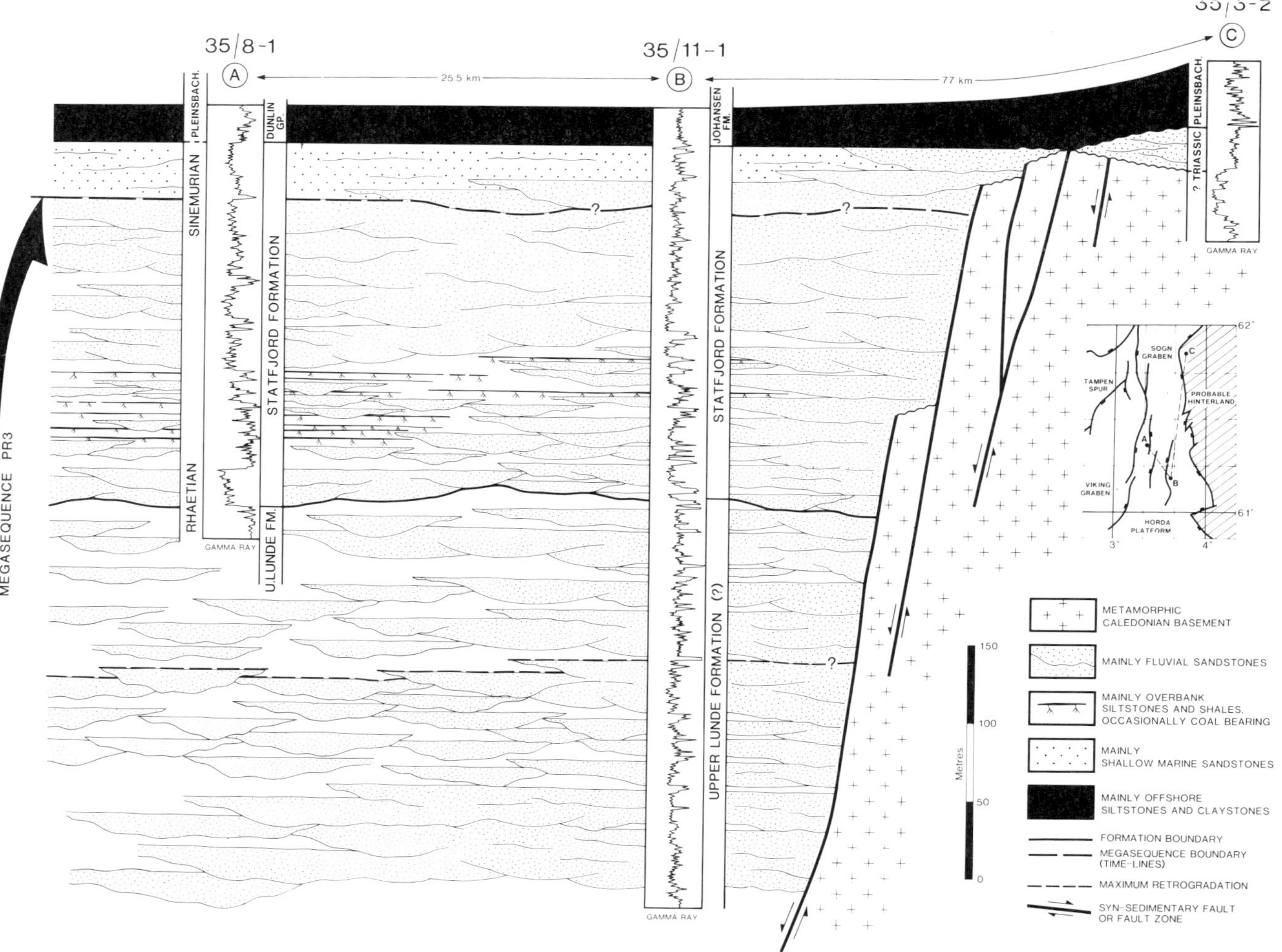

Fig. 16. Correlation panel for megasequence PR3 in the region north of the Horda Platform.

Horda Platform were headed northwards, and shed sediment towards the Sogn Graben.

The Statfjord Formation ranges from 130 m to 220 m on the Tampen Spur, and contains about 55% to 60% fluvial sandstone. To the east of the Tampen Spur alignment the thickness increases to approximately 230 m, accompanied by a slight increase in the fluvial sandstone content (65% in well E, Fig. 14). The uniformly high density of fluvial sandstone across the Tampen Spur and early Viking Graben (55–65%) suggests that major streams spread across the entire area east of the East Shetland Basin during deposition of the fluvial parts of the Statfjord Formation.

On the Horda Platform (Fig. 15), the Statfjord Formation thickens substantially from about 40 m along the eastern parts to more than 500 m close to the graben axis, indicating significant syn-depositional differential subsidence. Maximum fluvial sandstone content of approximately 65% is reached in the Bergen High area (wells B, C, D; Fig. 15), and tends to decrease both basinwards and landwards (40% and 48% in wells A and E, respectively). This alluvial architecture shows that streams existed across the entire eastern margin at this stage, and reflects a high preservation potential for overbank fines in areas of strong subsidence, as well as preferential funnelling of avulsing streams into such sites (Allen 1978; Bridge & Leeder 1979; Blakey & Gubitosa 1984).

Megasequence PR3 shows a general westward expansion across the Horda Platform (Fig. 9), with the most pronounced thickening taking place across the fault zone separating the Horda Platform from the Viking Graben (lineament 'X', Fig. 15). This thickness pattern is characteristic for a post-rift sequence, in which basinwards sequence expansions are being accomodated by planar, normal faults (Badley *et al.*, 1988). The expansion is pronounced for the Statfjord Formation part of the sequence, which suggests that differential subsidence was still significant during the later stages of sequence evolution.

To the north of the Horda Platform, megasequence PR3 apparently terminates abruptly against a boundary fault system (Fig. 16), to the east of which lower Jurassic (Pliensbachian and younger) strata onlap Caledonian basement rocks, and locally, a thin, undated (Triassic or older) red-bed unit (Fig. 16). This contrasts with the gradual expansion of the Statfjord Formation across the Horda Platform, and suggests a change in structural configuration south to north along the eastern margin of the early Jurassic basin.

Comparison of the Figs 14 & 15, suggests that certain reaches of the western and eastern graben margins were characterized, at times, by different modes of subsidence. This is particularly marked for the Statfjord Formation, which apparently is subjected to gradual, yet step-wise thickening into the graben across the Horda Platform (Fig. 15) as opposed to extremely abrupt expansion across the Alwyn–Ninian–Hutton alignment to the west. Accordingly, this latter tectonic alignment may have acted as a westerly master fault of the early Jurassic basin. The main depocenter was probably located close to the western margin south of 61°N. This is also supported by the asymmetric structure and westerly off-set of the graben axis of the South Viking Graben (e.g. Curtin & Ballestad 1986), and from the crustal half-graben structure with easterly dipping detachments interpreted from deep seismic sections (Beach *et al.* 1987; Gibbs 1987). A similar half-graben structure with end-Triassic listric movement along the Great Glen Fault has been invoked from the adjacent Moray Firth Basin (McQuillin *et al.* 1982; Frostick *et al.* 1988). The invoked north-south directed change in geometry along the eastern basin margin (Figs 14 & 16) does however suggest that the area north of the Horda Platform had a full-graben configuration.

Continental conditions dominated in the Viking Graben during the Rhaetian and Hettangian stages, despite some signs of weak marine influence in the Rhaetian muddy floodbasins. The distribution of sandy facies (Figs. 14–16) suggests that an axial drainage system was established during the late Rhaetian (Fig. 17A). During the Hettangian, major streams shifted across the entire area (Fig. 17B), with strong differential subsidence giving floodbasin deposits the best preservation potential along the graben axis. During the Sinemurian, marine waters entered the Viking Graben, probably from the north, with the establishment of fringing shoal-water deltas supplied from coastal plains on the graben flanks (Fig. 18).

The Rhaetian–Sinemurian interval was a period of general transgression in North West Europe (Ziegler 1982; Doré & Gage 1987), with marine waters encroaching from the Boreal (Arctic) and Tethyan domains. In the Norwegian–Danish Basin, the Rhaetian Vinding Formation and the Hettangian–Pleinsbachian Fjerritslev Formation accumulated in a brackish-marine basin into which deltas of the Gassum Formation (Rhaetian-Sinemurian) prograded from a northern source terrain (Fig. 18; see also Bertelsen 1978; Olsen & Strass 1982).

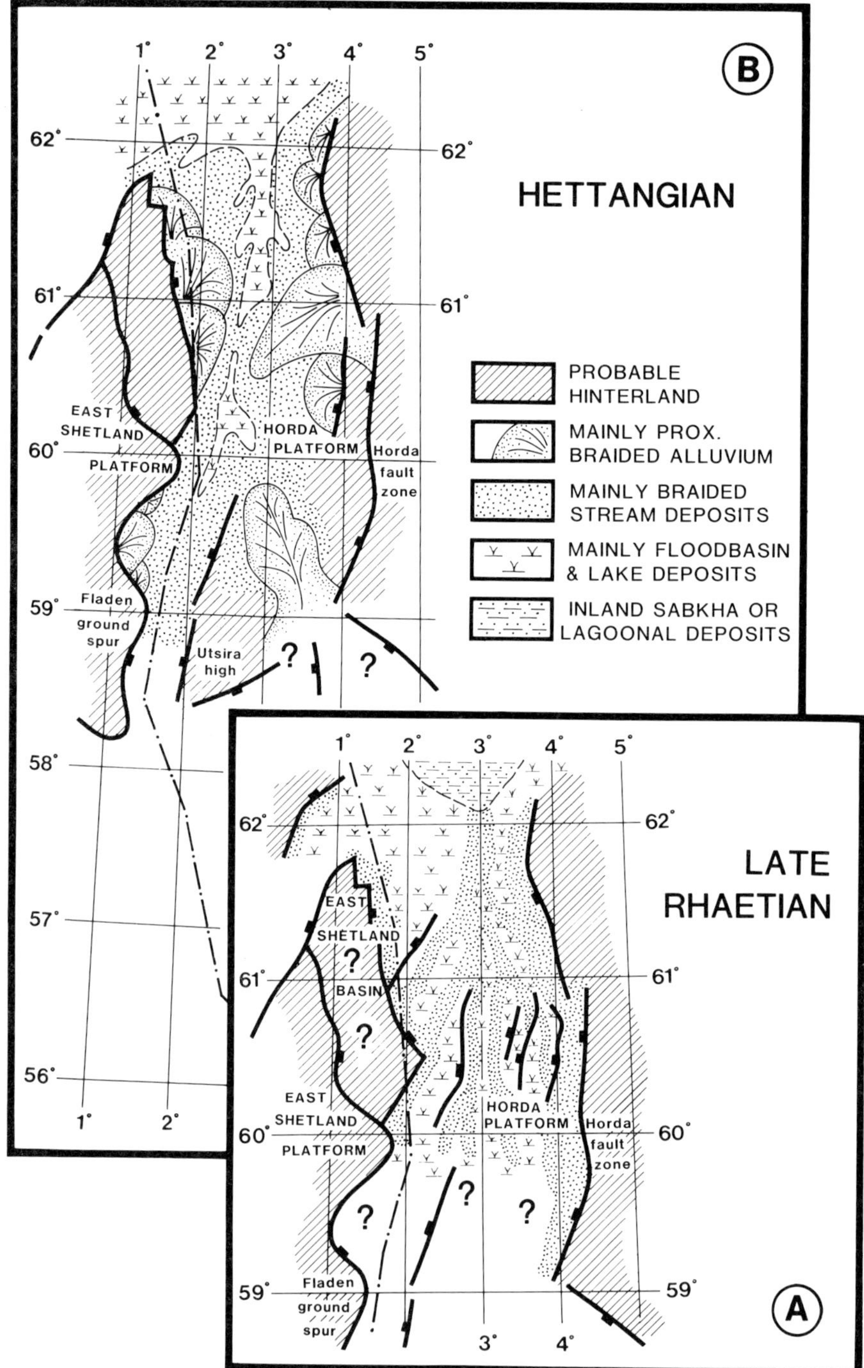

Fig. 17. Tentative and schematic palaeogeographic maps for megasequence PR3: (A) late Rhaetian and (B) Hettangian intervals in the northern North Sea.

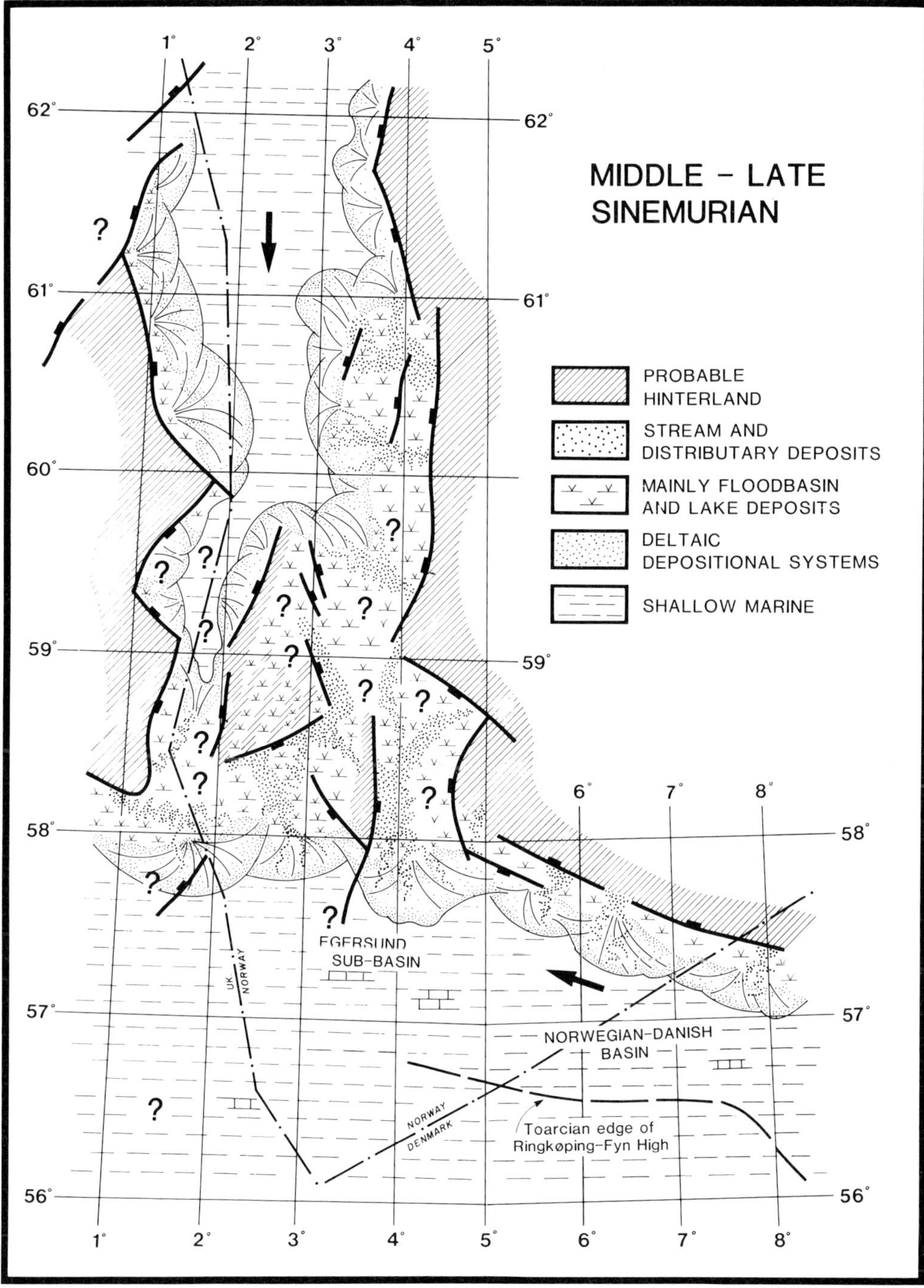

Fig. 18. Tentative palaeogeographic map for Sinemurian interval in northern and central North Sea region.

In the Hebrides Basin west of the British Isles, Steel (1977) documented a similar time-trend of sedimentation, though the sea encroached from the south with a pronounced diachroneity (Morton *et al.* 1987). Further north, in East Greenland, marginal marine facies in the Kap Stewart Formation (Rhaetian–Hettangian; Clemmensen 1976) may record incursions from the Arctic Sea, whereas continental conditions apparently persisted on the mid-Norwegian shelf until the Sinemurian (Gjelberg *et al.* 1987), in a depositional setting quite similar to the Viking Graben. This marked diachroneity in the regional distribution of marine strata may indicate that the Arctic and Tethyan seas did not link-up through the Viking Graben prior to the Sinemurian, as has been suggested from other regional studies (e.g. Ziegler 1982).

Conclusions

Sequence stratigraphy

Our choice of megasequence boundary at a stratigraphic level representing maximum progradation of the fluvial system, or showing maximum density of alluvial sandstones in basinal areas, leads to a definition of megasequences with a simple CU, or a composite FUCU character. In both cases the sandy intervals are separated by packages of sediment with a high content of fine-grained overbank, floodbasin or lacustrine deposits. These thick, fine-grained units imply a basin-marginwards retreat of the earlier sandy fluvial systms (as is confirmed by correlation panels) and a "drowning" of the region by floodbasin conditions. The occasional occurrence of marine strata within the finer intervals indicates that these units at least in some cases can be literally regarded as transgressive, and that they accumulated at stages of maximum flooding (Galloway 1989). Conversely, the megasequence boundaries can loosely be considered as levels of maximum regression. Consequently, our megasequence definition falls broadly in line with the practice of seismic stratigraphers (e.g. Haq *et al.* 1987), though we do not imply that each megasequence boundary represents an unconformity. The possibilty that fluvial channels may concentrate in basinal areas due to differential subsidence, so that dominantly fine material accumulates on large parts of marginal areas clearly complicates the definition of megasequences in alluvial basins. This seems to occur in some areas within PR3 (Rhaetian–Sinemurian; Fig. 14). Such variations in alluvial architecture require that sequences are not defined until a lateral correlation scheme has been established, and deviations from the general scheme of highest sandstone concentration in the proximal areas (Fig. 7) are incorporated.

On the basis of ages assigned to the post-rift megasequences, the latter have a longer duration (10–15 Ma) than the (eustatic) stratigraphic sequences of Haq *et al.* (1987), and we therefore suggest that factors other than eustacy controlled the large-scale cyclicity of the Triassic and early Juassic deposits. The possible cause of repeated megasequence development is of interest, and the following characteristics are relevant in this respect:

> The lowest levels of the megasequences are relatively sandy but show a general upwards trend to become mud-dominated.

As discussed above, this represents basin-marginwards retreat of the sandy alluial systems and was probably caused by a gradual increase in the rate of relative base level rise through time.

> The lower to middle levels of any megasequence are mudstone-rich, with isolated fluvial sand bodies.
>
> There are signs of marine influence at these lower and middle levels (notably in megasequences 2 and 3).

These features suggest higher rates of sediment accommodation (*sensu* Posamentier & Vail 1988) than below, with rate of relative base level rise exceeding rate of alluvial sediment input where marine influence is recorded.

> The upper levels of any megasequence are sand-rich, often with development of extensive and well connected fluvial sand bodies.
>
> Erosion surfaces are common, and a major hiatus can also occur preferentially within the sandy levels.

These features suggest lower rates of sediment accommodation, and lower rates of relative base level rise, or even some base level fall.

The above image of any megasequence is, of course, idealized. Megasequence PR2, in particular, deviates from the ideal. Here the CU-trend is limited to the lower part, with the upper sandy part being more irregular, or even tending to fine upwards in some areas (Nystuen *et al.* 1990). Nevertheless, the overall trend within the megasequences, is from an initial increase through a maximum and then to a decrease in the rate of relative base level rise. Because the megasequences cover a great basinal area, and sediment input was derived from various hinterlands, it is not likely that sandstone or mudstone dominance at different levels in the megasequences was caused by 'lithology' factors in the drainage areas. The changing relative base level, as referred to here, was therefore caused by increasing then decreasing eustatic sea level rise and/or by increasing, then decreasing rates of basinal subsidence. Modelling of sand body architecture in alluvial basins (Allen 1978) has shown clearly that decreasing rates of subsidence through time can produce increased stacking, reworking and amalgamation of sandbodies, resulting in a crudely coarsening upward sedimentary sequence. On the other hand, it is likely that the global sea level high-stand known to have occurred in Norian times (Haq *et al.* 1987) also contributed to the rising relative base level and to the generation of a marine influx near the base of megasequence 2. In contrast, a tendency to global sea level low stand (both short and long term) in early Rhaetian times (Haq *et al.* 1987) make it likely that the relative base level rise (and marine influx) reflected by megasequence PR3 is due to high rates of basinal subsidence rather than to eustatic causes.

An additional support of the hypothesis that variation in subsidence rate was generally more important than eustasy in the generation of megasequence stratigraphy is the observation of how the fine- or coarse-grained members of any megasequence thicken or thin against active, intra-basinal tectonic lineaments. Figure 19 shows two wells believed to have been located on a single structural block in Triassic times. Differential subsidence has caused the succession in the well at the downslope end of the block to thicken considerably compared to that at the upslope end. Note particularly that it was preferentially the fine-grained, lower parts of the megasequences which have thickened, suggesting maximum differential subsidence early in the accumulation history of each megasequence. This observation, linking maximum differential subsidence to intervals of maximum floodbasin development, suggests that regional subsidence rates may similarly have been at a maximum in these intervals. We conclude that the Triassic–early Jurassic megasequence stratigraphy of the northern North Sea basin points to an uneven, post-rift subsidence pattern, with subsidence rate maxima in late Scythian, late Carnian/early Norian and Rhaetian times.

Triassic tectonic activity and structure

Detailed reconstruction of Triassic fault activity (timing) and basin configuration is still difficult. Clearly, the seismic evidence from the eastern margin of the Viking Graben (Fig. 6; Badley *et al.* 1984, 1988) (see also Fig. 20) show beyond any doubt that an early (Permo-Triassic) stage of rifting affected this area. Likewise, thickening of Triassic sequences across faults bounding the Unst Basin, Sele High, Egersund Basin and the Horn Graben (e.g. Lervik *et al.* 1990) is evidence for Triassic fault activity. Swallow (1986) and Lervik *et al.* (1990) claim that no clear Triassic activity can be seen along the western margin of the Viking Graben, though Swallow (1986) stated that faulting had commenced by the early Jurassic. Our interpretation of post-rift Megasequence 3 (Rhaetian–Sinemurian) is clearly opposed to this view, because we see the termination of this unit against the Alwyn–Ninian–Hutton alignment as evidence for late Triassic–early Jurassic activity along this fault zone.

Other lines of evidence suggest to us that both the Alwyn–Ninian–Hutton aligment (Fig. 20) and other faults within the East Shetland Basin may have been active during the Triassic. Published material from the East Shetland Basin show that the thickness of Triassic strata between metamorphic and igneous basement rocks and the top of the Statfjord Formation is rather variable. Gray & Barnes (1981) reported approximately 1200 m of Triassic red-beds in well 2/5–4, whereas less than 460 m of Triassic deposits was found in well 211/21–1 (Lervik *et al.* 1990), which was drilled about 28 kilometres to the northeast. We find it unlikely that these thickness variations reflect a pre-Triassic (Caledonian ?) basement topography, and suggest that intra-Triassic faulting and possibly block rotation is a more likely cause. In addition, more than 2000 metres of Triassic strata have been drilled further east (correlated thickness in wells 33/12–2 and 33/12–5; Fig. 2). The most likely explanation for lateral thickness variations of this magnitude (2x–4x) is differential fault block movements and strong subsidence in the basin centre. Finally, the East Shetland Basin apparently underwent periods of erosion/non-deposition during the early and middle Triassic (Scythian–Ladinian; Fig. 10) and during the latest Triassic-earliest Jurassic (Rhaetian–Hettangian; Fig. 17), whereas deposition persisted to the east. Possibly, these events may relate to relative uplift of fault blocks along the right shoulder.

The inferred configuration of the 'Viking Gra-

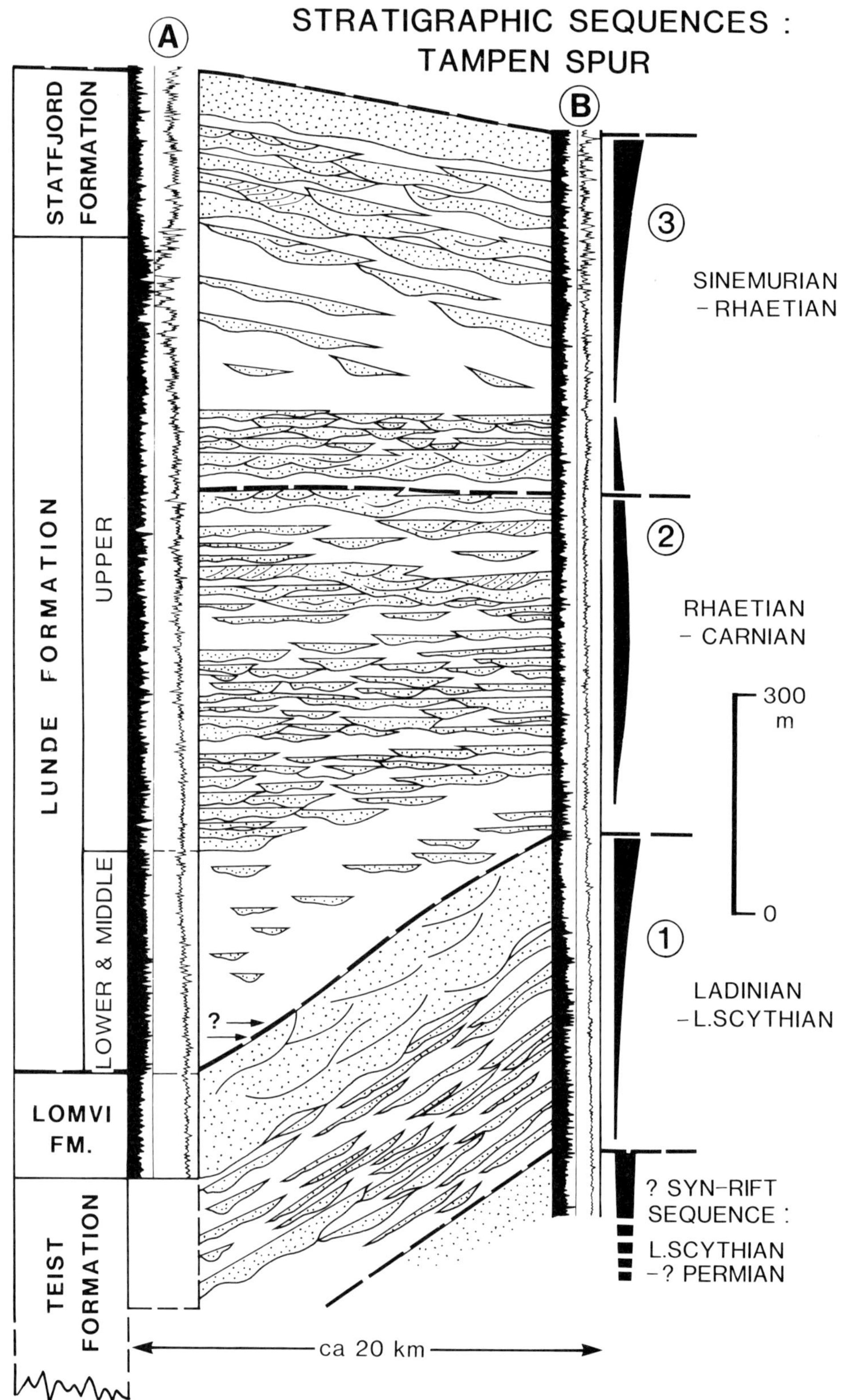

Fig. 19. Comparison of post-rift megasequences PR1–PR3 between wells 33/12–5 (A) and 34/10–13 (B). These 2 wells are believed to be located on downflank and upflank positions within the same Triassic, structural block. Note that the thickness expansion on megasequences PR2 and PR3 is largely taken up within the lower fine-grained intervals.

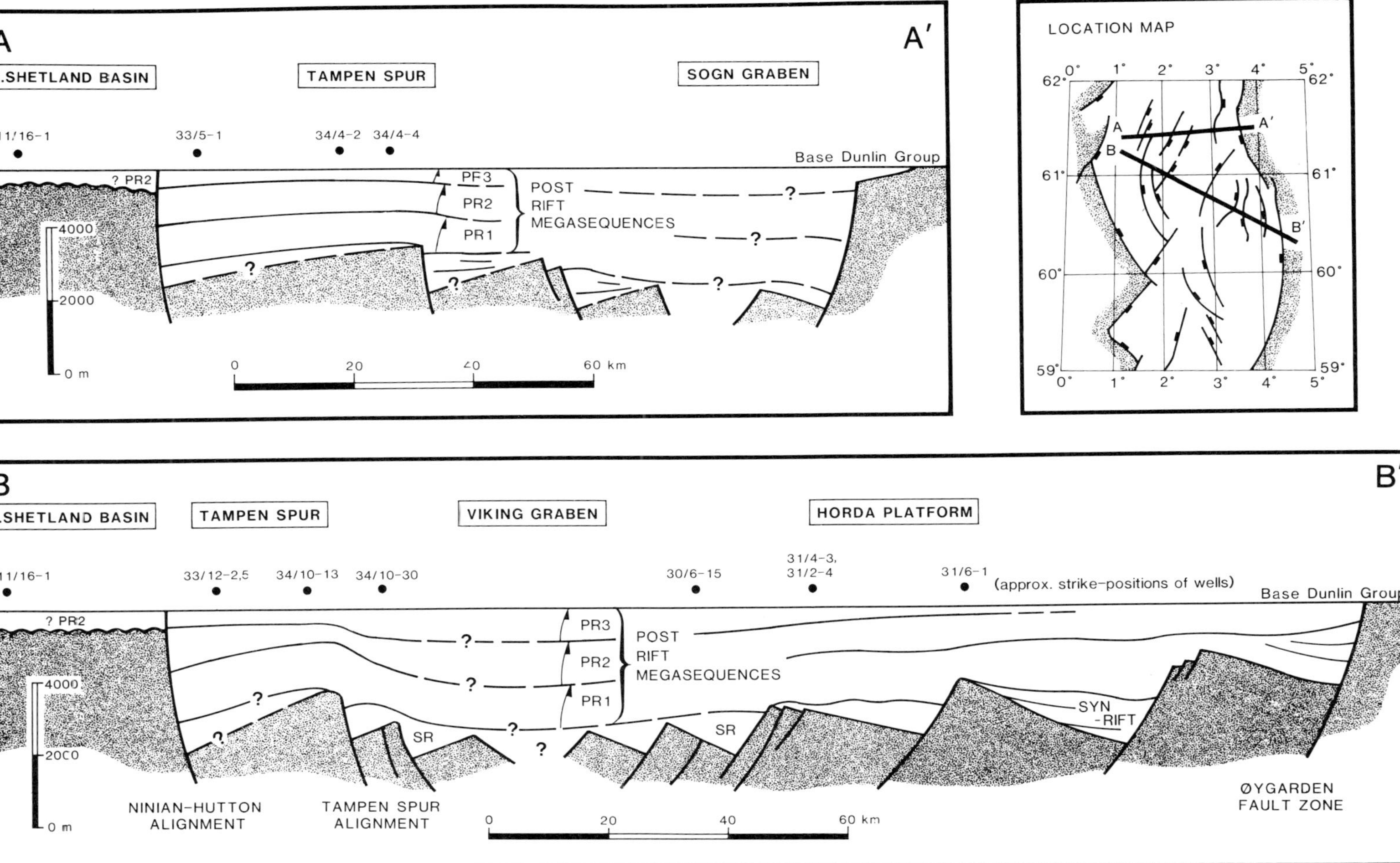

Fig. 20. Speculative and schematic cross sections showing the possible configuration of the rift and post-rift intervals, restored to a Sinemurian datum. The sections are based on well data (their approximate strike positions are shown) and on seismic data over the Horda Platform.

ben' during the Rhaetian–Sinemurian stage (PR3), comprising a half-graben structure and a western master fault system to the south, and a full-graben structure north of the Horda Platform, is clearly tentative, and will most likely require future modification. Half-graben structures alternating with limited full-graben areas and changes in basin polarity are common in modern continental rifts (e.g. Rosendahl *et al.* 1986; Frostick & Reid 1987) and should perhaps (?) be expected at early stages in the evolution of rift basins. To claim that this basin configuration persisted throughout the Triassic is very speculative, hence the reconstruction in Fig. 20 should be regarded merely as a working hypothesis.

We conclude that intra-Triassic tectonics in the northern North Sea can be read from the following.

(a) Early Triassic extensional structures.

(b) Differential subsidence across a number of fault lines throughout the Triassic, and particularly in latest Triassic and earliest Jurassic times.

(c) Signs of relatively high basinal subsidence rates at times within the middle and late Triassic post-rift phase, particularly in the late Scythian, late Carnian/early Norian and Rhaetian intervals.

We are indebted to B. Bakken, R. Faerseth and B. Noble for invaluable assistance at an early stage in the project and for help with the seismic interpretation and well tie (Fig. 6). A. M. Spencer, W. Helland Hansen and J. P. Nystuen read and made helpful suggestions on an earlier version of the paper. I. Neumann, N. Jansen and P. Gnanalingam are also thanked for assistance in the final stages of the work and with the figure drafting. Nopec is thanked for permission to use the seismic section in Fig. 6.

References

Allen, J. R. L. 1978. Studies in fluviatile sedimentation: an exploratory quantitative model for the architecture of avulsion-controlled alluvial suites, *Sedimentary Geology*, **21**, 129–147.

Badley, M. E., Egeberg, T. & Nipen, O. 1984. Development of rift basins illustrated by the structural evolution of the Oseberg feature, Block 30/6, offshore Norway, *Journal of the Geological Society, London*, **141**, 639–649.

——, Price, J. D., Rambech Dahl, C. & Agdestein, T. 1988. The structural evolution of the Viking Graben and its bearing upon extensional modes of basin formation, *Journal of the Geological Society, London*, **145**, 455–472.

Beach, A., Bird, T. & Gibbs, A. 1987. Extensional tectonics and crustal structure: deep seismic reflection data from the northern North Sea Viking graben. *In*: Coward, M. P., Dewey, J. F. & Hancock, P. L. (eds), *Continental Extensional Tectonics*, Geological Society, London, Special Publication **28**, 467–476.

Bertelson, F. 1978. The Upper Triassic-Lower Jurassic Vinding and Gassum Formations of the Norwegian-Danish Basin, *Geological Survey of Denmark*, Series B, No. 3.

Blakey, R. C. & Gubitosa, R. 1984. Controls of sandstone body geometry and architecture in the Chinle Formation (Upper Triassic), Colorado Plateau, *Sedimentary Geology*, **38**, 51–86.

Brennand, T. P. 1975. The Triassic of the North Sea, *In*: Woodland, A. W. (ed.), *Petroleum and the Continental Shelf of north-west Europe*, Applied Science, London, 295–311.

Brewer, J. A. & Smythe, D. K. 1984. MOIST and the continuity of crustal reflector geometry along the Caledonian-Appalachian orogen, *Journal of the Geological Society, London*, **141**, 105–120.

Bridge, J. S. & Leeder, M. R. 1979. A simulation model of alluvial stratigraphy, *Sedimentology*, **26**, 617–644.

Clemmensen, L. B. 1976. Tidally influenced deltaic sequences from the Kap Stewart Formation (Rhaetic-Liassic), Scoresby Land, East Greenland, *Bulletin Geological Society of Denmark*, **25**, 1–13.

——, Jacobsen, V. & Steel, R. 1980. Some aspects of Triassic sedimentation and basin development, East Greenland, North Sea, *The Sedimentation of the North Sea Reservoir Rocks*, Norwegian Petroleum Society, Geilo, Paper XVII.

Curtin, D. P. & Ballestad, S. 1986. South Viking Graben: habitat of Mesozoic hydrocarbons, *In* Spencer, A. M. (ed.), *Habitat of Hydrocarbons on the Norwegian Continental Shelf*, Norwegian Petroleum Society, Graham & Trotman, London, 153–157.

Deegan, C. E. & Scull, B. J. 1977. A standard lithostratigraphic nomenclature for the central and northern North Sea, *Bulletin Norwegian Petroleum Directorate*, No. 1.

Doré, A. G. & Gage, M. S. 1987. Crustal alignments and sedimentary domains in the evolution of the North Sea, North-east Atlantic Margin and Barents Shelf, *In* Brooks, J. & Glennie, K. (eds), *Petroleum Geology of North West Europe*, Graham & Trotman, London, 1131–1148.

Eide, F. 1990. Biostratigraphic correlation within the Triassic Lunde Formation in the Snorre area, in Collinson, J. D. (ed.), *Correlation in Hydrocarbon Exploration*, Norwegian Petroleum Society, Graham and Trotman, London.

Eynon, G. 1981. Basin development and sedimentation in the Middle Jurassic of the northern North Sea, *In* Illing, L. V. & Hobson, G. D. (eds) *Petroleum Geology of the continental Shelf of North West Europe*, Heyden, London, 196–204.

Fisher, M. J. 1984. Triassic, *In* Glennie, K. W. (ed.), *Introduction to the Petroleum Geology of the North Sea*, Blackwell, Oxford, 85–101.

Frost, R. E. 1987. The evolution of the Viking Graben tilted fault block structures: a compressional origin, *In* BROOKS, J. & GLENNIE, K. (eds), *Petroleum Geology of North West Europe*, Graham & Trotman, London, 1009–1024.

FROSTICK, L. E. & REID, I. 1987. Tectonic control of desert sediments in rift basins, ancient and modern, *In:* FROSTICK, L. & REID, I. (eds) *Desert Sediments: Ancient and Modern*, Geological Society, London, Special Publication, **35**, 53–68.

——, REID, I., JARVIS, J. & EARDLEY, H. 1988. Triassic sediments of the Inner Moray Firth, Scotland: early rift deposits, *Journal of the Geological Society, London*, **145**, 235–248.

GABRIELSEN, R. H., EKERN, O. F. & EDVARDSEN, A. 1986. Structural development of hydrocarbon traps, Block 2/2, Norway, *In* SPENCER, A. M. (ed.) *Habitat of Hydrocarbons on the Norwegian Continental Shelf*, Norwegian Petroleum Society, Graham & Trotman, London, 129–141.

GALLOWAY, W. E. 1989. Genetic stratigraphic sequences in basin analysis I: Architecture and genesis of flooding-surface bounded depositional units, *American Association of Petroleum Geologists Bulletin*, **73**, 125–142.

GIBBS, A. D. 1987. Deep seismic profiles in the northern North Sea. *In*: BROOKS, J. & GLENNIE, K. W., (eds), *Petroleum Geology of North West Europe*, Graham & Trotman, London, 1025–1028.

GILTNER, J. P. 1987. Application of extensional models to the northern Viking Graben, *Norsk Geologisk Tidsskrift*, **67**, 339–352.

GJELBERG, J., DREYER, T., HØIE, A., TJELLAND, T. & LILLENG, T. 1987. Late Triassic to Mid-Jurassic sandbody development on the Barents and Mid-Norwegian shelf, *In* BROOKS, J. & GLENNIE, K. W. (eds), *Petroleum Geology of North West Europe*, Graham & Trotman, London, 1105–1129.

GRAY, W. D. T. & BARNES, G. 1981. The Heather Oil Field, *In*: ILLING, L. V. & HOBSON, G. D. (eds) *Petroleum Geology of the continental shelf of North West Europe*, Heyden, London, 335–341.

HAQ, B. U., HARDENBOL, J. & VAIL, P. R. 1987. Chronology of fluctuating sea levels since the Triassic, *Science*, **235**, 1156–1166.

HAY, J. T. C. 1978. Structural development in the northern North Sea, *Journal of Petroleum Geology*, **1**, 65–77.

JOHNSON, A. & EYSSAUTIER, M. 1987. Alwyn North Field and its regional geological context, in *Petroleum Geology of North West Europe*, BROOKS, J. & GLENNIE, K. W. (eds), Graham and Trotman, London, 963–977.

JOHNSON, H. D. & STEWART, D. J. 1985. Role of clastic sedimentology in the exploration and production of oil and gas in the North Sea. *In*: BRENCHLEY, P. J. & WILLIAMS, B. P. J. (eds) *Sedimentology: Recent Developments and applied Aspects*, Geological Society, London, Special Publication **18**, 249–310.

LERVIK, K. S., SPENCER, A. M. & WARRINGTON, G. 1990. Outline of Triassic stratigraphy and structure in the central and northern North Sea. *In*: COLLINSON, J. D. (ed), *Correlation in Petroleum Exploration*, Norwegian Petroleum Society, Graham and Trotman, London.

MCQUILLIN, R., DONATO, J. A. & TULSTRUP, J. 1982. Development of basins in the Inner Moray Firth and the North Sea by crustal extension and dextral displacement of the Great Glen Fault, *Earth and Planetary Science Letters*, **60**, 127–139.

MORTON, N., SMITH, R. M., GOLDEN, M. & JAMES, A. V. 1987. Comparative stratigraphic study of Triassic-Jurassic sedimentation and basin evolution in the northern North Sea and north-west of the British Isles, *In* BROOKS, J. & GLENNIE, K. W. (eds), *Petroleum Geology of North West Europe*, Graham & Trotman, London, 697–709.

NEDERLANDSE AARDOLIE MAATSCHAPPIJ B. V. and RIJKS GEOLOGISCHE DIENST, 1980. Stratigraphic nomenclature of the Netherlands, *Verhandlingen van het Koninklijk Nederlands Geologisch Mijnbouwkundich Genootschap*, Deel 32.

NYSTUEN, J. P., KNARUD, R., JORDE, K. & STANLEY, K. O. 1990. Correlation of Triassic to lower Jurassic sequences, Snorre Field and adjacent areas, northern North Sea. *In* COLLINSON, J. D. (ed), *Correlation in Hydrocarbon Exploration*, Norwegian Petroleum Society, Graham and Trotman, London.

OLSEN, J. C. 1983. The structural outline of the Horn Graben, *Geologie en Mijnbouw*, **62**, 47–50.

OLSEN, R. C. & STRASS, I. F. 1982. The Norwegian-Danish Basin, *Norwegian Petroleum Directorate*, Paper No. 31.

POSAMENTIER, H. & VAIL, P. 1988. Eustatic controls on clastic deposition. *In* WILGUS, C. (ed.), *Sea-Level Changes – An integrated Approach*, Society of Economic Paleontologists and Mineralogists Special Publication **42**.

ROSENDAHL, B. R., REYNOLDS, D. J., LORBER, P. M., BURGESS, C. F., MCGILL, J., SCOTT, D., LAMBIASE, J. J. & DERKSEN, S. J. 1986. Structural expressions of rifting: lessons from Lake Tanganyika, Africa, *In* FROSTICK, L. *et al.* (eds) *Sedimentation in the Africa Rifts*, Geological Society, London, Speical Publication **25**, 29–43.

RØE, S–L. & STEEL, R. 1985. Sedimentation, sea-level rise and tectonics at the Triassic-Jurassic boundary (Statfjord Formation), Tampen Spur, northern North Sea, *Journal of Petroleum Geology*, **8**, 163–186.

SKJERVEN, J., RIIS, F., & KVALHEIM, J. E. 1983. Late Paleozoic to early Cenozoic structural development of the south-southeastern Norwegian North Sea, *Geologie en Mijnbouw*, **62**, 35–45.

SMYTHE, D. K., ROBINSON, A., MCQUILLIN, R., BREWER, J. A., MATTHEWS, D. H., BLUNDELL, D. J. & KELK, B. 1982. Deep structure of the Scottish Caledonides revealed by the MOIST reflection profile, *Nature*, **299**, 338–340.

STEEL, R. 1974. Cornstone (fossil caliche): its origin, stratigraphic and sedimentological importance in the New Red Sandstone, W Scotland, *Journal of Geology*, **82**, 351–369.

—— 1977. Triassic rift basins of north-west Scotland – their configuration, infilling and development, in FINSTAD, K. G. & SELLEY, R. C. (eds),

Proceedings Mesozoic Northern North Sea Symposium, Norwegian Petroleum Society, Oslo.

SWALLOW, J. L. 1986. The seismic expression of a low angle detatchment (sole fault) from the Beryl Embayment, central Viking Graben, *Scottish Journal of Geology*, **22**, 315–324.

TUNBRIDGE, I. P. 1981. Sandy high-energy flood sedimentation – some criteria for recognition, with an example from the Devonian of S. W. England, *Sedimentary Geology*, **28**, 79–95.

VOLLSET, J. & DORÉ, A. M. 1984. A revised Triassic and Jurassic lithostratigraphic nomenclature for the Norwegian North Sea, *Bulletin Norwegian Petroleum Directorate*, No. 3.

WOOD, R. & BARTON, P. 1983. Crustal thinning and subsidence in the North Sea, *Nature*, **304**, 561.

ZIEGLER, P. A. 1982. *Geological Atlas of Western and Central Europe*, Shell Internationale Petroleum Maatschappij BV., the Hague.

A review of Jurassic chronostratigraphy and age indicators for the UK

B. M. COX

Biostratigraphy Research Group, British Geological Survey, Keyworth, Nottingham, NG12 5GG, UK

Abstract: The basis of the chronostratigraphic subdivision of the Jurassic System is the sequence of ammonite faunas. At present, the 11 Jurassic stages (representing approximately 70 million years) can be divided into 145 ammonite-based zones or subzones. Methods of subdivision and correlation by various microfossil groups have also been developed for practical and economic reasons. Dating by fossils (biostratigraphy) underpins the broader-based approachs of event and sequence stratigraphy. It also supplies the primary stratigraphic control in the development of a standard magnetic polarity timescale (magnetostrâtigraphy) of global applicability. At present, the Jurassic System of the UK area is divided into 38 units (biozones and subzones) on the basis of dinoflagellate cysts and 23 on the basis of calcareous nannofossils. Coverage is less comprehensive using the benthonic groups (foraminifera and ostracoda) which are more affected by environmental controls. The Lower Jurassic is divided into 16 units on the basis of foraminifera, but higher in the System only local divisions can be recognised in the Bathonian and informal divisions in the Callovian and oldest Oxfordian. There is fuller but still incomplete coverage using ostracoda which have proved particularly useful in the non-marine, brackish and marginal marine sequences of the Bathonian and Portlandian. The recovery of radiolaria from Jurassic sediments in the UK area is a new and exciting development which allows division of Kimmeridgian and Portlandian strata in graben areas in the North Sea Basin.

The red-bed depositional environments which prevailed in the UK area during the Triassic were superseded by deposition in shallow shelf seas introduced by a regional marine transgression towards the end of that period. The climate was warm and equable, and the seas supported an abundant and varied fauna and flora (Hallam 1982; Sellwood 1978). This was the Jurassic Period which lasted for approximately 70 million years, from *c.* 205 to *c.* 135 million years ago. The marine deposits which accumulated at this time include sand, clay and limestone lithologies. In addition, nearshore facies (marginal marine, brackish and estuarine) also occur in the UK area. The strata deposited during the Jurassic Period constitute the Jurassic System, on which there is a very extensive literature; for the UK area, Arkell (1933) remains a classic.

Many workers involved with the Jurassic System are concerned with the relative age of deposits within it. This is the subject and application of chronostratigraphy — the classification of rock sequences into units on the basis of their age or time of origin. Absolute radiometric dating is the subject of geochronology and will not be considered in this paper except to say that, in the Jurassic, further advances and greater resolution will depend largely on the future recognition of petrologically suitable volcanic flows, bentonites, tuffs and glauconites within biostratigraphically controlled marine sequences (Odin 1984). Some recent datings for the Jurassic System are shown in Fig. 1.

The chronostratigraphic standard

The Jurassic strata of onshore Britain have played a key role in studies of the Jurassic System since they typify the fine degree of stratigraphic resolution that is possible (Arkell 1956). The basic chronostratigraphic subdivision of the Jurassic System is into 11 stages which are the 'lowest common denominator' for worldwide correlation (Fig. 1) and are the common language for all who work on Jurassic strata. The stages were conceived by d'Orbigny (1850) who gave for each one a selection of representative stratal units, localities and characteristic fossils. Only four changes have been made to d'Orbigny's original scheme and, apart from periodic controversy over the inclusion of the Aalenian (Sapunov 1964), the present set of stage names has been in use for 125 years (Arkell 1946). The Tithonian and Volgian stages — as alternatives for the terminal Jurassic stage — date from Oppel (1865) and Nikitin (1881) respectively. The base of each Stage is now defined at a particular point in a recommended

From HARDMAN, R. F. P. & BROOKS, J. (eds), 1990, *Tectonic Events Responsible for Britain's Oil and Gas Reserves*, Geological Society Special Publication No 55, pp 169–190.

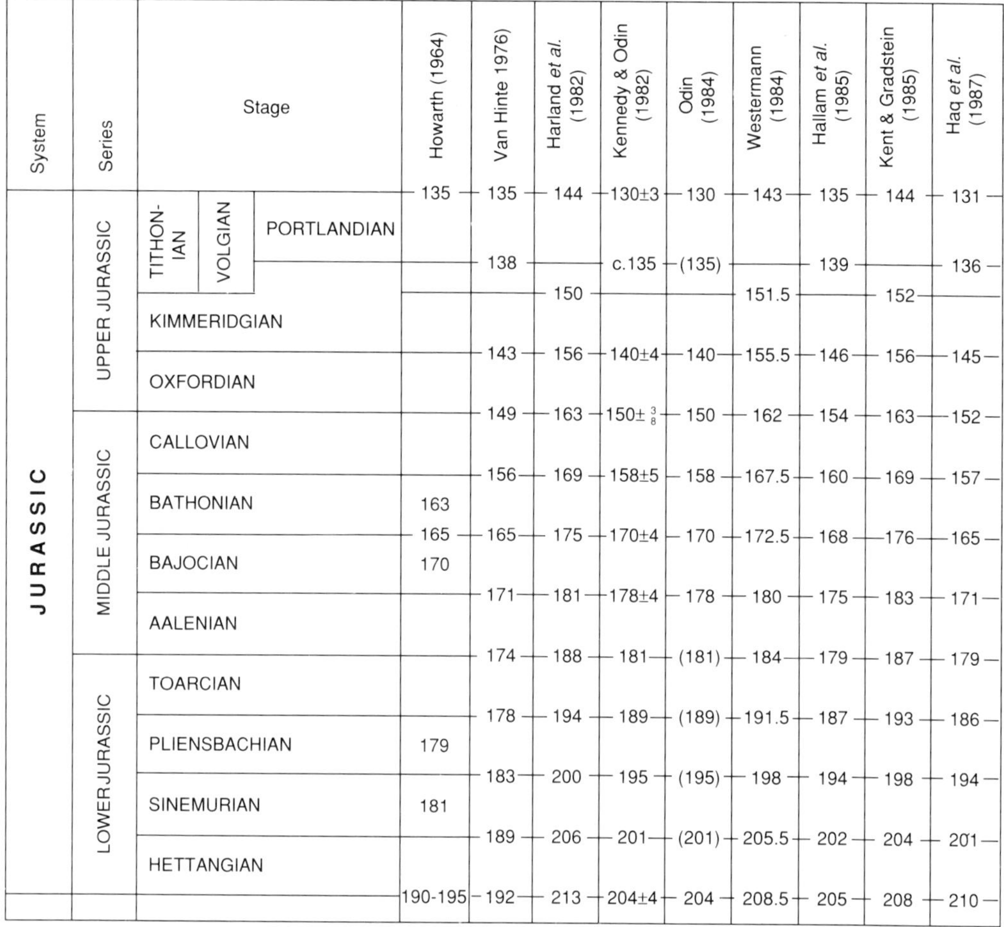

System	Series	Stage	Howarth (1964)	Van Hinte 1976)	Harland *et al.* (1982)	Kennedy & Odin (1982)	Odin (1984)	Westermann (1984)	Hallam *et al.* (1985)	Kent & Gradstein (1985)	Haq *et al.* (1987)
			135	135	144	130±3	130	143	135	144	131
JURASSIC	UPPER JURASSIC	TITHONIAN / VOLGIAN / PORTLANDIAN									
				138		c.135	(135)		139		136
					150			151.5		152	
		KIMMERIDGIAN									
				143	156	140±4	140	155.5	146	156	145
		OXFORDIAN									
				149	163	$150\pm^{3}_{8}$	150	162	154	163	152
	MIDDLE JURASSIC	CALLOVIAN									
				156	169	158±5	158	167.5	160	169	157
		BATHONIAN	163								
			165	165	175	170±4	170	172.5	168	176	165
		BAJOCIAN	170								
				171	181	178±4	178	180	175	183	171
		AALENIAN									
				174	188	181	(181)	184	179	187	179
	LOWER JURASSIC	TOARCIAN									
				178	194	189	(189)	191.5	187	193	186
		PLIENSBACHIAN	179								
				183	200	195	(195)	198	194	198	194
		SINEMURIAN	181								
				189	206	201	(201)	205.5	202	204	201
		HETTANGIAN									
			190-195	192	213	204±4	204	208.5	205	208	210

Fig. 1. Geochronology of the Jurassic System (figures indicate millions of years BP).

type section (the boundary stratotype). For many Jurassic stages, these type sections have been in England (Sinemurian, Toarcian, Bajocian, Callovian, Oxfordian, Kimmeridgian and Portlandian, with possible alternatives for the Hettangian and Pliensbachian) (Fig. 2). The intermediate groupings of the stages into series (Lower, Middle and Upper Jurassic) are shown in Fig. 1.

Ammonite-based zones

As well as the boundary stratotypes, the modern definition of each Stage is given in terms of the constituent (in particular basal) ammonite zones (George *et al.* 1969). The virtues of ammonites as stratigraphic 'guide fossils' and the status of ammonite zones has been discussed at length (viz. Kennedy & Cobban 1977; Torrens 1980b; Callomon 1984a, b) but, in order to demonstrate the nature of the primary standard scale or yardstick used in the Jurassic, a few basic facts concerning ammonite zones and so-called biostratigraphy are reiterated here. Biostratigraphic classification is the organisation of strata into units based on their fossil content. It is the most widely used and practical method of correlating rock sequences and is the usual means of getting from a purely lithology-based scheme (lithostratigraphy) to one having time significance (chronostratigraphy). The merits and philosophy of the latter are discussed by Holland (1986). Biozones are units of rock characterized by a particular fossil or group of fossils and whose boundaries are determined by the known lateral and vertical range of fossil taxa. They are, in concept, quite independent of lithology. They are based to a great extent on evolutionary

Stage (basal Zone)	**Recommended basal boundary stratotype**
PORTLANDIAN (Albani)	Houns-tout Cliff, Dorset: base of the Massive Bed (Arkell 1947)
KIMMERIDGIAN (Baylei)	Ringstead Bay, Dorset: base of the Inconstans Bed (Arkell 1947)
OXFORDIAN (Mariae)	Cornelian Bay, North Yorkshire: base of the Oxford Clay (Wright 1969)
CALLOVIAN (Herveyi)	Swan Inn Quarry, Long Handborough, Oxfordshire: base of Bed 4 of Douglas & Arkell (1928) (Page 1989) [inferred substitution for Sutton Bingham reservoir, near Yeovil, Somerset: base of Bed 4 of Arkell (1954)]*
BATHONIAN (Zigzag)	Bas Auran, Basses-Alps, France:base of Bed 23 of Sturani (1967)
BAJOCIAN (Discites)	Bradford Abbas railway cutting, Dorset: 0.2 m below the top of the Bradford Abbas [formerly Fossil] Bed (Parsons 1980)
AALENIAN (Opalinum)	a section in Swabia or Lower Saxony, Germany (pending)
TOARCIAN (Tenuicostatum)	Kettleness, Runswick Bay, NorthYorkshire: base of Bed 26 of Howarth (1955)
PLIENSBACHIAN (Jamesoni)	Charmouth, Dorset: base of Bed 105 of Lang *in* Lang et al. (1928) *OR* at Pliensbach, Württemberg, Germany (Geyer 1964)
SINEMURIAN (Bucklandi)	Seven Rock Point or Devonshire Head, Lyme Regis, Dorset: base of Bed 19 of Lang (1924)
HETTANGIAN (Planorbis)	Watchet, Somerset (pending)

Fig. 2. Recommended basal boundary stratotypes for the Jurassic stages (see Morton 1974). * To be revised.

changes in animals and plants and therefore they are strongly indicative of geological age, *but* the boundaries of a biozone are not necessarily time-planes and the relative stratigraphic ranges of biozones vary from place to place. In the Jurassic, however, the evolutionary rates of the ammonites were relatively so fast that it is widely believed that there is no practical difference between an ammonite-based biostratigraphy (zonation) and chronostratigraphy (Holland *et al.* 1978; Torrens 1980a). A scale based on ammonite biozones has therefore become the primary chronostratigraphic standard for the Jurassic and the standard ammonite-based zones are effectively chronozones. Their definition does not depend on fossil taxa, but their recognition certainly will (Callomon 1984a). The rapidity of ammonite evolution is the most important reason for their pre-eminence over other fossils in correlation.

Some palaeontologists and stratigraphers find difficulty in using ammonite-based zones as anything other than pure biozones. Nevertheless, it is not uncommon in practice to assign to a zone strata which contain no ammonite fauna or to identify a zone on the basis of characters other than the ammonite assemblage or index species – both of these practices do not follow strict biostratigraphic methodology. In ammonitiferous strata, there appears to be no dilemma when assigning a particular stratum to a particular zone (whether it be considered biozone or chronozone) but rather when pinpointing the zone boundaries. Purists would position the zone boundary at the lowest occurrence of a particular ammonite taxon. However, in continuously cored boreholes (and subsequently other types of section), the most sensible practice is to position boundaries between individual zones at some lithological change or marker (an 'event') at or close to the change in ammonite faunas that would mark the limit of a pure biozone. This practice seems entirely reasonable (and is also workable) because the occurrence of an ammonite (relatively large diameter) at a particular level in a core (relatively small diameter) is a chance phenomenon, dependent upon where the ammonite lay relative to the position of the drill bit. The lithological marker or 'event' is a more repeatable and correlatable feature (? a time plane) than a chance ammonite occurrence and would also be manifest on a geophysical log.

At present, there are 75 ammonite-based zones in use for the Jurassic of the UK area. The recommendation (see Holland *et al.* 1978; Callomon 1984a) that they should be labelled according to the ammonite species on which they are based and that this should be written in Roman fount with an initial capital (i.e. not italicized) is here strongly supported. In the Geological Society's Jurassic Correlation Chart (Cope *et al.* 1980a, b), the usage of ammonite zones as 'Standard Zones' was recognized, but the opportunity to abandon the use of italicized fossil names in the zonal notation was not taken.

Smaller divisions within the zones (subzones and horizons) are also recognisable. At present, 45 of the 75 ammonite-based zones are subdivided into a total of 115 subzones, giving a total of 145 constituent ammonite-based units for the Jurassic System. Furthermore, the standard 'Beds' described for the Oxfordian and Kimmeridgian strata of eastern England, but known to be recognizable further afield, are also small-scale chronostratigraphic units; these are recognised on the basis of combined lithological and macrofaunal characters (Gallois & Cox 1976, 1977; Cox & Gallois 1979). The 'standard units' established by Penn, Merriman & Wyatt (1979) for part of the Bathonian sequence of the Bath–Frome area, are recognized on a similar basis.

Figures 3–6 give reference sections and basal boundary stratotypes (where designated) for the Jurassic ammonite-based zones; these partly satisfy the requirement that formal chronostratigraphic units (in this case chronozones) should each have a designated type section (unit or boundary stratotypes), but also show the relative 'state of play' within the Jurassic System. Most chronostratigraphic units are defined by a basal boundary stratotype rather than a unit stratotype. This practice is deemed preferable because it allows a total stratigraphic scheme without gaps or overlaps but which is also sufficiently flexible to incorporate subsequent refinements whilst retaining stability (Callomon & Donovan 1974). Where the basal stratotypes of two succeeding zones or subzones appear in one continuous section, there is automatically, of course, a unit stratotype. Some Jurassic ammonite zones have had no formal statement of a stratotype of either sort but for many of these there are one or more obvious candidates which can be used, in the meantime, as reference sections. Others have a formally defined basal stratotype or, if this is not possible, a type area where a stratotype will eventually be established. For the Callovian and Oxfordian, the zones are defined in terms of the basal boundary stratotypes of their constituent subzones. Indeed, this is recognized as the ideal situation whereby a chronostratigraphic unit is defined in terms of the units of next lower rank that it contains, and only the lowest ranking units (subzones in this case) are defined typologically by time planes in sections (Callomon 1984a). For a handful of Middle Jurassic zones, not even a type area can be stated. There is no guarantee of permanence of availability or access for a type or reference section, particularly where these are in pits or quarries; some of those given in Figs 3–6 are already lost to view but they are well documented and generally well represented by collected material.

Fine-tuning of the ammonite-based standard is constantly taking place and, at Stage level and below, there are a number of more fundamental problems still to be overcome in connection with ammonite provincialism and adverse facies. Some decisions await ratification or further work and discussion by specialist groups. Nevertheless, for practical stratigraphic purposes, the present scheme of stages, ammonite-based zones and subzones provides the most usable key for unravelling the intricacies of Jurassic rock sequences and correlations.

Fig. 3. Reference sections for the Lower Jurassic ammonite-based zones. For zonation, see Dean, Donovan & Howarth (1961). Bed numbers follow [1]Lang (1924), [2]Palmer (1972), [3]Lang *in* Lang *et al.* (1923), [4]Lang *in* Lang *et al.* (1926), [5]Bairstow (1969) (see also Powell 1984), [6]Lang *in* Lang *et al.* (1928), [7]Lang (1936), [8]Howarth (1955) (see also Powell 1984), [9]Howarth (1973) (see also Powell 1984), [10]Howarth (1962) (see also Powell 1984), [11]Dean (1954) (see also Knox 1984).

Stage	Zone (subzones)		Reference Section(s)
TOARCIAN	Levesquei	(4)	Cliff and foreshore exposures at Blea Wyke Point, North Yorkshire: Blea Wyke Sandstone Formation and Whitby Mudstone Formation, Fox Cliff Siltstone Member (beds 63–82)[11]
	Thouarsense	(2)	Cliff and foreshore exposures between Ravenscar (Peak Fault) and Blea Wyke, North Yorkshire: Whitby Mudstone Formation, Fox Cliff Siltstone and Peak Mudstone members (beds 53–62)[11]
	Variabilis	(0)	Cliff and foreshore exposures between Ravenscar (Peak Fault) and Blea Wyke, North Yorkshire: Whitby Mudstone Formation, Peak Mudstone and Alum Shale members (beds 37–52)[11]
	Bifrons	(3)	Foreshore exposure between Whitby and Saltwick Nab: Whitby Mudstone Formation, Alum Shale Member (beds 49–72)[10] *or* between Ravenscar (Peak Fault) and Blea Wyke: Alum Shale Member (beds xvi–lvi)[10] (both North Yorkshire)
	Falciferum	(2)	Cliff and foreshore exposures south of Port Mulgrave (Rosedale Wyke to Lingrow Knock): Whitby Mudstone Formation, Jet Rock Member (beds 33–40)[10] *plus* south of Whitby (Saltwick Nab and Saltwick Bay): Jet Rock Member (beds 41–48)[10] (both North Yorkshire)
	Tenuicostatum	(4)	Cliff and foreshore exposures south of Staithes, NorthYorkshire; Brackenberry Wyke to Rosedale Wyke: Cleveland Ironstone Formation (beds 58–60)[8] and Whitby Mudstone Formation, Grey Shale Member (beds 1–32)[9] *or* Runswick Bay and Kettleness: Cleveland Ironstone (beds 26–28)[8] and Whitby Mudstone, Grey Shale Member (beds 1–32)[9]
PLIENSBACHIAN	Spinatum	(2)	Cliff and foreshore exposures near Staithes, North Yorkshire: Cleveland Ironstone Formation (beds 42–57)[8]
	Margaritatus	(3)	Foreshore exposures near Staithes, North Yorkshire: Staithes Sandstone Formation and Cleveland Ironstone Formation (beds 12–41)[8]
	Davoei	(3)	Robin Hood's Bay, North Yorkshire: Redcar Mudstone Formation, Ironstone/Pyritous Shales and Staithes Sandstone Formation (beds 581–600(v))[5] *or* Cliff section, east of Charmouth, Dorset: Green Ammonite Beds (beds 122–130)[7]
	Ibex	(3)	Robin Hood's Bay, North Yorkshire: Redcar Mudstone Formation, Ironstone/Pyritous Shales (beds 562–580)[5] *or* Cliff section, east of Charmouth, Dorset: Belemnite Marls (beds 118c–121)[6]
	Jamesoni	(4)	Robin Hood's Bay, North Yorkshire: Redcar Mudstone Formation, Ironstone/Pyritous Shales (beds 501.3(pars)–561)[5] *or* Cliff section, east of Charmouth, Dorset: Belemnite Marls (beds 105–118b)[6]
SINEMURIAN	Raricostatum	(4)	Robin Hood's Bay, North Yorkshire: Redcar Mudstone Formation, Siliceous Shales and Ironstone/ Pyritous Shales (beds 485–500)[5]
	Oxynotum	(2)	Robin Hood's Bay, North Yorkshire: Redcar Mudstone Formation, Siliceous Shales (beds 466–484)[5]
	Obtusum	(3)	Robin Hood's Bay, North Yorkshire: Redcar Mudstone Formation, Calcareous Shales and Siliceous Shales (beds 446.3–465)[5]
	Turneri	(2)	Robin Hood's Bay, North Yorkshire: Redcar Mudstone Formation, Calcareous Shales (beds 428–446.2)[5] *or* Cliff section, west of Lyme Regis, Dorset: Shales with Beef and Black Ven Marls (beds 73–83e)[3,4]
	Semicostatum	(3)	Cliff section, west of Lyme Regis, Dorset: Blue Lias and Shales with Beef (beds 47–72)[1,3]
	Bucklandi	(3)	Cliff section, west of Lyme Regis, Dorset: Blue Lias (beds 19–46)[1]
HETTANGIAN	Angulata	(2)	Cliff section, west of Lyme Regis, Dorset: Blue Lias (beds H84–18)[1] *or* St Audrie's Cliff, east of Watchet, Somerset: Blue Lias (beds C1–C100)[2]
	Liasicus	(2)	Cliff section, west of Lyme Regis, Dorset: Blue Lias (beds H57–H83)[1] *or* St Audrie's Cliff, east of Watchet, Somerset: St Audrie's Shales (beds B7–B15)[2]
	Planorbis	(2)	Cliff section, west of Lyme Regis, Dorset: Blue Lias (beds H25–H56)[1] *or* St Audrie's Cliff, east of Watchet, Somerset: Aldergrove Beds and St Audrie's Shales (beds A20–B6)[2]

Stage	Zone (subzones)		Reference Section(s)
BATHONIAN	Discus	(2)	Neue Tongrube Temme, near Hildesheim, NW Germany: levels *c.*11 m* to – 7.2 m (Westermann 1958: Lutze 1960)
	Orbis	(2)	Sengenthal/Opf., Franconian Alb, S Germany: Orbis Oolite (Dietl & Callomon 1988)
	Hodsoni	(0)	none designated; possibly in the Ain district of the southern Jura or Swabia, S Germany (Mangold 1984)
	Morrisi	(0)	Bruton railway cutting (south side), Somerset: Fullers Earth Rock (beds 2*–4) (Torrens 1974)
	Subcontractus	(0)	Troll Quarry, Thornford, Dorset: Fullers Earth Rock (beds 2*–21; top not seen) (Torrens 1974)
	Progracilis	(0)	none designated; possibly in Vendée, W France or northern Swabia, S Germany (Mangold 1984)
	Tenuiplicatus	(0)	none designated
	Zigzag	(3)	Bas Auran section, near Barrême, Basses-Alps, SE France: beds 23*–2 (Sturani 1967, emend. Torrens 1987)
BAJOCIAN	Parkinsoni	(2)	none designated
	Garantiana	(4)	none designated
	Subfurcatum	(3)	Département de la Moselle, NE France: Marnes à Longwy (pars) (Parsons 1976)
	Humphriesianum	(3)	Swabia, S Germany: beds between the Blaukalke and Subfurcatum Oolithe (Parsons 1976)
	Sauzei	(0)	Swabia, S Germany: Blaukalke (Parsons 1974)
	Laeviuscula	(2)	none designated
	Discites	(0)	Bradford Abbas railway cutting, Dorset: Bradford Abbas [formerly Fossil] Bed, upper part (Parsons 1980)
AALENIAN	Concavum	(0)	North Dorset: horizon not designated
	Murchisonae	(3)	none designated
	Opalinum	(2)	Swabia or Lower Saxony, Germany: horizon not designated (Morton 1974)

Fig. 4. Recommended type areas and sections for the Middle Jurassic ammonite-based zones. *base of bed = basal boundary stratotype for zone (see Torrens 1974; Mangold 1984).

Jurassic chronostratigraphy in practice

For obvious practical reasons, the close proximity of the North Sea Basin to the classic onshore areas does not make it any easier to apply the land-based stratigraphic standards in the North Sea Jurassic, although that is the endeavour. The primary chronostratigraphic scale cannot be determined directly in sequences of rock-cuttings: no core, no ammonite. Any ammonite recovered from an occasional cored interval is of great interest and considerable importance, particularly if several specimens are recovered over a run of core. The half and sliced cores from commercial wells that are deposited in the Department of Energy's Core Store at Edinburgh (managed by the British Geological Survey (BGS)) are destined to be preserved for posterity. Finding an ammonite is therefore entirely fortuitous and depends on a chance sighting on a broken core bedding surface. Consequently, most ammonite finds will depend on oil companies' vigilance, and the primary tool for the recognition of the chronostratigraphic units must necessarily be microfossils. For the Jurassic of the UK area, foraminifera, ostracoda, nannofossils (coccoliths) and palynomorphs (dinoflagellate cysts, acri-

Stage	Zone (subzones)		Reference Section(s)
OXFORDIAN	Rosenkrantzi	(0)	Staffin Bay, Isle of Skye: Staffin Shale Formation, Flodigarry Shale Member, Bed 35
	Regulare	(0)	Staffin Bay, Isle of Skye: Staffin Shale Formation, Flodigarry Shale Member, Bed 33 (pars) (7m below top*) – base Bed 35
	Serratum	(2)	Staffin Bay, Isle of Skye: Staffin Shale Formation, Flodigarry Shale Member, Bed 31 (pars) (4.5 m above base*) – Bed 33 (pars) (7m below top)
	Glosense	(2)	Staffin Bay, Isle of Skye: Staffin Shale Formation, Flodigarry Shale Member, Bed 31 (pars) (0.6m above base* – 4.5m above base)
	Tenuiserratum	(2)	Staffin Bay, Isle of Skye: Staffin Shale Formation, Glashvin Silt Member, Digg Siltstone Member and Flodigarry Shale Member, base Bed 22* – Bed 31 (pars) (0.6m above base)
	Densiplicatum	(2)	Staffin Bay, Isle of Skye: Staffin Shale Formation, Glashvin Siltstone Member, Bed 21 (pars) (5.4m above base* – base Bed 22)
	Cordatum	(3)	Staffin Bay, Isle of Skye: Staffin Shale Formation, Dunans Clay Member and Glashvin Silt Member, Bed 12 (pars) (0.8m above base) – Bed 21 (pars) (5.4m above base)
	Mariae	(2)	Cliff and foreshore exposures south of Scarborough, North Yorkshire (Cornelian Bay, Osgodby Nab, Cayton Bay, Red Cliff, Cunstone Nab, Gristhorpe Cliffs): Oxford Clay* and basal Lower Calcareous Grit (Wright 1983)
CALLOVIAN	Lamberti	(2)	Woodham brickpit, Buckinghamshire: Middle Oxford Clay (beds D2* and C) (Arkell 1939; Callomon 1968)
	Athleta	(3)	London brick Co.'s pit, Calvert, Buckinghamshire: Lower – Middle Oxford Clay (beds 10* – 13b) *plus* Woodham brickpit, Buckinghamshire: Middle Oxford Clay (beds E – D1) (Arkell 1939; Callomon 1968)
	Coronatum	(2)	London brick Co.'s pits, Peterborough, Cambridgeshire: Lower Oxford Clay (beds 14*–22c) (Callomon 1968)
	Jason	(2)	Kidlington, Oxfordshire: Lower Oxford Clay (beds 9*–26) (Callomon 1955) *or* London Brick Co.'s pits, Peterborough, Cambridgeshire: Lower Oxford Clay (beds 5–13) (Callomon 1968)
	Calloviense	(2)	Banks of the River Avon, south of Kellaways, Wiltshire: Kellaways Formation, Kellaways Sand Member* *plus* South Newbald Quarry, Humberside: topmost Kellaways Sand Member (above base of lower "oyster" band) and Cave Rock Member (Page 1989)
	Koenigi	(3)	Chippenham, Wiltshire: Kellaways Formation, Kellaways Clay Member* (Tytherton No. 3 Borehole, 6.95 m–22.40 m; Cave & Cox 1975) (Page 1989)
	Herveyi	(3)	Swan Inn Quarry, Long Handborough, Oxfordshire: Abbotsbury Cornbrash Formation (beds 4*–5) (Douglas & Arkell 1928) *plus* Cayton Bay, near Scarborough, North Yorkshire: Abbotsbury Cornbrash Formation (beds $\alpha_1 - \alpha_3$) and Cayton Clay "Formation" (Wright 1977; Page 1989) [minor non-sequences present]

Fig. 5. Reference sections for the Callovian and Oxfordian ammonite-based zones. For Callovian, see Callomon (1964), Callomon & Sykes *in* Duff (1980) and Page (1989). For Oxfordian and Staffin Shale bed numbers, see Sykes & Callomon (1979) emend. Birkelund & Callomon (1985, pp. 16–17) and Callomon (1964); where discrepancies occur between Sykes and Callomon's text, figures and Appendix, thicknesses and bed numbers follow their Appendix. *base of bed = basal boundary stratotype for zone.

tarchs and miospores) have a proven biostratigraphic capability; radiolaria are also now making some impact. The dating of Jurassic rock sequences using these various fossil groups underpins so-called event stratigraphy and sequence stratigraphy which are prevalent approaches to chronostratigraphy in the oil industry today (Vail & Todd 1981; Vail *et al.* 1984; Haq *et al.* 1987). The recognition of widespread 'unconformities and correlative con-

Stage	Zone (subzones)	Reference Section(s)
PORTLANDIAN	Lamplughi (0)	Nettleton Top Barn Sandpit, Lincolnshire: Lower Spilsby Sandstone, Bed 6 (Casey 1973)
	Preplicomphalus (0)	Nettleton Top Barn Sandpit, Lincolnshire: Lower Spilsby Sandstone, beds 4–5 (Casey 1973)
	Primitivus (0)	Nettleton Top Barn Sandpit, Lincolnshire: Lower Spilsby Sandstone, beds 2–3 (Casey 1973)
	Oppressus (0)	Winspit, 'Isle' of Purbeck, Dorset: Portland Stone [Formation], Shrimp Bed (Bed V), upper part and Purbeck 'Beds' (?pars) *or* West Dereham Flood Relief Channel, Norfolk: Sandringham Sands Formation, Roxham Beds (beds 1–5) (Casey 1973; Casey & Gallois 1973)
	Anguiformis (0)	Freshwater Bay (south side), Isle of Portland, Dorset: Portland Stone [Formation], Cherty 'Series' (pars) (top 3m*) and Freestone 'Series' (?pars)
	Kerberus (0)	Freshwater Bay (north side), Isle of Portland, Dorset: Portland Stone [Formation], Basal Shell Bed* and Cherty 'Series' (pars)
	Okusensiš (0)	Old quarry, Coate Water, Swindon, Wiltshire: Portland Stone [Formation], "Cockly Bed" (beds 4*–7; Wimbledon 1976) *or* Emmetts Hill, 'Isle' of Purbeck, Dorset: Portland Sand [Formation], Black Sandstones (pars) and Portland Stone, Cherty 'Series' (pars) (beds A28 & B–H)
	Glaucolithus (0)	Houns-tout Cliff (and Emmetts Hill), 'Isle' of Purbeck, Dorset: Portland Sand [Formation], White Cementstone*, St Alban's Head Marls, Parallel Bands and Black Sandstones (pars) (beds 13–A27)
	Albani (0)	Houns-tout Cliff, 'Isle' of Purbeck, Dorset: Portland Sand [Formation], Massive Bed* and Emmit [sic] Hill Marls (beds 1–12)
KIMMERIDGIAN	Fittoni (0)	Cliff and foreshore exposures between Hobarrow Bay and Chapman's Pool, Dorset: Kimmeridge Clay Formation, Rhynchonella and Lingula Beds (undifferentiated; pars), Hounstout Clay and Hounstout Marl (Cope 1978)
	Rotunda (0)	Kimmeridge Clay Formation, beds 2–1c** and Rhynchonella and Lingula Beds (undifferentiated; pars) (Cope 1978)
	Pallasioides (0)	Kimmeridge Clay Formation, beds 4–3**
	Pectinatus (2)	Kimmeridge Clay Formation, beds 20 (White Stone Band) – 5**
	Hudlestoni (3)	Kimmeridge Clay Formation, top of Blackstone to base of White Stone Band (beds 26 pars – 21**)
	Wheatleyensis (2)	Kimmeridge Clay Formation, *c.*3m above Cattle Ledge Stone Band to top of Blackstone (beds 33 pars – 26 pars**)
	Scitulus (0)	Kimmeridge Clay Formation, base of Yellow Ledge Stone Band to *c.*3m above Cattle Ledge Stone Band (beds 36–33 pars**)
	Elegans (0)	Kimmeridge Clay Formation, beds 42–37** (base of Yellow Ledge Stone Band)
	Autissiodorensis (0)	Kimmeridge Clay Formation, top of Flats Stone Band to base Bed 42**
	Eudoxus (0)	upper part: Kimmeridge Clay Formation, beach level (Hobarrow Bay Stone Band) to top of Flats Stone Band lower part: Blue Circle Portland Cement Co.'s pit, Westbury, Wiltshire: Kimmeridge Clay Formation, beds E1–E7 (Birkelund *et al.* 1983) [sections do not quite overlap]
	Mutabilis (0)	Blue Circle Portland Cement Co.'s pit, Westbury, Wiltshire: Kimmeridge Clay Formation, beds M1–M21 (Birkelund *et al.* 1983)
	Cymodoce (0)	Low cliff sections west of Osmington Mills [SY 7342 8174] or at Black Head [SY 7239 8195], Dorset: Kimmeridge Clay Formation, Wyke Siltstone and overlying *c.*5m of strata (Cox & Gallois 1981; Birkelund *et al.* 1983)
	Baylei (0)	Cliff and foreshore exposures, Ringstead Bay and Wyke Regis, near Weymouth, Dorset: Kimmeridge Clay Formation, Inconstans Bed* to base Wyke Siltstone

formites' allows a relatively broad breakdown of rock sequences which is of practical value, particularly offshore. However, the timing of these events depends ultimately on dating by fossils which remain the primary age indicators. Fossils are playing a similar role in the development of a Jurassic magnetostratigraphy (see Conclusions).

Palynology

When the Jurassic of the North Sea was first explored in the late 1960s and early 1970s, it became apparent that calcareous microfossils were very difficult to recover, if preserved at all. Organic-walled microfossils (palynomorphs) were found to be more readily recoverable and much effort was devoted to palynology. Dinoflagellate cysts (dinocysts) became, and remain, the key fossils. They are the resistant resting cysts of dinoflagellates, a group of unicellular marine phytoplankton, and are from 25 to 250 μm in overall diameter. Extraction from rock is by mechanical and chemical means; after mounting on slides (see for example Sarjeant 1974, Appendix A), they are examined using optical microscopes.

In Jurassic sequences, dinocysts are potentially good stratigraphic indicators for they satisfy most of the requirements of a good 'guide fossil'. They have a wide geographical distribution, are common and generally easily identified, and occur in all marine facies; preservation is favoured in argillaceous sediments. However, compared to the ammonites, their relatively slow evolutionary rates gave rise to relatively long stratigraphic ranges and it is therefore unlikely that they will ever afford the same fine degree of stratigraphic resolution as the ammonites. Also, the detailed study of the stratigraphic distribution and systematics of the dinocysts is still at a relatively early stage and much work is done in commercial companies where confidentially prohibits release or publication of results.

There are a number of published dinocyst zonations (divisions of the rock column into units based on dinocysts) for the Jurassic. These are either well documented but cover only part of the sequence or are all embracing but not sufficiently well documented to be generally usable; others are based on sections with inadequate stratigraphic control or on a compilation of published data worldwide and hence are not completely meaningful for any given area or section. Most commercial concerns have their own in-house zonal schemes but, when referred to in print, these are insufficiently documented to be usable by others (e.g. Rawson & Riley 1982, Fig. 3). To overcome these inadequacies, a dinocyst zonation for the Jurassic founded on classic British sections with the best possible stratigraphic control was compiled by the BGS (Woollam & Riding 1983). The Kimmeridgian part of the zonation has recently been revised (Riding & Thomas 1988) and the Jurassic is now covered by 16 dinocyst zones, 12 of which are divided into a total of 34 subzones, giving a total of 38 constituent dinocyst-based units for the System (Fig. 7). The zones are mainly assemblage zones (i.e. intervals distinguished by their particular contained dinocyst assemblage), although various types of range zones (i.e. intervals distinguished by the total, partial, concurrent or local stratigraphic range(s) of named dinocyst taxa) are also included. Most of the subzones are interval zones (i.e. intervals between the first and/or last occurrences of named dinocyst taxa). This zonation (a true biozonation) is the most usable zonation available in our present state of knowledge and is the best stratigraphic breakdown of the Jurassic that can be achieved using dinocysts based on present sampling procedures. A much closer sampling interval would almost certainly identify at least local 'Acme' biozones which might be of some stratigraphic usefulness within limited geographical ranges. Further revision and refinement of the zonation is currently being undertaken at the BGS.

In sequences of rock-cuttings, it is necessary to work downhole and to use extinctions and range tops to establish the chronostratigraphic succession. In the Woollam & Riding (1983) zonation, extinctions of 21 dinocyst taxa were highlighted. Riding (1984) subsequently elaborated on these and identified 35 extinctions in the Jurassic (see also Riding & Thomas 1988). In a long run of cuttings through a thick Jurassic sequence, one could make some chronostratigraphic sense of the sequence using these datum points but for a single cuttings sample or only a few samples any recorded taxa can give only a minimum age.

Fig. 6. Reference sections for the Kimmeridgian and Portlandian ammonite-based zones. Dorset Portlandian nomenclature and bed notation follow Arkell (1935; 1947), but see Townson (1975) for lithostratigraphic revision. For Dorset Kimmeridge Clay sections, see Cox & Gallois (1981). *base of bed = basal boundary stratotype for zone (for Portlandian, see Wimbledon & Cope (1978)) **bed numbers of Blake (1875) emend. Cope (1967; 1978).

System/Series/ Stage/Substage			Ammonite-Based Zone	Dinocyst Zone		Sub-zone
CRETACEOUS						
UPPER JURASSIC	PORTLANDIAN		Lamplughi	*Gochteodinia villosa* (Gv)		b
			Preplicomphalus			
			Primitivus			
			Oppressus			a
			Anguiformis	*Dichadogonyaulax pannea* (Dp)		b
			Kerberus			
			Okusensis			
			Glaucolithus			
			Albani			a
	KIMMERIDGIAN	Upper	Fittoni	*Glossodinium dimorphum* (Gd)		e
			Rotunda			
			Pallasioides			d
			Pectinatus			c
			Hudlestoni			b
			Wheatleyensis			
			Scitulus			
			Elegans			a
		Lower	Autissiodorensis	*Endoscrinium luridum* (El)		c
			Eudoxus			
			Mutabilis			b
			Cymodoce			a
			Baylei	*Scriniodinium crystallinum* (Sc)		d
	OXFORDIAN	Upper	Rosenkrantzi			c
			Regulare			b
			Serratum			
			Glosense			a
		Mid.	Tenuiserratum			
			Densiplicatum	*Acanthaulax senta* (As)		b
		Low.	Cordatum			a
			Mariae	*Wanaea fimbriata* (Wf)		
MIDDLE JURASSIC	CALLOVIAN	Upp.	Lamberti	*Wanaea thysanota* (Wt)		b
			Athleta			a
		Mid.	Coronatum	*Ctenidodinium ornatum – C. continuum* (Cc/Ccn)		
			Jason			
		Lower	Calloviense			
			Koenigi			
			Herveyi	*Ctenidodinium combazii – C. sellwoodii* (Ccb/Cs)		b
	BATHONIAN	Upper	Discus			a
			Orbis			
		Mid.	Hodsoni			
			Morrisi			
			Subcontractus			
			Progracilis			
		Low.	Tenuiplicatus			
			Zigzag			
	BAJOCIAN	Upper	Parkinsoni	*Nannoceratopsis gracilis* [superzone]	*Acanthaulax crispa* (Acr)	
			Garantiana			
		Lower	Subfurcatum		*Nannoceratopsis gracilis* (Ng)	c
			Humphriesianum			
			Sauzei			b
			Laeviuscula			
			Discites			a
	AALENIAN		Concavum		*Mancodinium semitabulatum* (Ms)	e
			Murchisonae			
			Opalinum			d
LOWER JURASSIC	TOARCIAN	Upper	Levesquei			c
			Thouarsense			b
			Variabilis			
		Lower	Bifrons			a
			Falciferum			
			Tenuicostatum			
	PLIENSBACHIAN	Upp.	Spinatum		*Luehndea spinosa* (Ls)	
			Margaritatus			
		Lower	Davoei	*Liasidium variabile* (Lv)		b
			Ibex			
			Jamesoni			
	SINEMURIAN	Upper	Raricostatum			a
			Oxynotum			
			Obtusum			
		Lower	Turneri	*Dapcodinium priscum* (Dp)		b
			Semicostatum			
			Bucklandi			a
	HETTANGIAN		Angulata			
			Liasicus			
			Planorbis			
TRIASSIC				*Rhaetogonyaulax rhaetica* (Rr)		

Fig. 7. Jurassic dinoflagellate cyst zonation (for details, see Wollam & Riding (1983) and Riding & Thomas (1988)).

In order to relate back to the primary chronostratigraphic standard, dinocyst ranges (and the zones which they define) are often 'rounded up' to the nearest appropriate ammonite-based zone boundary. This perhaps throws the dinocysts into a more favourable light, as regards their potential for primary chronostratigraphic resolution, than is strictly fair. For instance, in the Jurassic, 35 (out of 76) ammonite-based zone boundaries can seemingly be identified on the basis of dinocysts (Fig. 7). Also, detailed work on the dinocysts of any individual section, which is then used as part of a synthesis and subsequent zonation, is only as good as the original ammonite-based stratigraphy supplied with that section; any problems with that original classification are inherited by the resultant composite dinocyst 'standard'. These comments also apply, of course, to other microfossil groups. Nevertheless, in practice, there is no doubt that dinocysts are the most important tool for chronostratigraphic subdivision in the North Sea Jurassic.

As well as dinocysts, Jurassic palynological samples also yield acritarchs, and pollen and spores (so-called miospores or sporomorphs). The latter are the products of continental vegetation which were transported into the marine environment. They are particularly common in nearshore, deltaic and estuarine deposits but, compared with dinocysts, show even slower evolutionary rates and are therefore of much less value as age indicators in Jurassic strata.

Calcareous micropalaeontology

The calcareous groups of microfossils (nannofossils, foraminifera and ostracoda) can be used in sequences of rock-cuttings or core where, as in the Jurassic of the onshore UK and Southern North Sea Basin, diagenetic processes have not destroyed them.

Calcareous nannofossils. Like dinocysts, calcareous nannofossils are the remains of marine phytoplankton. They are a heterogeneous group of very small fossils (usually less than 15 μm in size) which represent the calcified plates of unicellular algae (Lord & Taylor 1982). During the Late Triassic and Early Jurassic, the coccolithophorid algae became an increasingly important component of the marine phytoplankton, and coccoliths are the dominant calcareous nannofossil group in Jurassic rocks (Bown 1987).

They are found most abundantly in silts and clays that have a calcareous component and were deposited in open ocean to nearshore environments. Clays appear to favour the preservation of the fine detail of coccolith structures (Hay 1977; Haq 1978). The planktonic habit of these organisms and their relatively rapid dispersal over large areas enhances their usefulness as correlative tools and, during the past 20 years, they have become increasingly important in Mesozoic biostratigraphy because of their great abundance and ease of preparation (see Taylor & Hamilton 1982). Ideally, both optical and scanning electron microscopes should be used but routine biostratigraphic service work is usually carried out with only an optical microscope, although not all specimens can be positively identified by this means (Thierstein 1976; Medd 1982). In practice, the small size of coccoliths means that they are easily reworked into younger sediments without necessarily showing signs of wear and, in boreholes, they can be recirculated with drilling mud relatively easily. Contamination of samples, however collected, is therefore a serious consideration.

Nannoplankton evolutionary rates were relatively slow in the Jurassic compared with the later Mesozoic and Cenozoic, and zonations based on coccoliths are less refined in the Jurassic compared with these later intervals (Haq 1978). Indeed, calcareous nannofossils of the Jurassic have received less study than those of other periods. One of the problems for coccolith workers trying to establish a zonation is the state of specimen preservation. Diagenetic effects can cause dissolution (the partial or total destruction of delicate central structures) and overgrowth or recrystallization (calcitic strengthening of other features which obscure the morphology). Taxonomic confusion may result if these effects, which appear to affect assemblages selectively, are not recognized. Taxonomy is commonly inconsistent and incoherent because classification has generally been purely morphological but without a standard evaluation of morphological characters (Bown 1987). These difficulties have led to poorly defined first and last occurrences of taxa which, together with limited numbers of samples and sections studied, and gaps due to facies limitations, have tended to impede attempts to zone the Jurassic on the basis of coccoliths. Bown *et al.* (1988) highlight these facts in introducing their most recent zonation. By selecting solution-resistant, commonly occurring and mostly well known species as key taxa, they hope to have brought some stability to nannofossil biostratigraphy for the European Jurassic. The Jurassic is covered by 17 nannofossil interval zones, six of which are each divided into two subzones, giving a total of 23 constitutent nannofossil-based units for the Jurassic System. Associated species which help identify the zones are cited, together with reference sections (those for 13 out of the 17 zones are in the UK) and the relationship to the ammonite-based standard. Crux (1987) reported 23 nannofossil 'events' (first or last occurrences of named taxa) in the Lower Jurassic of southern Britain, southwest Germany and the North Sea (see Fig. 8). Nonetheless, the realization of Medd's (1979) optimistic forecast that future developments in sample examination techniques would lead to greater accuracy and confidence in identification, and a consequent scheme of coccolith zones as refined as those based on ammonites, appears remote.

An investigation into Late Jurassic nannofossil provincialism (Cooper 1989) has shown a Boreal–Tethyan distribution pattern in which there are differences in the dominant nannofossil families and important genera. However, over 75% of all species recorded were found to occur in both realms. No detailed work has been published on variations within the Boreal Realm which might influence biostratigraphic results in the UK area.

Foraminifera. Foraminifera are one of the best known and most comprehensively studied groups of calcareous microfossils. They are unicellular protists with a test generally less than 1 mm in length, and are abundant, diverse and relatively easy to study. They can be extracted, quickly and cheaply, by simple mechanical and chemical processes from almost all post-Triassic marine sediments which have not been severely leached or become acidic. Most foraminifera are studied using an optical microscope with reflected light, although transmitted light is better for examining internal chamber arrangements which are taxonomically important in some groups. Foraminifera in indurated limestones are generally studied in thin-section. Scanning electron micrographs are usually used for illustrative purposes although photographs in reflected light and, to a lesser extent, transmitted light are sometimes given to aid identification (Brasier 1980).

As with other planktonic groups, such as the dinocysts and calcareous nannofossils, planktonic foraminifera are useful for biostratigraphic purposes. However, these types did not evolve until the latter half of the Jurassic Period and are relatively rare in Jurassic microfaunal assemblages (Boersma 1978). Not until the

System/Series	Stage	Substage	Ammonite-Based Zone	*	Nannofossil Zone	NJ	Nannofossil Subzone	NJ
CRETACEOUS								
UPPER JURASSIC	PORTLANDIAN		Lamplughi		*Stephanolithion atmetros*	NJ 17		
			Preplicomphalus					
			Primitivus					
			Oppressus					
			Anguiformis					
			Kerberus					
			Okusensis					
			Glaucolithus					
			Albani					
	KIMMERIDGIAN	Upper	Fittoni					
			Rotunda					
			Pallasioides					
			Pectinatus		*Stephanolithion helotatus*	NJ 16		
			Hudlestoni					
			Wheatleyensis					
			Scitulus					
			Elegans					
		Lower	Autissiodorensis					
			Eudoxus		*Cyclagelosphaera margaretii*	NJ 15	*Hexapodorhabdus cuvillieri*	NJ 15 b
			Mutabilis					
			Cymodoce					
			Baylei					
	OXFORDIAN	Upper	Rosenkrantzi					
			Regulare				*Lotharingius crucicentralis*	NJ 15 a
			Serratum					
			Glosense					
		Mid.	Tenuiserratum					
			Densiplicatum					
		Low.	Cordatum		*Stephano. bigotii max.*	NJ 14		
			Mariae					
MIDDLE JURASSIC	CALLOVIAN	Upp.	Lamberti		*Stephanolithion bigotii bigotii*	NJ 13		
			Athleta					
		Mid.	Coronatum					
			Jason					
		Lower	Calloviense		*Ansubsphaera helvetica*	NJ 12	*Watznaueria deflandrei*	NJ 12b
			Koenigi					
			Herveyi				*Stephanolithion hexum*	NJ 12 a
	BATHONIAN	Upper	Discus					
			Orbis					
		Mid.	Hodsoni		*Pseudoconus enigma*	NJ 11		
			Morrisi					
			Subcontractus					
			Progracilis					
		Low.	Tenuiplicatus					
			Zigzag					
	BAJOCIAN	Upper	Parkinsoni		*Stephanolithion speciosum*	NJ 10		
			Garantiana					
			Subfurcatum					
		Lower	Humphriesianum					
			Sauzei		*Watznaueria britannica*	NJ 9		
			Laeviuscula					
			Discites		*Biscutum intermedium*	NJ 8	*Lotharingius contractus*	NJ 8b
	AALENIAN		Concavum					
			Murchisonae				*Retecapsa incompta*	NJ 8a
			Opalinum					
LOWER JURASSIC	TOARCIAN	Upper	Levesquei		*Discorhabdus striatus*	NJ 7		
			Thouarsense					
			Variabilis					
		Lower	Bifrons	*				
			Falciferum	5	*Carinolithus superbus*	NJ 6		
			Tenuicostatum		*Lotharingius hauffii*	NJ 5	*Crepidolithus cavus*	NJ 5b
	PLIENSBACHIAN	Upp.	Spinatum	2			*Bisc. finchii*	NJ5a
			Margaritatus	2	*Biscutum novum*	NJ 4	*Crepidolithus granulatus*	NJ 4b
		Lower	Davoei	3				
			Ibex	1			*C. pliensbachensis*	NJ4a
			Jamesoni	3				
	SINEMURIAN	Upper	Raricostatum		*Crepidolithus crassus*	NJ 3		
			Oxynotum					
			Obtusum	1	*Parhabdolithus liasicus*	NJ 2	*Mitrolithus elegans*	NJ 2b
		Lower	Turneri					
			Semicostatum	1			*Parhabdo. marthae*	NJ 2a
			Bucklandi	4				
	HETTANGIAN		Angulata		*Schizosphaerella punctulata*	NJ 1		
			Liasicus					
			Planorbis	1				
TRIASSIC				1				

Fig. 8. Jurassic nannofossil zonation (for details, see Bown *et al.* (1988)). *= number of nannofossil events of Crux (1987).

Cretaceous did the planktonic foraminifera evolve and diversify sufficiently to become an important biostratigraphic tool. In the Jurassic, foraminiferal faunas were dominated instead by calcareous and agglutinated benthonic forms. As with any benthonic fauna, their potential as general environmental indicators is often as important as that for biostratigraphy.

There is no published composite foraminiferal zonation for the Jurassic of the UK area. For the Lower Jurassic, the zonation of Copestake & Johnson (1989), shown in Fig. 9, utilizes the most consistently occurring short-ranging and widespread species. It consists of 16 zones (without index species), three of which are subdivided into two subzones each, giving a total of 19 constituent foraminifera-based units for the British Lower Jurassic. The zonation can be applied using the 24 taxa-range bases (either of common occurrence or absolute), the 19 range tops and the 20 other foraminifera species 'events' which are given. All the zone/subzone boundaries coincide with ammonite zone or subzone boundaries; for reasons of space and legibility, the ammonite subzones are not shown in Fig. 9.

In contrast, the Middle Jurassic is poorly documented. The majority of foraminifera species are long-ranging and display complex facies associations, but local extinctions and appearances could be used as biostratigraphic markers within individual basins. The Middle Jurassic species discussed and illustrated by Morris & Coleman (1989) either have restricted stratigraphic ranges within the British succession, have reliably defined appearances or extinctions of correlative value, or constitute an important component of assemblages at one or more horizons although having overall a long total range; the last named will be important in the characterization of assemblages in any future zonation of the Middle Jurassic. For the UK area, the Aalenian–Bajocian interval has been almost totally neglected. For the Bathonian, only the localized zones ('faunules') first recognized by Cifelli (1959) are published (Coleman 1982); details of these units are reiterated by Morris & Coleman (1989) and are believed to relate to regional events. For the Callovian, three assemblage zones are recognised in the Oxford Clay of the English Midlands; the boundaries of each coincide with lithostratigraphic boundaries and change of

System/Series/ Stage/Substage			Ammonite-Based Zone	Foraminifera Zonation		
CRETACEOUS						
UPPER JURASSIC	PORTLANDIAN		Lamplughi	No independent zonation available for Upper Jurassic		
			Preplicomphalus			
			Primitivus			
			Oppressus			
			Anguiformis			
			Kerberus			
			Okusensis			
			Glaucolithus			
			Albani			
	KIMMERIDGIAN	Upper	Fittoni			
			Rotunda			
			Pallasioides			
			Pectinatus			
			Hudlestoni			
			Wheatleyensis			
			Scitulus			
			Elegans			
		Lower	Autissiodorensis			
			Eudoxus			
			Mutabilis			
			Cymodoce			
			Baylei			
	OXFORDIAN	Upper	Rosenkrantzi			
			Regulare			
			Serratum			
			Glosense			3
		Mid.	Tenuiserratum			
			Densiplicatum			
		Low.	Cordatum			
			Mariae			
MIDDLE JURASSIC	CALLOVIAN	Upp.	Lamberti	3	assemblage zones for the Oxford Clay of the English Midlands	assemblage zones for the Heather Formation of the Magnus Field (Northern North Sea)
			Athleta			
		Mid.	Coronatum	— ? —		2
			Jason	2		
		Lower	Calloviense			1
			Koenigi	1		
			Herveyi			
	BATHONIAN	Upper	Discus	E	"faunules" for Dorset and Bath area	
			Orbis	D		
			Hodsoni	C (1 – 3)		
		Md.	Morrisi			
			Subcontractus			
		Low.	Progracilis	B (1 – 2)		
			Tenuiplicatus			
			Zigzag	A		
	BAJOCIAN	Upper	Parkinsoni	No independent zonation available for Aalenian and Bajocian		
			Garantiana			
			Subfurcatum			
		Lower	Humphriesianum			
			Sauzei			
			Laeviuscula			
			Discites			
	AALENIAN		Concavum			
			Murchisonae			
			Opalinum			
				Zone		Subzone
LOWER JURASSIC	TOARCIAN	Upper	Levesquei	JF 16		
			Thouarsense			
			Variabilis	JF 15		
		Lower	Bifrons	JF 14		
			Falciferum	JF 13		
			Tenuicostatum	JF 12		b
	PLIENSBACHIAN	Upp.	Spinatum			a
			Margaritatus	JF 11		
		Lower	Davoei	JF 10		b
			Ibex	JF 9		a
			Jamesoni			
	SINEMURIAN	Upper	Raricostatum	JF 8		
			Oxynotum	JF 7		
			Obtusum	JF 6		
		Lower	Turneri	JF 5		
			Semicostatum	JF 4		b
			Bucklandi			a
	HETTANGIAN		Angulata	JF 3		
			Liasicus	JF 2		
			Planorbis	JF 1		
TRIASSIC						

Fig. 9. Jurassic foraminiferal zonation (for details of authorship, see text).

lithology. Morris & Coleman also detail three assemblages spanning the Mid and Late Bathonian to Mid Oxfordian interval of the Magnus Field (Northern North Sea) and present a less formal breakdown of the Bathonian-Callovian foraminiferal sequence in the Moray Firth Basin. There is clearly much work to be done.

In the Upper Jurassic, foraminiferal assemblages are very uniform; changes can be related to environment and facies rather than evolutionary causes. An exceptional foraminifera 'event' is usually geographically limited and still relates to major changes in the sea-floor environment (Shipp 1989). Apparently, onshore localities yield richer and more varied calcareous benthonic forms than offshore sections which represent deeper water depositional environments; in graben areas for instance, agglutinating forms generally predominate. Recovery in these areas is, in any case, limited. The Upper Jurassic species discussed and illustrated by Shipp (1989) include the rare short-ranging forms and the common long-ranging forms, chosen to give an overall impression of Upper Jurassic assemblages.

The geographical distribution pattern (Boreal/Tethyan) for Jurassic foraminifera is well known (Gordon 1970) and must be taken into account if a detailed biostratigraphy for the Middle/Upper Jurassic of the UK area is to be achieved. For the Upper Jurassic, Shipp (1989) notes Arctic assemblages dominated by simple agglutinating forms, and mixed shelf assemblages of simple agglutinates and various calcareous benthonic forms in a broad belt, including northwest Europe, north of the Tethyan region.

Ostracoda. Ostracoda are minute crustacea generally 0.2 to 1.5 mm in overall length. They have a bivalved, calcified shell and a well documented fossil record from the early Cambrian to the present day (Pokorny 1978). With foraminifera, they are a major constituent of shallow marine benthonic fossil faunas; more significantly, they occur in brackish and non-marine sediments where other groups are poorly represented (see below). As well as having biostratigraphic potential, they are particularly important as indicators of ancient shorelines, salinities and relative sea-floor depths. They are extracted and prepared in the same way as foraminifera (see above). Unlike foraminifera, thin-sectioning is not an appropriate method for identifying ostracods in indurated lithologies. Routine service work is carried out using optical microscopes, but for illustration and detailed study a scanning electron microscope is usually used.

Ostracoda tend to be restricted to a particular benthonic ecological niche and their distribution is very much related to the depositional environment. Indeed, it is for their environmental data that ostracods are valued in the oil industry (Oertli *et al.* 1983). Some ostracoda may have been attached to aquatic vegetation and not therefore so directly related to substrate type (Bate 1978). For the UK area, they have had most biostratigraphic application in the Bathonian (marginal marine and non-marine facies) and the Portlandian (brackish and non-marine Purbeck facies), where other biostratigraphically useful groups are wanting. Species may have relatively short vertical ranges but cannot be used for long distance correlations and only local zonal schemes have been developed (Neale 1988). According to Christensen (1988), an ostracod zone is often represented in only one or a few samples in a particular section and these occurrences relate to a slight change in the palaeoenvironment of that section; finds of useful ostracod faunas in commercial wells are, in any case, relatively uncommon. Ostracod zones for the Jurassic of the UK area are shown in Fig. 10. The basis of the zonation was summarized by Bate & Robinson (1978); a number of intervals have subsequently been refined or amended but coverage is patchy.

Early Jurassic ostracod faunas became increasingly rich and diverse with time except in the Sinemurian when there was low diversity. Distributions can be related in broad terms to gross lithofacies (and ultimately environmental) patterns (Lord 1978). Working on sections in England and the Netherlands, Park (1984) established four interval zones for the Hettangian to Lower Pliensbachian; three of these were each subdivided into two subzones. Unfortunately, this zonation, established for the area around the Southern North Sea Basin, and partly overlapping that of Michelsen (1975), remains unpublished. There are three published zones for the middle part of the Toarcian based on boreholes in the English Midlands (Bate & Coleman 1975).

In the Middle Jurassic, ostracod faunas are fairly constant throughout but much work still needs to be done (Bate 1978). There is no published ostracod zonation for the Aalenian or Upper Bajocian. The Lower Bajocian is incompletely covered by three zones, one of which is further divided into two subzones; these are based on sections in the English Midlands, Humberside and North Yorkshire (Bate 1978). For the Bathonian, there are five zones, one of which is subdivided into three subzones. This zonation was based on the ostracod faunas of southern England and northern France; zonal boundaries are fixed by first stratigraphic appearances of index species (Sheppard 1981). In the Early Callovian, ostracod diversity began to increase and continued to a maximum in the Late Oxfordian but thereafter decreased throughout the remainder of the Jurassic Period. No zonation is yet published for the Lower–Middle Callovian but Kilenyi (1978) listed three zones for the interval from the Middle Callovian to the Lower Kimmeridgian. Detailed zonations for the Kimmeridgian have been published for Dorset (Christensen & Kilenyi 1970; Kilenyi 1978) and eastern England (Wilkinson 1983a, b) (Fig. 10). Among Kimmeridgian microfossils, ostracods are proving biostratigraphically reliable but are absent from the oil-shales and more 'bituminous' mudstones. According to Kilenyi (1978), the marine Portlandian of southern England is covered by a single ostracod zone. The youngest Portlandian, in so-called Purbeck facies, is covered by two 'freshwater' ostracod zones, four 'assemblages' and 21 'faunicycles' (see Anderson 1985). The marine sandstones of this age in eastern England have not yielded ostracod faunas.

Radiolaria

Like foraminifera, radiolaria are a group of marine protists but they differ in being exclusively planktonic and having siliceous skeletons. They live at all water depths in the open ocean, especially just seaward of the continental slope (Brasier 1980). The fossils are 100–2000 μm in overall diameter and are extracted from the host rock by various mechanical and chemical processes depending on lithology (Pessagno 1977) (but see below for UK material). The most abundant and diverse faunas are found in deep water flysch and other sedimentary facies associated with subduction zones and spreading centres. Such facies of Jurassic age occur in Japan, western North America and Alpine/Tethyan Europe, and it is from these areas that most Jurassic radiolarian data is available. The group has a long geological history, rapid evolution, worldwide distribution and great diversity, and its biostratigraphic potential has long been recognized (Torrens 1980b). Relatively shallow water epicontinental sequences, which do not yield radiolaria, predominate in the Jurassic deposits of the UK area and deposits of more

System/Series/ Stage/Substage			Ammonite-Based Zone	Ostracoda Zonation	
CRETACEOUS				Zone *Cypridea granulosa*	ASSEMBLAGE 4; FAUNICYCLE 22 - 25
UPPER JURASSIC	PORTLANDIAN		Lamplughi	*Cypridea granulosa*	3; 16 - 21
			Preplicomphalus		
			Primitivus	*Cypridea dunkeri*	2, 1; 1 - 15
			Oppressus	S. ENGLAND	E. ENGLAND
			Anguiformis		
			Kerberus	*Macrodentina retirugata*	?
			Okusensis		
			Glaucolithus		
			Albani		
	KIMMERIDGIAN	Upper	Fittoni	*Gall. polita*	
			Rotunda	*Gall. spinosa*	
			Pallasioides		
			Pectinatus	*Mandelstamia maculata*	*un-named*
			Hudlestoni		*Gall. spinosa Eocyth. aquitanum*
			Wheatleyensis		*Mandelstamia maculata*
			Scitulus	?	*Mandelstamia horrida*
			Elegans		*un-named*
		Lower	Autissiodorensis		*Galliae. elongata* / *Macro. steghausi*
			Eudoxus	*Gall. elongata*	*Gall. kilenyii*
			Mutabilis		*Macrodentina proclivis*
			Cymodoce	*Galliaecytheridea dissimilis*	
			Baylei		
	OXFORDIAN	Upper	Rosenkrantzi		
			Regulare		
			Serratum		
			Glosense		
		Mid.	Tenuiserratum	*Nophrecythere cruciata oxfordiana*	
			Densiplicatum		
		Low.	Cordatum		
			Mariae		
MIDDLE JURASSIC	CALLOVIAN	Upp.	Lamberti	*Lophocythere interrupta* s.s.	
			Athleta		
		Mid.	Coronatum	?	
			Jason		
		Lower	Calloviense		
			Koenigi		
			Herveyi		
	BATHONIAN	Upper	Discus	*Micropneumatocythere falcata*	
			Orbis	*Fossaterquemula blakeana*	
			Hodsoni	*Progonocythere polonica*	
		Mid.	Morrisi	*Praeschuleridea confossa*	
			Subcontractus		
			Progracilis		
		Low.	Tenuiplicatus	*Nophrecythere rimosa*	*Merocythere postangusta*
			Zigzag		*Eosch. batei*
	BAJOCIAN	Upper	Parkinsoni		*Nop. rimosa*
			Garantiana	?	
			Subfurcatum		
		Lower	Humphriesianum	*Glyptocythere scitula*	
			Sauzei	*Glyptocythere polita*	
			Laeviuscula	*Kinkelinella (Ekt.) triangula*	*Prog. carinata*
			Discites		*Prog. reticulata*
	AALENIAN		Concavum	?	
			Murchisonae		
			Opalinum		
LOWER JURASSIC	TOARCIAN	Upper	Levesquei		
			Thouarsense	*Kink. debilis & Camp. toarciana*	
			Variabilis		
		Lower	Bifrons	*King. persica*	
			Falciferum	*Kink. sermoisensis & Kink. intrepida*	
			Tenuicostatum	?	
	PLIENSBACHIAN	Upp.	Spinatum		
			Margaritatus		
		Lower	Davoei	UNPUBLISHED ZONATION OF Park (1984)	
			Ibex		
			Jamesoni		
	SINEMURIAN	Upper	Raricostatum		
			Oxynotum		
			Obtusum		
		Lower	Turneri		
			Semicostatum		
			Bucklandi		
	HETTANGIAN		Angulata		
			Liasicus		
			Planorbis		
TRIASSIC				?	?

Fig. 10. Jurassic ostracod zonation (for details of authorship, see text).

favourable oceanic environments do not occur in the North Sea Jurassic. However, radiolaria have now been recovered from graben areas in the Northern North Sea Basin and onshore coastal localities in the Moray Firth (Kimmeridge Clay Formation) (Dyer & Copestake 1989). In most cases, the opaline silica of the radiolarian skeleton has been replaced by calcite or pyrite. The fossils are extracted using standard processing techniques with hydrogen peroxide but it is necessary to pick the fine fractions (63 μm and 125 μm residues). Dyer & Copestake (1990) give details of a more successful technique using hydrofluoric acid, which removes all the argillaceous component and concentrates the radiolaria in the insoluble residue. For routine investigations, all diagnostic features can be recognised using reflected light stereomicroscopes. For illustrative purposes or detailed examination, use of a scanning electron microscope is preferable.

Some of the practical difficulties of working on these Jurassic radiolaria are that the material is often badly preserved and careful processing methods are required for successful extraction. Also, there is no onshore standard for reference. The advantages are that the radiolaria occur in sediments where other microfossil groups such as dinocysts have been adversely affected by high thermal maturity or the drilling process, and they can be extracted from dysaerobic/anaerobic sediments devoid of benthonic forms such as foraminifera and ostracoda, or where calcareous microfossils have suffered dissolution.

Though Dyer & Copestake's biostratigraphic results are couched in confidentiality, their recognition of 13 radiolarian 'events' in the Upper Jurassic to Lower Cretaceous succession marks a significant milestone for the UK Jurassic, albeit for only a small part. The stratigraphic positions of the 13 'events' are shown in Fig. 11. Although the precise datum for each 'event' is not divulged, they apparently allow the recognition of most stage and substage boundaries in the Kimmeridgian–Ryazanian (Lower Cretaceous) interval. These 'events' are traced throughout the Northern North Sea Basin and west of Shetland and are based on assemblage changes including extinctions, appearances and the lower and upper limits of abundance of particular species; abundance limits are thought to have the greatest potential for establishing a biozonation. The 'events' have been related to

System/Series/ Stage/Substage			Ammonite-Based Zone	North Sea Radiolarian "Events"
CRETACEOUS				
	RYAZAN-IAN	Upper	Albidum	
			Stenomphalus	1
			Icenii	2
		Low.	Kochi	
			Runctoni	3
UPPER JURASSIC	PORTLANDIAN		Lamplughi	
			Preplicomphalus	
			Primitivus	4
			Oppressus	5
			Anguiformis	6
			Kerberus	?
			Okusensis	
			Glaucolithus	
			Albani	
	KIMMERIDGIAN	Upper	Fittoni	
			Rotunda	
			Pallasioides	7
			Pectinatus	8, 9
			Hudlestoni	
			Wheatleyensis	
			Scitulus	10
			Elegans	11
		Lower	Autissiodorensis	
			Eudoxus	
			Mutabilis	12
			Cymodoce	13
			Baylei	
	OXFOR-DIAN	Upper	Rosenkrantzi	
			Regulare	
			Serratum	
			Glosense	

Fig. 11. Stratigraphic position of radiolarian events in the Upper Jurassic (and Lower Cretaceous) of the Northern North Sea Basin (Dyer & Copestake 1989).

the ammonite-based standard using Woollam & Riding's (1983) dinocyst zonation. They have proved most useful in the classification and correlation of the Middle Volgian [Kimmeridgian Pallasioides Zone to Portlandian Oppressus Zone inclusive] where laterally impersistent sands are interbedded with organic-rich claystones and siltstones and with local non-sequences. This interval can be divided into up to five units on the basis of radiolaria.

Variations in abundance and consistent recognition of acmes and absences amongst radiolaria are thought to relate to variations in oceanic influence caused by eustatic and/or tectonically controlled sea level changes. These can be linked to phases of sand generation and can provide the basis for sequence stratigraphy analysis.

Four latitudinal provinces are recognised in the Jurassic radiolaria worldwide (Pessagno *et al.* 1984). Assemblages in the Jurassic of the North Sea, which are low in diversity and have many previously undescribed taxa, are of northern Boreal type. According to Dyer & Copestake (1989), it will be necessary to develop parallel zonal schemes for the Boreal and Tethyan areas although sufficient taxa are believed to be common to the various provinces to allow some degree of correlation.

Conclusions

The ammonites have been and remain the basis of the primary chronostratigraphic standard for the Jurassic System and it is important to ensure that there are always personnel who are sufficiently trained and expert to sustain and maintain the ammonite based chronstratigraphica standard. There is no cause for complacency for there are still many problems to solve and gaps to fill in the published documentation.

Commercial pressures and practical demands have meant that for some 25 years, stratigraphic resolution in the Jurassic by means of fossils other than ammonites has been sought, and all the microfossil groups have undergone a surge of investigation; to some extent, the present paper is a review of the progress of that effort.

Ammonite workers have sometimes adopted a rather patronising tone in defending the 'supremo' role of these fossils and, likewise, microfossil workers have been equally vehement in their own assertions and defence. These attitudes seem inappropriate for each group makes its own contribution which, depending on circumstances, will meet the shortfall of another. No fossil group is infallible, nor are the subjective judgements of palaeontologists working on scant and poor quality material. Stratigraphic resolution and correlation is a challenge not a contest. Working-groups on microfossils, under the auspices of the International Subcommission on Jurassic Stratigraphy, now exist alongside those on the individual Jurassic stages (Michelsen & Zeiss 1984).

For the Jurassic, it is not possible to advocate a single microfossil group as the best stratigraphic indicator because recovery of the various different groups is not consistent or uniform. However, in our present state of knowledge, and assuming adequate recovery of all groups, the best resolution will be achieved in the Lower Jurassic by foraminifera, with nannofossils, ostracoda and dinocysts following close behind. In the Middle Jurassic, ostracoda, foraminifera, dinocysts and nannofossils appear to afford equal resolution. In the Upper Jurassic, dinocysts are pre-eminent but radiolaria offer an exciting new area of study. A simplistic comparison of the time resolution possible with the different fossil groups is shown in Fig. 12 (based on data in Figs 1 and 7–10). The value of such diagrams is, however, open to doubt because of the gross approximations inevitably included

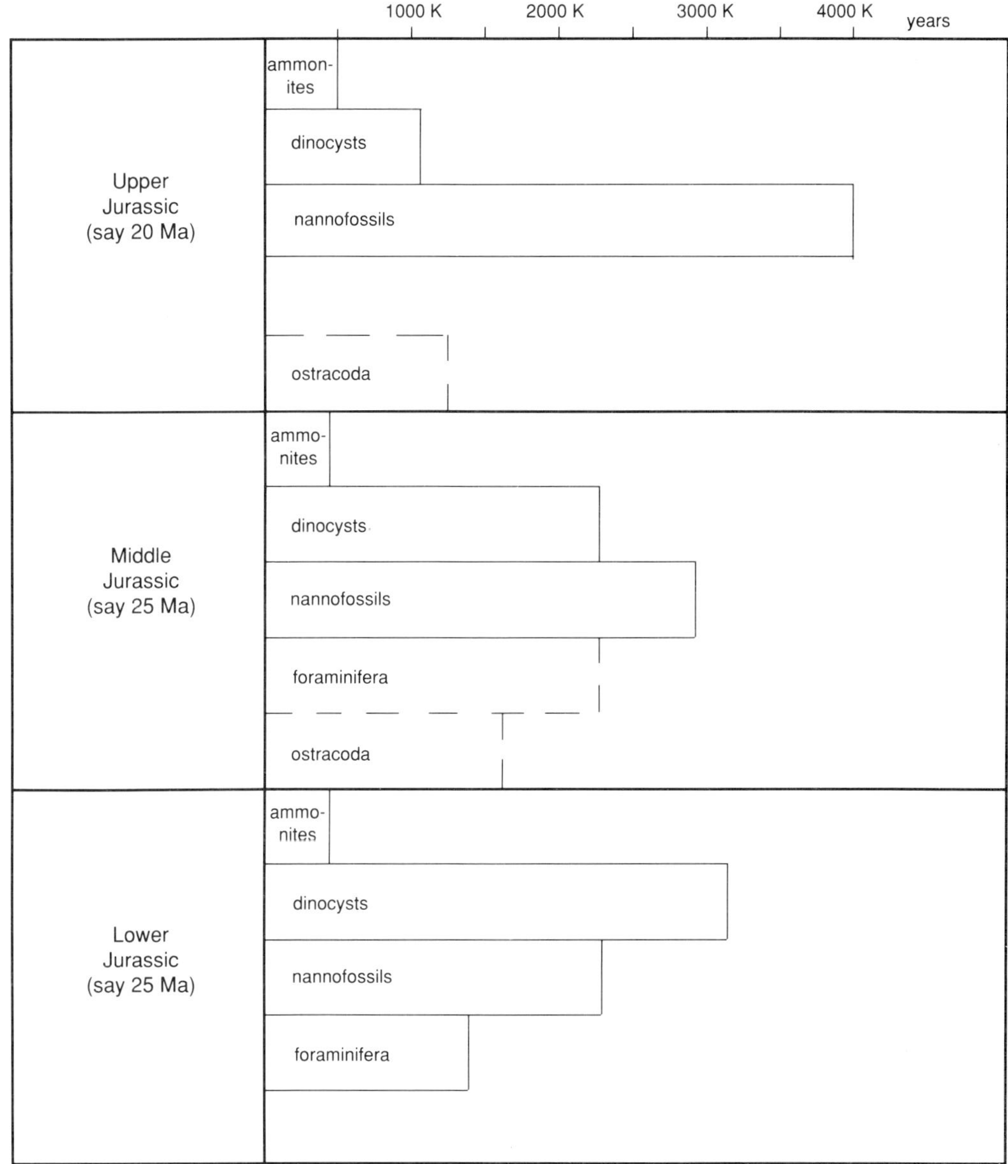

Fig. 12. Simplistic comparison of time resolution in the Jurassic using different fossil groups. Bars represent average duration of zonal/subzonal units for each fossil group.

and the false sense of accuracy consequently imparted.

In preparing this contribution, I have been made aware of the considerable number of recently published papers on Jurassic microfossil biostratigraphy which must indicate a healthy and encouraging 'state-of-the-art'. The publication of composite biostratigraphies based on several microfossil groups is also to be welcomed (e.g. Ainsworth *et al.* 1987).

In the future, magnetostratigraphy will undoubtedly play an increasingly important role in Jurassic chronostratigraphy but, at present, it is not a viable tool for the UK Jurassic. A global standard scale relating geomagnetic polarity reversals to stage boundaries has been partly achieved (Ogg & Steiner 1984) but local standards for the UK Jurassic need to be established. For practical purposes, the geomagnetic reversals need to have occurred at intervals which, relative to the sediment thickness, allow breakdown of the rock column into units of manageable scale. Rapid repeated reversals associated with relatively thin rock successions are likely to be difficult or impossible to correlate (Channell 1982). Recent published accounts of Jurassic sequences elsewhere in Europe indicate average reversal intervals of 260 Ka for the Bajocian–Bathonian and 215 Ka for the Callovian–Oxfordian, although intervals as small as 130 Ka may occur in the Oxfordian (and probably also the Pliensbachian) (Steiner *et al.* 1987). In the Toarcian and Kimmeridgian–Portlandian, the average reversal intervals were longer (maybe 350 Ka); these stages are likely, in practice, to have the best potential for magnetostratigraphic resolution (Ogg & Steiner 1984; Hallam *et al.* 1985).

I thank P. Copestake (BP Petroleum Development Ltd, Britoil plc), R. Dyer (GeoStrat Ltd), P. H. Morris (GeoStrat Ltd) and D. J. Shipp (The Robertson Group plc) for access to page proofs or preprints of their *in press* papers, and colleagues at the BGS, in particular, H. C. Ivimey-Cook, R. W. O'B. Knox, S. G. Molyneux, M. G. Sumbler, J. E. Thomas and I. P. Wilkinson, for their comments. This paper is published with the approval of the Director, British Geological Survey (NERC).

References

Ainsworth, N. R., O'Neill, M. O., Rutherford, M. M., Clayton, G., Horton, N. F. & Penney, R. A. 1987. Biostratigraphy of the Lower Cretaceous, Jurassic and uppermost Triassic of the North Celtic Sea and Fastnet Basins. *In*: Brooks, J. & Glennie, K. (eds) *Petroleum Geology of North West Europe*, Graham & Trotman, London, 611–622.

Anderson, F. W. 1985. Ostracod faunas in the Purbeck and Wealden of England. *Journal of Micropalaeontology*, **4**, 1–68.

Arkell, W. J. 1933. *The Jurassic System in Great Britain*. Oxford University Press.

—— 1935. The Portland Beds of the Dorset Mainland. *Proceedings of the Geologists' Association*, **46**, 301–347.

—— 1939. The ammonite succession at the Woodham Brick Company's pit, Akeman Street Station, Buckinghamshire, and its bearing on the classification of the Oxford Clay. *Quarterly Journal of the Geological, Society of London*, **95**, 135–222.

—— 1946. Standard of the European Jurassic. *Bulletin of the Geological Society of America*, **57**, 1–34.

—— 1947. The geology of the country around Weymouth, Swanage, Corfe & Lulworth. *Memoir of the Geological Survey of Great Britain*.

—— 1954. Three complete sections of the Cornbrash. *Proceedings of the Geologists' Association*, **65**, 115–122.

—— 1956. *Jurassic geology of the World*. Oliver & Boyd.

Bairstow, L. 1969. The Lower Lias of Robin Hood's Bay. *In*: Hemingway, J. E., Wright, J. K. & Torrens, H. S. (eds) *International Field Symposium on the British Jurassic. Excursion No 3. Guide for North East Yorkshire*, C24–C37. Geology Department, Keele University.

Bate, R. H. 1978. The Jurassic Part II – Aalenian to Bathonian. *In*: Bate, R. H. & Robinson, E. (eds) *A stratigraphical index of British Ostracoda*, 213–258. Seel House Press.

—— & Coleman, B. E. 1975. Upper Lias Ostracoda from Rutland and Huntingdonshire. *Bulletin of the Institute of Geological Sciences*, **55**, 1–42.

—— & Robinson, E. (eds). 1978. *A stratigraphical index of British Ostracoda*. Seel House Press (Geological Journal Special Issue No. 8).

Birkelund, T. & Callomon, J. H. 1985. The Kimmeridgian ammonite faunas of Milne Land, central East Greenland. *Bulletin Grønlands Geologiske Undersøgelse*, **153**.

——, Callomon, J. H., Clausen, C. K., Nøhr Hansen, H. & Salinas, I. 1983. The Lower Kimmeridge Clay at Westbury, Wiltshire, England. *Proceedings of the Geologists' Association*, **94**, 289–309.

Blake, J. F. 1875. On the Kimmeridge Clay of England. *Quarterly Journal of the Geological Society of London*, **31**, 196–233.

Boersma, A. 1978. Foraminifera. *In*: Haq, B. U. & Boersma, A. (eds) *Introduction to marine micropalaeontology*, 19–77. Elsevier.

Bown, P. R. 1987. Taxonomy, evolution, and biostratigraphy of Late Triassic–Early Jurassic calcareous nannofossils. *Special Papers in Palaeontology*, **38**, 1–118.

——, Cooper, M. K. E. & Lord, A. R. 1988. A calcareous nannofossil biozonation scheme for the early to mid Mesozoic. *Newsletters on Stratigraphy*, **20**, 91–114.

Brasier, M. D. 1980. *Microfossils*. Allen & Unwin.

Callomon, J. H. 1955. The ammonite succession in the Lower Oxford Clay and Kellaways Beds at Kidlington, Oxfordshire, and the zones of the Callovian Stage. *Philosophical Transactions of the Royal Society of London*, **239B**, 215–264.

—— 1964. Notes on the Callovian and Oxfordian stages. *In*: Maubeuge, P. L. (ed.) *Comptes rendus et Mémoires Colloque du Jurassique à Luxembourg 1962*, Institute grand-ducal, Section des Sciences naturelles, physiques et mathématiques, 269–291.

—— 1968. The Kellaways Beds and the Oxford Clay. *In*: Sylvester-Bradley, P. C. & Ford, T. D. (eds) *The geology of the East Midlands*, Leicester University Press, 264–290.

—— 1984a. Biostratigraphy, chronostratigraphy and all that – again! *In*: Michelsen, O. & Zeiss, A. (eds) *International Symposium on Jurassic stratigraphy*. Symposium volume 3, Geological Survey of Denmark, 612–624.

—— 1984b. The measurement of geological time. *Proceedings of the Royal Institution of Great*

Britain, **56**, 65–99.

—— & DONOVAN, D. T. 1974. A code of Mesozoic stratigraphical nomenclature, *Mémoires du Bureau de Recherches Géologiques et Minières*, **75** (for 1971), 75–81.

CASEY, R. 1973. The ammonite succession at the Jurassic-Cretaceous boundary in eastern England. *In:* CASEY, R. & RAWSON, P. F. (eds) *The Boreal Lower Cretaceous*, 193–266. Seel House Press.

—— & GALLOIS, R. W. 1973. The Sandringham Sands of Norfolk. *Proceedings of the Yorkshire Geological Society*, **40**, 1–22.

CAVE, R. & COX, B. M. 1975. The Kellaways Beds of the area between Chippenham and Malmesbury, Wiltshire. *Bulletin of the Geological Survey of Great Britain*, **54**, 41–66.

CHANNELL, J. E. T. 1982. Palaeomagnetic stratigraphy as a correlation technique. *In:* ODIN, G. S. (ed.) *Numerical dating in stratigraphy*, John Wiley & Sons, 81–103.

CHRISTENSEN, O. B. 1988. Ostracod zones and dispersion of Mesozoic fossils in the Scandinavian North Sea area. *In:* HANAI, T., IKEYA, N. & ISHIZAKI, K. (eds) *Evolutionary biology of Ostracoda,* Elsevier, Amsterdam, 1269–1281.

—— & KILENYI, T. I. 1970. Ostracod biostratigraphy of the Kimmeridgian in northern and western Europe. *Geological Survey of Denmark*. Series II, **95**.

CIFELLI, R. 1959. Bathonian foraminifera of England. *Bulletin of the Museum of Comparative Zoology at Harvard College*, **121**, 265–368.

COLEMAN, B. E. 1982. Lower and Middle Jurassic foraminifera in the Winterborne Kingston borehole, Dorset. *In:* RHYS, G. H., LOTT, G. K. & CALVER, M. A. (eds) *The Winterborne Kingston borehole, Dorset, England.* Report of the Institute of Geological Sciences **81/3**, 82–88.

COOPER, M. K. E. 1989. Nannofossil provincialism in the Late Jurassic–Early Cretaceous (Kimmeridgian to Valanginian) Period. *In*: CRUX, J. A. & VAN HECK, S. E. (eds) *Nannofossils and their applications*, Ellis Horwood (for British Micro palaeontological Society), 223–246.

COPE, J. C. W. 1967. The palaeontology and stratigraphy of the lower part of the Upper Kimmeridge Clay of Dorset. *Bulletin of the British Museum (Natural History), Geology*, **15**, No. 1.

—— 1978. The ammonite faunas and stratigraphy of the upper part of the Upper Kimmeridge Clay of Dorset. *Palaeontology*, **21**, 469–533.

——, DUFF, K. L., PARSONS, C. F., TORRENS, H. S., WIMBLEDON, W. A. & WRIGHT, J. K. 1980a. *A correlation of Jurassic rocks in the British Isles. Part Two: Middle and Upper Jurassic*. Geological Society, London, Special Report **15**.

——, GETTY, T. A., HOWARTH, M. K., MORTON, N. & TORRENS, H. S. 1980b. *A correlation of Jurassic rocks in the British Isles. Part One: Introduction and Lower Jurassic.* Geological Society, London, Special Report **14**.

COPESTAKE, P. & JOHNSON, B. 1989. The Hettangian to Toarcian (Lower Jurassic). *In*: JENKINS, D. G. & MURRAY, J. W. (eds) *Stratigraphical Atlas of Fossil Foraminifera*, 2nd edition, Ellis Horwood (for British Micropalaeontological Society), 129–188.

COX, B. M. & GALLOIS, R. W. 1979. *Description of the standard stratigraphical sequences of the Upper Kimmeridge Clay, Ampthill Clay and West Walton Beds.* Report of the Institute of Geological Sciences **78/19**, 68–72.

—— & GALLOIS, R. W. 1981. *The stratigraphy of the Kimmeridge Clay of the Dorset type area and its correlation with some other Kimmeridgian sequences*. Report of the Institute of Geological Sciences **80/4**, 1–44.

CRUX, J. A. 1987. Early Jurassic calcareous nannofossil biostratigraphic events. *Newsletters on Stratigraphy*, **17**, 79–100.

DEAN, W. T. 1954. Notes on part of the Upper Lias succession at Blea Wyke, Yorkshire. *Proceedings of the Yorkshire Geological Society*, **29**, 161–179.

——, DONOVAN, D. T. & HOWARTH, M. K. 1961. The Liassic ammonite zones and subzones of the north-west European province. *Bulletin of the British Museum (Natural History), Geology,* **4**, No. 10.

DIETL, G. & CALLOMON, J. H. 1988. Der Orbis-Oolith (Ober-Bathonium, Mittl. Jura) von Sengenthal/Opf., Fränk. Alb, und seine Bedeutung für die Korrelation und Gliederung der Orbis-Zone. *Stuttgarter Beiträge zur Naturkunde*, Serie B, **142**, 1–31.

DOUGLAS, J. A. & ARKELL, W. J. 1928. The stratigraphical distribution of the Cornbrash: I. The south-western area. *Quarterly Journal of the Geological Society of London*, **84**, 117–178.

DUFF, K. L. 1980. Callovian correlation chart. *In*: COPE, J. C. W., DUFF, K. L. *et al. A correlation of Jurassic rocks in the British Isles. Part Two: Middle and Upper Jurassic*. Geological Society, London, Special Report **15**, 45–60.

DYER, R. & COPESTAKE, P. 1989. A review of Late Jurassic to earliest Cretaceous radiolaria and their biostratigraphic potential to petroleum exploration in the North Sea. *In*: BATTEN, D. J. & KEEN, M. C. (eds) *Northwest European micropalaeontology and palynology*, Ellis Horwood (for British Micropalaeontogical Society), 214–235.

GALLOIS, R. W. & COX, B. M. 1976. The stratigraphy of the Lower Kimmeridge Clay of eastern England. *Proceedings of the Yorkshire Geological Society*, **41**, 13–26.

—— & COX, B. M. 1977. The stratigraphy of the Middle and Upper Oxfordian sediments of Fenland. *Proceedings of the Geologists' Association*, **88**, 207–228.

GEORGE, T. N., HARLAND, W. B., AGER, D. V., BALL, H. W., BLOW, W. H., CASEY, R., HOLLAND, C. H., HUGHES, N. F., KELLAWAY, G. A., KENT, P. E., RAMSBOTTOM, W. H. C., STUBBLEFIELD, C. J. & WOODLAND, A. W. 1969. Recommendations on stratigraphical usage. *Proceedings of the Geological Society of London*, **1656**, 139–166.

GEYER, O. F. 1964. Die Typuslokalität des Pliensbachium Württemberg Südwestdeutschland. *In*: MAUBEUGE, P. L. (ed.) *Comptes rendus et Mémoires Colloque du Jurassique à Luxembourg 1962*, Institute grand-ducal, Section des Sciences naturelles, physiques et mathématiques, 161–167.

GORDON, W. A. 1970. Biogeography of Jurassic foraminifera. *Bulletin of the Geological Society of America*, **81**, 1689–1704.

HALLAM, A. 1982. The Jurassic climate. *In: Climate in Earth History*, National Academy Press, 159–163.

——, HANCOCK, J. M., LABRECQUE, J. L., LOWRIE, W. & CHANNELL, J. E. T. 1985. Jurassic and Cretaceous geochronology and Jurassic to Paleogene magnetostratigraphy. *In*: SNELLING, N. J. (ed.) *The chronology of the geological record*, Geological Society, London, Memoir **10**, 118–140.

HAQ, B. U. 1978. Calcareous nannoplankton. *In*: HAQ, B. U. & BOERSMA, A. (eds) *Introduction to marine micropalaeontology*, Elsevier, 79–107.

——, HARDENBOL, J. & VAIL, P. R. 1987. Chronology of fluctuating sea levels since the Triassic. *Science*, **235**, 1156–1167.

HARLAND, W. B., COX, A. V., LLEWELLYN, P. G., PICKTON, C. A. G., SMITH, A. G. & WALTERS, R. 1982. *A geological time scale*. Cambridge University Press.

HAY, W. W. 1977. Calcareous Nannofossils. *In*: RAMSAY, A. T. S., (ed.) *Oceanic Micropalaeontology,* Academic, New York, 1055–1200.

HOLLAND, C. H. 1986. Does the golden spike still glitter? *Journal of the Geological Society, London*, **143**, 3–21.

——, AUDLEY-CHARLES, M. G., BASSETT, M. G., COWIE, J. W., CURRY, D., FITCH, F. J., HANCOCK, J. M., HOUSE, M. R., INGHAM, J. K., KENT, P. E., MORTON, N., RAMSBOTTOM, W. H. C., RAWSON, P. F., SMITH, D. B., STUBBLEFIELD, C. J., TORRENS, H. S., WALLACE, P. & WOODLAND, A. W. 1978. *A guide to stratigraphical procedure.* Geological Society, London, Special Report, **11**.

HOWARTH, M. K. 1955. Domerian of the Yorkshire coast. *Proceedings of the Yorkshire Geological Society*, **30**, 147–175.

—— 1962. The Jet Rock Series and the Alum Shale Series of the Yorkshire coast. *Proceedings of the Yorkshire Geological Society*, **33**, 381–422.

—— 1964. The Jurassic Period. *Quarterly Journal of the Geological Society of London,* **120** Supplement, 203–205.

—— 1973. The stratigraphy and ammonite fauna of the Upper Liassic Grey Shales of the Yorkshire coast. *Bulletin of the British Museum (Natural History), Geology,* **24**, No. 4.

KENNEDY, W. J. & COBBAN, W. A. 1977. The role of ammonites in biostratigraphy. *In*: KAUFFMAN, E. G. & HAZEL, J. E. (eds) *Concepts and methods of biostratigraphy*, Dowden, Hutchinson & Ross, 309–320.

—— & ODIN, G. S. 1982. The Jurassic and Cretaceous time scale in 1981. *In*: ODIN, G. S. (ed.) *Numerical dating in stratigraphy*, Wiley, New York, 557–592.

KENT, D. V. & GRADSTEIN, F. M. 1985. A Cretaceous and Jurassic geochronology. *Bulletin of the Geological Society of America*, **96**, 1419–1427.

KILENYI, T. 1978. The Jurassic Part III Callovian-Portlandian. *In*: BATE, R. H. & ROBINSON, E. (eds) *A stratigraphical index of British Ostracoda*, Seel House Press, 259–298.

KNOX, R. W. O'B. 1984. Lithostratigraphy and depositional history of the late Toarcian sequence at Ravenscar, Yorkshire. *Proceedings of the Yorkshire Geological Society*, **45**, 99–108.

LANG, W. D. 1924. The Blue Lias of the Devon and Dorset coasts. *Proceedings of the Geologists' Association*, **35**, 169–185.

—— 1936. The Green Ammonite Beds of the Dorset Lias. *Quarterly Journal of the Geological Society of London*, **92**, 423–437, 485–487.

——, SPATH, L. F., COX, L. R. & MUIR-WOOD, H. M. 1926. The Black Marl of Black Ven and Stonebarrow, in the Lias of the Dorset coast. *Quarterly Journal of the Geological Society of London*, **82**, 144–187.

——, SPATH, L. F., COX, L. R. & MUIR-WOOD, H. M., 1928. The Belemnite Marls of Charmouth, a series in the Lias of the Dorset coast. *Quarterly Journal of the Geological Society of London*, **84**, 179–257.

——, SPATH, L. F. & RICHARDSON, W. A. 1923. Shales-with-'Beef', a sequence in the Lower Lias of the Dorset coast. *Quarterly Journal of the Geological Society of London*, **79**, 47–99.

LORD, A. R. 1978. The Jurassic Part I (Hettangian-Toarcian). *In*: BATE, R. H. & ROBINSON, E. (eds) *A stratigraphical index of British Ostracoda*, Seel House Press, 189–212.

—— & TAYLOR, R. J. 1982. Introduction. *In*: LORD, A. R. (ed.) *A stratigraphical index of calcareous nannofossils*, Ellis Horwood (for British Micropalaeontological Society), 7–15.

LUTZE, G. F. 1960. Zur Stratigraphie und Paläontologie des Callovien und Oxfordien in Nordwest-Deutschland. *Geologisches Jahrbuch*, **77**, 391–531.

MANGOLD, C. 1984. Report of the Bathonian Working Group. *In*: Michelsen, O. & Zeiss, A. (eds) *International Symposium on Jurassic stratigraphy*. Symposium volume 1, 67–75. Geological Survey of Denmark.

MEDD, A. W. 1979. The Upper Jurassic coccoliths from the Haddenham and Gamlingay boreholes (Cambridgeshire, England). *Eclogae geologicae Helvetiae*, **72**, 19–109.

—— 1982. Nannofossil zonation of the English Middle and Upper Jurassic. *Marine Micropalaeontology*, **7**, 73–95.

MICHELSEN, O. 1975. *Lower Jurassic biostratigraphy and ostracods of the Danish Embayment.* Geological Survey of Denmark. Series II, **104**.

—— & ZEISS, A. (eds) 1984. *International Symposium on Jurassic stratigraphy*. Symposium volumes I–III. Geological Survey of Denmark.

MORRIS, P. H. & COLEMAN, B. E. 1989. The Aalenian to Callovian (Middle Jurassic). *In*: JENKINS, D. G. & MURRAY, J. W. (eds) *Stratigraphical Atlas of Fossil Foraminifera*, 2nd edition, 189–

236. Ellis Horwood (for British Micropalaeontological Society).

MORTON, N. (ed.) 1974. The definition of standard Jurassic stages. *Mémoires du Bureau de Recherches Géologiques et Minières*, **75** (for 1971), 83–93.

NEALE, J. W. 1988. Ostracoda — a historical perspective. *In*: HANAI, T., IKEYA, N. & ISHIZAKI, K. (eds) *Evolutionary biology of Ostracoda*, 3–15. Elsevier, Amsterdam.

NIKITIN, S. 1881. Die Jura-Ablagerungen zwischen Rybinsk, Mologa and Myschkin an der oberen Wolga. *Mémoires de l'Academie Imperiale des Sciences de St Pétersbourg*, Series 7, **28**, number 5.

ODIN, G. S. 1984. Geochronology of the Jurassic time: status in 1984. *In*: MICHELSEN, O. & ZEISS, A. (eds) *International Symposium on Jurassic stratigraphy*. Symposium volume 3, 768–776. Geological Survey of Denmark.

OERTLI, H. J., GROSDIDIER, E. & LE FEVRE, J. 1983. Ostracoda in petroleum exploration. *In*: MADDOCKS, R. F. (ed.) *Applications of Ostracoda*, 19–34. University of Houston.

OGG, J. G. & STEINER, M. B. 1984. Jurassic magnetic polarity time scale: current status and compilation. *In*: MICHELSEN, O. & ZEISS, A. (eds) *International Symposium on Jurassic stratigraphy. Symposium volume 3*, 778–794. Geological Survey of Denmark.

OPPEL, A. 1865. Die tithonische Etage. *Zeitschrift der Deutschen geologischen Gesellschaft*, **17**, 535–558.

ORBIGNY, A.d'. 1850. Résumé géologique: Division des terrains jurassiques en étages. *Paléontologie Française. Terrain jurassique. I Céphalopodes*, 600–611. Victor Masson.

PAGE, K. N. 1989. A stratigraphical revision for the English Lower Callovian. *Proceedings of the Geologists' Association*, **100**, 363–382.

PALMER, C. P. 1972. The Lower Lias (Lower Jurassic) between Watchet and Lilstock in North Somerset (United Kingdom). *Newsletters on Stratigraphy*, **2**, 1–30.

PARK, SE-MOON 1984. Lower Jurassic (Hettangian-Lower Pliensbachian) ostracoda from around the Southern North Sea Basin. *PhD thesis*, University College, London.

PARSONS, C. F. 1974. The *sauzei* and 'so-called' *sowerbyi* Zones of the Lower Bajocian. *Newsletters on Stratigraphy*, **3**, 153–180.

—— 1976. A stratigraphic revision of the *humphresianum*/*subfurcatum* Zone rocks (Bajocian Stage, Middle Jurassic) of southern England. *Newsletters on Stratigraphy*, **5**, 114–142.

—— 1980. Aalenian and Bajocian correlation chart. *In*: COPE, J. C. W., DUFF, K. L. *et al. A correlation of Jurassic rocks in the British Isles. Part Two: Middle and Upper Jurassic.* Geological Society, London, Special Report, **15**, 3–21.

PENN, I. E., MERRIMAN, R. J. & WYATT, R. J. 1979. *The Bathonian strata of the Bath–Frome area*. Report of the Institute of Geological Sciences, **78/22.**

PESSAGNO, E. A. 1977. Radiolaria in Mesozoic stratigraphy. *In*: RAMSAY, A. T. S., (ed.) *Oceanic Micropalaeontology*, Academic, New York, 913–950.

——, BLOME, C. D. & LONGORIA, J. F. 1984. A revised radiolarian zonation for the Upper Jurassic of western North America. *Bulletins of American Palaeontology*, **87**, 1–51.

POKORNY, V. 1978. Ostracodes. *In*: HAQ, B. U. & BOERSMA, A. (eds) *Introduction to marine micropalaeontology*, Elsevier, Amsterdam, 109–149.

POWELL, J. H. 1984. Lithostratigraphical nomenclature of the Lias Group in the Yorkshire Basin. *Proceedings of the Yorkshire Geological Society*, **45**, 51–57.

RAWSON, P. F. & RILEY, L. A. 1982. Latest Jurassic–Early Cretaceous Events and the 'Late Cimmerian Unconformity' in North Sea area. *Bulletin of the American Association of Petroleum Geologists*, **66**, 2628–2648.

RIDING, J. B. 1984. Dinoflagellate cyst range-top biostratigraphy of the uppermost Triassic to lowermost Cretaceous of northwest Europe. *Palynology*, **8**, 195–210.

—— & THOMAS, J. E. 1988. Dinoflagellate cyst stratigraphy of the Kimmeridge Clay (Upper Jurassic) from the Dorset coast, southern England. *Palynology*, **12**, 65–88.

SAPUNOV, I. G. 1964. Notes on the boundary between the Lower and Middle Jurassic and on the stage term Aalenian. *In*: MAUBEUGE, P. L. (ed.) *Comptes rendus et Mémoires Colloque du Jurassique à Luxembourg 1962*. Institut grand-ducal, Section des Sciences naturelles, physiques et mathématiques, 221–228.

SARJEANT, W. A. S. 1974. *Fossil and living dinoflagellates*. Academic, New York.

SELLWOOD, B. W. 1978. Jurassic, *In*: MCKERROW, W. S. (ed.) *The ecology of fossils*, Duckworth, 204–279.

SHEPPARD, L. M. 1981. Bathonian ostracod correlation north and south of the English Channel with the description of two new Bathonian ostracods. *In*: NEALE, J. W. & BRASIER, M. D. (eds) *Microfossils from Recent and fossil shelf seas*, Ellis Horwood (for British Micropalaeontological Society), 73 89.

SHIPP, D. J. 1989. The Oxfordian to Portlandian. *In*: JENKINS, D. G. & MURRAY, J. W. (eds) *Stratigraphical Atlas of Fossil Foraminifera*, 2nd edition, Ellis Horwood (for British Micropalaeontological Society), 237–262.

STEINER, M., OGG, J. & SANDOVAL, J. 1987. Jurassic magnetostratigraphy, 3. Bathonian-Bajocian of Carcabuey, Sierra Harana and Campillo de Arenas (Subbetic Cordillera, southern Spain). *Earth and Planetary Science Letters*, **82**, 357–372.

STURANI, C. 1967. Ammonites and stratigraphy of the Bathonian in the Digne-Barrême area (south-eastern France, dept. Basses-Alpes). *Bollettino della Società Paleontologia Italiana*, **5**, 3–57.

SYKES, R. M. & CALLOMON, J. H. 1979. The *Amoeboceras* zonation of the Boreal Upper Oxfordian. *Palaeontology*, **22**, 839–903.

TAYLOR, R. J. & HAMILTON, G. B. 1982. Techniques. *In*: LORD, A. R. (ed.) *A stratigraphical index of calcareous nannofossils*, Ellis Horwood (for British Micropalaeontological Society), 11–15.

Thierstein, H. R. 1976. Mesozoic calcareous nannoplankton biostratigraphy of marine sediments. *Marine Micropalaeontology*, **1**, 325–362.

Torrens, H. S. 1974. Standard zones of the Bathonian. *Mémoires du Bureau de Recherches Géologiques et Minières*, **75** (for 1971), 581–604.

—— 1980a. *Jurassic stratigraphy in practice.* Geological Society, London, Special Report **14**, 2–4.

—— 1980b. *Means of correlation of Jurassic rocks.* Geological Society, London, Special Report **14**, 4–16.

—— 1987. Ammonites and stratigraphy of the Bathonian rocks in the Digne-Barrême area (South-Eastern France, Dept. Alpes de Haute Provence). *Bollettino della Società Palaeontologica Italiana*, **26**, 93–108.

Townson, W. G. 1975. Lithostratigraphy and deposition of the type Portlandian. *Journal of the Geological Society, London*, **131**, 619–638.

Vail, P. R., Hardenbol, J. & Todd, R. G. 1984. Jurassic unconformities, chronostratigraphy, and sea-level changes from seismic stratigraphy and biostratigraphy. *In*: Schlee, J. S. (ed.) *Interregional unconformities and hydrocarbon accumulation*, American Association of Petroleum Geologists Memoir **36**, 129–144.

—— & Todd, R. G. 1981. Northern North Sea Jurassic unconformities, chronostratigraphy and sea-level changes from seismic stratigraphy. *In:* Illing, L. V. & Hobson, G. D. (eds) *Petroleum geology of the Continental Shelf of North-West Europe*, Heyden, 216–235.

Van Hinte, J. E. 1976. A Jurassic time-scale. *Bulletin of the American Association of Petroleum Geologists*, **60**, 489–497.

Westermann, G. E. G. 1958. Ammoniten-Fauna und Stratigraphie des Bathonien N.W. Deutschlands. *Beihefte zum Geologischen Jahrbuch*, **32**, 1–103.

—— 1984. Gauging the duration of Stages: a new approach for the Jurassic. *Episodes*, **7**, 26–28.

Wilkinson, I. P. 1983a. Kimmeridge Clay Ostracoda of the North Wootton Borehole, Norfolk, England. *Journal of Micropalaeontology*, **2**, 17–29.

—— 1983b. Biostratigraphical and environmental aspects of Ostracoda from the Upper Kimmeridgian of eastern England. *In*: Maddocks, R. F. (ed.) *Applications of Ostracoda*, University of Houston, 165–181.

Wimbledon, W. A. 1976. The Portland Beds (Upper Jurassic) of Wiltshire. *The Wiltshire Natural History Magazine*, **71**, 3–11.

—— & Cope, J. C. W. 1978. The ammonite faunas of the English Portland Beds and the zones of the Portlandian Stage. *Journal of the Geological Society of London*, **135**, 183–190.

Woollam, R. & Riding, J. B. 1983. *Dinoflagellate cyst zonation of the English Jurassic.* Report of the Institute of Geological Sciences, **83/2**.

Wright, J. K. 1969. Callovian. *In:* Hemingway, J. E., Wright, J. K. & Torrens, H. S. (eds) *International Field Symposium on the British Jurassic. Excursion No. 3. Guide for North East Yorkshire,* C10–C17. Geology Department, Keele University.

—— 1977. The Cornbrash Formation (Callovian) in North Yorkshire and Cleveland. *Proceedings of the Yorkshire Geological Society*, **41**, 325–346.

—— 1983. The Lower Oxfordian of North Yorkshire. *Proceedings of the Yorkshire Geological Society*, **44**, 249–281.

Note added in proof. The Proceedings of the 2nd International Symposium on Jurassic stratigraphy held in Lisboa in 1987 were not available to the author. The formal publication date for the two volumes (edited by R. B. Rocha and A. F. Soares) was May 1989 but distribution has been erratic. They include some up-dated information on basal boundary stratotypes for the Jurassic stages.

The early to mid-Jurassic evolution of the northern North Sea

PHILIP C. RICHARDS

British Geological Survey, 19 Grange Terrace, Edinburgh, EH9 2LF, UK

Abstract: The early to mid-Jurassic history of the northern North Sea is summarized as a complex transgressive–regressive–transgressive cycle controlled by variations in 'long-term' eustatic sea level rise, basin subsidence/margin uplift and sediment supply. The first of the transgressions occurred in the Hettangian when the Boreal Ocean spread southwards during a phase of fluctuating but generally increasing sea level rise. Offshore marine conditions developed in the north of the area during the Hettangian to Pliensbachian, while coastal to non-marine conditions developed around the ocean margins in the Unst Basin, the Beryl Embayment, the central Viking Graben and over parts of the Horda Platform. The coastal systems in the Beryl Embayment and central Viking Graben were drowned in the late Pliensbachian as a result of continued sea level rise. A widespread marine link was established with the Tethys Ocean to the south at this time. Sea level fall and/or basin margin uplift and erosion in the Aalenian led to the shedding of coarse clastic sediment transversely into the East Shetland Basin and the northern end of the Viking Graben. The main reservoir unit of the area, the Middle Jurassic Brent Group, formed in this way. A marine connection was maintained into the central Viking Graben area throughout the transverse progradation further north. Drowning of the progradational clastics and the start of the third major transgression began in the late Bajocian, but offshore conditions did not become fully established throughout the area until the Callovian or early Oxfordian.

Most of Britain's offshore oil reserves are found in Jurassic sandstones in the East Shetland Basin (Department of Energy 1988), an area of tilted fault blocks along part of the western margin of the Viking Graben in the northern North Sea (Fig. 1). The graben is the main structural element of the northern North Sea and is a 1000 km long, NNE–SSW trending rift which cross-cuts the Caledonian structural grain. It was probably initiated in the late Permian (Badley *et al.* 1988) and forms part of a trilete rift system which includes the Witch Ground and Central Grabens.

One often quoted model for the early and mid-Jurassic evolution of the northern North Sea envisages deposition of marine, Lower Jurassic strata throughout the graben area. This is then thought to have been followed by thermal doming at or near the confluence of the three grabens, with the consequent erosion of Lower Jurassic strata, and the disruption of a long established link between the Tethys and Boreal Oceans. Erosional products of this thermal uplift are usually thought to have been shed radially away from the domed area. The Middle Jurassic reservoir sandstones of the northern North Sea were supposedly derived in this way, and prograded northwards down the palaeoslope, by the migration of concentric facies belts along the Viking Graben (Budding & Inglin 1981; Eynon 1981; Ziegler 1982; Johnson & Stewart 1985; Graué *et al.* 1987).

This paper discusses the depositional patterns of the Lower and Middle Jurassic sediments of the northern North Sea and relates these patterns to the effects of basin subsidence, domal uplift and sea level change, to produce a new model for the early and mid-Jurassic evolution of the area. The model is illustrated by a new series of palaeogeographic maps based on over 300 wells. One important facet of the study is the inclusion of data from the Beryl Embayment and central Viking Graben (Fig. 1). Previous attempts to model the evolution of the basin have largely ignored data from this area but the nature of the sequence there indicates the limitations of models derived without reference to the area.

The early Jurassic

The Lower Jurassic sediments of the northern North Sea are referred to the Dunlin Group (Deegan & Scull 1977). The Group occurs over much of the area, attaining thicknesses of over 600 m according to Skarpnes *et al.* (1980) or 1000 m according to Badley *et al.* (1988) in the axial parts of the Viking Graben. The nature of the Group in various parts of the basin is reviewed briefly, followed by an account of early Jurassic palaeogeography.

From Hardman, R. F. P. & Brooks, J. (eds), 1990, *Tectonic Events Responsible for Britain's Oil and Gas Reserves*, Geological Society Special Publication No 55, pp 191–205.

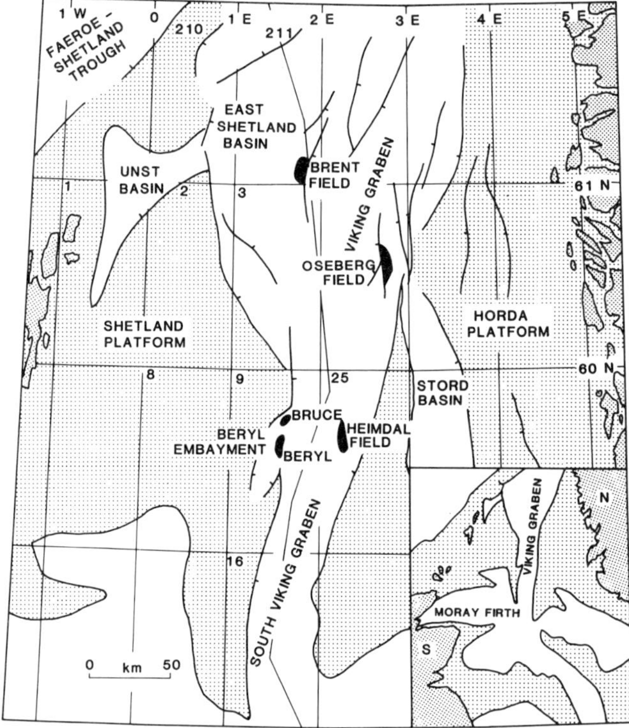

Fig. 1. Major structural elements of the northern North Sea.

Lower Jurassic sediments in the East Shetland Basin

Almost all of the 240 non-confidential exploration and appraisal wells in the East Shetland Basin penetrate Dunlin Group sediments, although the sequence is absent in a small number of wells on the crests of tilted fault blocks (Fig. 2).

The Group here consists of four siltstone or mudstone dominated formations overlying Rhaetian to Hettangian Statfjord Formation fan delta sediments (Roe & Steel 1985).

The Amundsen Formation at the base of the Group (Fig. 3) is probably of Hettangian to Sinemurian age. It is up to 109 m thick in this area and is composed predominantly of grey, bioturbated siltstones often with *Planolites* and *Teichichnus* burrows. Silty sandstones with *Terebellina* burrows, and sharp-based, coarser-grained sandstones occur at some levels. The sandstones occur over most of the basin, and their thickest trend appears to run sub-parallel to the basin margin, crossing the structural grain (Fig. 4a). Sandstones at the base of the formation are similar to those that comprise the underlying Stafjord Formation fan delta.

Hay (1978) and Skarpnes *et al.* (1980) interpreted the Amundsen Formation as a transgressive sequence deposited as a result of the southwards expansion of the Boreal Ocean. The sandy Nansen Member at the top of the underlying Statfjord Formation is of shallow marine origin (Roe & Steel 1985) and probably marks the actual start of the transgression, while the Amundsen Formation represents the first establishment of fully marine conditions. The generally fine grain size suggests deposition largely from suspension, and the bioturbation types indicate offshore conditions.

The overlying Burton Formation is probably of Sinemurian to Pliensbachian age and is over 65 m thick in the East Shetland Basin. Although not cored, the formation appears from cuttings and downhole log responses to be lithologically similar to the Amundsen Formation. The two formations probably have a similar offshore marine origin.

The Cook Formation is of Pliensbachian age. Like the underlying units, most of the formation in the East Shetland Basin is composed of grey siltstones, but thin beds of very fine to coarse grained sandstone comprise up to 20% of the succession in places, especially in the east of the

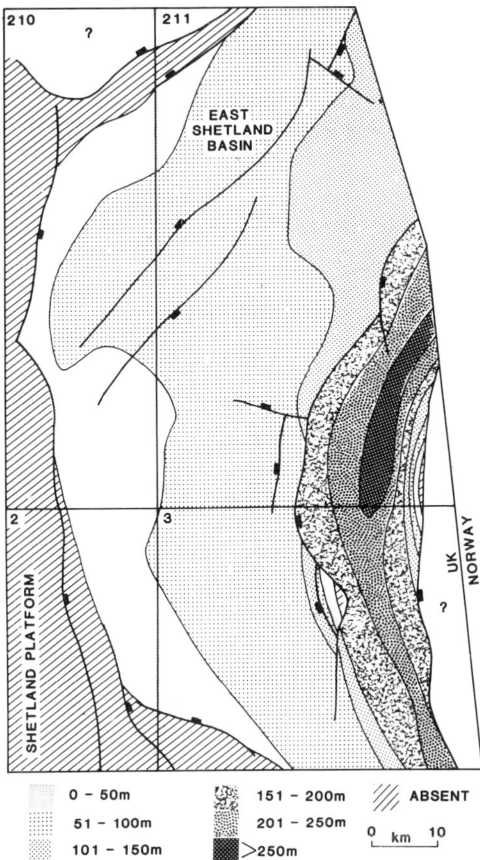

Fig. 2. Distribution of the Lower Jurassic Dunlin Group in the East Shetland Basin. The Group thickens eastwards, towards the axis of the Viking Graben, but thins locally onto some intra-basin faulted highs. All four Dunlin Group formations have the same area of distribution in the East Shetland Basin.

area (Fig. 4b). There appears to be no areal or vertical pattern to the distribution of grain size trends within the sandstones, which are often argillaceous, micaceous, chloritic and cemented by calcite.

The Cook Formation is probably also of offshore marine origin in the East Shetland Basin, and Vollset & Doré (1984) have interpreted the thin sandstones in the east of the area as turbidites derived from the east or southeast. Skarpnes *et al.* (1980) suggested that the Cook Formation is less widely distributed than the Burton Formation, and attributed this to the possibility that it was deposited during the minor, mid-Pliensbachian regressive event recognized by Vail *et al.* (1977). Although this is difficult to substantiate in the East Shetland Basin because the two formations there have similar distributions and the base of the Cook Formation cannot be dated accurately enough to link it to any specific short-term sea level fall, the Cook Formation does appear to correspond generally to the low point on the long-term eustatic sea level curve (Fig. 3) of Haq *et al.* (1987).

The Drake Formation at the top of the sequence is of late Pliensbachian to probably late Toarcian age and is up to 60 m thick in the East Shetland Basin. The formation is finer grained than underlying units and was termed the 'Shale Member' by Bowen (1975). There are no sandstones in this formation but the top of the sequence is interbedded over a few centimetres with the overlying, sandy Broom Formation (Richards & Brown 1987).

This formation is also probably of offshore marine origin. Skarpnes *et al.* (1980) have suggested it is much more widely distributed over the whole of the northern North Sea than other formations of the Dunlin Group. The finer grain size and absence of sandstones may suggest that it represents a slightly deeper water environment than underlying units. The formation may record the acme of the early Jurassic transgression of the northern North Sea, and corresponds to the high point of early Jurassic long-term sea level (Fig. 3) as defined by Haq *et al.* (1987).

Lower Jurassic sediments in the Norwegian area east of the East Shetland Basin

Although the Amundsen, Burton and Drake Formations are generally similar here to the East Shetland Basin, the Cook Formation is much sandier, and an additional sand dominated unit (the Johansen Formation of Vollset & Doré 1984) is also recognized (Fig. 3).

Cook Formation sandstones form minor reservoirs in the Gullfaks Field (Hazeu 1981), the Oseberg Field (Larsen *et al.* 1981), the Statfjord Field (Buza & Unneberg 1987) and the Snorre Field (Karlsson 1986), and have been interpreted as prograding shelf sands in the Horda Platform area and as marine shoal sands in the more basinal area to the west (Vollset & Doré 1984). Hazeu (1981) and Buza & Unneberg (1987) have also interpreted these sandstones as shallow marine or coastal deposits.

The Johansen Formation, found in the Horda Platform area (Fig. 1), is of Sinemurian to Pliensbachian age, and is equivalent to the Amundsen and Burton Formations (Fig. 3). Vollset & Doré (1984) describe the formation as a high-energy, shallow marine shelf deposit.

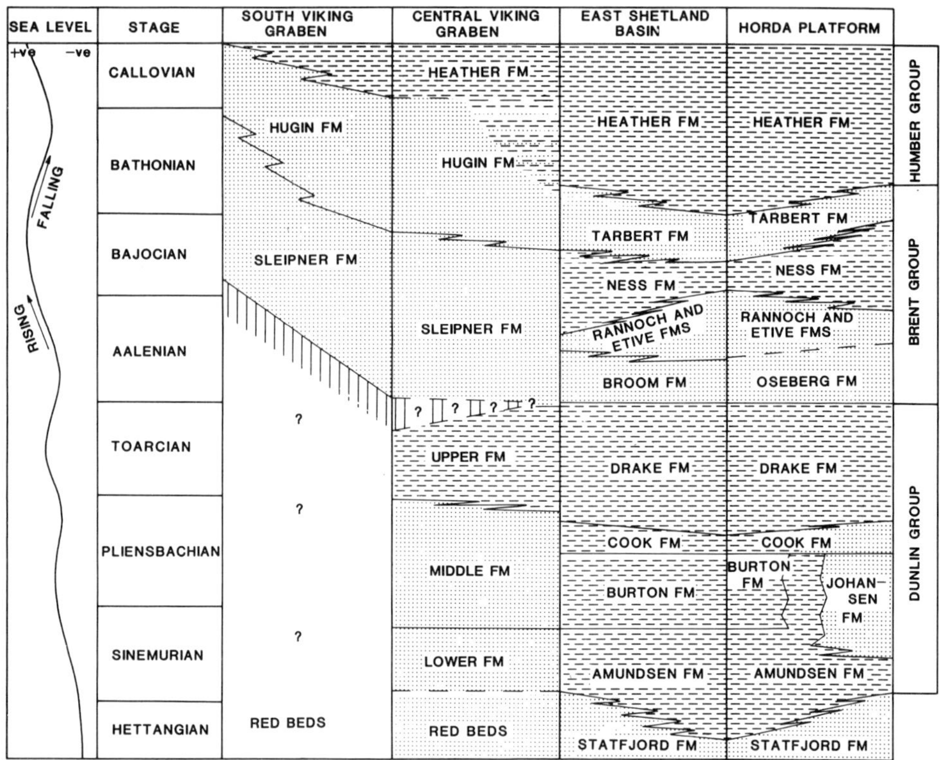

Fig. 3. Stratigraphic summary of Lower and Middle Jurassic strata in the northern North Sea. Ages of stratigraphic units based on present study, plus Deegan & Scull (1977), Vollset & Doré (1984), Graue *et al.* (1987) and Brown *et al.* (1987). The sea level curve shown is the 'long-term eustatic' curve of Haq *et al.* (1987).

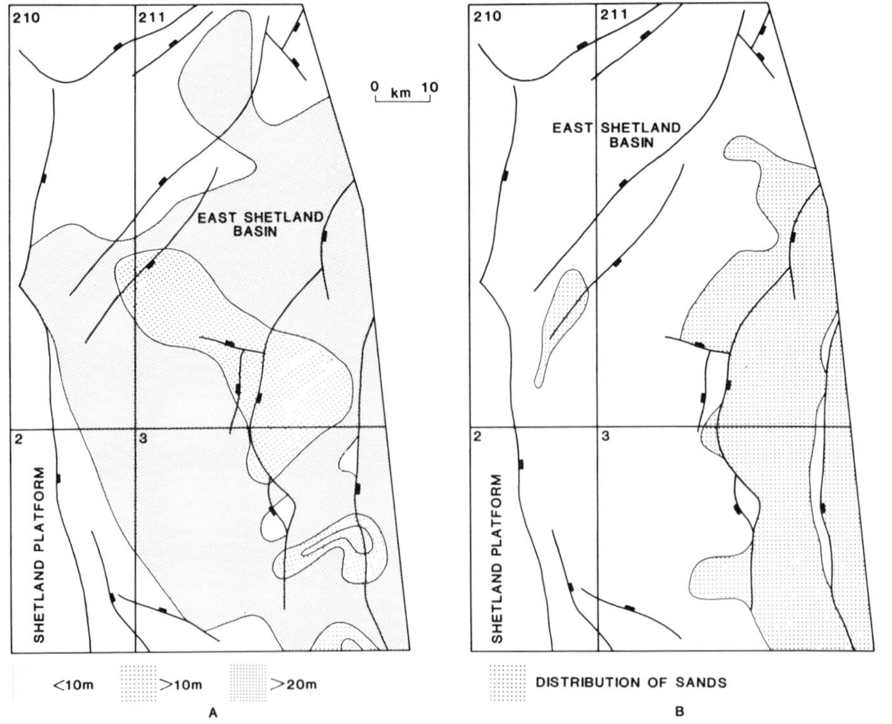

Lower Jurassic sediments in the Beryl Embayment and central Viking Graben

Over 20 non-confidential wells penetrate Lower Jurassic strata in the UK sector, proving a sequence significantly different to that in the East Shetland Basin. Three distinct units are recognized in the Dunlin Group in this area (Fig. 3).

The basal unit here is of Sinemurian age (J. B. Riding, pers. comm.) and is composed of interbedded sandstones, siltstones and minor coals (Fig. 5). It occurs over much of the Beryl Embayment and has also been proven in some wells in the graben area to the east of the Embayment. The sandstones often fine up in units up to 2.6 m thick above sharp bases, while mudstones range from blocky, with abundant root structures and fresh and brackish water palynomorphs, to fissile bedded, with abundant plant debris.

This basal unit is age equivalent to the offshore mudstones of the Amundsen Formation and the shallow marine sandstones of the Johansen Formation (Fig. 3). The fining-up, sharp-based sandstones probably represent fluvial channel deposits; the blocky, rooted siltstones probably represent marine influenced floodplain marsh deposits; the fissile mudstones lake deposits; and the coals vegetated swamps. The formation as a whole probably represents a marine influenced fluvial plain (Richards 1989).

The overlying unit is of Sinemurian to Pliensbachian age (J. B. Riding, pers. comm.) and consists dominantly of clean, fine to coarse grained, sometimes fining-up sandstones, and minor pebbly mudstones (Fig. 5) with marine palynomorphs and burrows. The unit has the same distribution as the underlying formation, occurring both within the Beryl Embayment and in the graben area to the east.

This sandy unit is partly age equivalent to the shallow marine to coastal sands of the Johansen and Cook Formations in the East Shetland Basin (Fig. 3). The unit is interpreted as a fan delta deposit by Richards (1989).

The unit at the top of the sequence is of late Pliensbachian to Toarcian age (J. B. Riding, pers. comm.). It is composed largely of grey, bioturbated siltstones with offshore-type burrows such as *Planolites* and *Terebellina*, and contains large numbers of marine dinoflagellate cysts. A number of thin sandstone beds occur throughout the unit.

This top unit is age equivalent to the relatively deep water Drake Formation elsewhere (Fig. 3) and is probably an offshore marine deposit representing drowning of the underlying fan delta during the acme of the early Jurassic transgression (Richards 1989).

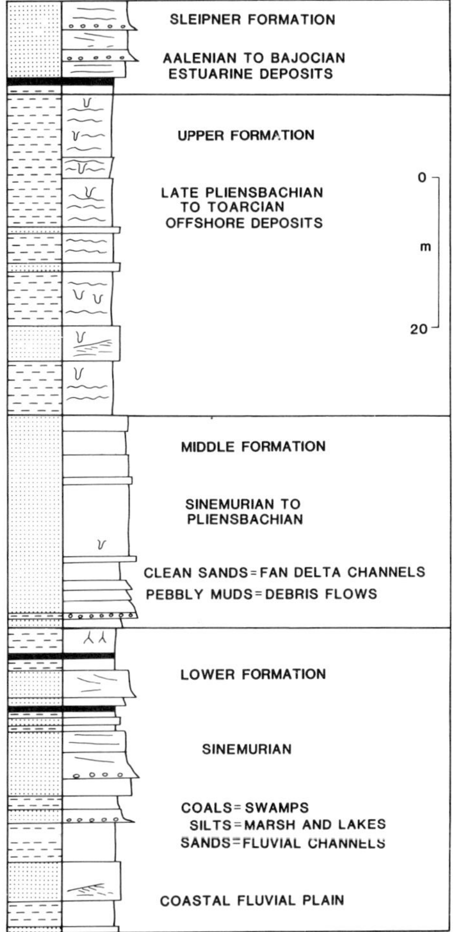

Fig. 5. Simplified core log of the Lower Jurassic Dunlin Group in the Beryl Embayment area. The two lower units represent coastal systems fringing the southern extension of the Boreal Ocean in the Viking Graben. The top unit is an offshore succession representing drowning of the underlying coarse, coastal clastics.

Fig. 4. (A) Distribution of Amundsen Formation sands in the East Shetland Basin. The thickest trend runs northwest to southeast, sub-parallel to the basin margin, and cross-cuts the structural grain. (B) Distribution of sands in the Cook Formation in the East Shetland Basin. Sands comprise up to 20% of the Formation, and are generally restricted to the east of the area.

A similar sequence of Lower Jurassic deposits has also been described by Thomas (1986) from the Heimdal Field on the eastern side of the central Viking Graben. Although the sequence there has been partly attributed to the Statfjord Formation by Thomas (1986), the three subdivisions described above can all be recognized, possibly suggesting that the depositional environments demonstrated in the Beryl Embayment spread across the central Viking Graben area.

Despite the fact that sediments similar to those in the Beryl Embayment are found in the graben area immediately to the east of the Embayment, no Lower Jurassic strata are known from the deepest, axial part of the graben between the Beryl and Heimdal fields, since most released wells there terminate in younger sediments. However, Norwegian well 24/12–1, situated some 25 km southeast of the Beryl Field, displays a 76 m thick succession of Pliensbachian siltstones and thin sandstones which may represent the upper unit of the Dunlin Group. Similar Lower Jurassic sediments have also been reported in UK well 16/28–3 in the south Viking Graben some 75 km south of Beryl, but the early Jurassic age of these siltstones has not yet been proven..

Lower Jurassic sediments in the Unst Basin

Johns & Andrews (1985) described an 81 m thick succession of grey-green coloured sandstones and subordinate siltstones in the Unst Basin as Dunlin Group. These sediments are actually undated, but their stratigraphic presence between Triassic red beds and Brent Group sandstones suggested to Johns & Andrews (1985) that they possibly represent a non-marine clastic equivalent to the marine Dunlin Group strata of the East Shetland Basin.

Early Jurassic palaeogeographic summary

Lower Jurassic sediments in the northern North Sea represent the deposits of a southern extension of the Boreal Ocean during the Hettangian to Toarcian. Ziegler (1982, 1988) has suggested that a full-scale marine connection was established between the Tethys and Boreal Oceans, along the Viking Graben, in the Sinemurian and persisted until thermal doming of the central North Sea area in the mid-Jurassic, when Lower Jurassic strata were eroded from the south Viking Graben. However, the stratigraphic evidence from parts of the northern North Sea is inconsistent with such a palaeogeographic model.

Offshore marine conditions were established in the East Shetland Basin and north Viking Graben by the Hettangian. While offshore marine deposits of the Amundsen, Burton and Cook Formations were accumulating in this northern part of the area during the Sinemurian to late Pliensbachian, sandstone dominated shallow marine to marginal and in some cases non-marine successions were being deposited in the Beryl Embayment, central Viking Graben, Horda Platform and Unst Basin areas (Fig. 6). The presence of coastal and non-marine clastics in the central Viking Graben area therefore prohibits a full-scale marine link with Tethys along the axis of the Viking Graben before the late Pliensbachian. A similar hypothesis has been postulated by Skarpnes *et al.* (1980). Note, however, that marine water may have extended southwards in places along local structural lows within the graben, and similarly also along the axis of the Stord Basin to the east, thereby facilitating a limited early marine link with Tethys, although such a connection cannot be proved from well data. It is likely that floodplain environments similar to those seen in the basal formation in the Beryl Embayment dominated in the south Viking Graben until the late Pliensbachian or Toarcian (Fig. 6).

The contiguous western limit of the Dunlin Group formations in the East Shetland Basin, together with the absence of a sandy, coastal facies in that area may suggest that the present-day western limit is an erosional one rather than an original depositional one. Marine deposition may have extended a short distance onto the Shetland Platform (Fig. 6), with coastal facies there having since been eroded.

Expansion of the marine depositional area during the late Pliensbachian and Toarcian (as a possible result of eustatic sea level rise) led to deposition of offshore argillaceous deposits over the coastal to shallow marine clastics in the Beryl Embayment, central Viking Graben and Horda Platform areas (Fig. 7). Skarpnes *et al.* (1980) suggested that it was this marine expansion event which resulted in the connection of the Tethys and Boreal Oceans for the first time via the Stord Basin. The presence of possible Lower Jurassic strata in Norwegian well 24/12–1 and UK well 16/28–3 possibly also indicates marine conditions in the south Viking Graben at this time. However, the extent of such a marine environment cannot be defined because of the sparsity of well data.

Badley *et al.* (1988) suggested that the early Jurassic was essentially a time of thermal subsidence in the northern North Sea, following a phase of rifting in the Permian to early Triassic.

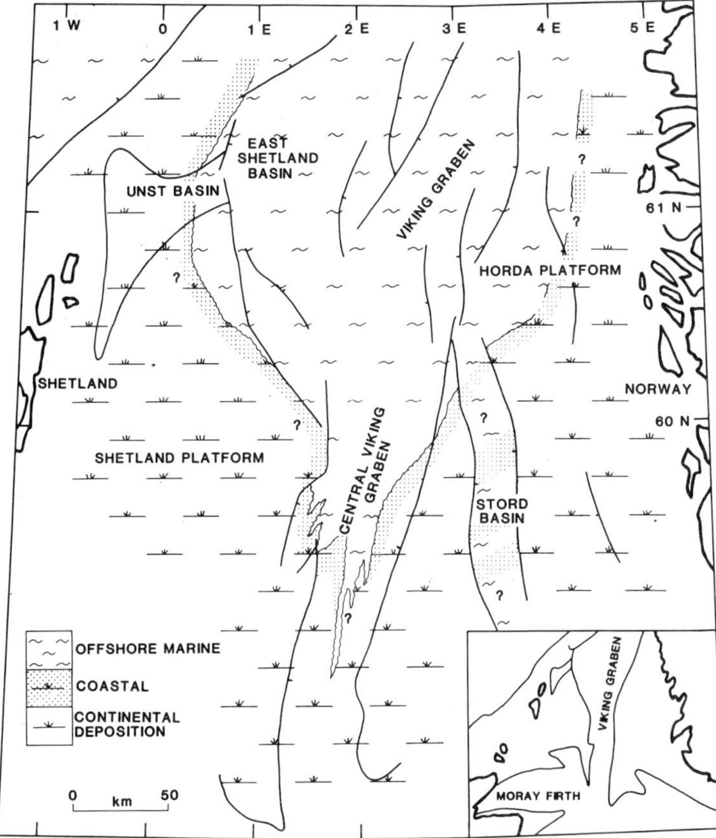

Fig. 6. Palaeogeographic reconstruction of the northern North Sea during the Sinemurian/late Pliensbachian. Offshore siltstones and gravity flow sandstones were deposited in the north of the area while continental and coastal deposits were accumulating in areas such as the Unst Basin and Beryl Embayment. Marine conditions may have extended some way south of the central Viking Graben along structural lows. No well data is available for the Stord Basin, so the nature of the succession there cannot be determined.

Although some thinning of the Dunlin Group is seen onto the crests of tilted fault blocks in the East Shetland Basin which may imply rift-associated tilting, Badley *et al.* (1988) explain such thinning as a result of footwall uplift without syn-depositional rifting. The inferred development noted above of Lower Jurassic marine sediments extending onto the Shetland Platform conforms to Badley *et al.*'s (1988) view that basin margin tectonism (and therefore relief) was negligible during deposition. However, the presence of Pliensbachian fan delta sediments in the Beryl Embayment suggests that basin margin faults were active in places during deposition.

The mid-Jurassic

Richards *et al.* (1988) have presented evidence which suggests that the progradational Middle Jurassic deltaic sediments of the northern North Sea were derived transversely, rather than axially along the graben after erosion of a thermally domed area to the south. A review of the evidence for transverse rather than axial drainage patterns is presented here, followed by a brief analysis of how the progradational sequence was drowned.

The progradational sequence in the East Shetland Basin

The Aalenian to Bajocian progradational part of the Brent Group comprises the Broom, Rannoch, Etive and part of the Ness Formations. Broom Formation sediments at the base of the Group are up to 48 m thick, and thin away from the basin margin fault, pinching out some 55 km

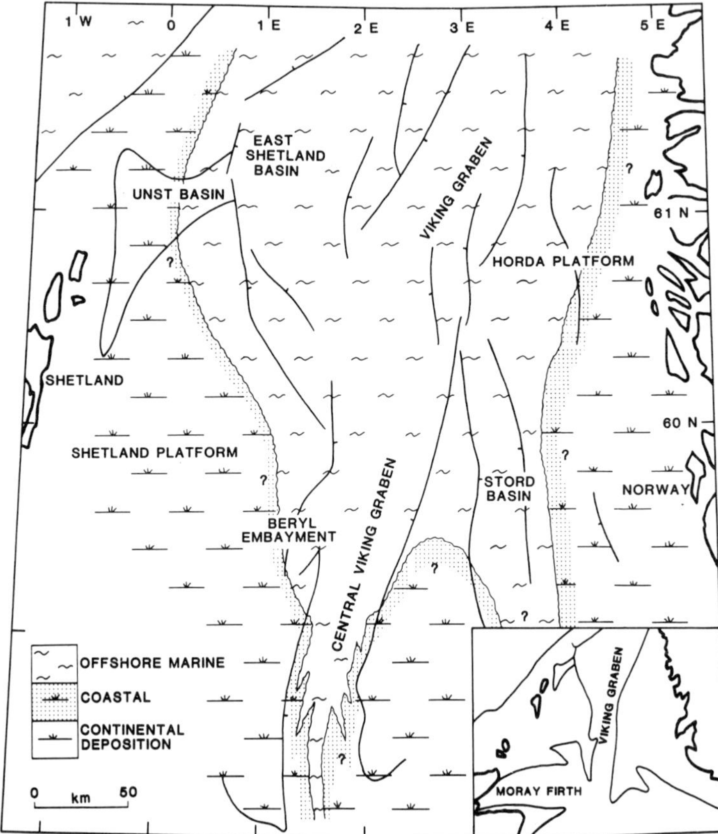

Fig. 7. Palaeogeographic reconstruction of the northern North Sea during the late Pliensbachian/Toarcian. Sea level rise during the late Pliensbachian resulted in expansion of the marine area compared with earlier times (Fig. 6). A full-scale marine link may have existed with the Tethys Ocean to the south via the Stord Basin (Skarpnes *et al.* 1980). A marine connection may also have existed along the axis of the Viking Graben at this time.

into the basin (Brown *et al.* 1987). The Formation is largely composed of medium to coarse grained subarkosic sandstone. In the southwest of the area it comprises cross bedded sets 1 m or so thick, while near its northeastern limit it comprises a thin bed of chamositic, oolitic sand. The textural and mineralogical immaturity of the sediment, and its thickness distribution suggests that it was derived from the Shetland Platform area, possibly as some form of fan delta system (Richards & Brown 1987) like the age equivalent Oseberg Formation (Graué *et al.* 1987) in Norwegian waters.

The Rannoch and Etive Formations, overlying the Broom Formation represent the main, marine progradational phase of the Brent Group in the East Shetland Basin. The Rannoch Formation is dominantly composed of very fine grained, laminated and hummocky cross stratified sandstones, although grain size is coarser in the southwest part of the basin. The Etive Formation is dominantly a fine to medium grained sandstone and occurs above the Rannoch Formation, either as the top part of a coarsening-up sequence, or in a fining-up unit with a sharp base.

Richards & Brown (1986) and Brown & Richards (1989) interpreted the Rannoch Formation as a storm-wave influenced, shoreface deposit and the Etive Formation as a composite barrier bar and channel system. Decrease of grain size away from the southwestern margin of the basin in the Rannoch Formation may indicate that, like the Broom Formation, these marine deposits prograded into the basin from the southwest.

An isopach map of the Rannoch and Etive Formations (Brown *et al.* 1987) shows a major thick zone trending northwest to southeast near the furthest limits of progradation defined by the limits of overlying, largely non-marine, delta top deposits. This thick development of shoreface and barrier deposits trends orthogonally to the direction of sediment derivation postulated above on grain size criteria, and the isopach pattern may therefore be consistent with sediment derivation from the southwest off the Shetland Platform. Budding & Inglin (1981) have also isopached the progradational marine sequence in the South Cormorant oilfield and have noted that the unit was prograding towards the northeast at that locality.

Ness Formation sediments probably accumulated in a 'delta top' setting, and record fluvial and lagoonal environments developed behind the barrier coast (Brown *et al.* 1987). Livera (1989) has demonstrated that channel sands in this formation in the Brent Field sometimes have an east–west orientation. These orientations may reflect the fact that the delta system was prograding towards the east or northeast.

Mineralogical evidence for transverse progradation off the Shetland Platform rather than from an upwarped dome to the south has been presented by Morton & Humphries (1983), Morton (1985) and Hamilton *et al.* (1987). Leeder (1983) has also suggested that the lack of basaltic detritus in the Brent Group sediments mitigates against an origin by erosion of the supposedly Triassic to mid-Jurassic volcanic material (Howitt *et al.* 1975) in a thermally domed region to the south, and has postulated that the Shetland Platform may have been uplifted in the mid Jurassic as a result of non-uniform lithospheric stretching, thereby providing a local source area. Furthermore, Latin *et al.* (this volume) have also suggested that the documented erosion of sediments in oil wells in the Forties area, often cited as evidence of domal uplift, may be no more than the effects of flank and fault-footwall uplift.

New dates for the igneous rocks in the supposed domed area (Ritchie *et al.* 1988) and reinterpretation of the dates of Howitt *et al.* (1975) on the new timescale of Hallam *et al.* (1985) suggests that igneous activity did not occur until the Callovian or Oxfordian, so basaltic detritus would not be expected to occur in the Aalenian to Bajocian Brent Group sediments in any case. Latin *et al.* (1990) have suggested that the Ritchie *et al.* (1988) dates may be the most reliable yet obtained, and point out that based on these dates the igneous rocks could be as young as Volgian or as old as Callovian (depending on the timescale used), but because the overlying sediments are of Oxfordian age they conclude the volcanics are of Callovian or Bathonian age.

Direct evidence that drainage systems on the Shetland Platform provided clastic detritus to the East Shetland Basin is rare because Jurassic sediments are missing over most of the Platform, and Tertiary deposits rest unconformably on ORS and Palaeozoic basement. However, Johns & Andrews (1985) have shown that the Unst Basin contains a coal-bearing Middle Jurassic sequence similar to the delta top deposits in the East Shetland Basin, but with no underlying progradational shoreface and barrier deposits. It is possible that these coal-bearing sediments represent a fluvial system that supplied sediment to the East Shetland Basin. Such a drainage system may originally have covered much of the Platform area (see evidence later for facies belt trends over the Platform).

The progradational sequence in the Norwegian sector

Graué *et al.* (1987) demonstrated that the lowest unit of the Brent Group in the northern Horda Platform and Viking Graben area is a westwards prograding fan delta system up to 60 m thick (the Oseberg Formation), derived from the Norwegian highlands. This formation is age equivalent to the Broom Formation in the East Shetland Basin (Fig. 3) and mirrors its transverse drainage into the basin.

Graué *et al.* (1987) suggested that the Oseberg Formation fan deltas were drowned in the Aalenian to early Bajocian before the northwards prograding Brent Group traversed the area. However, it is possible, considering the evidence presented below that deltaic progradation in the East Shetland Basin could not have been from the south, that the Rannoch and Etive Formation shoreface and barrier deposits above the Oseberg Formation developed in this region by continued progradation from the east following fan delta drowning. Indeed, Hurst & Morton (1988) have shown that the detrital garnet associations found in the Oseberg and Etive Formations in the Oseberg Field are very similar, and they have suggested that the two formations were derived from a similar source area, possibly from the east.

The progradational sequence in the Beryl Embayment and central Viking Graben

Aalenian to early Bajocian sediments in this area do not display the same range of facies as those in the Brent Group further north. The sequence here generally fines upwards from sand dominated units with marine burrows and sharp based, fining-up profiles near the base, through interbedded, laminated sandstones and lenticular bedded siltstones, up to siltstones and in situ coals. Brackish and marine palynomorphs (J. B. Riding, pers. comm.) and bivalves occur throughout, and probable herringbone cross stratification is also seen in places.

Richards (1989) interpreted these Aalenian to early Bajocian sediments as an estuarine succession. The fining-up and marine burrowed sandstones near the base of the sequence here are interpreted as estuarine channel and subtidal shoal deposits. Overlying lenticular bedded siltstones and laminated sandstones (sometimes with herringbones) are interpreted as estuarine tidal flat deposits, and interbedded, blocky siltstones and rooted coals at the top of the sequence are interpreted as supra-tidal marsh deposits. The estuary mouth probably lay towards the north.

Aalenian to early Bajocian palaeogeographic summary

The presence of a marine influenced estuarine succession in the central Viking Graben suggests that a marine connection was maintained into that area throughout deltaic deposition further north (cf. Richards *et al.* 1988). This marine connection was probably with the Boreal Ocean, along the structurally lowest, axial part of the Viking Graben. The presence of an estuarine sequence in the central Viking Graben, of age equivalence to the Brent Group further north, indicates that the Brent Group deltaic facies did not migrate along the Graben from the south. This is supported by evidence in the East Shetland Basin (and at least in the Oseberg Formation over the Horda Platform) that the Brent Group sediments were derived transversely, eroding highland and platform areas to the southwest and possibly the southeast (Fig. 8).

Evidence for transverse drainage in the north Viking Graben area, and marine connection into the central Viking Graben diminishes the importance of the supposed thermally uplifted area to the south as an influence on northern North Sea palaeogeography and sediment supply. The supposedly uplifted area to the south of the central Viking Graben may have been covered by fluvial environments similar to those on the Shetland Platform (Fig. 8). The area may have had a topographic relief of only a few tens of metres (Leeder 1983).

The observed changes at or about the Toarcian/Aalenian boundary from offshore marine deposition to regressive, sand dominated deposition coincides with the maxima of long-term eustatic sea level fall (Fig. 3) as defined by Haq *et al.* (1987). This eustatic sea level fall may have exposed large areas of platform to erosion, contributing to the regressive, transversely derived sedimentation in the basins in the Aalenian to early Bajocian.

Alternatively, sediment may have been derived from the basin margins as a result of tectonic uplift (Leeder 1983), although Badley *et al.* (1988) have suggested that the second rifting phase in the northern North Sea did not occur until the early Bathonian. Badley *et al.* (1988) do, however, acknowledge that thickness variations across faults cutting the Brent Group may represent the effects of precursors to the second rifting stage. In fact, the Broom and Oseberg Formations in particular display the classical elements of a syn-rift package as described by Badley *et al.* (1988) in so far as they are asymmetric packages that thicken into basin margin faults and thin to zero at the basin centre. They may, therefore, represent a discrete phase of rifting, with associated basin margin uplift continuing to supply sediment for the remainder of Brent Group times.

The transgressive sequence

Brent Group deltaic sediments in the East Shetland Basin were transgressed during the late Bajocian, coincident with a peak of long-term eustatic sea level rise (Fig. 3). Transgression resulted in deposition of the sandy Tarbert Formation in the north of the area and the sandy Hugin Formation in the central and southern Viking Graben. Tarbert Formation facies are, in part, somewhat similar to those in the shoreface Rannoch Formation (Brown *et al.* 1987) and the formation was deposited during stepped transgressive and regressive pulses (Ronning & Steel 1987; Graué *et al.* 1987). The lower part of the Hugin Formation in the central Viking Graben is laterally equivalent to the Tarbert Formation and is represented by shallowing-upwards, regressive deposits of barrier inlet and lagoonal origin (Richards 1989). Offshore marine siltstones (of the Heather Formation) may have been accumulating in the deepest parts of the Viking Graben while the Tarbert and lower Hugin Formation sands were accumulating around the margins of the depositional basin (Fig. 9).

Offshore marine conditions represented by Heather Formation siltstones became fully

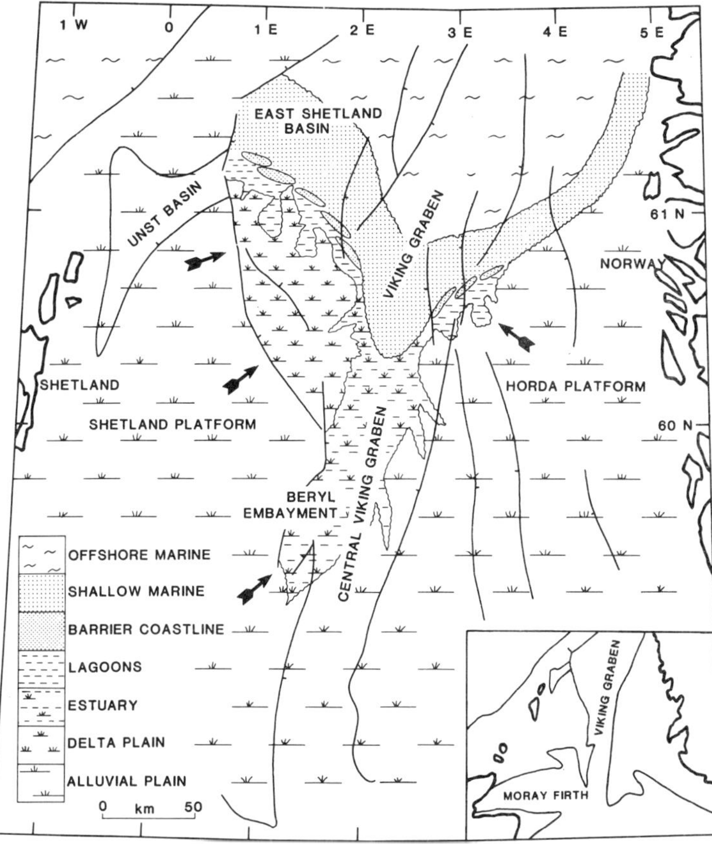

Fig. 8. Palaeogeographic reconstruction of the northern North Sea during the Aalenian/early Bajocian. Brent Group sediments probably prograded transversely into the basin from the southwest, rather than by the axial migration of concentric facies belts along the Viking Graben from the south. An estuary occupied the structurally low area between the transversely prograding 'deltaic' packages. The reconstruction of coastal orientation in the Norwegian sector is based partly on Graué *et al.* (1987).

established in the East Shetland Basin and north Viking Graben in the early Bathonian, but shallow marine to offshore sands and siltstones of the upper part of the Hugin Formation were still being deposited in the Beryl Embayment and southern Viking Graben at this time (Fig. 10). Offshore marine conditions did not become fully established in the central and southern Viking Graben until drowning of the Hugin Formation was achieved in the Callovian or early Oxfordian (Vollset & Doré 1984).

Also, fully marine offshore conditions may not have been established in the Unst Basin until the Callovian (Johns & Andrews 1985). Indeed, a downhole log correlation between the Unst Basin and central Viking Graben (Fig. 11) suggests a similar Middle Jurassic history for the two areas. The correlation suggests that facies belts in the UK sector may have trended sub-parallel to the western margin of the Viking Graben for much of the time, with similar broad environments over the Shetland Platform area between the Unst Basin and the Beryl Embayment (Figs 9 & 10).

Conclusions

During the Sinemurian to late Pliensbachian, a succession of offshore, marine siltstones and minor sandstones were deposited in the East Shetland Basin and north Viking Graben, while coarser, sometimes coal-bearing, fluvial to shallow marine deposits developed in the central Viking Graben and over parts of the adjacent platforms. There may have been a limited marine connection with the Tethys Ocean at

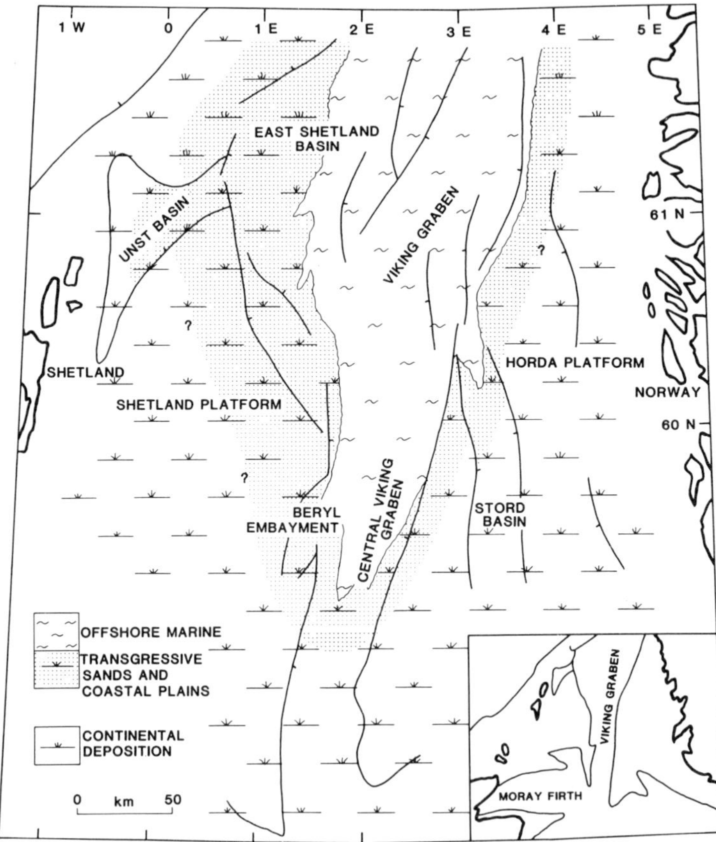

Fig. 9. Palaeogeographic reconstruction of the northern North Sea during the late Bajocian/early Bathonian (in part after Graué *et al.* 1987). The Brent deltas were transgressed as sea level rose, and marine shoreface to back-barrier sediments of the Tarbert and Hugin Formations were deposited over parts of the East Shetland Basin, Shetland Platform, central Viking Graben and Horda Platform while offshore marine conditions (represented by the Heather Formation) developed in the structurally lowest areas. Continental deposition with coals (the Sleipner Formation) continued over land areas surrounding the marine basin.

this time. An increase of sea level in the late Pliensbachian to Toarcian caused drowning of the coastal deposits in the central Viking Graben and Horda Platform areas, and possibly resulted in the establishment of a more widespread Boreal–Tethyan link through the North Sea.

The possible presence of terrestrial deposits over large areas to the south of the central Viking Graben during the earliest Jurassic implies the existence of some form of high ground to the south at the time. This high ground may be part of the thermally domed area envisaged by Ziegler (1982), although its inferred presence during the early Jurassic is earlier than previously postulated. Topographic relief on this 'high ground' may only have been of the order of a few tens of metres (Leeder 1983). Volcanic activity possibly first became established in the domed high ground area in the late Callovian to early Oxfordian (Ritchie *et al.* 1988), soon after initiation of the main rifting phase in the northern North Sea noted by Badley *et al.* (1988), and may have been related to rift-induced subsidence of the domed area.

Eustatic sea level fall and/or basin margin uplift during the Aalenian resulted in erosion of highland and platform areas, and the supply of coarse clastic material transversely into the basin during the Aalenian to early Bajocian. Throughout this time a marine connection was maintained into the central Viking Graben.

The prograding, transversely derived deltaic

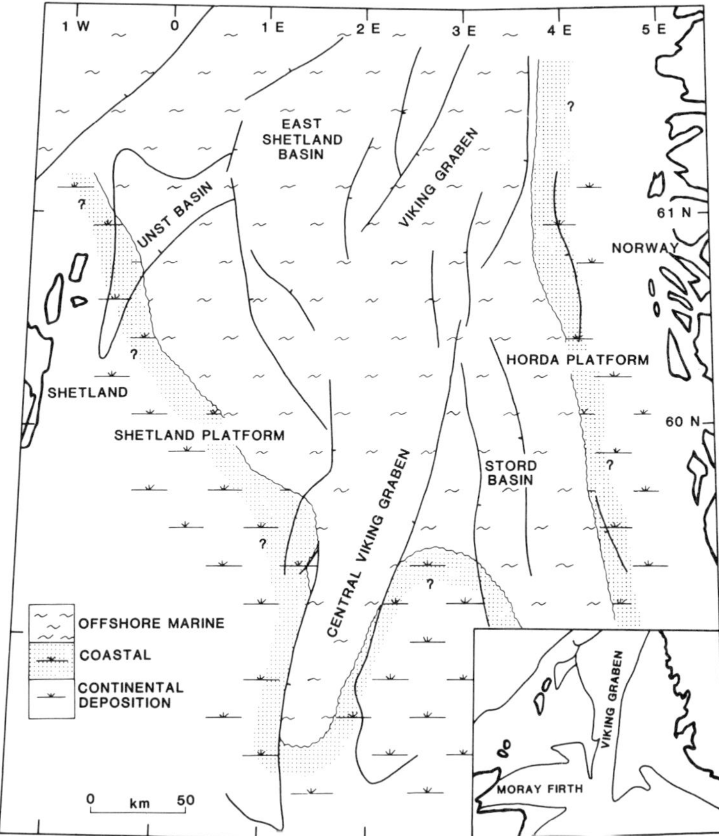

Fig. 10. Palaeogeographic reconstruction of the northern North Sea during the Bathonian/Callovian. Offshore marine conditions (represented by the Heather Formation) had become fully established in the north of the area by this time as a result of generally high global sea levels (Haq *et al.* 1987) and basin subsidence (Badley *et al.* 1988). Coastal sands may have fringed the marine basin over the Shetland Platform and Horda Platform areas but have since been largely removed by erosion. Hugin Formation shallow marine sandstones and Sleipner Formation fluvial sediments were still being deposited in the south of the area at this time.

system in the East Shetland Basin was transgressed during the late Bajocian to early Bathonian, and marine conditions spread down the graben. Although sea level was generally falling during the Bathonian (Haq *et al.* 1987), increased subsidence at this time (Badley *et al.* 1988) facilitated continued transgression. Offshore marine conditions became fully established throughout the Viking Graben by the Callovian or early Oxfordian.

There is no evidence to support the previously favoured model which envisaged deposition of marine strata throughout the Viking Graben from the Sinemurian to the Aalenian, followed by mid Jurassic thermal uplift and erosion near the confluence of the North Sea rifts, and the subsequent shedding of sediment northwards down the graben.

This paper forms part of a PhD project supervised by Drs. Roger Anderton and Stewart Brown, who are thanked for critically reviewing the original manuscript. British Petroleum Development Ltd., Britoil plc, Hamilton Brothers Oil and Gas Ltd., Total Oil Marine plc and their respective partners kindly allowed the use of unreleased well data for the original research project. The UK Department of Energy are thanked for encouraging the production of this paper. Publication is by permission of the Department of Energy and the Director, British Geological Survey (NERC).

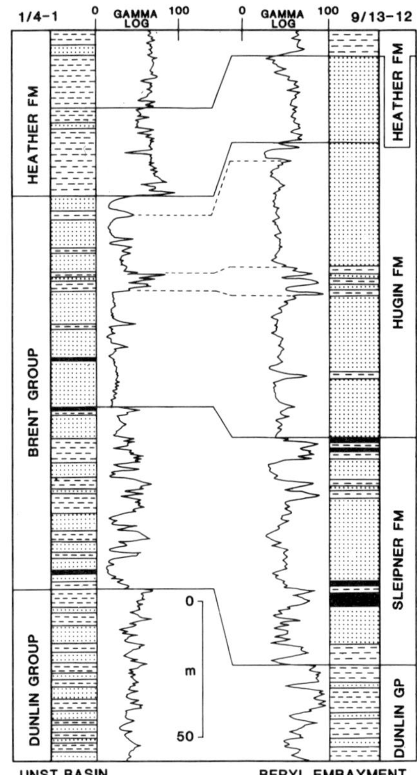

Fig. 11. Correlation of the Lower and Middle Jurassic successions in the Beryl Embayment and Unst Basin. The similarity of succession between the two areas suggests that the two basins had a similar depositional history, and may imply that facies belts ran across the Shetland Platform between the two basins.

References

BADLEY, M. E., PRICE, J. D., DAHL, C. R. & AGDESTEIN, T. 1988. The structural evolution of the northern Viking Graben and its bearing upon extensional models of basin formation. *Journal of the Geological Society, London*, **145**, 455–472.

BOWEN, J. M. 1975. The Brent oilfield. *In*: WOODLAND, A. W. (ed.) *Petroleum and the Continental Shelf of Northwest Europe*. Applied Science, Barking, 353–360.

BROWN, S. & RICHARDS, P. C. 1989. Facies and development of the mid Jurassic Brent delta near the northern limit of its progradation, UK North Sea. *In*: PICKERING, K. J. & WHATELY, M., (eds) *Deltas: sites and traps for fossil fuels*. Geological Society, London, Special Publication, **41**, 253–268.

——, —— & THOMSON, A. R. 1987. Patterns in the deposition of the Brent Group (Middle Jurassic) UK North Sea. *In*: BROOKS, J. & GLENNIE, K. W. (eds) *Petroleum Geology of North West Europe*. Graham & Trotman, London, 899–913.

BUDDING, M. C. & INGLIN, H. F. 1981. A reservoir geological model of the Brent sands in Southern Cormorant. *In*: ILLING, L. V. & HOBSON, G. D. (eds) *Petroleum Geology of the Continental Shelf of North-West Europe*. Heyden, London, 326–334.

BUZA, J. W. & UNNEBERG, A. 1987. Geological and reservoir engineering aspects of the Statfjord field. *In*: *North Sea Oil and Gas Reservoirs*. Graham & Trotman, London, 23–38.

DEEGAN, C. E. & SCULL, B. J. 1977. *A standard lithostratigraphic nomenclature for the Central and Northern North Sea*. Report Institute of Geological Sciences, 77/25.

DEPARTMENT OF ENERGY 1988. *Development of the Oil and Gas resources of the United Kingdom*. HMSO, London.

EYNON, G. 1981. Basin development and sedimentation in the Middle Jurassic of the northern North Sea. *In*: ILLING, L. V. & HOBSON, G. D. (eds) *Petroleum Geology of the Continental Shelf of North-West Europe*. Heyden, London, 196–204.

GRAUÉ, E., HELLAND-HANSEN, W., JOHNSON, J., LXMO, L., NXTTVEDT, A., RXNNING, K., RYSETH, K. & STEEL, R. J. 1987. Advance and retreat of Brent Delta system, Norwegian North Sea. *In*: BROOKS, J. & GLENNIE, K. W. (eds) *Petroleum Geology of North West Europe*. Graham & Trotman, London, 915–937.

HALLAM, A., HANCOCK, J. M., BREQUE, J. L., LOWRIE, W. & CHANNELL, J. E. T. 1985. Jurassic to Palaeogene: part 1. Jurassic and Cretaceous geochronology and Jurassic to Palaeogene magnetostratigraphy. *In*: SNELLING, N. J. (ed.) *The Chronology of the Geological Record*. Geological Society, London, Memoir **10**, 118–140.

HAMILTON, P. J., FALLICK, A. E., MACINTYRE, R. M. & ELLIOTT, S. 1987. Isotope tracing of the provenance and diagenesis of Lower Brent Group sands, North Sea. *In*: BROOKS, J. & GLENNIE, K. W. (eds) *Petroleum Geology of North West Europe*. Graham & Trotman, London, 939–950.

HAQ, U. B., HARDENBOL, J. & VAIL, P. R. 1987. Chronology of fluctuating sea levels since the Triassic. *Science*, **235**, 1156–66.

HAY, J. T. C. 1978. Structural development in the northern North Sea. *Journal of Petroleum Geology*, **1**, 65–77.

HAZEU, G. J. A. 1981. 34/10 Delta structure, geological evaluation and appraisal. *In*: *Norwegian Symposium on Exploration*. Norsk Petroleumsforening, Oslo, 13/1–36.

HOWITT, F., ASTON, E. R. & JACQUI, M. 1975. The occurrence of Jurassic volcanics in the North Sea. *In*: WOODLAND, A. W. (ed.) *Petroleum and the Continental Shelf of Northwest Europe*. Applied Science, Barking, 379–387.

HURST, A. & MORTON, A. C. 1988. An application of heavy-mineral analysis to lithostratigraphy and reservoir modelling in the Oseberg field, northern North Sea. *Marine and Petroleum Geology*, **5**, 157–169.

Johns, C. R. & Andrews, I. J. 1985. The petroleum geology of the Unst Basin, North Sea. *Marine and Petroleum Geology*, **2**, 361–372.

Johnson, H. D. & Stewart, D. J. 1985. Role of clastic sedimentology in the exploration and production of oil and gas in the North Sea. *In*: Brenchley, P. J. & Williams, B. P. J. (eds) *Sedimentology: recent developments and applied aspects*. Geological Society, London, Special Publication, **18**, 249–310.

Karlsson, W. 1986. The Snorre, Statfjord and Gullfaks oilfields and the habitat of hydrocarbons on the Tampen Spur, offshore Norway. *In: Habitat of Hydrocarbons on the Norwegian Continental Shelf*. Graham & Trotman, London, 181–196.

Larsen, V. Assheim, S. M. & Masset, J. M. 1981. 30/6 Alpha structure, a field case study in the Silver Block. *In: Norwegian Symposium on Exploration*. Norsk Petroleumsforening, Oslo, 14/1–34.

Latin, D. M., Dixon, J. E. & Fitton, J. G. 1990. Rift-related magmatism in the North Sea Basin. *In:* Blundell, D. J. & Gibbs, A. (eds) Tectonic Evolution of the North Sea Rifts.

Leeder, M. R. 1983. Lithostratigraphic stretching and North Sea Jurassic clastic sourcelands. *Nature*, **305**, 510–513.

Livera, S. E. 1989. Facies associations and sand body geometries in the Ness Formation of the Brent Group, Brent field. *In*: Pickering, K. J. & Whately, M. (eds) *Deltas: sites and traps for fossil fuels*, Geological Society, London, Special Publication, **41**, 269–286.

Morton, A. C. 1985. A new approach to provenance studies: electron microprobe analysis of detrital garnets from Middle Jurassic sandstones of the northern North Sea. *Sedimentology*, **32**, 553–566.

—— & Humphries, B. 1983. The petrology of the Middle Jurassic sandstones from the Murchison field, North Sea. *Journal of Petroleum Geology*, **5**, 245–260.

Richards, P. C. 1989. *Lower and Middle Jurassic sedimentology of the Beryl Embayment, and implications for the evolution of the northern North Sea*. PhD. thesis, University of Strathclyde.

—— & Brown, S. 1986. Shoreface storm deposits in the Rannoch Formation (Middle Jurassic), North West Hutton oilfield. *Scottish Journal of Geology*, **22**, 367–375.

—— & —— 1987. The nature of the Brent Delta, North Sea: a core workshop. *British Geological Survey, Open File Report* 87/17.

——, ——, Dean, J. M. & Anderton, R. 1988. A new palaeogeographic reconstruction for the Middle Jurassic of the northern North Sea. *Journal of the Geological Society, London*, **145**, 883–886.

Ritchie, J. D., Swallow, J. L., Mitchell, J. G. & Morton, A. C. 1988. Jurassic ages from intrusives and extrusives within the Forties Igneous Province. *Scottish Journal of Geology*, **24**, 81–88.

Roe, S.-L. & Steel, R. J. 1985. Sedimentation, sea-level rise and tectonics at the Triassic-Jurassic boundary (Statfjord Formation), Tampen Spur, northern North Sea. *Journal of Petroleum Geology*, **8**, 163–186.

Ronning, K. & Steel, R. J. 1987. Depositional sequences within a "transgressive" reservoir sandstone unit: the Middle Jurassic Tarbert Formation, Hild area, northern North Sea. *In*: *North Sea Oil and Gas Reservoirs*. Graham & Trotman, London, 169–176.

Skarpnes, O., Hamar, G. P., Jakobson, K. H. & Ormaasen, D. E. 1980. Regional Jurassic setting of the North Sea north of the central highs. *In*: *The Sedimentation of the North Sea Reservoir Rocks*. Norwegian Petroleum Society, 13/1–8.

Thomas, M. J. 1986. Diagenetic sequences and K/Ar dating in Jurassic sandstones, central Viking Graben: effects on reservoir properties. *Clay Minerals*, **21**, 695–670.

Vail, P. R., Mitchum, R. M. & Thompsen, S. 1977. Seismic Stratigraphy and Global Changes of Sea Level, Part 4: Global Cycles of Relative Changes of Sea Level. *In*: Payton, C. E. (ed.) *Seismic stratigraphy: applications to hydrocarbon exploration*. American Association of Petroleum Geologists, Memoir, **26.**

Vollset, J. & Doré, A. G. 1984. A revised Triassic and Jurassic lithostratigraphic nomenclature for the Norwegian North Sea. *Norwegian Petroleum Directorate – Bulletin* **3**.

Ziegler, P. A. 1982. *Geological Atlas of Western and Central Europe*. Shell, The Hague.

—— 1988. Post-Hercynian plate reconstruction in the Tethys and Arctic-North Atlantic domains. *In*: Manspeizer, W. (ed.) *Triassic-Jurassic Rifting: continental breakup and the origin of the Atlantic Ocean and passive margins*. Part B. Elsevier, Amsterdam, 711–755.

Mesozoic magmatic activity in the North Sea Basin: implications for stretching history

DAVID M. LATIN[1,2], JOHN E. DIXON[1], J. GODFREY FITTON & NICKY WHITE[2]

[1] *Department of Geology and Geophysics, University of Edinburgh, Edinburgh EH9 3JW, UK*

[2] *Bullard Laboratories, University of Cambridge, Cambridge CB3 OEZ, UK*

Abstract: Recent developments in the theory underlying magma generation allow the volumes and compositions of melt generated during extension of the lithosphere to be used to make quantitative statements about basin evolution. In the Mesozoic North Sea Basin there is a well defined relationship between magma chemistry and the degree of lithospheric attenuation. The largest extents of melting (<2%) resulted in the alkali basalts of the Forties province which occur in an area where β (the stretching factor) may be as high as 2, and which erupted within 30 million years of the onset of extension. In off-axis areas of the rift the degrees of melting are smaller and melt compositions are more undersaturated and extreme (nephelinites and ultrapotassic rocks). In the Forties province the observed melt compositions may be reconciled with melting on the dry peridotite solidus at a normal (1280°C) potential temperature (T_p) assuming a relatively thin mechanical boundary layer (MBL) of 70 km prior to rifting. Alternatively, a T_p of 1380°C will produce the alkali basalts from dry peridotite at the observed extension factors with an initial MBL thickness of 100 km. A T_p as high as 1480°C, which would give rise to regional pre- and syn-rift uplift, can be ruled out because the observed extent of melting is too small, the amounts of extension too large, and the rift episode too short. Off-axis magmatism may be explained by melting within a MBL with an initial thickness of ~100 km or more at a T_p of 1280°C. Models for extension of the lithosphere by simple shear along an initially planar detachment fault fail to account for either the existence or the location of North Sea magmatism, with or without elevated potential temperatures if the dry solidus governs melting. This conclusion is independent of the initial dip of the detachment and also holds when the initial dip in the lithospheric mantle is double that in the crust. The timing, nature, and location of magmatic activity can be used to provide information about the onset of rifting, the rates of extension, the amount of strain, and the thermal history of a basin. When linked with studies of faulting and subsidence this approach may be of some importance for hydrocarbon exploration.

This paper builds on two recent papers concerning the relationship between rifting and magmatic activity in the North Sea Basin (Latin *et al.* 1990) and the generation of melt by extension of the lithosphere (Latin & White 1990). Here the results of the two papers are summarized, and are used to make inferences about the nature of the rifting process in the North Sea during the Mesozoic.

The last eleven years have seen the development and consolidation of the theoretical framework explaining the formation of extensional sedimentary basins. In 1978, McKenzie proposed the uniform stretching model in which the lithosphere deforms by bulk pure shear. The uniform stretching model, adapted for finite rates of extension (Jarvis & McKenzie 1980), has since been successfully applied to explain the subsidence history of the North Sea (e.g. Sclater & Christie 1980; Barton & Wood 1984; Klemperer 1988; N. J. White 1989, 1990) as well as other basins and passive margins (e.g. Sclater *et al.* 1980; Le Pichon & Sibuet 1981; Sawyer *et al.* 1982, Royden *et al.* 1983).

Magma generation often accompanies basin development. Dixon *et al.* (1981) suggested that differences in the degree and style of extension may exert an important control on magma compositions. More recently the uniform stretching model has been used to estimate the volumes (Foucher *et al.* 1982; McKenzie & Bickle 1988) and compositions (McKenzie & Bickle 1988) of melts generated from the asthenosphere during adiabatic decompression.

In this paper we show how the timing, location, volumes and compositions of magmas produced during the Mesozoic extensional episode in the North Sea may be used to provide information about the stretching history and the nature of lithospheric extension. First we give a brief introduction to the theory of magma production as a result of adiabatic upwelling of the

From HARDMAN, R. F. P. & BROOKS, J. (eds), 1990, *Tectonic Events Responsible for Britain's Oil and Gas Reserves*, Geological Society Special Publication No 55, pp 207–227.

asthenosphere (McKenzie & Bickle 1988). We then consider the important parameters which control the volumes and compositions of melts produced by extension of the lithosphere. The nature of Mesozoic igneous activity is summarized and observed melt compositions are used to show that, at least for some of the North Sea magmas, the asthenosphere underwent melting but that the extent of asthenospheric melting is small. Because the amount of stretching (β), the initial thickness of the mechanical boundary layer, and the duration of rifting are, within limits, known, observations of magma character can be used to constrain the potential temperature at the time of stretching. The location and composition of melt can be used to discriminate between the uniform stretching model and the lithospheric simple shear model (Wernicke 1981, 1985) which has also been applied to the North Sea (e.g. Beach 1986; Gibbs 1987). The magmatic activity occurs in well defined temporal and spatial domains which may relate to the timing, strain rate and propagation of extensional rifts within the Mesozoic North Sea, and to the tectonic history of the North Sea lithosphere. The well defined relationships observed between extension and magmatism in the North Sea may be useful in the exploration of less well studied basins.

Production of melt during extension

Extensional sedimentary basins, such as the North Sea, which occur within continental plates, can be viewed as part of a continuum of stretching which may ultimately lead to the formation of an ocean basin and spreading centre. The production of magma beneath ocean ridges and its subsequent extrusion as MORB (mid-ocean ridge basalt) is the end result of the within-plate process. At an ocean ridge where there is no overlying lithosphere, and the process is essentially one of time-invariant steady-state upwelling, there are far fewer parameters governing magma production than in extensional basins. Quantitative models for magma production are therefore based on the ocean ridge system, because of its simplicity.

Melt production at ocean ridges

It has long been believed that large quantities of melt are generated at ocean ridges as a response to adiabatic upwelling of the asthenosphere (Verhoogen 1954; Green & Ringwood 1967). In a recent paper McKenzie & Bickle (1988) parameterized all the available melting experiments on dry mantle peridotite — thought to represent the source of MORB—to constrain the extent of melting at any temperature and pressure. Because extent of melting controls the volume and composition of melt the parameterization is able to predict these values for magmas generated under different $P-T$ conditions.

In the case of ocean ridges the most important control on the extent of melting is the potential temperature (T_p = the temperature on the adiabatic gradient projected to atmospheric pressure) of the upwelling asthenosphere. Figure 1 shows how the T_p determines the depth at which melting of the asthenosphere starts (where the solidus is intersected) and therefore the extent of melting it has attained by upwelling to the surface where the melt forms the oceanic crust. McKenzie & Bickle (1988) noted that the generally uniform, 6–7 km thickness of oceanic crust, regardless of spreading rate, implies that the asthenospheric temperature must be everywhere uniform at a given depth. Upwelling of asthenosphere with T_p = 1280°C gives the correct extent of melting (~20%) to produce 6–7 km of melt and is therefore inferred to be the 'normal' T_p. Anomalously thick crust in regions such as Iceland (27 km) is most readily explained by potential temperatures some 200°C higher than 'normal' in the asthenosphere below the ridge, causing the solidus to be encountered at a much greater depth in the upwelling path. Such regions of high T_p are due to the presence of a mantle plume (Courtney & White 1986; R. S. White *et al.* 1987; R. S. White & McKenzie 1989).

McKenzie & Bickle (1988) were able to predict the volumes and compositions of MORB, from the extent of melting. The constancy of MORB composition reflects its generation by between 15% and 20% partial melting of dry homogeneous asthenosphere. It follows that because continental rifts like ocean ridges are regions of adiabatic upwelling then we might also be able to predict the volumes and compositions of melts generated by decompression of the asthenosphere below continental rifts.

Melt production in continental rifts

The most important difference between ocean ridges and continental rifts is the presence of lithosphere in the latter. Figure 2 shows a thermal definition for the lithosphere prior to rifting (McKenzie & Bickle 1988). The upper part of the lithosphere (the 'plate' itself, or mechanical boundary layer — MBL) has an approximately linear temperature gradient

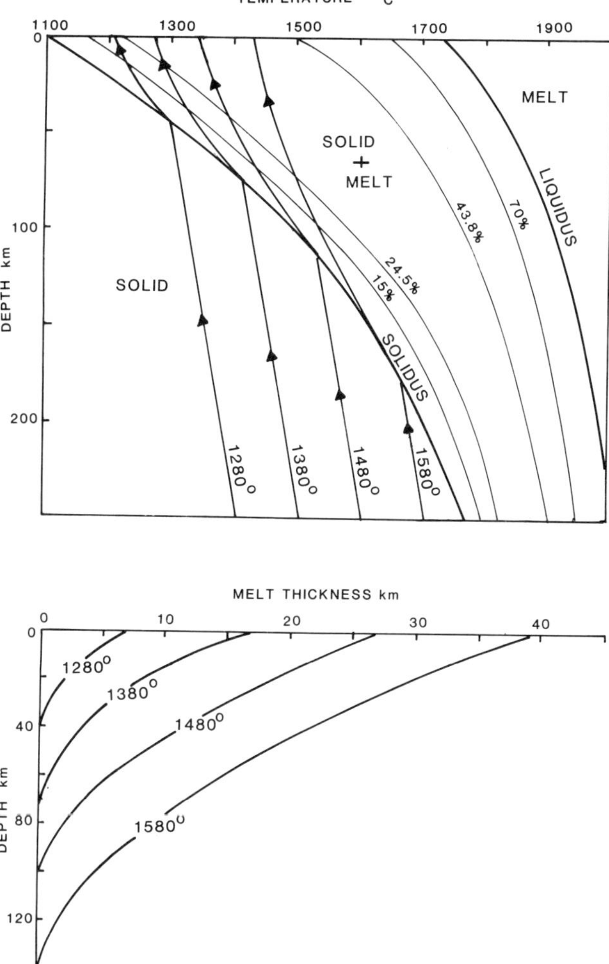

Fig. 1. Upper diagram shows adiabatic decompression paths for different potential temperatures. Curves between solidus and liquidus are labelled with % of melting of dry peridotite. Lower diagram shows total thickness of melt generated below a given depth plotted as a function of depth and potential temperature. Both diagrams from McKenzie & Bickle (1988).

governed by conductive heat loss. The MBL must overlie asthenosphere in which heat transport is advective and which therefore has an adiabatic geotherm. Between these two regions lies a thermal boundary layer (TBL) in which heat is transported both by conduction and by advection and which is inferred to be periodically/slowly incorporated into the asthenospheric circulation (Parsons & McKenzie 1978). The conductive and advective geotherms intersect at some point within the TBL, and this can be used to define the base of the lithosphere (McKenzie & Bickle 1988).

The equilibrium geotherm is a long way below the solidus for dry peridotite and so if this solidus controls melting of the asthenosphere, it will not occur (Fig. 2). For melting to occur the system must be perturbed in at least one of three ways: (a) by changing the position of the solidus due, for example, to the presence of volatile components; (b) by increasing the T_p of the asthenosphere; (c) by thinning the lithosphere causing adiabatic upwelling of the asthenosphere below. In the case of ocean ridges only T_p is important. The lithosphere has been thinned to zero and the volatile content of

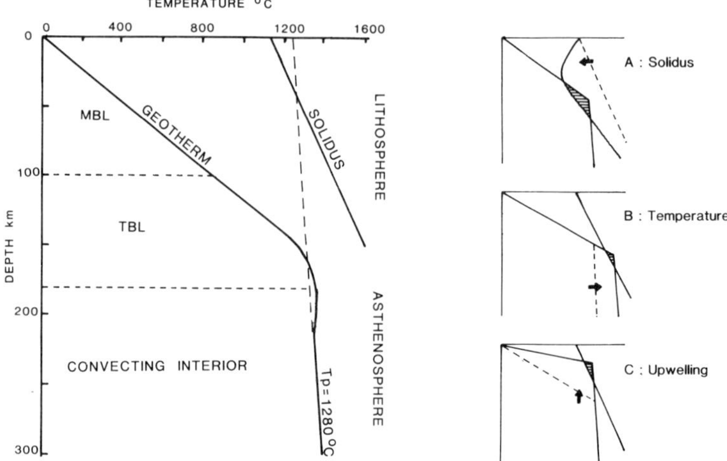

Fig. 2. Horizontally averaged thermal structure of lithosphere for potential temperature of 1280°C, mechanical boundary layer thickness of 100 km, and interior viscosity of $2 \times 10^{17} m^2 s^{-1}$ following McKenzie & Bickle (1988). Corresponding adiabatic upwelling curve (assuming no melting) is shown dashed. Solidus position for dry peridotite is also indicated. MBL, mechanical boundary layer; TBL, thermal boundary layer (see discussion in text). Cartoons to the right of the main diagram show three mechanisms for producing melt: (a) melting due to a change in solidus position, (b) melting by raising the potential temperature, and (c) melting by adiabatic upwelling.

MORB primary magmas is so low (Michael & Chase 1987) that the solidus is likely to be that of dry peridotite and thus has not changed. In continental rifts, however, these three controls on melting may operate together and can change in their relative importance depending on the specific situation.

In continental rifts the amount of extension is finite and thus controls the degree of upwelling of mantle for any T_p. However, the rate at which extension takes place is likely to be of critical importance because large amounts of heat can be lost by conduction if the duration of the rift phase is $>60/\beta^2$ Ma (Jarvis & McKenzie 1980). Slow rates of extension will reduce the extent of melting when thermal conduction can keep pace with upwelling, causing the asthenosphere to become lithosphere and follow a conductive cooling path. The initial thickness of the MBL, prior to stretching, is also important. For thinner initial thicknesses the asthenosphere is already closer to the solidus and so for a given amount of extension will not need to rise so far before it starts to melt.

The composition of a primary magma is essentially governed by the extent of melting of mantle of a particular starting composition — the magma *source*. A knowledge of source composition is required if we are to make estimates about extents of melting from observed magma compositions. McKenzie & Bickle (1988) implicitly assume that dry peridotite, depleted in incompatible elements and LILE (large ion lithophile elements) relative to bulk earth, is the source of MORB. However, isotopic data on continental rift (CRB) and ocean island basalts (OIB) imply a source that has not been depleted in incompatible elements and LILE to the same extent as the source of MORB (e.g. Fitton & Dunlop 1985; Allègre 1987). The sources of CRB and OIB are isotopically distinct from that of the MORB source, and if the MORB source represents bulk upper asthenosphere then these enriched (or less depleted) sources require physical separation from the bulk asthenosphere. There is little agreement in the literature about the location of the isotopically distinct, LILE-enriched, mantle material which acts as a source for CRB and OIB. Some authors ascribe enriched signatures to material brought up from depth by mantle plumes (e.g. Hofmann & White, 1982; Allègre 1982; McKenzie & O'Nions 1983) while others invoke a heterogeneous asthenosphere (e.g. Sleep 1984; Fitton & Dunlop 1985; Allègre 1987). It may well be that the first mechanism provides the heterogeneities for the second. Material may also be added to the asthenosphere at subduction zones. Alternative ideas involve enriched mantle residing within the MBL of the lithosphere (e.g. Bailey 1982; Harte 1983; Wyllie 1987; McKenzie 1989).

It is of critical importance to determine whether the source of the magmas was enriched in volatiles such as CO_2 and H_2O and other low melting point (fusible) components which *would change the position of the solidus* (e.g. Green 1970, 1973). Extreme enrichment of magmas in CO_2 or H_2O would imply a source in the MBL (or directly from a mantle plume) because only these parts of the system are separated from convection and the evidence from MORB is that the asthenosphere is effectively dry.

The melt once generated has to be extracted from the residue and make its way to the surface. The nature and efficiency of the extraction process may have implications for melt composition and quantity (McKenzie 1984). Three important assumptions are made in the following discussion. First, that extraction of melt poses no problem, even for very small amounts of melt (e.g. 10^{-3}%; M. Cheadle pers. comm., 1989; McKenzie 1989). Second, that the melt observed at a particular locality is the product of lithospheric attenuation directly below and has migrated vertically upwards (R. S. White & McKenzie 1989). Third, that the melts produced reflect something close to the maximum amount of lithospheric extension at any locality at a given time during the extensional phase.

In summary, the amounts and compositions of partial melts produced by upwelling are controlled by the following parameters: (1) The T_p of the convecting mantle, (2) the amount of extension, β, (3) the position of the solidus, (4) the composition of the magma souce, (5) the rates of extension, (6) the initial thickness of the MBL, and (7) the nature of the extraction process. After reviewing the observations concerning magmatism in the North Sea we assign known, or inferred, values to some of the above parameters in order to place constraints on other unknown parameter values.

The nature of Mesozoic magmas in the North Sea

This paper concentrates on igneous rocks produced during the well documented Mid-Jurassic to Early Cretaceous extensional event. Triassic extension is widely believed to have taken place in parts of the North Sea (e.g. Ziegler 1982; Badley *et al.* 1988; N. J. White 1990) but there is so far little evidence for Triassic magmatism. If this absence is real then it may support the view that the amount of extension in the Triassic was rather smaller than in the Mid-Jurassic to Early Cretaceous. Igneous rocks of Permian age are found in the central parts of the North Sea, notably in the Danish sector (Dixon *et al.* 1981), but little quantitative information is available about the tectonics at that time. The Mesozoic basaltic rocks, similar to those found in many other rift zones, vary from mildly to highly alkaline in character. The petrographic and chemical character of the rocks has most recently been reviewed in detail by Latin *et al.* (1990) – see also Dixon *et al.* (1981). Here we summarize the main features.

Four Mesozoic igneous provinces have been recognised in the North Sea (Fig. 3). These provinces differ from each other on the basis of age, location, and geochemical character. Many of the rocks have been severely altered so restricting much of the comparative geochemistry to ratios of immobile trace elements (eg. Ce/Y and Zr/Nb; Latin *et al.* 1990).

The Forties Province

The relatively large outpouring of alkali basalt in the area of the triple junction (Fig. 3) is the most striking feature of the magmatism during the Mesozoic. The Forties rocks have been described in detail by a number of authors (eg. Howitt *et al.* 1975; Woodhall & Knox 1979; Fall *et al.* 1982; Latin *et al.* 1990) since they were first discovered in 1970. The precise age of the basalts is still uncertain but recent $^{40}Ar/^{39}Ar$ dates (Ritchie *et al.* 1988) coupled with palynological data from interbedded sediments (Howitt *et al.* 1975) suggest a Callovian or Bathonian age (Fig. 4). Discrepancies between the radiometric and stratigraphic ages may be attributable to poor resolution in this part of the time scale.

The total amount of igneous rock in the province is still uncertain. Crude estimates of volume from the isopach map complied by Woodhall & Knox (1979) suggest that there may be as much as 9000 km^3 of igneous rocks. When this amount is averaged over the total area in which igneous rocks are found (>12 000 km^2) it suggests a maximum thickness of melt of 0.5 km. However, such estimates ignore any intruded or underplated material and probably represent a minimum melt thickness.

The dominant rock type is an extrusive, mildly undersaturated, porphyritic alkali basalt. Typically these basalts (ankaramites) contain abundant olivine and titaniferous augite phenocrysts. More evolved hawaiites and mugearites are also observed in a few wells. Abundant ocelli rich in hydrous phases (e.g. kaersutite and biotite) attest to some enrichment in both H_2O and K_2O in the parental Forties magmas. The majority of Forties rocks appear to be extrusive

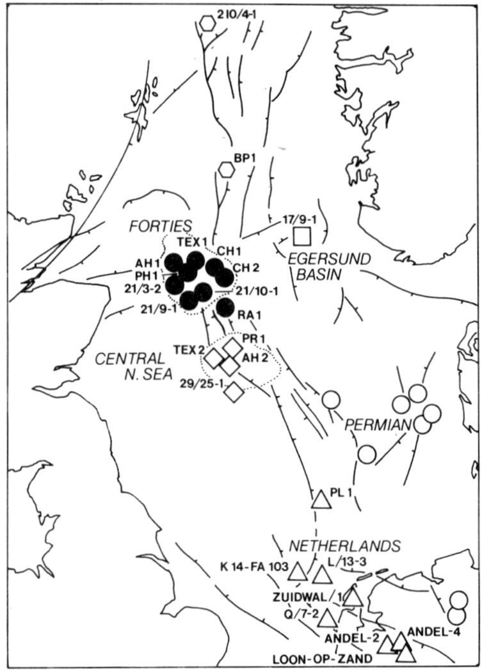

Fig. 3. Location of Mesozoic and Lower Permian igneous rocks in North Sea rift system. Labels correspond to well locations. Unreleased wells are given code names. Not all sites are discussed in text – see Latin *et al.* (1990) for details.

but there is no textural evidence for eruption into large bodies of water.

Chemically, all of the rocks are members of the alkaline series. On the TiO_2 against P_2O_5 diagram (Fig. 5), Forties rocks with > 4 wt% MgO plot as a distinct group with lower values than most of the other Mesozoic rocks. The spread within the group is probably a combination of variable melting at source, low-pressure fractionation (and crystal accumulation) of olivine and clinopyroxene, and alteration effects. On the incompatible element ratio plot (Fig. 6), which effectively removes the effects of low-pressure fractionation and alteration (Latin *et al.* 1990), the Forties basalts are more tightly grouped. The implications of their position on this diagram are discussed in the next section.

The Central North Sea province

The rocks in this province are restricted to quadrants 29 and 30 in the Central Graben area. The province has yet to be fully investigated but appears to have very distinct differences from Forties (Dixon *et al.* 1981; Latin *et al.* 1990). The rocks in this area are both intrusive (into Zechstein) and extrusive in the Upper Jurassic to Early Cretaceous. Radiometric dates on two suites of samples, one from a dyke from well 29/25–1 (Shell/Esso; $^{40}Ar/^{39}Ar$ date) and the other from two flows in the area of the Auk

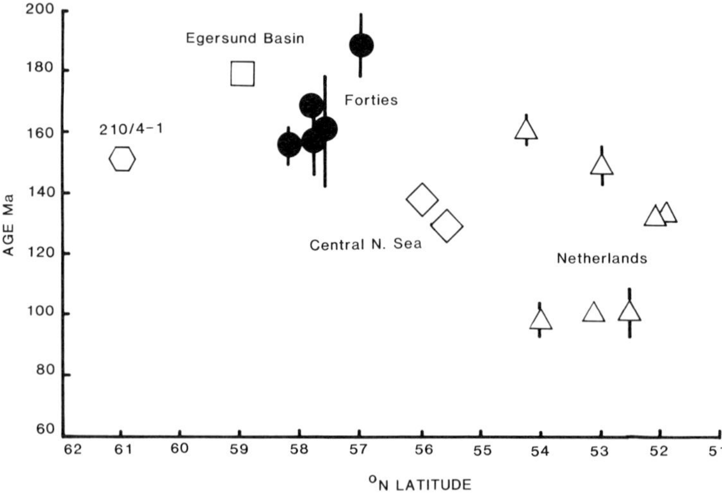

Fig. 4. Radiometric age data, both K/Ar and $^{40}Ar/^{39}Ar$, for North Sea rocks plotted against distance (as degrees of latitude) down the rift system from north to south. Where uncertainties in the date exceed the dimensions of the symbol error bars are shown. In some cases more than one age determination is included within single symbol and error bar (see Latin *et al.* 1990).

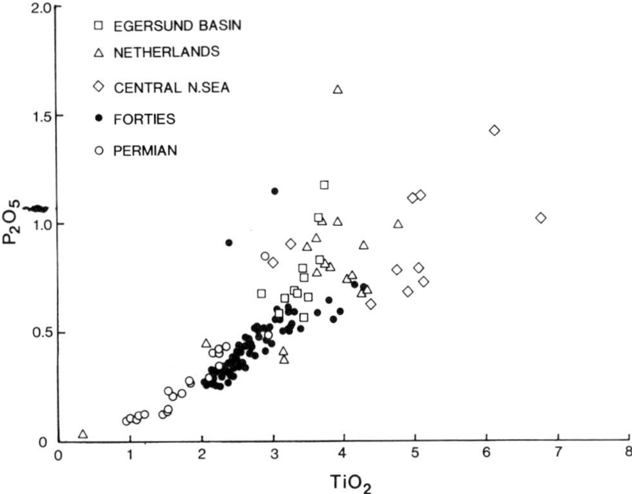

Fig. 5. TiO_2 against P_2O_5 (wt%) in Mesozoic and Lower Permian basic igneous rocks (> 4 wt% MgO) from the North Sea.

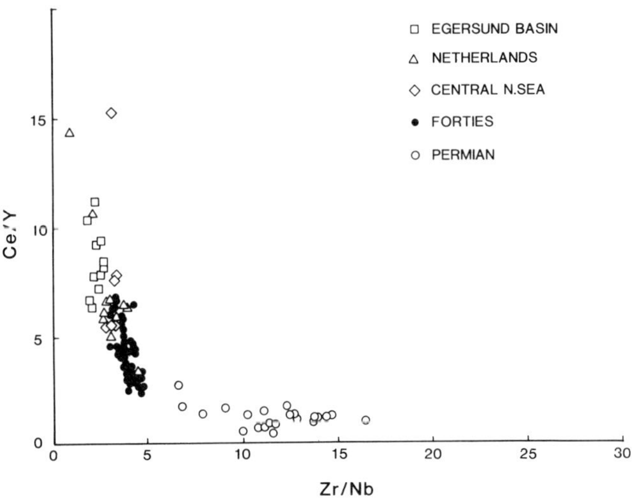

Fig. 6. Zr/Nb against Ce/Y in Mesozoic and Lower Permian basic igneous rocks (> 4 wt% MgO) from the North Sea.

field (K/Ar date) are similar between 140 and 130 Ma (Fig. 4). It is not yet known how much igneous material is present in this area but it is probably significantly less than in the Forties region.

The rocks studied so far are 'basaltic' and are more undersaturated and enriched in potassium and incompatible trace-elements than the Forties alkali basalts. They contain primary biotite and amphibole in the groundmass and as microphenocrysts, as well as large rounded phlogopite xenocrysts. The overall mineralogy (olivine, augite, alkali feldspar, leucite or sodalite, biotite +/− amphibole) leads to the term tephrite or mafic phonolite for the extrusive rocks and lamprophyre (monchiquite) for the dykes.

The rocks are altered but the high K_2O content is clearly original. Fresher samples are strongly *lc* normative (20+%). Incompatible elements are enriched relative to the Forties basalts in absolute terms (see Fig. 5) and match those in the fresh nephelinites from the Egersund Basin described below, with the exception of K_2O

which is up to ten times higher (5.24 wt% compared with 0.5 wt%). Although K is mobile during alteration, the distinctive mineralogy (e.g. the presence of phlogopite) suggests that this difference in K_2O content reflects that in the original magma. The Ce/Y and Zr/Nb ratios are generally intermediate between the Egersund nephelinites and the Forties basalts (Fig. 6).

The Egersund Basin

Igneous rocks have been encountered in several wells in this area (P. Ziegler, pers. comm.). Well 17/9–1 (Fig. 3; Esso) contains two sequences of igneous rocks. The upper sequence comprises 400 m of nephelinitic volcanic rocks which rest on Lower Jurassic shales. The nephelinites have been dated by conventional K/Ar methods (Furnes *et al.* 1982) and give Bajocian ages (177–180 Ma; see Fig. 4). The lower sequence is intrusive into Lower Jurassic and Triassic shales and sands and gives a similar age (177–178 Ma). There is no information to constrain volumes but they are likely to be small.

Textures in the lavas indicate very rapid cooling perhaps due to eruption into water. Prominent phenocrysts of titaniferous salite and large pseudomorphs after olivine are set in a groundmass of analcime (after nepheline), clinopyroxene and ore. The intrusive rocks are strikingly coarse grained aggregates of phlogopite, clinopyroxene and olivine, and the magma was presumably very rich in volatiles. Fresh samples are strongly *ne* normative. Absolute levels of incompatible trace-elements are comparable to those of other nephelinites from the literature, but some Ce/Y ratios are higher than those found in ocean islands (see the OIB field in Fig. 7).

The Netherlands Province

Igneous rocks occur in a number of locations in the Netherlands, both on-shore and off-shore (Fig. 3). The majority are intrusive rocks, with the notable exception of the Zuidwal volcano in the Waddensee near Texel island. There appear to be two distinct age groups within this province (Fig. 4). The on-shore occurrences (Loon-op-Zand, Andel, and Zuidwal) have older radiometric ages (150–130 Ma; Late Jurassic–Early Cretaceous) than the other rocks (all off-shore) which are all dated as close to 100 Ma.

The rocks comprising this province are, in detail, rather varied — especially those forming the Zuidwal edifice. However, an overall group-similarity is observed in terms of the strongly undersaturated character. The rock types found include basanites, tephrites, leucite basanites, phonolites, leucitites, olivine nephelinites, limburgites and lamprophyres. Particular attention is drawn to the similarities between the lamprophyres of PL1 (intruding the Zechstein and containing resorbed xenocrysts of amphibole) and those seen in 29/25–1 (Central North Sea) and between the nephelinites of this area (Loon-op-Zand and Andel) and those in the Egersund Basin.

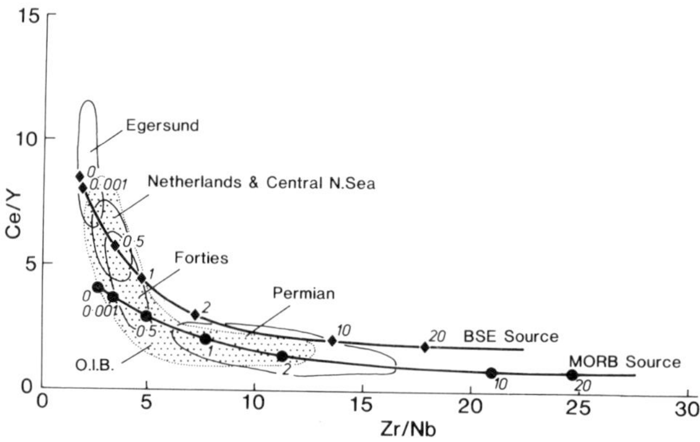

Fig. 7. Equilibrium partial melting (Shaw 1970) lines calculated for Zr/Nb against Ce/Y from a MORB source and a BSE source. Melting lines are marked for % melting and were calculated assuming simple single stage equilibrium partial melting using bulk distribution coefficients calculated by Fitton & Dunlop (1985) (see Latin *et al.* 1989 for details). Fields for basic igneous rocks (> 4 wt% MgO) from the North Sea and OIB (stippled; unpublished data of J. G. Fitton and D. James) are also outlined.

On both the TiO_2 against P_2O_5 and the Ce/Y against Zr/Nb diagrams (Figs 5 & 6) the Netherlands field overlaps that of the Central Graben rocks with which they have been grouped in later discussions. The key feature of Zuidwal is the implied, but not necessarily large, variability in the ratios of Na_2O to K_2O and of total alkalis to alumina and silica in the parental basaltic undersaturated magmas. Low pressure fractionation operating on these parents has then exploited what may be small bulk-chemical differences to generate a wide range of evolved products. Such implied variation in magma character from source is difficult to reconcile with melting of a uniform, well mixed asthenosphere.

Magma-source, solidus, and extent of melting

In order to make any quantitative predictions about tectonics and mantle dynamics using melt chemistry it is first essential to identify the nature of the source of the magmas and to determine the position of the solidus controlling melt production. When source composition and solidus position are known, estimates can be made for the degrees of partial melting represented by the primary magmas.

The chemical and isotopic composition of basic igneous rocks is controlled by five important variables: (1) the degree of partial melting of the mantle source, (2) the mineralogical, chemical and isotopic composition of the source, (3) the history of fractional crystallisation, (4) changes resulting from alteration, (5) and any effects due to crustal contamination. Only the first two variables are of interest in the present context. Latin *et al.* (1990) have demonstrated that for mafic basaltic rocks (i.e. rocks with 4 wt% MgO) the effects of variables other than extent of melting and source composition can be negated by use of ratios of immobile trace elements such as Ce/Y and Zr/Nb which are fractionated during partial melting events but which remain intact during crystallization. The values observed for such ratios thus reflect only the characteristics of the primary magma and may thus be related purely to the composition of the source and degree of partial melting.

In terms of magma source, solidus position, and extent of melting, the observations made by Latin *et al.* (1990) can be summarized in five parts as follows.

(1) Major- and trace-element compositions combined with the petrographic evidence for the importance of volatiles (particularly H_2O) suggest that it is unlikely that the production of *any* of the North Sea rocks, with the possible exception of the alkali basalts, will be governed by the solidus for dry peridotite, and none can be produced by melting of depleted MORB source asthenosphere alone.

(2) With the exception of the ultrapotassic rocks of the Central Graben and Netherlands Provinces the North Sea rocks have major-element compositions similar to OIB, suggesting that the same source may apply.

(3) The major-element compositions of many of the freshest Forties alkali basalts (e.g. A and B in Table 1) can be produced by the smallest extents of melting (<10%) in the parameterization (C in Table 1) of McKenzie & Bickle

Table 1. *Observed alkali basalt compositions from the Forties province compared with experimental data*

	A	B	C	D	E
SiO_2	47.81	45.00	45.53	47.80	49.00
Al_2O_3	11.90	15.07	17.26	15.60	14.50
FeO	9.19*	9.48*	9.78	9.91	9.50
MgO	9.63	8.99	10.71	11.20	11.60
CaO	12.43	11.10	12.05	9.71	9.10
Na_2O	2.02	2.19	3.24	3.33	3.10
K_2O	0.90	1.03	0.26	1.00	0.70
TiO_2	2.19	3.26	1.77	1.55	3.20

A: Ankaramitic alkali basalt from well AH1, Forties Province.
B: Alkali Basalt from well 21/9–1 (BP), Forties Province.
C: The first-melt calculated using the parameterisation of experimental data of McKenzie & Bickle (1988; p. 678, Table A1(a)) at a pressure of 2.0 GPa.
D: Experimental Run#43 of Takahashi & Kushiro (1983, p. 869, Table 3).
E: Experimental Run#1 of Jaques & Green (1980, p. 299, Table 4)
* Total iron originally calculated as Fe_2O_{3T} (Latin *et al.* 1990; Table 1) has been recalculated to generate FeO assuming $Fe_2O_3/(FeO + Fe_2O_3) = 0.2$.
Note: Major element oxides are given as wt%.

(1988, p. 678, Table A1(a)), and are also in good agreement with the melts produced close to the solidus in melting experiments of Takahashi & Kushiro (1983; D in Table 1). The compositions A to D are not as similar as they might be because A and B have both undergone a considerable amount of crystal fractionation of olivine and clinopyroxene. The values for K_2O and TiO_2 are generally higher (particularly TiO_2) in the Forties alkali basalts than those predicted from melting of dry peridotite (Table 1C). Peridotites that are relatively enriched in incompatible elements and LILE compared to the source of MORB, such as were used in Takahashi & Kushiro's (1983) experiments and more notably the Hawaiian pyrolite used by Jaques & Green (1980; E in Table 1 which is a tholeiitic melt), produce similar values for K_2O and TiO_2 to those observed in the Forties basalts. From the major-element compositions it can be concluded that the Forties alkali basalts can be produced by melting, in the less than 10% region, of dry peridotite, but that some input in terms of K_2O and TiO_2 is required from a more enriched source.

(4) Simple equilibrium melting calculations using trace element ratios can be used to estimate the extents of melting represented by the North Sea rocks when produced from sources of different composition (Fig. 7; see Latin *et al.* 1990 for details). The Forties basalts represent the largest degree melts in the Mesozoic. The calculations suggest that the extent of melting is small, *probably less than 2%* (excluding the Permian rocks) even if the source is relatively enriched compared to that of MORB. If the magmas were produced from dry peridotite of the type that gives rise to MORB on melting by 20% then the extents of melting are less than 0.5%. It is important to note that most of the North Sea rocks as well as OIB cannot be produced by melting of a depleted MORB source. However, at least some of the Forties basalts can be produced by melting of MORB source peridotite which is governed by the dry solidus. Use of trace elements to estimate extents of melting in this way gives a more precise figure than the less-than-10% value provided by the parameterization of McKenzie & Bickle (1988).

(5) The alkali basalts show petrographic evidence, in the form of ocelli and groundmass hydrous phases, for the presence of H_2O in the primary magma. However the volatile content is not likely to have been great and was probably less than 2 wt% at most. Does the presence of a small amount of volatiles make a source in the supposedly dry asthenospheric mantle unlikely? Michael & Chase (1987) consider that primary MORB magmas contain between 0.01 and 0.045 wt% H_2O. The asthenosphere that produces MORB is therefore probably *not* completely dry. If MORB represents a 20% partial melt and if all of the H_2O enters the melt then the source would have contained 0.002 to 0.009 wt% of H_2O. The trace element estimates for the extents of melting represented by the alkali basalts (<0.5% from a MORB source) would suggest that the primary alkali basalt magma could easily have contained as much as 1 wt% H_2O. Therefore, although the asthenosphere may be viewed as dry when the scale of melting is large, as is the case with MORB, on a scale of 1% melting or less, 1 wt% of H_2O in the melt may not be unreasonable. As we show later 1 wt% of H_2O can dramatically reduce the liquidus temperature of alkali basalt to a point where it is *molten at temperatures and pressures which are below those for the dry solidus* (Fig. 9; Green 1969, 1973). However, if the observed H_2O entered the magma while it resided in crustal magma chambers prior to eruption or during passage of the melts through the lithosphere, the dry solidus would still apply.

The alkali basalts of the Forties province and most of the other North Sea rocks are compositionally similar to OIB and most are more enriched in LILE than can be produced by melting depleted MORB source. Because the oceanic lithosphere is relatively young then this enriched signature, observed in OIB is not as likely to reside in the oceanic MBL as it is to be held in the form of enriched 'blobs' or 'streaks' in a more depleted host asthenosphere (Sleep 1984; Fitton & Dunlop 1985; Allègre 1987). In the North Sea, for the Forties magmas, we suggest such a source in heterogeneous, or 'streaky', asthenosphere, rather than one in the MBL may be more likely for the following reasons.

(1) In order for the relatively cold MBL to melt before there is any melting of the asthenosphere, the peridotite must be virtually saturated in C–H–O volatile species. Throughout most of the lithosphere the (pressures < 3 GPa) the water-saturated solidus curve will not apply (see Fig. 9). Amphibole and other minor phases (e.g. carbonate) can accommodate all available C–H–O volatiles without significant lowering of the dry peridotite solidus (see the pyrolite with 0.2 wt% H_2O curve in Fig. 9). The volatile content of the Forties alkali basalts was evidently not high — probably less than 1 wt% H_2O. They are rarely vesicular and fragmental and they only contain minor amounts of volatile phases, in the groundmass, unlike some of the

other North Sea rocks.

(2) The MBL in the area of the major Mesozoic rift structure in the North Sea is likely to be *thermally young*. The lithosphere in this area has been subject to repeated 'events', in the Carboniferous, Permian and Triassic, since the Caledonian tectonism (e.g. Ziegler 1982). Any or all of these events may have caused decompressive or thermal perturbations large enough to liberate any melt easily obtained from the MBL. In this case it seems unlikely that the MBL in the area of the main rifts will have had enough time to become enriched in the period between the Triassic and the Early to Middle Jurassic. It should be noted, however, that the rate at which such enrichment may occur is still not known (McKenzie 1989).

(3) At the onset of Jurassic rifting the lithosphere was probably still cooling from the remanent effects of the Triassic event. The thickness of the MBL was thus likely to be rather thinner than the existing estimates of ~100 km (see the discussion in the next section). We suggest that the MBL was probably closer to 70 km thick at commencement of rifting in the Forties area. Upon decompression such a thin MBL would be too cold to intersect an amphibole lherzolite solidus, considered appropriate for the solidus in the lithosphere (Green 1973), of the type shown in Fig. 9. If this is the case, *no melt* would be liberated from the lithosphere during extension. This argument will not hold for off-axis areas and areas not involved in the Triassic stretching event, where the MBL was likely to have been thicker in the Mid-Jurassic.

In conclusion, we assume that the alkali basalts are produced from the asthenosphere, but that the asthenospheric source is more enriched in K_2O and TiO_2 compared to the MORB source peridotites in the experiments parameterized by McKenzie & Bickle. It is likely that alkali basalts produced from 'dry' asthenosphere containing enriched streaks will still only be generated if the dry solidus has been overstepped to the same extent or to a greater extent than would be required if the source were more depleted. The question remains as to whether the melting was truly dry. The H_2O content of the alkali basalts, *if primary*, suggests that melting may have commenced at slightly lower temperatures and pressures than those of the dry solidus. In the remaining part of the paper we use two different solidus positions, one assumes a dry source, the other assumes a source with enough water to produce a melt containing 1 wt% H_2O. It is suggested that the North Sea rocks other than the alkali basalts are produced by smaller extents of melting of the streaky asthenosphere source or the a source within the MBL at temperatures and pressures considerably below those of the dry solidus. The ultrapotassic rocks, which have no oceanic equivalents and contain large amounts of volatile-rich phases, must have been produced in the MBL and may be controlled by something like the pyrolite with 0.2% H_2O solidus as shown in Fig. 9 (Green 1973).

The complete range of North Sea rocks can be explained by input from 2 different mantle sources. (1) 'Streaky' asthenosphere whose melting characteristics are essentially those of dry peridotite except for the first very smallest melt fractions where the melts may contain some H_2O. (2) Metasomatized MBL in areas which had not been tectonically reworked during the period immediately prior to the Jurassic rifting event.

Observed melt compositions and stretching history

It is now possible to combine the above observations concerning the extent of melting, solidus position and likely source of North Sea magmas with parameters such as β, the thickness of the MBL prior to rifting and the duration of rifting in order to make inferences about (A) the potential temperature, (B) the stretching mechanism, (C) the relations between on-axis and off-axis magmatism, β, crustal lineaments and the thermal age of the MBL, and (D) the relationship between locations of magmatic activity and hydrocarbon source rock maturity. The Forties basalts are the only on-axis Mesozoic magmas in the North Sea rift system. In most stretching models on-axis areas would be considered to be the areas of greatest lithospheric attenuation in the system, and so this is where the greatest degrees of melting are to be expected. In the North Sea the on-axis areas are also sites of repeated tectonism (e.g. Ziegler 1982) and so the magmas are unlikely to have their source in the lithosphere. The other localities of magmatic activity (see Fig. 3) are all situated in off-axis areas where lithospheric attenuation will have been considerably less than in the Forties area and where an input from a magma source in the MBL is perhaps to be expected.

Estimates of extension (β)

The amount of stretching in the triple junction is not well known, although estimates from the subsidence history suggest that β = 1.4 to 1.6 (Sclater & Christie 1980; Barton & Wood 1984).

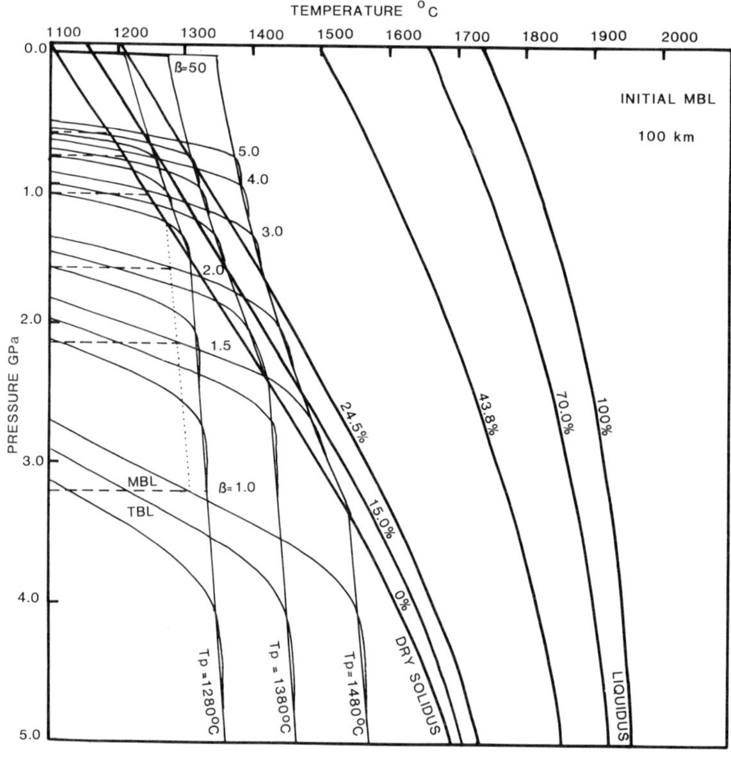
TEMPERATURE °C
1100
1200
1300
1400
1500
1600
1700
1800
1900
2000
0.0
1.0
2.0
3.0
4.0
5.0
PRESSURE GPa
INITIAL MBL
100 km
ß=50
5.0
4.0
3.0
2.0
1.5
ß=1.0
MBL
TBL
24.5%
15.0%
0%
DRY SOLIDUS
43.8%
70.0%
100%
LIQUIDUS
Tp=1280°C
Tp=1380°C
Tp=1480°C

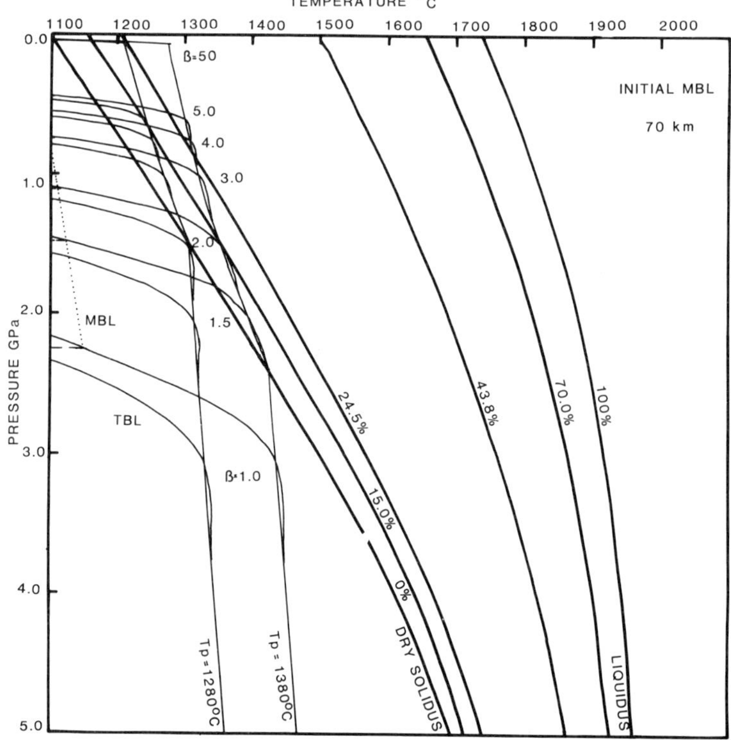
TEMPERATURE °C
1100
1200
1300
1400
1500
1600
1700
1800
1900
2000
0.0
1.0
2.0
3.0
4.0
5.0
PRESSURE GPa
INITIAL MBL
70 km
ß=50
5.0
4.0
3.0
2.0
1.5
ß=1.0
MBL
TBL
24.5%
15.0%
0%
DRY SOLIDUS
43.8%
70.0%
100%
LIQUIDUS
Tp=1280°C
Tp=1380°C

Similar values have been obtained for the Viking Graben (N. J. White 1988; 1990) and Central Graben (Wood 1982). The lack of well data for the deepest parts of the rift system (e.g. the Fisher Bank basin) is, however, a serious drawback in such studies and the β values in such regions may be somewhat higher. It has been suggested (Latin *et al.* 1990; Latin & White 1990) that the geometry of the rift system at the triple junction in the Forties area may give rise to a β value of between 2 and 2.5 as this is the square of the values in the rift arms. In off-axis areas the extension calculated from the subsidence is less. In the Egersund Basin theoretical (water-loaded basement) subsidence curves fit the decompacted data (well 17/9–1, unpublished data) extremely well with β = 1.21–1.23. On the flanks of the Central Graben β = 1.2–1.3 for the Mesozoic event. In the Netherlands values for β are less well known and estimates from the subsidence data are made difficult because of widespread inversion. However, melt compositions may suggest that the degree of extension is similar to that seen in the Egersund Basin.

Relating compositions to β

The existing theoretical framework only allows quantitative relationships to be determined for magmas caused by melting on the solidus for dry peridotite. These specific requirements may only be met at oceanic spreading centres, although even here the solidus requires some adjustment for the first small amounts of melt (see McKenzie 1989). Much of the following discussion is concerned with the Forties alkali basalts which are thought to represent small degree melts (< 2%) from a source assumed to be asthenospheric peridotite either with or without a small amount of H_2O (< 1 wt%). Most of the other magmas in the North Sea show a lithospheric component. In terms of Ce/Y values (Fig. 7) the Egersund Basin nephelinites are probably the most extreme example of asthenospheric 'streak' melts. In terms of minor elements such as TiO_2 and K_2O, the very high values shown by some of the Netherlands and Central North Sea rocks can *only* be reconciled with melting of the lithosphere. The most important point is that melts such as these can be produced by much less stretching (at a given potential temperature) because they are controlled by a different, much lower, solidus curve (Fig. 9).

MBL thickness prior to rifting

The extent of melting, particularly at small amounts of extension, is highly sensitive to the initial thickness of the MBL (Fig. 8) whatever the T_p. The source of magmas, as was discussed earlier, may also depend on the MBL thickness. Two estimates exist for the thickness of the *lithosphere* in the North Sea. Barton & Wood (1984) suggest that the similarity between the thermal time constant for basin subsidence both in the North Sea and in the oceans is evidence for a similar lithospheric thickness prior to extension – i.e. ~125 km (Parsons & Sclater 1977). A lithosphere 125 km thick would have a MBL thickness of ~100 km. Rayleigh wave dispersion studies (Stuart 1978) suggest a minimum present day lithosphere thickness of 70 km which would constitute a MBL thickness of ~60 km. A value somewhat less than 100 km for the MBL is suggested by the backstripped subsidence data. A thinner MBL would also be supported if the thermal effects of Triassic extension had also not completely decayed away by the Early Jurassic. We consider that a MBL thickness closer to 70 km (than 100 km) in the area of Forties and the main rift is most likely. However, in off-axis areas the thickness may have been closer to, or even greater than, 100 km.

Rates of extension and rift propagation

The rate at which extension takes place will have an important controlling influence on the amount of melt generated (and its composition) for given values of β and T_p. Rates of extension must also control when the maximum amount of upwelling occurs. If stretching is instantaneous then the maximum upwelling occurs immediately. However, when the stretching is protracted then the period of maximum upwelling will be controlled by an interplay between loss of heat transported by conduction opposed to that due to advection.

Fig. 8. Adiabatic upwelling due to stretching of a convective geotherm generated from a mechanical boundary layer (MBL) with an initial thickness of 100 km (a) and 70 km (b), and an internal viscosity of $4 \times 10^{15} m^2 s^{-1}$, for different values of β at different potential temperatures (T_p). Curves between solidus and liquidus are labelled for the % of melting. Boundary between MBL and TBL (thermal boundary layer) at different values of β is shown by dashed line and its path during extension (until MBL intersects the dry solidus in (a)) is shown by dotted line.

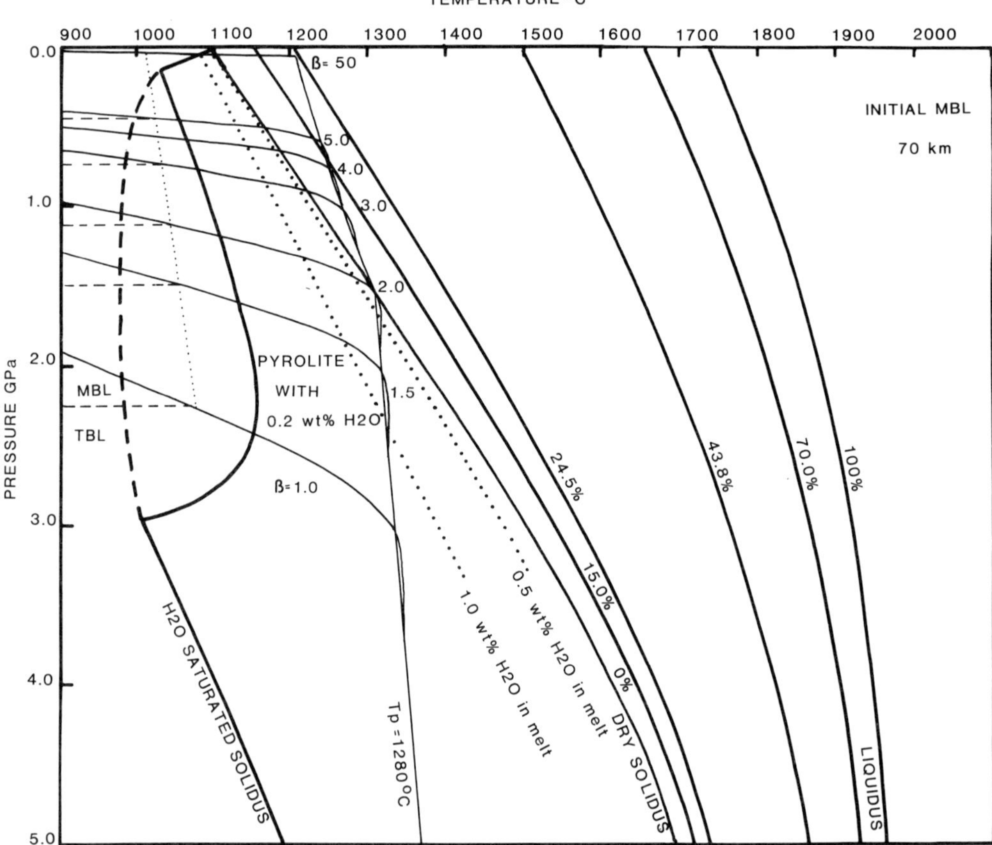

Fig. 9. Adiabatic upwelling due to stretching of a convective geotherm generated from a mechanical boundary layer (MBL) with an initial thickness of 70 km, an internal viscosity of $4 \times 10^{15} m^2 s^{-1}$, and a potential temperature of 1280°C, for different values of β. Dry peridotite solidus and liquidus with melt % intervals as in Fig. 8. Solidus for pyrolite containing 0.2 wt% H_2O, held in amphibole, is shown by the thick curved line from 0 to ~3 GPa (after Green 1973). At pressures > 3GPa amphibole is not stable and so breaks down releasing the water and creating water saturated conditions (indicated by the dashed curve). Liquidus curves for alkali basalts containing between 0.5 and 1.0 wt% water are shown dotted (after Green 1970).

Some indication of the rates of lithospheric thinning may therefore be indicated by the relationship between the initiation of rifting and the timing of magmatism. If rifting started in the Lower Jurassic or Early Middle Jurassic in the triple junction area, then the eruption of basalts so soon after the onset of rifting (< 30 Ma) suggests relatively rapid rates of extension early on. The Egersund Basin started extending in the Lower Jurassic and again the onset of magmatism is soon after the initiation of rifting (180 Ma).

The Viking Graben and Central Graben, on the other hand, are typically assumed to have a protracted, 60 Ma, rifting episode from Middle–Upper Jurassic to Early Cretaceous (Giltner 1988; Badley *et al.* 1988; N. J. White 1988, 1990). The conspicuous absence of volcanic rocks in the Viking Graben even at β values of 1.5 (N. J. White 1988, 1990) may be related to the slow rates of extension or may merely reflect a lack of well data in the deeper parts. On the basis of gravity and deep seismic data Holliger & Klemperer (1989) suggest that melt may have been underplated in some parts.

The magmatic activity in the Northern Viking Graben, the Central Graben and Netherlands all occurs later than in the Forties (see Fig. 4). The timing of magmatic activity appears therefore to young both to the north and to the south, away from the triple junction. It is interesting that the Central Graben magmas are found in a kink in the trend of the rift, similar to that observed in Forties — perhaps another region of rapid extension, but at a later date. Much of the magmatic activity in the Netherlands may be related to rapid thinning in wrench-bounded basins (P. A. Ziegler pers. comm., 1989). These observations may tentatively support a model of rift initiation and propagation away from the triple junction and from other structural nodal points, although much work remains to be done. The rifting event may well have been a complex affair with bursts of rapid extension in some areas and more gradual accommodation of strain elsewhere. The timing of the magmatism would at least support a model of later initiation of rifting in the southern North Sea (Upper Jurassic to Early Cretaceous) than in the area of the triple junction in the northern North Sea (Lower to Middle Jurassic).

Potential temperature: a North Sea mantle-plume?

Figure 8 shows the results of the one-dimensional calculations by McKenzie & Bickle (1988) for uniform stretching of the lithosphere over asthenosphere of different T_p. The degree of partial melting of 'dry' asthenosphere was calculated for a range of β values. Stretching is assumed to be instantaneous, or to occur within a period which is short compared to $60/\beta^2$ Ma (the thermal time constant for the lithosphere; Jarvis & McKenzie 1980), an assumption which may hold for the Forties province. The MBL is assumed to have an initial thickness of 100 km in Fig. 8a, which for a upper mantle kinematic viscosity of $4 \times 10^{15} m^2 s^{-1}$ represents an effective plate thickness of 120–125 km. In Fig. 8b the initial thickness of the MBL is assumed to be 70 km with the same viscosity. Larger upper mantle viscosities and greater thicknesses of the MBL will decrease the degree of partial melting resulting from given values of β and T_p.

Figure 8 shows that for a given value of β the amount of melting (if any) will be greater for higher T_p. A high T_p (1480°C) is inferred for the centres of mantle plumes such as Cape Verde, Iceland, and Hawaii (Courtney & White 1986; R. S. White & McKenzie 1989). A T_p of 1480°C combined with the known β values in the Forties area (β = 1.4–1.6) would for the uniform stretching model (Fig. 8a) produce 10–20% partial melting of asthenosphere with a dry solidus. Such large degrees of melting would give rise to up to 5 km of magma (McKenzie & Bickle 1988), rather than the observed 0.5 km, with tholeiitic rather than alkali basalt compositions. This problem is exaggerated if, as we suspect from the triple junction geometry, β is somewhat larger (2 to 2.5) in the Forties area. The compositions observed in Forties are produced by minimal intersection (if any) of the dry solidus. Conceivably, slow rates of extension (i.e. duration $> 60/\beta^2$ Ma) might reduce the extents of melting to the desired amounts, but this is not in agreement with the observed ages of the rocks with respect to the onset of extension nor does it agree with the subsidence data.

The amounts and compositions of melt in the Forties area thus appear to rule out plume-type elevated potential temperatures (1480°C). However, it should be noted that it is also difficult to explain the presence of melts from 'dry' asthenosphere at Forties if normal (1280°C) $\cdot T_p$'s are used. For uniform stretching, with T_p = 1280°C, a β value of 2.0 (Fig. 8b) is required to induce melting when the initial MBL thickness is 70 km and even larger β values (> 2.5) are required for a thicker MBL prior to rifting (Fig. 8a). Although these β values may not be unreasonable the observed extent of melting in the Forties province can be explained for lower β values if the liquidus position of alkali basalt is lowered by ~100°C (Fig. 9). Figure 9 shows the liquidus temperatures of alkali basalt containing 0.5 to 1 wt% H_2O. It is immediately clear that the observed β factors of 1.4 to 1.6 combined with a T_p of 1280°C, a MBL thickness of 70 km, and the presence of 1 wt% H_2O in the melt easily account for the production of alkali basalt in Forties.

As yet, a low temperature plume (T_p = 1380°C) cannot be ruled out for the North Sea. The observed amounts of extension over these temperatures could perfectly well explain the Forties alkali basalts. The predicted extents of melting, even with 1 wt% H_2O in the melt, are not too large so long as an initial MBL thickness of 100 km (Fig. 8a), and not 70 km (Fig. 8b) is assumed. We have already suggested that the 70 km value for the MBL thickness is more likely in the Forties area, and also, we feel that there must be unequivocal evidence for widespread (over a region 1000 km diameter) pre-rift uplift in the order of 0.5 to 1 km, before invoking a mantle plume. The documented erosion of pre- and syn-rift sediments observed in oil wells across the region, often cited as evidence for updoming (e.g. Ziegler & Van Hoorn 1989), may be no more than the effects of flank and fault-footwall uplift. However, the evidence for updoming from the distribution of sedimentary facies (e.g. Eynon 1981; Ziegler 1982; Richards, this volume) is more difficult to refute. The deeper parts of the rift system are not often drilled and so well-ties to the seismic reflection data in these areas are problematic. The true occurrence of pre-rift erosion throughout the area cannot yet be proven, and further work yet remains to be done.

In conclusion the model that fits best for the Forties alkali basalts in terms of potential temperature is the 1280°C one shown in Fig. 9 and not one of a major plume (1480°C). The presence of a lower temperature mantle-plume, or volatile components in the asthenosphere causing a slight lowering of the solidus cannot be ruled out but neither are necessary. The most likely model involves near normal potential temperatures, an initial MBL thickness of 70 km and β values close to 2.

Stretching mechanism: pure or simple shear?

Over the last eight years there has been considerable controversy regarding the nature of lithospheric extension. The ideas presented above all concern the uniform stretching model of McKenzie (1978) which argues that the lithosphere deforms by bulk pure shear. More recently, Wernicke (1981, 1985) argued that lithospheric extension could be accommodated by slip along an initially planar low-angle (10–30°) normal fault which penetrates the entire lithosphere (Fig. 11a). Both models have been widely applied to the formation of extensional sedimentary basins (e.g. Sclater & Christie 1980; Sclater *et al.* 1980; Barton & Wood 1984; Beach 1986; Gibbs 1987) and a variety of arguments have been used to try to distinguish between the two. These arguments include evidence from seismological data in areas of active extension such as the Aegean, East Africa and Suez (Jackson 1987), patterns of initial and thermal subsidence in older basins such as the North Sea (N. J. White 1989) and from the study of deep seismic reflection lines (Klemperer & White 1990). Below we discuss the quantitative implications of the lithospheric simple shear model for melt generation.

Figure 10a shows the lithosphere after extension along a planar detachment fault before the section has been isostatically balanced (as it is in Fig. 11a). The effective crustal thinning (β_c) is related to the dip of the detachment (θ_c) and to the horizontal separation of the two plates (ϵ) by

$$\beta_c = t_c/(t_c - \epsilon\tan\theta_c) \qquad (1)$$

where t_c is the thickness of the crust prior to extension and $\epsilon\tan\theta_c$ is equal to the vertical separation of the two plates (y_c in Fig. 10a).

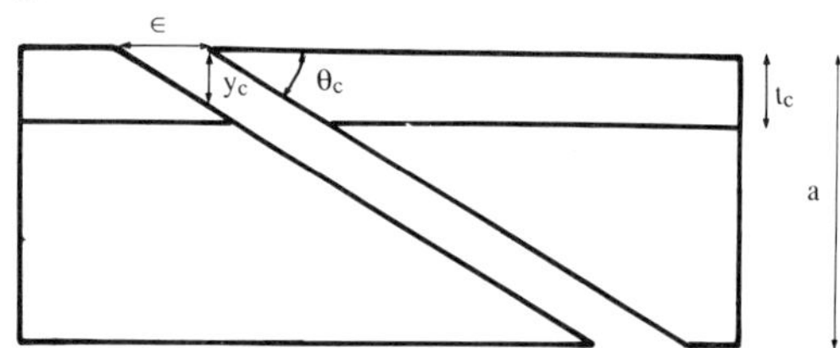

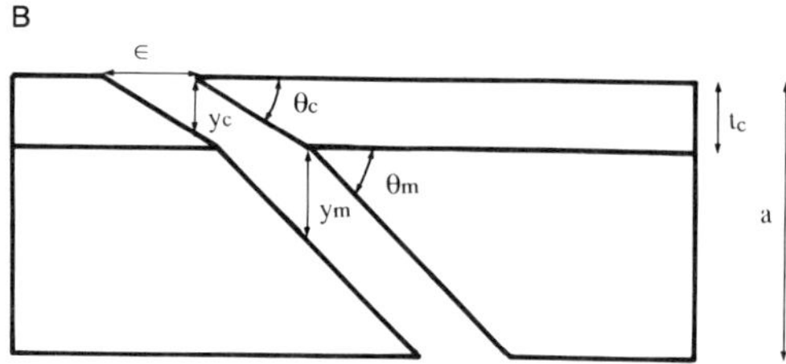

Fig. 10. Extension by lithospheric simple shear. (a) Cartoon shows lithosphere after extension along an initially planar detachment before the section is isostatically balanced. (b) Extension along an initially non-planar detachment fault, again before isostatic balancing. ϵ, horizontal separation of plates; y_c, vertical separation in crust; y_m, vertical separation in mantle; t_c, crustal thickness; a, lithospheric thickness; θ_c and θ_m, dip of detachment fault in crust and mantle respectively. See text for equations.

Figure 10b shows the situation, again before isostatic balancing, when the dip of the detachment fault in the crust (θ_c) is different to that in the mantle (θ_m), i.e. the fault is not initially planar. Rearranging equation (1) gives

$$\epsilon = t_c\,(1 - 1/\beta_c)/\tan\theta_c \qquad (2)$$

which when substituted into

$$\beta_m = (a - t_c)/(a - t_c - \epsilon\tan\theta_m) \qquad (3)$$

where a is the thickness of the lithosphere and $\epsilon\tan\theta_m$ is equal to the vertical separation of the two plates in the mantle (y_m in Fig. 10b), gives an expression for apparent thinning of the lithospheric mantle in terms of effective crustal thinning and detachment dip.

$$\beta_m = \frac{(a - t_c)}{[a - t_c - t_c\,(1 - 1/\beta_c)\tan\theta_m/\tan\theta_c]}. \qquad (4)$$

The results of calculations, similar to those used for Fig. 8, for the lithospheric simple shear model, are shown in Fig. 11. It is assumed that extension takes place along an *initially planar fault* (i.e. $\theta_c = \theta_m$; Fig. 10a), although after extension the dip of the detachment has in-

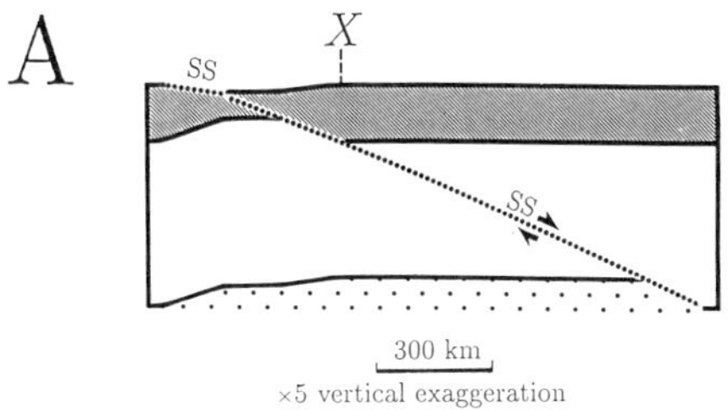

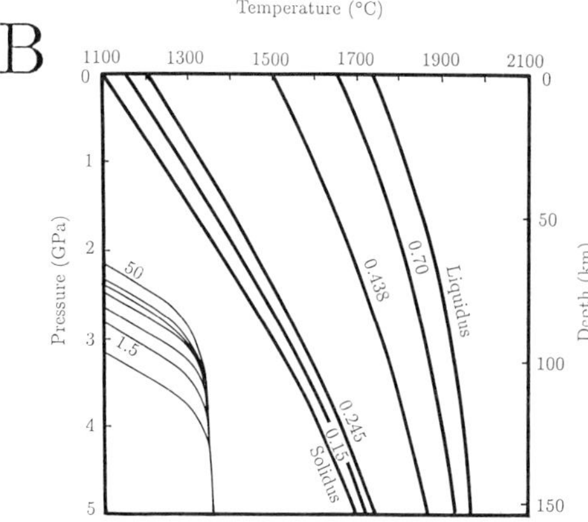

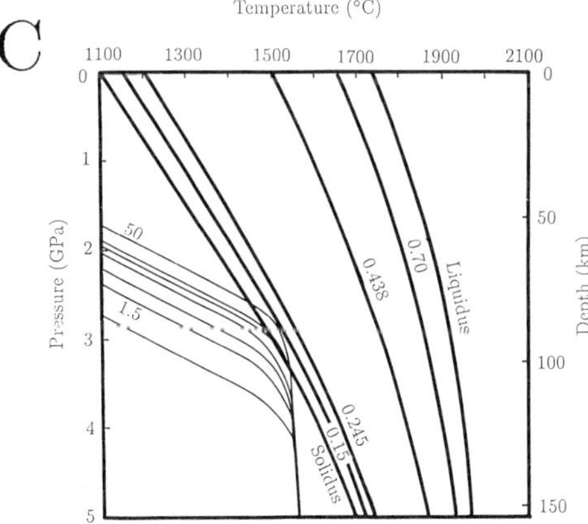

Fig. 11. Lithospheric Simple Shear Model (ss). (a) Cartoon illustrating lithosphere geometry produced by extension on a detachment fault dipping at 5° (effective $\beta_c = 2$). Crust shaded. X marks the position of maximum upwelling. (b) Adiabatic upwelling due to stretching of a convective geotherm, with an initial mechanical boundary layer 100 km thick and potential temperature of 1280°C, by different values of effective crustal β_c (other parameters as in Fig. 8). (c) As for (b) with an interior potential temperature of 1480°C. Effects of latent heat ignored. Note: *for an initially planar fault the amount of upwelling is independent of the dip of the detachment* (see text). From Latin & White (1990).

creased considerably with depth due to isostatic rebound (Wernicke 1981; 1985). Flexural rigidity is assumed to be negligible, simplifying the calculations and maximizing melt production. The initial thickness of the MBL is taken to be 100 km, which for an upper mantle kinematic viscosity of $4 \times 10^{15} m^2 s^{-1}$ represents an effective plate thickness of about 120 km. Only the region of maximum asthenospheric upwelling (X in Fig. 11a) was considered in the calculations and the curves drawn for the perturbed geotherm (Fig. 11b, c) are labelled for crustal thinning (β_c). From equation (1) it is clear that if β_c is fixed then $\epsilon \tan\theta$ is constant and therefore the amount of upwelling will be completely independent of the dip of the detachment. The perturbed geotherms shown in Fig. 11b and c overestimate the extent of melting because, unlike those in Fig. 8, they do not take account of the change in the adiabat due to the loss of latent heat upon melting.

For a T_p of 1280°C (Fig. 11b) and an initially planar fault it is clear that the perturbed geotherm will not intersect the dry solidus even when the values of crustal thinning are very large (e.g. $\beta_c = 50$). No melt is generated because although β_c is large, the effective thinning of the lithospheric mantle is too small (from equation (4) $\beta_m = 1.5$). Unlike the uniform stretching model, the lithospheric simple shear model requires that β_c be large while β_m is still very small. If the detachment fault is *not initially planar* ($\theta_c < \theta_m$; Fig. 10b) melting is accomplished more easily. However, equation (4) shows that even if the dip of the detachment is initially 15° in the crust and it steepens to 30° in the lithospheric mantle no melt will be generated from the dry solidus until $\beta_c >> 3$. Such values for β_c are clearly too large for the North Sea.

In the presence of a mantle plume (Fig. 11c; T_p = 1480°C) the lithospheric simple shear model predicts no melting until $\beta_c > 3$ (Fig. 11c). This value of β is still much larger than those estimated for the triple junction area. If the fault is not initially planar and if β_c has a value of 1.5 (close to the minimum estimate for the Forties area), assuming $\theta_c = 15°$ and $\theta_m = 30°$, β_m is 1.39 and some melting occurs. For the same dips, a value for β_c as large as 2.5 results in $\beta_m = 1.55$ and almost 15% partial melting of the asthenosphere.

Melting of the asthenosphere occurs principally within the triple junction area of the North Sea. The central location of this magmatic activity is of critical importance. Fig. 12 shows the predicted distribution of melting (assuming vertical melt migration) if rifting had occurred

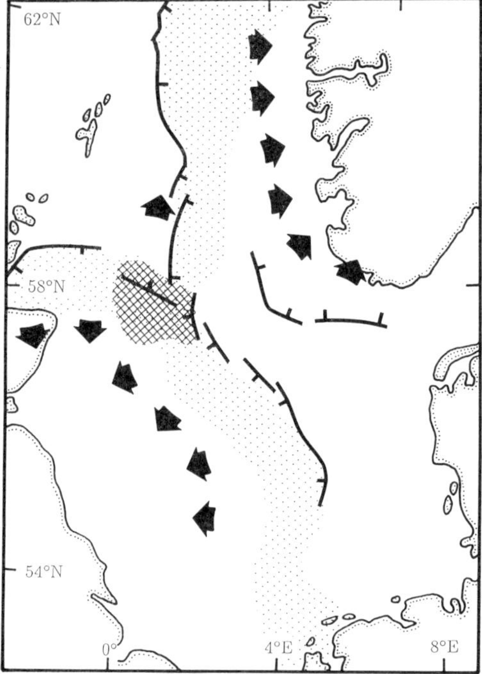

Fig. 12. Simplified tectonic map of the North Sea rift system. Light stippling, trilete rift system. Cross hatched shading, location of Forties province. Thick lines with tick marks, detachment faults at pre-rift surface as proposed by Beach (1986) & Gibbs (1987). Arrows point in directions away from the rift beyond which any vertically migrating melt would be expected to surface according to the lithospheric simple shear model. From Latin & White (1990).

by lithospheric simple shear, assuming it was controlled by the major 'detachments' identified by Beach (1986) and Gibbs (1987). The expected and observed distribution disagree. Even if magmas migrated upwards along very shallowly dipping detachment faults, the observed distribution of asthenospheric melt is difficult to explain.

In summary, for the estimated β values (1.4 to 2.5) Fig. 11 suggests that no melt is to be expected from the lithospheric simple shear model in the triple junction when T_p = 1280°C (Fig. 11b). If the T_p was 1480°C (a plume, Fig. 11c) and if the initial dip of the detachment as much as doubled as it passed from crust to mantle then alkali basalts might be produced, but they would probably be in the wrong place (Fig. 12).

On-axis and off-axis magmatism

Why are melts observed in some off axis locations, and yet not in others? Why is the triple junction the only on-axis area of magmatic activity? These questions may partly reflect the sampling problem. However, some answers may be found to lie in three separate arguments:

(1) the way in which the geometry of the triple junction is related to the amount and rate of extension;

(2) the ways in which long-lived crustal lineaments may control magma plumbing;

(3) the thermal history of the lithosphere and how it may control the production of magma from the MBL.

The triple junction may be the only on-axis area of magmatic activity because the amounts of extension are relatively large and the rates of extension were rapid enough for cooling by conduction to be negligible. In the Viking and Central Grabens the amounts of extension may be rather less (Latin *et al.* 1990). In the Viking Graben a rifting duration of 60 Ma is commonly mentioned (e.g. Wood 1982; Giltner 1988; Badley *et al.* 1988; N. J. White 1990) and this relatively slow rate may retard melting of the asthenosphere. Perhaps melts were not produced from the MBL in these areas because it had already been depleted by earlier Permian and Triassic events or it was too thin to intersect the amphibole-controlled solidus (shown in Fig. 9) upon decompression. An alternative explanation for the absence of melts in these areas may be provided by the underplating hypothesis of Holliger & Klemperer (1989) or may come down to sampling.

Why melts should underplate in some areas and erupt in others is unclear but it may be explained by crustal features. The coincidence between locations of magmatic activity and major lines of weakness within rift zones is well known (e.g. Bosworth 1987). The triple junction in the North Sea is not only the intersection of three major crustal rifts but it also appears to coincide with the extrapolated position of the Highland Boundary fault. The Central North Sea province may be related to the intersection of the Iapetus Suture (Klemperer & Hurich 1989) with the Mid-North Sea High while the Egersund Basin lies at the western end of the Tornquist line.

The reason why volcanic rocks are observed in the Central North Sea and the Netherlands may be that in these areas the graben intersects pre-existing and long-lived stable highs, the Mid-North Sea High, and Texel-Isselmeer High and the London Brabant-Massif (see Ziegler

1982). The MBL in these areas may have been relatively undisturbed by prior stretching events and so retained some enrichment and was able to undergo melting upon decompression. The observed ultrapotassic and volatile-enriched magmas lend support to this hypothesis.

Location of magmatism and source rock maturity

Consideration of the thermal maturity of hydrocarbon source rocks in the North Sea shows that the most thermally mature Mesozoic source rocks (MacKenzie & McKenzie 1983; Thorne & Watts 1989) coincide with areas of magmatic activity (notably in Forties and the Central Graben). Calculated predictions of source rock maturity based on predicted heat flow (from β factors) and rate constants for isomerization and aromatization reactions agree well with the observations in the North Sea (Mackenzie & McKenzie 1983) except where there has been Tertiary uplift (aromatization lower than predicted) or in areas of magmatic activity (aromatization higher than predicted). It is worth noting that in the Viking Graben, where volcanic rocks appear absent and there was little Tertiary uplift, the observed and predicted source rock maturity agree well (MacKenzie *et al.* 1984). The relationship between magmatism and maturity is yet to be fully examined but it is probably connected to the elevation of thermal gradients due to advection of heat into the upper crust by melts, and to more rapid rates of extension and subsidence than elsewhere. When degrees of extension are all relatively small then the absolute amounts of extension (in terms of β) and total sediment thickness may not be as important as the rates of extension and the presence of magmas.

Summary and conclusions

Although the quantitative study of rift-related magmatism is still in its infancy a number of important semi-quantitative relationships have been identified in the North Sea. Whether or not a melt composition can be attributed to easily melted enriched lithosphere, or to less easily melted asthenosphere, places limits on the amounts of extension in an area. Lithosphere-dominated melts occur in the North Sea when $\beta = 1.2$ to 1.3 and the MBL is thermally old, whereas in areas of higher extensional strain rates and thinner initial MBL thicknesses (e.g. Forties) input from the asthenosphere becomes more important at higher values of β. The rate of extension may be a critical factor in determining whether or not melt is produced. Slow strain rates may explain the absence of magmatic activity in much of the Viking and Central grabens. The timing of magmatic activity must be related to the onset of rifting (but will probably occur later) and to the rate of extension. In the North Sea the increase in the age of the activity towards the triple junction suggests that the rift system may have propagated from that point. This may be reasonably well constrained by crustal thinning and subsidence analysis.

In rift systems where the degree of extension is known then melt compositions and volumes can be used to place limits on the potential temperature of the asthenosphere at the time of rifting. Evidence from the Forties alkali basalts suggests that in the North Sea a 'hot' mantle plume ($T_p = 1480°C$) can be ruled out and that extension is by pure shear stretching rather than by lithospheric simple shear. The most likely combination of parameter values for the triple junction involves an initial MBL thickness of 70 km, a normal T_p (1280°C), rapid rates of extension ($< 60/\beta^2$ Ma), a value of β close to 2.0 (or more), and $< 2\%$ melting of a heterogeneous (streaky) asthenosphere source. Identification of asthenosphere potential temperature has very important implications for modelling palaeoheat-flow and the thermal evolution of a basin. Even without a knowledge of β the melt chemistry can place limits on the combinations of β and T_p that are possible. Taking the observations presented in this paper it is clear that useful information about the stretching history can be provided by rift-related igneous rocks found in extensional sedimentary basins. It would therefore be advantageous to core and study such material when it is encountered in exploratory wells.

We are indebted to the following oil companies for provision of the igneous material used in this study as well as for additional information and discussion: Amerada Hess Ltd, Amoco U.K. Exploration Co., Britoil, BP Exploration Co. Ltd, Chevron Exploration North Sea Ltd, Elf U.K., Esso (Norway), Nederlandse Aardolie M.B.V., Occidental Petroleum Ltd, Petroland (Elf), Phillips Petroleum Co. U.K. Ltd, Premier Oil Exploration Ltd, Ranger Oil, Shell International M.B.V., Shell U.K. Exploration and Production Ltd, Texaco Ltd, Total Oil-Marine Plc. We are grateful to Dan McKenzie for providing preprints of his work, computer programs and for discussion. We thank H. Nicholson and M. Wilding for their help during the early stages of this paper. DML acknowledges a N.E.R.C. studentship. NW is supported by a N.E.R.C. grant to the BIRPS group.

References

Allègre, C. J. 1982. Chemical geodynamics. *Tectonophysics*, **81**, 109–132.

—— 1987. Isotope geodynamics. *Earth and Planetary Science Letters*, **86**, 175–203.

Badley, M. E., Price, J. D., Rambech Dahl, C., & Agdestein, T. 1988. The structural evolution of the Viking Graben and its bearing on the extensional modes of basin formation. *Journal of the Geological Society of London*, **145**, 455–472.

Bailey, D. K. 1982. Mantle metasomatism – continuing chemical change within the Earth. *Nature*, **296**, 525–530.

Barton, P. & Wood, R. 1984. Tectonic evolution of the North Sea Basin: Crustal stretching and subsidence. *Geophysical Journal of the Royal Astronomical Society*, **79**, 987–1922.

Beach, A. 1986. Some comments on sedimentary basin development in the northern North Sea. *Scottish Journal of Geology*, **22**, 1–20.

Bosworth, W. 1987. Geometry of propagating continental rifts. *Nature*, **316**, 625–627.

Courtney, R. C. & White, R. S. 1986. Anomalous heat flow and geoid across the Cape Verde Rise: evidence for dynamic support from a thermal plume in the mantle. *Geophysical Journal of the Royal Astronomical Society*, **87**, 815–867.

Dixon, J. E., Fitton, J. G. & Frost, R. T. C. 1981. The tectonic significance of post-Carboniferous igneous activity in the North Sea Basin. *In*: Illing, L. V. & Hobson, G. D. (eds) *Petroleum Geology of the Continental Shelf of North-west Europe*, Heyden, London, 121–137.

Eynon, G. 1981. Basin development and sedimentation in the Middle Jurassic of the northern North Sea. *In*: Illing, L. V. & Hobson, G. D. (eds) *Petroleum Geology of the Continental Shelf of North-west Europe*, Heyden, London, 196–204.

Fall, H. G., Gibb, F. G. F. & Kanaris-Sotiriou, R. 1982. Jurassic volcanic rocks of the northern North Sea. *Journal of the Geological Society of London*, **139**, 277–292.

Fitton, J. G. & Dunlop, H. M. 1985. The Cameroon Line, West Africa, and its bearing on the origin of oceanic and continental alkali basalt. *Earth and Planetary Science Letters*, **72**, 23–38.

Foucher, J. P., Le Pichon, X. & Sibuet, J. C. 1982. The Ocean-Continent transition in the uniform lithosphere stretching model: role of partial melting in the mantle. *Philosophical Transactions of the Royal Society of London*, **A305**, 27–43.

Furnes, H., Elvsborg, A. & Malm, O. A. 1982. Lower and Middle Jurassic alkaline magmatism in the Egersund sub-basin, North Sea. *Marine Geology*, **46**, 53–69.

Gibbs, A. D. 1987. Deep seismic profiles in the northern North Sea. *In*: Brooks, J. & Glennie, K. W. (eds) *Petroleum Geology of North-west Europe*, Graham & Trotman, London, 1025–1028.

Giltner, J. P. 1988. Application of extensional models to the northern Viking Graben. *Norsk Geologisk Tiddskrift*, **67**, 339–352.

Green, D. H. 1970. The origin of basaltic and nephelinitic magmas. *Transactions of the Leicester Literary and Philosophical Society*, **64**, 28–54.

—— 1973. Experimental melting studies on a model upper mantle composition at high pressure under water saturated and water undersaturated conditions. *Earth and Planetary Science Letters*, **19**, 37–53.

—— & Ringwood, A. E. 1967. The genesis of basaltic magmas. *Contributions to Mineralogy and Petrology*, **15**, 103–190.

Harte, B. 1983. Mantle peridotites and processes — the kimberlite sample. *In*: Hawkesworth, C. J. & Norry, M. J. (eds) *Continental Basalts and Mantle Xenoliths*, Shiva, 46–91.

Hofmann, A. W. & White, W. M. 1982. Mantle plumes from ancient oceanic crust. *Earth and Planetary Science Letters*, **57**, 421–436.

Holliger, K. & Klemperer, S. L. 1989. A comparison of the Moho interpreted from gravity data and from deep seismic reflection data in the northern North Sea. *Geophysical Journal of the Royal Astronomical Society*, **97**, 247–258.

Howitt, F., Aston, E. R. & Jacque, M. 1975. The occurrence of Jurassic volcanics in the North Sea. *In*: Woodland, A. W. (ed.) *Petroleum and Continental Shelf of North-west Europe*, Applied Science, Barking, 379–388.

Jackson, J. A. 1987. Active normal faulting and crustal extension. *In*: Coward, M. P., Dewey, J. F. & Hancock, P. L. (eds) *Continental Extensional Tectonics*, Geological Society, London, Special Publication, **28**, 3–18.

Jaques, A. L. & Green, D. H. 1980. Anhydrous melting of peridotite at 0–15 kb pressure and the genesis of tholeiitic basalts. *Contributions to Mineralogy and Petrology*, **73**, 287–310.

Jarvis, G. T. & McKenzie, D. P. 1980. Sedimentary basin formation with finite extension rates. *Earth and Planetary Science Letters*, **48**, 42–52.

Klemperer, S. L. 1988. Crustal thinning and nature of extension in the northern North Sea from deep seismic reflection profiling. *Tectonics*, **7**, 803–821.

—— & Hurich, C. A. 1990. Lithospheric structure of the North Sea from deep seismic reflection profiling. *In*: Blundell, D. K. & Gibbs, A. (eds) *Tectonic Evolution of the North Sea Rifts*, Oxford University Press, Oxford.

Klemperer, S. L. & White, N. 1990. Coaxial stretching or lithospheric simple shear in the North Sea? Evidence from deep seismic profiling. *In*: Tankard, A. J. & Balkwill, H. R. (eds) *Extensional Tectonics and Stratigraphy of the North Atlantic Margins*. American Association of Petroleum Geologists Memoir, **46**, 511–522.

Latin, D. & White, N. 1990. Generating melt during lithospheric extension: pure shear vs simple shear. *Geology*, **18**, 327–331.

——, Dixon, J. E. & Fitton, J. G. 1989. Rift-Related Magmatism in the North Sea Basin. *In*: Blundell, D. J. & Gibbs, A. (eds) *Tectonic Evolution of the North Sea Rifts*, Oxford University Press, Oxford, 96–140.

Le Pichon, X. & Sibuet, J. C. 1981. Passive Margins:

A model of formation. *Journal of Geophysical Research*, **86**, 3708–3720.

MACKENZIE, A. S. & MCKENZIE, D. P. 1983. Isomerization and aromatization of hydrocarbons in sedimentary basins formed by extension. *Geological Magazine*, **120**, 417–528.

——, BEAUMONT, C. & MCKENZIE, D. P. 1984. Estimation of the kinetics of geochemical reactions with geophysical models of sedimentary basins and applications. *Organic Geochemistry*, **6**, 875–884.

MCKENZIE, D. P. 1978. Some remarks on the development of sedimentary basins. *Earth and Planetary Science Letters*, **40**, 25–32.

—— 1989. Some remarks on the movement of small melt fractions in the mantle. *Earth and Planetary Science Letters*, **95**, 53–72.

—— & O'NIONS, R. K. 1983. Mantle reservoirs and ocean island basalts. *Nature*, **301**, 229–301.

—— & BICKLE, M. J. 1988. The volume and composition of melt generated by extension of the lithosphere. *Journal of Petrology*, **29**, 625–679.

MICHAEL, P. J. & CHASE, R. L. 1987. The influence of primary magma composition, H_2O and pressure on mid-ocean ridge basalt differentiation. *Contributions to Mineralogy and Petrology*, **96**, 245–263.

PARSONS, B. & MCKENZIE, D. P. 1978. Mantle convection and the thermal structure of the plates. *Journal of Geophysical Research*, **83**, 4485–4496.

—— & SCLATER, J. G. 1977. An analysis of the variation of ocean floor bathymetry and heat flow with age. *Journal of Geophysical Research*, **82**, 803–827.

RITCHIE, J. D., SWALLOW, J. L., MITCHELL, J. G. & MORTON, A. C. 1988. Jurassic ages from intrusives and extrusives within the Forties igneous province. *Scottish Journal of Geology*, **24**, 81–88.

ROYDEN, L., HORVATH, F., NAGYMAROSY, A. & STEGENA, L. 1983. Evolution of the Pannonian Basin System 2. Subsidence and thermal history. *Tectonics*, **2**, 91–137.

SAWYER, D. S., TOKSÖZ, M. N., SCLATER, J. G. & SWIFT, B. A. 1982. Thermal evolution of the Baltimore Canyon Trough and Georges Bank Basin. *In*: WATKINS, J. S. & DRAKE, C. L. (eds) *Studies in Continental Margin Geology*. *American Association of Petroleum Geologists Memoir*, **34**, 743–762.

SCLATER, J. G. & CHRISTIE, P. A. F. 1980. Continental Stretching: An explanation of the post mid-Cretaceous subsidence of the Central North Sea Basin. *Journal of Geophysical Research*, **85**, 3711–3739.

——, ROYDEN, L., HORVATH, F., BURCHFIEL, B. C., SEMKEN, S. & STEGENA, L. 1980. The formation of the intra-Carpathian basins as determined from subsidence data. *Earth and Planetary Science Letters*, **51**, 139–162.

SHAW, D. M. 1970. Trace element fractionation during anatexis. *Geochimica et Cosmochimica Acta*, **34**, 237–243.

SLEEP, N. H. 1984. Tapping of magmas from ubiquitous mantle heterogeneities: an alternative to mantle plumes? *Journal of Geophysical Research*, **89**, 10029–10041.

STUART, G. W. 1978. The Upper Mantle structure of the North Sea from Rayleigh Wave dispersion. *Geophysical Journal of the Royal Astronomical Society*, **52**, 367–382.

TAKAHASHI, E. & KUSHIRO, I. 1983. Melting of a dry peridotite at high pressures and basalt genesis. *American Mineralogist*, **68**, 859–879.

THORNE, J. A. & WATTS, A. B. 1989. Quantitative analysis of North Sea subsidence. *American Association of Petroleum Geologists Bulletin*, **73**, 88–116.

VERHOOGEN, J. 1954. Petrological evidence on temperature distribution in the mantle of the earth. *Transactions of the American Geophysical Union*, **35**, 85–92.

WERNICKE, B. 1981. Low-angle normal faults in the Basin and Range province: Nappe tectonics in an extending orogen. *Nature*, **291**, 645–648.

—— 1985. Uniform sense simple shear of the continental lithosphere. *Canadian Journal of Earth Sciences*, **22**, 108–125.

WHITE, N. J. 1988. *Extension and subsidence of the continental lithosphere*. PhD Thesis, University of Cambridge (unpublished).

—— 1989. The nature of lithospheric extension in the North Sea. *Geology*, **17**, 111–114.

—— 1990. Does the uniform stretching model work in the North Sea? *In*: BLUNDELL, D. J. & GIBBS, A. (eds) *Tectonic Evolution of the North Sea Rifts*, Oxford University Press, Oxford, 213–235.

WHITE, R. S., FOWLER, S. R., MCKENZIE, D. P., WESTBROOK, G. K. & BOWEN, A. N. 1987. Magmatism at rifted continental margins. *Nature*, **330**, 439–444.

—— & MCKENZIE, D. P. 1989. Magmatism at rift zones: The generation of volcanic continental margins and flood basalts. *Journal of Geophysical Research*, **94**, 7685–7729.

WOOD, R. J. 1982. *Subsidence in the North Sea*. PhD Thesis, University of Cambridge.

WOODHALL, D. & KNOX, R. W. 1979. Mesozoic volcanism in the northern North Sea and adjacent areas. *Bulletin of the Geological Survey of Great Britain*, **70**, 34–56.

WYLLIE, P. J. 1987. Transfer of subcratonic carbon into kimberlites and rare earth carbonatites. *In*: MYSEN, B. O. (ed.) *Magmatic Processes: Physicochemical Principles*. The Geochemical Society, Special Publication, **1**, 107–119.

ZIEGLER, P. A. 1982. *Geological Atlas of Western Europe*. Shell Internationale Maatschappij B.V., The Hague.

—— & VAN HOORN, B. 1990. Evolution of the North Sea rift system. *In*: TANKARD, A. J. & BALKWILL, H. R. (eds) *Extensional tectonics and stratigraphy of the North Atlantic margins*. American Association of Petroleum Geologists Memoir, **46**, 471–500.

Late Jurassic half-graben control on the siting and structure of hydrocarbon accumulations: UK/Norwegian Central Graben

ALAN M. ROBERTS[1], JOHN D. PRICE[1] & TERKEL SVAVA OLSEN[2]

[1]*Badley Ashton & Associates Ltd, Winceby House, Winceby, Horncastle, Lincolnshire LN9 6PB, UK*

[2]*Statoil, Forus, N-4001, Stavanger, Norway*

Abstract: The late Jurassic structure of the Central Graben is discussed, both in terms of the extensional features developed at this time and their subsequent controls on younger inversion structures. The regional kinematics of the North Sea rift suggest that during the late Jurassic the Central Graben opened approximately orthogonally to the main NNW–SSE fault-trend within the basin. Some evidence that Triassic extensional faults were reactivated at this time is seen towards the basin margins, and a significant Triassic extensional history is inferred. The late Jurassic basin comprises the axial *c.*100 km of the Central Graben (*s.l.*). The basin axis is offset progressively to the SE by both transfer zones and discrete transfer faults. Footwall uplift of the basin margins during extension provided the source for the Upper Jurassic reservoir sands encountered by exploration drilling. These sands were probably preferentially fed into the basin at transfer zones in the margins. Salt structures are ubiquitous within the Central Graben, and their siting is everywhere controlled by the underlying fault-block topography. Salt structures form preferentially towards the crests of half-graben or buttressed against their boundary faults. Inversion structures in the Norwegian sector reactivated, with a reverse sense of displacement, late Jurassic extensional faults. A model is proposed which shows the two main inversion structures (Feda Graben, Norway; Tail End Graben, Denmark) to be detached on antithetic reverse faults riding in the salt. These faults terminate at tip-folds within major salt walls. The geometry of the inversion structures can be understood only when considered together with the geometry of the underlying late Jurassic half-graben.

The Central Graben is the NNW–SSE-trending, southern arm of the North Sea rift. From its intersection with the Outer Moray Firth Basin and Viking Graben, at about 58°N, it extends southwards *c.*500 km, through the UK, Norwegian, Danish, West German and Dutch sectors of the North Sea. Active extension of the Central Graben occurred during both the Triassic and late Jurassic, passive thermal subsidence ensuing in the early Cretaceous. It is not, however, our intention to describe here the complete structural history of the Central Graben, as to do so with justice to the complex tectonic history would require discussion well beyond the length of this paper. Instead we focus on the part that late Jurassic faulting played in controlling the siting and structure of hydrocarbon accumulations within the UK and Norwegian sectors. In pursuing such a discussion it is impossible to ignore the role of mobile Zechstein salt and thus much of our account emphasises the controls exerted by late Jurassic 'basement' structure on the siting of salt pillows and diapirs.

The main discussion comprises three sections: an outline kinematic and structural model for late Jurassic extension of the Central Graben, a discussion of late Jurassic structural controls on hydrocarbon accumulations in the UK Central Graben, and a discussion of late Jurassic half-graben control on selected Chalk-reservoir fields from the Norwegian sector.

A kinematic and structural framework for the late Jurassic Central Graben

The extension vector

Estimating the magnitude of late Jurassic extension across the Central Graben is a task fraught with difficulty (e.g. Ziegler 1983). Stretching estimates from seismic refraction data (e.g. Barton & Wood 1984) provide a figure for **finite**, whole-crustal strain, but do not indicate when such strain was imposed. Subsidence calculations (e.g. Sclater & Christie 1980; Barton & Wood 1984, Sclater *et al.* 1986; Hellinger *et al.* 1989) attempt to surmount this problem by relating stratigraphic data to the lithospheric stretching model (McKenzie 1978; Jarvis & McKenzie 1980). In a two-stage rift, such as the Central Graben, however, assessing the contribution of second stage (i.e. late Jurassic) stretch-

From Hardman, R. F. P. & Brooks, J. (eds), 1990, *Tectonic Events Responsible for Britain's Oil and Gas Reserves*, Geological Society Special Publication No 55, pp 229–257.

ing to basin subsidence is critically dependent on the timing, duration and magnitude of first-stage (i.e. Triassic) stretching. Thus, because knowledge of the Triassic event is at best partially-complete, Sclater & Christie (1980), Barton & Wood (1984), Sclater *et al.* (1986) and Hellinger *et al.* (1989) have each obtained a different estimate of Triassic extension from the same area of the Central Graben.

An alternative method of estimating basin extension is to sum the displacements on seismically-imaged fault surfaces (e.g. Ziegler 1983; Sclater & Shorey 1989). This approach is also, however, largely unsuccessful in delimiting the late Jurassic extension in the Central Graben, for two main reasons. (1) Attempts to measure meaningful fault displacement at stratigraphic levels above the Zechstein salt are rendered impossible by the common occurrence of salt diapirs adjacent to the main faults. (2) Measurements made at sub-salt (top Rotliegend) levels record the total post-Rotliegend extension, not just the late Jurassic displacement.

In an attempt to surmount these problems Roberts *et al.* (1990) constructed a vector triangle for the kinematically-linked Viking Graben, Moray Firth, Central Graben, late Jurassic rift system. Using more reliably constrained extension estimates from seismic data in the Viking Graben and Moray Firth a vector of *c.*20 km magnitude, trending ENE–WSW, was obtained for the likely seismically-resolvable extension in the Central Graben (Fig. 1). In this account we follow Roberts *et al.* and consider late Jurassic extension to have been of at least *c.*20 km magnitude across the Central Graben. Furthermore we acknowledge that if upper crustal extension below the limit of seismic resolution (Childs *et al.* 1990) were incorporated into the Viking Graben and Moray Firth extension estimates, an estimate of *c.*30 km extension for the Central Graben would perhaps be most likely. These figures are compatible with Sclater *et al.*'s (1986) estimate, from subsidence data, of 20–30 km late Jurassic extension across the axial 100 km of the graben. It is only within this axial region that significant late Jurassic extension is recorded.

Triassic extension

Seismic refraction data suggest that in total *c.*100 km of extension has occurred in the Central Graben, while we argue that only *c.*30 km of this is of late Jurassic age. Perhaps, therefore, evidence for as much as 70 km of Triassic extension should be sought. The Triassic history of the axial (late Jurassic) zone is too heavily over printed by subsequent deformation to allow a realistic assessment of early extension in this area. Large areas of the eastern Central Graben, however, *eg.* the Egersund and Søgne Basins (Norwegian sector) and Horn Graben (Danish sector) record Triassic extension outwith the zone of late Jurassic deformation. Regional seismic data show that the Triassic rift was areally much more extensive than its late Jurassic successor. Attributing appreciable extension to the Triassic is not therefore unrealistic.

Timing of extension and the significance of Jurassic volcanics

The effects of halokinesis on sediment thicknesses, combined with the problems of accounting for palaeobathymetric control on stratigraphic growth (Bertram & Milton 1989), make timing the onset and cessation of late Jurassic extension difficult, if not impossible, from seismic data alone. It has recently been argued, however, (Latin *et al.* 1990 and this volume) that the geochemistry of igneous rocks from the Forties province of the Central Graben shows them to have an origin synchronous with lithospheric stretching, rather than being the product of a pre-stretching mantle plume (*contra* Eynon 1981, Ziegler 1981). Perhaps therefore the most precise time markers of extension in the Central Graben are these igneous rocks. The most recent, and probably most reliable, ages for the rocks of this province are those derived by Ritchie *et al.* (1988), which, depending on the timescale against which they are calibrated, suggest a Callovian–Volgian age. The oldest rocks which overlie Forties volcanics are, however, Oxfordian shales of the Heather Formation which can be dated biostratigraphically. Thus, in agreement with Latin *et al.* (1990), we believe a Callovian, perhaps earliest Oxfordian at the youngest, age is likely for the volcanics. The majority of Jurassic extension may therefore have been gained very early in the late Jurassic, or even at the close of the Middle Jurassic. Much of the Humber Group (Heather and Kimmeridge Clay Formation) which overlies the volcanics, and might be thought of as syn-rift to late Jurassic extension, may in fact have post-dated most of the extension and simply filled in a deep and variable palaeobathymetry (see also Bertram & Milton 1989). Likewise the late Jurassic sands derived from footwall highs (see below) may also largely post-date active extension, being simply the erosion products of long-lived, upstanding fault-block crests. Care should be exercised in equating the presence of sand with necessarily-synchronous faulting.

Dip-slip faulting

It is frequently considered that strike-slip movement of varying magnitudes and orientations has controlled the opening of the late Jurassic Central Graben (e.g. Gowers & Sæbøe 1985, Gibbs 1985, J. C. Olsen 1987, Vejbæk & Anderson 1987). It is our suggestion, however, that such interpretations commonly derive from a misunderstanding of the rôle of halokinesis in controlling local structural geometries within the graben.

We suggest an alternative interpretation, following Roberts *et al.* (1990), which derives from superimposing the suggested ENE–WSW extension vector onto a fault pattern of the UK/Norwegian/Danish Central Graben (Fig. 1). This suggests that all of the major faults within the Central Graben are dip-slip structures, trending NNW–SSE. Any obliquity to pure orthogonal extension is only *c.*10° or less, and as such is far too small to manifest oblique-slip geometries within the deformed basin-fill (McCoss 1986). We therefore consider the late Jurassic Central Graben to be a basin dominated by NNW–SSE-trending extensional faults, linked kinematically by transfer zones or transfer faults at high angles to this trend.

Transfer faults and transfer zones

Although individual faults within the Central Graben strike NNW–SSE the external bounding envelope to the axial zone of late Jurassic faulting has a trend oblique to this, lying more NW–SE (Fig. 1). This arises because, unlike much of the Viking Graben, the margins of the Central Graben (with the exception of the Coffee Soil Fault in the Danish sector) are not defined by laterally-continuous fault segments. Rather both the margins and axial depocentre of the basin step progressively south and east (Fig. 1) in an en échelon fashion. Thus, although laterally offset from each other, the UK Central Graben, Norwegian Feda Graben and Danish Tail End Graben (Fig. 1) each define the local axis of maximum basin subsidence.

Southeastwards-stepping of the graben is

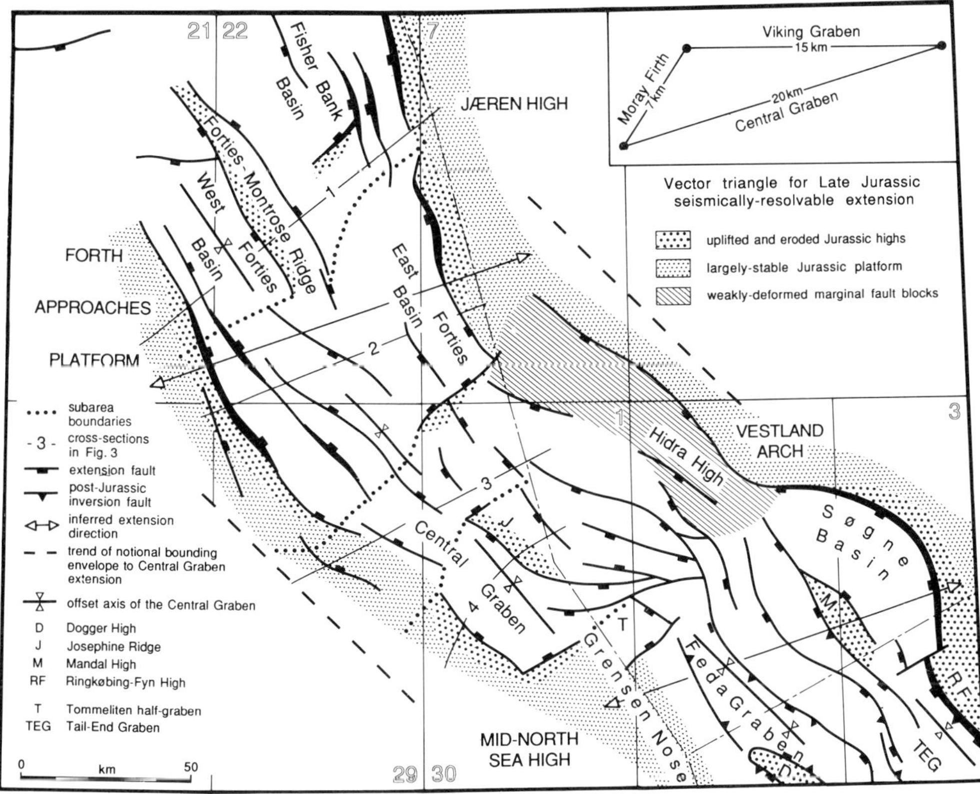

Fig. 1. Late Jurassic tectonic elements of the Central Graben showing tectonic subareas within the UKCS, separated by transfer fault/zones. Inset shows vector triangle from Roberts *et al.* (1990).

achieved by displacement partitioning between dip-slip faults, either along discrete transfer faults, such as those in the Tommeliten area (Fig. 1, see below), or across en échelon transfer zones, such as link the Feda and Tail End Graben (Fig. 1). The relationship between transfer zones and transfer faults, and the way in which the former may develop into the latter, is illustrated in Fig. 2. Not only have structures such as these partitioned the Central Graben into discrete sub-basins, they have probably also exerted a considerable influence on the derivation and distribution of Upper Jurassic sands within the graben (see below). A common misconception is that transfer faults are by definition strike-slip faults. Figure 2b demonstrates that this is not so. Transfer faults possess both a dip-slip and strike-slip component of displacement. If a transfer fault links two dip-slip faults inclined at >45° then its dip-slip displacement will be greater than the strike-slip displacement. Thus by recognizing transfer faults in the Central Graben it is not our intention also to imply significant lateral displacement on these structures.

The lack of evidence for 'Caledonian' controls on Mesozoic structures

The UK/Norwegian Central Graben lies between the Caledonian mountain belts of Scotland and Norway. Our suggested extension direction across the graben (ENE–WSW) lies close to the onshore Caledonian strike, and thus onshore structures have an orientation similar to the offshore structures which we identify as transfer zones and transfer faults. A simplistic approach, based on the correlation of similar-trending linear elements, would suggest a Caledonian influence on these Mesozoic structures (e.g. Johnson & Dingwall 1981; Threlfall 1981; Gage & Doré 1986). There is in reality, however, little factual evidence to support such an hypothesis.

Recent Caledonian plate reconstructions (e.g. McKerrow 1988; Soper 1988) suggest that prior to suturing, Scotland, England and Scandinavia were situated on separate plates, the UK being separated from Scandinavia by the Tornquist's Sea. Thus there should be no structural continuity at basement level between the UK and Scandinavian margins of the Central Graben. Reconstructions such as those presented by Gage & Doré (1986) and Grønlie & Roberts (1989), which link linear elements across the North Sea, do not honour the Tornquist's Sea hypothesis.

At a more detailed level, individual transfer faults within the graben are generally only a few kilometres in length, they are not fundamental structures dissecting the basin. There is little need to appeal to basement control on the siting of such local structures.

Onshore Caledonian structures in Scotland, such as the Highland Boundary Fault, can indeed be traced offshore for some distance using gravity and aeromagnetic data. The trend of these structures is, however, truncated within the Central Graben at the western margin of Mesozoic faulting, and although they are undoubtedly present at depth below the Mesozoic graben system, there is no evidence that they control the siting of younger structures.

We thus suggest that any NE–SW–(Caledonian)-trending faults within the Central Graben are entirely Mesozoic structures formed by the collapse of discontinuous faults (transfer zones) into transfer faults (Fig. 2), and endorse the comments of Klemperer & Hurich (1990) that neither the Highland Boundary Fault nor Tornquist Zone appear to have had any direct influence on the Mesozoic extensional history of the Central Graben.

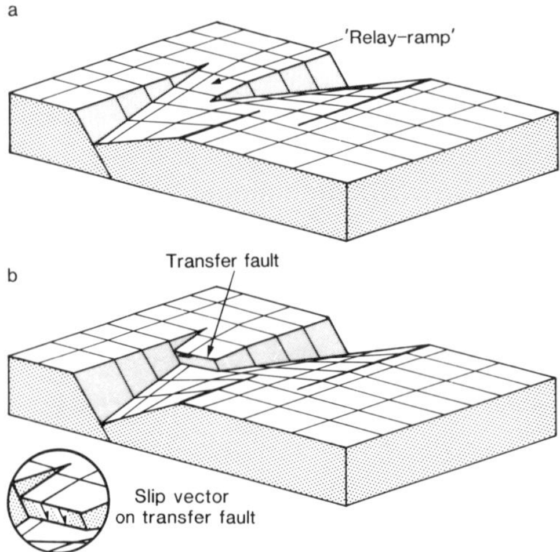

Fig. 2. Block diagrams showing (a) 'relay ramp' developed between two en-échelon normal faults, (b) collapse of relay ramp into a transfer fault after increased extension.

Late Jurassic fault control on hydrocarbon traps in the UK Central Graben

Since the close of the Permian the Central Graben has experienced two periods of extensional tectonism, Triassic and late Jurassic in age. Each has played a part in the formation of hydrocarbon accumulations. Our knowledge of the role of Triassic rifting is, however, limited, and so attention here is concentrated on the late Jurassic.

Characterization of hydrocarbon accumulations traditionally involves a description of: trapping style, the mode of deposition and distribution of the reservoir, and assessment of the mechanism and timing of hydrocarbon generation. Here we discuss only the tectonic controls on the first two of these topics. The third topic has been well covered by Cayley (1987). Trapping style and reservoir characteristics were both largely controlled by faulting and related halokenesis. An outline structural framework for the UK Central Graben is therefore presented before the more detailed discussion of hydrocarbon entrapment.

Structural framework

Late Jurassic extension in the Central Graben resulted in the formation of a series of largely ENE-dipping tilted blocks, bounded by NNW–SSE-orientated faults (Fig. 1). However, mapping a syn-rift reflector (i.e. Intra-Late Jurassic), the Base Cretaceous marker, or a reflector in the Triassic–Middle Jurassic, would not yield the simple structural pattern shown in Fig. 1. The reasons for this are linked to the presence of a thick sequence of Zechstein evaporites lying beneath the Triassic. These have been mobile since the Triassic, and have exercised significant control on the structural geometry of overlying sediments. Only by looking at a marker which is unaffected by halokinesis can the major basement-involved faults be identified. The most suitable marker is the Top Rotliegend/Base Zechstein, and it was by mapping of this reflector that the fault pattern in Fig. 1 has been deduced.

Subdivision of the UK Central Graben. The UK Central Graben is, in general, an asymmetric basin, with the majority of internal fault blocks dipping to the NE. The bounding faults to the graben are not everywhere clearly defined, and from NW to SE, are offset progressively to the SW by a series of transfer zones or transfer faults. Within the UK sector, four structural subareas can be recognized, separated by such transfer faults/zones. Cross sections through each of these are shown in Fig. 3, which illustrate how the internal structure of the graben changes progressively from north to south.

The northernmost subarea, bounded to the east by the Jæren High, and to the north by the complex 'triple junction' lying at the convergence of the Moray Firth, Viking Graben, and the Central Graben, is simply an easterly-dipping half-graben, bounded to the west by a broad anticline (of flexural origin?) rather than a discrete fault. The median Forties–Montrose Ridge is simply the crest of a subsidiary half-graben, the down-dip part of which forms the Fisher Bank Basin. This half-graben plunges and widens northwards as throw on the graben margin fault increases. The half-graben to the west of the ridge, known as the West Forties Basin, has a complementary relationship, widening and plunging southwards, as the western margin zone changes from a broad flexure, to a distinct fault.

The subarea lying immediately south of the Forties–Montrose Ridge comprises, from west to east, a series of three narrow (10 km wide) westerly-dipping terraces/half-graben, a true 'central' graben, and a broad (35 km wide) half-graben forming the SE Forties Basin, the crest of which is, essentially, the offset continuation of the Forties–Montrose Ridge.

Running between the two northern subareas is a complex series of transfer zones and faults. Offset of the graben margin is most marked in the east, where it amounts to approximately 10 km. North of the transfer, displacement in the boundary zone is partitioned between four faults which define a series of terraces (Fig. 4). South of the transfer, displacement is relayed into a single fault, which forms the boundary to the SE Forties Basin. To the east of the marginal fault lies the extensive Norwegian–Danish salt platform.

Further south, two more structural subareas can be recognized, each lying partly within the UK Sector. These both have a complicated internal structure, poorly defined by regional seismic data. It is within the southernmost subarea that the well documented Auk, Fulmar (Johnson *et al.* 1986) and Clyde (Smith 1987) Fields lie.

Transfer faults. The geometry of transfer faults was briefly described in the first part of this paper (Fig. 2). Rarely, however, do they assume the simple pattern shown there. More typically they serve to partition the throw between a series of faults, and for an example we turn to real data.

The NE part of Quadrant 22 is structurally complex (Fig. 1). Within it lie the Jæren High, the Fisher Bank Basin, the terraces transitional between these two structures, and the transfer zone between this part of the basin and the East Forties Basin to the south. The principal faults

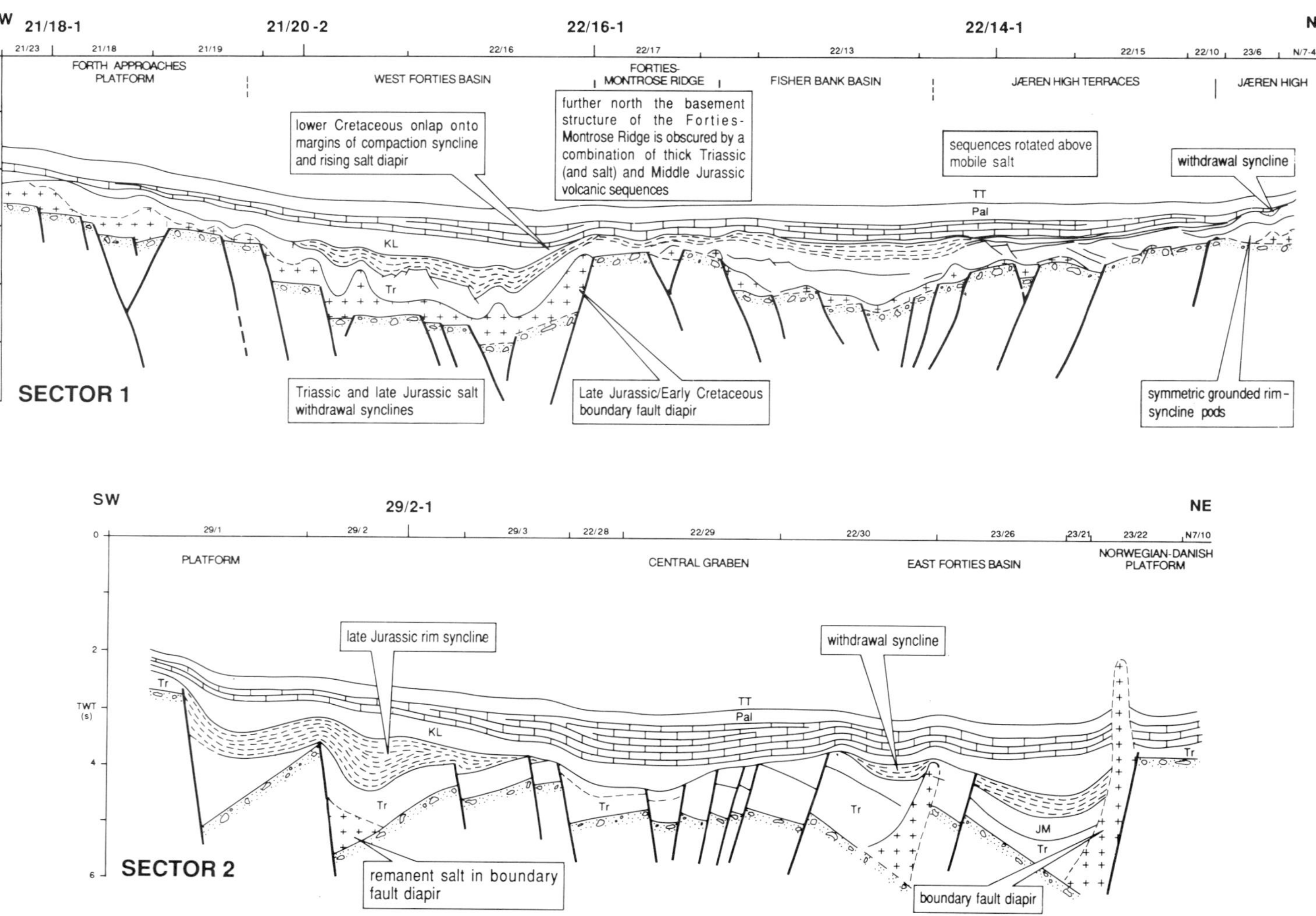
SW
21/18-1
21/20-2
22/16-1
22/14-1
NE
21/23
21/18
21/19
22/16
22/17
22/13
22/15
22/10
23/6
N/7-4
FORTH APPROACHES PLATFORM
WEST FORTIES BASIN
FORTIES-MONTROSE RIDGE
FISHER BANK BASIN
JÆREN HIGH TERRACES
JÆREN HIGH
lower Cretaceous onlap onto margins of compaction syncline and rising salt diapir
further north the basement structure of the Forties-Montrose Ridge is obscured by a combination of thick Triassic (and salt) and Middle Jurassic volcanic sequences
sequences rotated above mobile salt
withdrawal syncline
TT
Pal
KL
Tr
TWT (s)
SECTOR 1
Triassic and late Jurassic salt withdrawal synclines
Late Jurassic/Early Cretaceous boundary fault diapir
symmetric grounded rim-syncline pods
SW
29/2-1
NE
29/1
29/2
29/3
22/28
22/29
22/30
23/26
23/21
23/22
N7/10
PLATFORM
CENTRAL GRABEN
EAST FORTIES BASIN
NORWEGIAN-DANISH PLATFORM
late Jurassic rim syncline
withdrawal syncline
TT
Pal
KL
Tr
JM
TWT (s)
SECTOR 2
remanent salt in boundary fault diapir
boundary fault diapir

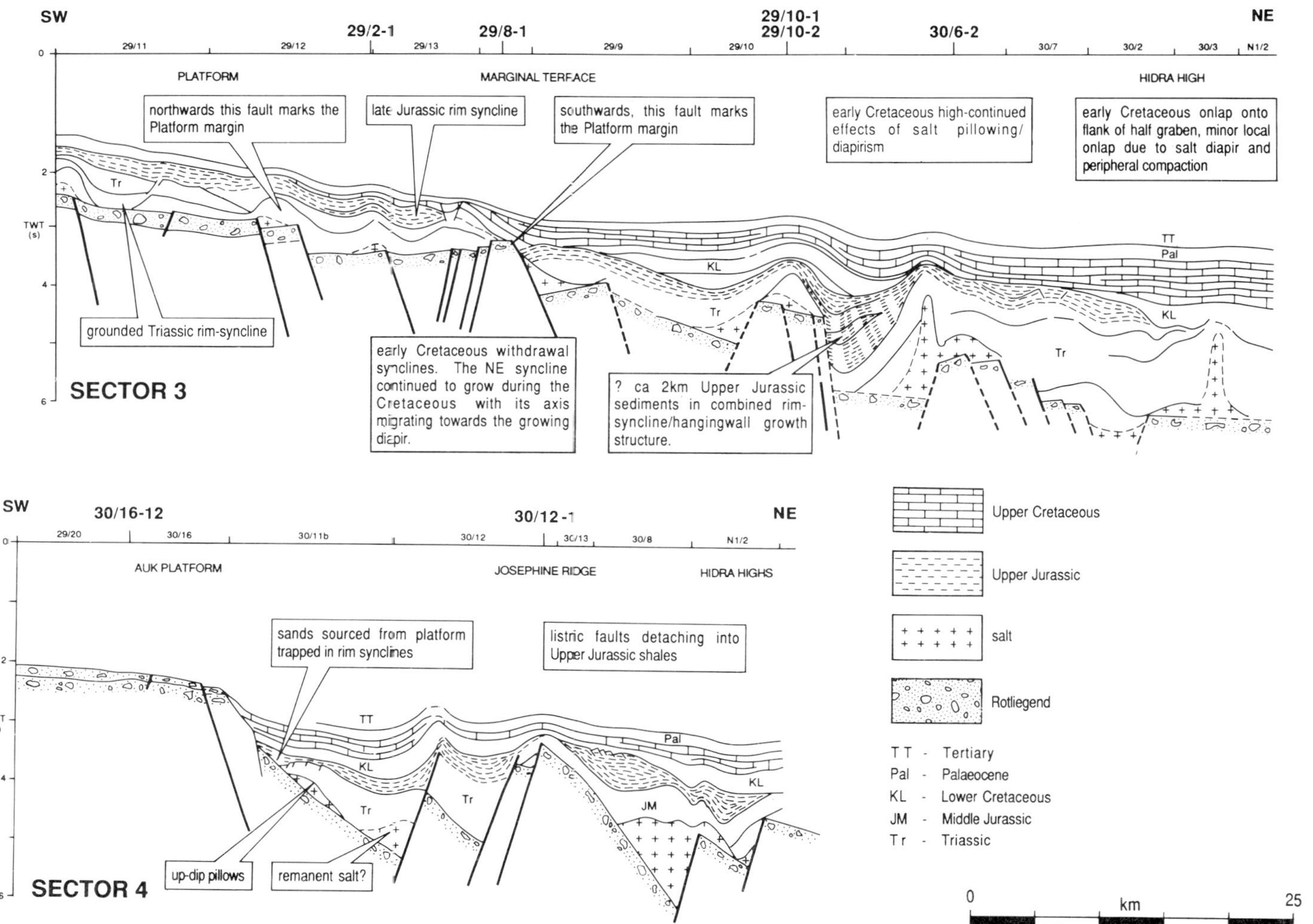

Fig. 3. Cross sections through the four tectonic subareas identified in the UK Central Graben.

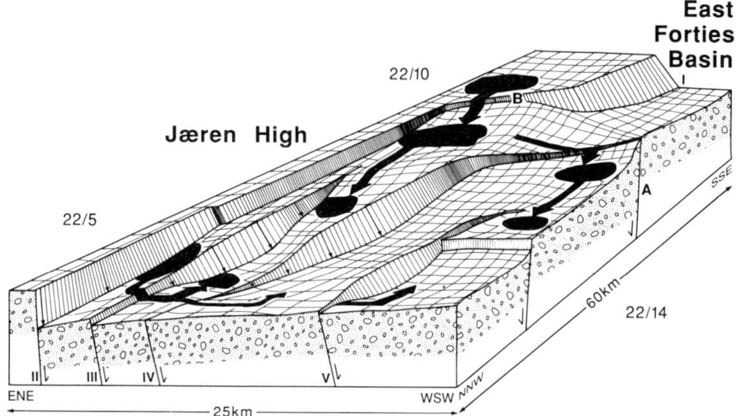

Fig. 4. Schematic block diagram showing the fault pattern and displacement style at Base Zechstein level, marginal to the Jæren High in the NE part of Quadrant 22. Shaded areas indicate sites of possible syn-rift sand accumulation and the direction of sediment transport.

in this zone are shown in a block diagram (Fig. 4). In the SE a single large fault (I) bounds the Central Graben. Further north displacement is partitioned between four faults (II–V) which define three terraces, the displacement being transferred between them via a series of relay-ramps. By contrast, displacement transfer at the southern ends of faults II, IV and V takes place through transfer faults (A and B). That these are unlikely to be single faults is clear from Fig. 5a, a seismic line passing through fault A, on which the transfer 'fault' is in fact a fault zone nearly 10 km wide.

Trap formation

Within the Central Graben potential hydrocarbon traps are to be found at the crests of tilted blocks, rotated during late Jurassic extension, above salt pillows and diapirs, and in the axes of grounded rim-synclines.

Fault-block rotation. Amongst the simplest of traps that have yielded hydrocarbons in the Central Graben are the crests of tilted fault blocks. The Josephine Field (Fig. 3) will produce from Jurassic sandstones at the crest of a structure often interpreted as a horst, one of the Central Highs of the graben (cf. Glennie 1986). The Josephine structure can, however, simply be regarded as the deeply-eroded footwall of a large easterly-dipping tilted fault block, the degree of erosion being attributable to the magnitude of the throw on the fault and the width of the fault block (Barr 1987). In contrast to the Viking Graben and Moray Firth, areally-large closure in this type of trap is rare in the Central Graben. Most examples are confined to Rotliegend/Jurassic sequences in the footwalls of platform margin faults (e.g. Auk).

The restricted development (or recognition) of hydrocarbon-bearing tilted fault blocks, should not be taken to indicate, however, that late Jurassic faulting exerted only a minor influence on the formation of traps within the graben. The most significant effects of faulting, from the point of view of trap formation, relate to the widespread triggering or acceleration of salt movement.

Salt mobility. To understand the formation of potential hydrocarbon traps resulting from salt movement (halokenesis), it is first necessary to understand how it moves within half-graben. A new conceptual model synthesising previous theoretical studies (Jackson & Talbot 1986) is proposed here (Fig. 6).

Salt movement, and subsequent disruption of sedimentary stratification, takes place as a result of density contrasts between salt and sediments, induced either by burial or loading.

Under normal burial conditions, the density contrast between salt and any overlying clastic sediments becomes sufficient for salt to migrate upwards, driven by buoyancy, at a burial depth of approximately 600–900 m (Jackson & Talbot 1986). Models of salt movement, involving three-phase rim-syncline development, synchronous with pillowing, diapirism, salt escape and sediment collapse, are well known (Trusheim 1960, Seni & Jackson 1983). Depocentres migrate, following the progressive and systematic withdrawal of the salt.

Above a flat, homogeneous substrate diapirs

form with a characteristic wavelength distribution, in a manner similar to that predicted by centrifuge modelling (Jackson & Talbot 1989). Rarely, however, is the sub-salt surface truly horizontal, and minor faults in the basement have long been known to exert a control on the sites of pillow formation.

'Up-dip' and 'boundary-fault' pillows. Much stronger controls on the siting of rising pillows are exerted when salt movement takes place above tilted surfaces, such as those seen in the Central Graben. Soon after burial (at <600–900 m), salt will flow **down** the dip of a half-graben (Fig. 6), its movement impeded only by faults in the underlying surface. Movement is halted when salt becomes buttressed against the half-graben boundary-fault. This initial movement of salt enhances any primary depositional thickening of the salt towards the deepest part of the half-graben.

As a result of continued subsidence a 'burial front' passes progressively upwards through the deepening half-graben, below which salt is now less dense than the overlying, partially-compacted sediments. Salt will now **rise** through buoyancy. Salt pillows begin to form. These lie against the boundary fault, and at the shallowest depth to which salt can rise by buoyancy (i.e. the burial front). Above the front salt may in addition still be moving down-dip from the shallower part of the half graben. At the point of convergence of the two trends, an up-dip pillow is formed.

Nearly all salt structures (e.g. Fig. 5b) in the Central Graben appear to conform with this simple 'up-dip' and 'boundary-fault' model, and probably began to form synchronous with block-tilting during the Triassic. The renewal of faulting during the late Jurassic may have either accelerated, or initiated further, salt movement, which is continuing to the present day.

It must be emphasized that salt withdrawal only amplifies or buffers the thicknesses of sediments accumulating in a basin, being responsible for local variations in the subsidence rate. Salt withdrawal does not drive tectonic basin subsidence. In general tectonic subsidence will accompany salt-induced subsidence.

Differential loading. Differential sediment-loading, such as that produced by the progradation of a clastic sequence, can allow salt-pillowing to occur more readily than buoyancy alone. It can even be initiated at burial depths below 600 m. Jackson & Talbot (1986) have argued that a depositional relief of only 16 m on the sea-bed can cause differential loading sufficient to induce salt movement. The differential stress required to mobilize salt by buoyancy effects alone is ten times greater than that required to mobilize it under the effects of differential loading.

Hospers *et al.* (1986) have argued that sediment loading, resulting from sediment progradation, triggered salt buoyancy throughout most of the Norwegian–Danish Platform. Differential loading is inevitable above salt sequences deposited in asymmetric half-graben, and the rate at which the differential load develops will be greatest during active faulting. This is particularly so if significant bathymetric relief allows the deposition of prograding/aggrading sequences.

Closure above diapirs and pillows. Salt movement in the Central Graben has resulted in the development of numerous, potential trapping configurations, either by directly inducing structural closure (e.g. a number of the Norwegian Chalk Fields, see later), or as a result of structural closure arising in sediments draped over incompactible salt.

Where salt removal is extreme, or has taken place on a tilted/faulted surface, blocks of overburden sediment can readily rotate, detached above mobile salt. Figure 5b shows, at its western end, a series of rotated blocks dipping to the west, a direction opposed to the dip of the underlying top Rotliegend, which marks the floor of the half-graben. The presence of salt in this half-graben is unequivocal, and a boundary-fault diapir/salt-wall rises at its eastern margin. Collapse-structures point to the presence of remanent salt pillows, and salt is probably present at the western end, in a boundary-fault pillow adjacent to the Forties–Montrose High. The rotated blocks have probably detached above the salt, and have slid 'down-hill', down the dip of the half-graben. Similar features have been modelled by Vendeville & Cobbold (1987) and their formation has been described by Jackson & Talbot (1986) as a modified form of buoyancy halokinesis, initiated after normal faulting in the cover sequence has produced differential loading.

Grounded rim-synclines. Sediments deposited synchronously with salt movement characteristically thicken above sites of salt withdrawal (primary rim-synclines), and thin above rising diapirs (Fig. 7). Subsequent evacuation of the diapirs/pillows allows thick sediments to be deposited in their place (secondary rim synclines), and compaction effects permit the development of thickening peripheral to the

(a)

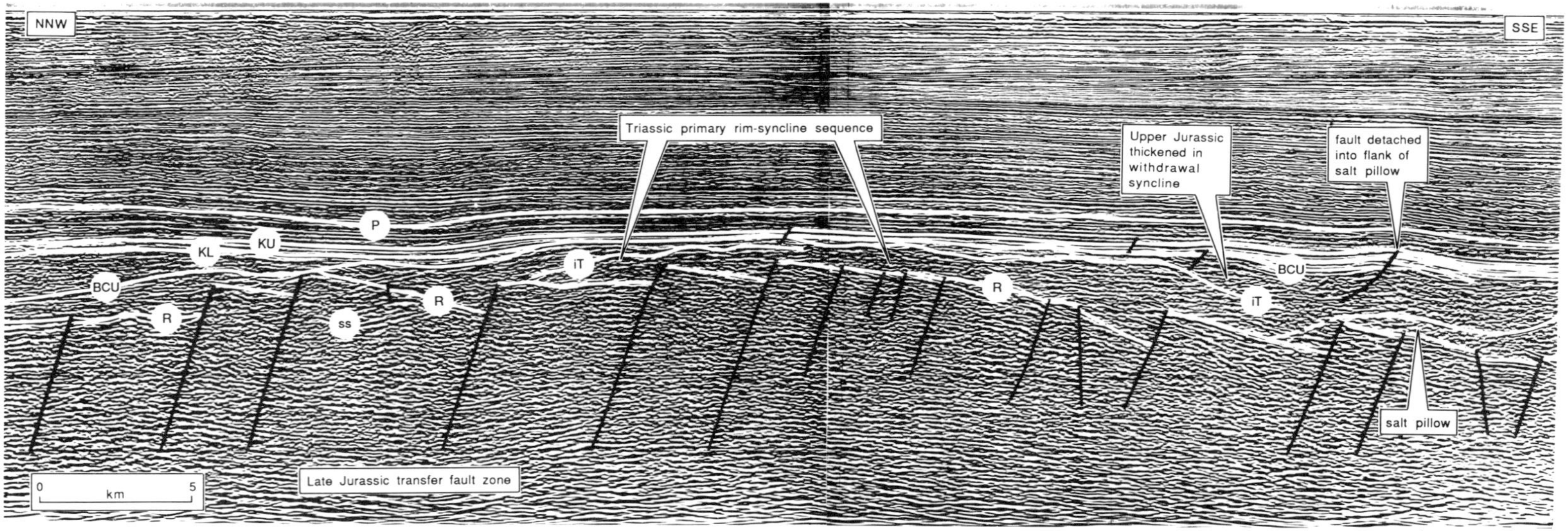

Fig. 5a. Seismic line through a NE–SW-trending transfer zone in Blocks 22/9 & 22/14. The Jurassic and Lower Cretaceous are thin or absent over the crest of the structure, which forms the footwall to the transfer zone. Key to intervals and tops (and Figs 8 & 9). P Palaeocene, KU Upper Cretaceous (Chalk Gp), KL Lower Cretaceous, JU Upper Jurassic, JM Middle Jurassic, TR Triassic, iT intra Triassic, R Rotliegend, BCU Base Cretaceous Unconformity, ss out-of-plane reflections (Courtesy GECO/NOPEC).

(b)

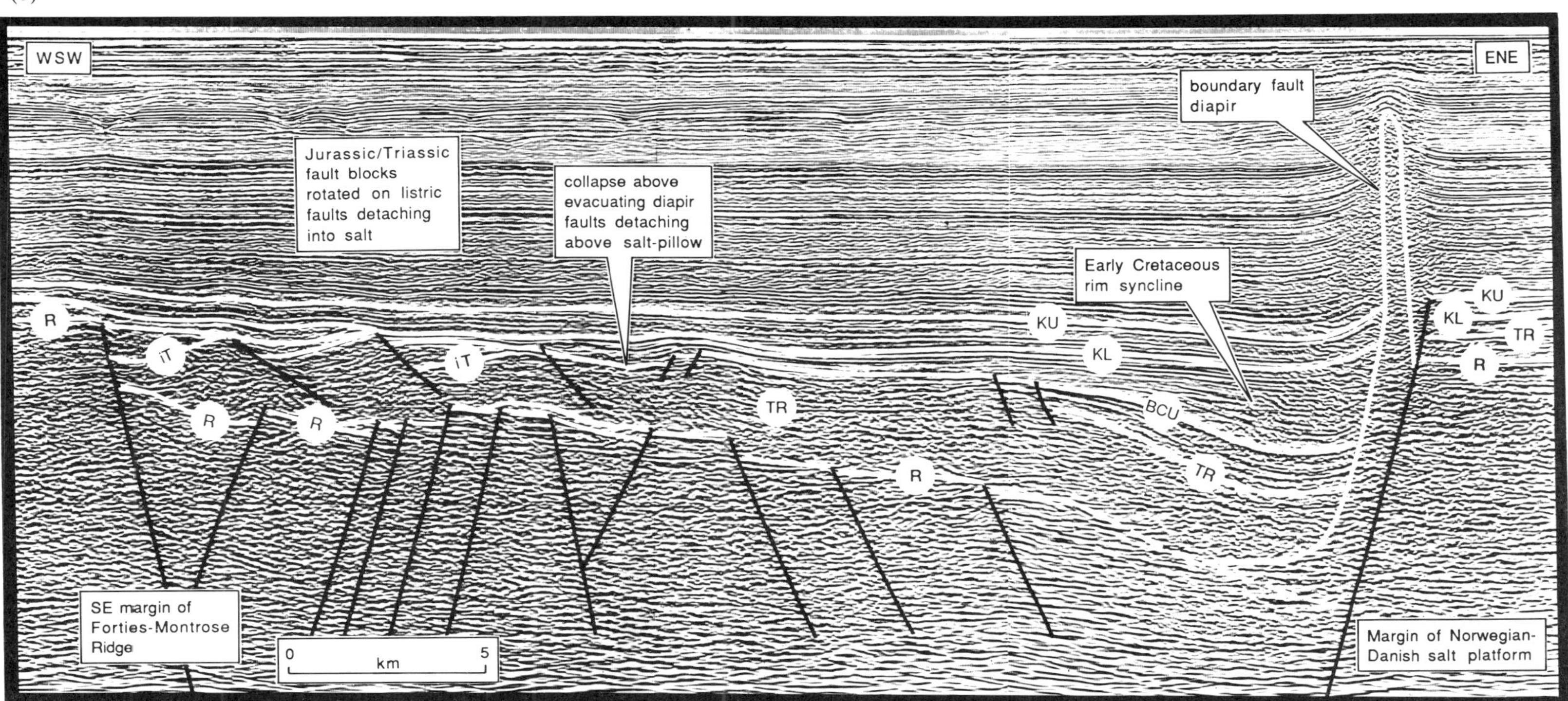

Fig. 5b. Seismic line across the northern part of the East Forties Basin, showing the boundary-fault diapir against the Norwegian–Danish salt salt platform, up-dip pillows, rim synclines, and block rotation resulting from salt withdrawal (Courtesy GECO/NOPEC). See Fig. 5a for key.

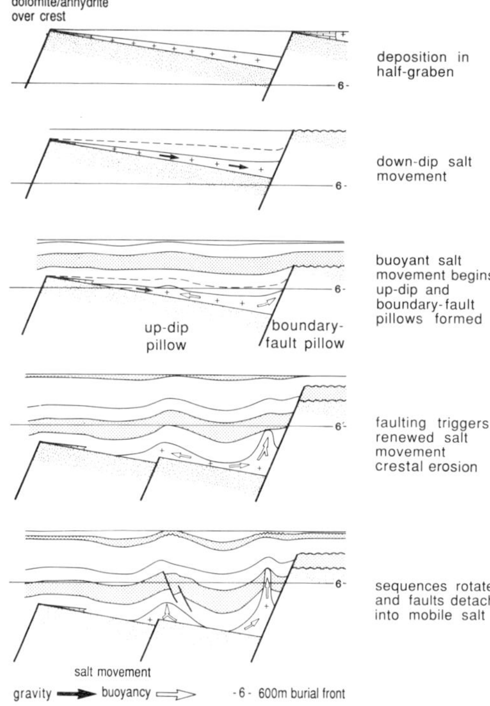

Fig. 6. Model for the migration of salt in a half-graben, illustrating the interplay between salt sinking and salt rising. Salt migration results both from the direct effects of gravity and from the effects of differential sediment-loading.

withdrawn diapir (tertiary rim-synclines). Complete evacuation of previously-formed diapirs results in the apparent structural inversion of the primary rim-syncline, which now assumes a flat base, and a curved upper surface, and so produces structural closure (Fig. 8). These features are variously referred to as grounded rim-synclines, turtle-backs, or more simply, as 'pods'. Closure has developed in the Tertiary sands of the Nelson Field, probably because they are draped over grounded rim-synclines (Fig. 9).

Reservoir formation

Sand source. Our concern here is primarily with the role of late Jurassic tectonism in the generation and deposition of sand-grade sediments. Topography related to extensional faults probably exerted a strong control on deposition in the largely terrestrial Triassic basin, but our knowledge of the basin geometry is too rudimentary, obscured by subsequent faulting and halokinesis, for a reliable predictive model yet to be established.

Most sediment introduced into the Central Graben during the late Jurassic was probably derived from the erosion of areas uplifted during rifting. In addition to such purely tectonic controls, erosion may also have occurred above the crests of rising salt structures.

Erosion of footwall crests. During normal faulting, subsidence of the hangingwall block is accompanied by subordinate uplift of the footwall (e.g. Vening Meinesz 1950; Jackson & McKenzie 1983; Barr 1987; Barnett *et al.* 1987; Kusznir *et al.* 1988; Roberts & Yielding 1991). This is principally because the footwall has been isostatically unloaded by the lateral motion of the hangingwall, and thus rises. The amount of uplift is directly related to throw on the fault, and the wavelength of the area affected by uplift is largely dependent on the effective elastic thickness of the crust (and to a lesser extent on the density contrast between the faulted rocks, and the hangingwall fill, normally water or sediment).

The effects of footwall uplift and subsequent erosion resulting from late Jurassic faulting, are best seen on the largest fault in the UK Central Graben, that forming the western boundary to the Jæren High. Here, Jurassic sediments are largely absent, and the Triassic is deeply eroded across a distance of *c.*25 km into the footwall (Sclater *et al.* 1986, fig. 2). Similar erosion is recorded adjacent to other major faults which bound the Central Graben (e.g. the Coffee Soil Fault, Roberts & Yielding 1991), and on the major intra-basinal half-graben and transfer faults (Forties–Montrose Ridge, Josephine High; Fig. 3). The Fulmar Formation represents the erosion products of this footwall uplift, and the distribution of these sediments is discussed below. Note, however, that the age of the Fulmar Formation (Oxfordian–Kimmeridgian) provides only a minimum age for fault activity. It is quite possible that rapid footwall emergence preceded a more sedate degradation of uplifted block crests.

Erosion of sediments above diapirs. Sediments can be uplifted above the core of a rising diapir (Fig. 6). Typically the result is an area of non-deposition above the diapir, subsidence being countered by the rise of the salt. When the rate of rise exceeds subsidence, uplift occurs, and if sediments are brought above wave-base, erosion may occur. Although salt diapirism is an obvious mechanism for generating uplift and erosion, the products are volumetrically insignificant and

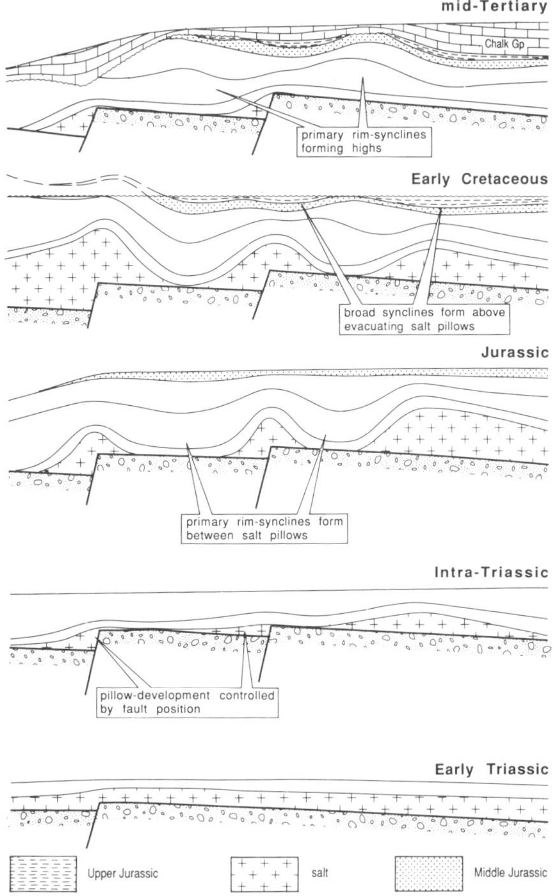

Fig. 7. Model for the Permian–Early Cretaceous evolution of grounded rim synclines (pod-structures) over the Jæren High.

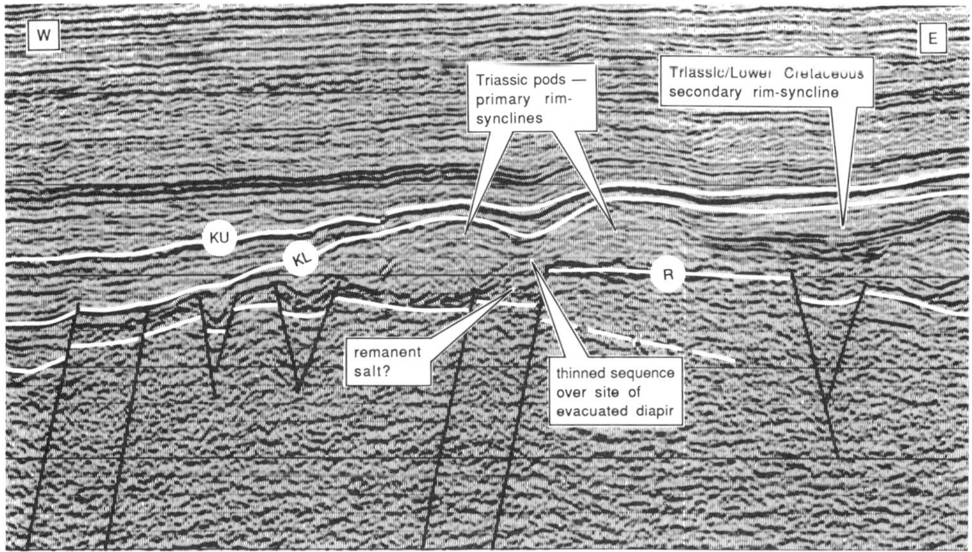

Fig. 8. Seismic line showing grounded Triassic primary rim synclines over the Jæren High. (See Fig. 5a for key) (Courtesy GECO/NOPEC).

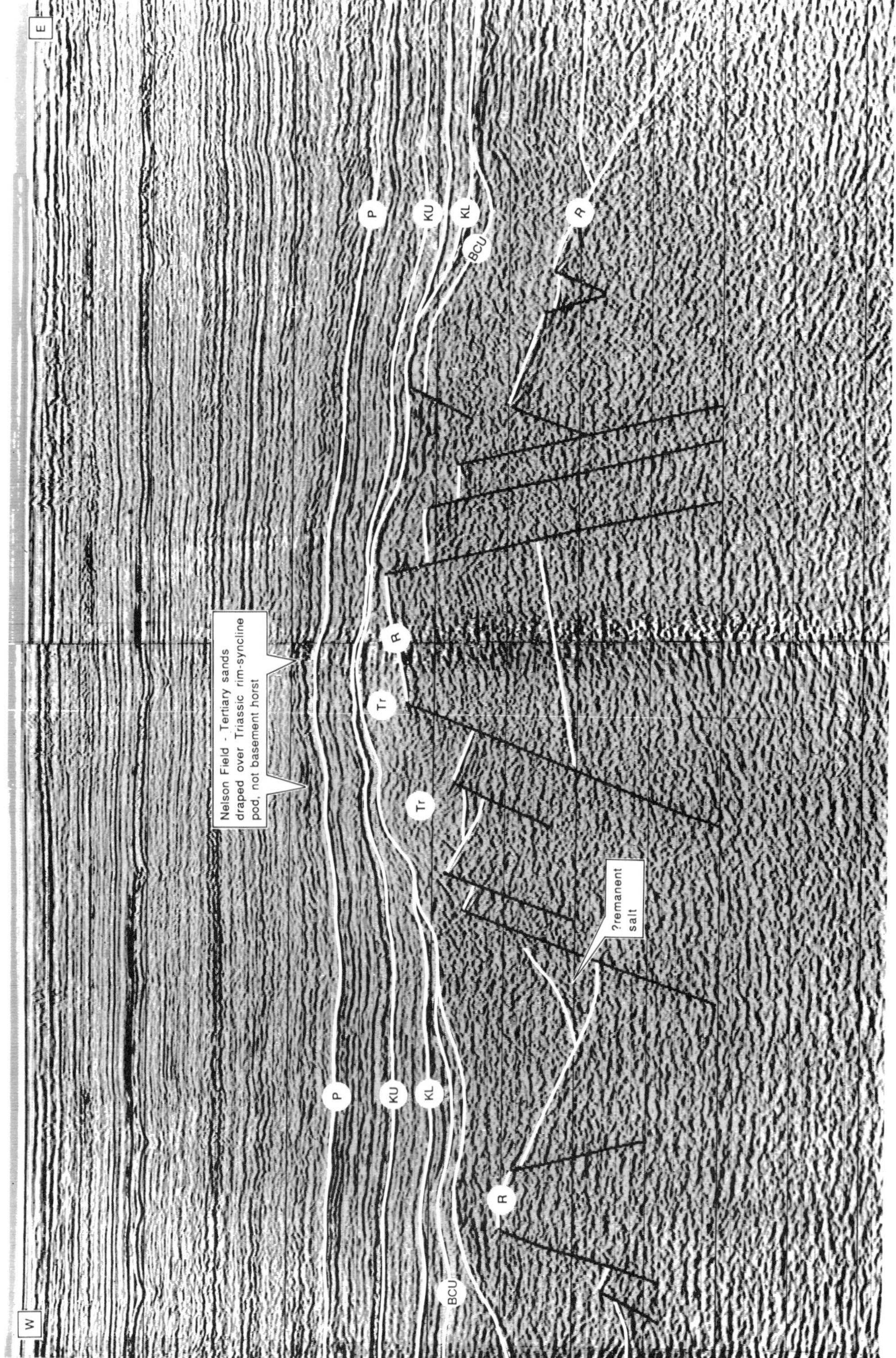
W
E
Nelson Field - Tertiary sands draped over Triassic rim-syncline pod, not basement horst
?remanent salt
P
KU
KL
BCU
R
Tr
P
KU
KL
BCU
R
R
Tr

will have only local effect, providing detritus to fill synchronously-subsiding rim-synclines. Rising diapirs have, however, controlled not the supply of late Cretaceous Chalk, but its facies around locally-developed highs (Nielsen *et al.* 1990).

Nature of eroded sediments. The late Jurassic subcrop in the Central Graben and surrounding platform areas comprises dominantly Triassic sediments. These, deposited largely in a distal alluvial/lacustrine setting, are dominated by mudrocks, and have provided low volumes of sand for deposition in the late Jurassic basin. In the platform areas and locally within the basin (e.g. Josephine High), Devonian rocks were eroded. Middle Jurassic rocks were probably eroded from areas peripheral to the basin, although they were most likely only thinly developed over the platforms, and in the northern part of the Central Graben comprise largely sand-poor volcaniclastic sequences. Sediment being introduced into the late Jurassic basin was therefore probably, in general, sand-poor, and the concentration of it to form viable hydrocarbon reservoirs was highly dependent on depositional processes. The patchy distribution of late Jurassic sands within the basin may therefore not reflect periodic fault activity, as might first seem likely, but rather the progressive exposure and erosion of sand-prone footwall sequences.

Depositional sites. The inception of rifting during the late Jurassic resulted in block rotation and regional subsidence, with local uplift and erosion of footwall crests. A marine basin was formed, at the latest by the early Oxfordian (and possibly during the Callovian; although there are no released wells drilled in the structurally-deep hanging-wall sites which might prove this). The accelerated rate of basin subsidence, combined with the effects of active faulting, triggered renewed salt movement. Late Jurassic tectonism thus exerted a dual control on the depositional sites of Upper Jurassic reservoirs.

Deposition close to active faults. It has already been argued that erosion of material for deposition in the Central Graben took place close to active faults, principally at the margins of the basin, and that the eroded sediments were predominantly mudstones. The observed distribution of Upper Jurassic sandstones fits this model well, for they are principally to be found fringing the basin, adjacent to the marginal faults. Absence of sandstones from some such areas is attributed to a failure in sediment supply. Locally, at the sites of particular wells, erosion/non-deposition may be attributable to halokenesis.

Deposition in areas tectonically active at the present day is greatly controlled by structural trend and fault-controlled topography. In terrestrial settings, drainage patterns commonly run parallel with fault strike, only deviating where individual fault segments terminate, namely in the transfer zones described above (*cf.* Leeder & Gawthorpe 1987, fig. 4; Roberts & Jackson 1991). Thus, in modern settings (Aegean, Turkey, Gulf of Suez), although alluvial fans do form at the feet of active fault scarps, far more important are the fans formed in the topographic lows between en échelon active fault strands (Roberts & Jackson 1991). Within the Upper Jurassic of the Central Graben, there is a good correspondence between sand deposition at sites relatively distant from the active marginal faults and the presence of transfer zones.

Deposition in rim-synclines. The migration of depocentres coincident with salt migration has already been described above. During the late Jurassic (and Cretaceous-early Tertiary) rim-synclines acted as sinks which allowed the accumulation of thick sedimentary sequences. Where the rim-synclines developed in areas of active sand supply, potential reservoir sequences were deposited, of greater thickness than similar sequences developed in areas undergoing only tectonic subsidence. Such sequences may, in addition, be thicker than would be expected for their depositional setting.

Distribution of the Fulmar Formation. Most non-coal-bearing Upper Jurassic sands encountered in the Central Graben are referred to the Fulmar Formation, which characteristically comprises thick (150–350 m) marine, sand-prone sequences, interpreted as having been deposited in a prograding shore-face, or as migrating storm-dominated, shelf sandsheets (Johnson *et al.* 1986).

Characteristics of the Fulmar sands include:

Fig. 9. Seismic line across the Nelson Field showing closure in Tertiary sands probably developed by drape over Triassic grounded rim-synclines, rather than over the basement horst (See Fig. 5a for key) (Courtesy GECO/NOPEC).

(1) their large thickness and massive nature; (2) their restriction to the basin margins (belts less than 10 km in width); (3) their localised occurrence, passing rapidly laterally and basinwards into mudstones; (4) their mineralogical immaturity, with a prevalence of terrigenous carbonaceous and woody material. These factors imply local sediment supply, best provided by a point-sourced, fluvio-deltaic input. The location of the input points is controlled by the fault transfer zones. A currently unreleased well on the western margin of the basin, penetrates a proximal sequence probably deposited in such a setting.

The Fulmar Formation sandstones range in age from early Oxfordian to Kimmeridgian (Eudoxus Zone). Within the Central Graben they are widely either barren of microfaunas and palynofloras, or buried to depths at which the palynofloras have been destroyed by thermal maturation. Recent advances in the understanding of the micropalaeontology of these sequences has, however, enabled a biostratigraphical subdivision to be made. Three crude time intervals are represented within the Fulmar Formation, and their recognition allows a temporal variation in the distribution of the Fulmar Formation to be observed.

Maps (Fig. 10) showing the extent of Fulmar Formation sand deposition chart the widening of the basin during the Oxfordian, particularly of the West Forties Basin, and a consequent reduction in the sub-aerially exposed part of the Forties–Montrose High. Widening of the basin (probably the result of basinwide, thermal? subsidence, rather than direct fault control) did not automatically result in an increase to the area of sand deposition, perhaps because of the limited volume of sand supply. Although the basin continued to widen during the Kimmeridgian, sand deposition was restricted to the western margin, and in particular to the Fulmar–Clyde area.

Summary of late Jurassic tectonic controls in the UK sector

Factors determining the deposition of potential Upper Jurassic hydrocarbon reservoirs and the development of trapping geometries, can be summarized as follows.

1. Active faulting, which resulted in the emergence of footwall crests, (particularly at the basin margin) and the supply of sediment into the basin.
2. Concentration of sediment transport into the basin at transfer zones, the areas between active fault strands.
3. Restriction of sand supply, dependent on the nature of the eroded sediments, resulting in deposition at proximal sites, close to the active faults at the margins of the basin.
4. Salt movement, controlled by active faulting, in turn creating salt-withdrawal synclines, which acted as sediment traps.
5. Movement of overburden above mobile salt, forming potential traps in rotated blocks, above diapirs/pillows, or at the axes of grounded rim synclines.
6. Faulting, accompanied by block rotation of pre-rift reservoirs, which produced simple tilted fault block traps.

It is clear that the formation of Upper Jurassic

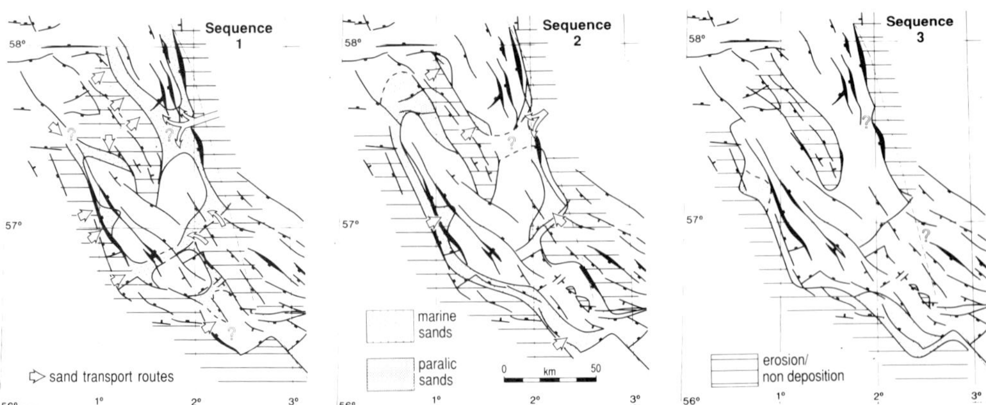

Fig. 10. Fulmar Formation sand distribution maps for three successive time intervals during the Oxfordian–Kimmeridgian.

reservoirs in intrabasinal settings is less likely than at the margins, unless a local source for sand supply can be demonstrated, such as, for example, the eroded crest of an intra-basinal fault block. Alternatively the introduction of sand into the basin via a transfer zone,, perhaps transported by turbidity currents, may locally have carried sediment beyond the marginal zone.

Chalk fields of the Feda graben, Norway

The preceding account has dealt largely with the wide-ranging effects of late Jurassic tectonism on contemporary Upper Jurassic reservoirs in the UK Central Graben. Together with Palaeocene turbiditic sands these have formed the primary exploration target in the UK sector. In the Norwegian sector, however, the most successful exploration target has been the late Cretaceous/Palaeocene Chalk, with more than ten commercial discoveries made in the axial part of the graben. The structural setting of the Chalk fields is generally attributed to salt tectonics, 'wrench'-driven inversion or a combination of both (e.g. Brewster & Dangerfield 1984; Gowers & Sæbøe 1985; D'Heur *et al.* 1985; Skjerven *et al.* 1983; Norbury 1987; Watts 1983). A satisfactory model linking together basement (i.e. Jurassic half-graben) structure, inversion features and the siting of salt pillows and diapirs has, however, yet to be established. We attempt here to establish such a model with reference to the six Chalk fields of the Feda Graben (Figs 1 & 11), namely Tommeliten Alpha, Tommeliten Gamma, Edda, Eldfisk, Valhall and Hod. Of paramount importance in understanding the Chalk fields is the need to understand the older, deeper late Jurassic structure.

The Feda Graben and Lindesnes Ridge, published models

The area under discussion is illustrated in Fig. 11. This is the deep, axial part of the Norwegian

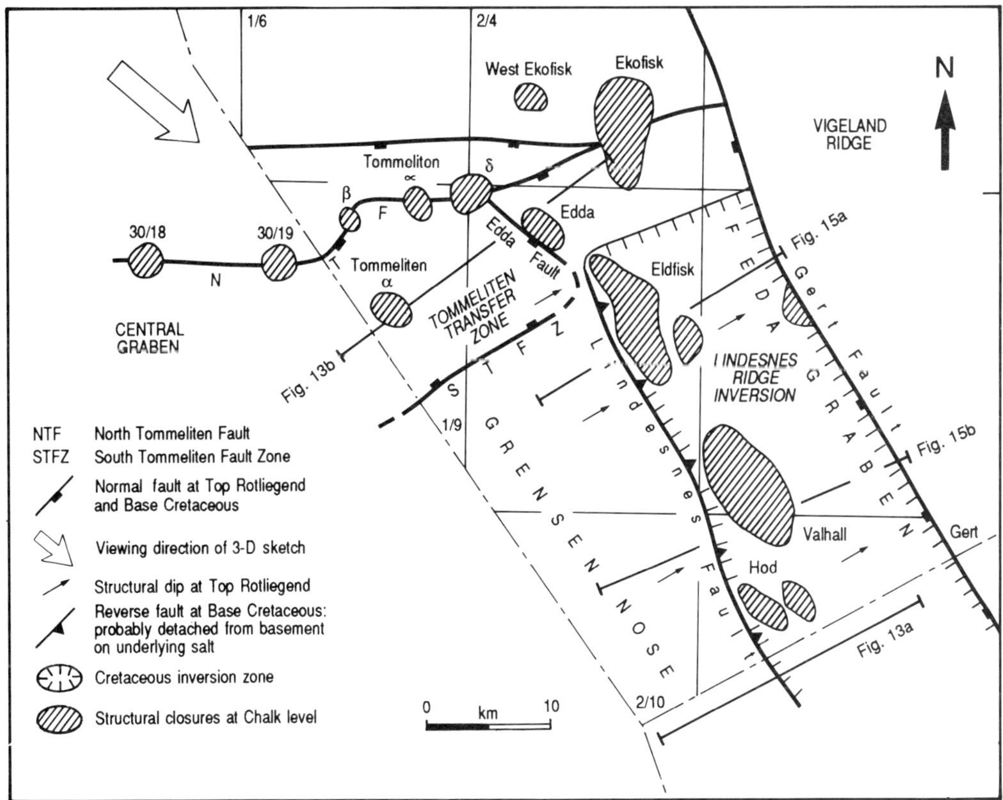

Fig. 11. Tectonic elements of the Feda Graben area (Norway), showing the location of seismic lines illustrated in Figs 13 & 15.

Central Graben, known as the Feda Graben. Although top basement (= top Rotliegend) in the Feda Graben is structurally very deep, approaching 8 km sub-sea in some areas, much of the graben at Chalk reservoir level is dominated by a structural high, known as the Lindesnes Ridge (Fig. 11). Many of the Chalk fields lie on the Lindesnes Ridge trend. All published models describing the structure of the Lindesnes Ridge and Feda Graben (e.g. Skjerven *et al.* 1983; Brewster & Dangerfield 1984; D'Heur *et al.* 1985; Gowers & Sæbøe 1985; Munns 1985; Møller 1986; Cartwright 1989) share the following essential features.

1. The graben at 'basement' level is an asymmetric (late Jurassic) structure, **dipping to the west.**
2. A basement-involved, down-to-the-east late Jurassic fault (the Lindesnes Fault) underlies the Lindesnes Ridge.
3. The Lindesnes Fault is interpreted as the master fault to the Feda Graben during late Jurassic extension and was reactivated as a compressional, oblique-slip structure during the late Cretaceous-early Tertiary. This compressional reactivation generated the Lindesnes Ridge.

The existence of a basement-involved fault below the Lindesnes Ridge cannot, however, be proven because of the poor quality of seismic data in this area. We would suggest that the poor seismic response is the result of ray-path distortion through a salt-wall underlying the ridge (Fig. 12).

We believe there are problems in invoking the late Jurassic master fault to the Feda Graben to lie below the Lindesnes Ridge. These can be summarized as follows.

1. Section balance. The Grensen Nose (Figs 11 & 12) marks the footwall to the putative Lindesnes fault. At top Rotliegend level the Grensen Nose dips to the east (Fig. 11; Skjerven *et al.* 1983; Møller 1986), yet if it lies in the footwall to a west-dipping half-graben it too, having been rotated by footwall uplift (Jackson & McKenzie 1983, Barr 1987, Barnett *et al.* 1987, Roberts & Yielding 1991), should dip to the west.
2. The Edda Fault. Within the Feda Graben the only seismically-imaged basement-fault

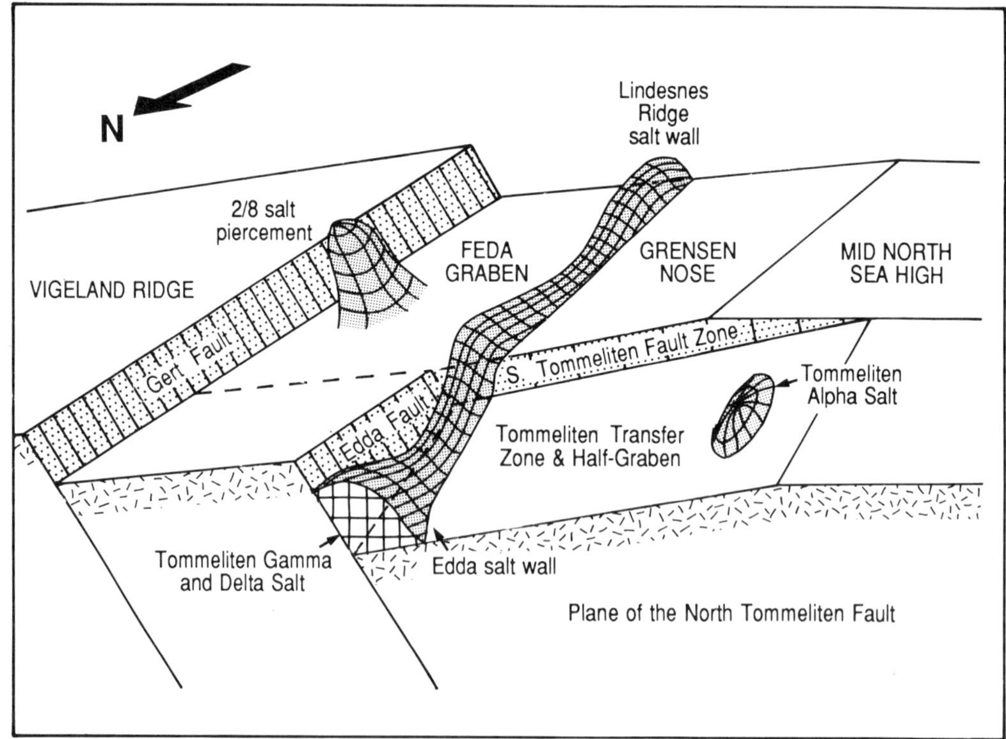

Fig. 12. Three-dimensional cartoon of the salt and sub-salt structure in the Feda Graben area. Note that the basement structure below the Lindesnes Ridge (Feda Graben) is interpreted as an east-dipping half-graben.

on trend with the Lindesnes Ridge lies to the west of the Edda Field, (Figs 11 & 12) and throws down-to-the-west (Nielsen *et al.* 1990, and later discussion of Edda). It therefore has a displacement sense opposite to the putative Lindesnes basement-fault.

3. Upper Jurassic growth synclines. All seismic lines across the Feda Graben show a clearly-defined Jurassic growth-syncline within the graben and below the Lindesnes Ridge (e.g. Møller 1986, fig 9; Norbury 1987, plate 5; Brewster & Dangerfield 1984, fig 2b). The published models of the Feda Graben/ Lindesnes Ridge do not explain the significance and persistence of this structure.

A new model for the Lindesnes Ridge and Feda Graben (Late Jurassic half-graben control on Late Cretaceous inversion)

In the light of the problems raised above we suggest a new model for the structure of the Lindesnes Ridge, which links the asymmetry of the late Jurassic Feda Graben, the siting of salt structures and the generation of inversion features. This model is discussed with reference to Fig. 12 and a seismic line across the Feda Graben/Lindesnes Ridge from near the Norwegian/Danish border (Fig. 13a). We also show below how this model has controlled the siting of hydrocarbon accumulations.

Although the model is described with particular reference to the Lindesnes Ridge, we believe it to be equally applicable to the inversion of the Tail End Graben in the Danish sector, a structure directly analogous to the Feda Graben.

Our mapping of the Feda Graben suggests it, as shown in previously published examples, to be bounded to the east by a down-to-the-east normal fault. This fault we call the Gert Fault, since it forms the western margin to the Norwegian/Danish Gert structure (e.g. Møller 1986; Cartwright 1989). We believe, however, that this fault is the master fault to the Feda Graben, and that the Feda Graben **dips to the east** at top basement level, into this fault.

We do not recognize a basement fault directly beneath the Lindesnes Ridge and suggest that the Grensen Nose is in direct structural continuity with the Feda Graben, dipping continually eastwards towards the Gert Fault.

Mobile salt is known to lie below the Lindesnes Ridge where it has been penetrated at a high level, below Upper Jurassic, in well 2/7–3(N). Clear evidence of salt diapirism against the Gert Fault can also be seen in block 2/8 (Figs 11 & 12). We suggest that the central late Jurassic/early Cretaceous growth-syncline within the Feda Graben is a halokinetic rim-syncline, related to salt movement westwards up-dip to the west towards the Grensen Nose (e.g. 2/7–3) and down-dip towards the Gert Fault (block 2/8). Thus the syncline is not directly related to any underlying basement topography, although the trigger for salt moving up-dip to the west was probably the formation of the east-dipping half-graben during the late Jurassic.

Figure 13a illustrates anticlinal culminations below the western margin of the Lindesnes Ridge which we interpret to be salt pillows. Similar structures, already attributed to local salt buoyancy, lie below the Hod (Norbury 1987), Valhall (Munns 1985) and Eldfisk (Brewster & Dangerfield 1985) fields. No features indicative of the presence of salt lie on the Grensen Nose (Fig. 11). Well penetrations on the Grensen Nose confirm an absence of salt. It is therefore suggested that the inversion fault which marks the western margin of the Lindesnes Ridge is coincident with a major salt wall.

We link these observations in a model similar to that tentatively proposed by Cartwright (1989) for the Danish Tail End Graben, and also to that proposed by Hayward & Graham (1989) for the Broad Fourteens Basin (Netherlands). Subsequent to formation of the **east-dipping** Feda (half-) Graben in the late Jurassic salt migrated up-dip into a major salt wall at the future western margin of the Lindesnes Ridge, perhaps terminating at a halite-dolomite/anhydrite facies transition within the Zechstein. During the late Cretaceous and Palaeocene the Gert Fault was reactivated as a reverse structure. Some of the displacement at basement-level was, however, transferred into an antithetic detachment fault (*cf.* Cooper *et al.* 1989) riding within the salt of the Feda Graben. The shortening associated with this antithetic fault was ultimately accommodated in a tip-fold developed at the site of the pre-existing salt wall, thus the detached inversion structure within the half-graben did not propagate onto the Grensen Nose. In contrast, therefore, to all previously published models for the Lindesnes Ridge, we believe the Lindesnes Fault to be a salt-detached reverse fault, with no expression in the underlying basement. The basic geometries of the previous models and our new model are compared in Fig. 14.

The kinematics of our inversion model are very similar to those observed in sandbox inversion models (Koopman *et al.* 1987). In these models a very clear antithetic backthrust is

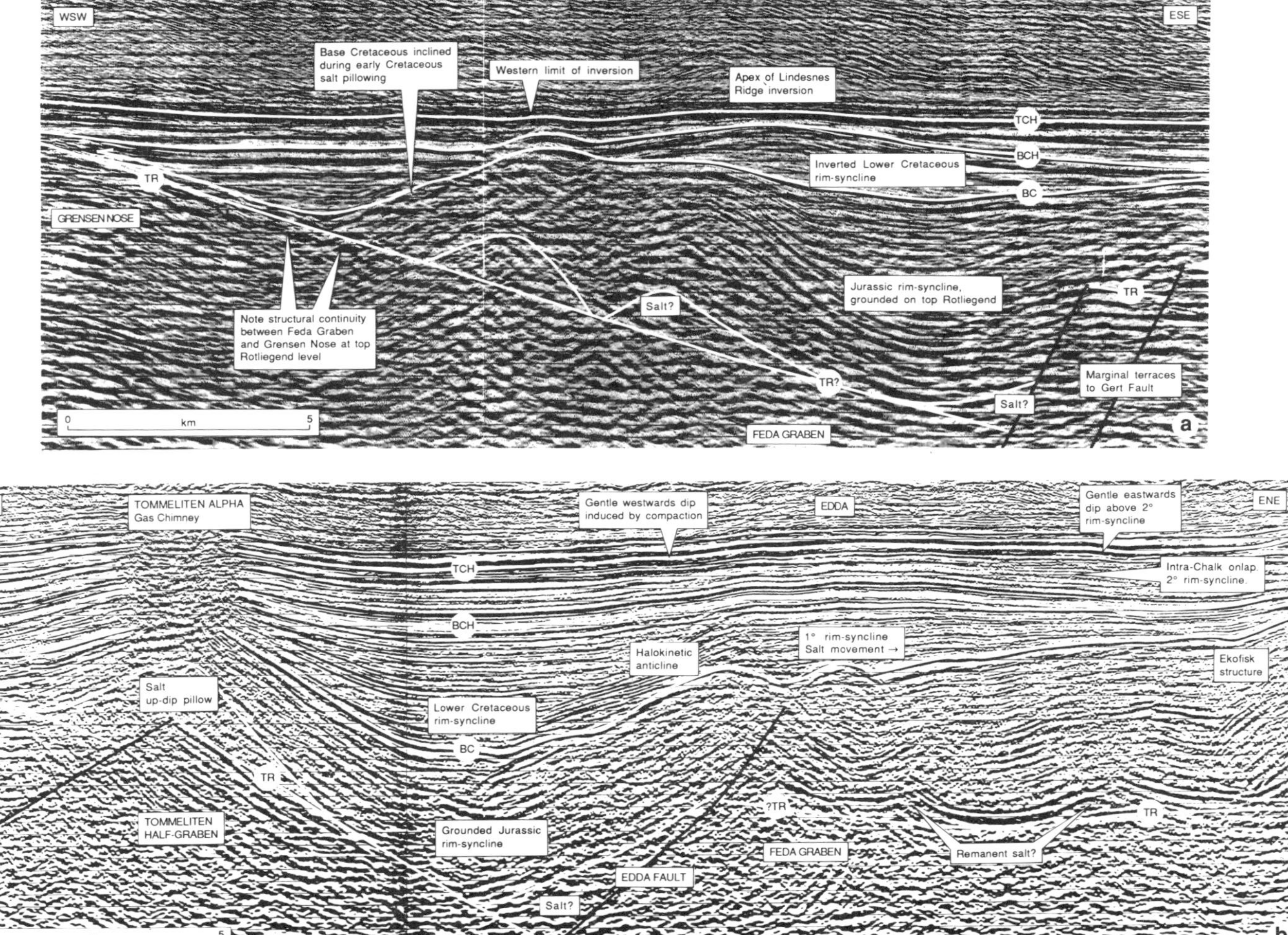

Fig. 13. (a) Interpreted seismic line across the Feda Graben/Lindesnes Ridge from close to the Norwegian/Danish border. (b) Interpreted seismic line across the Tommeliten Alpha and Edda structures. TCH = top Chalk, BCH = base Chalk, BC = Base Cretaceous, TR = top Rotliegend. See Fig. 11 for locations.

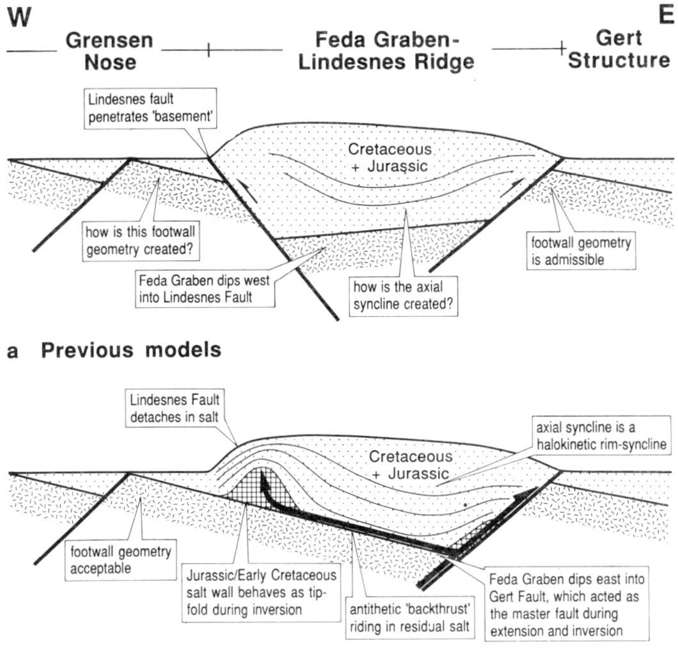

Fig. 14. (a) A summary of the geometric properties of previous models describing the Lindesnes Ridge inversion. (b) Our new model for the structure of the Lindesnes ridge, showing the inversion anticline to be detached on a reverse fault within the salt, antithetic to the master Gert Fault. The inversion strain terminates westwards in a salt-cored tip-fold.

produced, defining a 'pop-up' within the half-graben, similar to our model (Fig. 14). The models of Koopman *et al.* were constructed using an homogenous 'basin fill'. Koopman *et al.* concluded in addition, however, that the presence within the basin fill of a weak horizon, such as salt, would probably control the siting of the antithetic backthrusts developed during inversion. Such salt-detached faults, it was suggested, could transfer inversion strain some distance away from the master fault. This conjecture is borne out by our observations of structural geometry in the Feda Graben.

The Tommeliten transfer zone and half-graben

Having established that the axial part of the Norwegian Central Graben (*ie.* the Feda Graben) is an east-dipping late Jurassic half-graben (Figs 11 & 12), subsequently inverted in the late Cretaceous and early Tertiary, we can now suggest how this structure may be kinematically linked to the axial part of the UK Central Graben, which lies *c.*50 km to the NW. Displacement transfer between the offset axial zones was accomplished via the Tommeliten half-graben and its three bounding faults (Nielsen *et al.* 1990). The Tommeliten half-graben is bounded to the east by the extensional Edda Fault, to the north by the oblique-slip North Tommeliten Fault and to the south by the oblique-slip South Tommeliten Fault Zone. To the west the half-graben opens into the UK Central Graben (Fig. 11). Together this combination of faults serves to act as a transfer zone (Fig. 2) between the en échelon basin axes.

The internal structure of the half-graben, at top Rotliegend level, is similar to that of the adjacent Feda Graben. It is an east-dipping structure, bounded to the east by a major extensional fault. Within the half-graben Zechstein salt has been mobile, deforming the overlying basin-fill. There has, however, been no post-extensional inversion of the Tommeliten half-graben.

Structural controls on the Chalk fields

Previously all the Chalk fields of the Feda Graben area have been simply described as part of the Lindesnes Ridge inversion trend (D'Heur 1984, Brewster & Dangerfield 1984). Having established a new regional structural framework, characterized by east-dipping, late Jurassic half-graben, we would, however, suggest that the six fields of the 'Lindesnes Ridge' trend are the product of five different, but related, structural processes. A regional map of the top-reservoir (top Chalk) structure in this area has previously been published by Brewster & Dangerfield (1984) and the reader is referred to this in conjunction with the illustrations provided here.

Tommeliten Alpha (Fig. 13b). This is structurally the simplest of the fields under discussion. The closure is defined by a dome, although no wells have yet penetrated deep enough to confirm the presence of salt. The salt pillow lies towards the crest of the east-dipping Tommeliten half-graben, and its source-area rim-syncline lies immediately down-dip to the east. The detailed movement history of salt into the Alpha structure will be discussed by T. S. Olsen *et al.* in a future paper. In summary, however, detailed mapping shows that up-dip salt migration began in the late Jurassic (perhaps intra-Volgian) and continued through the early Cretaceous. By the end of the early Cretaceous the Alpha structure was largely established, and was passively buried and compacted subsequent to this.

The prime control on the siting of Tommeliten Alpha is its position within, and towards the crest of, a Jurassic half-graben. Movement of salt into the Alpha structure can be attributed to simple gravitational buoyancy alone.

Tommeliten Alpha is entirely remote from the Lindesnes Ridge inversion, although there is a strong similarity, however, between both the basement and halokinetic structure of the two areas (Fig. 13). The timing and siting of salt movement in particular are similar in both half-grabens.

Tommeliten Gamma. The geology of Tommeliten Gamma has recently been described in detail by Nielsen *et al.* (1990). It has previously been described as part of the Lindesnes Ridge inversion trend by D'Heur (1984) and D'Heur & Pekot (1987). We believe this not to be the case. Seismic and isopach data across the Gamma structure do not support the view that inversion extended north to this area (Nielsen *et al.* 1990).

The Gamma structure sits directly above the North Tommeliten (transfer) Fault, and it is the buttressing of a salt diapir against the footwall to this fault which has generated the structural closure. Several wells on the structure have penetrated high-level salt, which in some areas has risen diapirically to intra-Chalk level. Salt movement into the Gamma structure began in the early Cretaceous and extended through the late Cretaceous, affecting reservoir facies and distribution. The salt has been sourced largely from the adjacent Tommeliten half-graben.

Thus Tommeliten Gamma, like Tommeliten Alpha, is a halokinetic structure whose siting is intimately controlled by the underlying basement structure, imposed during the late Jurassic.

Edda (Fig. 13b). The Edda structure has been described as part of the Lindesnes Ridge inversion (D'Heur *et al.* 1985). It is, however, a far more subtle, low-relief structure than the Eldfisk, Valhall and Hod structures, which lie on the ridge further south. It appears to a require a different explanation.

Buttressed against the Edda Fault (Figs 11 & 12) is a salt wall, which perhaps connects the Tommeliten Gamma salt to the Lindesnes Ridge salt. At Base Cretaceous level there is a clear anticlinal culmination above this salt wall (Fig. 13b). This culmination, however, is offset from the overlying Edda closure at Chalk level. Edda is not therefore a 'simple' salt-induced culmination like the Tommeliten structures.

We suggest that two separate mechanisms have combined to generate the Edda structure. The western limb of the closure overlies both the Edda Fault and the rim-syncline created by salt withdrawal into Tommeliten Alpha and the Edda salt wall (Fig. 11). Simple differential compaction across this fault, the salt wall and the rim-syncline may be responsible for imposing the gentle westwards dip on the Chalk sequence (Fig. 13b).

The eastern limb of the Edda structure is more complex (Fig. 13b). At Chalk level there is a gentle eastwards dip. The underlying Base Cretaceous, however, dips to the west, while the top of the Lower Cretaceous is approximately flat. We suggest that salt withdrawal from below the Edda structure, towards the Ekofisk structure to the east, began in the early Cretaceous and that this imposed the initial dip to the west of the Base Cretaceous. The Lower Cretaceous below Edda was deposited within the primary rim-syncline associated with this salt withdrawal. The secondary rim-syncline to the Ekofisk structure is a more subtle feature, but is recognizable from onlap relationships and sequence-thickening within the Chalk, between Edda and Ekofisk. This secondary, more proximal salt withdrawal towards Ekofisk has imparted an eastwards dip to the Chalk, above the west-dipping Base Cretaceous, and thus created the eastern limb of the Edda structure.

We therefore suggest that the Edda structure formed by a combination of differential compaction over basement and halokinetic topography, and salt withdrawal on the footwall block to the Edda Fault. No inversion or underlying halokinesis is responsible for the Edda culmination, hence the subtlety of its expression by comparison with the surrounding hydrocarbon-bearing structures.

Eldfisk (Fig. 15a) and Hod. Inspection of seismic data along the strike of the Lindesnes Ridge, and of structural maps in this area (e.g. Brewster & Dangerfield 1984, fig. 2), shows the Lindesnes Ridge inversion to terminate northwards between Edda and Eldfisk, perhaps controlled by an eastwards extension of the South Tommeliten Fault Zone (Fig. 11).

Eldfisk and Hod both lie within the broad anticline, extending across the Feda Graben, which defines the inversion structure. The structural culminations defining the two fields are, however, localised sharper anticlines superimposed on the inversion structure. Following our model for generation of the Lindesnes Ridge (Fig. 14) we suggest that these culminations (Fig. 15a) overlie the salt-wall tip-folds at the western margin of the inversion. Support for a model involving a salt-core to these localised culminations is provided by well 2/7–3(N), which penetrated salt below the Upper Jurassic. Reflection geometries adjacent to the 2/7–3 location (Fig. 15a) and to the east of Hod (Fig. 13a; Norbury 1987, plate 5), suggest that salt movement commenced during the late Jurassic and continued into the early Cretaceous. The passively-induced salt-cored anticlines were then further enhanced during late Cretaceous/early Tertiary inversion, when they behaved as actively deforming tip-folds. Relief on the structures has been accentuated by differential compaction across the salt wall.

The siting of Eldfisk and Hod is thus thought to have been controlled initially by salt movement up the dip of basement topography, accentuated subsequently by inversion detached on the salt, which exploited the pre-existing structure.

Valhall (Fig. 15b). Valhall lies between the Eldfisk and Hod structures, at the apex of the Lindesnes Ridge. It shows, however, a style of closure distinct from its two neighbours. The closure lies at the centre of the Lindesnes Ridge, offset to the east from Eldfisk and Hod (Brewster & Dangerfield 1984, fig. 2). It is a much broader culmination than these neighbouring closures.

Below and to the west of the apex of the Chalk closure lie two smaller anticlines at Base Cretaceous level. These are possibly the equivalent structures to the salt-cored anticlines at Eldfisk and Hod. As at Edda (Fig. 13b), however, the Base Cretaceous structure at Valhall exerts no direct control on Chalk-level closures. Valhall is not therefore interpreted as a broad, salt-cored anticline, but as the apex of the inversion anticline, unsupported by salt. Reflection geometries below Valhall are extremely complex (Fig. 15b) and open to many interpretations. We believe, however, that they are readily interpreted in a manner consistent with our east-dipping half-graben hypothesis (Figs 13a & 14), and that the steep internal dips of the half-graben fill reflect detached halokinetic deformation of Jurassic and early Cretaceous age rather than deformation in direct response to basement-involved faulting.

Although the Valhall structure lies at the crest of an inversion anticline its siting is directly controlled by the pre-existing, late Jurassic Feda Graben.

Summary of Chalk-reservoir play-types in the Feda Graben. This account has dealt with the structural controls on six superficially similar Chalk fields in the Feda Graben. All owe their origin to the initial development of late Jurassic half-graben structures in the area, and yet five distinct structural settings are recognized.

1. Tommeliten Alpha. Salt pillow lying towards the crest of a small half-graben within a late Jurassic transfer zone.
2. Tommeliten Gamma. Salt diapir buttressed against the footwall of a major late Jurassic fault.
3. Edda. Gentle structure produced by the combination of compaction across a major fault and adjacent salt wall, together with salt redistribution on the footwall block to this fault.
4. Eldfisk and Hod. Salt-cored anticlines, similar to Tommeliten Alpha, but accentuated when exploited as active tip-folds during late Cretaceous/Tertiary inversion of a late Jurassic half graben.
5. Valhall. The apex of a broad inversion anticline detached from its underlying basement on an antithetic reverse fault riding in salt.

This list is not exhaustive, even in the local context of the Norwegian Central Graben. Other Chalk fields such as Albuskjell, Tor and Ekofisk may be different from those described. The importance of the late Jurassic extensional

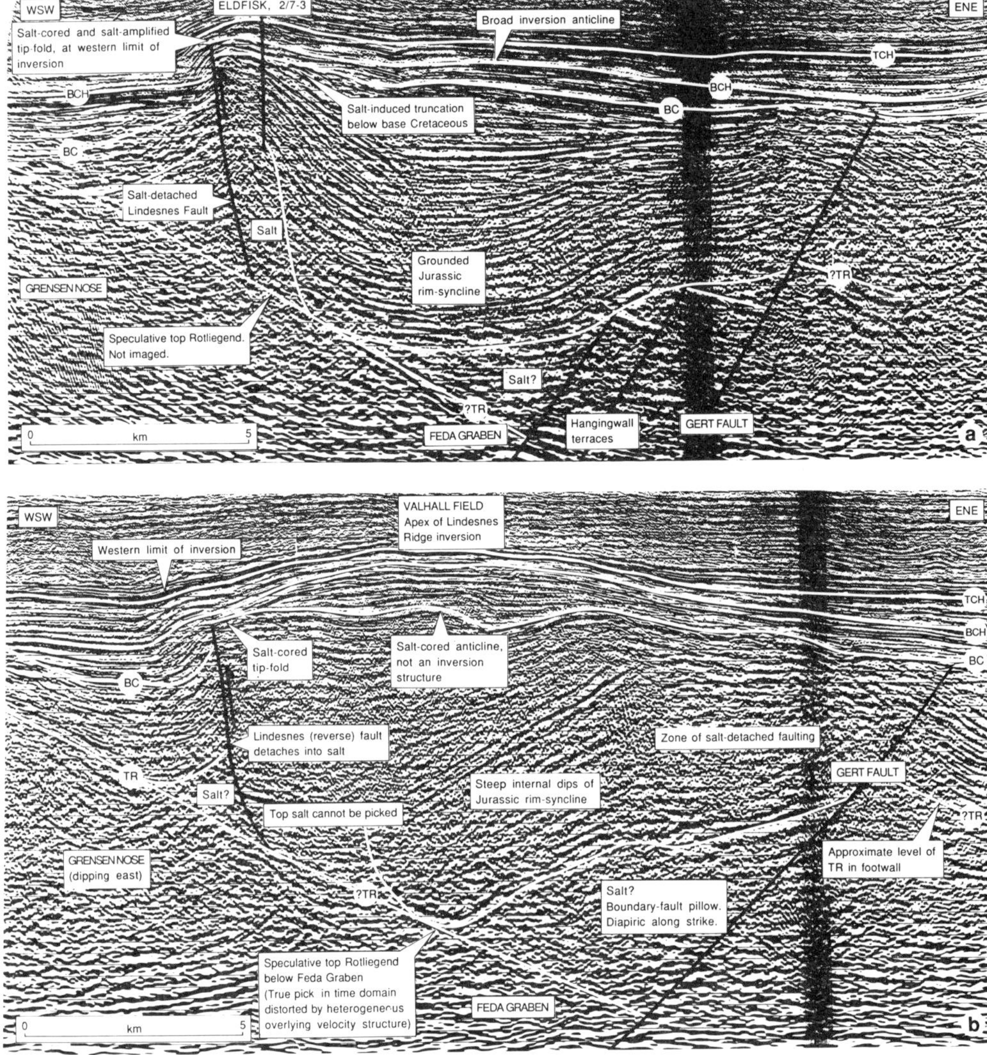

Fig. 15. (a) Interpreted seismic line across the Eldfisk structure and Lindesnes Ridge inversion. (b) Interpreted seismic line across the Valhall structure and Lindesnes Ridge inversion. TCH, top Chalk; BCH, base Chalk; B, Base Cretaceous; TR, top Rotliegend. See Fig. 11 for locations.

episode on hydrocarbon accumulations in much younger reservoirs is demonstrated by this review.

Conclusions

With emphasis on the late Jurassic, the main conclusions of our study of UK and Norwegian Central Graben are as follows.

1. Extension of the Central Graben occurred during the Triassic and late Jurassic. Late Jurassic deformation is confined to the axial *c.* 100 km of the graben and probably involved *c.* 30 km of extension orthogonal to the main NNW–SSE-trending fault set.
2. Transfer zones and transfer faults, both within and at the margins of the graben, are responsible for a southeastwards en échelon stepping of the basin axis.
3. Late Jurassic sands were preferentially introduced into the basin at or via these transfer zones, subsequent to their initial derivation from eroded footwall highs.
4. Late Jurassic sands are preferentially con-

centrated close to the basin margins, perhaps as a result of the relative paucity of sand in the largely Triassic source areas.

5. Salt structures define many trapping geometries within the basin. The location of salt structures is intimately related to basement, *ie*. fault-induced, topography of late Jurassic and Triassic age. Salt pillows and diapirs occur preferentially either towards the crest of half-graben, or buttressed against half-graben bounding faults.
6. In the Norwegian sector late Jurassic faulting has controlled the subsequent siting of all the Chalk fields in the Feda Graben. A variety of structurally-controlled play-types is recognized.
7. A new model for the generation of the Lindesnes Ridge inversion has been suggested which links together late Jurassic half-graben morphology, structurally-controlled halokinesis and the position of Chalk-reservoir hydrocarbon fields upon the inversion ridge.

Foremost, for allowing us to publish our ideas, we thank Amerada Hess (UK) and their UK 11th Round Central Graben partners (Unocal, Goal, Coalite and Aran), together with Statoil and their Tommeliten partners (Fina and Agip). We also thank Richard Hardman, Erik Nielsen and Jan Høyen with whom we undertook much of the work, and Mike Badley, Mike Ashton and Graham Yielding for help in the geological analysis for this paper. Paul Watson of GeoStrat kindly provided the biostratigraphical data, with the courtesy of Amerada Hess, which enabled us to draw the Fulmar Formation sand distribution maps, and Steve Roberts and Simon Price explained to us their ideas on sediment deposition in currently-extending basins in the Aegean and Turkey. Finally, our thanks go to David Kemp and Alethea Saville for their help in preparing the figures.

References

BARNETT, J. A. M., MORTIMER, J., RIPPON, J. H., WALSH, J. J. & WATTERSON, J. 1987. Displacement Geometry In the Volume Containing A Single Normal Fault. *American Association of Petroleum Geologists Bulletin*, **71**, 925–937.

BARR, D. 1987. Structural/stratigraphic models for extensional basins of half-graben type. *Journal of Structural Geology*, **9**, 491–500.

BARTON, P. & WOOD, R. 1984. Tectonic evolution of the North Sea basin: crustal stretching and subsidence. *Geophysical Journal of the Royal Astronomical Society*, **79**, 987–1022.

BERTRAM, G. T. & MILTON, N. 1989. Reconstructing Basin Evolution from Sedimentary Thickness; the Importance of Palaeobathymetric Control, with Reference to the North Sea. *Basin Research*, **1**, 247–257.

BREWSTER, J. & DANGERFIELD, J. A. 1984. Chalk fields along the Lindesnes Ridge, Eldfisk. *Marine and Petroleum Geology*, **1**, 239–278.

CARWRIGHT, J. A. 1989. The kinematics of inversion in the Danish Central Graben. *In:* COOPER, M. A. & WILLIAMS, G. D. (eds). *Inversion Tectonics*. Geological Society, London, Special Publication, **44**, 153–176.

CAYLEY, G. T. 1987. Hydrocarbon migration in the Central North Sea. *In*: BROOKS, J. & GLENNIE, K. W. (eds). *Petroleum Geology of North West Europe*, Graham & Trotman, London, 549–556.

CHILDS, C., WALSH, J. J. & WATTERSON, J. 1990. A method for estimating the density of fault displacements below the limit of seismic resolution in reservoir formations. *In: North Sea Oil and Gas Reservoirs* (May 1989), Norwegian Petroleum Society.

COOPER, M. A. *et al.* 1989. Inversion tectonics — a discussion. *In:* COOPER, M. A. & WILLIAMS, G. D. (eds). *Inversion Tectonics*. Geological Society, London, Special Publication, **44**, 335–347.

D'HEUR, M. 1984. Porosity and hydrocarbon distribution in the North Sea chalk reservoirs. *Marine and Petroleum Geology*, **1**, 211–238.

——, de WALQUE, L. & MICHAUD, F. 1985. Geology of the Edda Field Reservoir. *Marine and Petroleum Geology*, **2**, 327–340.

—— & PEKOT, L. J. 1987. Tommeliten. *In:* SPENCER, A. M. *et al.* (eds). *Geology of the Norwegian Oil and Gas Fields*, 117–128.

EYNON, G. 1981. Basin development and sedimentation in the Middle Jurassic of the northern North Sea. *In*: ILLING, L. V. & HOBSON, G. D. (eds). *Petroleum Geology of the Continental Shelf of North West Europe*, Heyden, London, 196–204.

GAGE, M. S. & DORÉ, A. G. 1986. A regional geological perspective of the Norwegian offshore exploration provinces. *In*: SPENCER, A. M. *et al.* (eds) *Habitat of Hydrocarbons on the Norwegian Continental Shelf*, Norwegian Petroleum Society, 21–38.

GIBBS, A. D. 1985. Discussion on the structural evolution of extensional basin margins. *Journal of the Geological Society, London*, **142**, 941–2.

GLENNIE, K. W. (ed.) 1986. *Introduction to the Petroleum Geology of the North Sea*, Blackwell, Oxford.

GOWERS, M. B. & SÆBØE, A. 1985. On the structural evolution of the Central Trough in the Norwegian and Danish sectors of the North Sea. *Marine and Petroleum Geology*, **2**, 298–318.

GRØNLIE, A. & ROBERTS, D. 1989. Resurgent strike-slip duplex development along the Hitra-Snåsa and Verran Faults, Møre Trøndelag Fault Zone, Central Norway. *Journal of Structural Geology*, **11**, 295–306.

HAYWARD, A. B. & GRAHAM, R. H. 1989. Some geometrical characteristics of inversion. *In:* COOPER, M. A. & WILLIAMS, G. D. (eds). *Inversion Tectonics*. Geological Society, London, Special Publication, **44**, 17–40.

HELLINGER, S. J., SCLATER, J. G. & GILTNER, J. 1989. Mid-Jurassic through mid-Cretaceous extension in the Central Graben of the North Sea — part 1: estimates from subsidence. *Basin Research*, **1**, 191–200.

HOSPERS, J., RATHORE, J. S., JIANHUA, F. & FINNSTRØM, E. G. 1986. Thickness of pre-Zechstein salt Palaeozoic sediments in the southern part of the Norwegian sector of the North Sea. *Norsk Geologisk Tiddsskrift*, **66**, 295–303.

JACKSON, J. & MCKENZIE, D. 1983. The geometrical evolution of normal fault systems. *Journal of Structural Geology*, **5**, 471–482.

JACKSON, M. P. A. & TALBOT, C. J. 1986. External shapes, strain rates, and dynamics of salt structures. *Geological Society of America Bulletin*, **97**, 305–323.

—— & —— 1989. Anatomy of mushroom-shaped diapirs. *Journal of Structural Geology*, **11**, 211–230.

JARVIS, G. T. & MCKENZIE, D. P. 1980. Sedimentary Basin Formation With Finite Extension Rates. *Earth and Planetary Science Letters*, **48**, 42–52.

JOHNSON, R. J. & DINGWALL, R. G. 1981. The Caledonides: their influence on the stratigraphy of the North-West European Continent Shelf. *In*: ILLING, L. V. & HOBSON, G. D. (eds). *Petroleum Geology of the Continental Shelf of North West Europe*, Heyden, London, 85–97.

JOHNSON, H. D., MACKAY, T. A. & STEWART, D. J. 1986. The Fulmar Oil-field (Central North Sea): geological aspects of its discovery, appraisal and development. *Marine and Petroleum Geology*, **3**, 99–125.

KLEMPERER, S. L. & HURICH, C. A. 1990. Lithospheric Structure of the North Sea from Deep Seismic Profiling. *In*: BLUNDELL, D. J. & GIBBS, A. D. (eds). *Tectonic Evolution of the North Sea Rifts.* Oxford University Press.

KOOPMAN, A., SPEKSNIJDER, A. & HORSFIELD, W. T. 1987. Sandbox model studies of inversion tectonics. *Tectonophysics,* **137**, 379–388.

KUSZNIR, N. J., MARSDEN, G. & EGAN, S. 1988. Fault block rotation during continental lithosphere extension: a flexural cantilever model. (Abstract). *Geophysical Journal*, **92**, 546.

LATIN, D. M., DIXON, J. E. & FITTON, J. G. 1990. Rift-related magmatism in the North Sea Basin. *In*: BLUNDELL, D. J. & GIBBS, A. D. (eds). *Tectonic evolution of the North Sea Rifts,* Oxford University Press, Oxford.

LEEDER, M. R. & GAWTHORPE, R. L. 1987. Sedimentary models for extensional tilt-block/half-graben basins. *In*: COWARD, M. P., DEWEY, J. F. & HANCOCK, P. L. (eds). *Continental Extensional Tectonics*, Geological Society, London, Special Publication, **28**, 139–152.

MCCOSS, A. M. 1986. Simple constructions for deformation in transpression/transtension zones. *Journal of Structural Geology*, **6**, 715–719.

MCKENZIE, D. 1978. Some remarks on the development of sedimentary basins. *Earth and Planetary Science Letters*, **40**, 25–32.

MCKERROW, W. S. 1988. The development of the Iapetus Ocean from the Arenig to the Wenlock. *In*: HARRIS, A. L. & FETTES, D. J. (eds). *The Caledonian-Appalachian Orogen*. Geological Society, London, Special Publication, **38**, 405–412.

MUNNS, J. W. 1985. The Valhall Field: a geological overview. *Marine and Petroleum Geology*, **2**, 23–43.

MØLLER, J. J. 1986. Seismic structural mapping of the Middle and Upper Jurassic in the Danish Central Trough. *Danmarks Geologiske Undersøgelse.* Serie A Nr. 13.

NIELSEN, E. B., FRASELLE, G. & VIANELLO, L. 1990. Tommeliten Gamma Field, Norway-North Sea. *In*: *Atlas of oil and gas fields*, American Association of Petroleum Geologists.

NORBURY, I. 1987. Hod. *In*: SPENCER, A. M. *et al.* (eds). *Geology of the Norwegian Oil and Gas Fields.* 107–116.

OLSEN, J. C. 1987. Tectonic evolution of the North Sea region. *In*: BROOKS, J. & GLENNIE, K. W. (eds). *Petroleum Geology of North West Europe*, Graham & Trotman, London, 398–402.

RITCHIE, J. D., SWALLOW, J. L., MITCHELL, J. G. & MORTON, A. C. 1988. Jurassic ages from intrusive and extrusives within the Forties Igneous Province. *Scottish Journal of Geology*, **24**, 81–88.

ROBERTS, A. M., YIELDING, G. & BADLEY, M. E. 1990. A kinematic model for the orthogonal opening of the Late Jurassic North Sea Rift System, Denmark-Mid Norway. In: BLUNDELL, D. J. & GIBBS, A. D. (eds). *Tectonic Evolution of the North Sea Rifts*, Oxford University Press.

—— & —— 1991. Deformation around basin-margin faults in the North Sea/Norwegian rift. *In*: ROBERTS, A. M., YIELDING, G. & FREEMAN, B. (eds). *The Geometry of Normal Faults.* Geological Society, London, Special Publication, **56**, 61–78.

ROBERTS, S. C. & JACKSON, J. A. 1991. Active normal faulting in Central Greece: an overview. *In*: ROBERTS, A. M., YIELDING, G. & FREEMAN, B. (eds). *The Geometry of Normal Faults.* Geological Society, London, Special Publication, **56**, 125–142.

SCLATER, J. G. & CHRISTIE, P. A. F. 1980. Continental Stretching: an explanation of the post mid-Cretaceous subsidence of the Central North Sea Basin. *Journal of Geophysical Research*, **85**, 3711–3739.

——, HELLINGER, S. J. & SHOREY, M. 1986. An analysis of the importance of extension in accounting for the post-Carboniferous subsidence of the North Sea basin. *University of Texas Institute for Geophysics Internal Report.*

—— & SHOREY, M. D. 1989. Mid-Jurassic through mid-Cretaceous extension in the Central Graben of the North Sea — part 2: estimates from faulting observed on a seismic reflection line. *Basin Research*, **1**, 201–216.

SENI, S. J. & JACKSON, M. P. A. 1983. Evolution of salt structures, East Texas Diapir Province, Part 1: Sedimentary Record of Halokinesis. *American Association of Petroleum Geologists Bulletin*, **68**, 1219–1244.

SKJERVEN, J., RIJS, R. & KALHEIM, J. E. 1983. Late Palaeozoic to early Cenozoic structural development of the south-southeastern Norwegian North Sea. *Geologie en Mijnbouw*, **62**, 35–46.
SMITH, R. L. 1987. The structural development of the Clyde Field. *In*: BROOKS, J. & GLENNIE, K. W. (eds). *Petroleum Geology of North West Europe*, Graham & Trotman, 523–532.
SOPER, N. J. 1988. Timing and geometry of collision, terrane accretion and sinistral strike-slip events in the British Caledonides. *In*: HARRIS, A. L. & FETTES, D. J. (eds). *The Caledonian-Appalachian Orogen*. Geological Society, London, Special Publication, **38**, 481–492.
THRELFALL, W. F. 1981. Structural framework of the central and northern North Sea. *In*: ILLING, L. V. & HOBSON, G. B. (eds). *Petroleum Geology of the Continental Shelf of North West Europe*, Heyden, London, 98–103.
TRUSHEIM, F. 1960. Mechanism of salt migration in Northern Germany. *American Association of Petroleum Geologists Bulletin*, **44**, 1519–1540.
VEJBAEK, O. V. & ANDERSON, C. 1987. Cretaceous-Early Tertiary inversion tectonism in the Danish Central Trough. *Tectonophysics*, **137**, 221–218.
VENDEVILLE, B. & COBBOLD, P. R. 1987. Glissements gravitaires synsédimentaires et failles normales listriques: modèles expérimentaux. *Comptes Rendus de l'Académie des Sciences Paris*, **305**, 1313–1319.
VENING MEINESZ, F. A. 1950. Les graben africains, resultat de compression ou de tension dans la croute terrestre? *Inst. r. colonial belge Bull.*, **21**, 539–552.
WATTS, N. L. 1983. Microfaults in Chalks of Albuskjell Field, Norwegian Sector, North Sea: Possible Origin and Distribution. *American Association of Petroleum Geologists Bulletin*, **67**, 201–234.
ZIEGLER, P. A. 1981. Evolution of sedimentary basins in North-West Europe. *In*: ILLING, L. V. & HOBSON, G. D. (eds). *Petroleum Geology of the Continental Shelf of North West Europe*. Heyden, London, 3–39.
—— reply by WOOD, R. & BARTON, P. 1983. Discussion on: Crustal thinning and subsidence in the North Sea. *Nature*, **304**, 561.

Question by J. Clemson (Elf UK Limited)

You state that the sand distribution of the Upper Jurassic is controlled by the faulting along the boundaries of the Central Graben and associated transfer faults, and that sand is sourced from the footwalls of these major faults.

Could not the distribution of the sand be directly related to zones of salt dissolution and withdrawal? Where there was a relatively thin overburden during Fulmar Formation times, penetrative salt occurred along the Forth Approaches Platform and Jæren High, where contemporaneous dissolution has occurred. During these times there was a subdued shallow marine topography with areas of subsidence over the zones of salt dissolution and along the flanks of Triassic turtlebacks in deeper parts of the Central Graben.

In this sea, sands mainly derived from the Mid-North Sea High, Rynkøbing–Fyn High and Jæren High were transported across this open shelf as isolated sand bars and storm-generated sheets and deposited in the topographic lows created by salt withdrawal and dissolution. Continued interstitial water circulation in the sands of these dissolution zones allowed salt withdrawal to continue. This salt dissolution and contemporaneous withdrawal along the flanks of the turtlebacks resulted in the accumulation of the stacked sandbar sequences of the Fulmar Formation.

Sand deposition ceased with late Jurassic transgression and shale deposition. These shales 'sealed' the sands and protected the salt from dissolution and therefore subsidence ceased over these salt features.

This would explain sand distribution without the need for local and derivation and stresses the importance of contemporaneous salt mobility.

Comment by R. Gawthorpe (University of Durham) & R. Collier (University of Leeds)

We feel it necessary to clarify the character of basin margin-derived fans (whether alluvial, deltaic or submarine) following the contribution of Roberts, Price & Olsen.

(i) Transfer zones may act as routes for sediment movement from the rear of a footwall crest to the hangingwall. However, in many cases fault–echelon accommodation zones or relay transfer zones act as conduits from hinterland drainage areas and the related fans may be larger than contemporaneous footwall fans formed along the adjacent border fault segments by drainage of the footwall scarp alone. The size of a transfer fan will be related to the area of their associated drainage basin. The sediment derived directly from footwall uplift and related erosion generally forms a minor component to these fan systems, as exemplified in active extensional basins in the Aegean and Basin and Range.
(ii) The size/location relationships of footwall-derived fans is further complicated by the antecedent drainage system which may have a large drainage basin area. If an antecedent river at high angle to the basin margin fault trend can incise at an equal rate to footwall uplift, then the antecedent drainage plan behind the fault footwall crest may be maintained. Anonomously large footwall fans may thus result.

Question by J. M. Peters (Shell Exploration UK)

What is your view on the role of earlier tectonic phases in the Central Graben on the subsurface distribution of Zechstein salts, and the localisation and style of late Jurassic faulting?

Reply to J. Clemson

As covered in the text of the paper, we readily acknowledge that salt-withdrawal and salt-dissolution have created sediment traps, within which sands have preferentially accumulated. This does not, however, account for the original nature of the sand source. To say, simply, that sands are derived from erosion of the Ringkøbing–Fyn and Mid-North-Sea Highs, and then transported 100–200 km northwards, is to misunderstand both the nature of these structures, and that of the basin interior, during the Late Jurassic.

The Ringkøbing–Fyn and Mid-North-Sea Highs are structures very similar to the Jæren High. They sit in the footwalls of major extensional faults, and during extension on these faults were subject to flexural footwall-uplift and consequent erosion (Roberts & Yielding 1991). Certainly these eroded footwall crests shed sand into the basin (e.g. Clyde, Fulmar, Gert, Ula), but there is no evidence that these sands were transported far from their source areas. The Jæren High also suffered flexural footwall uplift and erosion during extension, as did other parts of the basin margin. It is therefore simpler to appeal to a local sand derivation close to these structures, rather than to derivation from an 'inexhaustible' sand supply many tens of kilometres to the south.

This also leads us to ask how, in a basin whose topography was dominated by a complex network of tilted fault blocks, sediment could have been transported unimpeded across the basin floor. We know from the local distribution of the Brae, Magnus and Munin sands in the Viking Graben, that sands derived from uplifted footwall crests accumulate in the closest topographic lows, they are not distributed across the basin. When erecting depositional models for such Late Jurassic sands in the North Sea the fault block topography must be considered. A sedimentary model which takes no account of this is inherently flawed.

Reply to R. Gawthorpe and R. Collier

We thank these authors for their comments, knowing as we do that they are based on observational data from recent sediments. Clearly models of sub-surface geology can be greatly refined from such work.

We take no issue with what they say and simply draw attention to the text of the paper, in which we recognize the low volumes of sand within the basin, suggesting little hinterland drainage into the basin and possibly sand derivation largely from footwall areas alone.

One possible reason for this may be the direct consequence of the regional drainage slope **away** from the basin, established when major footwall-uplift occurs at basin margins (Roberts & Yielding 1991). The imposition a regional hinterland tilt may in fact result in sediment accumulation on the footwall block, within the flexural low peripheral to the much larger uplift, unless of course ready access to a transfer zone results in bypass of the hinterland dip. An example of this type of footwall sand accumulation may be the Mid-Norway Draugen field (Ellenor & Mozetic 1986). Perhaps such sands may also be present on, for example, the Jæren and Ringkøbing–Fyn Highs, adjacent to the Central Graben.

Reply to J. Peters

As discussed briefly in the introductory section of our paper we readily recognize that the Central Graben had an important pre-Jurassic structural history. In order to do justice to the complexity of this earlier history, however, a full separate publication is required, and with this in mind we were invited to discuss specifically only the influence of Late Jurassic extension in this contribution. We would, however, make the following points in summary.

1. Triassic extension is recognizable within the Central Graben and throughout the North Sea rift. In a number of other recent publications we have attempted to draw attention to this (e.g. Badley *et al.* 1988; Roberts *et al.* 1990a, b; Marsden *et al.* 1990; Yielding *et al.* 1991).

2. Structures on the Norwegian/Danish shelf, such as the Stord, Egersund, Søgne Basins and the Horn Graben are almost entirely of Triassic age with little subsequent overprint.

3. Other structures, however, such as the East Troll and Ninian/Hutton Faults in the Viking Graben, are probable Triassic structures reactivated during the Late Jurassic (Yielding *et al.* 1990). In the axial Central Graben the Late Jurassic overprint is in general too strong for the Triassic structure to be discerned in detail. We believe, however, that it is likely that the major Late Jurassic faults of the Central Graben may, in many instances, have a Triassic extensional history.

4. Elucidation of the pre-Triassic history in the Central Graben is even more difficult. Very rare seismic evidence of Permian(?) rotational faulting is seen in marginal areas. In addition it is also possible that the distribution of Zechstein facies alludes to contemporary structural topography. Wells on the Grensen Nose in the Norwegian sector contain Zechstein of dolomite/anhydrite facies, with no halite present. It is possible that up-dip salt structures, such as the Lindesnes salt wall and Tommeliten Alpha salt pillow, formed at the original Zechstein facies boundary between halite (down-structure) and anhydrite/dolomite (up-structure). This would suggest a late Permian history for some of our Jurassic half-graben.

Additional References

Badley, M. E., Price, J. D., Rambech Dahl, C. & Agdestein, T. 1988. The structural evolution of the northern Viking Graben and its bearing upon extensional modes of basin formation. *Journal of the Geological Society, London*, **45**, 455–472.

Ellenor, D. W. & Mozetic, A. 1986. The Draugen oil discovery. *In*: Spencer, A. M. *et al.* (eds). Habitat of Hydrocarbons on the Norwegian Continental Shelf, 313–316.

Marsden, G., Yielding, G., Roberts, A. M. & Kusznir, N. J. 1990. Application of a flexural cantilever simple-shear/pure-shear model of continental extensional to the formation of the North Sea Basin. *In*: BLUNDELL, D. J. & GIBBS, A. D. (eds). *Tectonic evolution of the North Sea Rifts*, Oxford University Press, Oxford.

ROBERTS, A. M., YIELDING, G. & BADLEY, M. E. 1990a. A kinematic model for the orthogonal opening of the Late Jurassic North Sea Rift System, Denmark-Mid Norway. *In*: BLUNDELL, D. J. & GIBBS, A. D. (eds). *Tectonic evolution of the North Sea Rifts*, Oxford University Press, Oxford.

——, BADLEY, M. E., PRICE, J. D. & HUCK, I. W. 1990b. The structural history of a transtensional basin, Inner Moray Firth, NE Scotland. *Journal of the Geological Society London*, **147**, 87–103.

ROBERTS, A. M. & YIELDING, G. 1991. Deformation around basin-margin faults in the North Sea/Norwegian rift. *In*: ROBERTS, A. M., YIELDING, G. & FREEMAN, B. (eds). *The Geometry of Normal Faults*. Geological Society, London, Special Publication, **56**, 61–78.

YIELDING, G., BADLEY, M. E. & FREEMAN, B. 1991. Seismic reflections from normal faults in the northern North Sea. *In*: ROBERTS, A. M., YIELDING, G. & FREEMAN, B. (eds). *The Geometry of Normal Faults*. Geological Society, London, Special Publication, **56**, 79–89.

Timing, nature and sedimentary result of Jurassic tectonism in the Outer Moray Firth

S. A. R. BOLDY[1] & S. BREALEY[2]

[1]*Amerada Hess Limited, 2 Stephen Street, London W1P 1PL, UK*
[2]*University College London, Gower Street, London, UK*

Abstract: Jurassic sedimentation in the Outer Moray Firth took place under a changing tectonic regime, accompanied by regional transgression in the late Jurassic. An early phase of tectonism is recognised, accompanying the collapse of the Central North Sea Dome. This Mid-Cimmerian phase of tectonism is characterized by North–South, Viking trend, faults and reactivated Northeast–Southwest Caledonian faults. These predominantly Bathonian age faults controlled subsidence and deposition of the Middle Jurassic Rattray Volcanics Formation. The Rattray Volcanics are unconformably overlain by the Sgiath and Piper Formations, of Oxfordian to early Kimmeridgian age, deposits of a northward prograding delta that continued to be affected by movement on Mid-Cimmerian faults.

A dramatic change in tectonic regime occurred during the Kimmeridgian, possibly concurrent with the major Eudoxus Zone transgression, that heralded the onset of Kimmeridge Clay deposition. During this Late Cimmerian rift-phase, sedimentation took place under conditions of active extension controlled by northwest–southeast, Witch Ground Graben trend, faults. The Kimmeridge Clay Formation is a typical syn-rift sequence with sands deposited on the downthrown sides of rotational fault blocks, whose crests were commonly eroded. Studies of ammonites recovered from cored sequences have refined the interpretation of timing of tectonic and transgressive events and have also highlighted anomalies in correlation between ammonite and dinocyst zonation schemes.

The Moray Firth Basin is the term given to the complex series of fault blocks and grabens, which displays overall east–west trend, extending offshore from the Moray Firth (Fig. 1). This basinal area has commonly been considered to form the third arm of a trilete rift system or triple junction (Whiteman *et al.* 1975; Woodhall & Knox 1979; Ziegler 1981), the other rift axes being the Viking and Central Grabens (Fig. 1).

On the basis of structural style and stratigraphical succession, the overall Moray Firth Basin is sub-divided into an Inner and an Outer Basin, across an axis that trends north–south through the central part of Quadrant 13 (Barr 1985). The stratigraphy, tectonic evolution and palaeogeography have been well documented in a number of recent papers (Turner *et al.* 1984, Andrews & Brown 1987, Harker *et al.* 1987 and Boote & Gustav 1987).

Exploration in the Outer Moray Firth has resulted in the delineation of a number of major oil fields, the principal ones being Piper, Claymore, Tartan and Scott. The most important reservoir rocks occur in the Upper Jurassic, in the shallow-marine to deltaic Piper Sands and in the deeper water submarine-fan Claymore Sands. Total recoverable reserves anticipated from fields in production and under development amount to 1.93 billion barrels (Department of Energy 1989), highlighting the economic importance of the Outer Moray Firth.

This paper presents the results of regional and detailed stratigraphical and structural evaluation carried out during the exploration and appraisal of Block 15/21 where two fields, Ivanhoe and Rob Roy have recently entered production. Both these fields produce from Upper Jurassic Piper Sandstone reservoirs. A further major field has recently been delineated in the eastern part of Block 15/21; it extends into Block 15/22 and has been named Scott.

The results of these studies have been to record an important change in tectonic activity during the Jurassic. During Bathonian to Early Kimmeridgian times, northeast–southwest trending Caledonian faults and north–south Mid-Cimmerian 'Viking' trending faults were dominant, exerting a strong control on isopach patterns. However, during the Kimmeridgian a change in the dominant fault trend occurred and northwest–southeast faulting controlled isopach patterns. These faults parallel the major structural feature in the Outer Moray Firth, the Witch Ground Graben and are therefore termed 'Witch Ground' trend. These faults were formed in response to Late Cimmerian rifting which was the major phase of extensional tectonism affecting the Outer Moray Firth.

From HARDMAN, R. F. P. & BROOKS, J. (eds), 1990, *Tectonic Events Responsible for Britain's Oil and Gas Reserves*, Geological Society Special Publication No 55, pp 259–279.

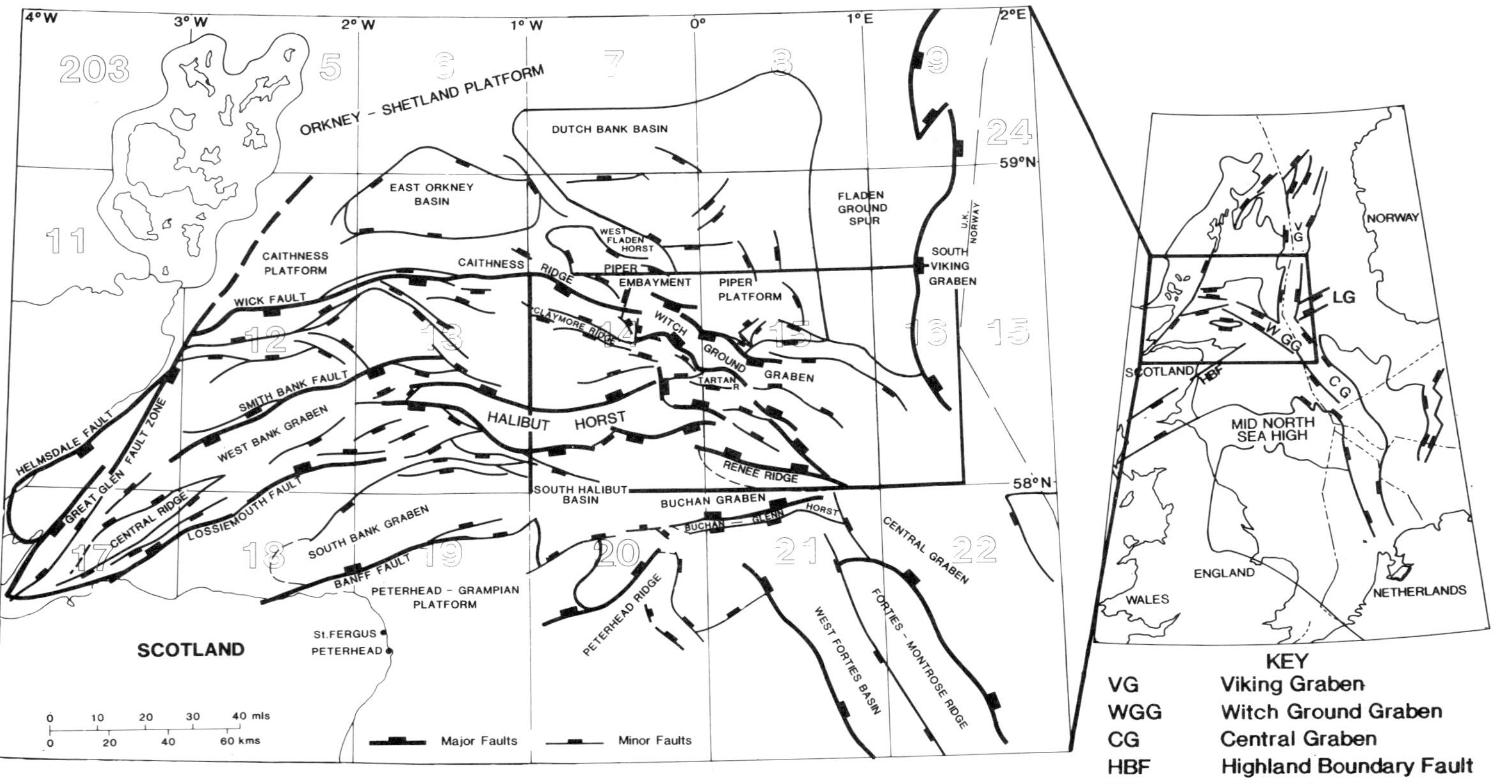

Fig. 1. Moray Firth: regional structural elements.

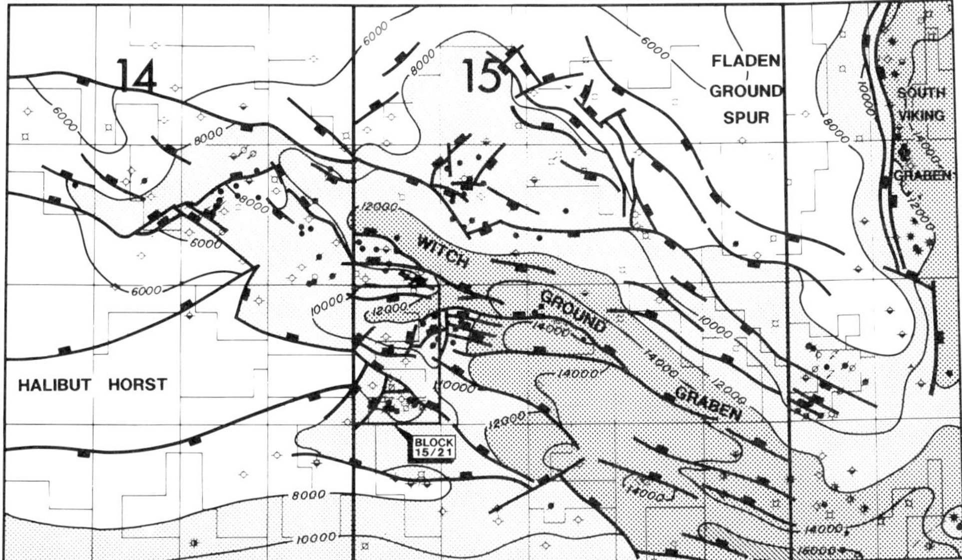

Fig. 2. Outer Moray Firth: depth structure map at base Cretaceous, highlighting major structural elements. Depth contours in feet subsea.

Ammonites recovered from cored sequences in wells in Block 15/21 provide detailed biostratigraphical zonation and allow the relationship between tectonic movements and sedimentary response to be documented more clearly.

Regional structural elements and pre-Jurassic stratigraphy

The regional structural elements map of the Moray Firth (Fig. 1) and the more local structure map at Base Cretaceous of the Outer Moray Firth area (Fig. 2), highlight the interaction of a number of fault trends.

Caledonian basement

Caledonian faults trend northeast-southwest and the principal faults are the Great Glen Fault Zone and the Highland Boundary Fault. The Great Glen Fault Zone has had a complex post-Caledonian history (Rogers *et al.* 1989) and has played a part in basin formation in the Inner Moray Firth from Devonian to Cretaceous times.

The Highland Boundary Fault System can be traced offshore in a northeastward direction to the northeast–southwest trending Ling Graben in the Norwegian sector (Doré & Gage 1987). This lineation, which has been postulated to represent a major crustal discontinuity (Doré & Gage 1987), marks the southern boundary of the Outer Moray Firth province.

The structure map at Base Cretaceous level (Fig. 2) shows that a number of northeast–southwest Caledonian trend faults traverse the Outer Moray Firth, and have influenced basin development in the Mesozoic. One prominent fault set along this trend traverses the Claymore Field in Block 14/19 and parallel faults have also been documented in the Piper Field (Maher 1980) and in Block 15/21.

Upper Palaeozoic sedimentation and Hercynian tectonism

The Moray Firth is underlain by a major Devonian basin, believed to have been initiated by transtensional movement along the Great Glen Fault Zone (McQuillin *et al.* 1982; Rogers *et al.* 1989). Within this basin, thick sequences of continental clastics accumulated. In the Outer Moray Firth, the Devonian is succeeded by Lower Carboniferous strata (Harker *et al.* 1987), believed to have been deposited during an extensional tectonic regime (Leeder & Boldy 1990), similar to that described in other North British Basins (Leeder 1982, 1987). An east–west orientation of the Outer Moray Firth Carboniferous Basin is suggested by the pre-Mesozoic sub-crop and isopach pattern and this is parallel to other well documented Carboniferous basins such as the Northumberland–Solway Basin.

It is likely that major east-west faults were initiated during this Upper Palaeozoic extensional phase and faults of this trend include the Wick Fault and the faults bounding the Halibut Horst. However, the Hercynian Orogeny, with its overall north–south compression, resulted in widespread uplift and erosion of the Outer Moray Firth area.

Early Cimmerian tectonism

There is a regional unconformity in the Outer Moray Firth between the Carboniferous of Dinantian age, and overlying Permian strata (Fig. 3). The Rotliegend is only locally developed, but Zechstein carbonates and evaporites are widespread. Triassic strata conformably overlie the Zechstein and consist of red siltstones and shales of the Smith Bank Formation, overlain in the western part of the Outer Moray Firth by more arenaceous sequences of the Skagerrak Formation (Fig. 3).

It is difficult to discern any evidence during the Permo-Triassic depositional episode of active fault control on sedimentation. It appears more likely that deposition under conditions of regional subsidence took place with the sinking of a basement composed of Caledonian and Upper Palaeozoic rocks, melded during the Hercynian Orogeny.

Jurassic sedimentation and tectonics

Episode 1: Mid-Cimmerian tectonism and Mid-Jurassic deposition

Throughout the Outer Moray Firth area there is an unconformity between Jurassic Strata and the underlying Triassic or older section. The oldest Jurassic rocks present are the volcanic sequences of the Rattray Volcanic Formation and the time-equivalent paralic sediments of the Pentland Formation (Fig. 3). These two formations are together classified as the Fladen Group (Deegan & Scull 1977).

Dating of the volcanics has indicated ages no older than Bajocian, with most dating indicating a Bathonian age (Howitt *et al.* 1975). Recent work using Ar/Ar dating on intrusives and volcanics has yielded an age of 153 ± 4 Ma, within the Callovian (Ritchie *et al.* 1988). However, biostratigraphical analysis of Pentland Formation sequences and of sediments interbedded with the volcanics of the Rattray Formation, supports a Bathonian age.

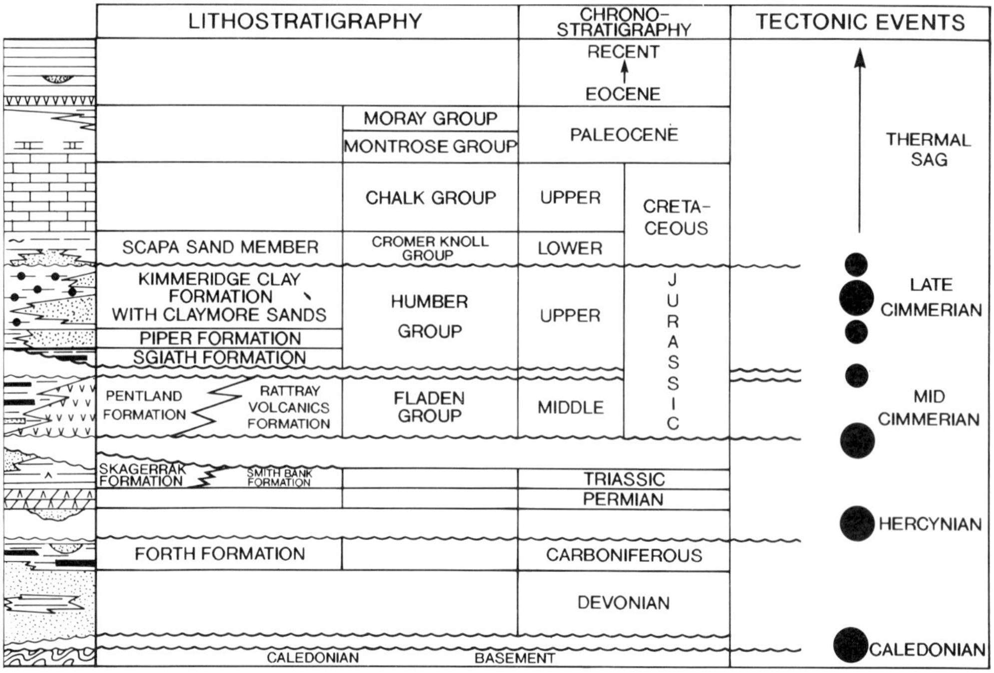

Fig. 3. Outer Moray Firth: stratigraphic column.

There are therefore no early Jurassic sediments known to be preserved in the Outer Moray Firth and this has been attributed to their removal during the pre-rift updoming of the Central North Sea area (Eynon 1981). Elsewhere, within the North Sea and throughout much of North West Europe, the early Jurassic records an important transgressive pulse, drowning the low-lying Triassic hinterland and leading to the deposition of shallow marine clastics. It is impossible to decipher the extent of early Jurassic deposition within the Outer Moray Firth, but it is generally agreed that the onset of uplift did not occur until the Aalenian to Toarcian (Eynon 1981).

The uplift and subsequent collapse of the Central North Sea Dome are the response in the Outer Moray Firth to a widespread tectonic event, generally termed the Mid-Cimmerian. The volcanics and sedimentary sequences of the Fladen Group were deposited during extensional rifting, associated with collapse of the dome. An isochore map of this sequence in the Outer Moray Firth (Fig. 4) displays a strong north–south pattern, which is attributed to control by north–south and northeast–southwest faults. These faults are termed the 'Viking' trend as they parallel the faults that control the Viking Graben system to the east.

Also apparent from the isochore map, is the absence of the Middle Jurassic Fladen Group in the western part of the Outer Moray Firth in Quadrant 14 and to the east, where the Middle Jurassic strata overstep the Triassic and Zechstein to onlap the Fladen Ground Spur, which acted as a major positive structure throughout the Jurassic. It seems likely that a positive north–south fault block, similar to the Fladen Ground Spur, extended through Quadrant 14, but the effects of later tectonism obscure this relationship.

The thickest sequences of the Fladen Group

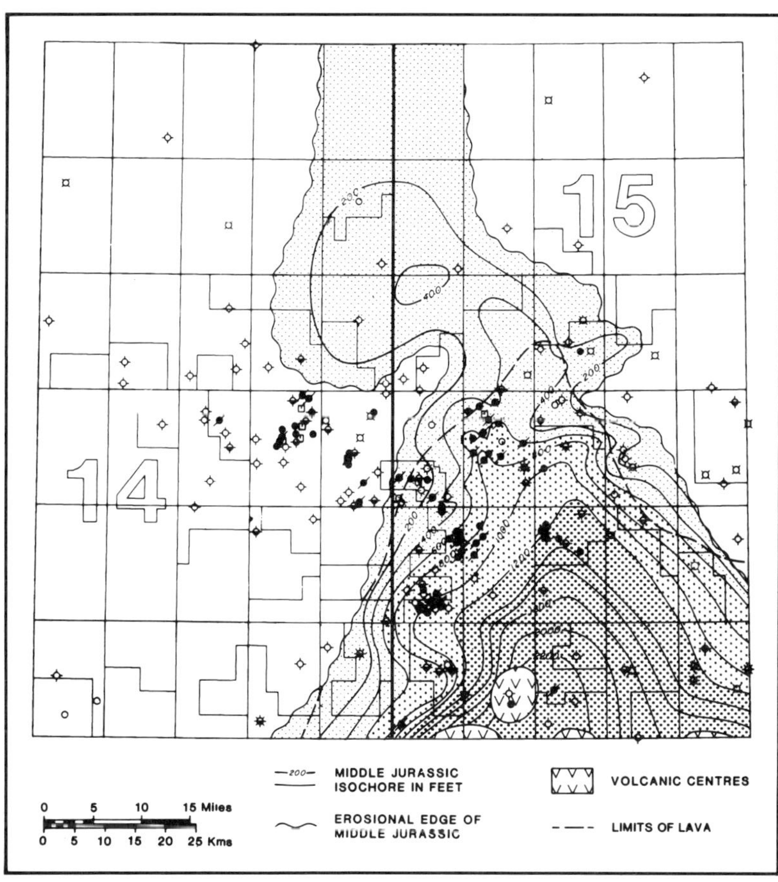

Fig. 4. Quadrants 14 and 15: Middle Jurassic Fladen Group distribution and thickness.

occur in the southern part of the Outer Moray Firth, in the south of Quadrant 15, where several volcanic vents have been postulated on geophysical evidence (Howitt *et al.* 1975). Several wells have proven thicknesses of volcanics of 1000 to 2000 ft; whilst the thickest recorded section occurs in the 15/27–2 well, which terminated having drilled 3691 ft of volcanics. Woodhall & Knox (1979) have suggested, from interpretation of marine magnetic data, that thicknesses in excess of 10000 ft may occur in the southern part of Quadrant 15.

Although locally extremely thick, the Middle Jurassic volcanics are restricted areally to the junction between the Outer Moray Firth, South Viking Graben and Central Graben. The location of the volcanics at the junction of these three basins has lead to speculation that this represents a plume generated triple junction (Whiteman *et al.* 1975; Woodhall & Knox 1979; Ziegler 1981), of the type described by Burke & Dewey (1974). However, the triple junction geometry consists of a combination of north-south 'Viking' Trend structures, younger WNW–ESE 'Witch Ground' Trend structures and northwest-southeast trending structures of the Central Graben that postdate the Middle Jurassic Volcanics. Furthermore, Latin *et al.* (this volume) have shown that the geochemistry of the Middle Jurassic Volcanics indicates that if they result from melting of normal aesthenosphere (i.e. that which produces MORB, Mid-Ocean Ridge Basalt), then only a small degree of melting has taken place which may be incompatible with a mantle plume origin.

Episode 2: Upper Jurassic: Sgiath and Piper Formation deposition

The oldest sediments overlying the Middle Jurassic Fladen Group in the Outer Moray Firth are the paralic coal-bearing sequences assigned to the Sgiath Formation by Harker *et al.* (1987) and these are considered to be no older than Mid-Oxfordian (Harker *et al.* 1987). There is therefore a significant unconformity and period of non-deposition between the Bathonian–?Callovian Fladen Group and the overlying Sgiath Formation of Upper Jurassic age (Fig. 3).

The Upper Jurassic was a period of progressive rise in sea level (Vail & Todd 1981; Rawson & Riley 1982; Haq *et al.* 1987) and this is recorded in the sedimentary succession in the Outer Moray Firth, with basal paralics of the Sgiath Formation passing upward into the shallow marine/shoreface sequences of the Piper Formation, which in turn are succeeded by deeper marine organic shales of the Kimmeridge Clay Formation.

The Upper Jurassic strata can be subdivided into two sequences (Fig. 3). An early, essentially pre-rift sequence, comprises the Sgiath and Piper Formations, whose isopach pattern mirrors that of the preceding Middle Jurassic and thus relates to the dying effects of Mid-Cimmerian tectonism. The Kimmeridge Clay Formation displays a different isochore pattern, with much more rapid thickness variation. The Kimmeridge Clay Formation forms the first part of the Late Cimmerian syn-rift sequence, that also includes Lower Cretaceous strata; these were deposited during the major phase of extension affecting the Outer Moray Firth.

The isochore map for the combined Sgiath and Piper Formations (Fig. 5) displays an overall north-south control with the major depocentre trending along the junction between Quadrants 14 and 15. This pattern is attributed to movement on syndepositional north–south 'Viking' trend faults. One of the thickest sequences of Sgiath and Piper drilled to date occurs in the 15/21–2 well where some 1285 ft of section was drilled, with the well terminating within the Sgiath Formation.

The combined Sgiath/Piper interval thins eastwards onto the Fladen Ground Spur. This may reflect a combination of both depositional thinning and later erosional truncation. The Piper and Sgiath Formations are absent in the west, over much of Quadrant 14 and over the Halibut Horst. It is considered likely that the Halibut Horst was originally covered by sequences of Sgiath and Piper but these have been subsequently eroded, yielding abundant clastic supply to the Witch Ground Graben during the Late Cimmerian extension, with the Piper Sands being reworked into the Claymore Member Sands of the Kimmeridge Clay Formation.

Overprinted upon the north-south depositional control of the Piper–Sgiath isochore pattern are the effects of Late Cimmerian extensional tectonism. Many of the tilted fault block structures of the Outer Moray Firth display erosion of the Sgiath and Piper Formations from the crestal parts of fault blocks formed by west–northwest trending (Witch Ground) normal faults. Such crestal erosion has been documented from the Claymore Field (Maher & Harker 1987) and Piper Field (Maher 1980), and also occurs in the Rob Roy and Galley Fields.

In summary, although there was activity on north–south Viking Trend faults during the

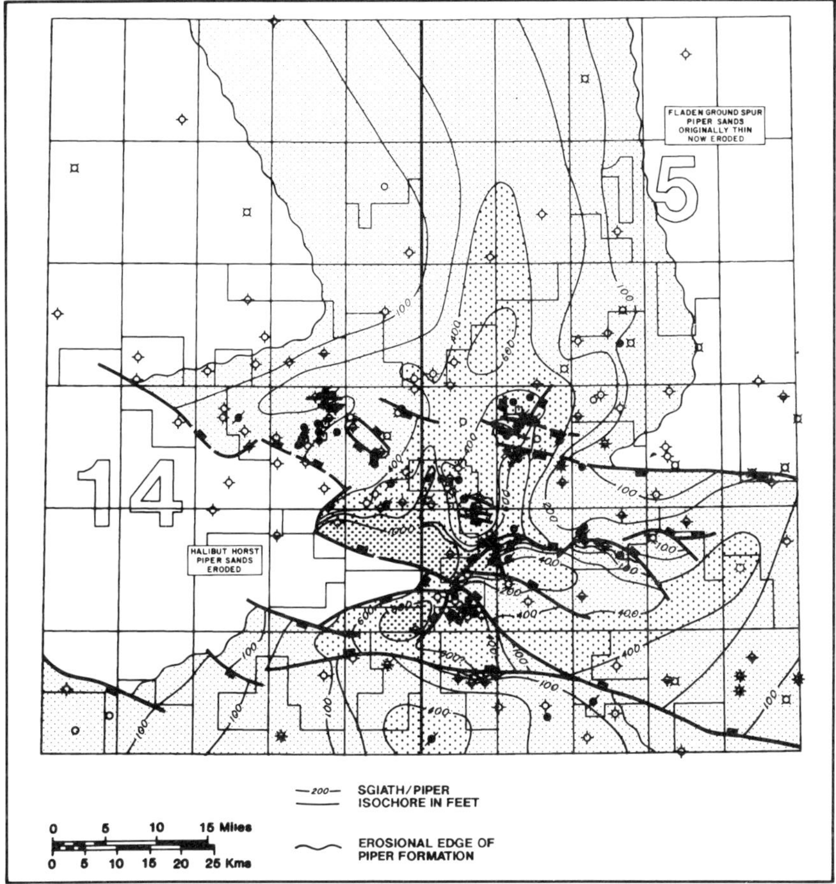

Fig. 5. Quadrants 14 and 15: Upper Jurassic Sgiath and Piper Formations distribution and thickness.

deposition of the Sgiath and Piper Formation, these sequences were deposited during the late stages of Mid-Cimmerian tectonism and can be considered as a pre-rift succession to the succeeding Late Cimmerian syn-rift sequence.

Episode 3: Upper Jurassic: Kimmeridge Clay Formation deposition

The regional isochore map of the Kimmeridge Clay Formation (Fig. 6), displays clearly the change in subsidence and sedimentation patterns associated with the opening of the Witch Ground Graben. Faults defining the graben are oriented WNW–ESE and the isochore pattern is parallel to these faults.

The Kimmeridge Clay Formation, of Kimmeridgian to early Ryazanian age (Harker *et al.* 1987), consists of typical organically rich black shales, but also contains several significant sandstone units, deposited by mass flow mechanisms down fault controlled palaeoslopes. The Claymore Sand Member of the Claymore Field is the only one of these sand units to have been formally defined and until the lithostratigraphical nomenclature is formalized further, all sands within the Kimmeridge Clay Formation are referred to here as Claymore Sands, although they are not all of the same age as those seen in the Claymore Field.

The isochore map of the Kimmeridge Clay Formation (Fig. 6) displays great variability over short distances, reflecting the strong tectonic control on this syn-rift sequence. In the deepest parts of the Witch Ground Graben thicknesses in excess of 3000 ft have been proven by drilling. However, there are a number of positive tectonic features over which the Kimmeridge Clay is absent. In the eastern part of Quadrant 15, the Fladen Ground Spur continued as a high area during Kimmeridge Clay

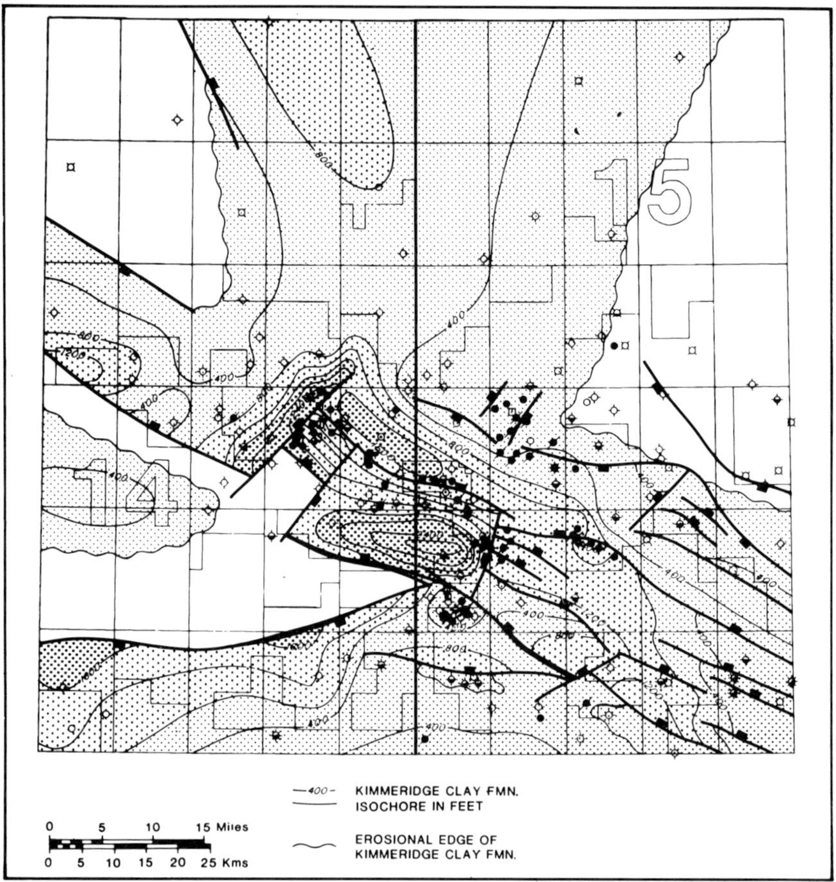

Fig. 6. Quadrants 14 and 15: Upper Jurassic Kimmeridge Clay Formation distribution and thickness.

deposition and the absence of Kimmeridge Clay here may be a function of both non-deposition and later erosion. A similar scenario can be envisaged for the Halibut Horst which is interpreted to have emerged as a positive block during Kimmeridge Clay times and undergone intense erosion, shedding coarse clastics to form the Claymore Sands. The mineralogical maturity of the Claymore Sands suggests strongly that they represent redeposition of Piper Sands, eroded from areas such as the Halibut Horst and other intrabasinal fault blocks.

Within the Witch Ground Graben itself, the variation in thickness of the Kimmeridge Clay Formation (Fig. 6) reflects the infilling of the tilted fault block topography that was formed by Late Cimmerian extensional tectonics. Well data from Block 15/21 has shown that there is often a hiatus above the Piper Formation, with a considerable part of the earliest Kimmeridge Clay Formation absent. This marks the onset of Late Cimmerian tectonics and is dated to be within the Kimmeridgian, probably occurring within the Eudoxus Zone.

Jurassic tectonics and sedimentation in block 15/21

Structural elements

The major structural elements in Block 15/21 are shown in Fig. 7. This structural configuration has been delineated utilizing a very extensive database, comprising two 3D seismic surveys, several 2D seismic datasets and 25 exploration and appraisal wells.

The southern flank of the Tartan Ridge extends in a east–west direction across the northernmost part of Block 15/21. The other major positive structural feature is the Halibut Horst Spur (Fig. 7). The Halibut Horst itself is a very

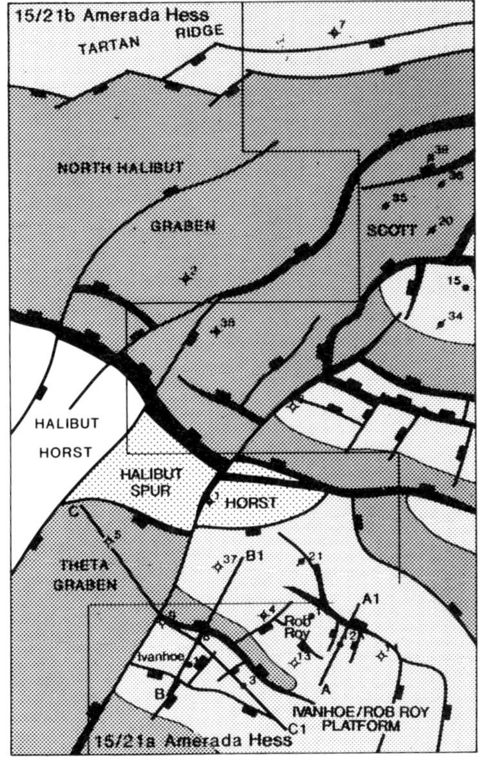

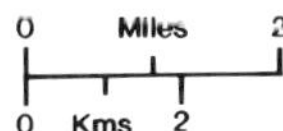

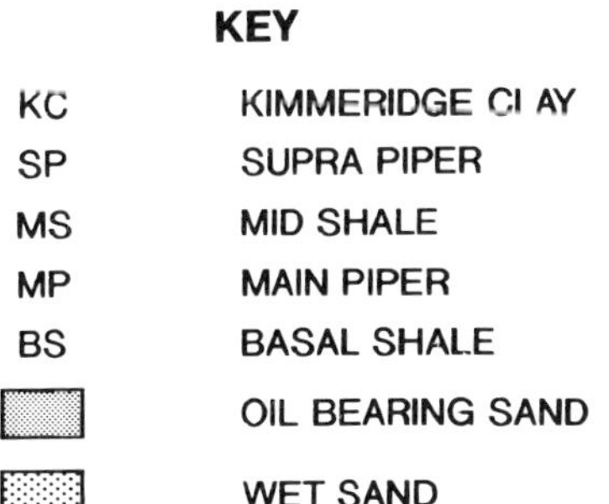

Fig. 7. Block 15/21: Structural Elements. Location of structural sections illustrated in Figs 12, 13 & 15 are annotated A–A[1], B–B[1] and C–C[1] respectively.

shallow feature with little or no Mesozoic section preserved, whilst the Halibut Horst Spur retains some Jurassic section, albeit severely truncated.

In the southernmost part of Block 15/21 the Ivanhoe and Rob Roy Fields lie in an area of intermediate structural relief termed the Ivanhoe/Rob Roy Platform (Fig. 7).

Two grabens traverse Block 15/21 and these reflect the two different fault trends recognised in the regional review. A graben bounded by northeasterly trending faults separates the Halibut Horst from the Ivanhoe/Rob Roy Platform (Fig. 7) and this feature is termed the Theta Graben. It is defined most clearly in the southern part of block 15/21, but continues at depth to the northeast, where it interacts with a graben traversing the northern half of Block 15/21 that is defined by east–west and northwest–southeast faults of the Witch Ground trend. This graben, called the North Halibut Graben, separates the Tartan Ridge to the north, from the Halibut Horst, to the South (Fig. 7).

Jurassic lithostratigraphy

The Jurassic lithostratigraphic nomenclature utilized in Block 15/21 is shown in Fig. 8, against the electric logs from the 15/21a–11 Rob Roy discovery well. A major unconformity separates the Middle Jurassic Rattray Volcanics Formation of the Fladen Group, from the overlying Upper Jurassic Humber Group.

Historically, in Block 15/21 the Humber Group has been divided into two constituent formations: the Piper Formation and the overlying Kimmeridge Clay Formation. The Piper Formation was further subdivided into four members, from the base upward these are: the Basal Shale Member, Main Piper Sand Member, Mid-Shale Member and Supra Piper Sand Member.

However, Harker *et al.* (1987), has defined a new unit, the Sgiath Formation, as comprising a paralic sequence of sandstones, coals carbonaceous mudstones and siltstones. The top of the Sgiath Formation and base of the overlying Piper Formation is defined by the change in log character from low gamma ray sands to high gamma ray mudstones of the marine 'I' Shale (Harker *et al.* 1987).

Considerable uncertainty remains in correlating the 'I' shale from the type section in the Piper Field southwestward into Block 15/21. However, in Block 15/21 a tripartite division of the basal shale sequence into a Marine Unit, a Paralic Unit and a Coal Unit can readily be recognized (Fig. 8). The lower two units are assigned to the Sgiath Formation as they represent non-marine sedimentation. The Marine Shale Unit is considered to form the lowermost part of the Piper Formation and represents a

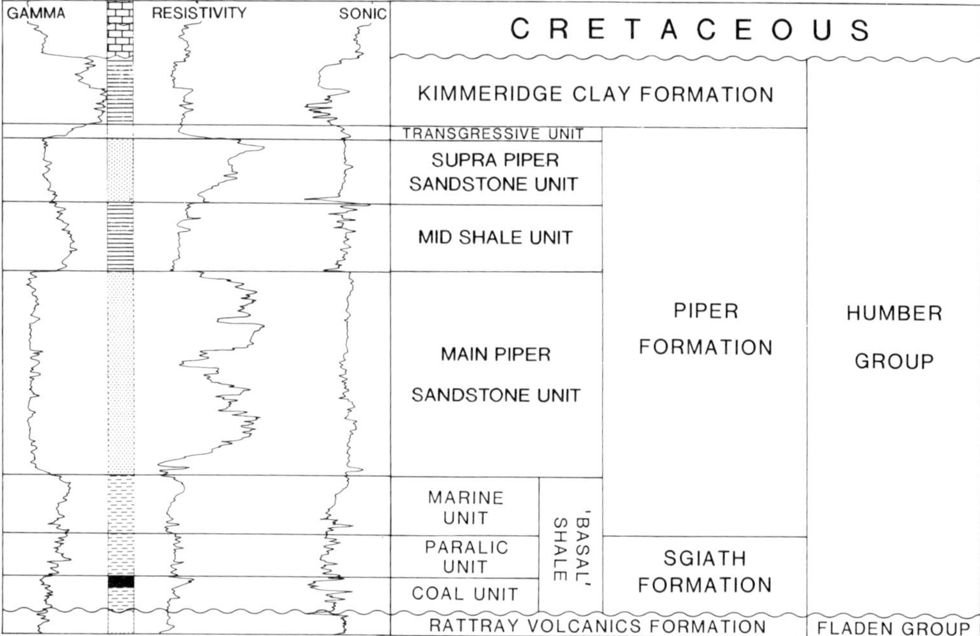

Fig. 8. Block 15/21: Jurassic lithostratigraphy.

major transgression which resulted in a fully marine depositional environment.

The Marine Shale Unit coarsens upward and passes gradationally into the overlying Main Piper Sandstone Unit (Fig. 8), which consists almost entirely of medium to coarse grained quartzose sandstones. Another major marine transgression terminated deposition of the Main Piper Sands and resulted in the accumulation of the Mid-Shale under fully marine conditions.

A second major coarsening-upward regressive sequence is recognized, with the Mid-Shale passing upwards into the Supra Piper Sandstone (Fig. 8). This is composed of medium-fine grained sandstones, significantly richer in detrital felspar grains in comparison to the Main Piper Sands.

The uppermost unit recognised within the Piper Formation is a thin fining-upward transgressive sequence. This is overlain by highly radioactive shales of the Kimmeridge Clay Formation (Fig. 8). There is a marked velocity contrast between the relatively low velocity organic shales of the Kimmeridge Clay formation and the higher velocity 'cold' shales of the Transgressive Unit of the Piper Formation. This contrast gives rise to a seismic event of variable intensity, but mappable over much of the area.

Furthermore, the boundary between the Kimmeridge Clay and Piper Formations is often a significant hiatus, with the oldest Kimmeridge Clay condensed or even absent altogether. The great variation in thickness of this early Kimmeridge Clay is strong evidence of significant faulting, concomittant with the major transgression that heralded the onset of Kimmeridge Clay deposition.

Lithologies within the Kimmeridge Clay Formation are highly variable within Block 15/21. In places it contains very thick sandstone sequences, assigned to the Claymore Member.

Tectonic controls on Sgiath and Piper Formation deposition

In the regional review of Jurassic sedimentation and tectonics, attention was drawn to the change in the isochore pattern between the north–south 'Viking' trend of the Middle Jurassic and Sgiath/Piper interval; and the northwest–southeast 'Witch Ground' Trend of the Kimmeridge Clay Formation (Figs 4, 5 & 6). This change in tectonic control is highlighted at a local level within Block 15/21 (Figs 9 & 10).

The isochore map for the combined Sgiath and Piper Formations has been composed using both well and seismic data. There is clear evidence of north–northeasterly trending depositional axes, the most pronounced of these being the Theta Graben, where a much thicker Sgiath/

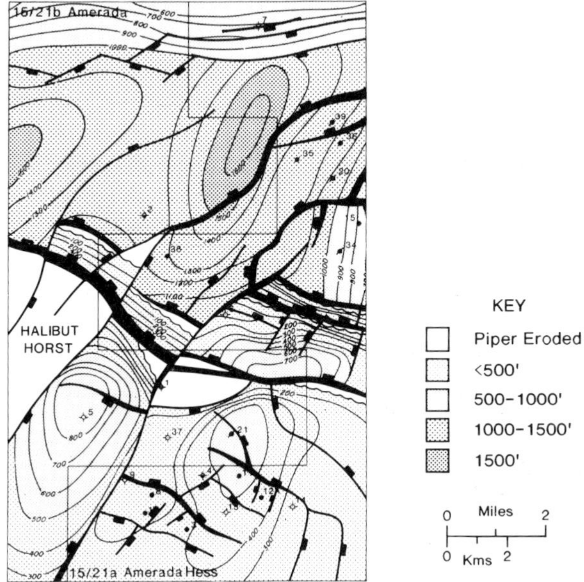

Fig. 9. Block 15/21: isochore map of Upper Jurassic Sgiath and Piper Formations.

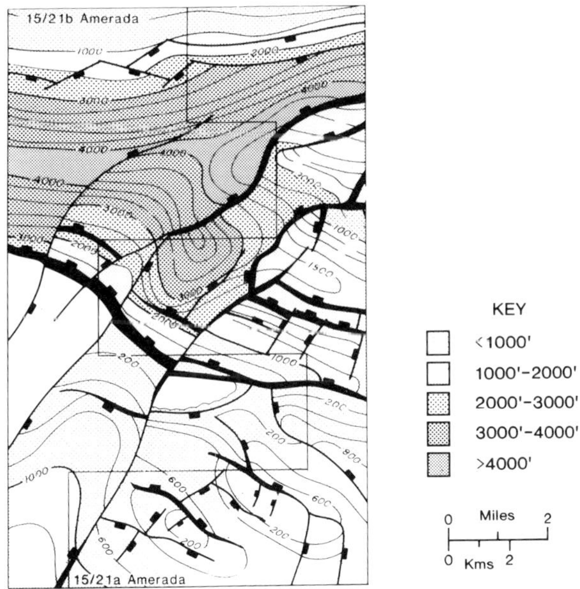

Fig. 10. Block 15/21: isochore map of Upper Jurassic Kimmeridge Clay Formation.

Piper sequence was penetrated in the 15/21–5 well (800 ft +) than is seen in wells on the Ivanhoe/Rob Roy Platform (Fig. 9). The maximum drilled thickness was encountered in the 15/21–2 well, where the combined Sgiath/Piper interval totals 1285 ft. Even thicker sequences are postulated within the northeasterly extension of the Theta Graben where up to 1600 ft may be present.

A parallel depositional axis is recognised trending north–northeasterly through the eastern part of Ivanhoe Field and the western part

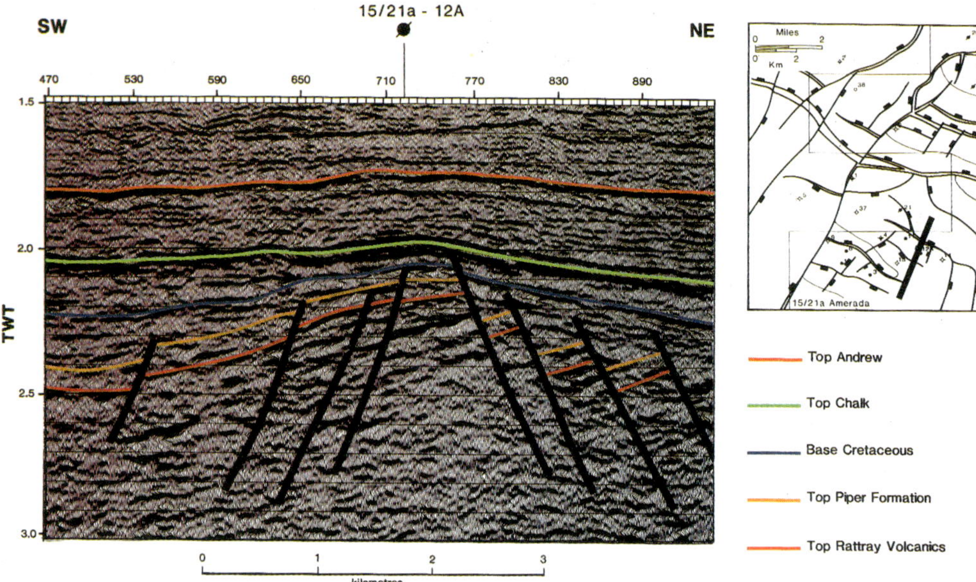

Fig. 11. Block 15/21: seismic line through the Rob Roy Field.

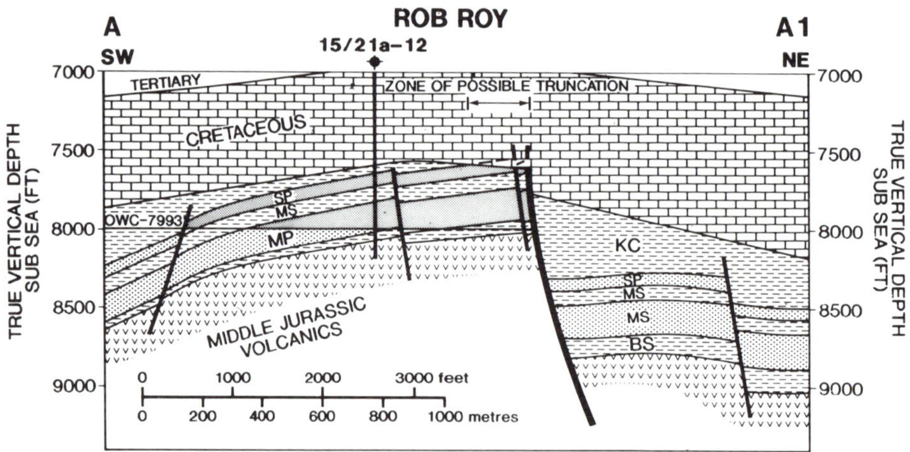

Fig. 12. Block 15/21: Structural cross section through the Rob Roy Field.

of the Rob Roy Field (Fig. 9). Here a maximum thickness of more than 500 ft of Piper–Sgiath has been proven by drilling, with thinning occurring both eastward towards Rob Roy and westward towards Ivanhoe, where the combined Sgiath and Piper interval is less than 300 ft thick.

The WNW–ESE Witch Ground trend faults also display strong control on the isochore pattern (Fig. 9). However, these faults are not considered to have been active during deposition of the Sgiath and Piper Formations, with the thinning seen along the crests of west–northwest trending fault blocks being a function of crestal erosion of rotating fault blocks, that occurred during the deposition of the Kimmeridge Clay. Several wells, such as 15/21–1 and 15/21–6 contain thick paralic and coal-bearing Sgiath sequences indicating that these areas were structurally low during the initial

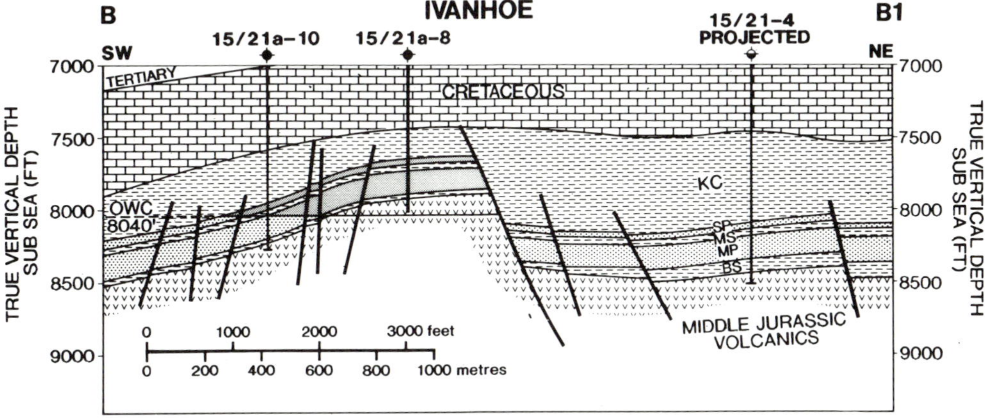

Fig. 13. Block 15/21: Structural cross section through the Ivanhoe Field.

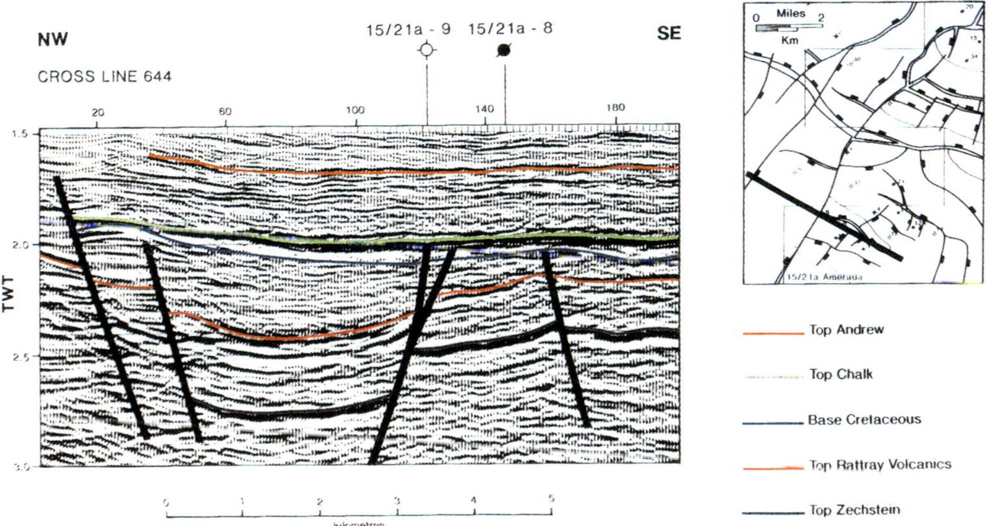

Fig. 14. Block 15/21: Seismic line illustrating the Theta Graben.

phase of Upper Jurassic sedimentation. These truncated sequences are unconformably overlain by Cretaceous Chalk and Kimmeridge Clay, respectively, in the 15/21–1 and 15/21–6 wells, testifying to the major phase of uplift and erosion which commenced in the late Jurassic and continued into Cretaceous times.

Tectonic controls on Kimmeridge Clay Formation deposition

The isochore map of the Kimmeridge Clay Formation in Block 15/21 (Fig. 10) displays marked contrast to the pattern illustrated for the Sgiath/Piper interval. The major depocentre lies within the axial part of the North Halibut Graben and is oriented in an east-west direction. This change in thickness pattern is related to the onset of significant movement on the west-northwest 'Witch Ground Trend' faults, connected with the opening of the Witch Ground Graben. The thickest sequence of the Kimmeridge Clay drilled to date on Block 15/21 occurs in the 15/21–2 well, where more than 3000 ft of section was found.

Within the North Halibut Graben a number

of tilted fault blocks are developed, commonly backtilted to the southwest. Fault blocks of this type form the structural traps in the Scott Field in the eastern part of Block 15/21 (Fig. 10), where the Kimmeridge Clay Formation displays marked thinning over the crests of individual fault blocks.

The Halibut Horst and its eastward continuation, the Halibut Horst Spur, are interpreted to have formed during the deposition of the Kimmeridge Clay Formation. On the Halibut Horst itself and on much of the Halibut Horst Spur, there is no Kimmeridge Clay Formation preserved. Intense erosion of these areas during deposition of the Kimmeridge Clay led to the removal of the Piper Sands, which were redeposited as submarine fan-sands of the Claymore Member. These Claymore Sands have been found in wells both to the north (15/21−2) and to the south (15/21−5) of the Halibut Horst.

Over the Ivanhoe/Rob Roy Platform, relatively thin sequences of the Kimmeridge Clay Formation occur and range up to about 800 ft (Fig. 10). The Ivanhoe and Rob Roy Fields are tilted fault block traps defined by westnorthwest Witch Ground Trend normal faults. These fault blocks are also backtilted to the southwest and the Kimmeridge Clay thins over the crests, with less than 100 ft preserved on the crest of Rob Roy Field.

The northerly and northeasterly Viking Trend faults, dominant during deposition of the Sgiath and Piper Formations, continued to effect the depositional pattern of the Kimmeridge Clay Formation. In the 15/21−5 well, within the Theta Graben, more than 800 ft of Kimmeridge Clay is present and seismic evidence indicates further thickening to the southwest. It seems likely that pre-existing Viking Trend faults were exploited as transfer faults, between the major Witch Ground Trend fault blocks, during the Late Cimmerian phase of extension and block rotation.

Seismic and structural cross sections illustrating Jurassic tectonism

Seismic lines and structural cross sections through the Ivanhoe and Rob Roy area illustrate Jurassic tectonism. A northeast−southwest seismic line through the Rob Roy Field (Fig. 11) clearly displays the tilted fault block nature of the structure. The most prominent seismic events are at Top Chalk, Base Cretaceous and Top Middle Jurassic Volcanics. The Top Piper Event is mappable over parts of the field area, but is often masked by Base Cretaceous reflections in the crestal part of the structure. In a gross structural sense the Piper Formation is concordant with the underlying Middle Jurassic Volcanics event, and formed a 'pre-rift' sequence, that is overlain by the 'syn-rift' Kimmeridge Clay Formation.

The structural cross section through the 15/21a−12 Rob Roy well (Fig. 12) illustrates the marked thinning of the Kimmeridge Clay Formation over the crest of the structure. On the crest of the Rob Roy Field footwall, adjacent to the major bounding normal fault, erosion of the uppermost Piper Formation has been postulated from seismic interpretation and has recently been proved in development drilling. Usually a thin veneer of Kimmeridge Clay is present, but fault movement continued into the early Cretaceous and in places the Kimmeridge Clay may be entirely eroded. Thick sequences of Kimmeridge Clay Formation occur to the northeast of the Rob Roy Field, in the hanging wall block.

This variation in thickness of the Kimmeridge Clay Formation is not seen in the underlying Piper/Sgiath sequence. As is apparent on the structural cross section (Fig. 12) this interval thickens from southwest to northeast, toward the crest of the Rob Roy Field, indicating that the structure post-dates deposition of this sequence.

A northwest−southeast structural section through the Ivanhoe Field, intersecting the 15/21a−8 and 10 wells (Fig. 13) also illustrates thinning of the Kimmeridge Clay Formation onto the crest of the structure, but activity on the major bounding fault appears to have died out during the Jurassic, as there is no displacement seen at Base Cretaceous.

The northeasterly 'Viking' trend faults that define the Theta Graben are illustrated on the seismic line shown in Fig. 14. This line is oriented WNW−ESE, parallel to the major Witch Ground trending faults. In the eastern part of the line the Ivanhoe Field is evident as a major high fault block covered by a relatively thin Jurassic sequence. The major down to the west fault that forms the eastern boundary to the Theta Graben is visible on Fig. 14 to the west of the 15/21a−9 well. Across this fault there is marked thickening of the Upper Jurassic, as defined by the interval between the Base Cretaceous and the Top Middle Jurassic Volcanics. However, there is also marked thickening in the deeper section, between the Top Middle Jurassic and the Top Zechstein, indicating earlier, pre-Upper Jurassic movement on this fault. This is highlighted by the thickening of the Middle Jurassic Volcanics package, identified by high-amplitude reflectors, across the fault. These Viking trend faults were therefore certainly active in Mid-Jurassic times, but are likely to have been initiated earlier in the Mesozoic, probably during the Triassic.

A structural cross section across the Theta Graben (Fig. 15) parallel to the seismic line (Fig. 14), illustrates the rapid thickening of the Piper/Sgiath interval across the fault to the west of 15/21a–9. The thick sequence of Piper/Sgiath Formations proved by the 15/21–5 well, is restricted to the Theta Graben, with much reduced thicknesses penetrated in Ivanhoe Field wells to the east.

Jurassic tectonism: conclusions

From the foregoing examination of seismic and isochore data at both regional (Outer Moray Firth) and local (15/21) scale, a number of conclusions can be drawn concerning Jurassic tectonism and its effect upon sedimentation and structure.

(i) North–south and northeast–southwest 'Viking' Trend Faults were active in Mid-Jurassic times and form the principal control upon the isochore patterns of the Fladen Group and the Sgiath and Piper Formations.

(ii) If the assumption is made that the extension direction is orthogonal to the major normal faults, this would argue for an early phase of east–west extension in the Outer Moray Firth, which formed a north–south basin, parallel to the Viking Graben.

(iii) There is no evidence that the west-northwest Witch Ground Trend faults were active during the deposition of the Fladen Group, or the Sgiath or Piper Formations. Some faults of this orientation may have existed as transfer faults, but the major structural feature of the Witch Ground Graben, had not been formed. Therefore during this Mid Cimmerian phase of Jurassic tectonism the North Sea rift system did not display the triple-junction geometry as defined by the Viking, Witch Ground and Central Graben intersections.

(iv) A major change in tectonic regime occurred during the Kimmeridgian, with the onset of opening of the Witch Ground Graben along WNW–ESE 'Witch Ground' Trend faults. This Late Cimmerian phase of rifting was much more intense than the earlier tectonism.

(v) Once again, if the assumption is made that the extension direction is orthogonal to the major normal faults, this argues for a northeast–southwest orientated extensional regime, suggesting a change in extension direction during the later Jurassic.

(vi) The tilted fault block structural traps of the Outer Moray Firth are defined by Witch Ground Trending faults and were formed during Kimmeridgian to early Cretaceous times. The Ivanhoe, Rob Roy, Scott and Piper Fields are all tilted fault block traps of this type.

(vii) The Witch Ground Graben is a symmetrical feature with fault blocks backtilted both to the northeast, in the north, as shown by the Piper Field (Fig. 16) and to the southwest, in the south, as shown by the Ivanhoe, Rob Roy and Scott Fields.

(viii) The Witch Ground Trend faults are important in controlling the distribution of Claymore Sands, which accumulated as toe of slope submarine fans (Gustav & Boote 1987). These Witch Ground faults also effect the isochore pattern of the Sgiath and Piper Formations through widespread erosion of the crestal parts of tilted fault blocks, as seen in the Rob Roy Field.

(ix) Many of these conclusions are supported by Maher (1980) in his work upon the Piper Field (Fig. 16). Maher (1980) stated that only a few faults could be shown to have been active during deposition of the Piper Formation, the principal one of these being the 'D' fault which is orientated in a northeasterly direction. Furthermore, he demonstrated that erosion occurs along the crestal part of the footwall associated with the northwest trending faults.

Timing of Jurassic events

Dating the Jurassic sequences of the Outer Moray Firth has relied upon radiometric techniques for the Middle Jurassic Rattray Volcanics, and micropalaeontological zonation for the Upper Jurassic sequence (Harker *et al.* 1987; O'Driscoll *et al.* this Volume). Of particular importance has been the use of dinoflagellate cysts and several zonation schemes have been proposed utilising these palynomorphs. Wollam & Riding (1983) proposed 18 dinoflagellate cyst assemblage zones covering the interval from latest Triassic to earliest Cretaceous. The latest Oxfordian to earliest Port-

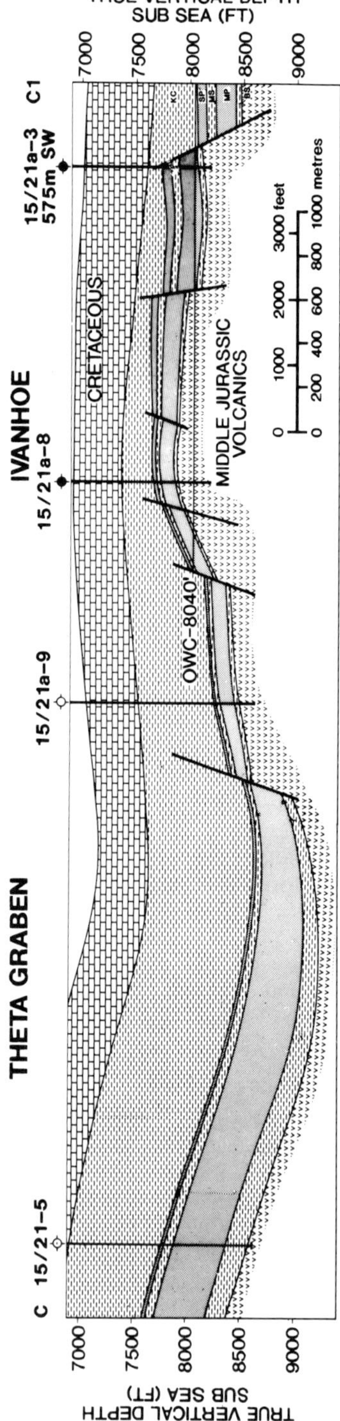

Fig. 15. Block 15/21: Structural cross section through the Theta Graben.

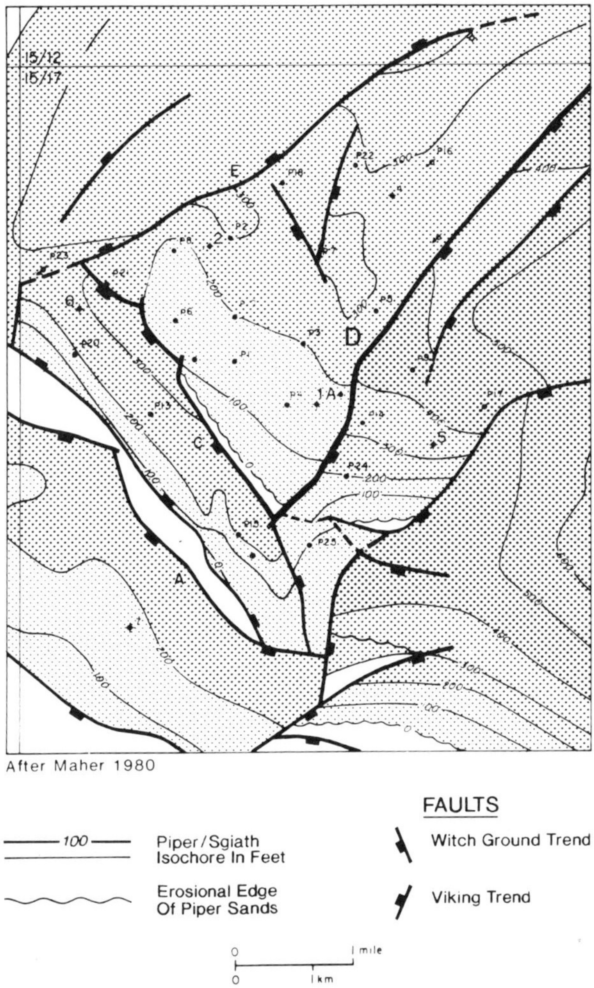

Fig. 16. Block 15/17: isopach map of the Piper and Sgiath Formations highlighting fault trends (after Maher 1980).

landian part of this zonation scheme has been revised by Riding & Thomas (1988).

Within Block 15/21 cores from several wells have yielded well preserved ammonites. Identification of these ammonites has allowed more precise definition of the timing of transgressive and tectonic events. Ammonites have most commonly been recovered from the Mid-Shale Unit of the Piper Formation, but occasional specimens have also been recovered from the lowermost part of the Kimmeridge Clay Formation. This distribution certainly displays sample bias, as far more core material has been obtained from the Mid-Shale Unit than from any of the other Upper Jurassic shale sequences. To date, no identifiable ammonites have been recovered from the Marine Shale Unit of the Piper Formation.

Ammonites identified from four wells in the Rob Roy Field are shown in Fig. 17. Identification of these ammonites has shown that the Mid-Shale Unit represents the lowermost part of the Kimmeridgian, with definitive *baylei*, *cymodoce* and *mutabilis* forms and assemblages recovered. The dating of the lowermost part of the Mid Shale Unit as *baylei* Zone, the lowermost zone within the Kimmeridgian, suggests strongly that Main Piper Sand deposition was terminated by the base Kimmeridgian transgression (Rawson & Riley 1982) and that the

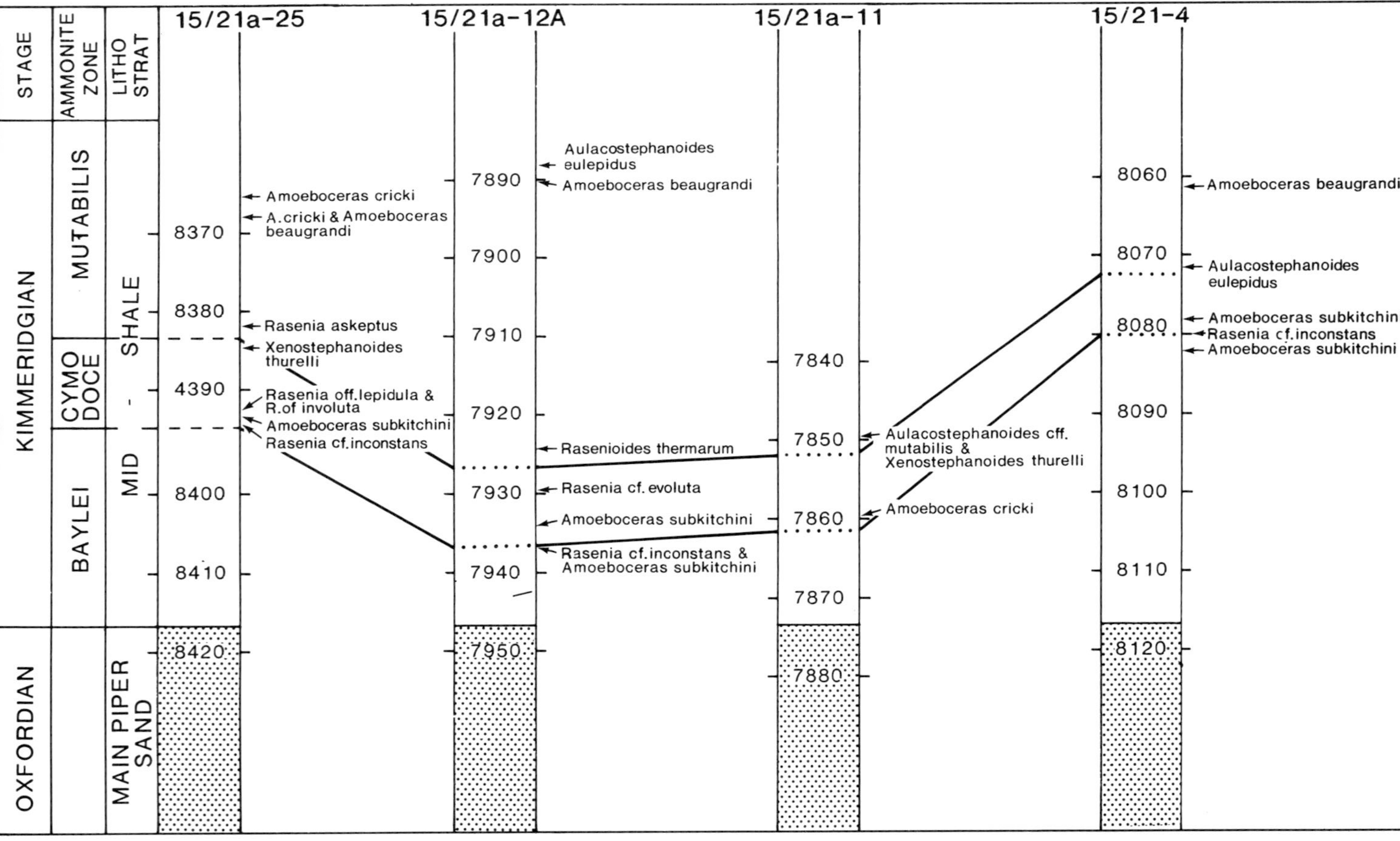

Fig. 17. Block 15/21: Rob Roy Field Ammonite Zonation.

junction between the Mid Shale and the Main Piper must be very close to the Oxfordian/ Kimmeridgian boundary (Fig. 18).

A zonal subdivision of the Lower Kimmeridgian stage *sensu anglico* was proposed first by Salfeld (1914), who created five ammonite zones: *baylei* (lowest); *cymodoce*; *mutabilis*; *yo* and *pseudomutabilis* (highest). Ziegler (1962) replaced the two highest with the *eudoxus* and *autissiodorensis* Zones. More precise definition of zonal boundaries is still needed and, as yet, no formal subzonal scheme has been applied to Northwest Europe. However, it appears from a number of richly fossiliferous sequences that three faunas can be recognised for the *cymodoce* Zone and three for the *mutabilis* Zone, allowing identification of lower, middle and upper parts of these zones (Fig. 19, Birkelund *et al.* 1978, 1983).

The ammonite fauna recovered from Rob Roy wells indicates that Mid Shale deposition continued at least until mid-*mutabilis* Zone times, with the Supra Piper Sand being of upper *mutabilis* Zone age (Fig. 18).

Ammonites recovered from the basal part of the Kimmeridge Clay Formation in the 15/ 21a–25 well indicate an earliest *eudoxus* Zone age, from the co-existence of *Amoeboceras kochi* and *Aulacostephanites eulepidus*. The final transgressive pulse that terminated Piper Sand

NW EUROPEAN AMMONITE ZONES	FAUNAL SUBDIVISIONS
Aulacostephanus autissiodorensis	
Aulacostephanus eudoxus	
Aulacostephanoides mutabilis	Aspidoceras orthocera
	Aulacostephanoides mutabilis
	Rasenioides askepta
Rasenia cymodoce	Rasenia evoluta
	Rasenia involuta
	Rasenia cymodoce
Pictonia baylei	Rasenia inconstans

Fig. 19. Lower Kimmeridgian Ammonite Zones and informal subdivisions.

deposition and heralded the onset of Kimmeridge Clay deposition, can therefore be dated as earliest *eudoxus* Zone (Fig. 18), as suggested by Harker *et al.* (1987).

Concomitant with the study of the ammonite fauna, samples have been prepared for palynological analysis and from these more than 50

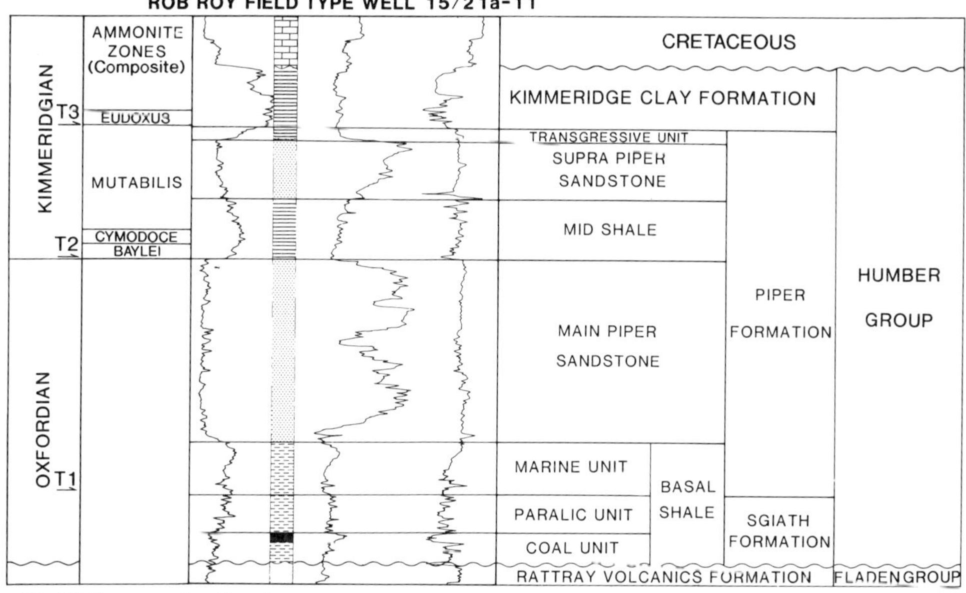

Fig. 18. Block 15/21: Jurassic Ammonite Biostratigraphy.

STAGE	RIDING & THOMAS 1988					BLOCK 15/21a
	AMMONITE ZONE	ZONE	SUB-ZONES	RANGE-TOPS	RANGE-BASES	AMMONITE ZONE
LOWER KIMMERIDGIAN	AUTISSIODORENSIS	ENDOSCRINIUM LURIDUM (E l)	c	E. luridum		
	EUDOXUS					EUDOXUS
	MUTABILIS		b	S. scarburghense A.dictyota pyra		MUTABILIS
	CYMODOCE		a	T.dangeardi E.galeritum	S?inaffecta S?paeminosum P.pannosum	
	BAYLEI	SCRINIODINIUM CRYSTALLINUM (Sc)	d	S.crystallinum S.orbis N.pellucida	C.longicorne O.patulum	CYMODOCE / BAYLEI
UPPER OXFORDIAN	ROSENKRANTZ1		c	C.ornatum		
	REGULARE		b		O.balia D.tuberosum	

Fig. 20. Comparison of Block 15/21 Ammonite and Dinocyst Zonation with Riding & Thomas (1988).

taxa of dinoflagellate cysts have been identified, including many of the critical taxa used in zonation by Riding & Thomas (1988). However, as can be seen in Fig. 20, the correlation of the dinocyst zones to the Boreal ammonite zonal sequence in Block 15/21 does not agree in detail with the scheme published by Riding & Thomas (1988) for the Dorset coastal sections. This anomaly is explained by the extended ranges in the 15/21 wells of *Scriniodinium crystallinum* and *Endoscrinium galeritum*, both these forms occurring in conjunction with *mutabilis* Zone ammonites (Fig. 20). Other key taxa utilized by Riding & Thomas (1988) are absent in the 15/21 sequences and these include *T. dangeardi, S. inaffectun, D. tuberosum* and *C. ornatum*. Work is continuing to clarify the anomaly between the ammonite and dinocyst biostratigraphy.

In conclusion, the identification of ammonites has allowed precise ages to be assigned to two of the three major transgressive events seen in the Upper Jurassic in Block 15/21. No ammonites have been recovered from the first transgressive event represented by the Marine Unit of the Piper Formation, but the Mid-Shale transgression has been demonstrated to be basal Kimmeridgian in age and the Kimmeridge Clay transgression has been dated as early *eudoxus* Zone. It seems most likely that the Kimmeridge Clay transgression accompanied the onset of Late Cimmerian tectonism within Block 15/21, suggesting a *eudoxus* Zone age for this tectonic event. In contrast, no tectonic control is apparent accompanying the Mid-Shale transgression, although such a relationship may become apparent with further evaluation.

The authors would like to thank the management of Amerada Hess Limited and their co-venturers in Block 15/21; Deminex UK Oil and Gas Limited, Kerr McGee Oil UK plc and Pict Petroleum plc, for permission to publish this paper. The research work being undertaken by Simon Brealey is financed by a research studentship from the 15/21 Group, which is also gratefully acknowledged.

We would also like to thank the numerous Amerada Hess geologists and geophysicists, who have contributed to the successful exploration of Block 15/21. In particular we acknowledge the efforts of Ron Krenzke, Alan Booth and Chris Thomas in interpreting the seismic data and Richard Hardman and Richard Warren for the long hours spent in discussion.

Finally, we are most grateful to Carol Tait for typing the manuscript and the Amerada Hess Limited Drawing Office for draughting the figures.

References

Andrews, I. J. & Brown, S. 1987. Stratigraphic evolution of the Jurassic, Moray Firth. *In*: Brooks, J. & Glennie, K. W. (eds). *Petroleum Geology of North West Europe*. Graham & Trotman, London, 785–795.

Barr, D. 1985. 3-D Palinspastic restoration of normal faults in the Inner Moray Firth: implications for extensional basin development. *Earth and Planetary Science Letters*, **75**, 191–203.

Birkelund, T., Thusu, B. & Vigran, J. 1978. Jurassic–Cretaceous biostratigraphy of Norway with comments on the British Cymodoce Zone *Palaeontology*, **21**(1), 31–63.

——, Callomon, J. H., Clausen, C. K., Hansen, M. N. & Salinas, I. 1983. The Lower Kimmeridge Clay at Westbury, Wiltshre, England. *Proceedings of the Geological Association*, **94**(4), 289–309.

BOOTE, D. R. D. & GUSTAV, S. H. 1987. Evolving depositional systems within an active rift, Witch Ground Graben, North Sea. *In*: BROOKS, J. & GLENNIE, K. W. (eds). *Petroleum Geology of North West Europe*. Graham & Trotman, London, 819–833.

BURKE, K. & DEWEY, J. F. 1974. Plume generated triple junctions: key indicator in applying Plate tectonics to old rocks. *Journal of Geology*, **81**, 406–433.

DEEGAN, C. E. & SCULL, B. J. 1977. (Compilers). *A standard lithostratigraphic nomenclature for the Central and Northern North Sea*. Report of the Institute of Geological Sciences, 77/25.

DEPARTMENT OF ENERGY, 1989. *Development of the Oil and Gas Resources of the United Kingdom*. HMSO, London.

DORE, A. S. & GAGE, M. S. 1987. Crustal alignments and sedimentary domains in the evolution of the North Sea, north-east Atlantic Margin and the Barents shelf. *In*: BROOKS, J. & GLENNIE, K. W. (eds). *Petroleum Geology of North West Europe*. Graham & Trotman, London, 1131–1149.

EYNON, G. 1981. Basin development and sedimentation in the Middle Jurassic of the Northern North Sea. *In:* ILLING, L. V. & HOBSON, G. D (eds). *Petroleum Geology of the Continental Shelf of North West Europe*. Heyden, London, 196–204.

HAQ, B. U., HARDENBOL, J. & VAIL, P. R. 1987. Chronology of Fluctuating Sea Levels Since the Triassic. *Science*, **235**, 1156–1167.

HARKER, S. D., GUSTAV, S. H. & RILEY, L. A. 1987. Triassic to Cenomanian stratigraphy of the Witch Ground Graben. *In*: BROOKS, J. & GLENNIE, K. W. (eds). *Petroleum Geology of North West Europe*. Graham & Trotman, London, 809–818.

HOWITT, F., ASTON, E. & JACQUE, M. 1975. The occurrence of Jurassic volcanics in the North Sea. *In*: WOODLAND, A. W. (ed.) *Petroleum and the Continental Shelf of North-West Europe*. Applied Science, Barking, 379–387.

LATIN, D. M., DIXON, J. E. & FITTON, J. G. 1990. Rift-Related Magmatism in the North Sea Basin. *In*: BLUNDELL, D. J. & GIBBS, A. (eds). *Tectonic Evolution of the North Sea Rifts*, Oxford University Press.

LEEDER, M. R. 1982. Upper Palaeozoic basins of the British Isles – Caledonide inheritance versus Hercynian plate margin processes. *Journal of the Geological Society, London*, **139**, 479–491.

—— 1987. Tectonic and palaeogeographic models for Lower Carboniferous Europe. *In*: MILLER, J. M., ADAMS, A. F. & WRIGHT, V. P. (eds). *European Dinantian Environment*. Wiley, Chichester, 1–20.

—— & BOLDY, S. A. R. 1990. The Carboniferous of the Outer Moray Firth Basin, Quadrants 14 and 15, Central North Sea.

MAHER, C. E. 1980. The Piper Oilfield. *In: Giant Oil and Gas Fields of the Decade: 1968–1978*. American Association of Petroleum Geologists Memoir, **30**, 131–172.

—— & HARKER, S. D. 1987. The Claymore Oilfield. *In*: BROOKS, J. & GLENNIE, K. W. (eds). *Petroleum Geology of North West Europe*. Graham & Trotman, London, 835–845.

MCQUILLIN, R., DONATO, J. A. & TULSTRUP, J. 1982. Development of basins in the Inner Moray Firth and North Sea by crustal extension and dextral displacement on the Great Glen Fault. *Earth and Planetary Science Letters*, **60**, 127–139.

RAWSON, P. F. & RILEY, L. A. 1982. Latest Jurassic–Early Cretaceous events and the "Late Cimmerian Unconformity" in North Sea Area. *Bulletin of the American Association of Petroleum Geologists*, **66**, 2628–2648.

RIDING, J. B. & THOMAS, J. E. 1988. Dinoflagellate cyst stratigraphy of the Kimmeridge Clay (Upper Jurassic) from the Dorset coast, Southern England. *Palynology*, **12**, 65–88.

RITCHIE, J. D., SWALLOW, J. L., MITCHELL, J. G. & MORTON, A. C. 1988. Jurassic ages from intrusives and extrusives within the Forties Igneous Province. *Scottish Journal of Geology*, **24**, 81–88.

ROGERS, D. A., MARSHALL, J. E. A. & ASTIN, T. R. 1989. Devonian and later movements on the Great Glen fault system, Scotland. *Journal of the Geological Society, London*, **146**, 369–373.

SALFELD, H. 1914. Die Gliederung des oberen Jura in Nordwest europa. *Neues. Jahrb. Miner. Geol. Palaeont. Beil. Bd.* **37**, 125–246.

TURNER, C. C., RICHARDS, P. C., SWALLOW, J. L. & GRIMSHAW, S. P. 1984. Upper Jurassic stratigraphy and sedimentary facies in the Central Outer Moray First Basin, North Sea. *Marine and Petroleum Geology*, **1**, 105–117.

VAIL, P. R. & TODD, R. G. 1981. Northern North Sea Unconformities, Chronostratigraphy and Sea-Level Changes from Seismic Stratigraphy. *In*: ILLING, L. V. & HOBSON, G. D. (eds). *Petroleum Geology of the Continental Shelf of North West Europe*. Heyden, London, 216–235.

WHITEMAN, A. J., REES, G., NAYLOR, D. & PEGRUM, R. M. 1975. North Sea troughs and plate tectonics. *In*: WHITEMAN, A. J., ROBERTS, D. & SELLEVOLE, M. A. (Eds.) *Petroleum Geology and Geology of the North Sea and NE Atlantic Continental Margin*, Bergen. Norg. geol. Unders, **316**, 137–162.

WOODHALL, D. & KNOX, R. W. O'B. 1979. Mesozoic volcanism in the northern North Sea and adjacent areas. *Bulletin Geological Survey G.B.* **70**, 34–56.

WOLLAM, R. & RIDING, J. B. 1983. Dinoflagellate cyst zonation of the English Jurassic. *Report of the Institute of Geological Sciences* 83/2.

ZIEGLER, B. 1962. Die Ammonite Gattung *Aulacostephanus* im Oberjura (Taxionomie, Stratigraphie, Biologie). *Palaeontographica*, A, **119**, 1–172.

ZIEGLER, P. A. 1981. Evolution of Sedimentary Basins in North-West Europe. *In*: ILLING, L. V. & HOBSON, G. D. (eds). *Petroleum Geology of the Continental Shelf of North West Europe*. Heyden, London, 3–39.

Fault traps in the Northern North Sea

A. M. SPENCER & V. B. LARSEN

Statoil, Forushagen, Postboks 300, 4001 Stavanger, Norway

Abstract: The 250 hydrocarbon finds in the northern North Sea have total resources of $8000 \times 10^6 Sm^3$ oil-equivalent and 70% are in fault block traps, all with Jurassic or older reservoirs. They can be classified into plays by reservoir age: pre-rift is pre-Jurassic and Lower–Middle Jurassic; syn-rift is Upper Jurassic. 40 finds have sufficient published information to allow analysis of the geological relationships, structural and stratigraphic, which have given rise to their faulted hydrocarbon traps. All of the pre-rift and some of the syn-rift finds are traps in footwall blocks. They form a series with respect to the amount of conformable versus unconformable cap rock. In many the up-dip seal is due to stratigraphic truncation of the reservoir below an unconformable cap rock: the hydrocarbon pool does not extend to the bounding fault. The amount of erosion on the fault blocks and the footwall uplift which occurred seem to be related to the magnitude of fault throw. Most of the other syn-rift finds are hanging wall traps with entirely conformable cap rocks. The Brae-trend finds are classic syn-rift traps, located in the hanging wall of a major fault, movement on which was responsible for the supply of the reservoir clastics. The traps in the Central Graben are the most varied: all are different, probably due to complications produced by the underlying Permian salt.

In the northern North Sea about 250 hydrocarbon finds have been made containing recoverable resources totalling of the order of $8000 \times 10^6 Sm^3$ of oil-equilvalent. Some 70% of these finds have been made in fault block traps all with Jurassic or older reservoir rocks. Many geological factors have contributed to creating this prolific set of finds: excellent source and reservoir rocks, the fault structures themselves giving traps and juxtaposing the source and reservoir rocks, maturation of the source and hydrocarbon migration into the reservoirs and traps.

Fault traps have been defined as 'traps for oil and gas in which the closure results from the presence of one or more faults' (Bates & Jackson 1980). Industry practice in the northern North Sea accords with this definition in classifying most of the traps we describe here as fault traps. In detail, however, the up-dip seal is frequently due to stratigraphic truncation and many of the traps have also been classified as subcrop traps below unconformities (North 1985, p 313).

Numerous topics are important in understanding the structure, history and genesis of fault blocks and are the subject of active research by many groups, e.g. Wernicke & Burchfiel (1982), Jackson & McKenzie (1983), Gibbs (1984), Mandle (1987) and Walsh & Watterson (1988). In the northern North Sea the tectonic evolution of the graben system has also been much studied, e.g. Ziegler (1982), Badley *et al.* (1988). And, in more detail, previous authors have reviewed the trapping styles of the northern North Sea hydrocarbon finds (Færseth *et al.* 1986), their structural styles in relation to graben formation (Harding 1984) and their structural interpretation (Harding & Tuminas 1989). Also, several topics are important to understanding the sealing of hydrocarbon pools, such as the controls on their spill points (Allan 1989) and the nature of the sealing beds (Downey 1984; Watts 1987). In this article we wish to examine the geological relationships, structural and stratigraphic, which have given rise to the faulted hydrocarbon traps in the northern North Sea. Our aim is to 'distill the essence of entrapment'.

Structural maps and profiles of about 40 of the finds have been published, principally in the proceedings of the main petroleum conferences (Woodland 1975; Illing & Hobson 1981; Spencer *et al.* 1986, 1987; Brooks & Glennie 1987). We have mainly used this published information but have redrawn maps and profiles to uniform scales so as to facilitate comparisons between the finds.

Hydrocarbon plays related to late Jurassic rifting

The key to understanding the Jurassic to Recent evolution of the northern North Sea is the late Jurassic to early Cretaceous rifting episode. The overall pattern of the rifts is shown by an isopach map of Upper Jurassic strata. It shows

From HARDMAN, R. F. P. & BROOKS, J. (eds), 1990, *Tectonic Events Responsible for Britain's Oil and Gas Reserves*, Geological Society Special Publication No 55, pp 281–298.

three converging rifts: the Viking, Moray Firth and Central Grabens. Also noteworthy is the asymmetry of the faulted troughs and the major erosion of the adjacent highs, which are the footwall blocks to the boundary faults (Fig. 1).

The hydrocarbon geology can be analysed in terms of plays grouped in relation to this rifting episode: pre-rift, pre-Jurassic and Lower–Middle Jurassic plays; syn-rift, Upper Jurassic and Lower Cretaceous plays; post-rift, Chalk and Paleogene plays (Pegrum & Spencer, 1990; Fjæran & Spencer 1990). The foundation for all the plays is the presence of organic-rich Upper Jurassic shales (source rocks) and the rift system itself. The rifting provided many of the structures (traps) and post-rift cooling caused the subsidence necessary for hydrocarbon generation. Fault traps are only important in the pre-Jurassic and Jurassic plays. The plays with younger reservoirs have traps formed by stratigraphic pinch-out, drape or halokinesis.

The pre-Jurassic play includes reservoirs of widely different ages ranging from Devonian sandstones to Permian sandstones and carbonates to Triassic sandstones (Fig. 2). The common factor in these finds is the stratigraphic relationships of their reservoirs: all lie in areas where the reservoir is unconformably overlain by Upper Jurassic or Cretaceous strata. They are located in the eroded highs which formed during the late Jurassic rifting episode. Also, they are close to areas with mature Upper Jurassic source rocks, indicating the importance of short migration routes for this play.

The Lower–Middle Jurassic play is of outstanding importance and contains many of the largest finds. The principal reservoirs are the widespread, blanket, Statfjord and Brent sandstones. Traps occur in the fault blocks created during the late Jurassic rifting, which are widely distributed throughout the northern North Sea (Fig. 1). The play is more restricted in geographical extent, largely to the Viking Graben region. Only there are the reservoirs present in traps which have short migration routes to mature source areas.

The Upper Jurassic play is the most varied of the three: almost every find is unique. The complicated distribution of some of the sandstone reservoir rocks is due to deposition contemporaneous with the rifting (e.g. the submarine fan sandstones of Fig. 3). Also, the structures of the traps are varied. Some are traps in hanging walls and others in footwalls. The most complicated traps are in the Central Graben where the fault structures have been much affected by movement of the underlying Zechstein salt layer.

Pre-rift hydrocarbon finds

The pre-Jurassic finds include Snorre, Buchan, Marnock, Auk, Argyll and finds on the Fladen Ground Spur (Fig. 2). Snorre is the largest field in area and reserves. It lies on the same fault trend as Gullfaks, but erosion has here cut down to mid-Triassic levels. The tilted fault block has been planed off, so that although over 1000 m of strata contribute to the reservoir, the oil column is less than 300 m high. The small Argyll field is located on the eroded high bounding the Central Graben and on the crest of the fault block Cretaceous strata rest unconformably on Devonian. The Auk field is similar in its location and stratigraphic relationships. The Buchan field is a horst structure with Cretaceous strata resting unconformably on the Devonian reservoir. At Marnock, Cretaceous strata rest unconformably on the Triassic reservoir.

The Lower–Middle Jurassic finds are a family of traps with similar reservoirs and structural appearances. The fault blocks range considerably in areal size (Statfjord to Thistle) and also in structural complexity (Ninian to Gullfaks) (Figs 4 & 5). With respect to the degree of erosion of the fault blocks, there is a continuous series from fields with none (Murchison, Thistle) to fields with major truncation (Gullfaks). The pre-Jurassic finds are simply a continuation of this series, where erosion has cut even deeper. The complex, undulating structure of Sleipner Vest is anomalous and is probably due to the presence of the underlying mobile Zechstein salt.

Syn-rift hydrocarbon finds

The Upper Jurassic finds are of several different types (Figs 6 & 7). The Piper field has an Oxfordian to Kimmeridgian age shallow marine sandstone reservoir which on the crest of the tilted fault block is unconformably overlain by Upper Cretaceous strata. The Magnus field is a generally similar tilted fault block but with a Kimmeridgian submarine fan sandstone reservoir. At Brage and Troll the reservoirs are shallow marine sandstones of Bathonian to Kimmeridgian age. All these finds are analogous to the Lower–Middle Jurassic finds, with traps in the footwalls of tilted fault blocks: their reservoirs pre-date the main rifting movements.

The Claymore field is more complex. The main reservoir is a Kimmeridgian to Volgian submarine fan sandstone deposited during the onset of rifting; minor hydrocarbons occur in the unconformably underlying Permian and

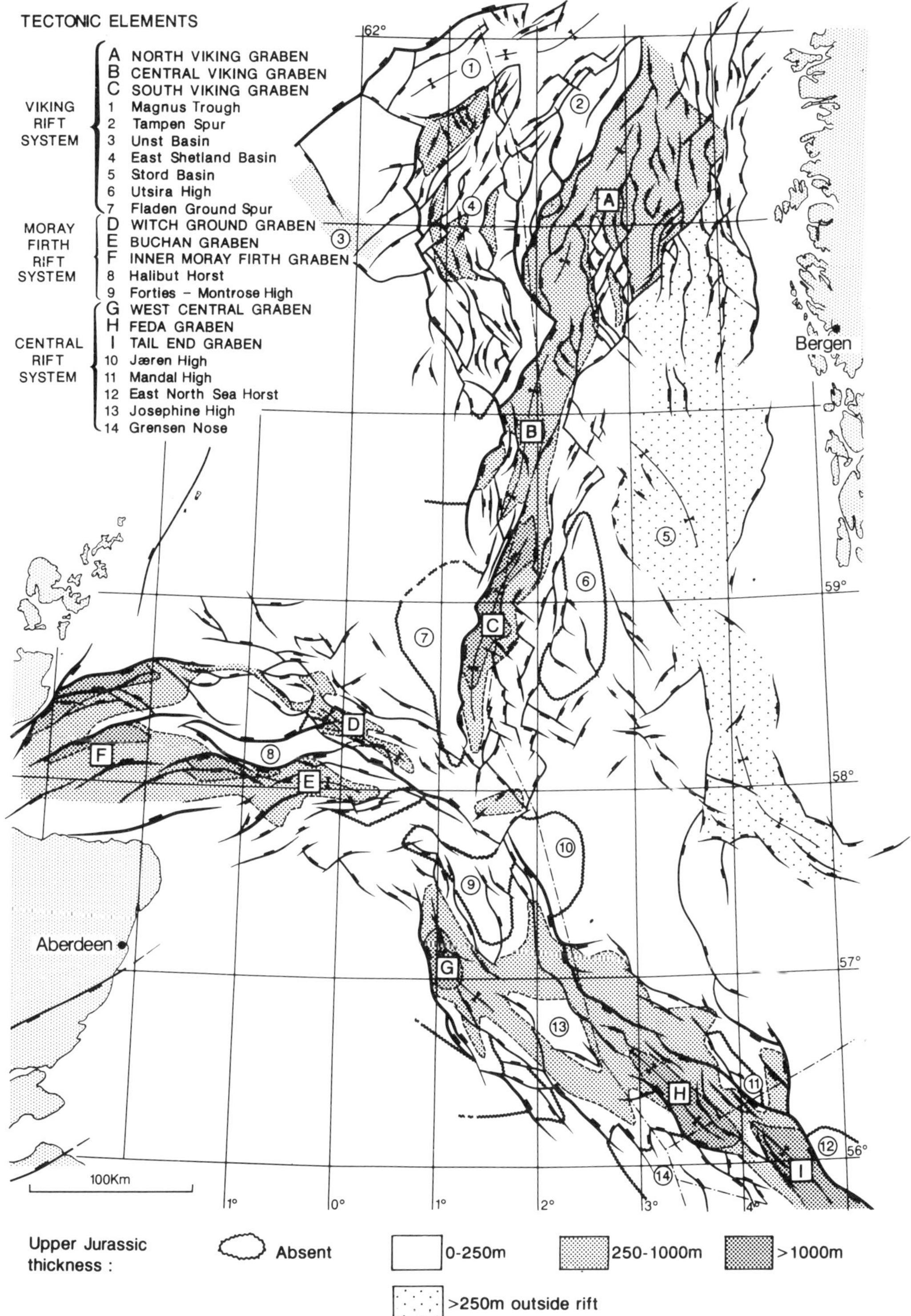

Fig. 1. Tectonic map of the late Jurassic to early Cretaceous rift system. Major subsidence occurred along the rifts, as shown by the Upper Jurassic isopach, which is based on drilled thicknesses in wells. Uplifted, eroded highs developed flanking the rifts (numbered 6 to 14). Modified from Pegrum & Spencer (1990).

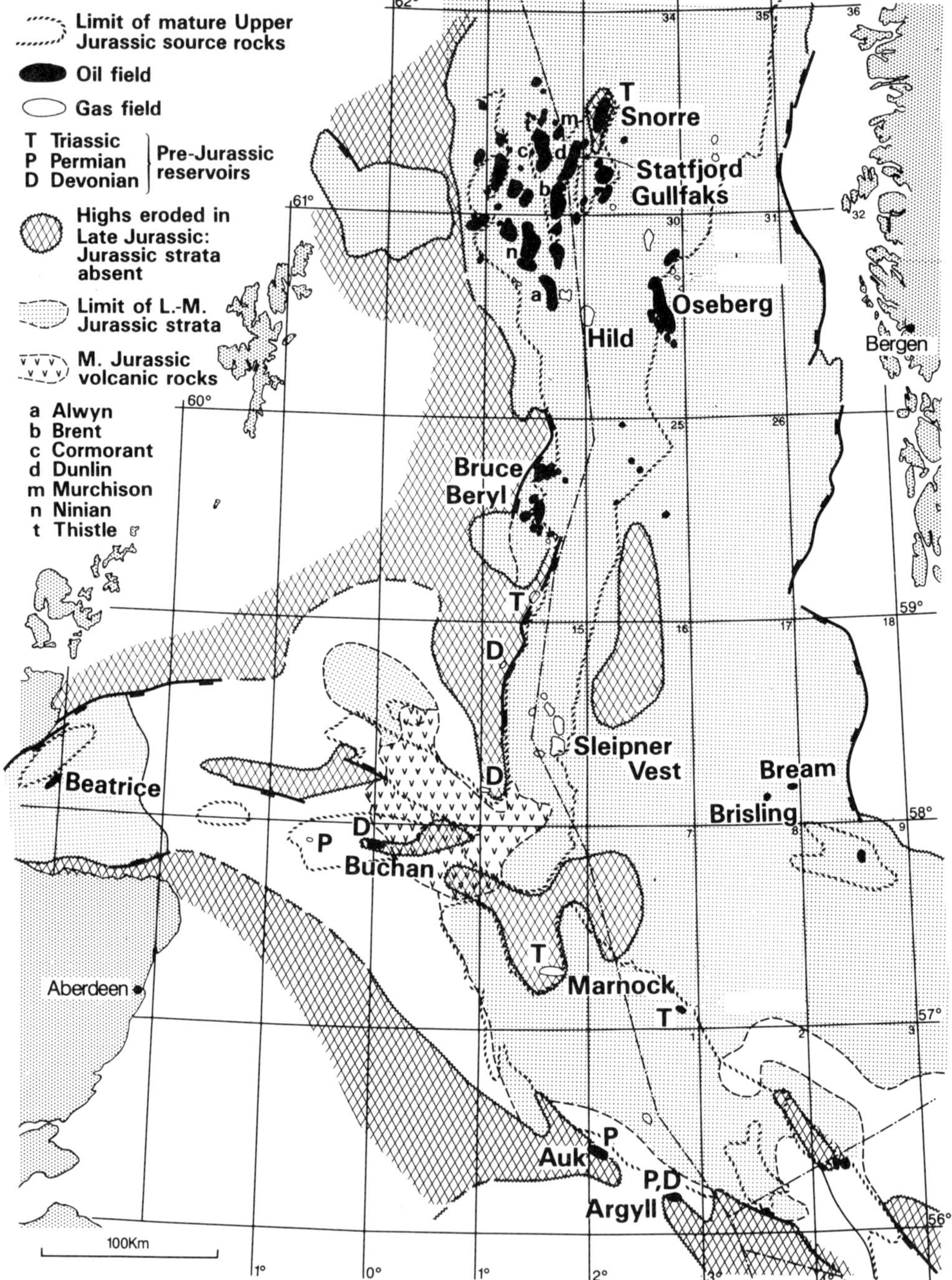

Fig. 2. Pre-rift play map. Fields with Triassic, Permian or Devonian reservoirs are identified by T, P and D. They are located on the highs eroded in late Jurassic times. The remaining fields all have Lower and/or Middle Jurassic reservoirs. Modified from Fjæran & Spencer (1990).

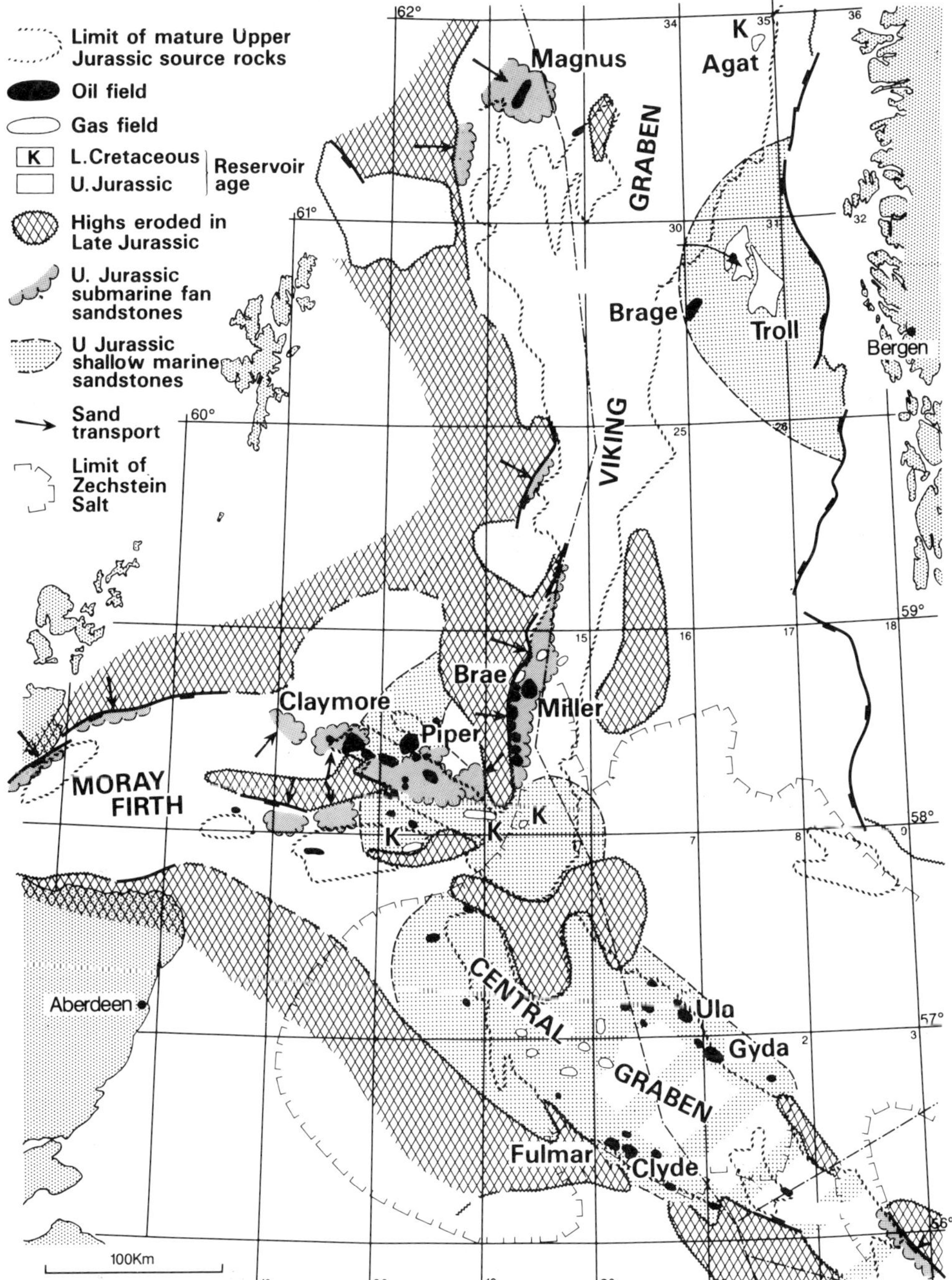

Fig. 3. Syn-rift play map. Fields with Lower Cretaceous reservoirs are present only in the Outer Moray Firth and at Agat. All other fields have Upper Jurassic reservoirs. Modified from Færan & Spencer (1990).

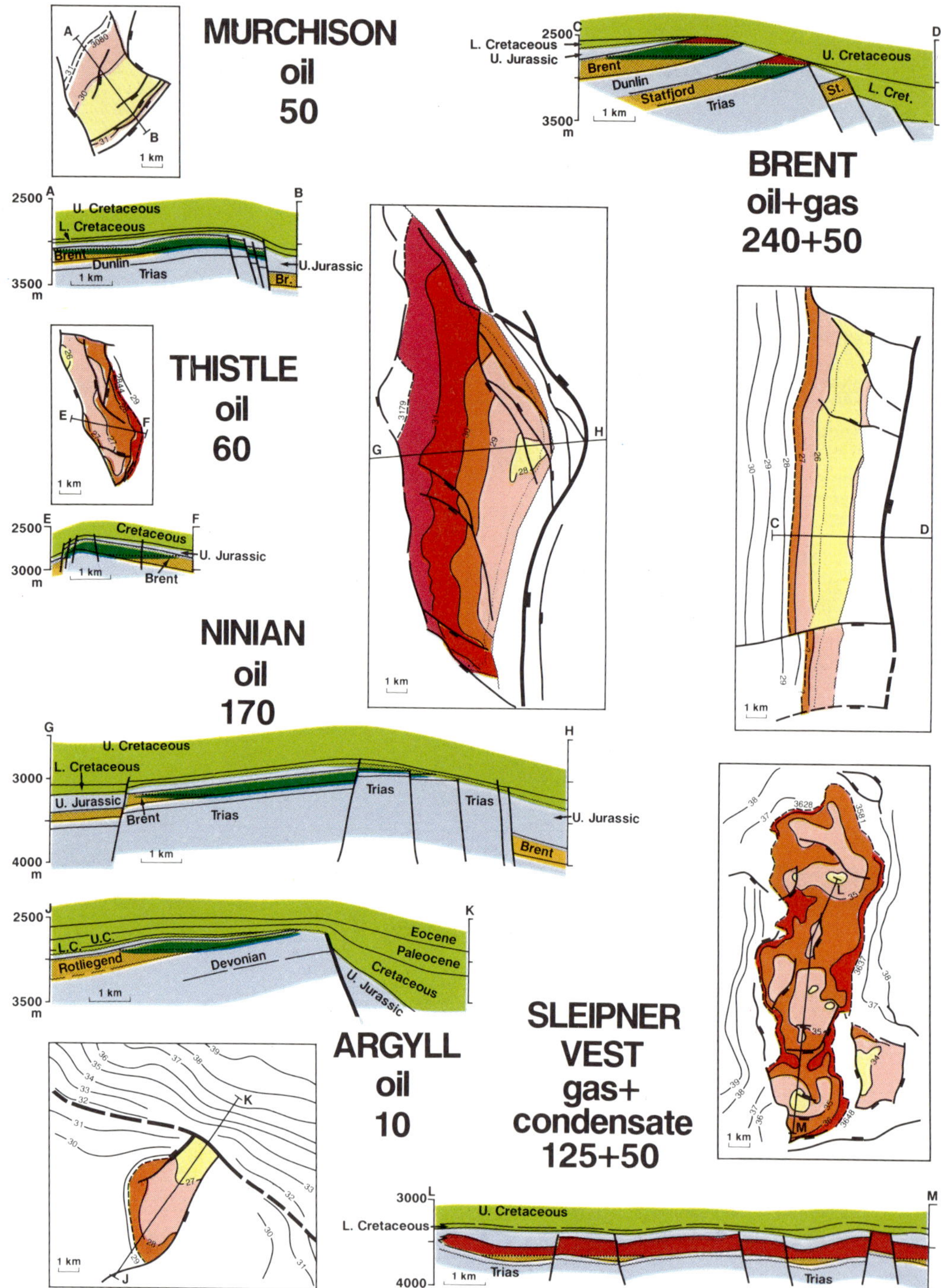

Fig. 4. Field examples from the pre-rift play. Sources: Murchison, Engelstad (1987); Thistle, Hallet (1981); Ninian, Albright *et al.* (1980); Argyll, after Pennington (1975); Brent, profile, Bowen (1975), map, Harding (1984); Sleipner Vest, Ranaweera (1987). Note that, for ease of comparison, Figs 4–7 and 9–12 have the same scales: all field maps are the same scale; all profiles have VE × 2 and have horizontal scales twice the vertical scale. Also, recoverable reserves of the fields are given in $\times 10^6 Sm^3$ oil or $\times 10^9 Sm^3$ gas. In Figs 4–7 the maps are contouredusing 100 m contour values and are layer coloured. The profiles are also uniformly coloured. Cretaceous and younger, light green; Jurassic and older, blue; reservoirs, yellow; oil, dark green; gas, red.

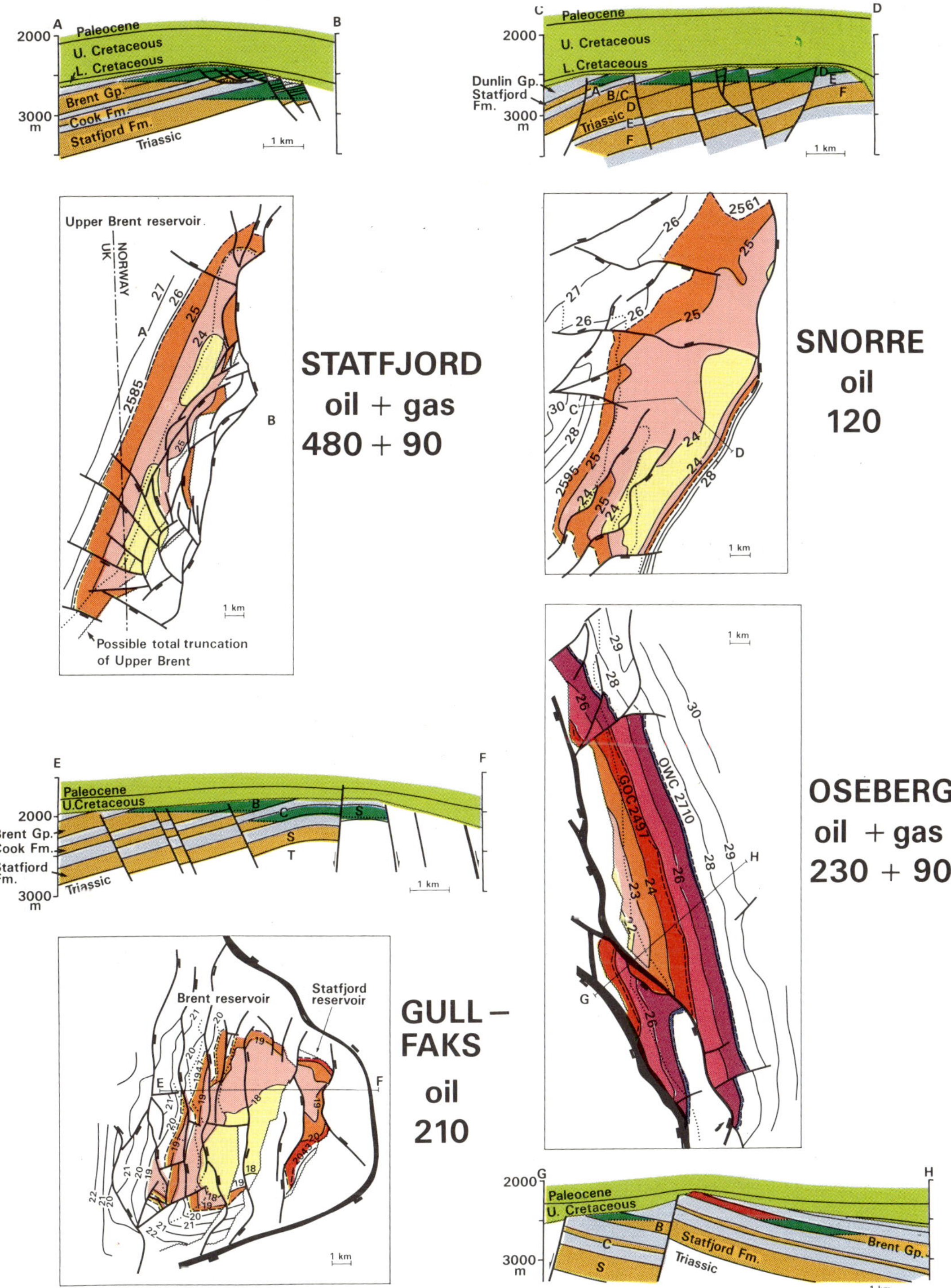

Fig. 5. Field examples from the pre-rift play. Sources: Statfjord, after Roberts *et al.* (1987); Snorre, Hollander (1987); Oseberg, profile after Nipen (1987). For further details, see caption to Fig. 4.

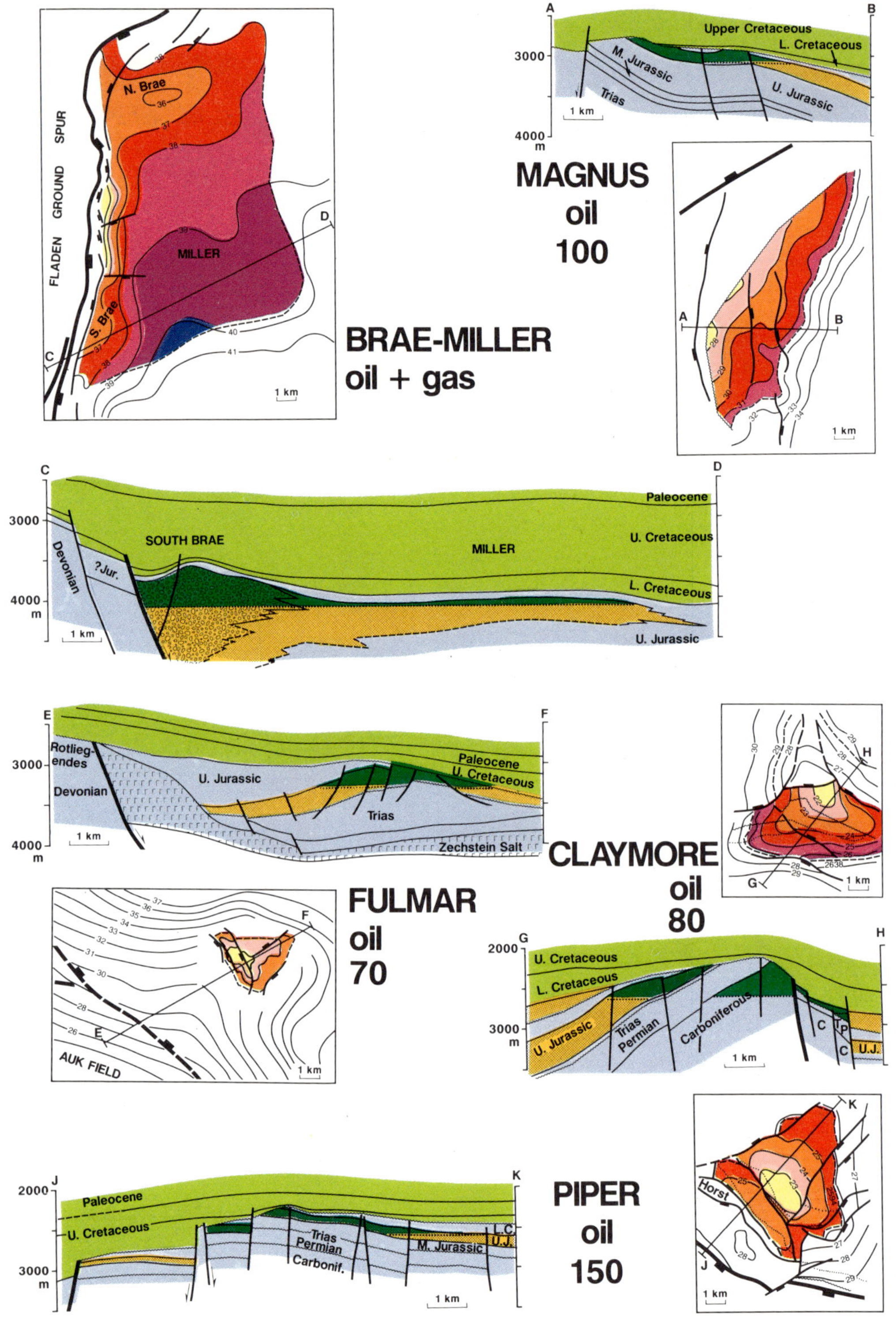

Fig. 6. Field examples from the syn-rift play. Sources: Brae-Miller, map, after Harms *et al.* (1981), profile, Turner *et al.* (1987); Fulmar, after Johnson *et al.* (1986); Magnus, De'Ath & Schuyleman (1981); Piper, Maher (1981); Claymore, Maher & Harker (1987). For further details see Fig. 4.

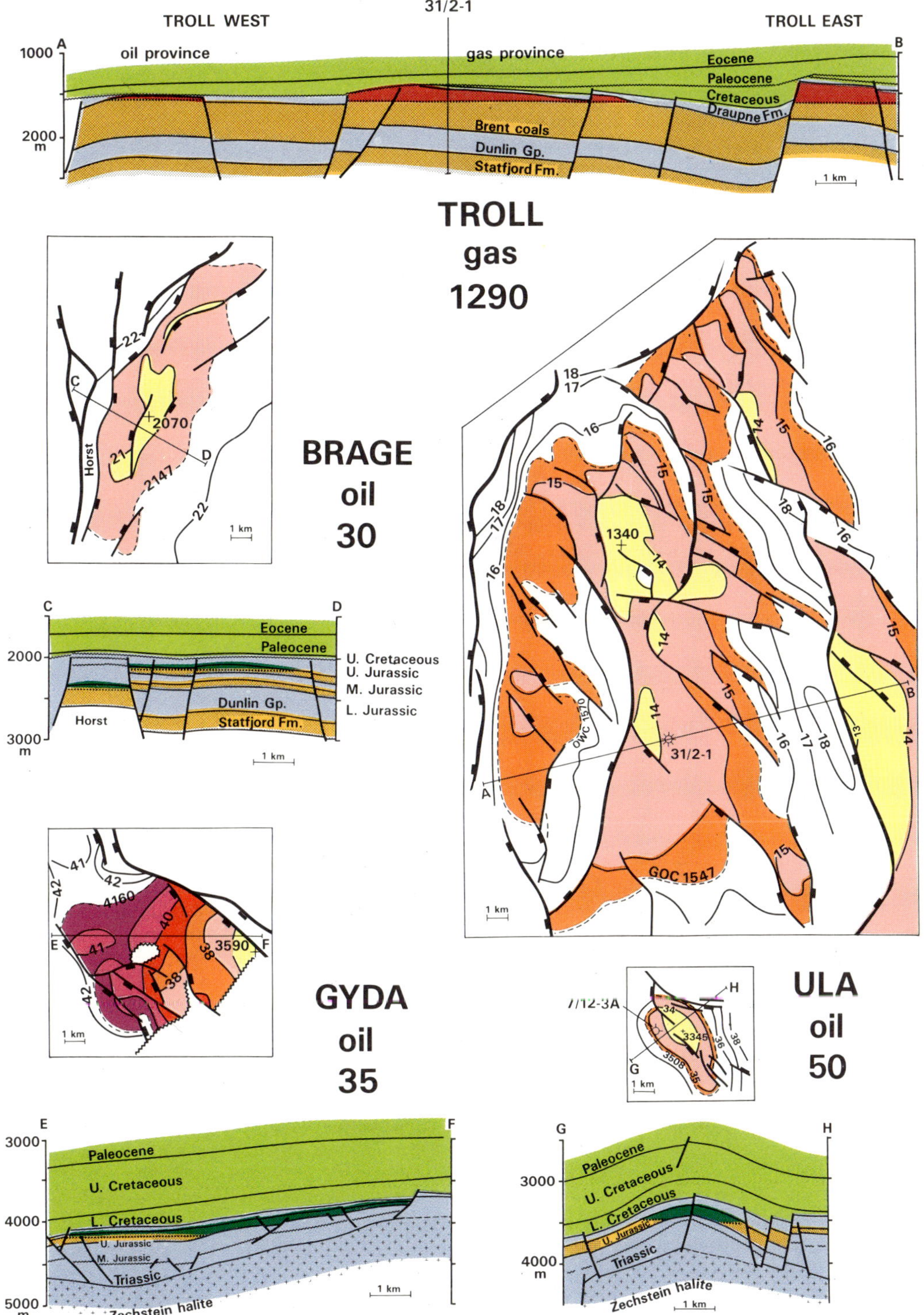

Fig. 7. Field examples from the syn-rift play. Sources: Troll, profile after Gray (1987), map, Birtles (1986); Brage, Hage *et al.* (1987); Gyda, Spencer *et al.* (1986); Ula, profile, Spencer *et al.* (1986), map, Home (1987). For further details see Fig. 4.

Carboniferous strata; and all these reservoirs lie in a tilted fault block which is unconformably covered by Cretaceous strata. Substantial hydrocarbons also occur in Lower Cretaceous submarine fan sandstones which unconformably drape the fault block.

The Brae trend includes ten finds along 100 km of the eastern, downfaulted side of the Fladen Ground Spur, the main bounding fault of the South Viking Graben. The reservoirs consist of conglomerates and sandstones of Volgian age. They formed as submarine fans containing material eroded from the Fladen Ground Spur, so that only 10–15 km east of the fault line they are replaced by basinal mudstones. These finds are classic syn-rift traps. They are trapped in the hanging wall of a major fault, movement on which was responsible for the supply of the reservoir clastics.

In the Central Graben the Fulmar Field has a mainly Oxfordian to Kimmeridgian shallow marine sandstone reservoir with a subordinate deepwater Volgian sandstone. The trap is domal in form, perhaps produced largely by salt withdrawal during late Jurassic times (Johnson *et al.* 1986). The Gyda and Fulmar Fields are analogous in that both are located close to the main bounding faults of the Central Graben. They are distinct from the Brae-type fields, for at Fulmar and Gyda the relationship between the graben bounding fault and the deposition of the reservoirs is not so clear. Also, the traps are not simple closures in the hanging wall of the graben bounding faults. These differences are the result of the presence in the Central Graben of the mobile underlying Zechstein salt layer.

Hydrocarbon entrapment

The hydrocarbon pools of the finds described here are sealed by six types of surfaces (Fig. 8, a–f). Three types of surfaces can define the top of the pool (a, b, c) and the other three underlie the hydrocarbon pool. Figures 9–12 show the finds analyzed according to these sealing elements.

The pre-rift finds described here (Figs 9 & 10) are all closures in footwall blocks. They vary in reservoir thickness and in dip angle. They form, however, one continuous series of traps with respect to the amount of conformable versus unconformable cap rock. Murchison and Thistle have cap rocks which are entirely conformable. Argyll has a small area of unconformable cap and Ninian, Oseberg, Brent, Statfjord, Gullfaks and Snorre have progressively increasing amounts of unconformable cap. Thus it is only at Murchison and Thistle that the up-dip sealing element is the fault (element c). For most of the other fields the up-dip seal is due to the stratigraphic truncation of the reservoir below the unconformable cap rock: the hydrocarbon pool does not extend as far as the bounding fault of the fault block. Cross faults are important sealing elements in many of the fields, for example at Murchison, Thistle, Ninian, Oseberg, Brent and Statfjord.

TRAP IN FOOTWALL BLOCK

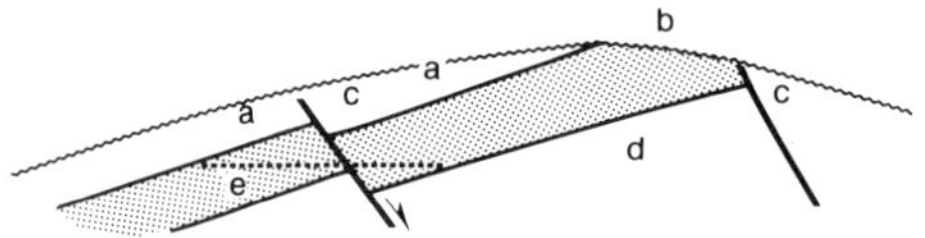

TRAP IN HANGING WALL

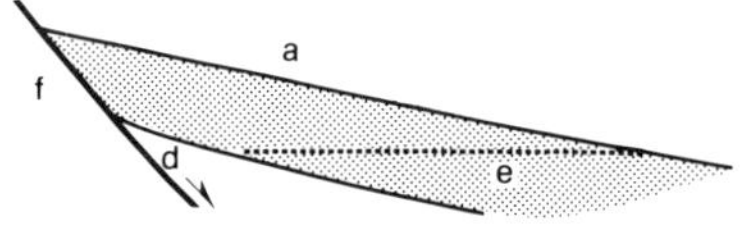

a Conformable cap rock
b Unconformable cap rock
c Cross-fault seal provided by hanging wall
d Conformable bottom seal
e Hydrocarbon-water contact
f Cross-fault seal provided by footwall

Fig. 8. Analysis of the seals of fault traps.

The syn-rift finds show greater variety in their sealing elements (Figs 11 & 12). One group are footwall closures and like the pre-rift finds show increasing amounts of unconformable cap: Piper, Claymore, Magnus and Troll (Fig. 13). Magnus is another good example of a hydrocarbon pool which does not reach the bounding fault because of up-dip stratigraphic truncation of the reservoir.

Of the other syn-rift finds Brae, Gyda and Brage belong to a quite different group — hanging wall traps in which the cap rock is entirely conformable. At Gyda and Brage conformable bottom seal (element d) is important, whilst at Brae the reservoir is so thick that the hydrocarbon-water contact reaches to the main fault zone. Fulmar is in a group on its own, having a partially unconformable cap rock, perhaps due to movements connected with the underlying mobile salt layer.

Figure 13 gives a comparison of the seal types in the finds. The footwall traps, with their increasing amounts of erosion, form a continuous series (Thistle/Murchison to Snorre). Troll lies

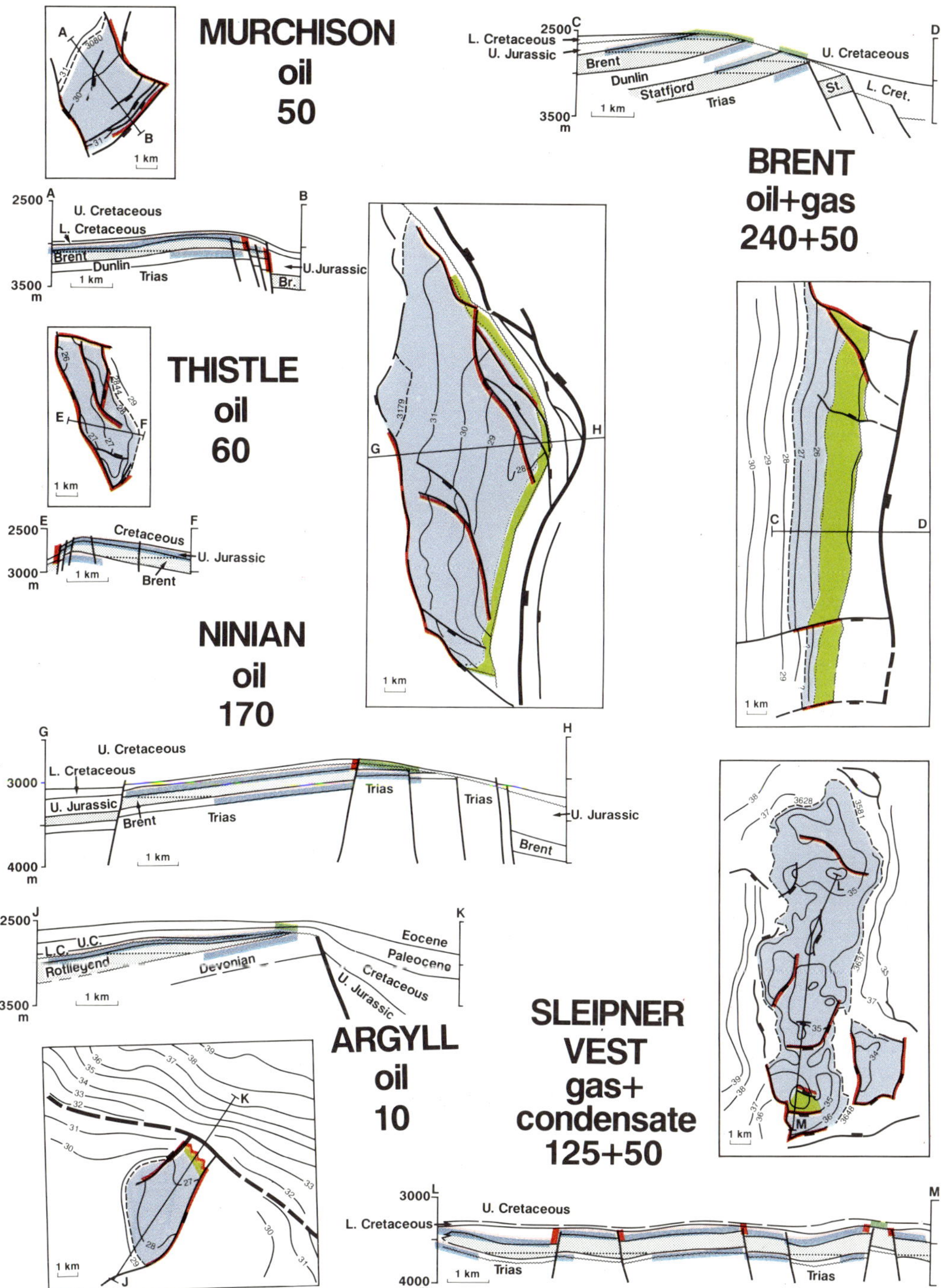

Fig. 9. Seal analysis of the pre-rift fields of Fig. 4. This analysis has been applied to Figs 9–12 using the colour scheme: a, d, light blue; b, light green; c, red; f, shaded red.

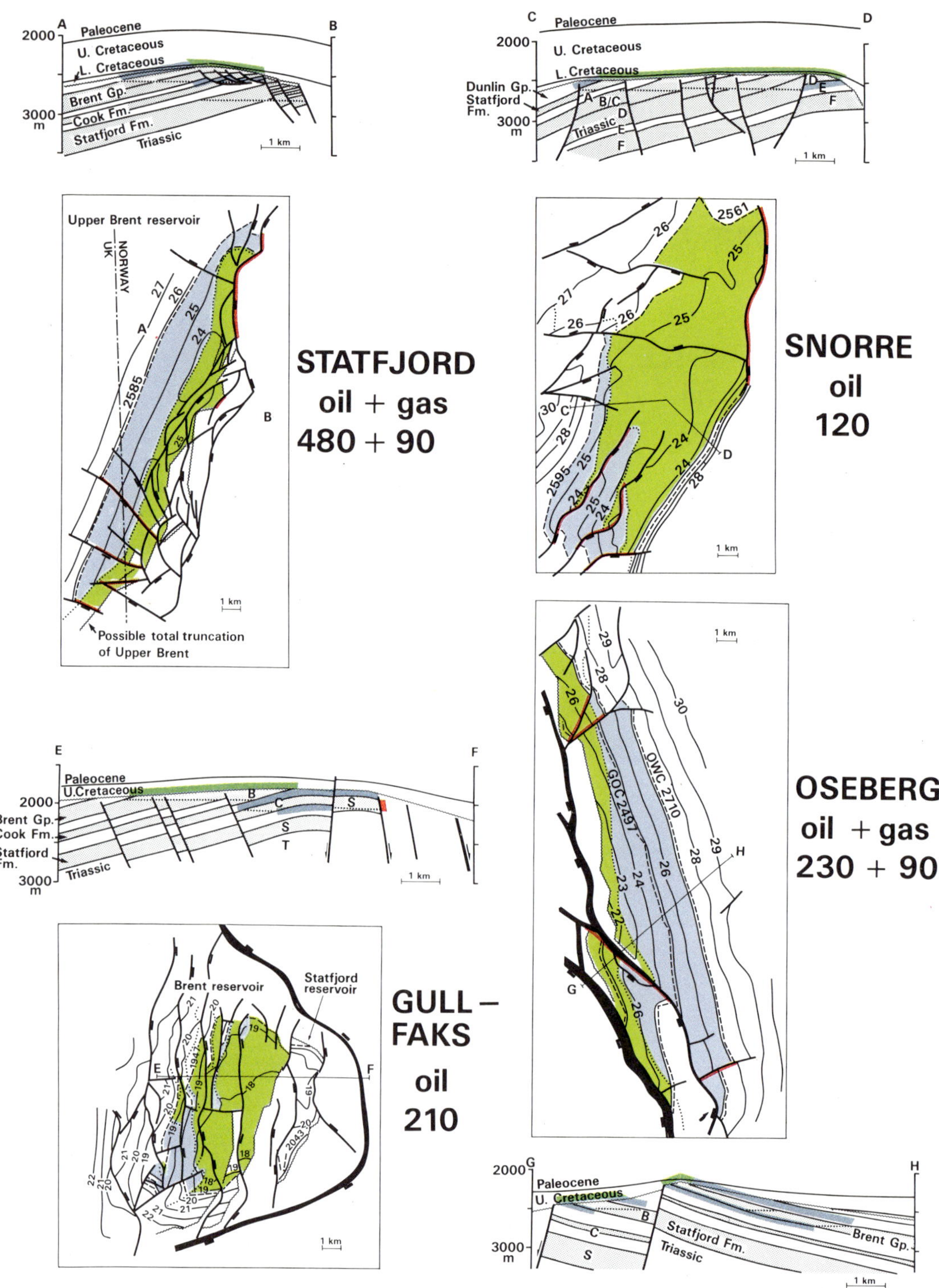

Fig. 10. Seal analysis of the pre-rift rields of Fig. 5.

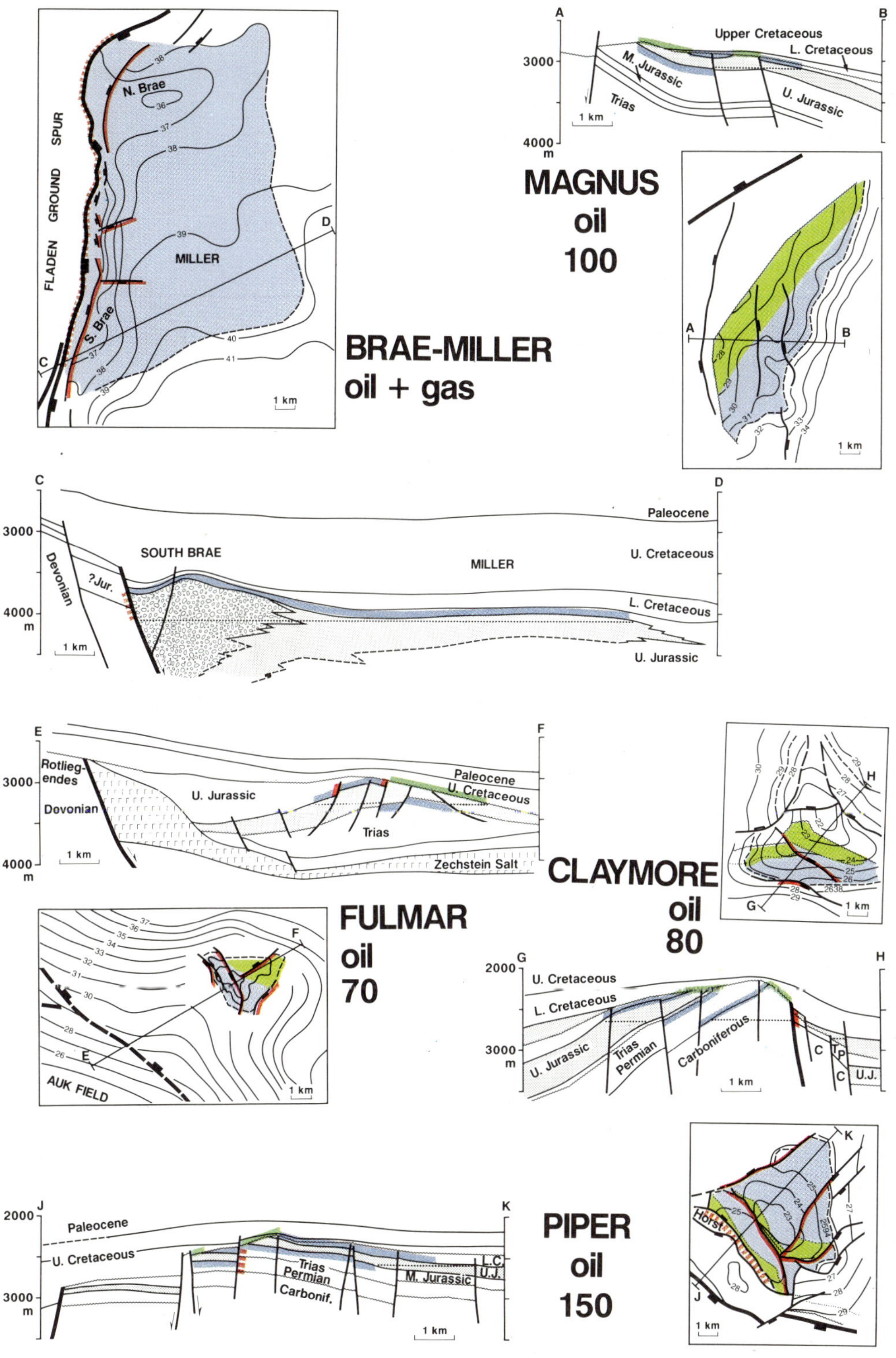

Fig. 11. Seal analysis of the syn-rift fields of Fig. 6. For further details, see caption to Fig. 9.

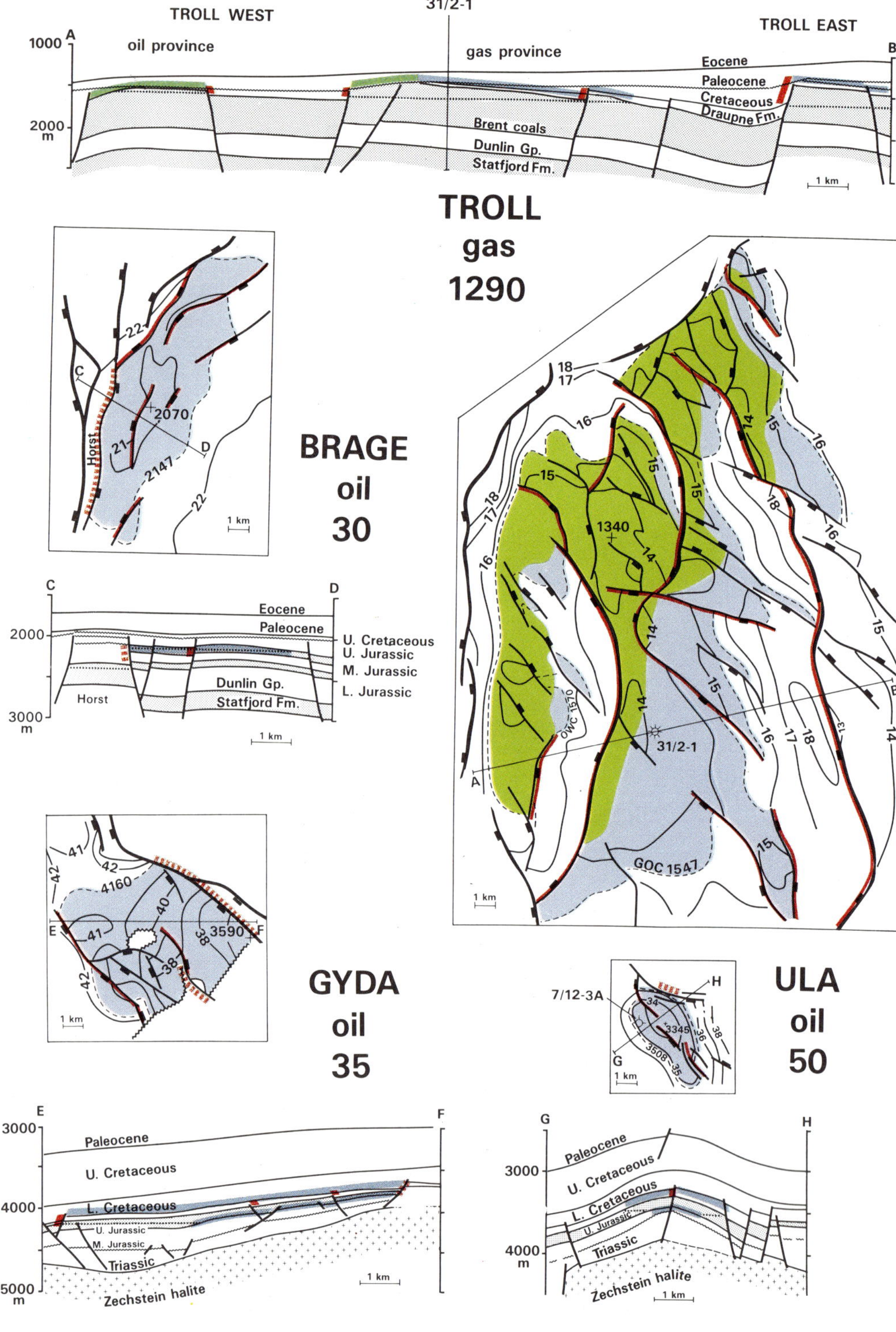

Fig. 12. Seal analysis of the syn-rift fields of Fig. 7. For further details, see caption to Fig. 9.

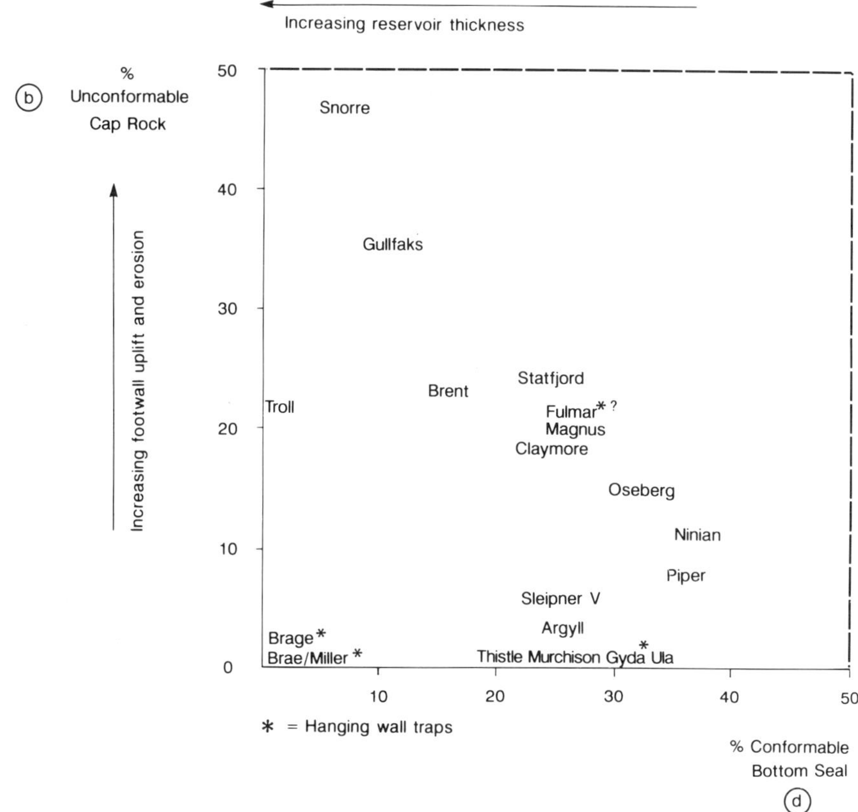

Fig. 13. Graph showing top seal versus bottom seal. The values were derived using the profiles of Figs 9–12, by measuring the horizontal component of each of the six sealing elements shown in Fig. 8. The values were expressed as percentages. Seal elements which overlie the hydrocarbon pool (a+b+c) thus total to 50%, as do those which underlie the pool (d+e+f). Because the horizontal component values related to faults amount to only a few per cent, the vertical axis principally shows increasing unconformable cap rock (decreasing conformable cap rock) and the horizontal axis shows increasing conformable bottom seal (decreasing amount of hydrocarbon/water contact).

anomalously off the trend of this series because of the thickness of the reservoir there. The syn-rift, hanging wall closures lie along the horizontal axis because of their conformable cap rocks; Fulmar is anomalous.

Footwall uplift

We have just seen that the footwall traps form a series with respect to the amount of unconformable cap rock. This is because the fault blocks they lie in form a series with respect to the magnitude of the erosion they have undergone (Table 1). Estimates of the depth erosion has cut down (i.e. the thickness of strata eroded) are uncertain. The stratal time gap measured in million years is better controlled, at the many available wells. It gives a measure of the erosive down-cutting plus the lack of deposition when the block stood relatively high. Table 1 is ordered according to the magnitude of throw on the main block boundary fault. The magnitude of erosion, whether assessed by depth of erosion or stratal time gap, seems to be related to the magnitude of the throw on the block bounding fault.

The erosion of these footwall blocks is due to the footwall uplift which occurred during the faulting. Table 1 shows, therefore, that the magnitude of footwall uplift is related to the overall throw of the fault.

Conclusions

We have analysed the fault traps for which adequate published information is available: a

Table 1. *Fault blocks and footwall uplift*

	Approximate magnitude of erosion			Approximate size of fault block	
	Width of eroded zone (km)	Depth erosion has cut (km)	Stratal time gap (Ma)	Width (km)	Overall fault throw (km)
Fladen Ground Spur boundary fault	40	?3	250	?40	5
Magnus	5	0.8+	150	15	4
Gullfaks	8	0.8	130	25	4
Snorre	6	1.0	120	12–30	3
Oseberg	2	0.3	100	15	1.5–3
Claymore	3	0.9	180	?10–15	2
Brent	3	0.8	120	20	2
Ninian	2	0.3	100	10–25	2
Statfjord	2	0.5	80	15	2
Piper	1	0.3	80	?15	2
W. Troll	6–8	0.2	90	8–10	1
Murchison	0	0	0	?12	0.4
Thistle	0	0	0	?7	0.4

Note: The Fladen Ground Spur boundary fault is included for comparison; it is the largest of the late Jurassic faults.

total of around 40 hydrocarbon finds. When the geological relationships which have given rise to these traps are studied these conclusions can be made.

(i) Traps in footwall blocks form a clear family. Many involve up-dip stratigraphic truncation of the reservoir below an unconformable cap rock. All of the pre-rift finds discussed are of this type, as are some of the syn-rift finds.

(ii) The family of footwall traps shows increasing amounts of unconformable cap rock, related to increasing erosion of the fault blocks. This correlates to the amount of footwall uplift, which seems to be related to the magnitude of the fault throw. These relationships have important practical effects: exploration wells located on the crests of the fault blocks may fail due to the absence of the reservoir bed.

(iii) Syn-rift fault traps are of several types. Some are footwall closures. The hanging wall Brae-trend finds are classic syn-rift trap types, the fault movement causing reservoir deposition and producing the trapping geometry; these traps have conformable cap rocks. The Central Graben traps are the most varied; all are different, probably due to complications produced by the underlying salt.

There is one group of fault traps which we have not investigated: pre-rift hanging wall closures. Little information has been published on them, except for an article by Hindle (1990).

The authors wish to thank Statoil for giving permission to publish this article. S. M. Aasheim and G. Yielding provided helpful comments. We thank E. Heitman, T. Oliversen & S. Todnem for preparing the drawings.

References

Abright, W. A., Turner, W. L. & Williamson, K. R. 1980. Ninian field, U.K. sector, North Sea. *In*: Halbouty, M. T. (ed.). *Giant Oil and gas fields of the decade 1968–1978.* Memoir of the American Association of Petroleum Geologists, **30**, 173–193.

Allan, U. S. 1989. Model for hydrocarbon migration and entrapment within faulted structures. *Bulletin of the American Association of Petroleum Geologists*, **73**, 803–811.

Badley, M. E., Price, J. D., Rambech Dahl, C. & Agdestien, T. 1988. The structural evolution of the northern Viking Graben and its bearing upon extensional models of basin formation. *Journal of the Geological Society, London*, **145**, 455–472.

Bates, R. L. & Jackson, J. A. 1980. *Glossary of Geology*. American Geological Institute, Virginia. 2nd edition.

Birtles, R. 1986. The seismic flatspot and the discovery and delineation of the Troll Field. *In*: Spencer, A. M. *et al.* (eds). *Habitat of Hydrocarbons on the Norwegian Continental Shelf.*

Graham & Trotman, London, 207–215.

Bowen, J. M. 1975. The Brent oil-field. *In:* Woodland, A. W. (ed.). *Petroleum and the Continental Shelf of North-west Europe*. Applied Science, Barking, 353–361.

Brooks, J. & Glennie, K. W. (eds). 1987. *Petroleum Geology of North West Europe*. Graham & Trotman, London.

De'ath, N. G. & Schuyleman, S. F. 1981. The geology of the Magnus oilfield. *In*: Illing, L. V. & Hobson, G. D. (eds). *Petroleum Geology of the Continental Shelf of North-west Europe*. Heyden, London, 342–351.

Downey, M. W. 1984. Evaluating seals for hydrocarbon accumulations. *Bulletin of the American Association of Petroleum Geologists*, **68**, 1752–1763.

Engelstad, N. 1987. Murchison. *In*: Spencer, A. M. *et al.* (eds). *Geology of the Norwegian Oil and Gas Fields*. Graham & Trotman, London, 295–305.

Fjæran, T. & Spencer, A. M. 1990. Proven hydrocarbon plays offshore Norway. *In*: *Proceedings of the First Conference of the European Association of Petroleum Geoscientists*, June 1989, Berlin.

Færseth, R. B., Oppebøen, K. A. & Sæbøe, A. 1986. Trapping styles and associated hydrocarbon potential in the Norwegian North Sea. *In*: Halbouty, M. T. (ed). Future petroleum provinces of the world. *Memoir of the American Association of Petroleum Geologists*, **40**, 585–597.

Gibbs, A. D. 1984. Structural evolution of extensional basin margins. *Journal of the Geological Society, London*, **141**, 609–620.

Gray, D. I. 1987. Troll. *In*: Spencer, A. M. *et al.* (eds). *Geology of the Norwegian Oil and gas fields*. Graham & Trotman, London, 389–401.

Hage, A., Bomstad, K. & Strand, J. E. 1987. Brage. *In*: Spencer, A. M. *et al.* (eds). *Geology of the Norwegian Oil and gas fields*. Graham & Trotman, London, 371–378.

Hallett, D. 1981. Refinement of the geological model of the Thistle Field. *In*: Illing, L. V. & Hobson, G. D. (eds). *Petroleum Geology of the Continental Shelf of North-west Europe*. Heyden, London, 315–325.

Harding, T. P. 1984. Graben hydrocarbon occurrences and structural style. *Bulletin of the American Association of Petroleum Geologists*, **68**, 333–362.

—— & Tuminas, A. C. 1989. Structural interpretation of hydrocarbon traps sealed by basement normal block faults at stable flank of foredeep basins and at rift basins. *Bulletin of the American Association of Petroleum Geologists*, **73**, 812–840.

Harms, J. C., Tackenberg, P., Pickles, E. & Pollack, R. E. 1981. The Brae oilfield area. *In*: Illing, L. V. & Hobson, G. D. (eds) *Petroleum Geology of the Continental shelf of North-west Europe*. Heyden, London, 352–357.

Hindle, A. D. 1990. Downthrown traps of the NW Witch Ground Graben. *Journal of Petroleum Geology*, **12**, 405–418.

Hollander, N. B. 1987. Snorre. *In*: Spencer, A. M. *et al.* (eds). *Geology of the Norwegian oil and gas fields*. Graham & Trotman, London, 307–318.

Home, P. C. 1987. Ula. *In*: Spencer, A. M. *et al.* (eds). *Geology of the Norwegian oil and gas fields*. Graham & Trotman, London, 143–151.

Illing, L. V. & Hobson, G. D. (eds). 1981. *Petroleum Geology of the Continental shelf of North-west Europe*. Heyden, London.

Jackson, J. & McKenzie, D. 1983. The geometrical evolution of normal fault systems. *Journal of Structural Geology*, **5**, 471–482.

Johnson, H. D., Mackay, T. A. & Stewart, D. J. 1986. The Fulmar oilfield (Central North Sea): geological aspects of its discovery, appraisal and development. *Marine and Petroleum Geology*, **3**, 99–125.

Maher, C. E. 1981. The Piper oilfield. *In*: Illing, L. V. & Hobson, G. D. (eds). *Petroleum Geology of the Continental Shelf of North-west Europe*. Heyden, London, 342–351.

—— & Harker, S. D. 1987. Claymore oil field. *In*: Brooks, J. & Glennie, K. (eds). *Petroleum Geology of North West Europe*. Graham & Trotman, London, 835–845.

Mandl, G. 1987. Tectonic deformation by rotating parallel faults: the "bookshelf" mechanism. *Tectonophysics*, **141**, 277–316.

Nipen, O. 1987. Oseberg. *In*: Spencer, A. M. *et al.* (eds). *Geology of the Norwegian Oil and Gas Fields*. Graham & Trotman, London, 379–387.

North, F. K. 1985. *Petroleum geology*. Allen & Unwin, London.

Pegrum, R. M. & Spencer, A. M. 1990. Hydrocarbon plays in the northern North Sea. *In*: Brooks, J. (ed). *Classic Petroleum Provinces*. Geological Society, London, Special Publication, **50**, 441–470.

Pennington, J. J. 1975. The geology of the Argyll field. *In*: Woodland, A. W. (ed). *Petroleum and the Continental Shelf of North-west Europe*. Applied Science, Barking, 285–291.

Ranaweera, H. K. A. 1987. Sleipner Vest. *In*: Spencer, A. M. *et al.* (eds). *Geology of the Norwegian Oil and Gas Fields*. Graham & Trotman, London, 253–264.

Roberts, J. D., Mathieson, A. S. & Hampson, J. M. 1987. Statfjord. *In*: Spencer, A. M. *et al.* (eds). *Geology of the Norwegian Oil and Gas Fields*. Graham & Trotman, London, 319–340.

Spencer, A. M., Home, P. C. & Wiik, V. 1986. Habitat of hydrocarbons in the Jurassic Ula trend, Central Graben, Norway. *In*: Spencer, A. M. *et al.* (eds). *Habitat of Hydrocarbons on the Norwegian continental shelf*. Graham & Trotman, London, 111–127.

Turner, C. C., Cohen, J. M., Connell, E. R. & Cooper, D. M. 1987. A depositional model for the South Brae oilfield. *In*: Brooks, J. & Glennie, K. W. (eds). *Petroleum Geology of North West Europe*. Graham & Trotman, London, 853–864.

Walsh, J. J. & Watterson, J. 1988. Dips of normal faults in British Coal Measures and other sedimentary sequences. *Journal of the Geological Society, London*, **145**, 859–873.

WATTS, N. L. 1987. Theoretical aspects of cap-rock and fault seals for single- and two-phase hydrocarbon columns. *Marine and Petroleum Geology*, **4**, 274–307.

WERNICKE, B. & BURCHFIEL, B. C. 1982. Modes of extensional tectonics. *Journal of Structural Geology*, **4**, 105–115.

WOODLAND, A. W. (ed.) 1975. *Petroleum and the Continental Shelf of North West Europe*. Applied Science, Barking.

ZIEGLER, P. A. 1982. *Geological Atlas of Western and Central Europe*. Shell Internationale Petroleum Maatschappij B.V., The Hague.

The structural controls on Upper Jurassic and Lower Cretaceous reservoir sandstones in the Witch Ground Graben, UK North Sea

D. O'DRISCOLL, A. D. HINDLE & D. C. LONG

Texaco Ltd, 1 Knightsbridge Green, London SW1X 7QJ, UK

Abstract: The Witch Ground Graben is a product of Mid- and Late Cimmerian extensional tectonism. The principal orientation of normal faulting was NNE–SSW during the Mid-Cimmerian (Bajocian to Early Kimmeridgian) and a combination of NW–SE and E–W during the Late Cimmerian (Late Kimmeridgian to Barremian). Block rotation on the dominant Late Cimmerian trends, combined with reactivation of older structural grains formed the fault block traps in the area.

Two major depositional systems form the reservoirs of the late Jurassic and early Cretaceous of the Witch Ground Graben, namely, the deltaic and shallow marine systems of the Sgiath and Piper Formations and the deeper water turbiditic sands of the Kimmeridge Clay and Valhall Formations. The transition between the systems is highly diachronous. The turbiditic sandstones are often enclosed within or onlap sealing strata, thus providing a stratigraphic element to many of the traps in the area.

A chronostratigraphic framework was established and used as the basis for understanding firstly, the timing of the tectonic episodes responsible for moulding the Witch Ground Graben and secondly, the importance of these events in determining reservoir sandstone distribution and trap development.

The Witch Ground Graben is a NW–SE trending Mesozoic rift located in the Outer Moray Firth Basin within the UK sector of the North Sea (Fig. 1). The graben is a prolific hydrocarbon province having produced some 1.25 billion barrels of oil from 6 fields up to the end of 1988, with numerous remaining exploration and development opportunities (Fig. 2). Texaco has interests in 12 whole or part blocks covering some 1870 km^2 in the area shown on Fig. 2, and has participated in 112 exploration and 120 development wells in Quadrants 14 and 15.

The stratigraphy of the Outer Moray Firth Basin records a number of episodes of basin formation with sequence boundaries being marked by major unconformities associated with regional tectonic events (Fig. 3). The NW–SE trend of the Witch Ground Graben is a product of intense rifting during Volgian to Barremian times, which overprints and obscures much of the earlier tectonism. The most important reservoirs of the Witch Ground Graben are the deltaic and shallow marine sequences of the Sgiath and Piper Formations. In addition an important suite of syn-tectonic clastic turbidite reservoirs were shed off the graben margins during rifting. These include the Claymore and Galley Sandstones of Volgian age and the Scapa Sandstone of early Cretaceous age. The Late Cimmerian tectonic phase (Ziegler 1988) generated most of the fault block traps present in the area.

A biostratigraphic framework for the Outer Moray Firth Basin has been developed by Deutsche Texaco AG using cored sections from 20 key wells in the Witch Ground Graben. Detailed studies of these cores were made during 1986–1988 using dinoflagellate cysts, foraminifera, radiolaria, ostracoda, pollen and spores (Dworatzek *et al.*, pers. comm.). The present study extrapolates the biostratigraphic control over Quadrants 14 and 15 and beyond, where necessary, to establish basin limits for the Upper Jurassic. Similar methods were used to study the Lower Cretaceous of four blocks in the centre of the area (14/18, 19, 20 and 15/16). All available wireline log, core sedimentology and petrography data have been integrated into the biostratigraphic framework. A total of 11 biostratigraphically defined time slices were studied: seven for the late Jurassic and four for the early Cretaceous. The sedimentological features of each interval are described and, where possible, related to regional or semi-regional tectonic events. Isopachs for selected time units were constructed to define the timing of structural phases within the graben's history.

Initially a discussion of the pre-late Jurassic regional history is necessary to identify earlier structural grains which influenced the depositional and structural development during the late Jurassic to early Cretaceous.

From Hardman, R. F. P. & Brooks, J. (eds), 1990, *Tectonic Events Responsible for Britain's Oil and Gas Reserves*, Geological Society Special Publication No 55, pp 299–323.

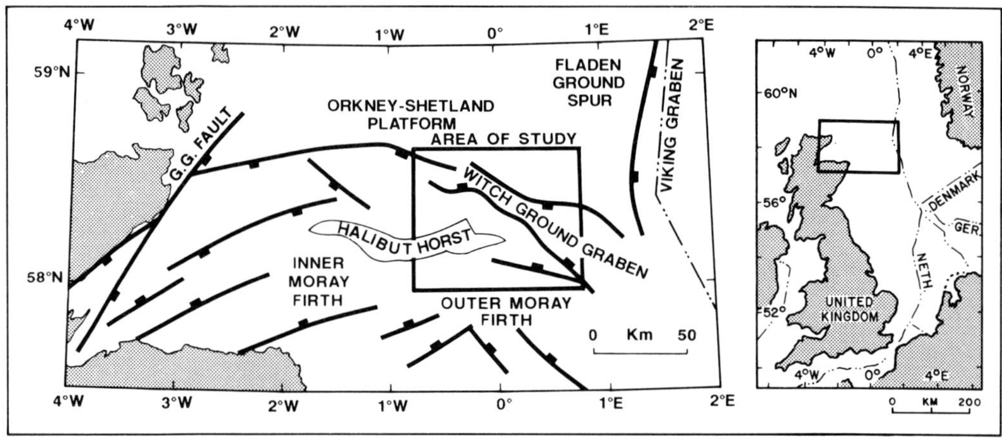

MORAY FIRTH LOCATION MAP

Fig. 1. Location Map of the Moray Firth illustrating area of study.

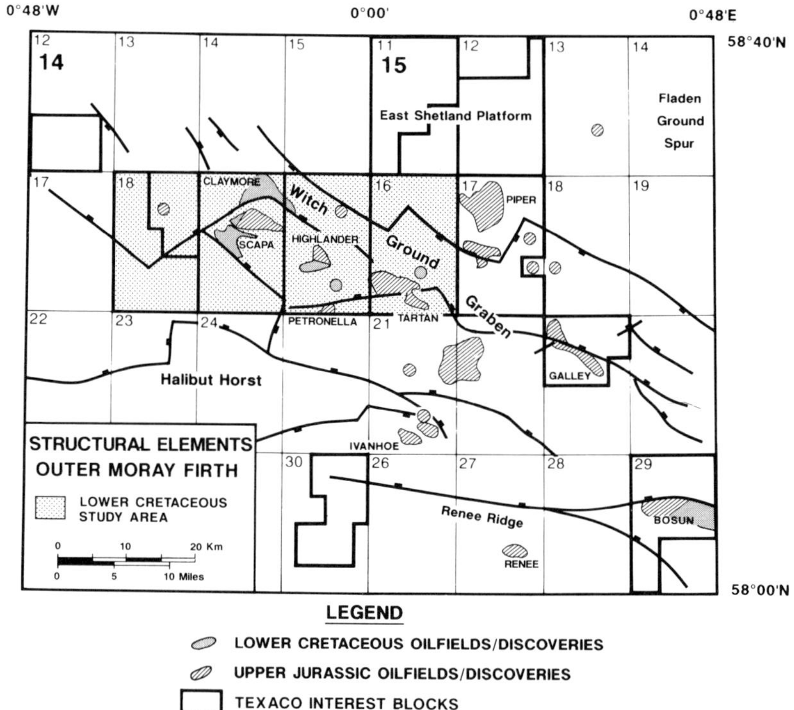

LEGEND

LOWER CRETACEOUS OILFIELDS/DISCOVERIES

UPPER JURASSIC OILFIELDS/DISCOVERIES

TEXACO INTEREST BLOCKS

Fig. 2. Structural elements of the Outer Moray Firth Basin.

Fig. 3. Stratigraphic column for the Outer Moray Firth Basin. BCU = Base Cretaceous Unconformity.

DEVONIAN TO EARLY CRETACEOUS STRATIGRAPHY OF THE WITCH GROUND GRABEN

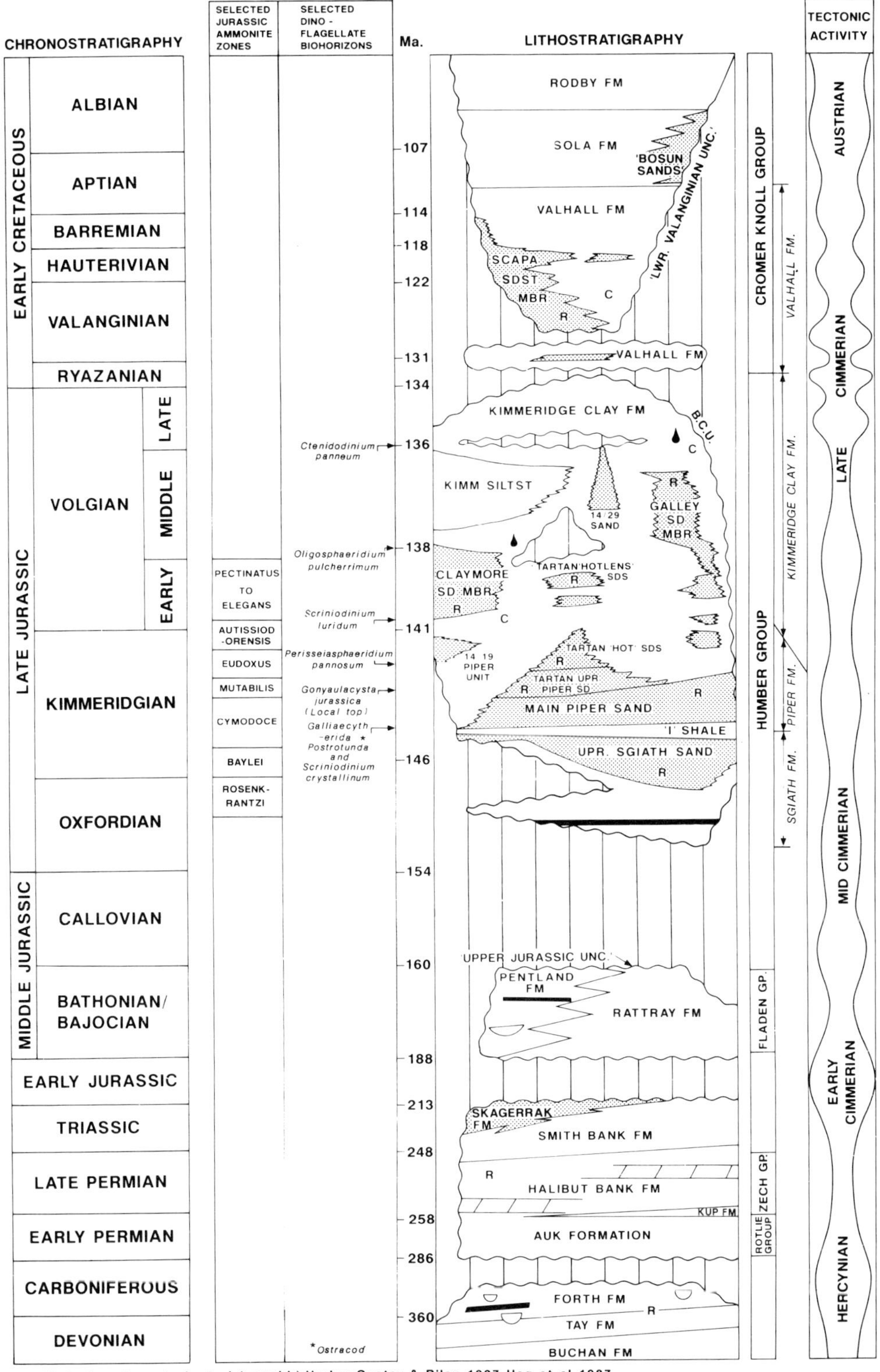

Modified after Dworatzek et al (unpubl.), Harker, Gustav & Riley 1987 Haq et al 1987.

Source Rock   R Reservoir Rock   C Cap Rock

Pre-late Jurassic tectonic history and structural elements

Pre-Hercynian tectonism

During the Devonian, a thick sequence of continental red beds was deposited over much of the Witch Ground Graben which, at that time, formed part of the Orcadian Basin. (Donovan *et al.* 1976). These sediments were derived from erosion of the surrounding mountain ranges such as the Scottish Caledonides to the west. A NE–SW structural grain developed from the oblique plate collision that caused the Caledonian orogen (Coward, this volume). The Great Glen Fault Zone has this orientation and has been active from the early Palaeozoic to the present day (McQuillan *et al.* 1982 and Rodgers *et al.* 1989) and forms the western boundary of the Inner Moray Firth Basin. Old Red Sandstone deposition came to an end in the earliest Carboniferous. With the development of a humid climate, coal-bearing fluvio-deltaic sequences were deposited in the Outer Moray Firth Basin over a developing E–W fault pattern. Sandstones of this age produce small quantities of oil in the Central Area of the Claymore Field (Maher & Harker 1987).

Hercynian and early/mid Cimmerian tectonism

A major regional unconformity associated with the Hercynian orogeny exists between Lower Carboniferous and the Lower Permian Rotliegend Group in the Moray Firth Basin. An E–W structural fabric is associated with late Hercynian compression and wrench faulting in the North Sea during the late Carboniferous and early Permian (Pegrum 1984; Lake & Karner 1987; Glennie, this volume). The subsequent Permo-Triassic basin fill in the Moray Firth Basin consists of arid red bed clastics in the west, grading to argillites, evaporites and carbonates in the east. Isopach and facies analysis of the Zechstein within the Outer Moray Firth Basin suggest evidence of a NW–SE tectonic control to deposition, which may be related to an extension of the well documented Tornquist Zone (Dore & Cage 1987; Cartwright 1987; Ziegler 1982, 1988). Zechstein dolomites produce small quantities of oil in the Central Area of the Claymore Field only where highly fractured (Maher & Harker 1987).

The Triassic to early Jurassic rifting in the N–S trending Viking Graben (Badley *et al.* 1988) and documented over much of the North Sea (Ziegler 1988) is referred to as Early Cimmerian tectonism. Lower Jurassic deltaic sequences are preserved in the Viking Graben and the Inner Moray Firth Basin, but similar aged rocks are absent across the Outer Moray Firth Basin and the Central Graben. No definitive fault trend linked with this early rifting phase exists in the Witch Ground Graben. The absence of this sequence may be a result of thermal updoming near the intersection of the Moray Firth Basin and the Central and Viking Grabens associated with anomously thin continental crust documented by Christie & Sclater (1980). Cooling of the dome was followed by collapse with subsequent rifting during the Middle Jurassic. This phase of rifting, termed Mid-Cimmerian tectonism, accentuated the Viking Graben and formed lesser sub-parallel structural lows in the Outer Moray Firth. A syn-rift suite of volcanics, the Rattray Formation, is found in the SE of the study area. It interdigitates with and is onlapped by paralic and deltaic sediments of the Pentland Formation to the north and east. These formations are grouped together as the Fladen Group and are dated as Bajocian to Bathonian (Deegan & Scull 1977).

Major tectonism and partial reactivation of previous tectonic episodes continued into the late Jurassic, resulting in complex structural inter-relationships.

Depositional systems of the Oxfordian to Barremian

There are two depositional systems in the Mid–late Jurassic and early Cretaceous of the Witch Ground Graben. These are the paralic, deltaic and shallow marine systems of the Sgiath and Piper Formations, deposited mainly under the influence of wave and tidal energy; and the Kimmeridge Clay and Valhall Formations which contain deeper water marine shales and marls, and sandstones deposited by turbidite flows. Previous studies (notably by Maher 1980, Turner *et al.* 1984, Andrews & Brown 1987 and Boote & Gustav 1987) have recognized this twofold division of the depositional systems of the area. The structural controls on the deposition of the later turbidite sands are well understood because of the absence of subsequent structural overprinting. In contrast, the distribution of the Sgiath and the Piper Formation sandstones is controlled by the Mid-Cimmerian tectonic episode. Structural controls at this level are less certain and different main provenance areas for the sands have been favoured by prior studies.

In the following sections, sedimentological discussion is limited to broad facies classifications. A series of seven time-interval isopach maps are presented for the mid–late Jurassic and two for early Cretaceous. The time intervals represented have been chosen both to illustrate the tectonic evolution of the Witch Ground Graben and to correspond to published and widely accepted biostratigraphic markers, shown in Fig. 3 (Haq *et al.* 1987, Harker *et al.* 1987).

Upper Jurassic: Sgiath Formation

Harker *et al.* (1987) defined the Sgiath Formation in the Witch Ground Graben as a sequence of 'medium to coarse grained quartzose sandstones, interbedded with massive coal beds and thin carbonaceous mudstones and siltstones of Oxfordian age'. The base of the formation rests unconformably on the Middle Jurassic Fladen Group in the SE Witch Ground Graben and in the NW, on older lithologies. In places the formation passes conformably upwards into the shallow marine sandstones and shales of the Piper Formation, the base of which is defined in Harker *et al.* (1987) as the 'I shale'. The type and reference wells used for the 'I shale' were 14/19–4 and 15/17–4. However, in this study, they are both recognized as containing condensed sequences, with thicker development of fully marine sand and shale sequences below the 'I shale' occurring in the south of block 15/17 and block 15/21. Thus, the Sgiath Formation contains in places a shallow marine sandstone, here informally termed the Upper Sgiath Unit sandstone, which has lithological affinities to the Piper Formation.

In the following section, the Sgiath Formation is used as defined by Harker *et al.* (1987), but it is informally sub-divided into a lower 'Paralic Unit' (reference well section 13,550–13,703 ft in 15/17–9) and an 'Upper Sgiath Unit' (reference well section 13,197–13,550 ft in 15/17–9 (Figs 10 & 11)).

Paralic Unit (Oxfordian). The unit consists of carbonaceous shales with occasional sands and thick coal seams deposited in a paralic to deltaic environment. The Paralic Unit is widespread in the Witch Ground Graben and an age-equivalent coal-bearing sandstone sequence occurs in the southern part of the Moray Firth Basin, south of the Halibut Horst. An unconformity exists between this unit and the underlying Middle Jurassic Fladen Group. The scale of the hiatus is unclear from biostratigraphic data but it appears that Callovian and early Oxfordian are absent over much of the basin.

The Upper Sgiath Unit (Oxfordian to Lowermost Kimmeridgian). A marine transgression in late Oxfordian times deposited a grey, organic-rich, pyritic shale which tends to coarsen upwards into lower shoreface and finally upper shoreface delta front sands. Figure 4 shows the isopach of the Upper Sgiath Unit sandstone, the oldest major reservoir unit in the area which demonstrates a well defined NNE–SSW channel axis and shale-out towards the south and west. Deposition is dominantly influenced by the basement control line shown on Fig. 4, where Upper Jurassic strata rest on Triassic rocks or older to the west of this line, whilst the Middle Jurassic section is preserved to the east. The basement high is linked to a reactivation of Caledonian fault trends which are associated with rifting along the Viking Graben at this time (Johnson & Dingwall 1981). The Sgiath thick axis contains up to 500 ft of coarse sandstone with erosively based fining-upward units interpreted to be tidally influenced distributary channel deposits together with medium to coarse grained, upper shoreface sandstones. The Upper Sgiath Unit sandstone, together with the succeeding Main Piper Sandstone were interpreted by Maher (1980) as having been sourced from the NE, on the basis of core palaeontology and sedimentology studies in the Piper Field.

Uplift and erosion of the Fladen Ground Spur in the late Oxfordian is suggested by the stratigraphy of the South Brae Field (Stow 1982, Turner *et al.* 1987). This would have provided a source area with a dominantly WSW-dipping slope. The Upper Sgiath Unit sandstone is quartz-arenitic and is believed to be derived from erosion on the Fladen Ground Spur which shed such material throughout the late Jurassic.

The Upper Sgiath Unit sandstone is interpreted here to be age equivalent to the Main Piper Sand Member of block 15/21. This forms the main reservoir unit of the Rob Roy and Ivanhoe fields, due on stream in 1989 (Boldy & Brealey, this volume). Distal argillaceous, coarsening-upwards sequences are formed to the south of the study area, in Quadrants 20 and 21.

Upper Jurassic: The Piper Formation

The Piper Formation consists of a sequence of clean, shallow marine sandstones with occasional interbedded marine shales, forming the main reservoir unit in the Piper, Tartan, High-

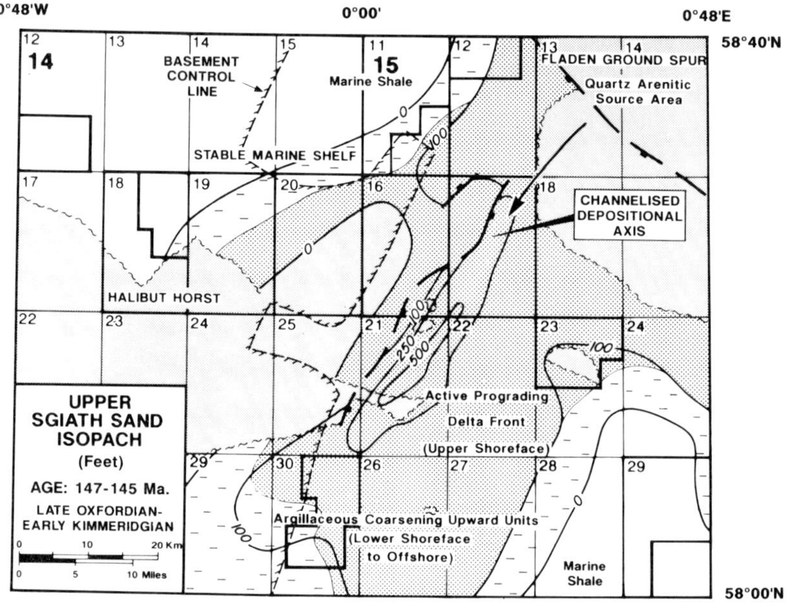

Fig. 4. Upper Sgiath Unit sand isopach, Sgiath Formation.

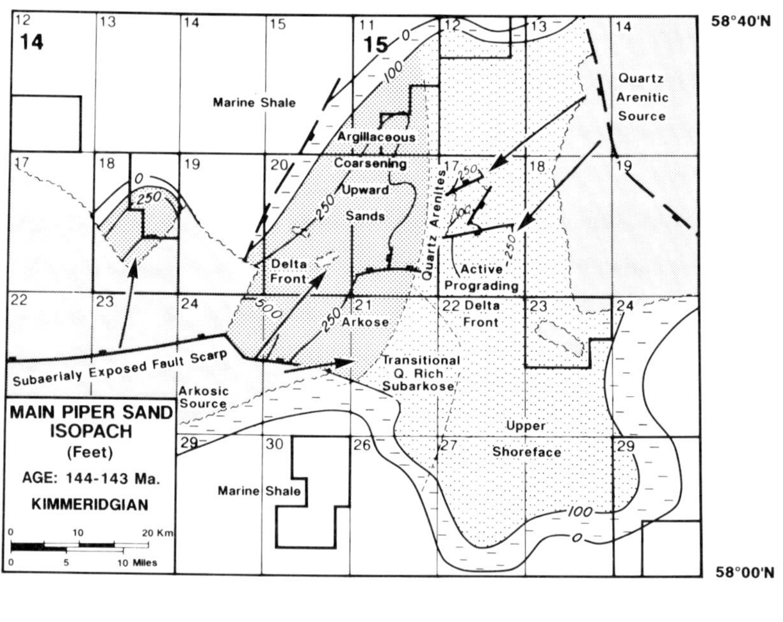

Fig. 5. Main Piper Sand isopach, Piper Formation.

lander and Petronella Fields. The Upper Sgiath Unit passes conformably upward into the Piper Formation. The base of the Piper Formation is marked by a marine transgression and the deposition of the 'I' shale (nomenclature of Maher 1980, of the Piper Field) dated as early Kimmeridgian. The 'I' shale is lithologically similar to the basal shale of the Upper Sgiath Unit and hence without firm biomarkers, confusion between the two can arise. The 'I' shale coincides with a marked downhole increase in the occurrence of the dinocyst *Gonyaulacysta jurassica* and the top occurrence of both the ostracod *Galliaecytheridea postrotunda* and the dinocyst *Scriniodinium crystallinum*. The 'I' shale thickens to the SW (Fig. 10), indicating a SW to NE marine transgression over the underlying Sgiath Unit.

Main Piper Sandstone (Lower–Middle Kimmeridgian). Deposition of the Main Piper Sandstone occurred under more complex tectonic controls than were evident during deposition of the Upper Sgiath Unit sandstone. Isopachs of the Lower to Middle Kimmeridgian combined with petrographic studies reveal that two depositional systems were active during this period (Fig. 5).

Following the marine incursion of the 'I' shale, a delta system once again prograded from NE to SW, supplying dominantly quartz-arenitic detritus across the Piper Field. Deposition in an upper shoreface environment subsequently formed the main reservoir units of the Piper Field. Early units shale out towards the south, being replaced by a thick "I" shale unit in Blocks 15/21 and 27. Renewed progradation during the Middle Kimmeridgian once again established upper shoreface conditions over these blocks producing the 'Supra Piper' sands of the Rob Roy and Ivanhoe fields, and the main reservoir of the Renee Field.

Figure 5 also shows a pronounced thickening in the opposite direction over blocks 14/20 and 15/16, to the west of the Sgiath depocentre and against the northern side of the Halibut Horst. Upper shoreface sandstones developed here, depositing the Main and Lower Sandstones of the Tartan and Petronella Fields. These 'Tartan-type' sandstones are petrographically distinct from the quartz arenites of the Piper Field, being subarkosic to lithic subarkoses (Fig. 6). The Tartan-type sandstones form coarsening-upward to massive units which pass laterally northward into more argillaceous, coarsening-upward units on block 15/11 (Figs 5, 10 & 11). The evidence indicates that the Halibut Horst to the SW was a separate source for the Tartan-type sandstones, resulting from a reactivation of the E–W faults bounding the northern margin of the horst. This relationship was recognized by Turner *et al.* (1984). To the south of the Halibut Horst, in Quadrants 20 and 21, time-equivalent rocks are the marine shales of the Renee Mudstone (Andrews & Brown 1987).

A subsidiary lobe over 300 ft thick is preserved to the west in block 14/18. Sparse palynological data suggest that it is age-equivalent to the Main Piper Sandstone. The top of this time inverval is defined by the top occurrence of the dinocyst *G. jurassica* in the Witch Ground Graben.

Depositional Model. The depositional processes for all the 'Piper-' and 'Tartan-type' sandstones are similar. Boote & Gustav (1987) proposed a wave-dominated delta environment for the Piper Formation Sandstones. The dominance of clean, high-energy sands and the rarity of lagoonal back barrier type sediments supports this interpretation. However, it is concluded in this study from isopach patterns that tidal processes were equally important. The occurrence of marine shale laterally along the coastline does not fit an entirely wave-dominated coastal model (Coleman & Prior 1980). The position of the study area between the emergent Fladen Ground Spur and the Halibut Horst (or a NNE–SSW trending basement high to the west), would have been ideal for concentrating tidal currents between marine basins to the NW and the SE resulting in the development of highly winnowed sands.

Upper Piper Sandstone (Upper Kimmeridgian). A continuation of the marine transgression during the late Kimmeridgian caused the Piper delta to retreat further to the NE. In the area to the east of the basement control line of Fig. 4, sand deposition is limited to offshore bars with local upper shoreface development on structurally high blocks. Over the area from the Tartan and Highlander Fields to block 15/11 sediment input kept pace with the rising sea level and a thick shale-free, but poorly sorted (bimodal) sand accumulated. The isopach trend, facies distribution and petrography are all similar to the preceding 'Tartan-type' sandstone in this area (Fig. 7).

This unit is an important reservoir in the Tartan, Highlander and Petronella Fields, where it is known as the '15/16–6' Sandstone Member. The base occurrence of the dinocyst *Perisseiasphaeridium pannosum* (previously *P. pannosum sp.A.*) defines the top of this time interval.

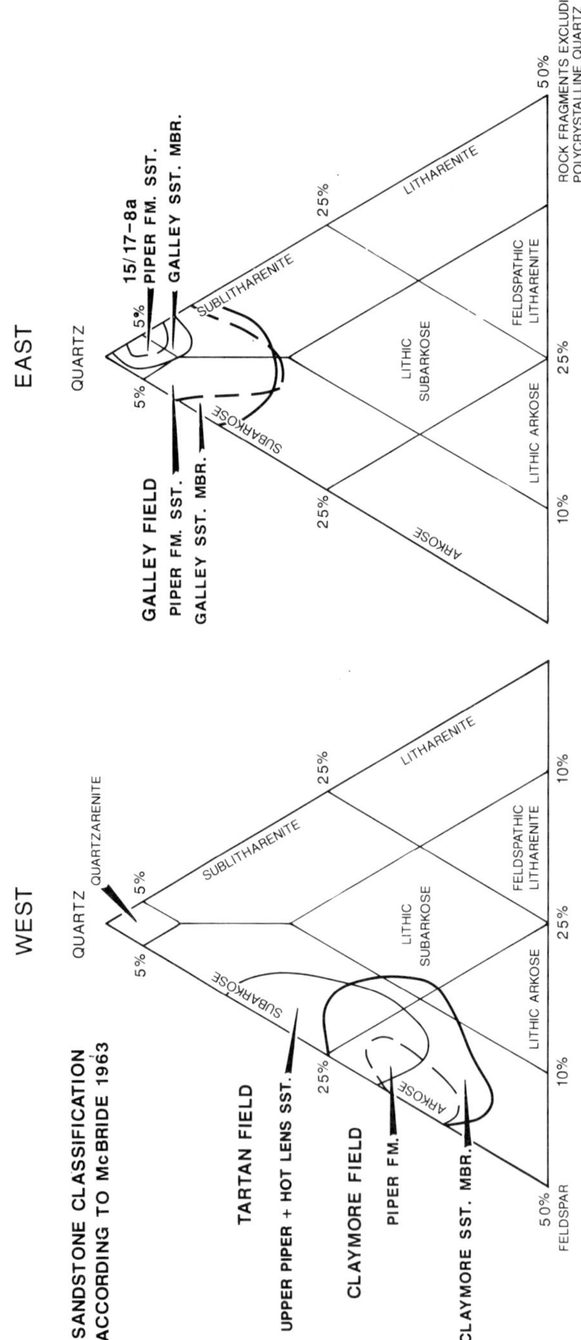

Fig. 6. Petrographic fields for Upper Jurassic sandstones in the Witch Ground Graben (after Dworatzek *et al.* 1988). See text for discussion.

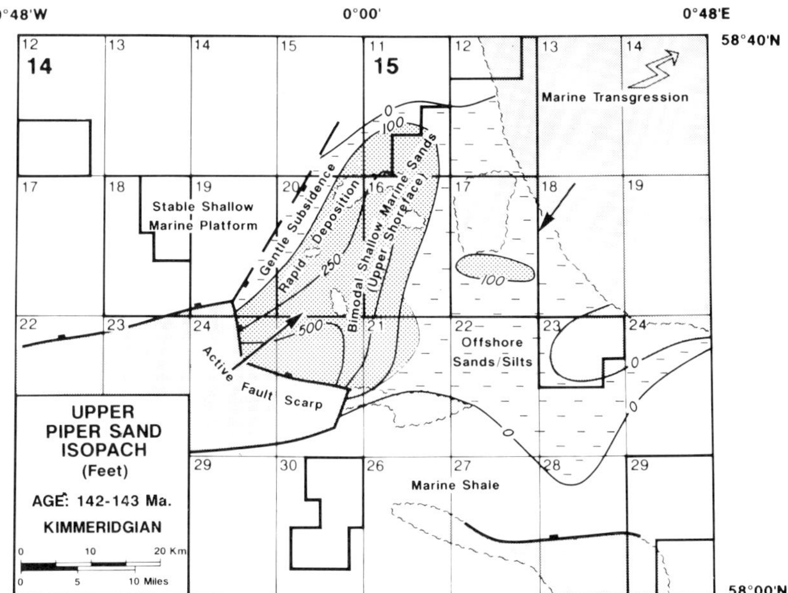

Fig. 7. Upper Piper Sand isopach, Piper Formation.

The 'Hot Sand' Unit (Uppermost Kimmeridgian–Lowermost Volgian). This time interval contains both shallow marine and turbiditic sandstones with distribution patterns being controlled by both mid- and late Cimmerian trends. As the marine transgression progressed during the late Kimmeridgian, upper shoreface facies persisted only in the area of well 14/24–2, close to the Halibut Horst source. In addition, a high gamma ray sandstone was desposited over the Tartan Field, called the 'Hot Sand' Member (Fig. 8). This sandstone is heavily bioturbated and is interpreted to have been deposited in an offshore shelf environment. It is therefore included within the Piper Formation.

The top of the 'Hot Sand' is of earliest Volgian age, correlating to the top occurrence of the dinocyst *S. luridum*. Almost everywhere else in the basin, Kimmeridge Clay facies were being deposited, except in the Claymore Field, where the Piper-equivalent unit consists of a coarsening-upward, siltstone–sandstone sequence (Fig. 11) of mainly lower shoreface facies.

With rising sea level, the northeast source area had now migrated to the flanks of the Fladen Ground Spur, from where, for the first time, channelized turbidities were supplied into parts of the NE Witch Ground Graben. The retreating Piper delta provided a source of unconsolidated sand which was transported to the graben throughout the early and middle Volgian. Such deposition may have been triggered by a resurgence of fault activity associated with Late Cimmerian tectonism along the Viking Graben. The E–W and NW–SE structural controls on the Tartan type sand deposition can now be seen over much of the Witch Ground Graben, although the interval isopach points to NNE–SSW control as being still dominant (Fig. 8).

Upper Jurassic to Lowermost Cretaceous: The Kimmeridge Clay Formation

Claymore Sandstone Member and equivalents (Lower Volgian). Early Volgian facies relationships are illustrated by an isopach of the Claymore Sandstone Member of the Kimmeridge Clay Formation (Fig. 12). The Claymore Sandstone contains a dinocyst assemblage of *P. pannosum* and *Oligosphaeridium pulcherrimum* sensu loannides *et al.* The top is diachronous, marked generally by a base occurrence of common *Muderongia* sp. A. Davey, but in the east ranging slightly older to the top occurrence of *O. pulcherrimum*.

Continued E–W and NW–SE faulting strongly influenced deposition during the early Volgian. In the west, wells in block 14/19 record the development of a major, subarkosic turbidite fan complex, thickening markedly towards

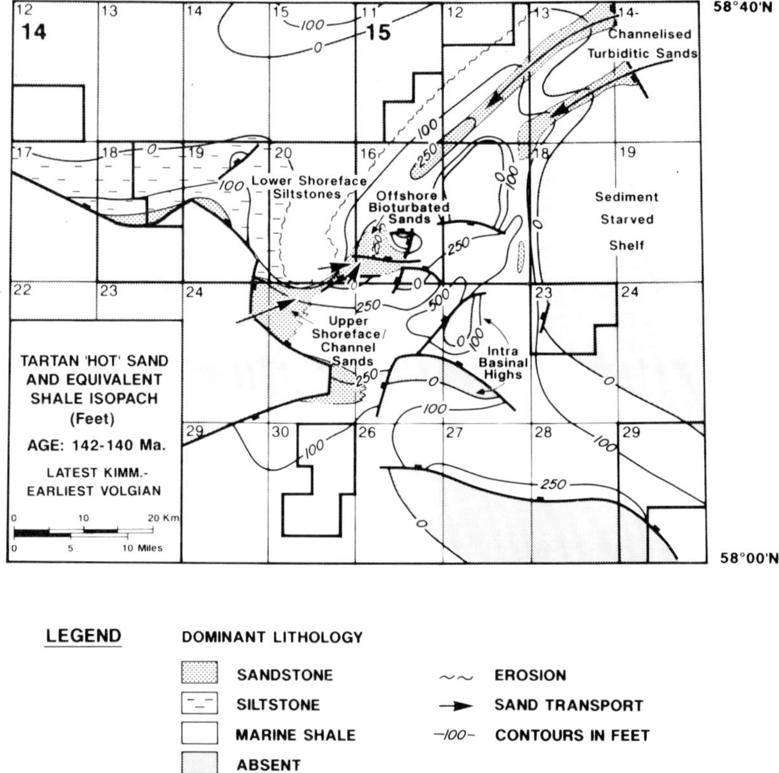

Fig. 8. Tartan 'Hot Sand' and equivalent shale isopach.

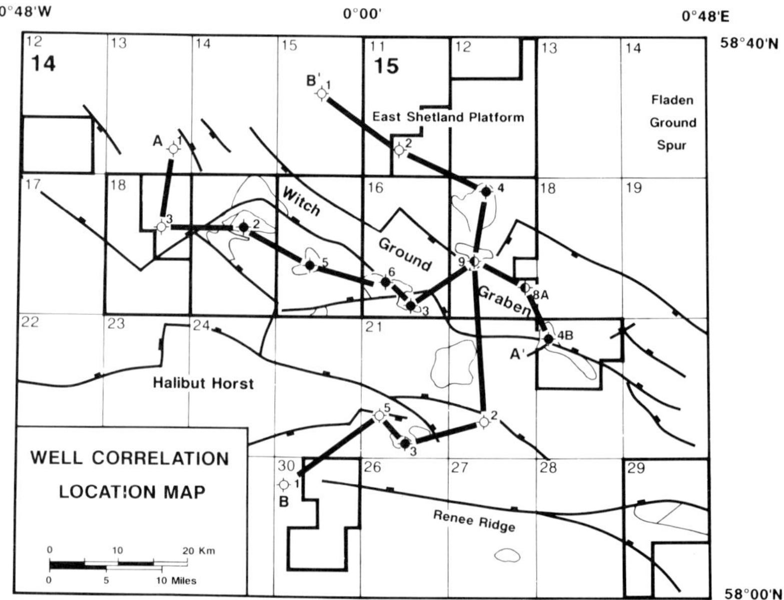

Fig. 9. Jurassic well correlation location map (see Figs 10, 11).

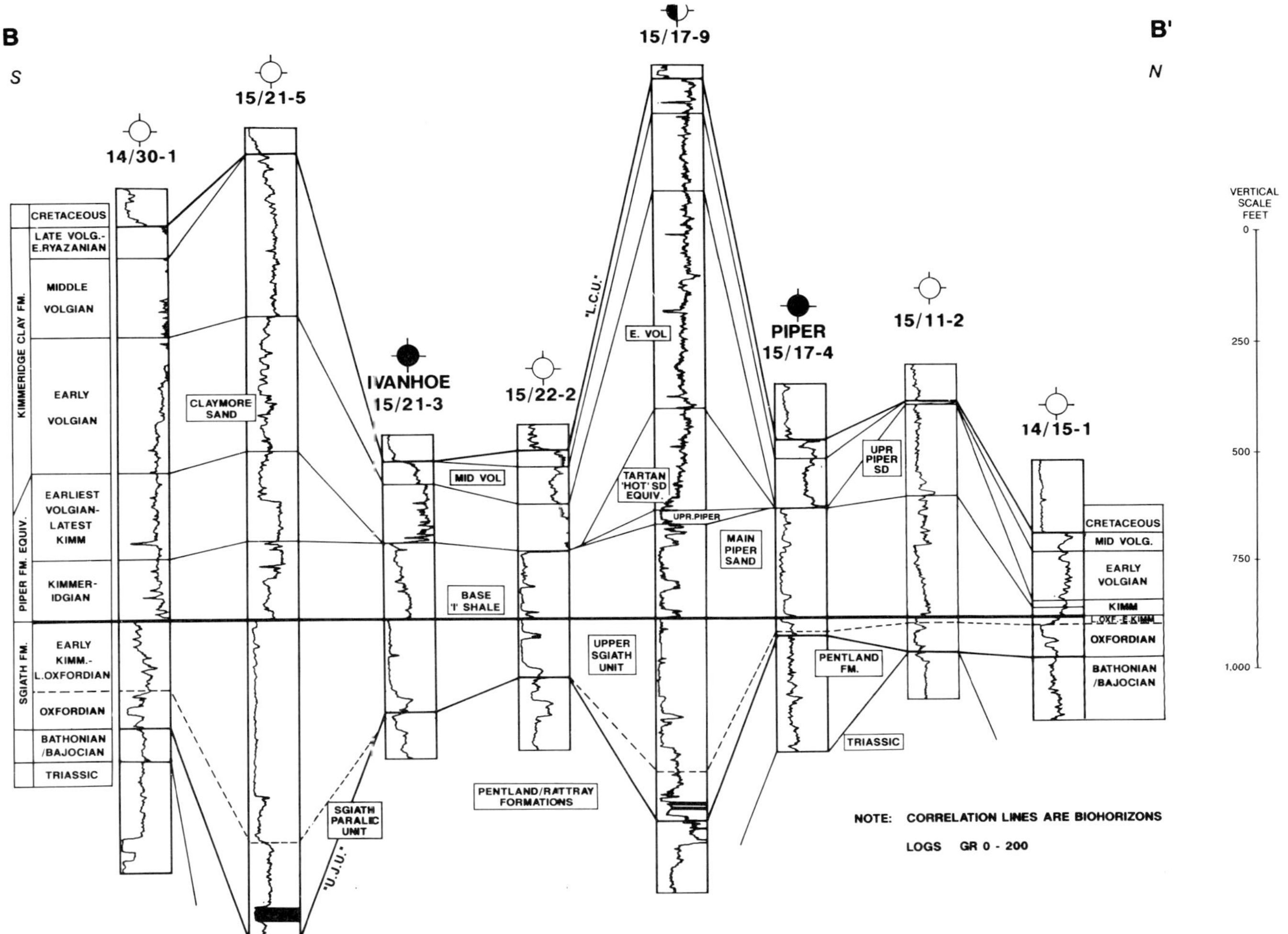

Fig. 10. North–south Jurassic Well correlation across the Witch Ground Graben (See Fig. 9 for location).

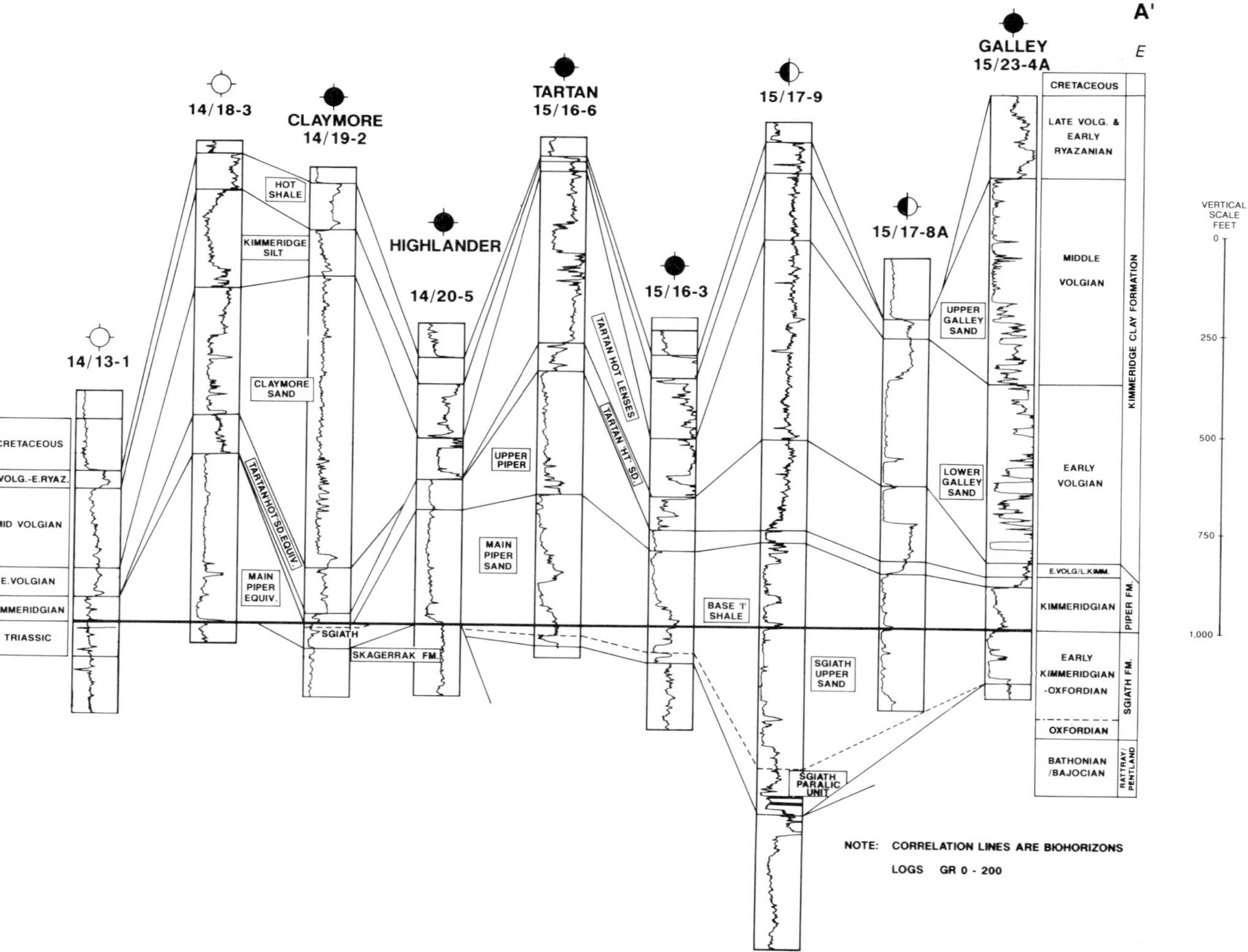

Fig. 11. East–west Jurassic Well correlation along the Witch Ground Graben (See Fig. 9 for location).

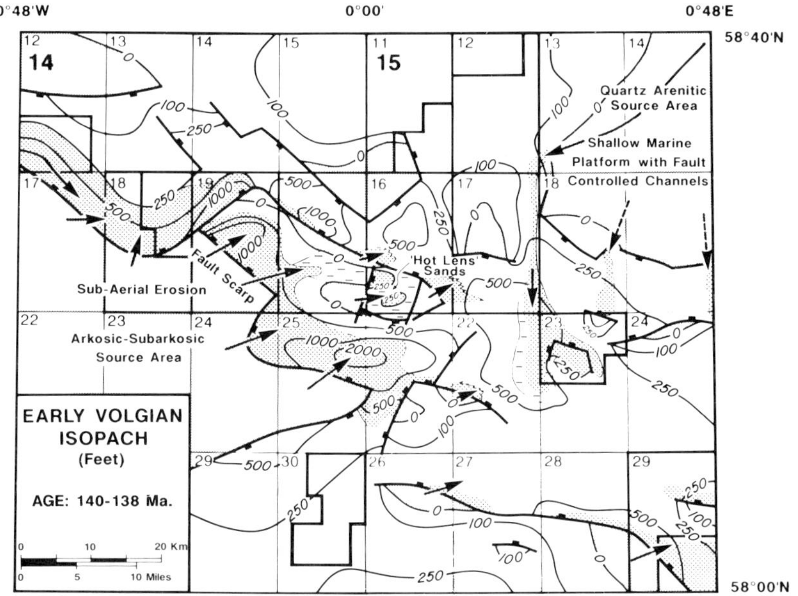

Fig. 12. Early Volgian isopach, Kimmeridge Clay Formation.

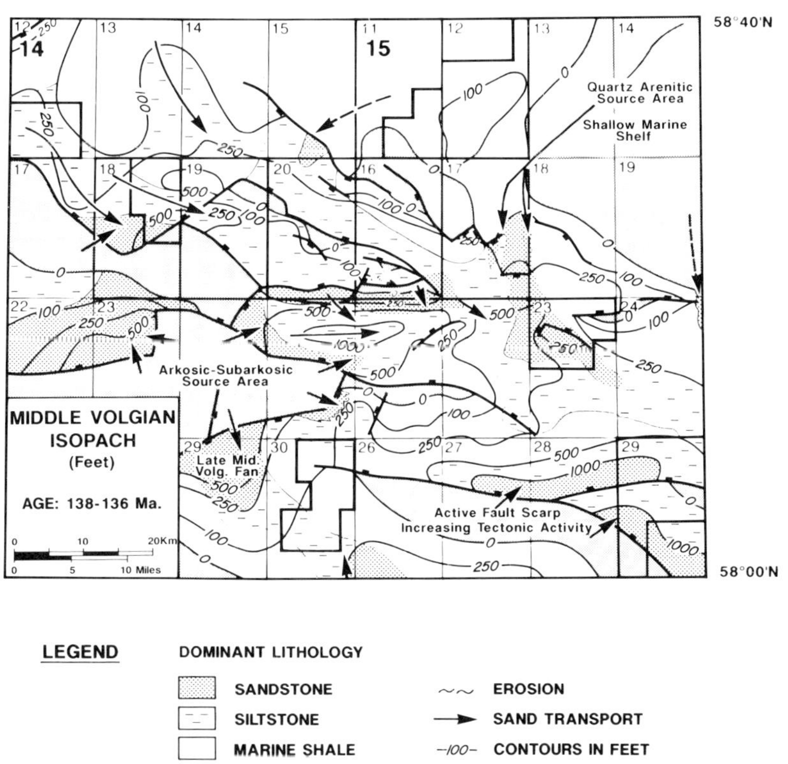

Fig. 13. Middle Volgian isopach, Kimmeridge Clay Formation.

and abutting against the Halibut Horst. The fan complex extends to the NW into block 14/18 and to the SE into blocks 14/24 and 14/25. More distal equivalents are found further north in the Highlander, Petronella and Tartan areas ('Hot Lens' Sandstone Members of the Tartan Field). Fault-normal sediment transport directions are commonly documented in the literature (Mutti 1985) and would be consistent with by depositional isopach patterns in block 14/18. Additionally, oblique, or even fault-parallel transport patterns (Leeder & Gawthorpe 1987) appear to have been caused by differential subsidence along syn-depositional faults. Such a deflection of fault-normal flows along the basin axis is seen in parts of block 14/19, 14/20 and 14/24. In addition, water depth may have been much shallower than inferred in the classical models of fan sedimentation and more akin to fan delta sedimentation resulting from a well sorted, fluviodeltaic input into a shallow marine basin (Stow 1985, Westcott & Etheridge 1983). Additional sediment input may have come from sources further removed to the west, in the Inner Moray Firth Basin. However, the much larger fan complexes developed here are Middle Volgian or younger in age.

Towards the centre of the Witch Ground Graben localized highs in the Tartan area supplied detritus over short distances into the surrounding lows. In the east, channelized quartz-arenitic turbidites continued to prograde south across the gentle slope from the Fladen Ground Spur shelf to form the lower reservoir units in the Galley area. Evidence suggests that the highly channelized nature of these turbidites in the north is replaced further south, in the basin plain, by elongate turbidite fans onlapping the Galley High (Fig. 14). This is analagous to Walker's (1978) general model of fan sedimentation and to aspects of Stow's (1985) elongate fan model. At this time, turbidite sedimentation commenced further SE off the northern margin of the Renee Ridge in block 15/29.

During the early Volgian, general basin subsidence and isopach thickness are governed by NW–SE and E–W faulting. In the west, transport directions are controlled by faults trending WNW–ESE. However, further to the east in the Galley area, transport directions are also influenced by the older NNE–SSW structural fabric (Fig. 12). Hence, the interaction of at least two fault trends, combined with rising sea level, produced a complex pattern of sedimentation in the Galley area. An increase in tectonic activity during the early Volgian is associated with a marked increase in the rate of differential subsidence.

Kimmeridge Siltstone and Galley Sandstone Member (Middle Volgian). Middle Volgian time is represented by the acme of *Muderongia* sp. A. Davey and an upper boundary marked by a top range of *Ctenidodinium panneum* (dinocysts). Deposition of turbidites continued in the Galley area during the middle Volgian, with periodic rapid avulsion of the feeder channels to the north, which still supplied material from the Fladen Ground Spur to elongate fans in the graben (Fig. 13). The system prograded south with time and deposition commenced first in the north during the earliest Volgian, followed by

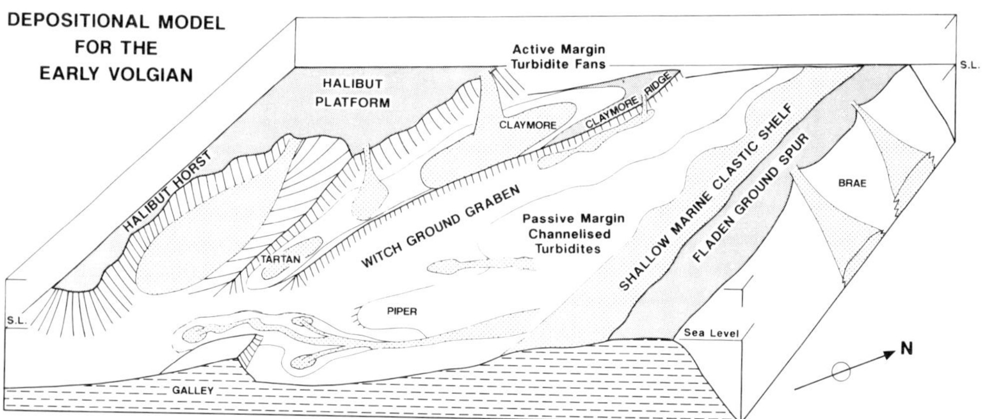

Fig. 14. Depositional model for the Early and Middle Volgian. A contrast in turbidite depositional style during this period between areally extensive fan-shaped turbidites and confined channelized turbidities is related to basin floor slope and position of the main active faults.

maximum development of the Galley turbidite system in the middle Volgian. The transition from channel to elongate fan occurred south of a significant change in slope associated with the northern graben margin (Fig. 14). Many of the Mid-Volgian Galley Sandstones are more feldspathic than those of the early Volgian. This may be due to a sediment input from the Halibut Horst and the Tartan Ridge which were being eroded at the time.

To the west, the Claymore fan had ceased to be active at the end of the early Volgian. The Middle Volgian is represented in the Claymore area by the Kimmeridge Siltstone which, as the basin subsided, extended well beyond the limits of the underlying Claymore Sandstone. Only relatively minor fans continued to deposit sediment in block 14/24, whilst southerly backcutting of the northern margin of the Halibut Horst resulted in the deposition of coarse detritus into the block 14/23 area. Further south, in the area of blocks 14/29 and 14/30, an uppermost Middle Volgian to Upper Volgian fan developed off the southern margin of the Halibut Horst. To the east, sedimentation continued off the northern margin of the Renee Ridge, with the general pattern of sedimentation controlled by a combination of structural trends, in a similar manner to that which existed during early Volgian times.

Hot Shale (Upper Volgian–Lower Ryazanian). These rocks were deposited during a period of gently falling sea level (Haq *et al.* 1987) and increased basin-wide subsidence. The intrabasinal clastic sources had been drowned during the middle Volgian and organic-rich, high gamma ray shales were deposited over most of the Outer Moray Firth Basin (Fig. 15). A regional hiatus at the boundary between mid and late Volgian is suggested on biostratigraphic data, which probably represents a condensed sequence rather than an actual unconformity. The main structural grain remained NW–SE and E–W, although active faulting seems to have decreased with intra-basinal highs being covered by condensed shale sequences. Clastic input was active only on the southern margin of the Halibut Horst during this period.

Lower Cretaceous: The Valhall Formation

The stratigraphic subdivision of the Lower Cretaceous into the youngest Ryazanian aged Kim-

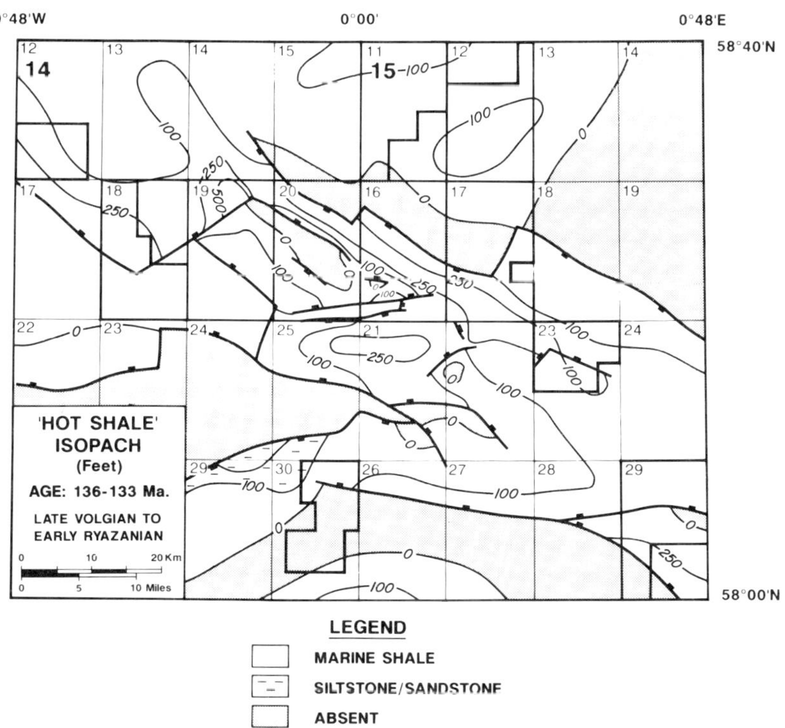

Fig. 15. Late Volgian 'Hot Shale' isopach, Kimmeridge Clay Formation.

meridge Clay Formation succeeded by the Valhall, Sola and Rodby Formations of the Cromer Knoll Group (Fig. 3) is well established for the Witch Ground Graben (Harker *et al.* 1987). Sandstones occur within the Valhall and Sola Formations, however the development of significant sandstone thickness within the dark grey mudstones of the Sola Formation is confined primarily to the SE Witch Ground Graben in the Bosun area (Bisewski, this volume). The only sandstones of the Lower Cretaceous in the NW Witch Ground Graben which are oil-bearing are those of the Valhall Formation (Fig. 2). Sandstones in the area of the Scapa Field in block 14/19 have been assigned to the Scapa Sandstone Member (Harker *et al.* 1987). The following section attempts to correlate the sandstones in the Scapa Field with those found elsewhere in the NW Witch Ground Graben. In this study, the Valhall is subdivided informally into four time intervals, the oldest is Interval 'S' passing into Intervals 'A', 'I' and 'L' (Fig. 16). These spell the word SAIL which was the name assigned to blocks 14/20 and 15/16 by Texaco during the UK 4th Round of Licensing.

Within the Lower Cretaceous, 13 markers have been identified and correlated over the area using biostratigraphic studies and seismic mapping in addition to well log data. The correlation section illustrated in Fig. 16 passes from the Halibut Horst through the Scapa Field to a Lower Cretaceous reservoir on the Main Claymore Field (Well 14/19–5) and on through the North Claymore and Highlander Fields to the north flank of the Tartan Field. Markers 1 and 2 are unconformities, the former marking the top of the Kimmeridge Clay Formation and is commonly referred to as the Base Cretaceous Unconformity. Marker 2 is referred to in this paper as the Lower Valanginian Unconformity which previously has not been recognized, and marks the top of Interval 'S'. In areas of sandstone deposition, it is usually located at the base of the sandstone unit, associated with a change in dip recorded on the dipmeter and a gamma-ray increase. There is little or no change on the sonic log. This fact, combined with its close proximity (often less than 100 ft., or 20 ms two-way-time) and mirroring of Marker 1, means that it is not easily identified on seismic sections (Fig. 17).

Markers 3–13 inclusive are biostratigraphic correlations, with Markers 4, 5, 7, 8 and 9 forming the chronostratigraphic boundaries: Top Early Valanginian, Late Valanginian, Early Hauterivian, Late Hauterivian and Early Barremian respectively. Marker 10 has distinctive sonic and gamma ray responses and is correlated to Occidental's informal 'Viper's Tongue Marker'. The correlation section in Fig. 16 is referred to this datum. Markers 11, 12 and 13 are the tops of the Valhall, Sola and Rodby Formations respectively, as used by Harker *et al.* (1987).

Significant abrupt angular changes on the dipmeter logs were recorded. These aided correlation and an assessment of tectonic activity (Fig. 16) during the Lower Cretaceous. The depositional systems within each of the four time intervals of the Valhall Formation are now discussed.

Interval 'S' (Late Ryazanian–Early Valanginian). Following the change from the poorly aerated seas with organic-rich shale deposition of the Kimmeridge Clay Formation to well aerated seas in Ryazanian times, a calcareous–argillaceous — sandy formation of usually no more than 200 ft preserved thickness was deposited (Fig. 16). The sediments within the interval are usually well bioturbated and the sands form beds generally not exceeding 2 ft in thickness which are impermeable due to calcite cementation. The interval is dated as late Ryazanian to early Valanginian. The small variation in thickness of the unit together with its wide distribution (Fig. 16) suggests that there was little differential subsidence at this time. The Base Cretaceous Unconformity at the base of Interval 'S' is indicated by biostratigraphic data and dipmeter logs, resulting in local truncation of the uppermost Kimmeridge Clay Formation.

Interval 'A' (Early Valanginian–Early Hauterivian). The base of Interval 'A' is marked by a significant unconformity termed here the Lower Valanginian Unconformity (Marker 2) and the top is defined as Marker 6 (Fig. 16). Renewed rifting at the beginning of Interval 'A' was very substantial and was largely responsible for generating, by fault block rotation, the basins and highs which were subsequently infilled and onlapped by Valhall Formation sediments of early Cretaceous age (Figs 16 & 17). This rifting generated a great fault scarp on the northern side of the Halibut Horst, against which vast thicknesses of conglomerates (locally greater than 2000 ft) accumulated, trapped in a narrow half graben which developed immediately to the north of the Horst. Such conglomerates are observed in Well 14/19–9 (Fig. 16). In time, the crest of the Halibut Horst was eroded. During the early Cretaceous a waning of tectonism together with a gradual eustatic rise in sea-level (Vail *et al.* 1977, Haq *et al.*

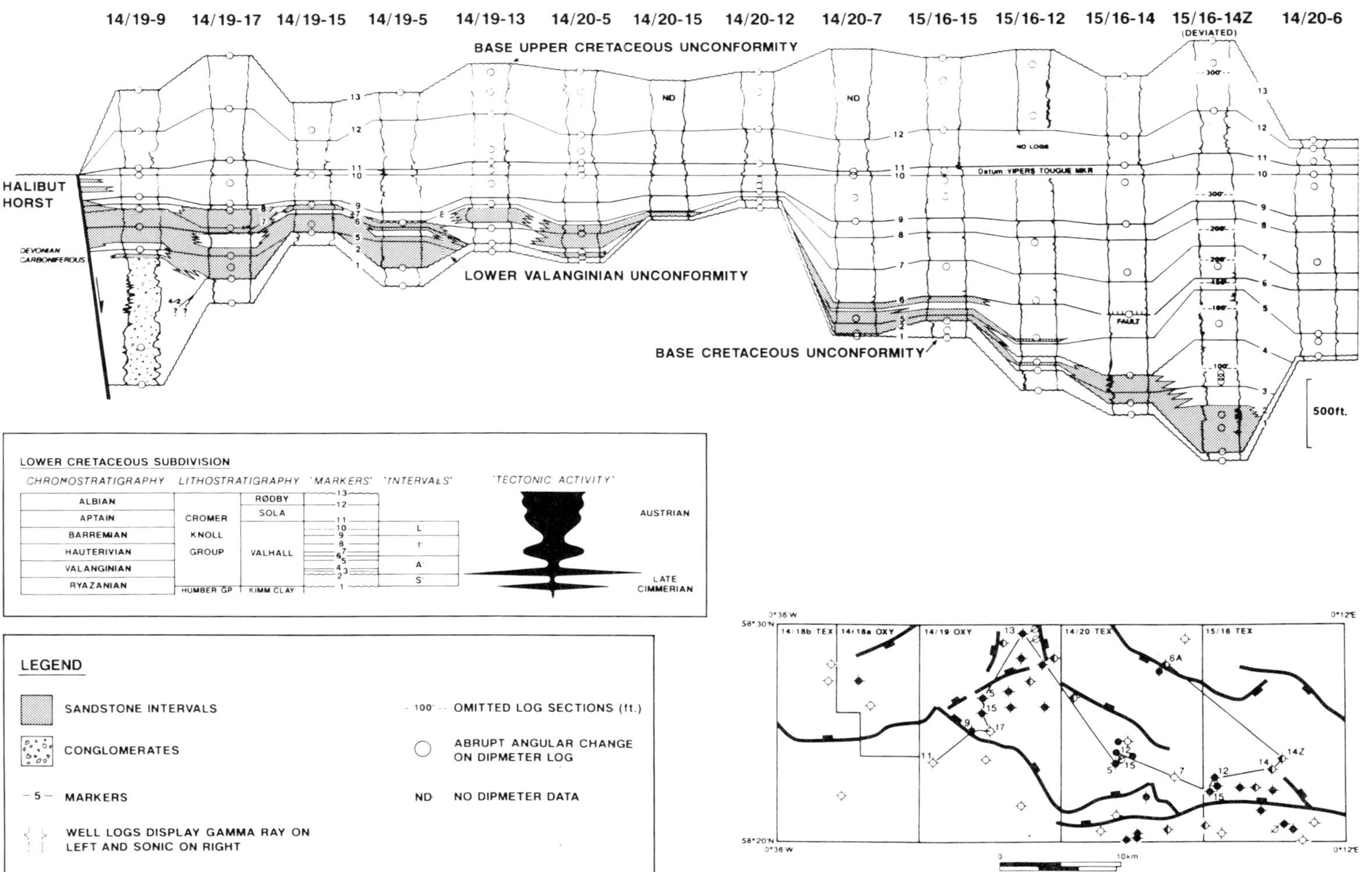

Fig. 16. Lower Cretaceous correlation section across the NW Witch Ground Graben.

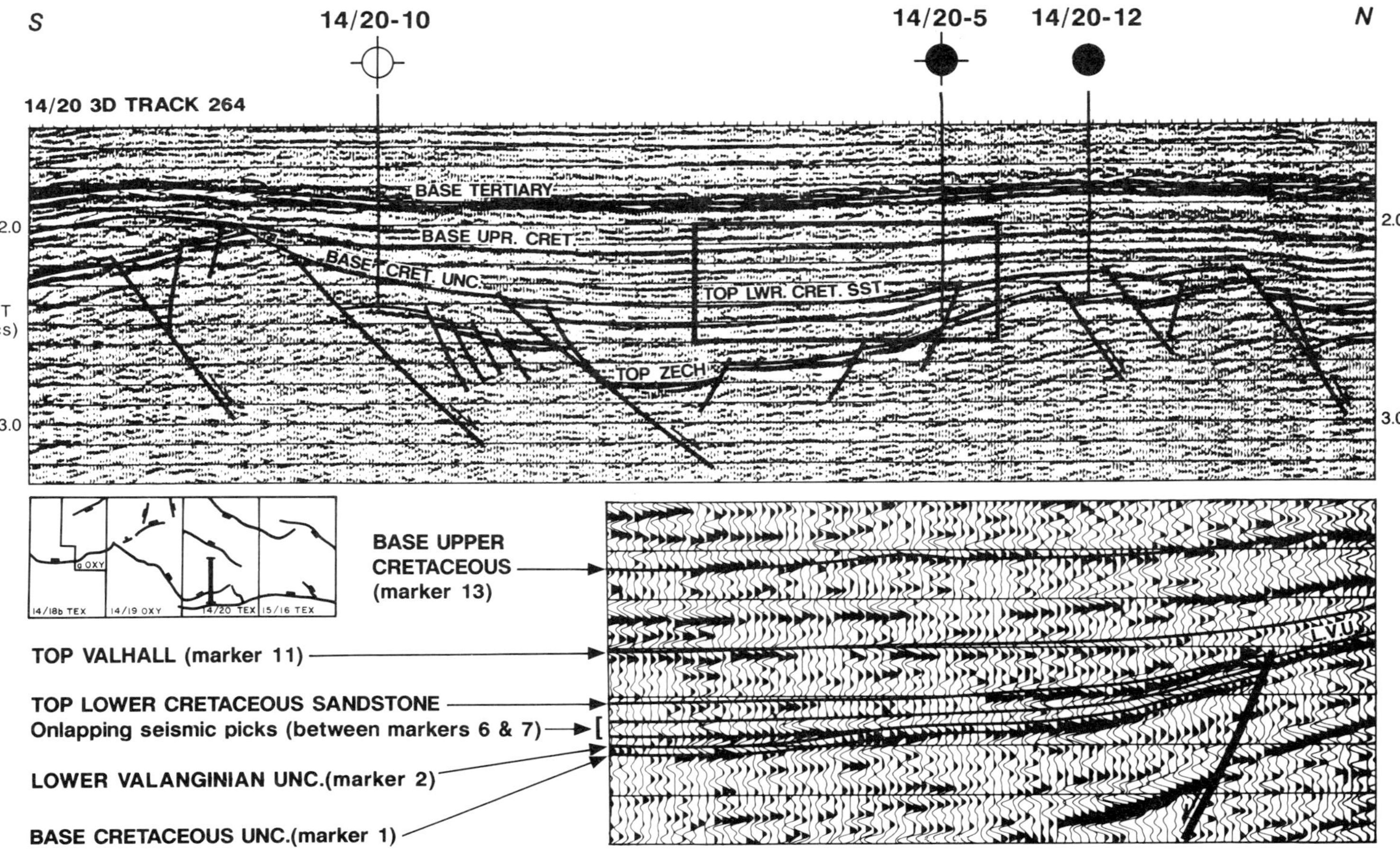

Fig. 17. A seismic section across the SW Highlander sub-basin illustrating the infill and onlap of Valhall Formation sandstones and mudstones onto the Lower Valanginian Unconformity.

1987) resulted in a reduced conglomeratic input and the accumulation of sand grade material on a narrow shelf on the Halibut Horst. This material ws subsequently re-deposited as sand-rich turbidites, which spread out northwards from the Halibut Horst area across the Scapa Syncline (Fig. 18). These sandstones onlap the Claymore–Highlander High to the north of the syncline in a similar fashion to the early Volgian sediments. The Claymore High and the Tartan Ridge also provided a significant sandstone input with possible minor contributions from high areas to the north and east of the graben. However, the sandstones deposited at this time in North Claymore and the northern flank of Tartan contain numerous Permo-Triassic and Carboniferous palynomorphs suggesting that the Halibut Horst provided the majority of the sandstone input. There are only limited areas of the Claymore High and Tartan Ridge where Carboniferous rocks subcrop Lower Cretaceous rocks.

The correlation section illustrated in Fig. 16 shows how the sandstone development in Interval 'A' in blocks 14/20 and 15/16 is diachronous, retreating through time towards the Halibut Horst, as the source areas are eroded closer to sea level.

Interval 'I' (Early Hauterivian–Early Barremian). The beginning of this interval is marked by a reduction in sandstone deposition, which suggests that the Halibut Horst and other surrounding high areas had been largely submerged. There was then a resumption of tectonic activity before Marker 7, the top of the early Hauterivian, resulting in further uplift and erosion of the Halibut Horst and continued sandstone deposition as turbidites. The central area of the Scapa Field was initially the main sandstone depocentre. However, a combination of uplift of this area (indicated by a local unconformity at the Marker 7 horizon such as seen in Well 14/19–5) together with a mounding of sediments resulted in a bypass of sands subsequently shed from the horst. These sands were transported further afield into the NW Claymore sub-basin and reached the Northern Area of the Claymore Field (NAC). They form a major component of the 'Massive Sand' which contains more than 90% of the NAC reserves (Maher & Harker 1987).

In blocks 14/18, 14/19 and 14/20, the general distribution of Interval 'I' sandstones is more extensive than that of sands deposited during Interval 'A'. This is largely because the basins generated by the Lower Valanginian tectonic episode had been progressively infilled and were now broader and consequently, the sandstones could onlap higher structural areas (Figs 16 & 17). Further east in block 15/16, sandstones of Interval 'I' are represented by beds of tight sandstone less than 10 ft thick.

In general, the sediments within the two intervals consist of laterally extensive, well-sorted sandstones abutting source and other structurally high areas. They pass distally into thick marl intervals with thin non-reservoir sandstone stringers in the main NW Witch Ground Graben (Fig. 18). Such thin stringers have not been shown on Fig. 16 for clarity. The transition from massive, high-quality reservoir sandstones into these non-reservoir sandstones in the main NW Witch Ground Graben is usually quite abrupt, coinciding approximately with the 100 ft gross sandstone thickness contour (Fig. 18). The conglomerates found close to the source areas, such as the Halibut Horst, are generally confined to the early part of Interval 'A', in the Valanginian and are not laterally equivalent to the reservoir sandstones of Intervals 'A' and 'I'. It is believed that the sandstones were well sorted on the Halibut Horst, on a narrow high energy intertidal shelf, prior to re-sedimentation as laterally extensive turbidites. This early Cretaceous depositional system is not adequately explained by many of the existing published turbidite models, but is probably best described by the models of Mutti (1985). Evidence suggests that sedimentation in both Intervals 'A' and 'I' evolved from a Type I to Type II system of Mutti's scheme.

Water depths can be estimated by assuming that the morphology of the basin into which the sands were deposited was mainly generated in the Valanginian and subsequently onlapped. Maximum water depths can be gauged by determining the elevation difference between the top of the Carboniferous on the crest of the Claymore High (which was an area of erosion throughout Intervals 'A' and 'I') and the Lower Valanginian Unconformity at the point of interest. This indicates a water depth of the order of 500 ft in the Scapa Field area, reaching greater than 1500 ft in the main NW Witch Ground Graben. Later adjustments on the basin margin faults have distorted these elevation relationships, but only to a minor degree.

By the time of Marker 8, the eustatic sea level rise had largely drowned the Halibut Horst and sandstone deposition continued only immediately adjacent to the Halibut Horst.

Interval 'L' (Early Barremian–Early Aptian). With the exception of a narrow sandstone fringe to the Halibut Horst, this interval is represented

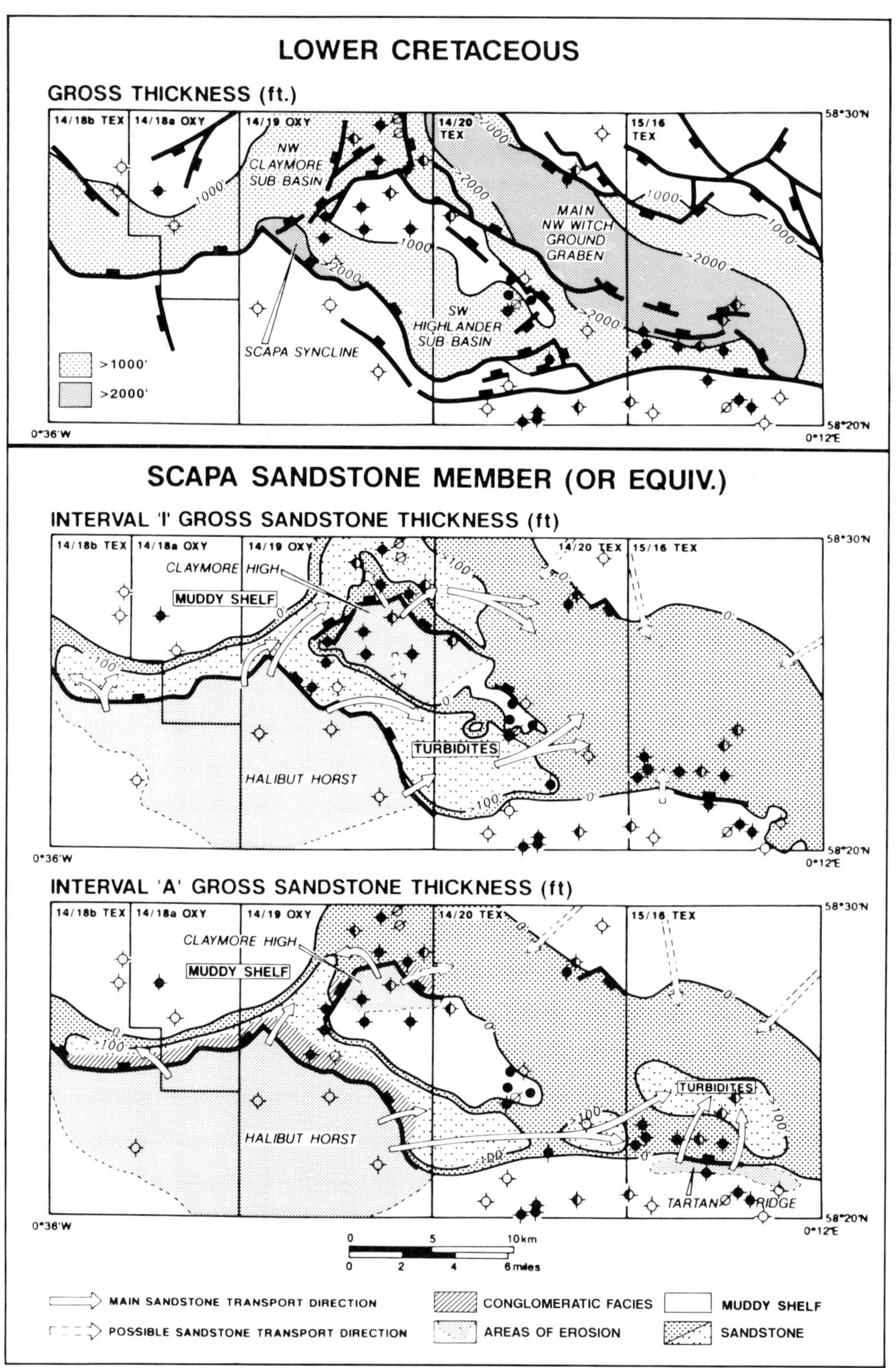

Fig. 18. (a) Gross isopach for the Lower Cretaceous in the NW Witch Ground Graben. (b) Gross sandstone isopachs for intervals 'A' and 'I' of the Lower Cretaceous in the NW Witch Ground Graben.

by argillaceous marine carbonate across the graben (Fig. 16).

Structural evolution and trap development

Most traps in the Witch Ground Graben can be assigned to one of two broad categories: 'upthrown traps' with four-way dip closure or tilted fault blocks, generally located on the flanks of the graben and 'downthrown traps', with closure against a major fault, located downthrown on the main graben fault systems (Hindle 1989). A number of these traps were initiated in the Volgian, but it was the extensional tectonic period of the Ryazanian–Valanginian, represented by the Base Cretaceous and Lower Valanginian Unconformities, that was largely responsible for generating the great number of traps that have made the area such an exploration success. The Witch Ground Graben is structually very complex and in detail split into a number of sub-basins separated by intra-graben highs. This complexity results from the interaction of the many inherited fault trends, detailed above, together with new faults which developed in response to the Ryazanian–Valanginian extensional forces. Many of these new faults generated terraces on the graben margins, resulting in a retreat of the crest of the margins from the basin centres. The traps were subsequently modified by Austrian tectonism (Fig. 3) and continued differential subsidence in the Tertiary, but these periods generally tightened the pre-existing structures and did not destroy them. Many of the traps with reservoirs of late Jurassic and early Cretaceous age deposited in turbidite systems have stratigraphic trapping elements where the sandstones are enclosed within or onlap sealing strata. The preceeding subdivision on a basinwide scale of the Sgiath, Piper, Kimmeridge Clay and Valhall Formations has been necessary to investigate accurately the timing of tectonic phases within the Witch Ground Graben. The following section discusses the structural evolution of the Witch Ground Graben from the Oxfordian to the Barremian.

The Jurassic

Deposition during the Sgiath Formation was largely controlled by a NNE–SSW trending basement high, which defines the boundary between preservation or erosion of the Pentland/Rattray Formation in the Outer Moray Firth Basin (Fig. 4). This indicates that the Sgiath Formation was controlled by such NNE–SSW faulting which defined the Mid-Jurassic basin. Sedimentation was able to keep pace with subsidence in the depocentre whereas on the eastern flank, up to two unconformities developed during late Callovian and earliest Kimmeridgian. Here subsidence was reduced and deltaic sedimentation infilled and onlapped an irregular topographic surface showing a NNE–SSW structural grain.

This fabric is sub-parallel to the Caledonian trend and is probably related to reactivation during the Early and Mid-Cimmerian tectonic episodes. Sands of the Upper Sgiath Unit prograded from the northeast along the same orientation (Fig. 4). These sands were generated by erosion on the Fladen Ground Spur, which had been uplifted during the Mid-Cimmerian tectonic episode.

During the Kimmeridgian, the emergence of the Halibut Horst provided another sediment source resulting in more widespread deposition of high-energy, shallow marine facies (Figs 5 & 7). The preceding structural fabric remained the dominant control, with sediment transport trending NNE–SSW. In latest Kimmeridgian to earliest Volgian times, sea level reached a new maximum, with shoreline migration to the NE and widespread deposition of shales in the east (Fig. 8). Faults trending E–W and NW–SE associated with emergence of the Halibut Horst, formed a marked block and basin regime and combined with the older trends to form isolated intrabasinal submarine highs. These trends represent a new tectonic phase, the Late Cimmerian tectonic episode, which is responsible for the formation of the present structural configuration of the Witch Ground Graben at the Base Cretaceous Unconformity level. The subsequent early Volgian time unit records deposition in the west of the graben controlled primarily by the NW–SE and E–W faulting and dominated by the rejuvenation of the Halibut Horst (Fig. 12). The large-scale mobilization of sediment from the Halibut Horst occurring in conjunction with a rise in sea level (Haq *et al.* 1987) indicates the dominance of tectonic control at this time, as source area uplift and erosion kept pace with rising sea level. To the east, general basin subsidence was controlled by similar fault trends, but the relic NNE–SSW structural fabric inherited from Sgiath and earlier times appears to have controlled the confinement of linear turbidite channels from the Fladen Ground Spur (Figs 12 & 14). This control continued into the middle Volgian with deposition of the Galley Sandstone Member (Fig. 13).

In the west during the middle Volgian, part of the Halibut Horst source area became submerged, but widespread subsidence allowed the deposition of the Kimmeridge Siltstone over the Highlander High and onlapping the Claymore Ridge (Fig. 13). The mid to late Volgian is characterized by a basinwide hiatus in deposition. Although the European sea level curves of Haq *et al.* (1987) depict a slight drop in sea level with a regional unconformity at this time, the break is more likely to represent a basinwide condensed sequence developed in response to a rising sea level. In the NW–SE trending basin centre, near complete sections are believed to exist onlapping onto structural highs such as the Piper Field, parts of the Tartan Field and the Claymore–Highlander Ridge. Subsequent deposition in the late Volgian is characterized by high gamma ray, hemipelagic mudstones (Fig. 15).

The Jurassic–Cretaceous boundary is marked by transition in environmental conditions, from the anaerobic bottom water deposition of organic-rich Kimmeridge Clay shale in the late Volgian, to the well aerated conditions in which calcareous clastics accumulated in the late Ryazanian.

Early Cretaceous

During the deposition of the Valhall Formation, the distribution of potential reservoir sands was intimately linked to tectonic episodes within the background of a pulsed, eustatically rising sea level. The majority of the sandstones were deposited in two broad intervals, Marker 2 to Marker 6 within Interval 'A' and Marker 6 to Marker 8 in Interval 'I' (Fig. 16). In each case, sandstone deposition was triggered by tectonic activity which resulted in rotation and emergence of the Halibut Horst, Claymore High and additionally, during Interval 'A', the Tartan Ridge. The sands eroded from these areas were deposited in turbidite systems which continued until rising sea level drowned the source areas.

The subject of Late Cimmerian tectonics in the North Sea at the boundary of the late Jurassic and early Cretaceous is controversial (see John 1975, Ziegler 1981, Fyfe *et al.* 1981, Vail & Todd 1981, Rawson & Riley 1982 and Badley *et al.* 1984 for discussions). Two major unconformities can be recognized in the earliest Cretaceous of this area, the Base Cretaceous and Lower Valanginian Unconformities (Markers 1 and 2 respectively). The former is the most dominant on seismic sections, but the Lower Valanginian Unconformity has the greater tectonic importance, representing a time of significant structural movement in the area and being largely responsible for generating the tectonic fabric into which subsequent Lower Cretaceous sands and marls where ponded and onlapped.

The greatest fault throws occurred on the NE flank of the Halibut Horst and in the main NW Witch Ground Graben, in particular along the W–E trending fault separating the upthrown and downthrown blocks of the Tartan Field where a throw in excess of 2000 ft can be demonstrated at this level. This indicates that the extension direction must have been approximately NNE–SSW during this tectonic period (Fig. 18). The inherited NE–SW Caledonide-oriented faults were still active at this time, but since they were approximately parallel to the extension direction, they probably experienced strike-slip movement and acted as transfer faults to the NW–SE and E–W trending faults (as suggested by Beach 1984).

The correlation section in Fig. 16 may be an over-simplification, for example Markers 3 to 7 inclusive are shown to onlap the Lower Valanginian Unconformity (Marker 2), but may be present in very condensed sequences not resolvable biostratigraphically or seismically over structural highs. Marls deposited from suspension are known to form condensed sequences over the intra-basinal highs and tend to mask sequence boundaries and the related tectonic episodes.

The time interval between Markers 6 and 8 is the next significant tectonic period, being a phase of further extension and block rotation across the area. This probably marks a peak of tectonism in a period of continuous differential subsidence throughout the early Cretaceous.

Differential subsidence continued in response to tectonic activity during the deposition of the Sola and Rodby Formations, associated with Austrian tectonics. Reservoir sandstone deposition had ceased, however, because the high stand in sea level (Vail *et al.* 1977, Haq *et al.* 1987) had drowned the sandstone source areas in the NW Witch Ground Graben. Seismic mapping of the top of the Halibut Horst indicates that it had been largely peneplained by this time. The seismic line in Fig. 17 indicates that, in addition to regional subsidence over the area, differential subsidence continued after the early Cretaceous between basin and high areas which is indicated by warping at the base of the Tertiary reflector. Only locally, such as around the Halibut Horst, was there continued upward fault propagation after the early Cretaceous.

The analysis of tectonic activity during the Lower Cretaceous illustrates the importance of not placing too much reliance on differences in sediment thicknesses between basins and highs to indicate periods of tectonic activity in an area where sediments are infilling a pre-existing submarine topography. Previous work has tended to identify the Hauterivian in the Witch Ground Graben as the most dominant tectonic period within the Lower Cretaceous using this principle (for example Beach 1984).

The distribution and timing of sandstone deposition in the early Cretaceous (Figs 16 & 18) can be adequately explained by local tectonic activity combined with a general, pulsed eustatic sea level rise. Note, however, that the onlapping early Cretaceous sediments did not rely on this sea level rise for their deposition, only their preservation. This relationship could have been generated by a still stand but the sandstone content and distribution would have been different. This itself could impact the eustatic sea level curves, such as those of Vail *et al.* (1977), who use the onlapping relationships, albeit on a larger basin scale, to indicate sea level rise. The importance of small sea level fluctuations driven by local or regional tectonics, or eustatic mechanisms, on individual sandstone pulses into the basins, cannot as yet be determined for the early Cretaceous.

Conclusions

A detailed biostratigraphic framework has been essential in understanding the distribution in time and space of Jurassic and Cretaceous reservoir sandstones in the Witch Ground Graben. The most important tectonic episodes have been recognised, and the morphology and structural fabric of the basins which were generated by such episodes have been determined. The Middle Jurassic to Lower Cretaceous period records the progressive switch in dominance from NNE–SSW faulting during the Mid-Cimmerian tectonic episode recorded in Oxfordian to Early Kimmeridgian sediments, to one dominated by NW–SE and W–E faulting during the late Cimmerian tectonic episode of the late Kimmeridgian to Barremian period. Many faults were inherited from previous tectonic phases during the Devonian to Middle Jurassic period and have suffered multiple reactivation in the late Jurassic and early Cretaceous.

Reservoir distribution in the Witch Ground Graben has been shown to be determined largely by tectonic activity, which generated the source areas of the sandstones and created the basin morphology into which the sandstones were deposited. However, sea level ultimately controls the surface area available for erosion and sorting of the sandstones. In the Late Volgian and the Barremian, rising sea level drowned source areas and resulted in reservoir units being covered by non-reservoir mudstones which provide important seals. The mudstones of the late Volgian are, in addition, a very effective oil source rock. These factors, combined with the generation of numerous fault block traps during the Late Cimmerian tectonic episode, and accentuated by Austrian tectonism in the mid Cretaceous, prior to oil generation and migration in the late Cretaceous, have made the Witch Ground Graben a very prolific area for hydrocarbons.

The Piper Formation sandstone reservoirs on the Tartan and Piper Fields overlap in age, but are generally younger on Tartan, where shallow marine deposition persisted until the earliest Volgian.

A significant unconformity, termed here the 'Lower Valanginian Unconformity' has been newly recognized. Although largely masked on seismic, it marks the most important tectonic event of the early Cretaceous of the Witch Ground Graben.

The management of Texaco are thanked for permission to publish this paper. The authors are indebted to M. Dworatzek and the staff at Deutsche Texaco AG for numerous biostratigraphy and sedimentology reports over the last 15 years and for establishing the detailed biostratigraphic framework. Our colleagues are thanked in Texaco's departments of Exploration and Development for comments during the preparation of this paper, in particular, J. Hart, N. Clark, C. Swarbrick, C. Abernethy, W. Moon, W. Stanworth, M. Whitehead, D. Beeden and M. Caulfield. Also, we are grateful to our reviewer for some very useful comments. Finally we are grateful to M. Sumner and the Drawing Office staff for the diagrams.

References

ANDREWS, I. J. & BROWN, S. 1987. Stratigraphic evolution of the Jurassic, Moray Firth. *In*: BROOKS, J. & GLENNIE, K. W. (eds) *Petroleum Geology of North West Europe*. Graham & Trotman, London, 785–796.

BEACH, A. 1984. Structural evolution of the Witch Ground Graben. *Journal of the Geological Society, London*, **141**, 621–628.

BADLEY, M. E., EGEBERG, T. & NIPEN, O. 1984. Development of rift basins illustrated by the structural evolution of the Oseberg Feature, Block 30/6, Offshore Norway. *Journal of the Geological Society, London*, **141**, 639–649.

——, PRICE, J. D., RAMBECH DAHL, C. & AGDESTEIN, T. 1988. The structural evolution of the northern Viking Graben and its bearing upon extensional modes of basin formation. *Journal of the Geological Society, London*, **145**, 455–472.

BOOTE, D. R. D. & GUSTAV, S. H. 1987. Evolving

depositional systems within an active rift, Witch Ground Graben, North Sea. *In*: Brooks, J. & Glennie, K. W. (eds) *Petroleum Geology of North West Europe*. Graham & Trotman, London, 819–833.

Christie, P. A. F. & Sclater, J. D. 1980. An extensional origin for the Buchan and Witchground Graben in the North Sea. *Nature*, **283**, 729–732.

Coleman, J. M. & Prior, D. B. 1980. *Deltaic Sand Bodies.* A 1980 Short Course Note Series **15**. American Association of Petroleum Geologists, Tulsa, Okla.

Deegan, C. E. & Scull, B. J. 1977. (Compilers). *A standard lithostratigraphic nomenclature for the Central and Northern North Sea.* Report of the Institute of Geological Sciences, 77/25.

Donovan, R. N., Archer, R., Turner, P. & Tarling, D. H. 1976. Devonian palaeogeography of the Orcadian Basin and the Great Glen Fault. *Nature*, **259**, 550–551.

Haq, B. U., Hardenbol, J. & Vail, P. R. 1987. Chronology of fluctuating sea levels since the Triassic. *Science*, **235**, 1156–1166.

Harker, S. D., Gustav, S. H. & Riley, L. A. 1987. Triassic to Cenomanian stratigraphy of the Witch Ground Graben. *In*: Brooks, J. & Glennie, K. W. (eds) *Petroleum Geology of North West Europe*. Graham & Trotman, London, 809–818.

Hindle, A. D. 1989. Downthrown Traps of the NW Witch Ground Graben, U.K. North Sea. *Journal of Petroleum Geology*, **12**, 405–418.

Johnson, R. J. & Dingwall, R. G. 1981. The Caledonides: their influence on the stratigraphy of the north-west European continental shelf. *In*: Illing, L. V. & Hobson, G. D. (eds). *Petroleum Geology of the Continental Shelf of North-west Europe.* Heyden, London, 85–97.

Lake, S. D. & Karner, G. D. 1987. The structure and evolution of the Wessex Basin, Southern England: an Example of inversion tectonics. *In*: Ziegler, P. A. (ed.) Compressional Intra-Plate Deformations in the Alpine Foreland. *Tectonophysics*, **137**, 347–378.

Leeder, M. R. & Gawthorpe, R. L. 1987. Sedimentary models for extensional tilt-block/half graben basins. *In*: Coward, M. P., Dewey, D. K. & Hancock, P. L. (eds) *Continental Extensional Tectonics.* Geological Society, London, Special Publication, **28**, 139–152.

Maher, C. E. 1980. The Piper Oilfield. *In: Giant Oil and Gas Fields of the Decade: 1968–1978.* American Association of Petroleum Geologists Memoir, **30**, 131–172.

—— & Harker, S. D. 1987. The Claymore Oilfield. *In*: Brooks, J. & Glennie, K. W. (eds). *Petroleum Geology of North West Europe*. Graham & Trotman, London, 835–845.

McBride, E. F. 1963. A classification of sandstones. *Journal of Sedimentary Petrology*, **33**, 664–665.

McQuillan, R., Donato, J. A. & Tulstrup, J. 1982. Development of basins in the Inner Moray Firth and the North Sea by crustal extension and dextral displacement on the Great Glen Fault. *Earth and Planetary Science Letters*, **60**, 127–139.

Mutti, E. 1985. Turbidite systems and their relation to depositional sequences. *In*: Zuffa, G. G. (ed.) *Provenance of Arenites*. Reidel, Dordrecht, 65–93.

Pegrum, R. M. 1984. Structural development of the southwestern margin of the Russian-Fennoscandian Platform. *In*: Spencer, A. M. *et al.* (eds). *Petroleum Geology of the North European Margin*. Graham & Trotman, London, 359–369.

Rawson, P. F. & Riley, L. A. 1982. Latest Jurassic–Early Cretaceous events ahd the "Late Cimmerian Unconformity" in North Sea Area. *Bulletin of the American Association of Petroleum Geologists*, **66**, 2628–2648.

Rodgers, D. A., Marshall, J. E. A. & Astin, T. R. 1989. Devonian and later movements on the Great Glen fault system, Scotland. *Journal of the Geological Society, London*, **146**, 369–373.

Stow, D. A. V. 1982. Sedimentology of the Brae Oilfield, North Sea: Fan models and controls. *Journal of Petroleum Geology*, **5**, 129–148.

—— 1985. Deep-sea clastics: where are we and where are we going? *In*: Brenchley, P. J. & Williams, B. P. J. (eds). *Sedimentology – Recent Developments and Applied Aspects*. Geological Society London Special Publication, **18**, 67–94.

Surlyk, F. 1978. Submarine fan sedimentation along fault scarps on tilted fault blocks (Jurassic-Cretaceous boundary, East Greenland). *Bull. Gronlands geol. Unders*, **128**, 108.

Turner, C. C., Cohen, J. M., Conell, E. R. & Cooper, D. M. 1987. A depositional model for the South Brae Oilfield. *In*: Brooks, J. & Glennie, K. W. (eds). *Petroleum Geology of North West Europe*, Graham & Trotman, London, 853–864.

——, Richards, P. C., Swallow, J. L. & Grimshaw S. P. 1984. Upper Jurassic stratigraphy and sedimentary facies in the Central Outer Moray Firth Basin, North Sea. *Marine and Petroleum Geology*, **1**, 105–117.

Vail, P. R., Mitchum, R. M., Todd, R. G., Widmier, J. M., Thompson, S., Sangree, J. B., Bubb, J. N. & Hatledid, W. G. 1977. Seismic Stratigraphy and Global changes in sea-level. *In*: Payton, C. E. (ed.) *Seismic Stratigraphy–application to hydrocarbon exploration*. American Association of Petroleum Geologists Memoir, **26**, 49–212.

—— & Todd, R. G. 1981. North Sea Jurassic unconformities, chronostratigraphy and sea-level changes from seismic stratigraphy. *In*: Illing, L. V. & Hobson, G. D. (eds) *Petroleum Geology of the Continental Shelf of North-west Europe*. Heyden, London, 216–235.

Wescott, W. A. & Etheridge, F. G. 1983. Eocene fan delta submarine fan deposition in the Wagwater Trough, east-central Jamaica. *Sedimentology*, **30**, 235–247.

Ziegler, P. A. 1981. Evolution of sedimentary basins in North-West Europe. *In*: Illing, L. V. & Hobson, G. D. (eds). *Petroleum Geology of the Continental Shelf of North-west Europe.* Heyden & Son, London, 3–42.

—— 1982. *Geological Atlas of Western and Central Europe*. Elsevier, Amsterdam.
—— 1988. *Evolution of the Arctic–North Atlantic and the Western Tethys*. American Association of Petroleum Geologists Memoir **43**.

Occurrence and depositional environment of the Lower Cretaceous sands in the southern Witch Ground Graben

HANS BISEWSKI

Conoco (UK) Limited, Park House, 116 Park Street, London, W1Y 4NN, UK (Present address: Conoco (UK) Limited, Rubislaw House, Anderson Drive, Aberdeen AB2 4AZ, UK)

Abstract: Lower Cretaceous sands in the southern Witch Ground Graben are of Aptian–Albian age and are part of the Sola Formation. In the Witch Ground Graben the Sola Formation is generally argillaceous. However, in the southern Witch Ground Graben, thick sands are developed and the argillaceous Sola Formation is replaced, almost in its entirety, by the sandy Kopervik Formation, also called the 'Bosun Sands'. The increased clastic input in the upper part of the Lower Cretaceous is associated with the regression in the Central North Sea as a result of the Austrian tectonism. The sands of the Kopervik Formation are derived from sources on locally exposed and rotated fault blocks and emergent terraces and are deposited as submarine fan or turbiditic sands.

The Lower Cretaceous Aptian–Albian Kopervik Sandstone is found in the vicinity of the triple junction of South Viking, Central and Witch Ground Grabens and adjacent basins. Its presence is due to the Austrian orogenic phase which caused localized emergent areas and fan sand deposition into the surrounding depressions.

Various discoveries in the Kopervik Sandstone have been made and hydrocarbon types vary from medium gravity oil to gas/condensate. These discoveries concentrate on an area either side of the Renee Ridge and south of the Fladen Ground Spur comprising the southern most parts of Quadrants 15 and 16 and the northernmost parts of Quadrants 21 and 22 (Fig. 1).

The stratigraphic evolution of the Lower Cretaceous, the depositional environment, the distribution and the reservoir characteristics of the Aptian–Albian sands as well as the tectonic setting of the southern Witch Ground Graben are described in order to highlight the exploration significance.

Regional geological setting

The Witch Ground Graben is part of a major rift complex in the North Sea which developed during the Upper Jurassic after the collapse of the previously uplifted triple junction of the Central, South Viking and Witch Ground Grabens (Ziegler 1975). Vast amounts of sediment were removed from the rift shoulders and deposited into the grabenal systems forming the well known reservoirs such as the Hugin, Brae, Miller, Ula, Fulmar and Piper Formations. During the latest Jurassic stages the rift basins deepened and were filled by anoxic shales and intercalated deepwater turbidities. The Jurassic sands are confined to the grabens or their immediate flanks and form the producing hydrocarbon reservoirs for Piper, Claymore and Tartan in the Witch Ground Graben and Brae, Miller, the T-complex as well as the Birch and Oak discoveries in the South Viking Graben (Fig. 2).

The post-rift phase saw a time of quiescence in the Central North Sea with the exception of the northern Witch Ground Graben, adjacent to the Halibut Horst. Generally Neocomian and Barremian marls and shales blanket the Upper Jurassic sediments and transgressed onto formerly exposed areas. The onset of the Austrian orogenic phase in the Aptian resulted in localized uplifts and fan sand deposition especially in the area of the former triple junction. Regional subsidence replaced the active tectonic events in the middle Albian as the continuously rising sea level caused the inundation of the emerged areas like the Fladen Ground Spur and Halibut Horst (Figs 3 & 4).

Lower Cretaceous stratigraphy

In the Central North Sea the anoxic sediments of the Upper Jurassic Humber Group Kimmeridge Clay are overlain by the early Cretaceous Cromer Knoll Group which ranges in age from latest Ryazanian to Albian (Fig. 5). It is in turn overlain by the late Cretaceous Chalk Group, sedimentation of which commenced in the Cenomanian.

Early Cretaceous sedimentation within the southern Witch Ground Graben was controlled by eustatic/tectonic interaction and divides into three main phases. These are the transgressive Valhall Formation (late Ryazanian to early Aptian), the regionally regressive Sola Forma-

From Hardman, R. F. P. & Brooks, J. (eds), 1990, *Tectonic Events Responsible for Britain's Oil and Gas Reserves*, Geological Society Special Publication No 55, pp 325–338.

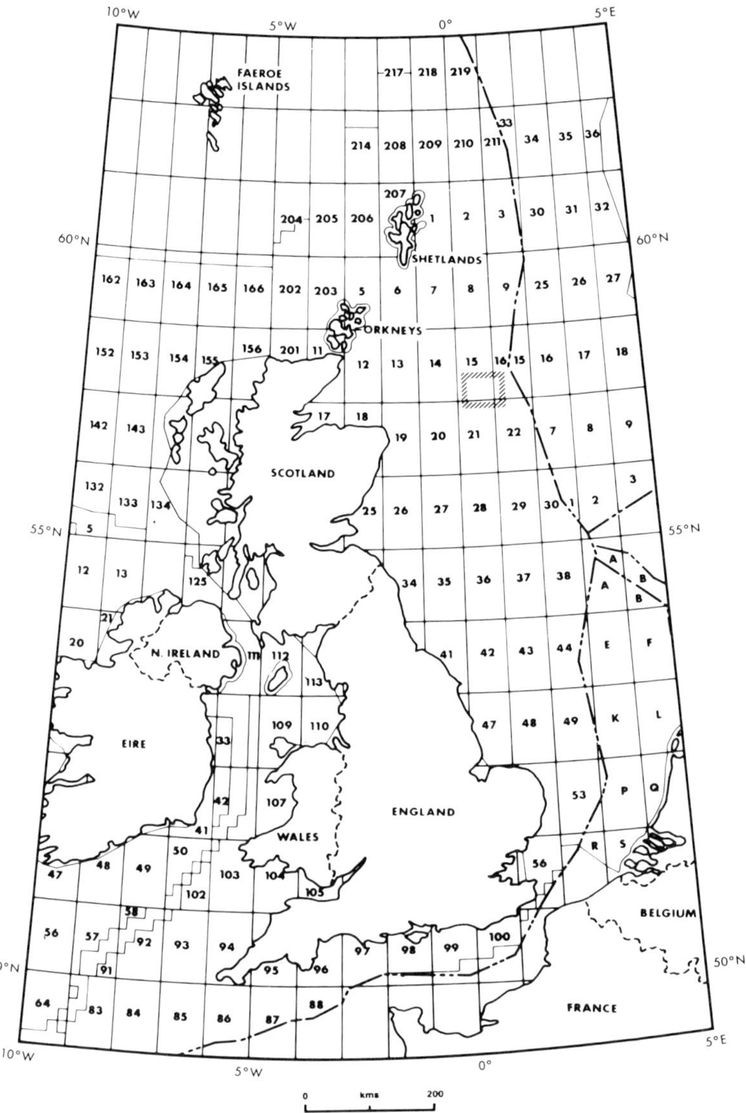

Fig. 1. Area of evaluation.

tion (early Aptian to early Albian) and the transgressive Rodby Formation (mid- to late Albian).

A major transgressive pulse in late Ryazanian flushed the anoxic basins and terminated the deposition of the Kimmeridge Clay Formation (Rawson & Riley (1987), stenomphalus transgression). This event initiated changes in water mass chemistry and a major lithological change but, within the basins, no major interruption in sedimentation. The Valhall Formation sediments are generally poor in terrigenous material and are characterized by a vertical increase in carbonates which is related to the overall transgressive nature of the sequence. Four transgressive pulses can be recognized in the southern Witch Ground Graben and fivefold subdivision of the Valhall Formation based on log character can be applied to wells in the basinal areas.

The lowermost part of the Valhall Formation (Unit I) is represented by highly calcareous claystones and limestones of latest Ryazanian to early Valanginian age. This unit is overlain by a series of uniform grey claystones (Unit II)

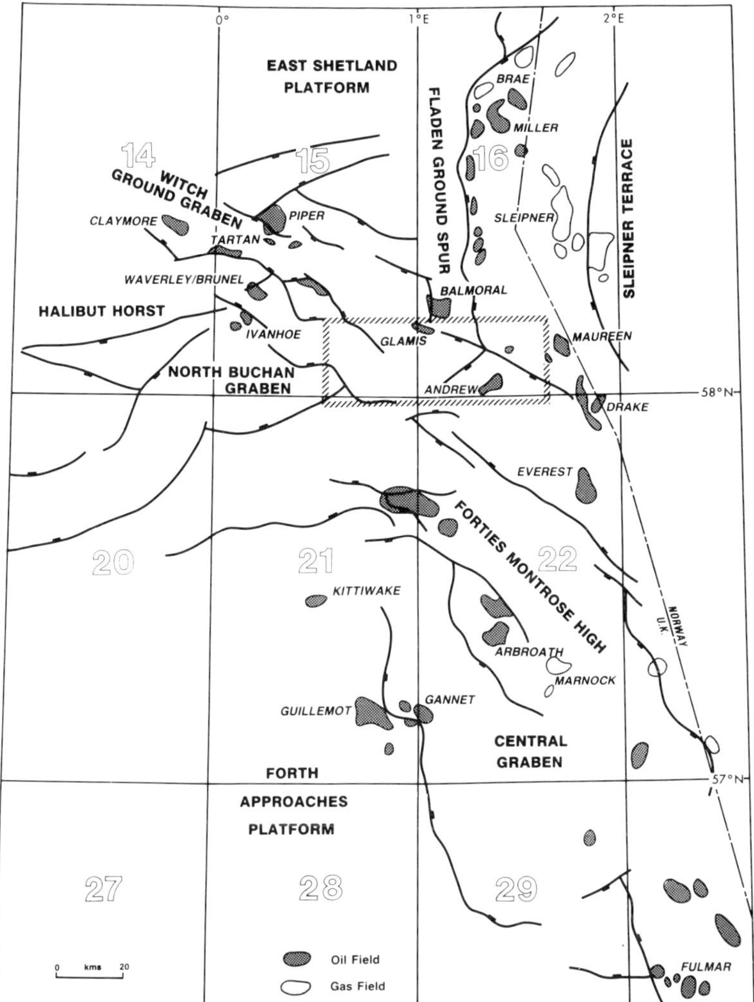

Fig. 2. Structural features and hydrocarbon distribution.

of late Valanginian to late Hauterivian age. The Unit I to Unit II contact appears to be related to the base late Valanginian transgressive event (dichotomites Zone). The overlying Unit III represents a period of increased calcareous deposition initiated as response to an intra-late Hauterivian transgressive pulse (gottschei Zone). Within Unit III, ranging in age from late Hauterivian to intra-mid Barremian, is a high-gamma, low-velocity marker which coincides approximately with the biostratigraphically defined early to middle Barremian boundary. Unit IV, the upper part of the middle Barremian, indicates an increased clastic component with respect to the underlying Unit III reflecting a regressive phase in the overall rising sea level.

Unit V comprises a high-gamma-ray shale at the base and a calcerous claystone at the top. The sapropelic character of the basal Unit V is considered to result from water mass stratification initiated during a questionable late Barremian to early Aptian transgression.

The upper part of the Unit V is dated as early Aptian. It appears that within the southern Witch Ground Graben this Unit does not persist and an eastward facies change to an arenaceous unit is apparent.

The overlying Sola Formation comprises an arenaceous basal unit, the Kopervik Sandstone Member, and an argillaceous upper unit. The Kopervik Sandstone Member exhibits an informal threefold subdivision, the 'Reservoir

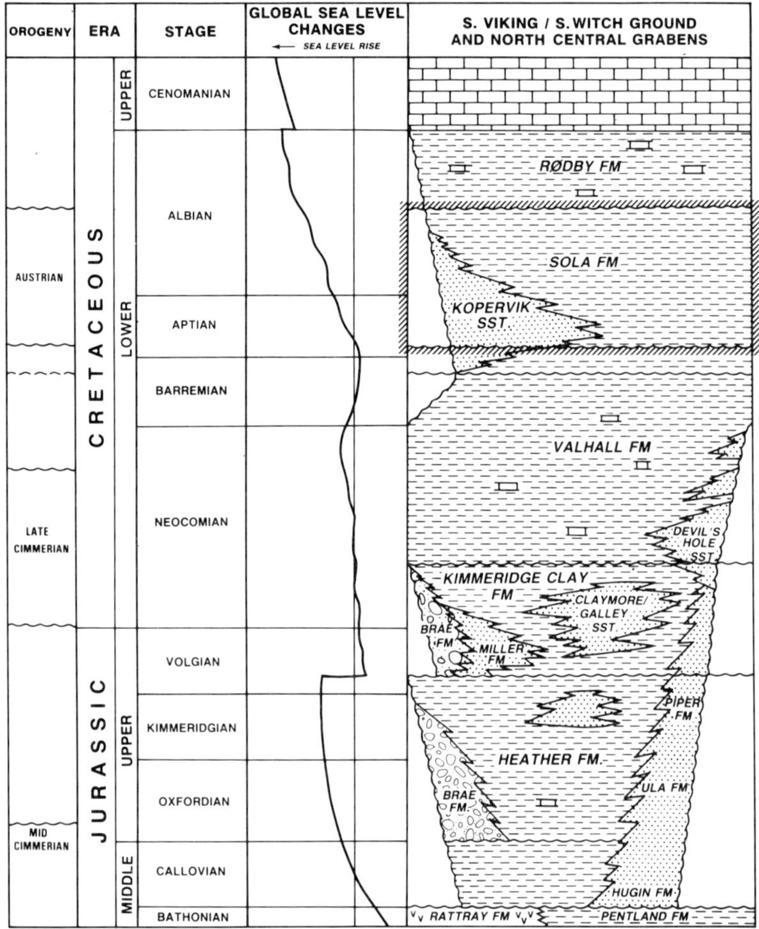

Fig. 3. Lower Cretaceous and Upper Jurassic lithostratigraphy.

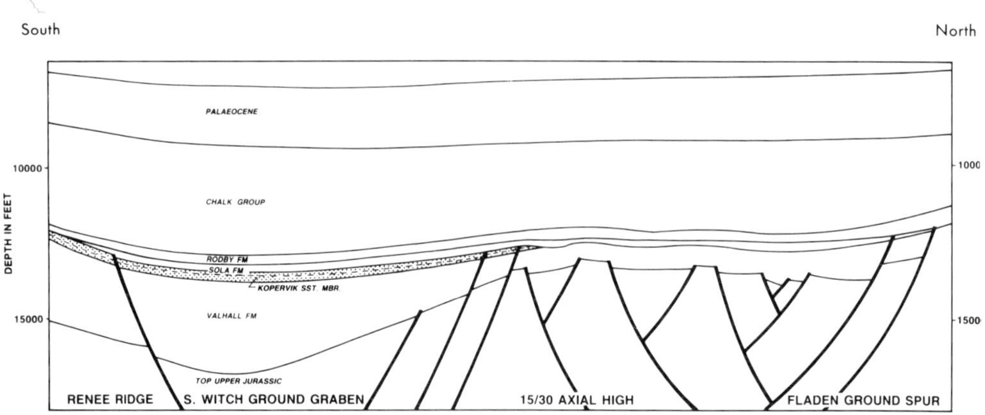

Fig. 4. Schematic cross section across the southern Witch Ground Graben.

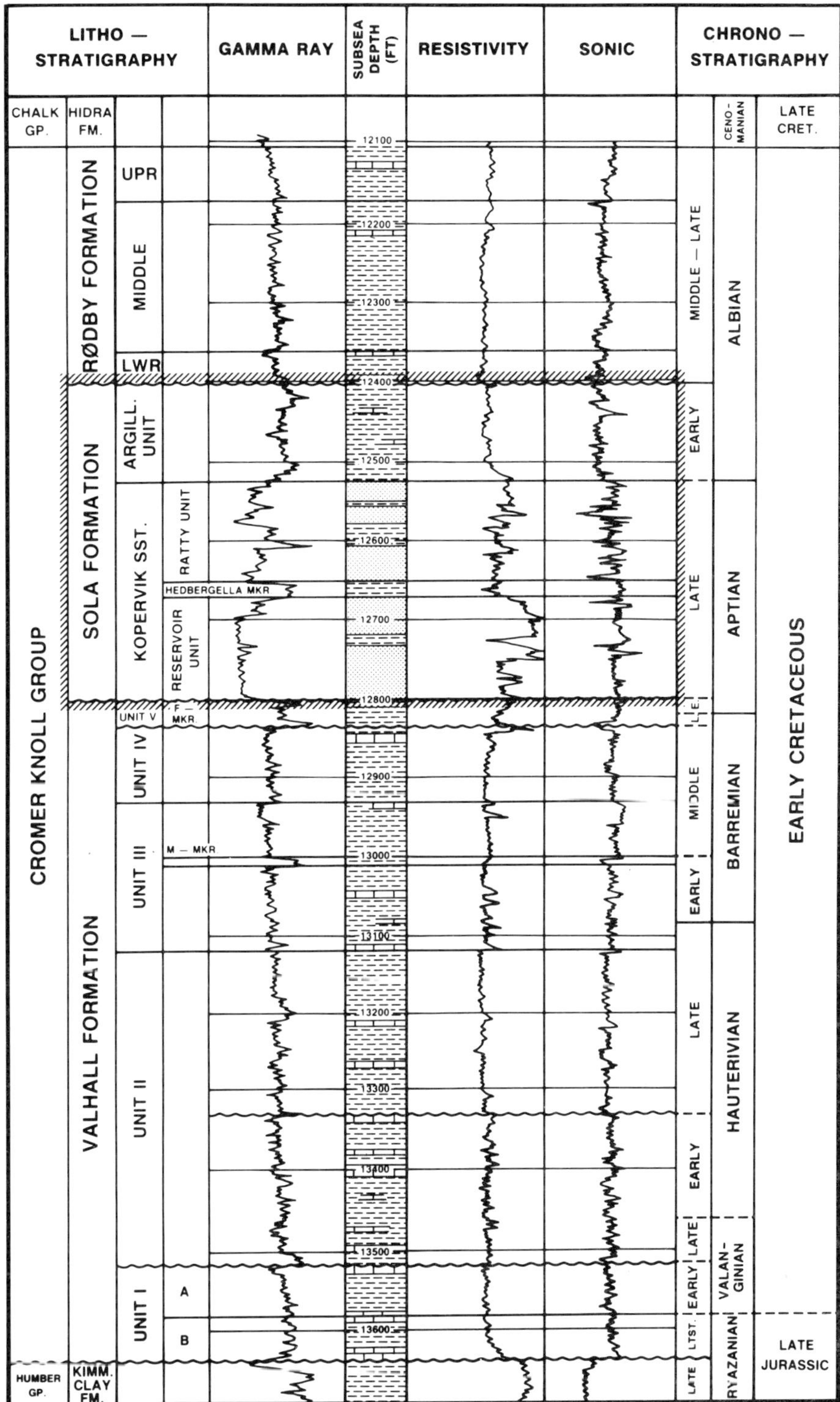

Fig. 5. Generalized section of the Lower Cretaceous interval, southern Witch Ground Graben.

Unit' at the base is overlain by a shale marker, the 'Hedbergella Marker', which in turn is succeeded by the Ratty Unit. The Sola Formation is thought to be essentially regressive in character as a result of the Austrian orogenic phase which initiated rifting in areas outside the North Sea and caused exposure of localized structural high areas and basin isolation in the North Sea. This tectonic activity overprinted a general eustatic rise in sea level. Variation of the age of onset of sand sedimentation and biostratigraphic evidence suggest an uncomformable contact between the Valhall and the Sola Formations. Within the Sola Formation tuffaceous material is found ranging in age from early Aptian to early Albian suggesting volcanic activity was associated with the Austrian orogenic event.

The sandstones are poorly sorted, fine to coarse grained and suggest a turbiditic or mass flow origin. The interbedded claystones exhibit the same lithological and wireline characteristics as those of the overlying Sola Formation Argillaceous Unit. The onset of deposition of the 'Argillaceous Unit' results from an early Albian transgressive event, the inundation of emerged areas and the cessation of coarse clastic sedimentation.

The youngest formation in the Cromer Knoll Group is the Rodby Formation, the deposition of which was initiated by the mid-Albian transgressive event continuing into the Turonian. The Rodby Formation sedimentary cycle terminated the overall Aptian to early Albian regressive regime of the Sola Formation. The eustatic transgressive nature of the mid Albian sequence is reflected by the remarkable lateral uniformity of the Rodby Formation in the Central North Sea. The Sola to Rodby Formation boundary seems to be conformable only in the basinal areas while over the basin margins it is marked by an unconformity. The Rodby Formation can be subdivided into three units which are a marly and limestone rich lower unit, an argillaceous middle unit and a calcareous upper unit which is slightly more transgressive and suggests deeper water in comparison to the middle unit.

Distribution and depositional environment

The Kopervik Sandstone unit in the Southern Witch Ground Graben is found in an area covering of 1600 km^2. The thickest sequence is penetrated south of the Fladen Ground Spur/ Andrew Ridge which was probably the depocentre of the Kopervik basin (Figs 6 & 7).

The presence of sands in the east, underlying the 'Reservoir Unit', suggests the initiation of the sand sedimentation to be earlier in the basin centre than its surrounding areas. Succeeding younger units extended further and enlarged the area covered indicating a basin filling sedimentation with little or no subsidence. The 'Hedbergella Marker' separating the Reservoir and the Ratty Units is traceable throughout the sedimentary basin (Fig. 8).

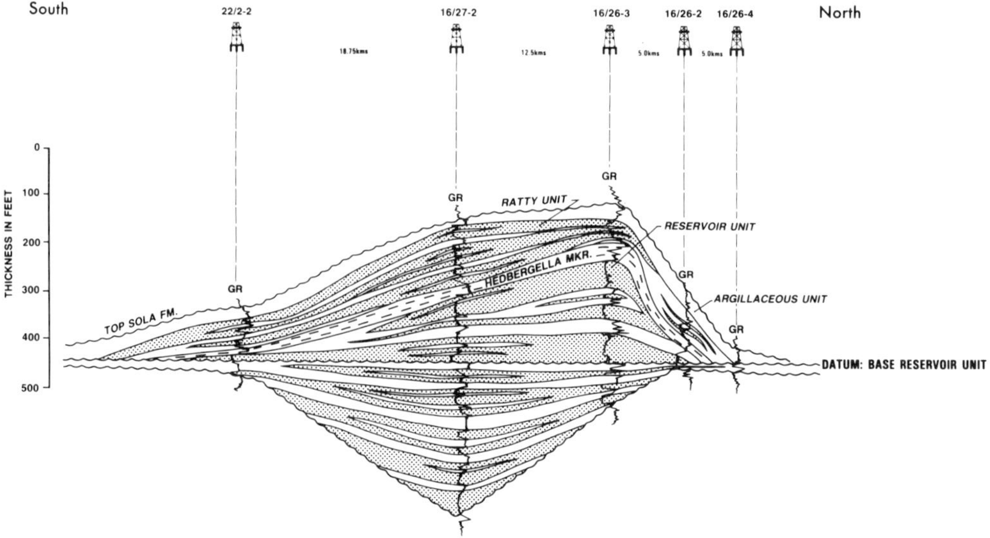

Fig. 6. Stratigraphic cross section of the Sola Fm. across the 16/26–16/27 sub-basin.

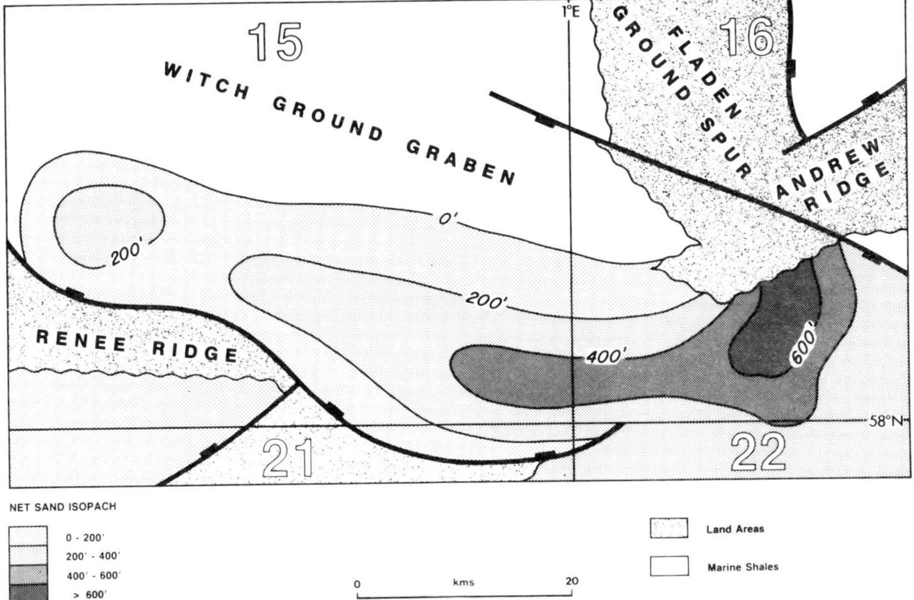

Fig. 7. Lower Cretaceous net sand isopach.

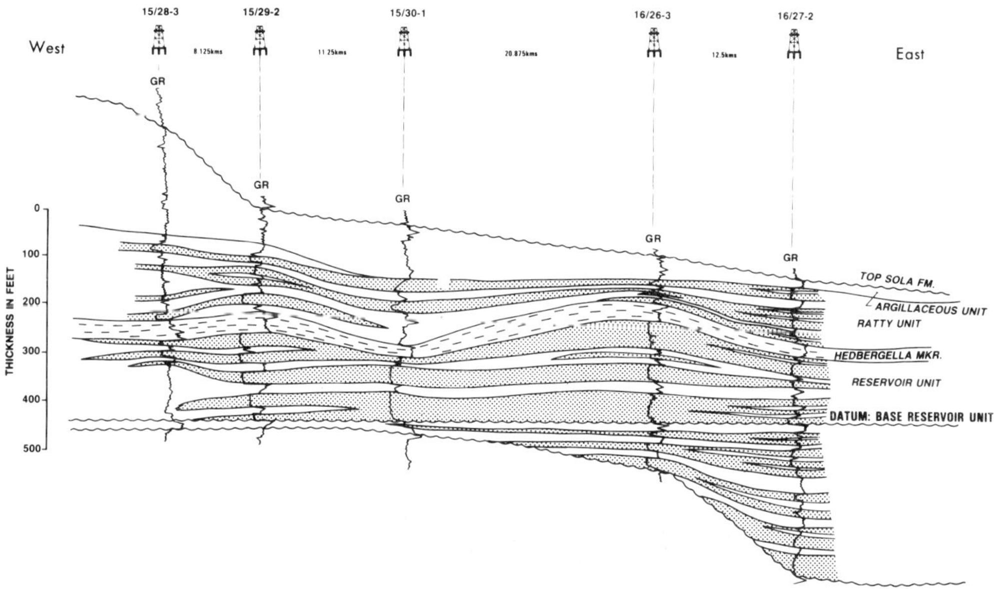

Fig. 8. Stratigraphic cross section of the Sola Fm. along the southern Witch Ground Graben axis.

The 'Ratty Unit' and especially the overlying Argillaceous Unit seem to onlap onto the basin margins. Older remaining structures within the basin seem to have a profound influence on the Kopervik Sandstone distribution and are overlain by only the younger units indicating the rising sea level during the sedimentation of the Sola Formation (Fig. 9).

The Kopervik sands are described as generally poorly sorted, pale coloured, mainly fine to

South
North
15/30-1
15/30-3
15/30-2
15/30-4
GR
TOP SOLA FM.
ARGILLACEOUS UNIT
RATTY UNIT
HEDBERGELLA MKR.
RESERVOIR UNIT
DATUM: BASE RESERVOIR UNIT
THICKNESS IN FEET
0
100
200
300
400
500

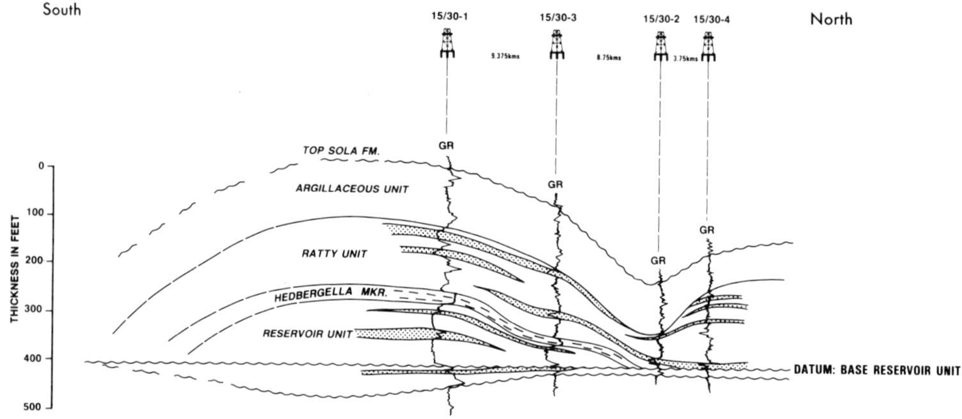

Fig. 9. Stratigraphic cross section of the Sola Fm. across the 15/30 sub-basin.

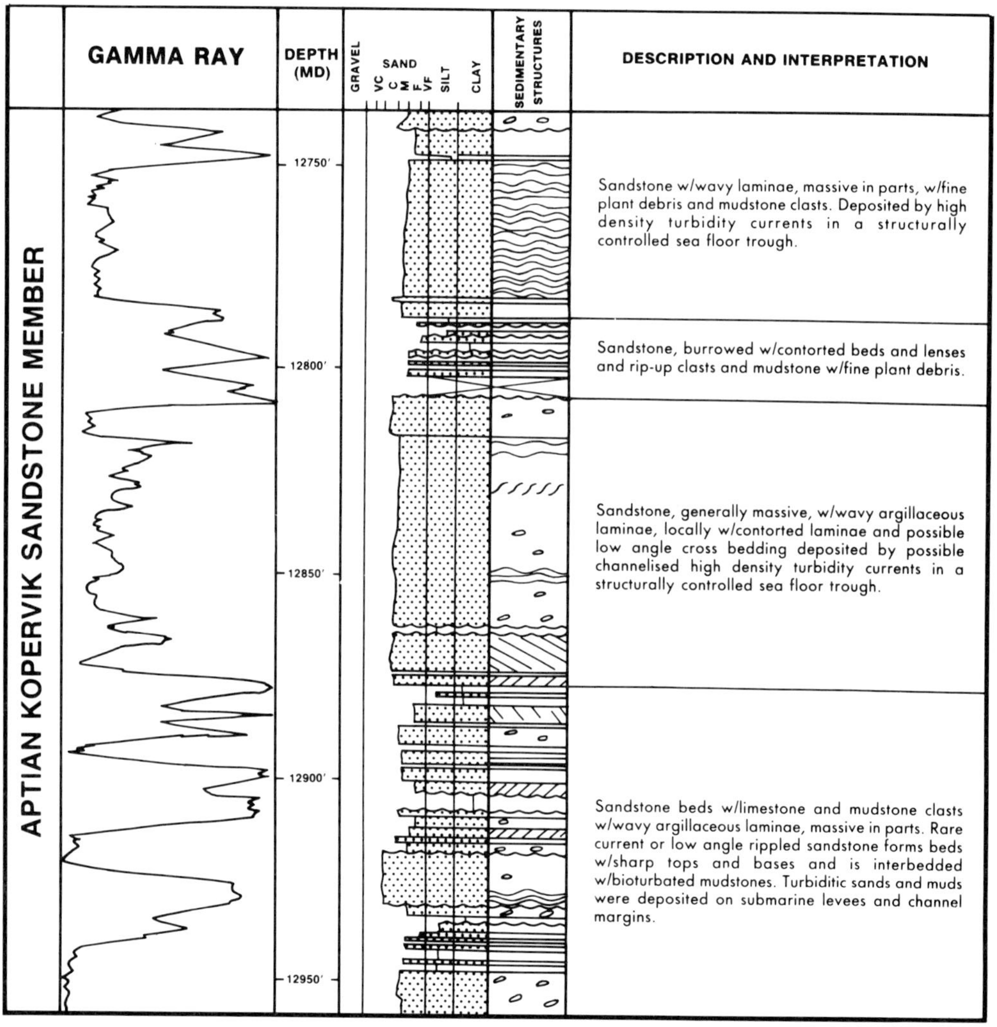

GAMMA RAY
DEPTH (MD)
GRAVEL
SAND
VC
C
M
F
VF
SILT
CLAY
SEDIMENTARY STRUCTURES
DESCRIPTION AND INTERPRETATION
APTIAN KOPERVIK SANDSTONE MEMBER
12750′
12800′
12850′
12900′
12950′
Sandstone w/wavy laminae, massive in parts, w/fine plant debris and mudstone clasts. Deposited by high density turbidity currents in a structurally controlled sea floor trough.
Sandstone, burrowed w/contorted beds and lenses and rip-up clasts and mudstone w/fine plant debris.
Sandstone, generally massive, w/wavy argillaceous laminae, locally w/contorted laminae and possible low angle cross bedding deposited by possible channelised high density turbidity currents in a structurally controlled sea floor trough.
Sandstone beds w/limestone and mudstone clasts w/wavy argillaceous laminae, massive in parts. Rare current or low angle rippled sandstone forms beds w/sharp tops and bases and is interbedded w/bioturbated mudstones. Turbiditic sands and muds were deposited on submarine levees and channel margins.

Fig. 10. Core log of a Lower Cretaceous Kopervik Sandstone section, southern Witch Ground Graben.

medium, partly coarse grained, embedded in dark grey, non-calcareous shales (Fig. 10). The log character of the individual sandstone beds generally display sharp bases and tops indicating erosional bases and rapid abandonment or sediment switching. Occasionally, however, both fining and coarsening upward sequences are observed. Sedimentary structures are generally massive or wavy laminated. Low-angle and contorted laminae are occasionally observed. Within the sands, rip-up clasts, plant debris as well as mudstone clasts are common. The intercalated mudstones as well as the thin sandstone beds indicate bioturbation.

The sands, most of them arenites or wackes, consist chiefly of quartz, minor feldspars which are partly degraded, and minor cements as carbonate or quartz overgrowths. The grains are cemented by sparry carbonate and little detrital clay is present in the sands. Some authigenic kaolinite and quartz overgrowth contribute to the cementation.

The first stage in the diagenetic sequence was initiated following deposition. Compaction, following burial, caused grain to grain suturing, deformation of glauconite grains and mudstone clasts, splitting of feldspar grains and pressure solution. Consequently the feldspar degraded to smectite, sometimes kaolinite, with the creation of some secondary porosity. The later quartz and feldspar overgrowths were succeeded by carbonate cementation and finally by the growth of authigenic clays such as kaolinite, illite–smectite and illite.

The diagenetic sequence has had a major influence on porosity distribution (Fig. 11). The original arenites and wackes had probably high porosities restricted only by the presence of significant quantities of clay matrix. Porosity distribution is primarily facies dependent with depth of burial acting as a secondary control. The proximal and mid-fan sands show higher porosities and better reservoir characteristics then the distal member of the same fan complex.

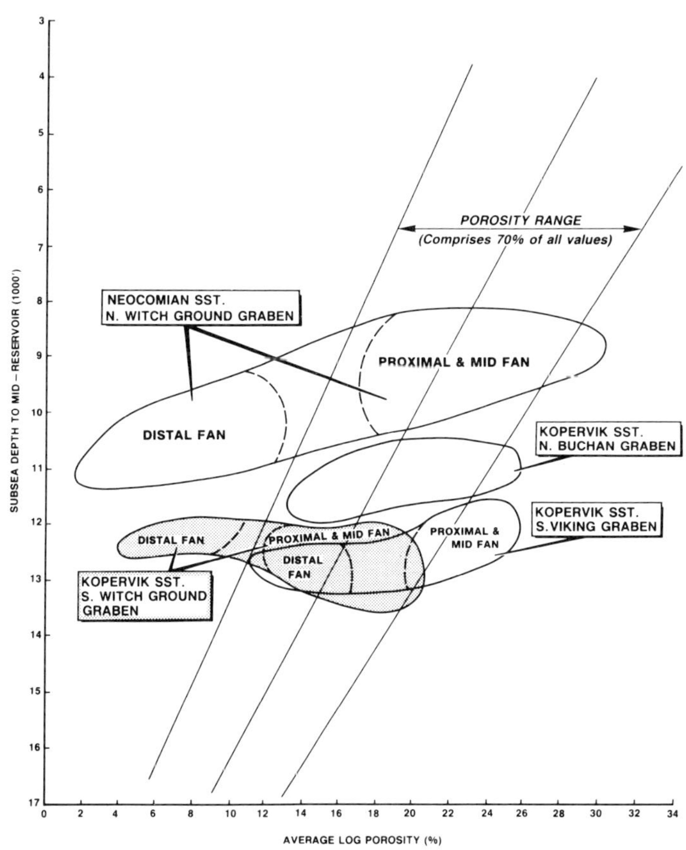

Fig. 11. Porosity distribution of the Lower Cretaceous Sandstones.

In addition, the formation of the diagenetic phases are inhibited by early emplacement of hydrocarbons resulting in less porosity occlusion in the hydrocarbon bearing zone and greatly enhanced reservoir quality of the sands.

The Austrian 'tectonic event'

The Central North Sea, situated between the Boreal Sea in the north and the Tethyan Sea in the south, was probably affected only indirectly by the Austrian tectonic event (Fig. 12). Connections with both oceans seems to have existed along the Viking Graben to the north and, in times, through the East Midland Shelf/Wessex Basin to the south (Rawson 1972). The Austrian tectonic event initiated the sea floor spreading in the Gulf of Biscay (Hesjedal & Hamar 1983) and the first phase of oceanic spreading in the Rockall Trough (Dore & Gage 1987). The tectonically induced volcanic centres seem to have been outside the Central North Sea area along the Fennoscandinan Border (Hesjedal & Hamar 1983) or within the Rockall Trough and tuffaceous material found in the Sola Formation was probably air-transported. Within the Central North Sea the tectonic effects of the mid Aptian orogenic phase were manifested as minor accommodating movements along the Jurassic North Sea rifts. In the Central North Sea the main effect of the Austrian Event was a drop in sea level due to water mass withdrawal which caused the change in lithology.

Despite an overall global sea level rise in Aptian times, the water withdrawal into the Gulf of Biscay and into the Rockall Trough overprinted this transgressive phase and in the basinal areas of the Central North Sea resulted in regressive sedimentation changing from very calcareous, greyish coloured claystone of the deeper, well connected basins to non-calcareous and darker coloured ones containing more terrigeneous material reflecting shallower and less well connected basins. Along the basin margins it caused exposure of the still unconsolidated Barremian and Neocomian and, on higher terraces, of older sediments.

In the southern Witch Ground Graben area,

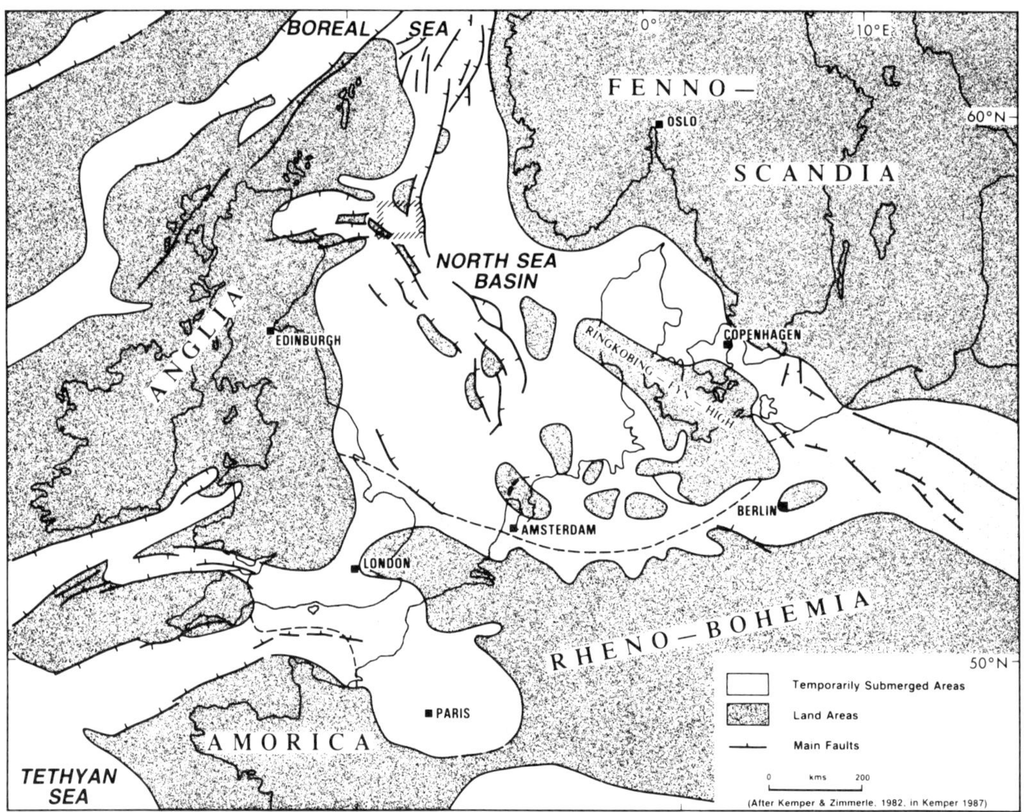

Fig. 12. Palaeogeography during Late Aptian (Lower Cretaceous) for Central and Northern Europe.

the Fladen Ground Spur and the East Shetland Platform in the north, the Halibut Horst in the west, the Renee Ridge and the Forties High in the south as well as the Jaeren High and the Sleipner Terrace were exposed to erosion. Additionally a ridge emerged as a result of halokinesis connecting the Fladen Ground Spur and the Sleipner Terrace. It intersected the connection along the South Viking to the Boreal Sea and caused the enclosure or part-enclosure of the southern Witch Ground Graben (Fig. 13).

Another interesting observation is that the organic carbon content of the Sola Formation is considerably higher than in the underlying Valhall Formation (Jensen & Buchardt 1987) which adds evidence to the theory of an enclosed basin. The Sola Formation was probably deposited in a restricted sea with reduced oxygen supply in the basin centres and restricted water circulation.

Prior to the mid Aptian sea level drop Neocomian and Barremian sediments onlapped or covered the Fladen Ground Spur and probably

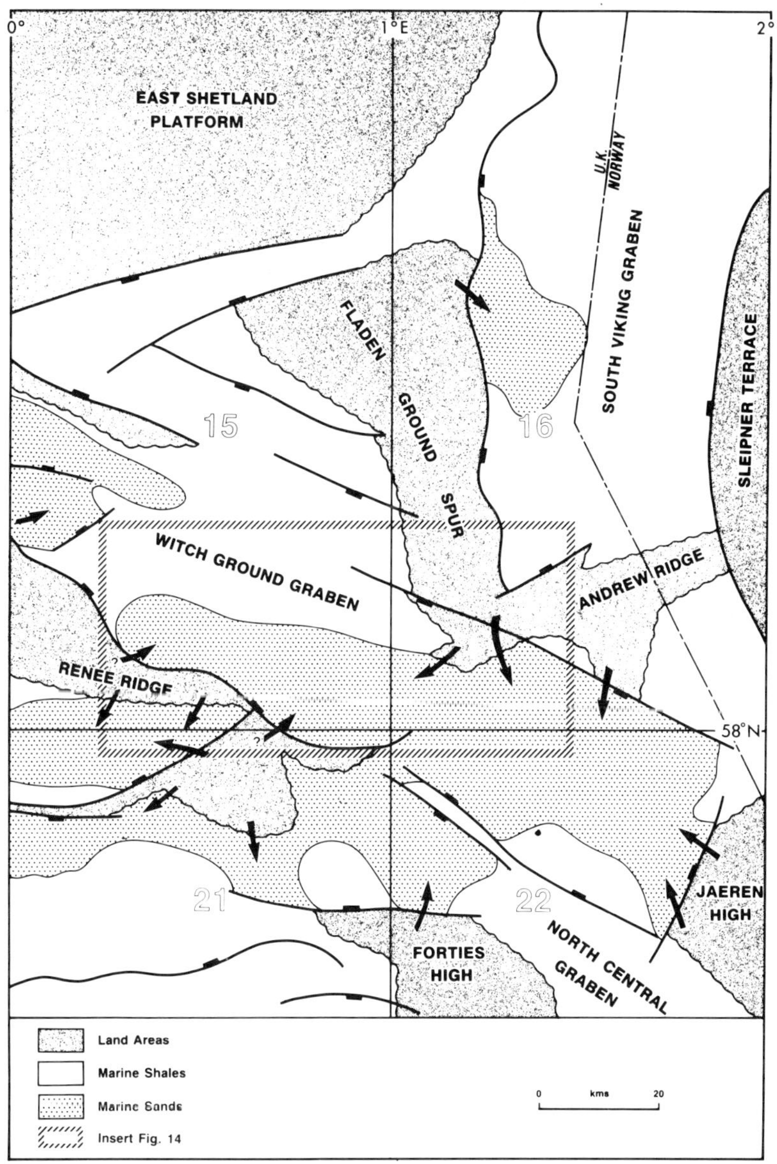

Fig. 13. Lower Cretaceous palaeogeography.

the Renee Ridge as well. These emergent features are overlain, if at all, by thin sediments of generally late Lower Cretaceous age. Positive submarine structures are known to have condensed the Lower Cretaceous sequence (Fig. 14).

With the onset of the Austrian orogenic phase the Fladen Ground Spur became exposed to erosion and was slightly tilted to the east. The Renee Ridge was probably subaerially exposed, as well as the Forties and Jaeren Highs and the Sleipner Terrace. Sands were removed from these areas and deposited in the Witch Ground Graben as turbidity currents or submarine gravity flows. These followed the sea bottom topography into the deepest parts of the basin (Fig. 15) forming an almost sand-sheet like geometry.

Basal Cretaceous as well as Upper Jurassic and Carboniferous reworking is described indi-

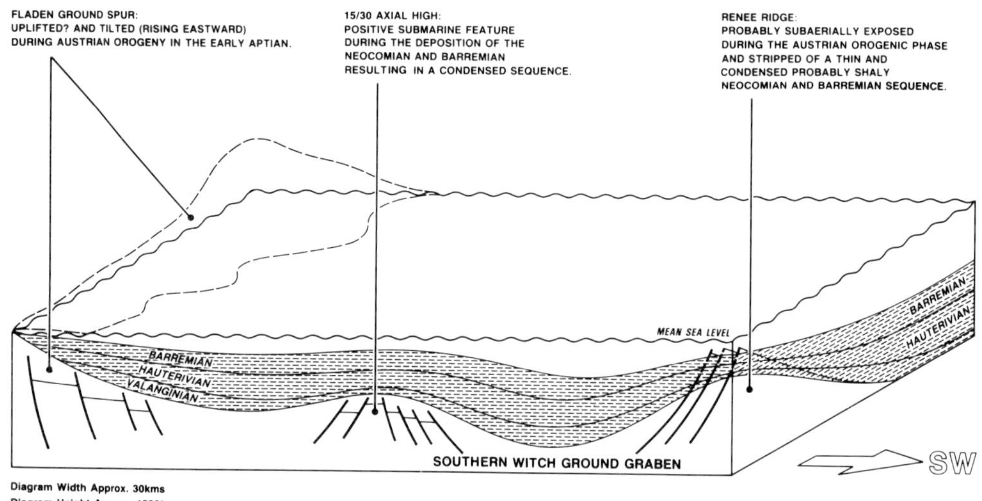

Fig. 14. Block diagram of the southern Witch Ground Graben prior to the Sola Formation deposition.

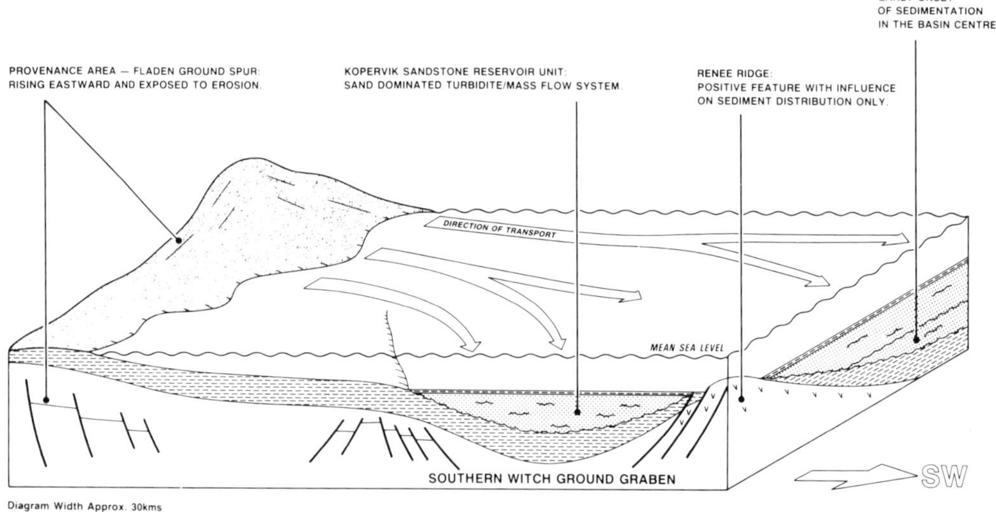

Fig. 15. Block diagram of the southern Witch Ground Graben prior to the deposition of the Kopervik Sandstone 'Ratty Unit'.

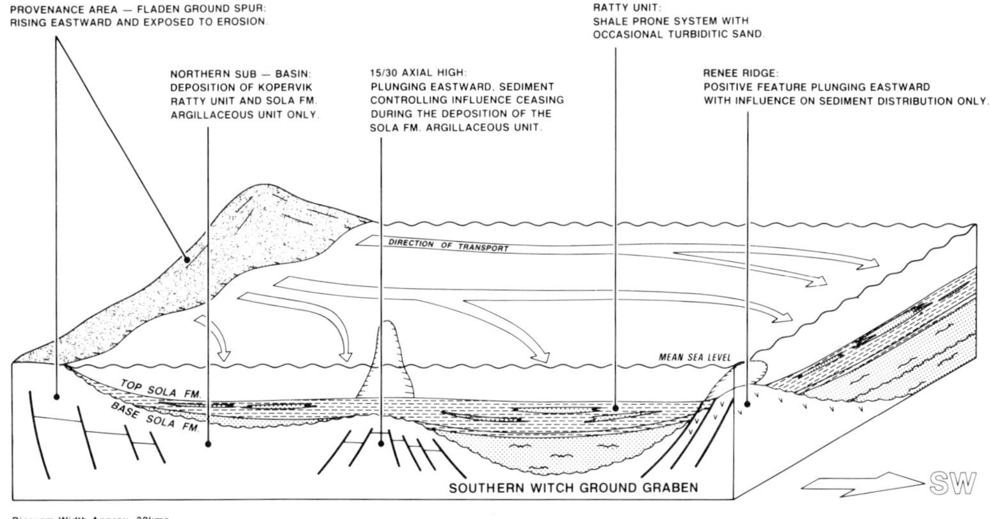

Fig. 16. Block diagram of the southern Witch Ground Graben after the Sola Formation deposition.

cating the exposure of Barremian, Neocomian and Upper Jurassic sediments. A gap in sand sedimentation appears to be represented by the shale marker overlying the Reservoir Unit which persists throughout the southern Witch Ground Graben. A renewed pulse of sand sedimentation saw the deposition of the Ratty Unit consisting of thin interbedded sands, silts and shales. This unit can be correlated throughout the southern Witch Ground Graben. The shale prone Ratty Unit with occasional probably channelized turbiditic sands indicate the continuous rising Aptian sea level and the levelling of the provenance areas (Fig. 16).

The overlying Argillaceous Unit, deposited in the same depositional environment as the Kopervik Sandstone unit, shows the cessation of the coarse clastic deposition. The Austrian orogenic influence on the sedimentation ceased in the early Albian with the deposition of the argillaceous unit. Its truncation in some parts of the southern Witch Ground Graben suggests some late tectonic activity, and younger sands might be found closer to the provenance areas.

Conclusions

The Sola Formation and its arenaceous member, the Kopervik Sandstone, is found in Aptian basins only. Its presence is due to the Austrian orogenic phase which heralded the rifting of the Gulf of Biscay and the Rockall Trough. The Central North Sea was affected by this tectonic event only indirectly. Due to water mass withdrawal, emergent areas were exposed to erosion providing the sediment source areas and compensating fault movement might have generated minor uplifts and block rotation.

Despite an overall global Aptian sea level rise the Kopervik Sandstone was deposited as a 'regressive' basin filling sediment. Younger units of the Kopervik Sandstone extended further and covered larger areas. By the end of the Sola Formation deposition the coarse clastic input ceased and the onset of the Rodby Formation saw a reversal of the depositional environment from enclosed basins to open marine conditions.

The Kopervik Sandstone Reservoir Unit was deposited as turbidites or submarine fan deposits forming almost sand sheet like features, while the Ratty Unit, a rather shale prone sequence with less clastic material, was deposited probably as channelized turbidites. The sedimentary structures are generally massive with minor wavy laminae. Most sandstone beds have an erosional base and sharp top. Porosities within the hydrocarbon bearing sands range from 12% to 24% with the distribution of the better quality sands dependent on depositional facies.

The author wishes to thank Conoco management for permission to publish this paper, the interpretations and conclusions of which are based partly on proprietary data. I wish to acknowledge the contributions of various colleagues of the Conoco Exploration Department and to thank the staff of the Drafting Department.

References

Dore, A. G. & Gage, M. S. 1987. Crustal alignment and sedimentary domain in the evolution of the North Sea Northeastern Atlantic Margin and Barents Shelf. *In*: Brooks, J. & Glennie, K. W. (eds) *Petroleum Geology of North West Europe*, Graham & Trotman, London, 1131–1148.

Hesjedal, A. & Hamar, G. P. 1983. Lower Cretaceous stratigraphy and tectonics of the south–southeastern Norwegian offshore. *Geologie en Mijubouw*, **62**, 135–144.

Jensen, T. F. & Buchardt, B. 1987. Sedimentology and geochemistry of the organic carbon-rich Lower Cretaceous Sola Formation (Barremian–Albian), Danish North Sea. *In*: Brooks, J. & Glennie, K. W. (eds) *Petroleum Geology of North West Europe*, Graham & Trotman, London, 431–440.

Kemper, E. 1987. Das Klima der Kreide-Zeit. *Geol. Jb.*, **A96**, 5–185.

Rawson, P. F. 1972. Lower Cretaceous (Ryazanian–Barremian) marine connections and cephalopod migration between the Tethyan and Boreal Realms. *In*: Casey, R. & Rawson, P. F. (eds) *The Boreal Lower Cretaceous*, Seel House, Liverpool, 131–144.

—— & Riley, L. A. 1982. Latest Jurassic–Early Cretaceous events and the 'Late Cimmerian Unconformity' in the North Sea. *American Association of Petroleum Geologists Bulletin*, **66**, 2628–2648.

Ziegler, P. A. 1975. Geologic evolution of the North Sea and its tectonic framework. *Bulletin American Association of Petroleum Geologists*, **59**, 1073–1097.

Early Palaeogene tectonics and sedimentation in the Central North Sea

N. J. MILTON, G. T. BERTRAM & I. R. VANN

BP Exploration, 301 St Vincent Street, Glasgow G2 5DD, UK

Abstract: The early Palaeogene deposits of the Central North Sea have been divided by Stewart into ten depositional units, on the basis of seismic stratigraphy. These are interpreted as the products of variations in relative sea level. The units may be traced from the shelf into the basin using wireline log markers correlated within a biostratigraphic framework. These are interpreted as the signature of transgressive maxima when basinal clastic supply was at a minimum. The attitude of coal or lignite beds within the shelfal areas can be used to demonstrate a period of Palaeogene net uplift, followed by tilting and sinking of the shelf-edge and basinal areas during the later depositional episodes. Other than by a mechanism of load-induced differential compaction around buried or partially buried Mesozoic features no significant rejuvenation of Mesozoic faults can be demonstrated. The area of uplift extended at least as far as the western limit of the Beauly Formation.

The products of Stewart's depositional episodes vary depending on the tectonic activity. During the uplift phase, thick massive basinal fans were deposited and little or no shelfal deposits are preserved. During the tilting/sinking phase thick progradational wedges were deposited, and basinal deposition was largely confined to shales with relatively small local sand systems.

The Palaeogene marks a period of considerable clastic input to the Central North Sea Basin. The pelagic chalks and marls of the Cretaceous were replaced by thick submarine gravity-flow sands and reworked tuffs, followed by prograding coastal wedges. The influx of clastics and the development of coastal systems is related to an episode of uplift affecting the Scottish Mainland and much of the northwestern North Sea Basin (Rochow 1981). This uplift was probably related to the early Tertiary opening of the North Atlantic (Bott 1975), and has since partially collapsed in the North Sea Basin. A similar influx of clastics into the Faeroes Basin to the west of the UK, occurred at the same time (Mudge & Rashid 1987). In this paper we investigate the timing of uplift and collapse with respect to the depositional episodes, and the effects of the uplift and collapse on the preserved deposits of those episodes.

Early Palaeogene sedimentation

The Palaeogene deposits of the Central North Sea Basin (Fig. 1) are well suited to seismic stratigraphic analysis. Parker (1975) first described the seismic expression of early Palaeogene sedimentary facies, recognizing an early turbiditic unit and a late 'deltaic' progradational unit. This was refined and elaborated by Rochow (1981), and integrated with the lithostratigraphy of Deegan & Scull (1977), who recognized a threefold subdivision of the turbiditic unit (Montrose Group) and a twofold subdivision of the 'deltaic' unit (Moray Group; Dornoch and Beauly Formations). Stewart (1987) was able to demonstrate a tenfold subdivision of the early Palaeogene, on the basis of seismic stratigraphy. He interpreted the cyclicity of sedimentation as a result of changes in relative sea level, and the bounding hemipelagic mudstones between the basinal fans as related to high-stand events.

Examination of Stewart's (1987) type well ties shows that the seismic sequence boundaries in a basinal setting coincide exactly, within the limits of seismic resolution, with consistent wireline log markers; namely cuspate high gamma readings (gamma peaks) and low sonic velocities (sonic troughs). These are illustrated on a type well from the Central North Sea in Fig. 2. The cuspate log markers are believed to represent pauses between pulses of submarine fan deposition, when basinal sedimentation was relatively condensed and restricted to pelagic muds. They therefore coincide with biostratigraphic hiatuses or condensation, and with the marine onlap/downlap surface below the subsequent fan system. They contain the bounding hiatal surfaces of Galloway (1989) when maximum transgression effectively cuts off fluvially-derived clastic input to the basin by raising base-level. Stewart's (1987) seismic sequences may therefore be equated with parts of Gallo-

From Hardman, R. F. P. & Brooks, J. (eds), 1990, *Tectonic Events Responsible for Britain's Oil and Gas Reserves*, Geological Society Special Publication No 55, pp 339–351.

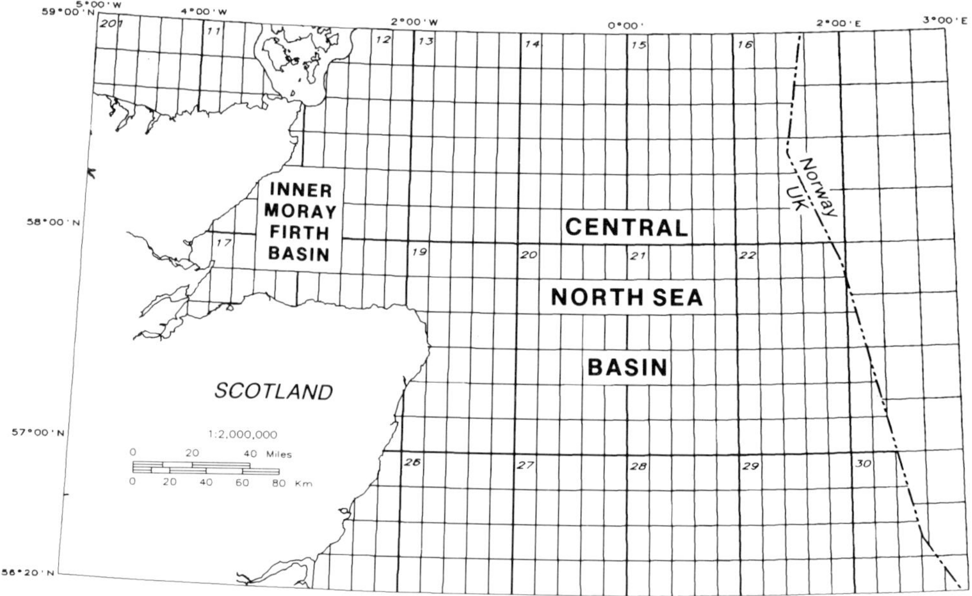

Fig. 1. Location of the study area.

way's (1989) genetic depositional episodes. In a shelf setting (Fig. 3), similar gamma maxima/sonic minima are recognised lying above the major progradational units. These are isochronous with the basinal hiatal markers, within the resolution of current biostratigraphic control. The gamma peak above Stewart's (1987) Sequence 9 is a particularly widespread marker, and has been used (with the aid of biostratigraphy) to correlate the basinal Sequence 9 with the Beauly Formation. The correlation between the basinal Sequence 10 of Stewart (1987) and the progradational unit shown in Fig. 3 is more tentative. These gamma peaks in a shelf setting represent the temporary establishment of marine conditions subsequent to drowning of the prograding units, and may be related directly with the maximum flooding surfaces of Galloway (1989).

The shelfal units (8, 9 & 10) are frequently capped by widespread coals or lignites. In the west of the area the sands at the top of the Dornoch Formation coarsening-up unit are frequently absent, and the coals lie on prodelta shales (Fig. 4). Any unconformity must therefore lie below the coal, and the coals can be ascribed to the retrogradational part of the episode, when relative sea-level rise led to aggradation of the coastal plain during a period of little or no sand input.

A chronostratigraphic representation of the early Palaeogene deposits of seismic Sequences 2 to 9 of Stewart (1987) is shown as Fig. 5 using the interpretation of the log markers (above) as maximum flooding markers to divide the stratigraphy into depositional episodes *sensu* Galloway (1989). The relationship between Stewart's (1987) seismic sequences and the lithostratigraphy of Deegan & Scull (1977) is shown in Table 1. In Fig. 5 the basinal submarine fans are shown as representing a later part of the depositional episode than the prograding wedges. This is a model as there is insufficient biostratigraphic resolution to determine the respective time relationships between fan and wedge in a single depositional episode.

Early Palaeogene tectonics

The nature of basal, early Palaeogene tectonics is difficult to determine. Except in those areas with preserved coals, there are no horizons which may be restored to flat and used to indicate relative subsidence and fault movement, or to derive any causal link between sedimentary pulses and increases and decreases in tectonic (i.e. compensated for loading) subsidence rate. Under these circumstances tectonic reconstruction is difficult (Bertram & Milton 1989). Published analyses of the early Palaeogene agree on the interpretation of an end-Cretaceous episode of uplift and tilting of the basin margins, and many authors also infer a simultaneous increase in subsidence of the basin and a rejuvenation of

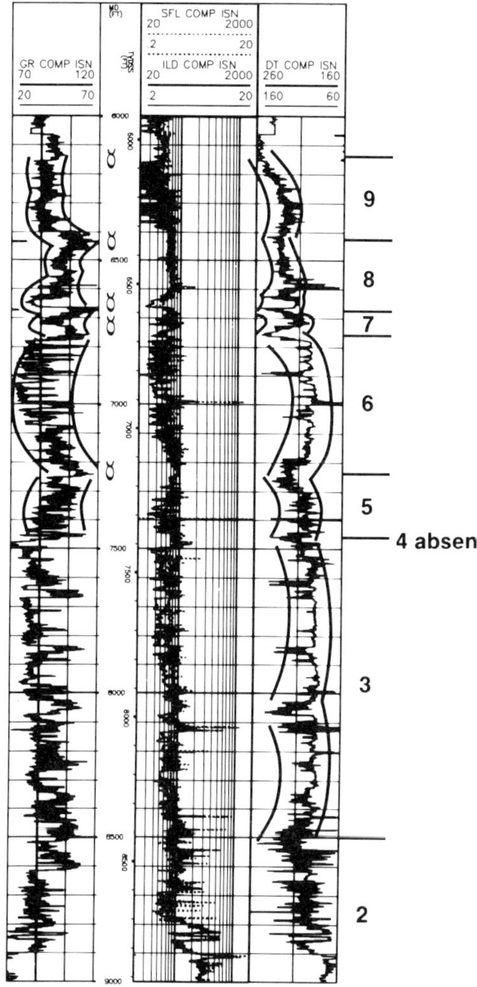

Fig. 2. Wireline log response of basinal early Palaeogene deposits.

the Mesozoic fault trends (e.g. Parker 1975; Rochow 1981; Morton 1982; Conort 1986).

The uplift of the basin margins is clearly demonstrated by the onlap of late Palaeogene coals onto pelagic Cretaceous chalk (Fig. 6), showing appreciable shallowing in this area. Uplift of a basin-margin source area may also be indirectly inferred from the sudden influx of clastics.

The interpretation of evidence for rejuvenation of Mesozoic faults is, however, more subjective. Most major faults in the Central North Sea show seismic geometries in the Palaeogene which can be explained entirely by differential compaction of a pre-Palaeogene sequence around a basement high, both during and after Palaeogene deposition (Fig. 7). Where the base Palaeogene is broken by a 'fault plane', it is probable that the basement high existed as a fault scarp at the sea-bed. There is no direct evidence for fault rejuvenation except as a result of loading-driven differential compaction (Bertram & Milton 1989).

The evidence for early Palaeogene subsidence of the grabens is also ambiguous. The coals of episodes 8, 9 & 10 are frequently tilted to the eastsoutheast, which would suggest post – episode 10 downwarp.

Backstripping this area shows that the current position of the Top Sequence 8 coals (which were probably deposited at or close to sea level) is incompatible with a simple post-Jurassic rift-related thermal decay model (Fig. 8). The coals are considerably deeper, and the ratio of net post-coal to net pre-coal subsidence considerably greater, than the simple model would predict. Thorne & Watts (1989) also note this apparent post-Palaeocene increase in subsidence rate.

Some or all of this tilting and apparent in-

Table 1. *A comparison of Stewart's sequences with published lithostratigraphy*

Sequence	Shelfal Unit	Basinal Sand	Basinal Shale
10	Unnamed	Frigg Fm	
9	Beauly Fm	Cod sands*	Balder Fm
8	Dornoch Fm	Unnamed	Sele Fm
7	Unnamed	Forties Fm	Sele Fm
6	Unnamed	Unnamed	Lista Fm
5	Not preserved?	Unnamed**	Lista Fm
4	Not preserved	Andrew Tuff	Lista Fm
3	Not preserved	Andrew Fm	Lista Fm
2	Not preserved	Maureen Fm	Marl Unit
1	Not preserved	Ekofisk Fm	Ekofisk Fm

* after Conort (1986).
** 'Montrose Group undifferentiated' of Deegan & Scull (1977).

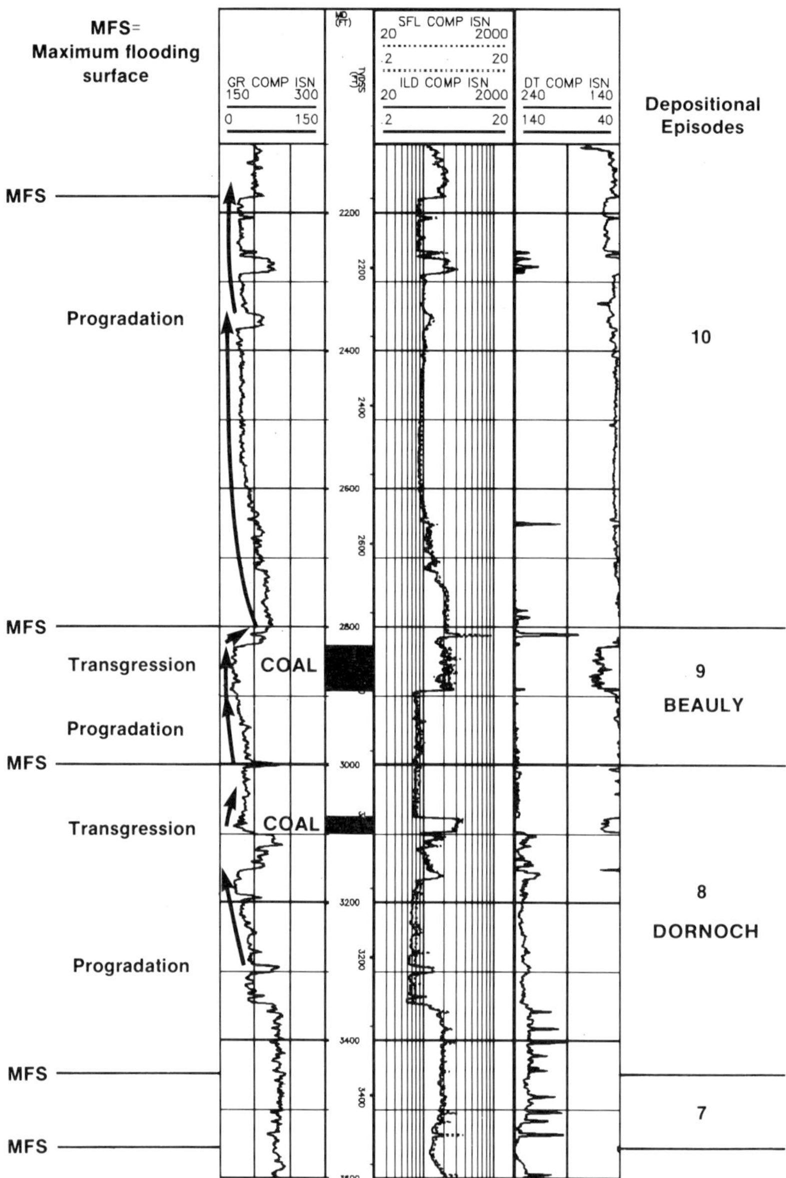

Fig. 3. Wireline log response of shelfal early Palaeocene deposits.

creased subsidence may be a result of flexural loading by the thick post-Eocene deposits of the Central North Sea. Alternatively the anomalous depth of the coals could be related to collapse of an early Palaeogene uplift period which reached its widest extent during deposition of sequence 8. The net pre-coal subsidence in this area would therefore be equal to the post-Jurassic thermal subsidence minus the net pre-coal uplift.

The geometry of the Beauly Formation (Deegan & Scull 1977), approximately equivalent to the coastal deposits of genetic depositional episode 9 (Fig. 9) is that of a double wedge thinning to the east and west, westwards by onlap onto the subcropping Dornoch Formation, and eastwards by basinward downlap and depositional thinning of a coastal wedge. Where coals are present above and below the Beauly Formation, the thickness of the formation may

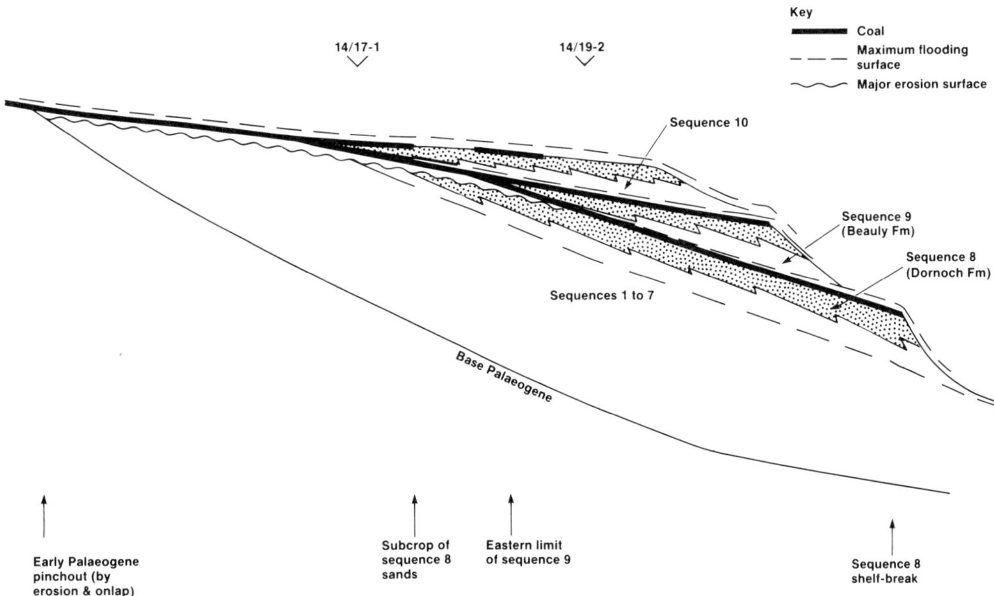

Fig. 4. Diagrammatic cross section of the early Palaeogene near latitude 58 30′ North, showing convergence of the coal/toplap horizons, and erosion below the Sequence 8 coal.

be directly equated to the total subsidence (plus any eustatic sea-level rise) during the period between deposition of the two coals. Where the coals converge the sum of subsidence and eustatic rise is zero. Subsidence during Beauly deposition therefore increased eastwards. West of the line of convergence there is erosion below the coal, which sits on sequence 7 marine muds, with units 8 and 9 absent, and by inference uplift increased westwards. The convergence of the coals marks the location of a tilting axis at a point in time.

Further west (Fig. 10) the coal on top of the sequence 10 wedge meets the top Beauly coal. Along this line there has been no net subsidence during the time of deposition of Unit 10.

So both Units 9 and 10 show evidence of contemporary tilting. The westwards shift of the tilting axis from Unit 9 to Unit 10 is evidence of the shrinking of the area of uplift, and the re-establishment of subsidence over this area of the North Sea during the early Eocene.

The depositional limits of progradational coastal wedges are partially constrained by relative sea-level, and are therefore influenced by uplift and subsidence (Bertram & Milton 1989). The limits of the coastal wedges of episodes 8, 9 & 10 (Fig. 11) are subparallel to the Orkney-Shetland axis, and show the outline of the uplifted area. By contrast the depositional limits of the submarine fans of the earlier episodes (see maps in Stewart 1987) are controlled by basin-floor topography, which itself is controlled by compaction into the buried Mesozoic Grabens. These 'compaction grabens' are persistent throughout the Palaeogene, and influence basinal deposition during episodes 2b, 3, 4, 5 and 7. Thus, although they may have been infilled and buried by any one episode of fan deposition (which will replace pre-existing basin-floor topography with a sedimentary fan profile), these compaction features must have become re-established during the inter-fan hiatuses as dewatering followed loading.

Early Palaeogene tectonics may be summarized as follows.

(i) Uplift of North West Britain at the end of the Cretaceous led to rejuvenation of source areas and the deposition of submarine clastics into the Central North Sea.

(ii) During deposition of the Dornoch and Beauly Formations (episodes 8 and 9) there was a period of tilting, with uplift in the west and subsidence in the east.

(iii) By the time of the deposition of the Beauly Formation, the eastern extent of net uplift reached as far as the line of convergence of the top and base Beauly Formation coals. It is probable that net uplift reached at least this far during deposition of the Dornoch Formation, and may have reached as far as the eastern limit of the Top Dornoch (sequence 8) coal.

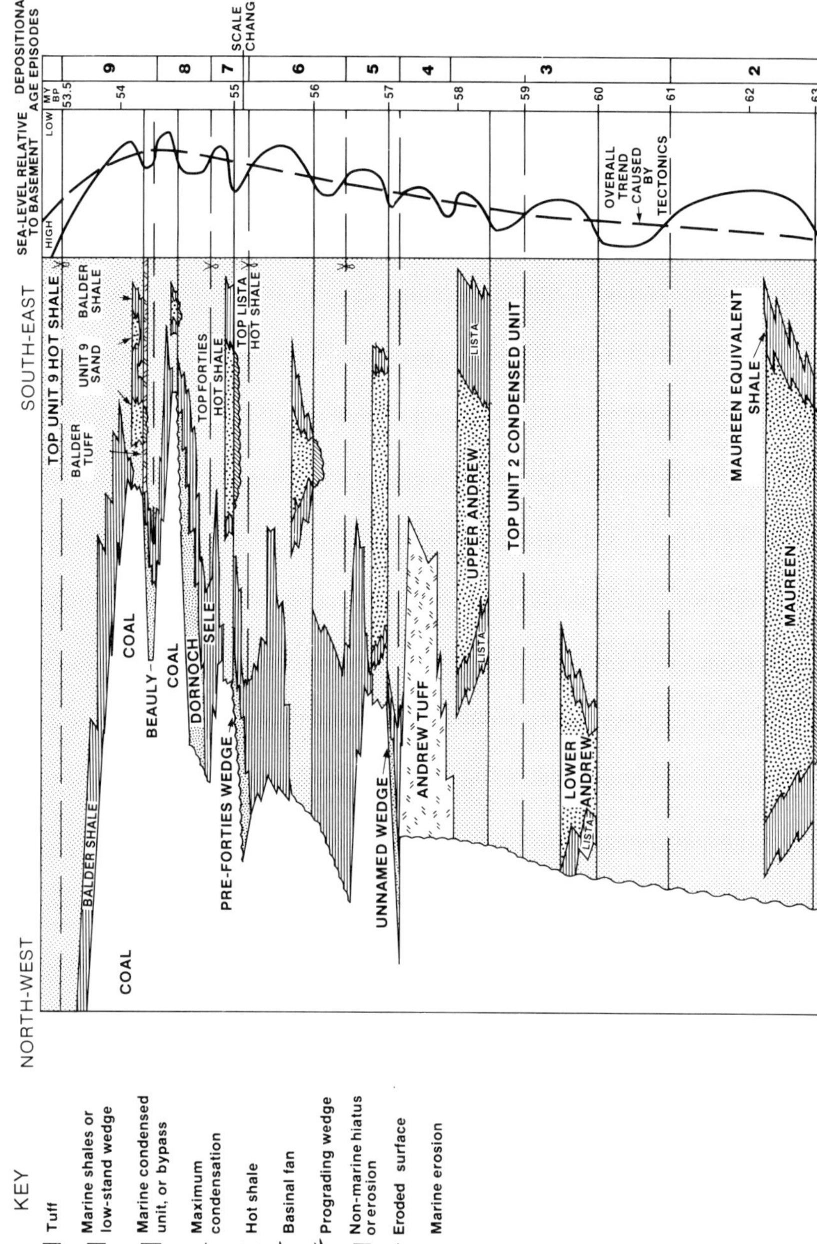

Fig. 5. Chronostratigraphic diagram of the early Palaeogene of the Central North Sea.

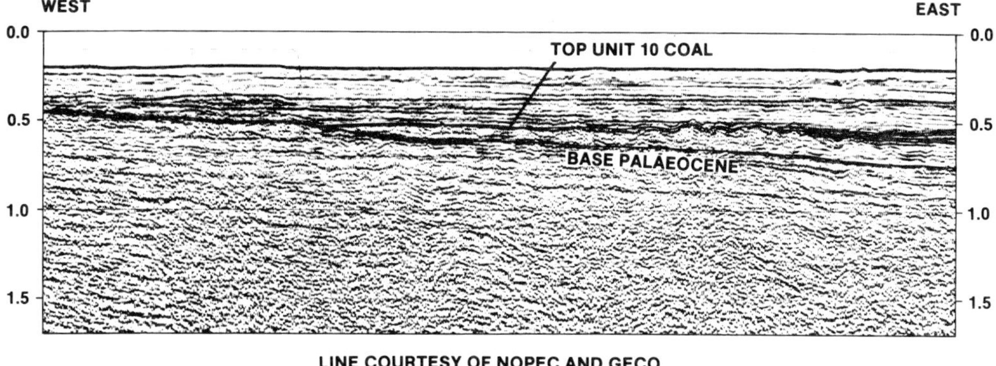

Fig. 6. Western limit of the early Palaeogene, where the coal/toplap horizon near the top of Unit 10 onlaps the base Palaeogene surface.

(iv) During depositional episode 10 (and probably 9) the zone of uplift retreated westwards as subsidence became more widespread. The tilting mentioned above possibly took place during the transition from uplift to subsidence.

(v) Thermal subsidence (the Post-Rift phase to the Mesozoic rifting) has now resumed over much of the North Sea. The Inner Moray Firth, Scotland and some of the Marginal Platform areas retain significant net uplift.

(vi) At no time during this cycle of uplift and collapse in the Central North Sea is there evidence of rejuvenation of Mesozoic faults. The cycle can be explained by loading – induced differential compaction or burial of sea-bed topography.

The control of tectonics on sedimentation

It is argued above that the Palaeocene sediments are the product of ten depositional episodes superimposed on one tectonic episode. The depositional episodes can be related to changes in relative sea-level (Stewart 1987), and the tectonic episode of uplift and subsidence results in a fall and rise of relative sea level. The mechanism causing the ten short-scale relative sea-level cycles is unknown; possibly eustatic sea-level variations, intra-plate stress or short-scale tectonic variations. The depositional products of the short-scale cycles depends on their position on the large-scale tectonic cycle; ie whether they are part of the uplift phase or sinking phase.

The uplift phase (depositional episodes 2 to 7) results in a period of relative sea-level fall. The deposits of smaller-scale cycles within the uplift phase comprise the Montrose Group, (Deegan & Scull 1977), an alternation of thick massive submarine fans with a sheet-like morphology, with background pelagic mudstones. The fans cover wide areas (see maps in Stewart 1987). No coals are preserved, and no significant shelfal sands are preserved. Earlier Palaeogene and older microfossils are extensively reworked. Log response is dominated by sonic-log bows, and the intra-depositional condensed gamma peaks are frequently poorly developed.

The tilting/sinking phase (depositional episodes 8 to 10) results in a transition from relative sea-level fall to relative sea-level rise.

The deposits of smaller-scale cycles within the tilting/sinking phase comprise the Moray Group (Deegan & Scull 1977), a series of progradational coastal/shelfal/deltaic wedges with thin basinal shale equivalents (the Rogaland Group) Toplap surfaces and transgressive (?) coals are frequently preserved. Basinal fans (the Frigg sands of McGovney & Radovich (1985), the 'Cod Sands' of Conort (1986) and Enjolras *et al.* (1986) and the Sequence 9 sands of Stewart (1987)) are relatively small in areal extent, frequently with a characteristic mounded geometry. Generally however the basinal deposits of the sinking phase consist of pelagic mudstones punctuated by widespread correlative gamma peaks. Reworking of Palaeogene microfossils is less common than during the uplift phase.

It should be noted that during this episode, sinking (or relative sea-level rise) may have been restricted to the basin centre. Certainly during deposition of Unit 9 the basin flanks were still rising. However the area of uplift was shrinking, and the areas of net relative sea-level rise during depositional episodes 8 to 10 include the shelf edges of the previous depositional episodes. Also the inferred relative sea-level rise refers to the background basement subsidence. Falls of

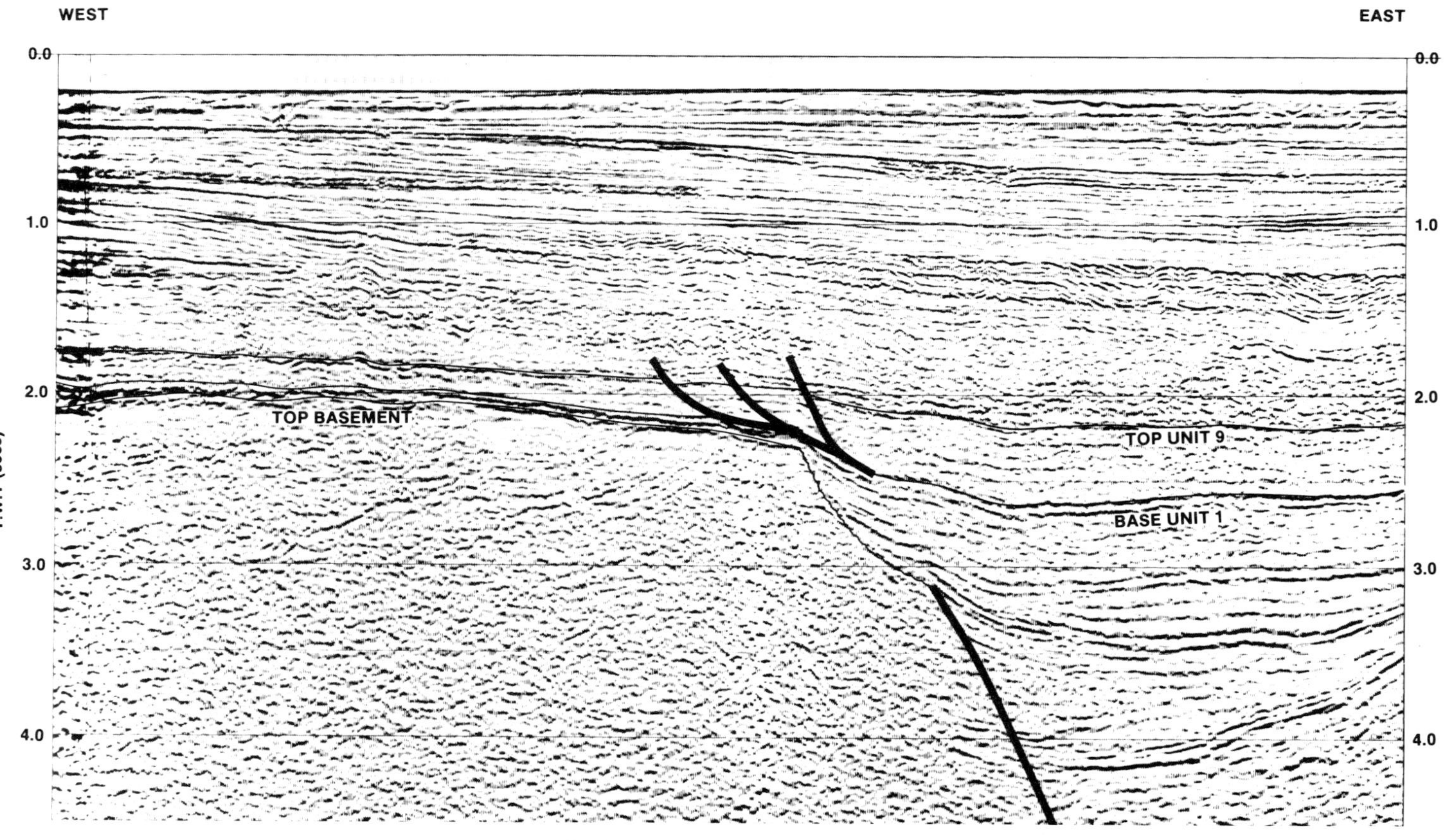

Fig. 7. The response of the Palaeogene to a major Central North Sea fault. Bedding geometries are compatible with differential compaction of the pre-Palaeogene sequence over a buried basement feature.

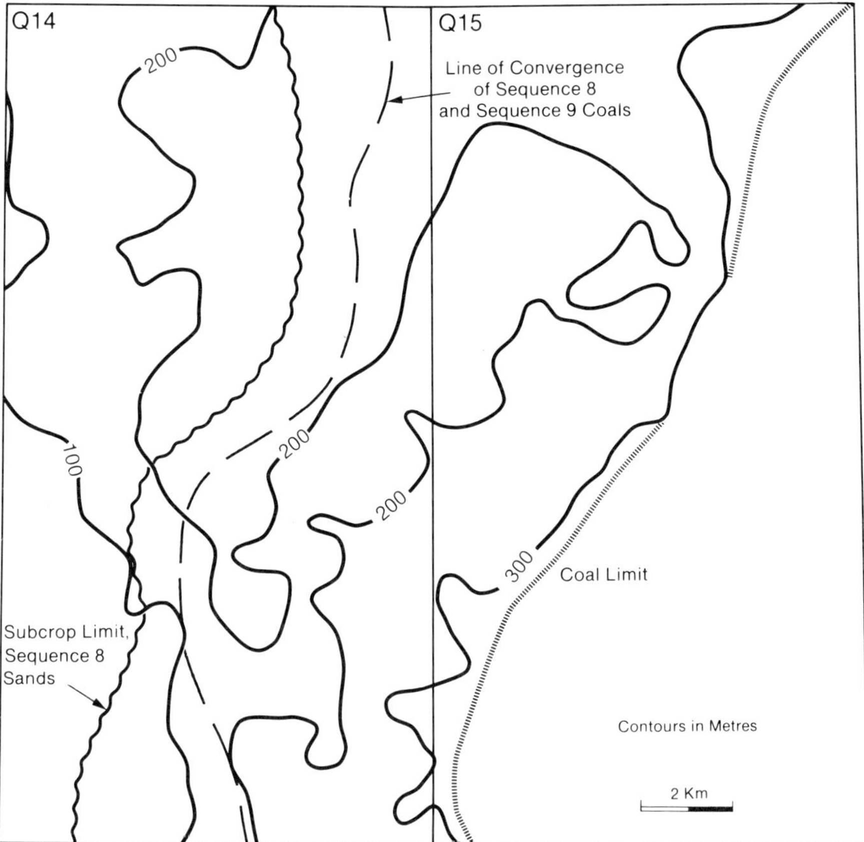

Fig. 8. Depositional palaeobathymetry of Top Sequence 8, reconstructed from a simple model of post-Jurassic thermal subsidence. A correction for compaction has been made, but not for differential loading. As this horizon is marked by a coal, this simple model is not sufficient.

relative sea level could have occurred on a smaller time scale.

Enjolras *et al.* (1986) identify the submarine sandy deposits of the tilting/sinking phase cycles with Type-II turbidites of Mutti (1985). They contrast these with the submarine fans of the uplift phase cycles which are more closely allied with Type-I turbidites. Mutti (1985) suggests that these types of turbidites were both deposited during relative lowering of sea-level and lowstand, and that they differ by the volume of the gravity flows (taken to be largely a function of relative sea-level position). If the Central North Sea submarine sands were to be related to periods of relative sea-level fall and lowstand during the ten small-scale cycles, then one could expect these to be amplified during the uplift phase and subdued during the tilting/sinking phase, resulting in a significant difference in gravity-flow volume.

Conclusions

The deposits of the early Palaeogene have been divided by Stewart (1987) into ten separate depositional units on the basis of seismic stratigraphy.

These may be interpreted as the products of episodes of shelfal or basinal sedimentation punctuated by transgression-related hiatuses, in the manner described by Galloway (1989).

The ten episodes of deposition are superimposed on a single episode of uplift and tilting/sinking (or tectonically-controlled fall and rise of relative sea-level).

The deposits of the uplift (relative sea-level fall) phase consist of massive basinal fans similar to Type-I turbidites of Mutti (1985, Enjolras *et al.* 1986) with little or no preserved shelf.

The deposits of the tilting/sinking (relative sea-level rise) phase consist of back-stepping

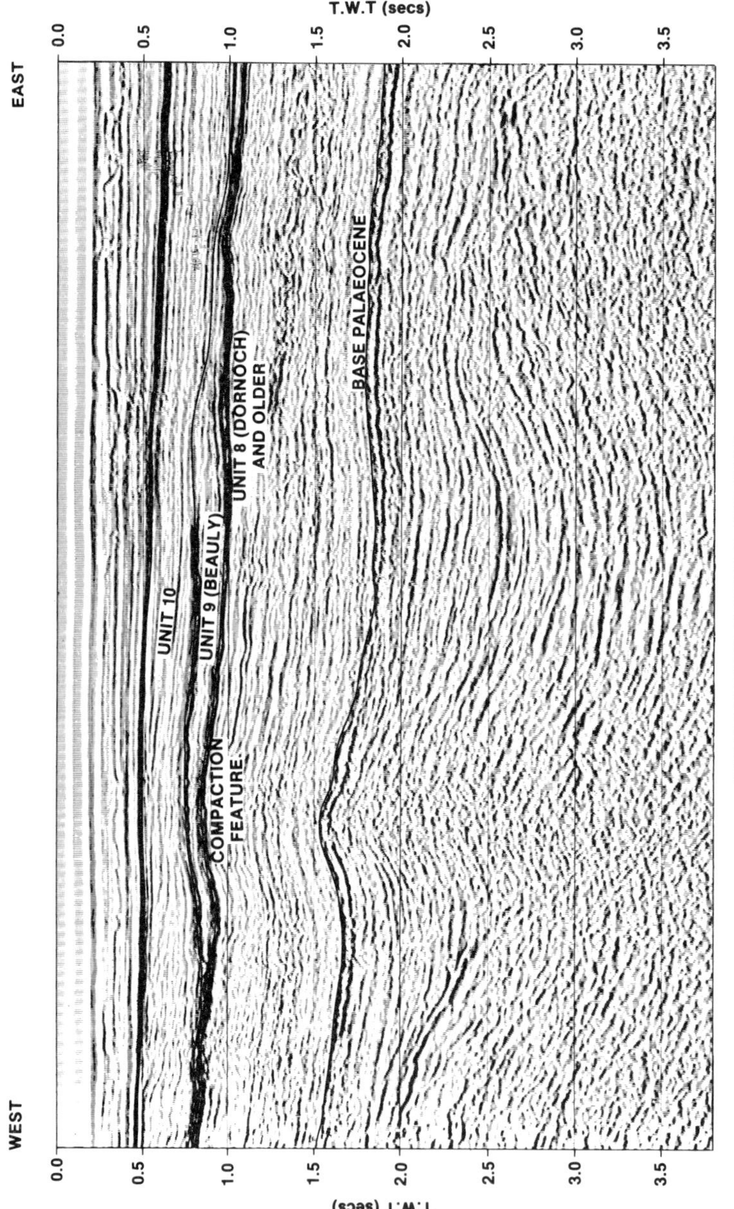

Fig. 9. The geometry of Unit 9. High-amplitude reflectors near the top and base of Unit 9 are coals which converge westwards.

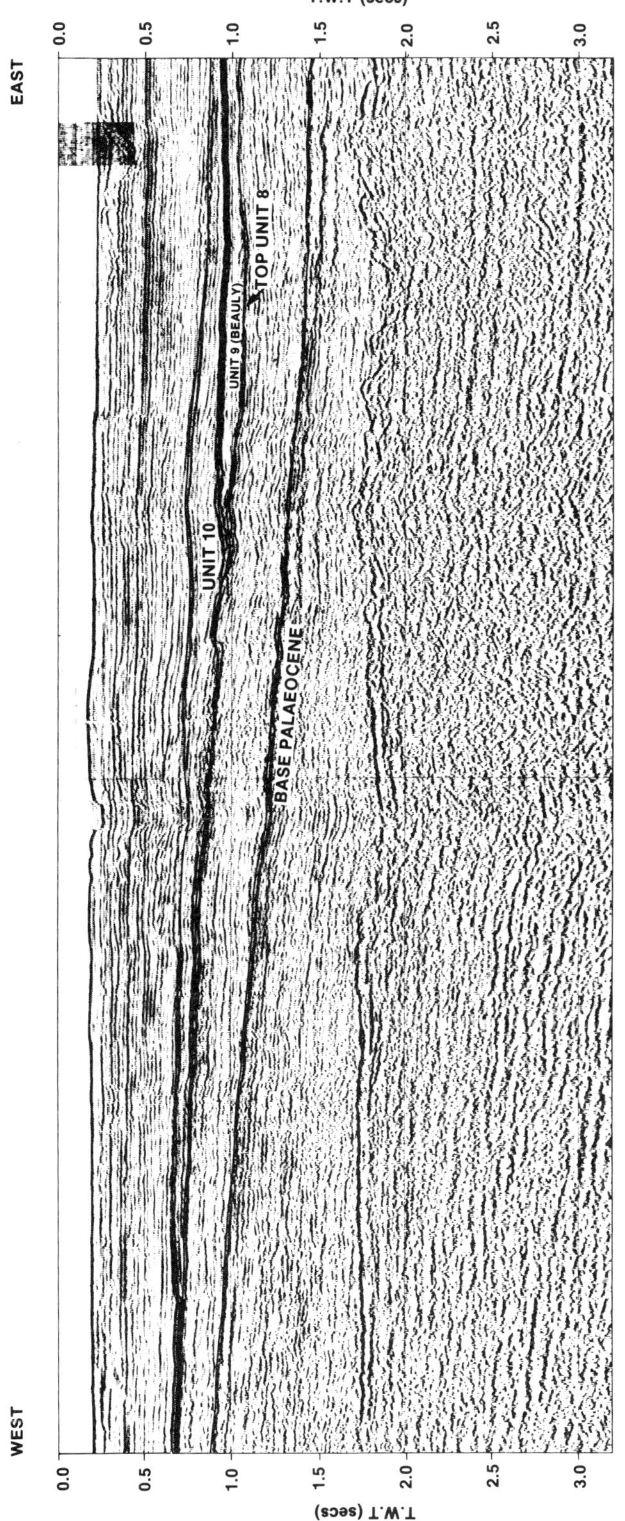

LINE COURTESY OF NOPEC AND GECO

Fig. 10. The geometry of Unit 10. This unit thins by internal onlap westwards and downlap eastwards. A possible coal at the top of the unit is indicated by high amplitudes to the west.

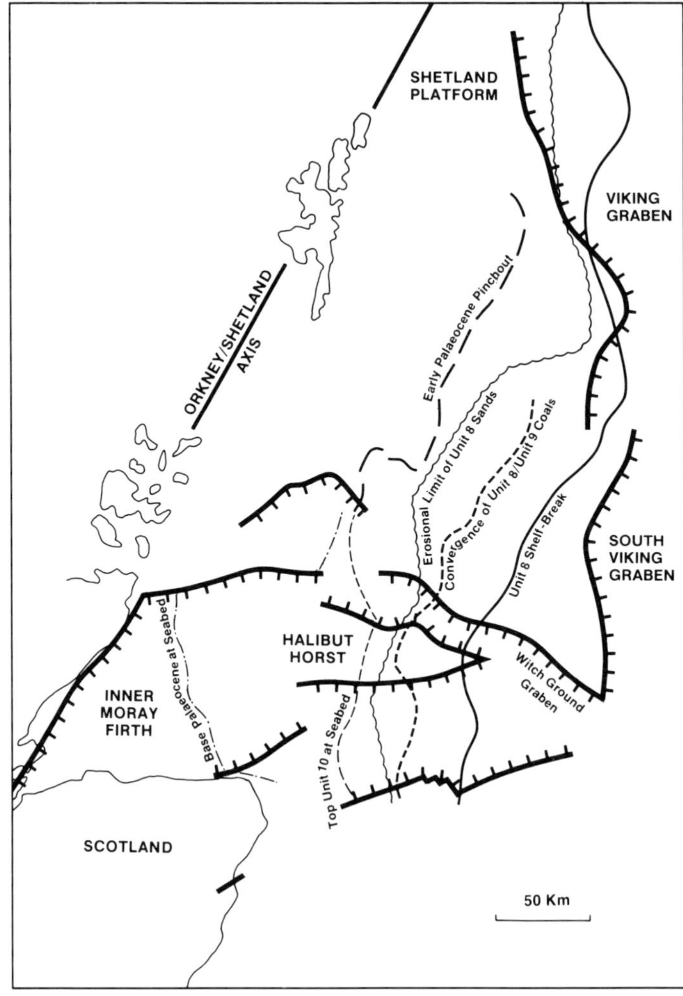

Fig. 11. Depositional and preservational limits of the shelfal early Palaeogene depositional units.

shelf/shoreline systems, with pelagic mudstones in the basin. Submarine fan systems are similar to those of Mutti (1985) Type II.

Over much of the Central North Sea the tectonic cycle of uplift and sinking resulted in the deposition in the basin of widespread reservoir sands followed by pelagic shale topseal. This combination, combined with compaction drape over Mesozoic highs, has resulted in a large number of oil and gas fields.

The authors wish to thank BP for permission to publish this paper, to acknowledge the support and assistance of their colleagues in BP, and to thank the backstripping group for permission to use Fig. 8.

References

Bertram, G. T. & Milton, N. J. 1989. Reconstructing basin evolution from sedimentary thickness; the importance of palaeobathymetric control, with reference to the North Sea. *Basin Research*, **1**, 247–257.

Bott, M. H. P. 1985. Structure and Evolution of the North Scottish Shelf, the Faeroe block and the intervening region. *In:* Woodland, A. W. (ed.) *Petroleum Geology and the Continental shelf of North-West Europe*, Applied Science, Barking, 105–116.

Conort, A. 1986. Habitat of Tertiary Hydrocarbons, South Viking Graben. *In: Habitat of Hydrocarbons on the Norwegian Continental shelf,*

N.P.S., Graham & Trotman, London, 159–170.

DEEGAN, C. E. & SCULL, B. J. 1977. *A standard lithostratigraphic nomenclature for the Central and northern North Sea.* Inst. Geol. Sci. Rept. 77/25; Bull. Norwegian Petrol. Direct. No 1.

ENJOLRAS, J. M., GOUADAIN, J., MUTTI, E. & PIZON, J. 1986. New Turbidite Model for the Lower Tertiary sands in the South Viking Graben. *In*: *Habitat of Hydrocarbons on the Norwegian continental shelf, N.P.S.*, Graham & Trotman, London, 171–178.

GALLOWAY, W. E. 1989. Genetic stratigraphic sequences in Basin Analysis I: Architecture and Genesis of Flooding — surface bounded deposition units. *American Association of Petroleum Geologists Bulletin*, **73**, 125–142.

McGOVNEY, J. E. & RADOVICH, B. J. 1985. Seismic Stratigraphic and Facies of the Frigg Fan Complex. *In*: BERG, O. & WOOLVERTON, D. (ed.) *Seismic Stratigraphy 2*. American Association of Petroleum Geologists Memoir 39, 139–154.

MORTON, A. C. 1982. Lower Tertiary Sand Development in the Viking Graben, North Sea. *American Association of Petroleum Geologists Bulletin*, **66**, 1542–1559.

MUDGE, D. C. & RASHID, B. 1987. The Geology of the Faeroes Basin. *In*: BROOKS, J. & GLENNIE, K. W. (eds) *Petroleum Geology of North-West Europe*. 751–763.

MUTTI, E. 1985. Turbidite systems and their relations to depositional sequences. *In*: ZUFFA, G. C. (ed.) *Provenance of Arenites,* Reidel, Dordrecht, 65–93.

PARKER, J. R. 1975. Lower Tertiary Sand Development in the Central North Sea. *In*: WOODLAND, A. W. (ed.) *Petroleum Geology and the Continental shelf of North-West Europe*. Applied Science, Barking, 447–453.

ROCHOW, K. A. 1981. Seismic stratigraphy of the North Sea 'Palaeocene' deposits. *In*: ILLING, L. V. & HOBSON, D. G. (eds) *Petroleum Geology of the Continental Shelf of North-West Europe*, 557–576.

STEWART, I. J. 1987. A revised stratigraphic interpretation of the Early Palaeogene of the Central North Sea. *In*: BROOKS, J. & GLENNIE, K. W. (eds) *Petroleum Geology of North-West Europe*, 557–576.

THORNE, J. A. & WATTS, A. B. 1989. Quantitative analysis of North Sea Subsidence. *American Association of Petroleum Geologists Bulletin*, **73**, 88–116.

Controls on Eocene submarine fan deposition in the Witch Ground Graben

A. W. HARDING[1], T. J. HUMPHREY[2], A. LATHAM[1], M. K. LUNSFORD[1] & M. H. STRIDER[3]

[1]*Chevron UK Limited, 2 Portman Street, London W1A 0AN, UK*
[2]*Formerly Chevron UK Ltd., now Chevron Overseas Petroleum Inc., San Ramon, USA*
[3]*Formerly Chevron UK Ltd., now Amoseas Indonesia, PO Box 2782/JKT Jakarta 10001, Indonesia*

Abstract: The framework of Eocene sedimentation in the Witch Ground Graben is described using an integrated analysis of seismic stratigraphy, biostratigraphy, and lithofacies mapping. Primary controls on the distribution of Eocene submarine fan sands are (1) Cimmerian tectonic elements, (2) palaeotopography created by deposition of Palaeocene sediments, and (3) variations in sea level.

The discovery of the Alba Field in 1984 by the Chevron-operated Sea Search Group opened up a new UK North Sea oil play, the Middle Eocene sands. Well 16/26–5 encountered 108 feet of oil pay in sands of Middle Eocene age. Subsequent appraisal wells of the Alba Field have confirmed the presence of a large stratigraphic trap, with over 1000 million barrels of oil in place. This paper will discuss the depositional environment of the Middle Eocene Alba sequence and its regional setting.

Regional stratigraphic setting

The Alba Field is located in UKCS Block 16/26, 140 miles northeast of Aberdeen (Fig. 1).

The Tertiary basin framework was shaped by Mesozoic tectonic events. Upper Jurassic to Lower Cretaceous rifting led to the development of a complex graben system in the Block 16/26 area as shown in Fig. 2. The Witch Ground Graben, with a northwest–southeast trend, and the South Viking Graben, with a north–south trend, are separated by the Fladen Ground Spur. Block 16/26 lies at the junction of these two graben systems. The Lower Cretaceous Epoch saw the termination of rifting and the development of a large thermal sag basin which continued into the Tertiary.

At the close of the Cretaceous, uplift and erosion (Ziegler 1982) of the Shetland Platform in the northwest provided large volumes of coarse clastic material which was transported by mass transport processes into a marine basin in excess of 1000 feet deep (Parker 1975). For a discussion of mass transport processes in deep water environments, see Middleton & Hampton (1976). This led to the development of (1) a southeastward prograding shelf in the northwest of the study area, (2) a slope striking generally southwest-northeast, and (3) large, sand-rich submarine fans of the Palaeocene Andrew Formation and the overlying Forties Formation.

This basin architecture persisted into the Eocene, but the continued denudation of the East Shetland Platform produced a finer-grained sediment supply to the shelf, changing basinal deposition from sand dominated facies in the Palaeocene to mud dominated facies in the Eocene. During the remainder of the Tertiary period, basinal sedimentation continued, with shale being the dominant lithology. At the end of the Middle Miocene, a phase of extensional faulting occurred, probably related to the continued sag of the North Sea Basin. The faults appear on seismic profiles to detach primarily within or above the Eocene section. The mechanism and geometry of the faulting are not, however, well understood at present and are the subject of continuing study.

Upper Cretaceous and Palaeocene

Figs 3a and 3b are a composite regional seismic line composed of the NOPEC/GECO line CNST86–16 and three Chevron group proprietary lines. The line runs from Block 15/13 in the northwest across the Alba Field in Block 16/26 to Block 22/1b in the southeast (Fig. 4) and is approximately orthogonal to regional strike of the Tertiary Basin. The northern boundary fault of the Witch Ground Graben can be seen in the left centre of Fig. 3a.

The lowest sequence of interest for this paper is the Upper Cretaceous Chalk. The post-rift

From Hardman, R. F. P. & Brooks, J. (eds), 1990, *Tectonic Events Responsible for Britain's Oil and Gas Reserves*, Geological Society Special Publication No 55, pp 353–367.

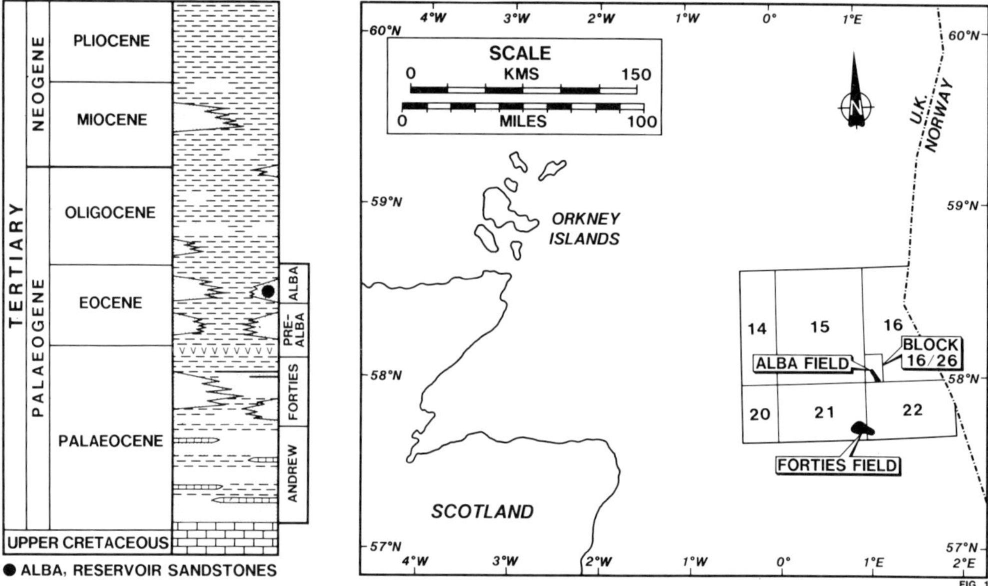

Fig. 1. Study area. Map of Central North Sea highlighting the study area. Column shows Tertiary stratigraphy and highlights the sequences discussed in this paper.

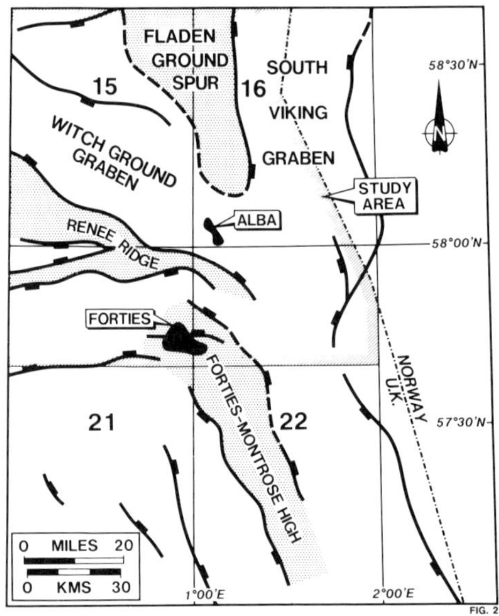

Fig. 2. Tectonic features. The main tectonic features at the termination of Late Cimmerian rifting.

sediments of the Chalk Group and the Lower Palaeocene buried the Mesozoic highs, but differential sediment compaction and the continued basin subsidence, produced variations in the basin floor topography which influenced Tertiary sedimentation by focusing sediment deposition in the area of the Witch Ground Graben.

Overlying the Chalk is the Andrew/Maureen sequence (Stewart 1987). Within the study area, the Palaeocene Andrew Formation consists of a series of submarine fan sandstones intercalated with minor claystone units.

The Forties sequence, consisting of the Forties Formation and its chronostratigraphic equivalents, overlies the Andrew Formation (Stewart 1987). From the northwest of the study area, the deltaic sands and shales of the Moray Group grade southeastwards through pro-delta slope sediments to the submarine fan sands of the Forties Formation. The regional southeastward thinning of the sequence is well demonstrated on Figs 3a and 3b. The Forties sequence slope runs across the study area from Block 21/1 in the south to Block 15/18 in the north (Fig. 5).

Pre-Alba sequence

The base of the Pre-Alba sequence is taken as the base of the Balder Formation. This is a well defined regional seismic reflector.

The upper boundary of the pre-Alba sequence is a good log marker and a recognisable seismic

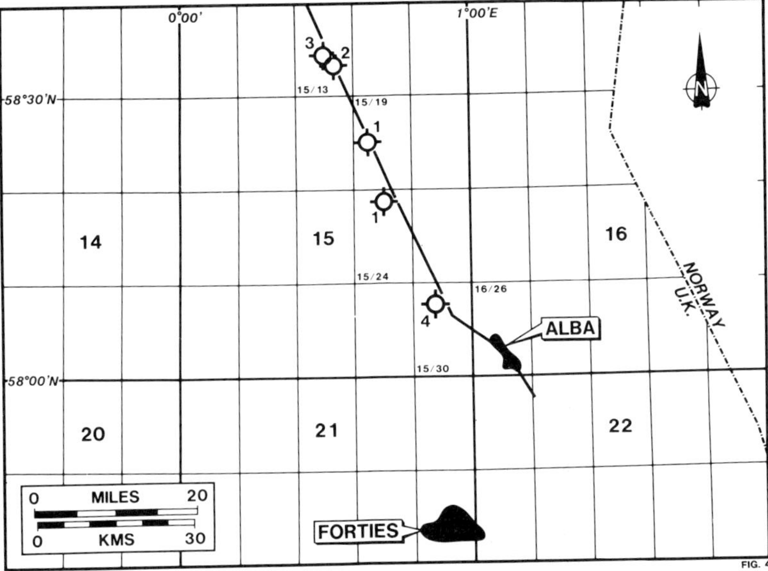

Fig. 4. Location map. The location of the regional seismic line (Fig. 3) and geological cross section (Fig. 8) is shown.

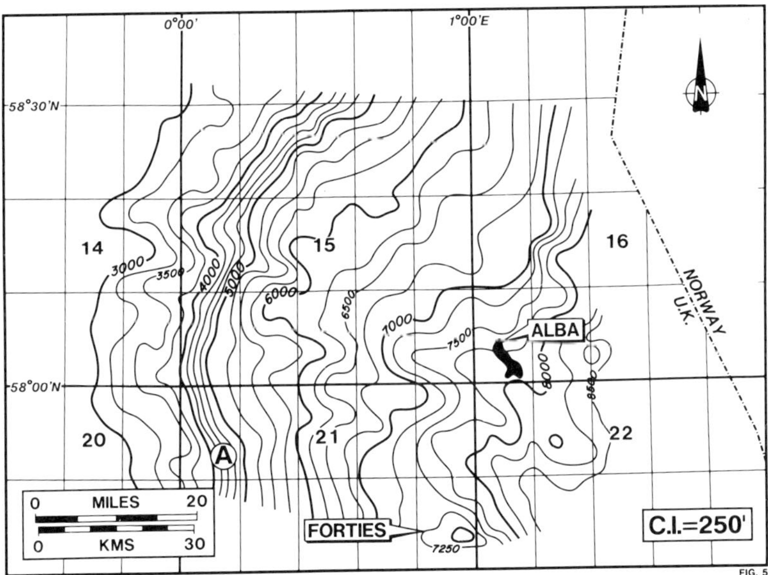

Fig. 5. Base Balder (equiv.) structure. Depth structure map at base Balder Formation (= base Pre-Alba Sequence) in feet. Moray Group Slope indicated (A). Contour interval 250 feet.

reflector referred to as the Blue Marker (Fig. 3). On logs the Blue Marker is identified as a downhole decrease in resistivity and density and an increase in sonic transit time. Seismically, the reflector is a low-frequency doublet and is mappable over most of the study area. The reflector is identified by biostratigraphic data as early Middle Eocene in age, within the lower part of palynofloral zone PE3c of the Robertson Group's biostratigraphic zonation

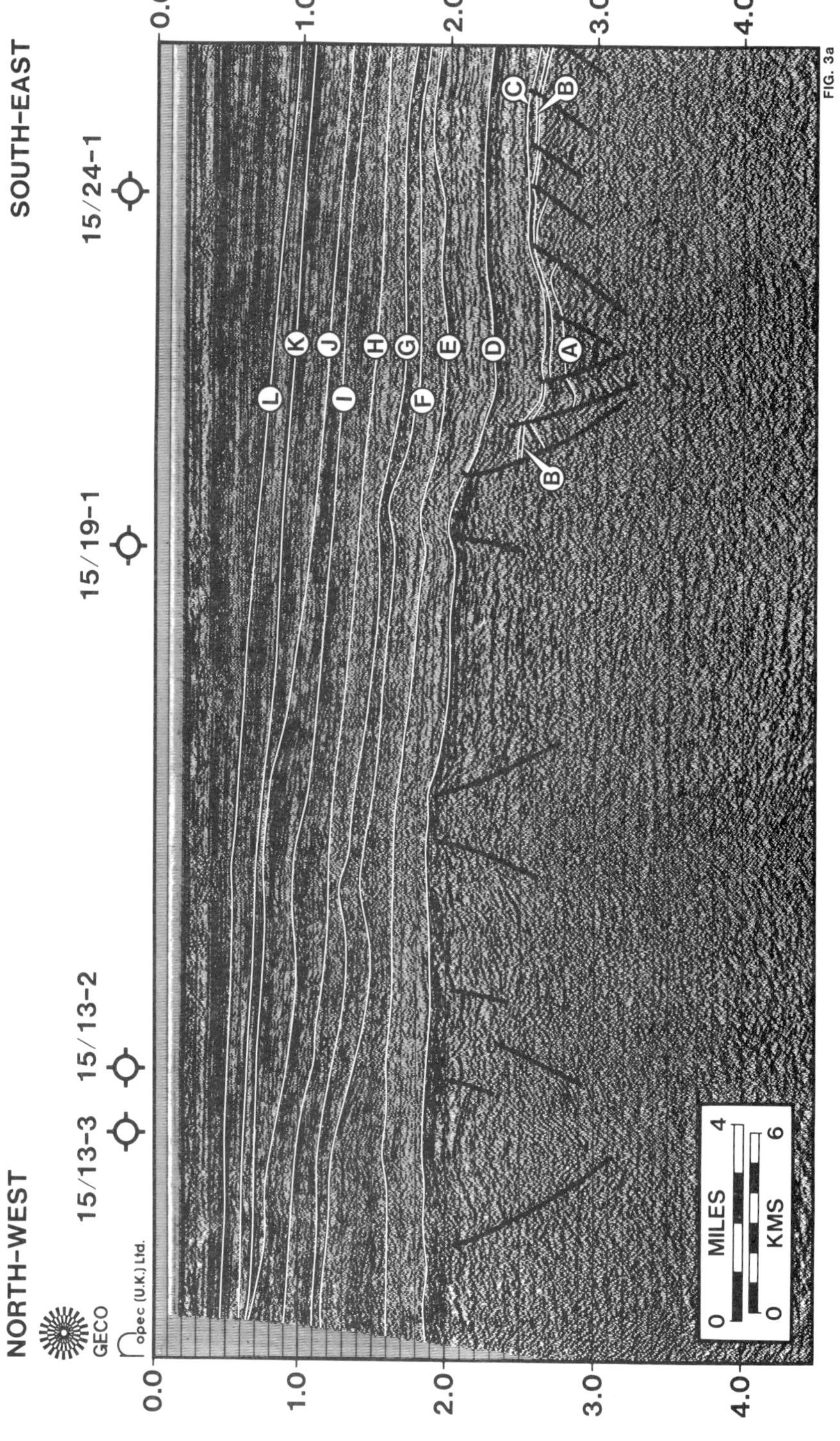
NORTH-WEST
SOUTH-EAST
15/13-3
15/13-2
15/19-1
15/24-1
GECO
opec (U.K.) Ltd.
0.0
1.0
2.0
3.0
4.0
MILES
KMS
0
4
6
FIG. 3a

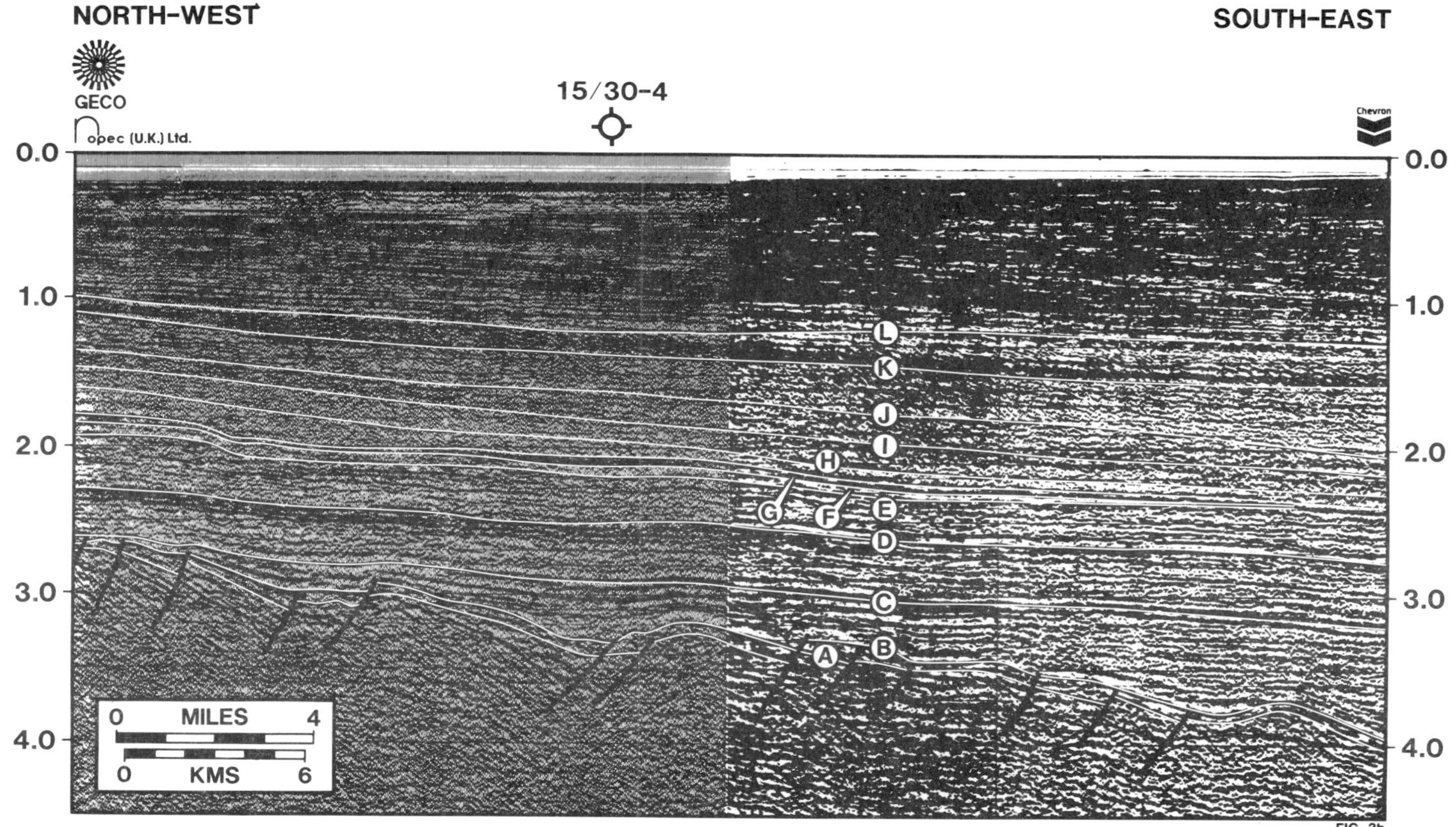

Fig. 3. (a) Regional seismic section North. (b) Regional seismic section South. Horizons shown are: A, Middle Jurassic volcanics; B, base Cretaceous unconformity; C, base Lower Cretaceous; D, top Chalk group; E, top Andrew Formation; F, base Balder Formation (= Base Pre-Alba Sequence); G, top Balder; H, top Lower Eocene; I, blue Marker (= Base Alba Sequence); J, Oligocene unconformity (= top Alba sequence); K, top Oligocene; L, Mid-Miocene unconformity. Vertical exaggeration is approximately 16 times.

scheme. The structure of the top of the sequence (Blue Marker) is shown in Fig. 6.

Within the pre-Alba sequence, the top Balder is a strong regional seismic reflector whereas the top Lower Eocene seismic event is weak. The upper part of the sequence is Middle Eocene in age and, in the southeast of the study area, is characterized by steeply dipping, cross-cutting reflectors which are interpreted to be submarine channel cuts. These channels are filled predominantly with shale except in the area of Quadrant 16.

The structure of the base of the Pre-Alba sequence and a map of its overall thickness are shown on Figs 5 & 7 respectively. In the northwest of the area, sediments consisting of sands interbedded with coals and shales are demonstrated by wells 15/13−3 and 15/13−2 (Fig. 8) to be characteristic of a deltaic environment. Sand thicknesses of greater than 600 feet are preserved in this area (Fig. 9). On the seismic line of Fig. 3a, the pre-Alba Eocene sequence is shown to prograde southeastward over the Forties sequence shelf and slope.

The Pre-Alba sequence has its maximum thickness in the slope area (Figs 7 & 8), indicating the development of a depocentre. Much of the sediment carried across the shelf was trapped in this depocentre. In the area of Blocks 16/17 and 16/22, sand thicknesses in excess of 800 feet are preserved. These sands are interpreted to be deep water sediments. Biostratigraphic evidence and the regional setting indicate water depths up to or exceeding 1200 feet. On Fig. 8, the basinal sediments are predominantly shales with sands appearing only in the far southeast. These sands are at the most southerly extent of the Viking Graben fan system.

Pre-Alba sequence palaeogeography is shown in Fig. 10 and a map showing the principal facies developed is shown in Fig. 11. The relationship of deltaic shelf, slope, and basin within the Pre-Alba Eocene sequence is similar to that within the Forties sequence but the features are offset in the southeast direction. A shelf area existed to the northwest of the study area where sands accumulated. These sands and their associated shales provided the source of sediment for gravity flow deposition. Sand-rich sediment was transported from the shelf predominantly into a basin area in Quadrant 16 to form the South Viking Graben fan. Elsewhere basinal sedimentation was dominated by hemipelagic shales and muddy turbidites with rare sandstones. The sands of the slope area in Block 15/22 are interpreted to be slope apron facies (Stow *et al.* 1984) deposited by mass transport processes induced by shelf edge and slope failure. The submarine fan system of the Pre-Alba sequence is interpreted to have been a response to a relative low stand of sea level in the early part of the Middle Eocene. A rise in

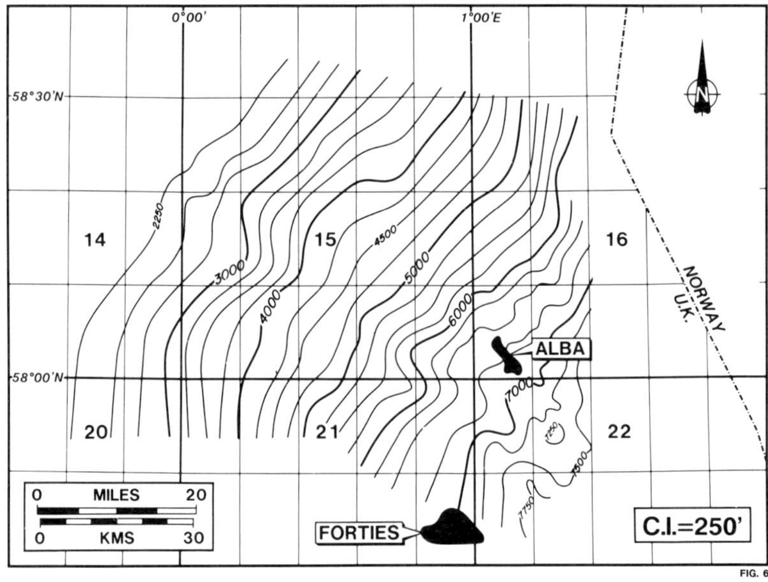

Fig. 6. Base Alba sequence structure. Depth structure map of Blue Marker (= base Alba sequence) in feet.

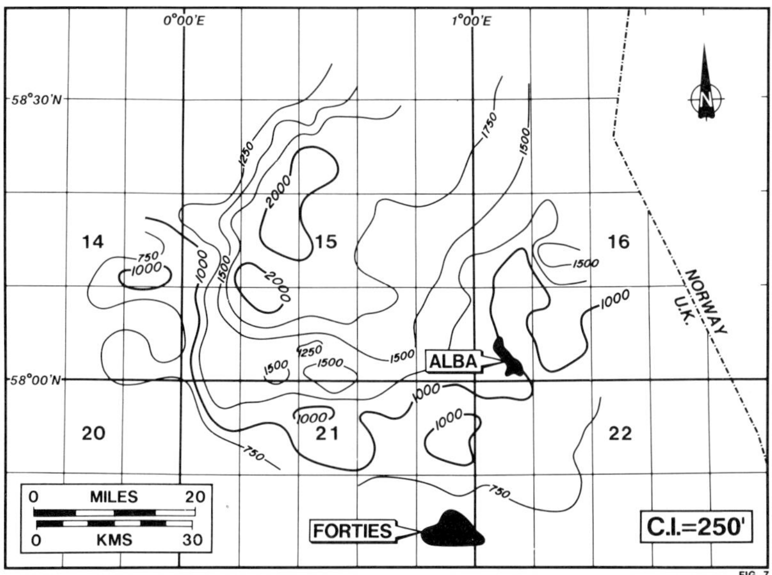

Fig. 7. Pre-Alba sequence isopach. Total isopach of Pre-Alba sequence in feet.

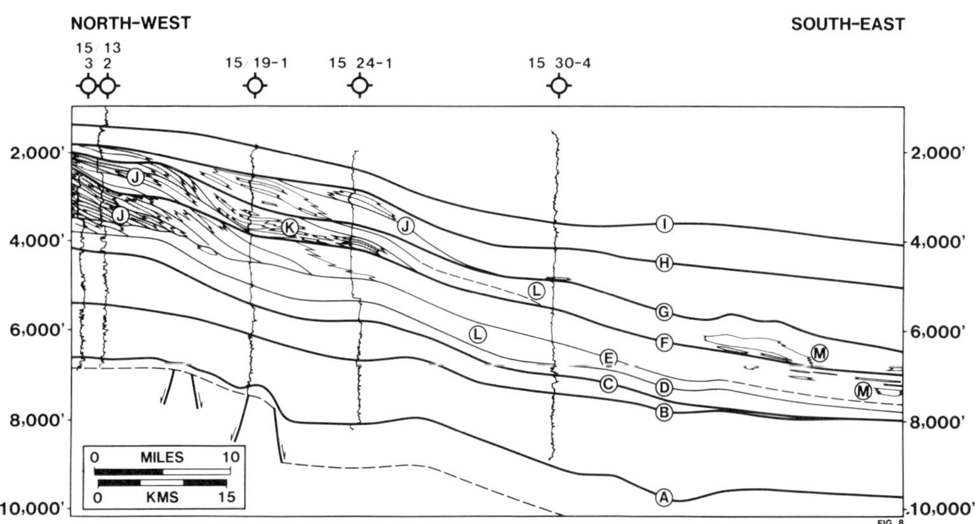

Fig. 8. Geological cross section. Geological cross section coincident with seismic line of Fig. 3. Horizons shown are: A, top Chalk Group; B, top Andrew Formation; C, base Balder Formation (= base Pre-Alba sequence); D, top Balder; E, top Lower Eocene; F, Blue Marker (= base Alba Sequence); G, Oligocene unconformity (= top Alba Sequence); H, top Oligocene; I, Mid-Miocene unconformity. Deltaic Shelf sediments (J), slope apron deposits (K), slopes shales (L) and deep water channel sands (M) are indicated. Sands are shown stippled.

relative sea level terminated these depositional processes and hemipelagic shale deposition became prevalent. This is indicated by the Blue Marker seismic and log event.

Sands within the upper part of the Pre-Alba sequence have not been found to be oil-bearing in the study area, probably because of interconnection of the sand bodies and their ultimate

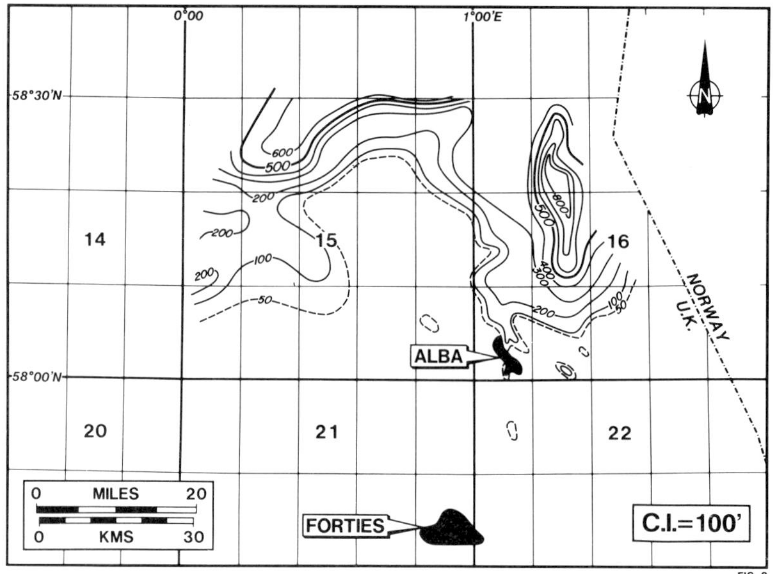

Fig. 9. Pre-Alba sequence sand isopach. Isopach of Pre-Alba sequence sand in feet. No zero contour has been drawn. A 50-foot contour is shown dashed. No sands of the Balder Formation are included.

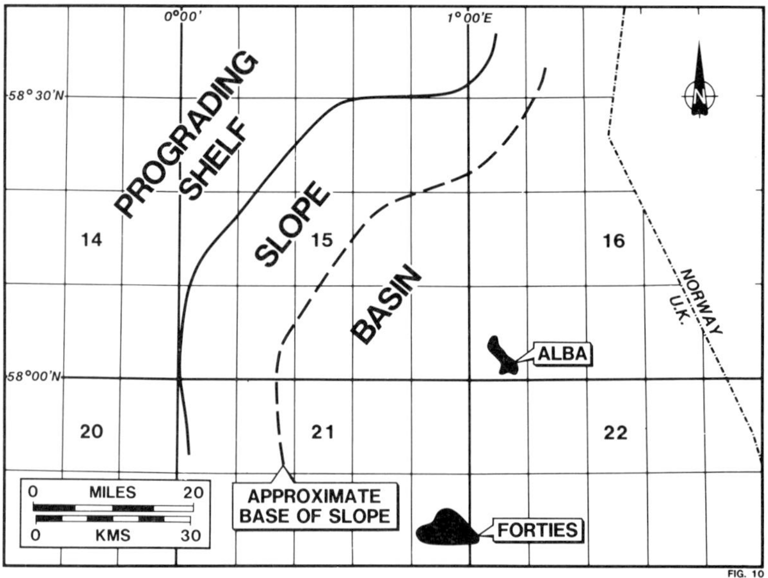

Fig. 10. Pre-Alba sequence palaeogeography. Shelf, slope and basin areas are indicated. The boundary between shelf and slope is based on biostratigraphic and seismic evidence. The boundary between slope and basin is interpretive, based on seismic evidence.

continuity with the sands of the South Viking Graben Fan. These sands may, however, have been a conduit through which oil migrated from the Palaeocene sands into the Middle Eocene Alba sequence sand system.

Alba sequence

The Alba sequence is defined at its base by the Blue Marker discussed above. The upper boundary is the Oligocene Unconformity (Figs

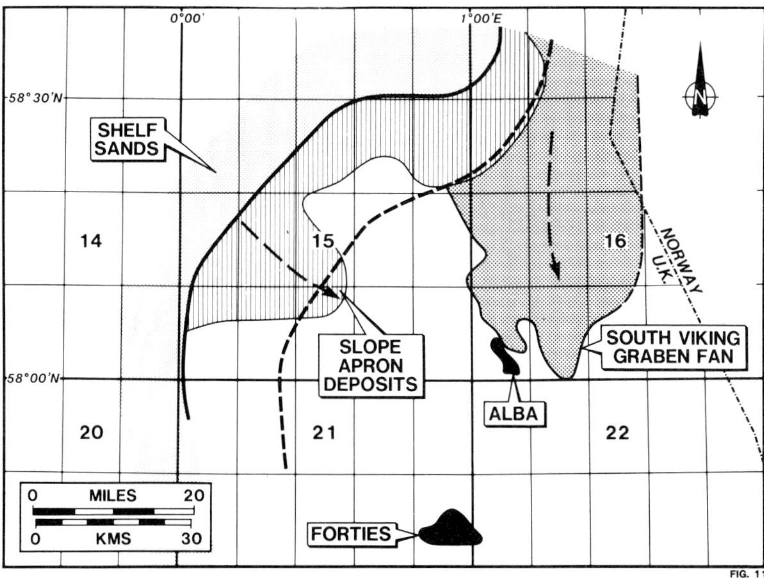

Fig. 11. Pre-Alba sequence facies. The principal facies are indicated, namely shelf sands, slope apron deposits and submarine fan basin deposits.

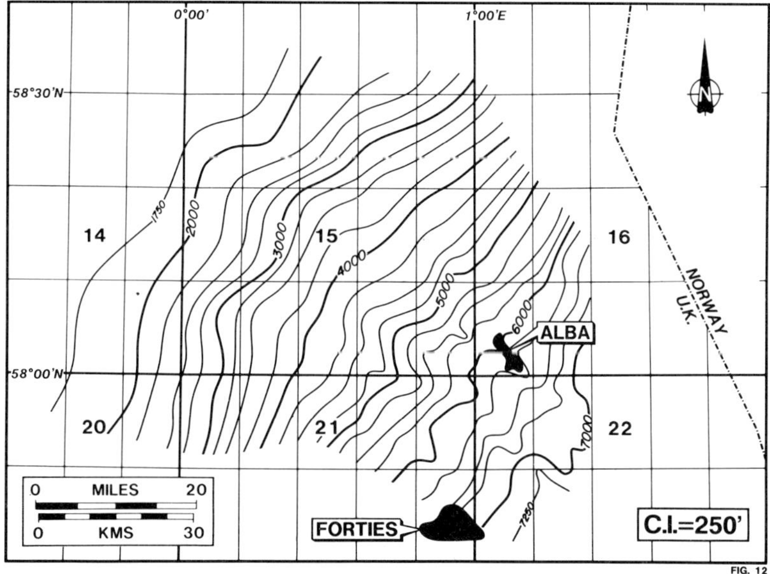

Fig. 12. Top Alba sequence structure. Depth Structure map of Oligocene Unconformity (= Top Alba sequence) in feet.

3 & 12). Biostratigraphic data show the Alba sequence to be of Middle Eocene age. Sediments of the Upper Eocene have not been recognised in the Witch Ground Graben, and the Middle Eocene sequence is overlain unconformably by shales of Early Oligocene age. No satisfactory mechanism explaining the lack of Upper Eocene section in this basinal setting has been suggested, and the question is still under investigation.

Seismically, the sequence is characterized in the northwest of the study area by low-angle

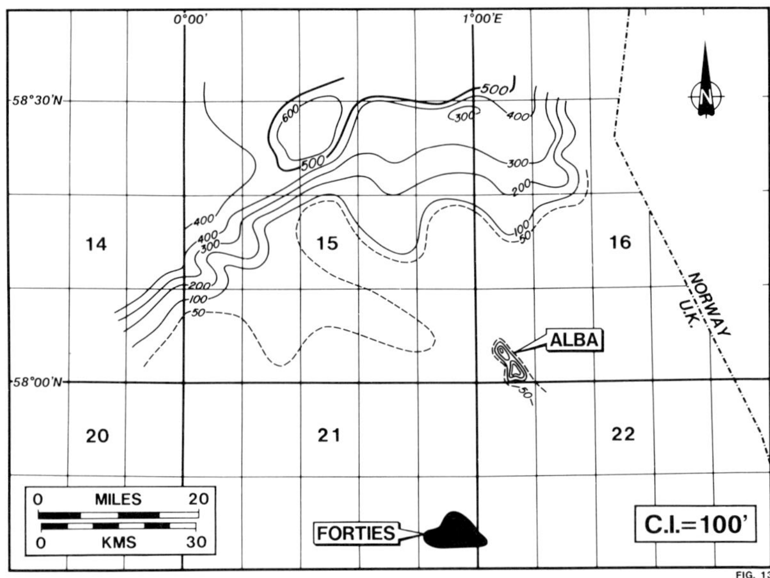

Fig. 13. Alba sequence sand isopach. Isopach of Alba sequence sand in feet. No zero contour has been drawn. A 50-foot contour is shown dashed.

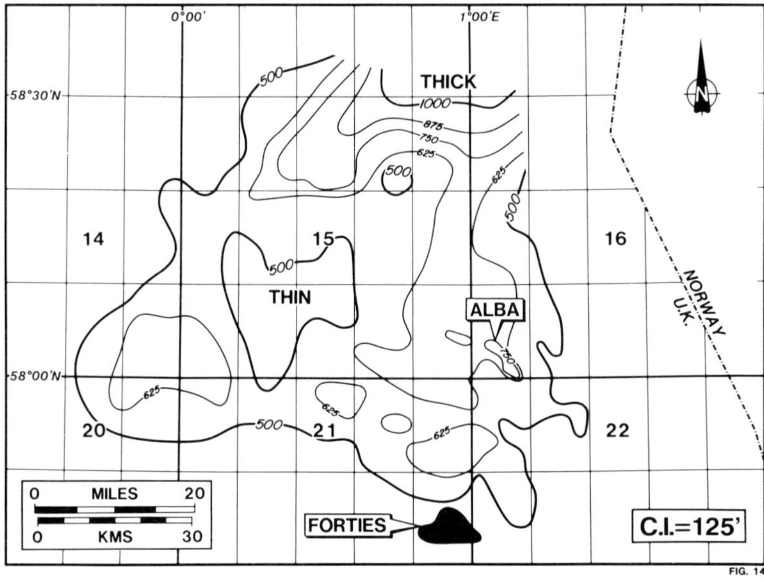

Fig. 14. Alba sequence isopach. Total Isopach of Alba sequence in feet.

prograding clinoforms, the top sets being a series of sub-parallel, high-amplitude reflectors. Well data (15/13–2 and 3 on Fig. 8) show that the high-amplitude reflectors correspond to interbedded deltaic sands and shales. An accumulation of sand with a present-day thickness in excess of 600 feet developed in this area, as shown by Fig. 13, the net sand isopach for the Alba sequence. The geological section (Fig. 8) suggests that the slope is characterized by shale deposition, with some sands shed from the shelf by mass transport processes. The total isopach of the Alba sequence (Fig. 14) shows the slope area to be relatively thin, indicating that this area received reduced sediment volumes relative to the Pre-Alba sequence slope (Fig. 7).

Beyond the base of slope, in the centre and southeast of the study area, the sequence is predominantly characterized by parallel seismic reflectors indicative of hemipelagic shale deposition, but there are anomalous areas with mounded reflection character. Within these mounds, the seismic data show bi-directional downlap, channel-cut features and seismic dead zones. The anomalies are approximately one mile wide and trend northwest-southeast. The Alba Field is an example of one such anomaly and, based on the extensive log and core information available in the field area, the seismic anomalies are interpreted to be submarine fan channel complexes.

The Alba Field reservoir section is divided into three sand units, as illustrated in Fig. 15. Of these, the Lower Sand is discontinuous and separated from the second unit, the Main Sand, by a shale section up to 50 feet thick.

The Main Sand is thick, homogeneous and has a sharp top and base. It is fine to very fine grained, is unconsolidated, and is extremely well sorted with, on average, only two percent clay. In core, dewatering structures are common but pebble horizons and rip-up clasts are rare. Amalgamation surfaces are interpreted in cores but fining upward sequences are not obvious. Average porosity is 33% and average permeability is approximately 2.8 darcies.

The third unit is the Upper Sands. These are a series of thin sands interbedded with shales. The petrophysical characteristics of these upper sands are similar to those of the Main Sand. Core evidence demonstrates that at least some of the individual upper sands result from sedimentary injection of sand from the thicker sand units into the surrounding shale section. Figure 16 shows the detailed seismic expression of the Alba Field. This line was shot transverse to the axis of the field and shows the Main Sand unit within the mounded facies. The Base Oligocene and Blue Marker reflectors are shown. The mound is flanked by features interpreted to be submarine channel levee deposits which form the lateral seals for the field. The complex of depositional sand-filled channel and flanking levees built up contemporaneously, producing a topographic anomaly on the area floor (Fig. 17a). The amplitude of this anomaly was enhanced subsequently by differential compaction between sand and shale. A rise in relative sea level towards the end of the Middle Eocene caused the sand sediment source to be inundated, the Alba channel system was abandoned and sand deposition in the basin ceased (Fig. 17b). Turbiditic and hemipelagic shales were deposited over the abandoned channel system thus providing the top seal for the Alba Field. Updip of the Alba Field, the channel system was erosional, or at least non-depositional in character. The subsequent shale deposition here formed the updip seal of the field.

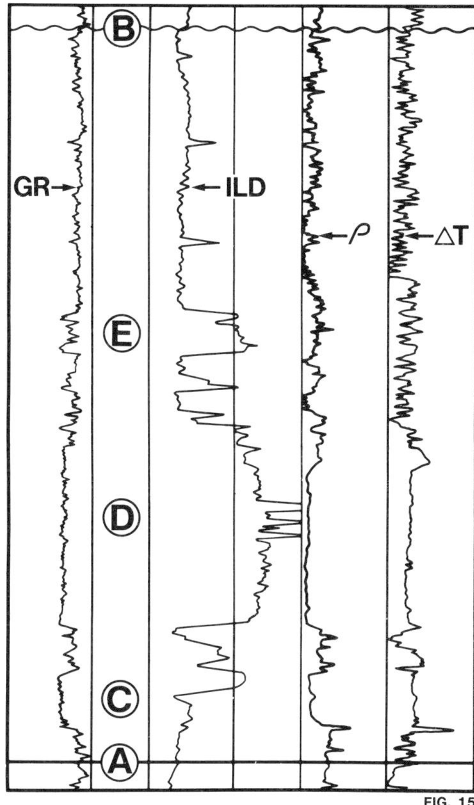

Fig. 15. Alba field schematic well. Typical well log response of the Alba Field sands. Blue Marker or base Alba sequence (A), Oligocene unconformity or top Alba sequence (B), Lower Sands (C), Main Sands (D) and Upper Sands (E) are indicated.

The biostratigraphic evidence (agglutinated foraminiferal assemblages) and the regional setting indicate that the channel complex was deposited in a deep water environment below storm wave base, with water depths up to or exceeding 1200 feet.

Figure 13 shows the Alba Field as an isolated sand body within the predominantly shaly Middle Eocene interval. This is, however, a simplified map and other isolated sand lobes are present in the study area. The isopach of the

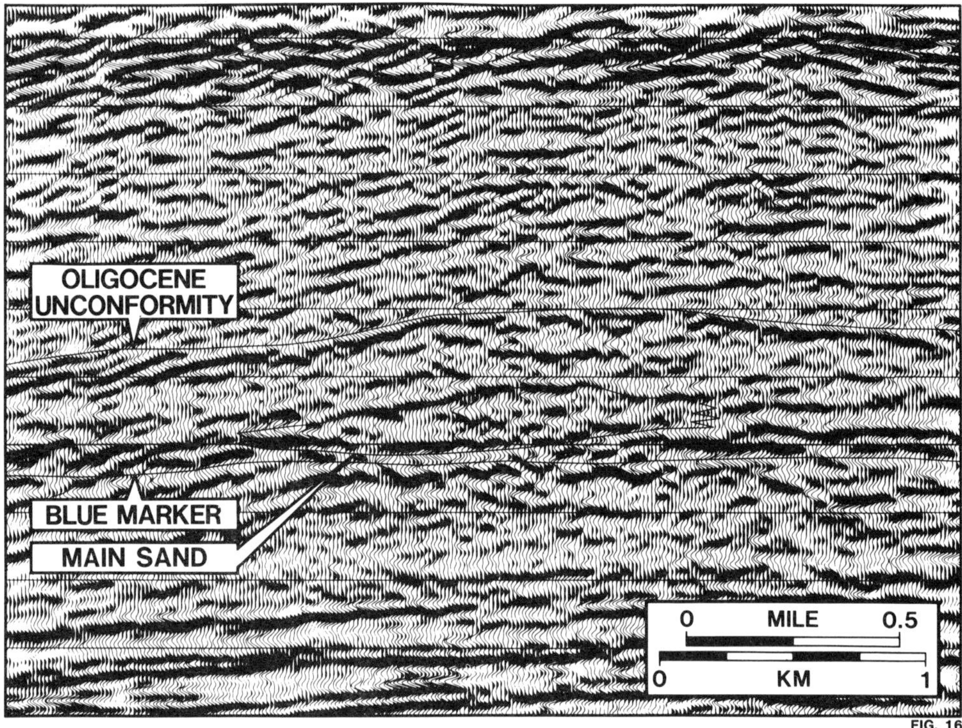

Fig. 16. Seismic line (detail). Seismic line over the Alba Field orientated southwest–northeast. The interpreted outline of the Main Sand is shown.

Alba sequence (Fig. 14) shows an anomalously thick area at the Alba Field but other lobes can be observed to the southwest, east and southeast of Alba. The 22/2a discovery lies in one of these lobes, and other lobes have been the objective of recent drilling. Exploration in this area will continue.

The palaeogeographic setting of the Alba sequence is illustrated in Fig. 18 and a map showing the principal facies developed is shown in Fig 19. In the northwest of the study area, a shelf edge prograded southeastward over the pre-Alba sequence slope. Sands accumulated on the shelf and were deposited over the slope area as slope apron deposits resulting from shelf edge and slope failure. Mud-rich turbidity flows were also active, transporting sediment from the shelf and slope areas for distances in excess of 30 miles into the basin. The Alba Fan system thus developed in this basinal environment. Channels within the fan deposited sands flanked by levee deposits. The submarine fans of the Alba sequence, like those of the Pre-Alba sequence, are believed to have been a response to a relative low stand of sea level. A rise in sea level at the end of the Middle Eocene terminated the deposition of the Alba fan system.

Post-Alba sequences

During the Oligocene, subsidence of the basin continued and the shelf area further prograded southeastwards over the Alba sequence shelf (Fig. 8). Well control shows the sediments to be predominantly shales.

Conclusion

Regional studies of seismic stratigraphy, biostratigraphy and lithofacies lead to the interpretation of a succession of shelf, slope and basinal deposits of Palaeocene to Oligocene age, prograding southeastwards and becoming progressively mud dominated. The Alba Field reservoir sands are interpreted to be a submarine fan channel deposit of Middle Eocene age within the basinal part of the sequence. The principal controls governing the deposition of submarine fan sands in the Middle Eocene were (1) the underlying basin topography, (2) the location and nature of the shelf area which provided the source of sediment for the basin, and (3) variations in relative sea level.

(a)

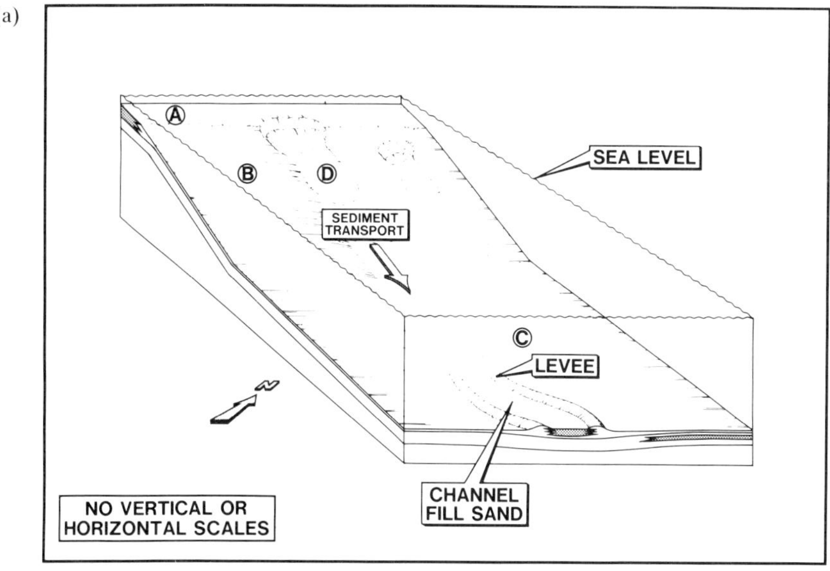

(b)

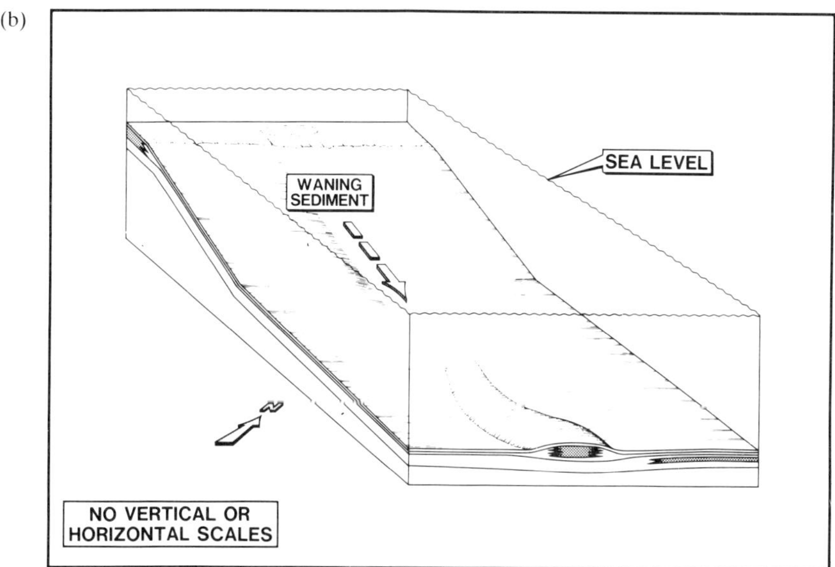

Fig. 17. Depositional history. Block Diagram showing a model for Eocene sedimentation. (a) Early Stage: sediment is transported by mass flow processes from the shelf area (A), down the slope (B) and deposited in the basin (C). Sand is deposited in a channel confined by levees. Slope apron deposits are indicated (D). (b) Late stage: sea level rise causes sediment to be trapped on the shelf. Waning sediment supply leads to the channel system to be abandoned and filled with hemipelagic shale. Figure 16 can be thought of as the appearance of the front face of the block diagram on a seismic section.

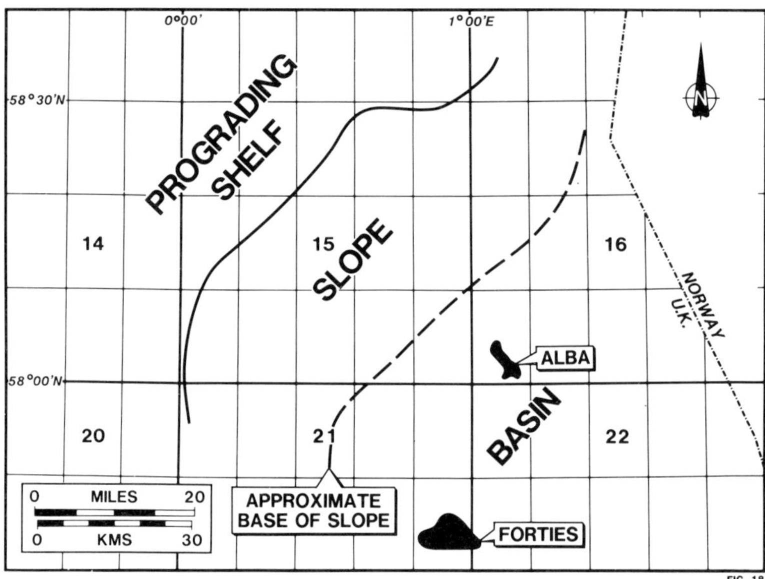

Fig. 18. Alba sequence palaeogeography. Shelf, slope and basin areas are indicated. The boundary between shelf and slope is based on biostratigraphic and seismic evidence. The boundary between slope and basin is interpretive, based on seismic evidence.

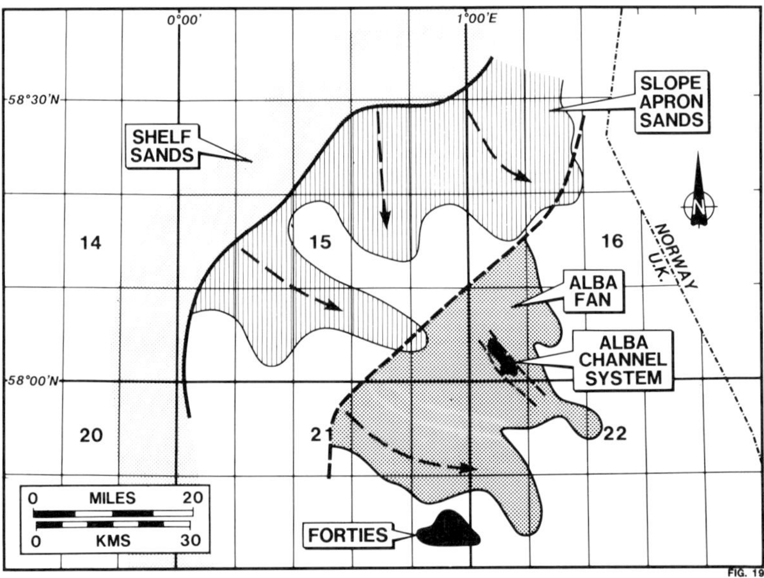

Fig. 19. Alba sequence facies. The principal facies are indicated, namely shelf sands, slope apron deposits and submarine fan basin deposits.

The authors wish to thank Chevron and the members of the UK Block 16/26 Partnership for permission to publish this paper. Nopec and Geco kindly gave permission for the use of their seismic line CNST86–16, as did Conoco, BP and Shell over whose Licence Blocks the line as shot. Special thanks are due to June Brown and Ed Wallis who drafted the figures.

References

MIDDLETON, G. V. & HAMPTON, M. A. 1976. Subaqueous sediment transport and deposition by sediment gravity flow. *In*: STANLEY, D. J. & SWIFT, D. J. P., (eds), *Marine Sediment Transport and Environmental Management,* Wiley, New York, 197–218.

PARKER, J. R. 1975. Lower Tertiary sand development in the central North Sea. *In*: WOODLAND, A. W. (ed.), *Petroleum and the Continental Shelf of Northwest Europe, 1, Geology*, Applied Science, Barking.

STEWART, I. J. 1987. A revised stratigraphic interpretation of the Early palaeogene of the central North Sea. *In*: BROOKS, J. & GLENNIE, K. W. (eds), *Petroleum Geology of North West Europe*, Graham & Trotman, London.

STOW, D. A. V., HOWELL, D. G. & NELSON, C. H. 1984. Sedimentary, Tectonic and Sea-Level Controls on Submarine Fan and Slope-Apron Turbidite Systems. *Geo-Marine Letters,* **3**, 57–64.

ZIEGLER, P. A. 1982. *Geological Atlas of Western and Central Europe*. Shell Internationale Petroleum Maatschappij B.V.

Cenozoic subsidence and uplift in the North Sea region: Implications for mechanisms of basin formation

RICHARD K. MORGAN

Robertson Group, Petroleum Division, Llandudno, Gwynedd, LL30 1SA, UK

Abstract: The scale on which flexural subsidence has taken place in the North Sea region from the Late Palaeocene is very much greater than the scale on which fault-controlled subsidence and flexural subsidence occurred during the Mesozoic. In addition, uplift of the Scottish Highlands occurred during the Cenozoic on a scale comparable with that of the subsidence in the North Sea region during this time.

The Moray Firth, Mesozoic to Tertiary basin is divided into two distinct areas as a consequence of its Tertiary subsidence history, the Inner Moray Firth being uplifted and eroded and the Outer Moray Firth undergoing apparently continuous subsidence. The geometries of the pre-erosional sequences in the Moray Firth area indicate a Cenozoic history of both subsidence and uplift, providing a link between the subsiding North Sea region and the uplifted Scottish Highlands. Using seismic mapping and cross sections constructed therefrom, across the Moray Firth and North Central Graben areas, the relative subsidence and uplift is displayed as part of a unified process of crustal flexuring.

It is suggested that the flexural subsidence and uplift in the North Sea region and Scottish Highlands during the Cenozoic is associated with the onset of sea-floor spreading of the North Atlantic to the west. Furthermore it is proposed that the magnitude of Cenozoic subsidence in the North Sea region does not relate to the magnitude of Mesozoic extension which preceded it, the two processes being driven by different mechanisms.

From HARDMAN, R. F. P. & BROOKS, J. (eds), 1990, *Tectonic Events Responsible for Britain's Oil and Gas Reserves*, Geological Society Special Publication No 55, p 369.

Tertiary structures and hydrocarbon entrapment in the Weald Basin of southern England

MALCOLM BUTLER[1] & CHRISTOPHER P. PULLAN[2]

[1] *Brabant Petroleum Limited, Sterling House, 150/152 High Street, Tonbridge, Kent TN9 1BB, UK*

[2] *Hunt United Corporation, William Blake House, 8 Marshall Street, London W1V 1LP, UK*

Abstract: The Weald Basin of southeast England was formed by rapid subsidence associated with thermal relaxation following early Mesozoic extensional block faulting. The basin appears initially to have taken the form of an easterly extension of the Wessex Basin but became the major depocentre during the Upper Jurassic and Lower Cretaceous, with associated active faulting. These movements appear to have ceased prior to Albian times and a full Upper Cretaceous cover is believed to have been deposited in a gentle downwarp which extended far beyond the confines of the Weald and Wessex Basins. Major inversion of the Weald Basin took place in the Tertiary, with both gentle regional uplift, which in the eastern part of the basin is estimated to have exceeded 5000 feet (1525 metres), and intense local uplift along pre-existing zones of weakness, which led to the formation of compressional features such as tight folds and reverse faults. Zones of Tertiary deformation appear to have been strongly influenced by underlying, particularly Hercynian, structural trends.

Lower Jurassic source rocks reached maturity in the early Cretaceous and initial migration occurred at this time, often over long distances, into traps closed by pre-Aptian faults. Tertiary tilting and uplift led to the breaching of many of these pre-existing traps and the formation of large folded closures. A second phase of hydrocarbon migration, particularly of gas, took place at this time, with significant vertical migration along fault zones. Major reservoirs located to date occur in Middle Jurassic carbonates and Upper Jurassic sandstones, but deep burial in the basin has caused considerable destruction of primary reservoir characteristics; changes in the temperature and pressure regimes and the mobilization of fluids within the basin resulting from the Tertiary uplift caused further diagenetic changes, particularly in the carbonate reservoirs.

Exploration of the Weald Basin remains at a very early stage, with a low drilling density to date. The more recent drilling has focussed on earlier structures, but traps formed or modified during the Tertiary movements represent important exploration objectives, although general deterioration in reservoir quality towards the centre and east of the basin makes large fold closures in these areas less attractive.

Hydrocarbon exploration of the Weald Basin entered its first major phase between the 1930s and the 1960s, when a number of wells were drilled on the basis of surface mapping, gravity and some early seismic data. Although significant shows of oil and gas were encountered in most wells and the small gasfield at Bletchingley (TQ 360480) was discovered, overall results were disappointing. The discovery of Wytch Farm oilfield (SY 980860) in 1973 led to a resurgence of interest in southern England and the introduction of new seismic techniques to the Weald Basin in the late 1970s led oil geologists to the realisation that the surface geology differed significantly from that at depth: the compressional features at the surface represented a Tertiary overprint on a Jurassic to Cretaceous extensional basin and, most importantly, because of inversion, the surface structures were offset from deeper closures.

Exploration activity in the late 1970s and early 1980s was concentrated on the older fault blocks of Jurassic and early Cretaceous age and led to the discovery of a number of oil accumulations in the Weald area. More recent discoveries have, however, indicated that structures which were formed or strongly modified as a result of the Tertiary movements still represent important exploration objectives in southern England.

Geological framework

The Weald Basin is located in southeast England (Fig. 1) and forms an easterly extension to the Wessex Basin (which here is considered to be

From Hardman, R. F. P. & Brooks, J. (eds), 1990, *Tectonic Events Responsible for Britain's Oil and Gas Reserves*, Geological Society Special Publication No 55, pp 371–391.

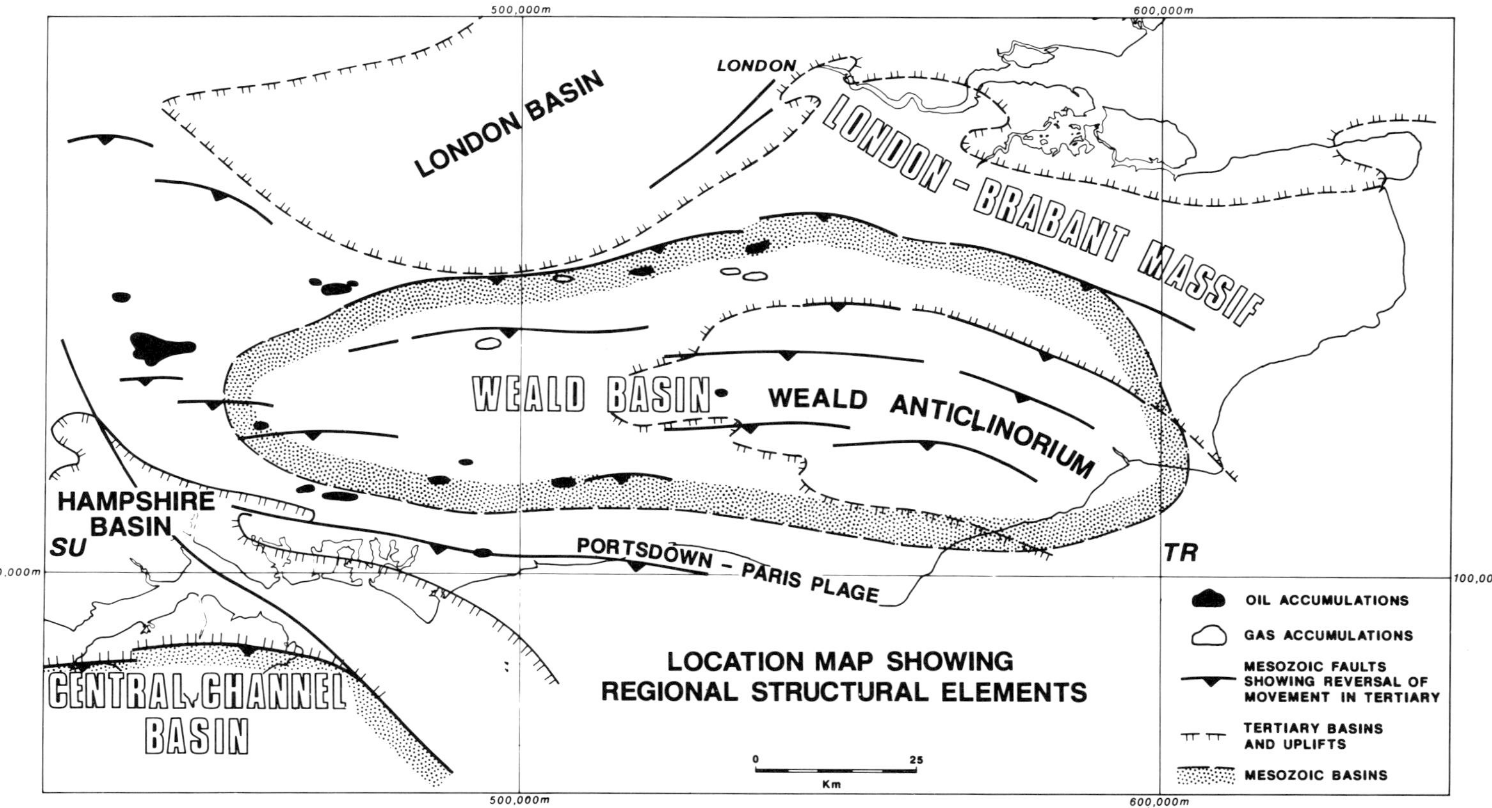

Fig. 1. Location map, showing structural elements.

restricted to the areas of thick Triassic and Lower Jurassic rocks underlying Dorset, Somerset and parts of Wiltshire). It is bounded to the north and east by the London–Brabant Massif and is separated from the Central Channel Basin and the Paris Basin by a regional arch, the Portsdown–Paris Plage ridge. The majority of outcrop within the basin area is of Lower Cretaceous age, but Upper Cretaceous Chalk outcrops around the margins and there are extensive areas of Tertiary outcrop in the London Basin, to the north, and the Hampshire Basin, to the south. Figure 2 demonstrates the general stratigraphic succession in southern England.

The geological history of the Weald Basin during the uppermost Jurassic and Lower Cretaceous has been studied in some detail over the years. However, knowledge of the earlier history is confined to seismic and borehole data (there have been few penetrations to the Palaeozoic in the deepest part of the basin) and the record of Upper Cretaceous and Tertiary deposition has been largely removed by erosion over the central part of the basin. Subsurface information indicates that the Alpine uplifts which caused the inversion of the Weald Basin were very much influenced by underlying, older structure and it is therefore important to understand the older geological framework when studying the effects of Tertiary movements.

The Weald of southern England is underlain at depth by Palaeozoic rocks of mainly Middle Devonian to Lower Carboniferous age. These underlying Palaeozoic rocks were deformed during the Hercynian orogeny, but show no signs of metamorphism in the Wealden area. Several authors have demonstrated evidence for deep-seated thrusting underlying Mesozoic rocks to the west of the Weald Basin (for example, Chadwick *et al.* 1983 and Smalley & Westbrook 1982) produced evidence for a thrust underlying the area of the Hog's Back (SU 900490), at the boundary of the London Platform and Weald Basin. The dominant Hercynian trends appear to be oriented east–west beneath the western and central Weald and there is some evidence that the Mesozoic development of the basin has also been influenced by northeast–southwest oriented trends of

STRATIGRAPHY			HYDROCARBON OCCURRENCES	RESERVOIR	SOURCE	SEAL	TECTONIC EPISODES	SEISMIC EVENTS
TERTIARY		MIOCENE					ALPINE INVERSION	
		OLIGOCENE						
		EOCENE						
CRETACEOUS	UPPER	CHALK						
	LWR	GREENSAND/GAULT					CIMMERIAN U/C	←
		WEALDEN	☼ HEATHFIELD, BOLNEY	√		√		←
		PURBECK	☼ ALBURY					←
JURASSIC	UPPER	PORTLAND	☼ GODLEY BRIDGE	√		√		
		KIMMERIDGE CLAY			√	√		←
		CORALLIAN	● PALMERS WOOD ☼ BLETCHINGLEY	√ √				←
		OXFORD CLAY			√	√		
		KELLAWAYS						
	MIDDLE	CORNBRASH/ FOREST MARBLE	● KIMMERIDGE	√				←
		GREAT OOLITE	✸ HUMBLY GROVE, HORNDEAN, GOODWORTH, STOCKBRIDGE, BAXTER'S COPSE, STORRINGTON, SINGLETON, LIDSEY	√				
		INFERIOR OOLITE				√		←
	LOWER	UPPER LIAS	● WYTCH FARM, ARNE, WAREHAM, STOBOROUGH, WADDOCK'S CROSS	√				
		MIDDLE LIAS		√	√	√		
		LOWER LIAS						
TRIASSIC		PENARTH GP	✸ HUMBLY GROVE	√				←
		MERCIA MUDSTONE GROUP			√?	√		
		SHERWOOD SANDSTONE GROUP	● WYTCH FARM	√				
PERMIAN		AYLESBEARE GROUP						
U.CARB. / BASEMENT		COAL MEASURES / CARBONIF'S AND OLDER		√?	√?		HERCYNIAN COMPRESSION	← ?

Fig. 2. Southern England stratigraphic column, showing hydrocarbon occurrences.

probable Caledonian age; faulting along this orientation occurs in the Tertiary and Upper Cretaceous rocks under South London and the abrupt northern margin of the Weald Basin at the Hog's Back changes to a broad shelf area along the projection of these faults. A pronounced change in structural grain, from east–west to northwest–southeast, occurs in the eastern Weald along a line which also has a northeast–southwest trend (Fig. 1). Information from boreholes suggests that the Palaeozoic subcrop underlying the northern part of the Weald is, in general, younger than that underlying the London Platform; this may be due in part to a reduction in depositional thicknesses of the Upper Palaeozoic over the London Platform, particularly of the Lower Carboniferous, but it may also indicate that there is no extensive major thrust complex dividing the two areas. The Weald area appears to have formed part of an exterior fold belt during late Hercynian time, lying north of the zone of major deformation which it is believed lay in the Channel. It is interesting to note that the area of the London Platform subsided during Upper Coal Measures time, with deposition of thick sequences of this age in the Reading and Oxford areas, and there are indications that the Devonian Old Red Sandstone sequence is also very thick under the London Platform, indicating that this area has undergone a series of inversions through time.

Little is known of the history of the area which was later to form the Weald Basin during the Permian and Triassic. Well penetrations in the east indicate a lack of Permo-Triassic sediments, other than a widely developed 'Rhaetic', but limited penetrations to the west indicate that there was some local deposition of both Sherwood Sandstone Group and Mercia Mudstone Group. Triassic deposition appears to have been controlled by active fault movements, giving rise to continental alluvial fan sequences which were reworked by a system of braided channels (Smith 1986). The distribution of reservoirs within the Triassic is poorly known, but good sands may have been developed locally where Old Red Sandstone and possibly Upper Coal Measures outcrop was being eroded.

The Weald Basin appears to have become a major depocentre during the Lower Lias, separated from the Wessex Basin by a saddle along a line from Southampton to Salisbury: although seismic definition is not good at this level, well evidence indicates significant variations in Lower Lias thickness across the basin, suggesting active faulting at this time. It is likely that this faulting was a continuation of that which began in the Triassic and is recognised in the western part of southern England. Chadwick (1985a) suggested that these movements were closely related to relaxation along the lines of underlying Hercynian thrust trends. For the remainder of the Lias and the Inferior Oolite, it appears from seismic and well evidence that fault movements became less marked and it is likely that the Weald Basin underwent regional thermal relaxation subsidence in response to a cooling lithosphere (Chadwick 1985a): outcrop evidence from the Cotswold region (Arkell 1933) may, however, indicate a renewal of minor fault activity during Inferior Oolite time. During the deposition of the Great Oolite Series and the overlying Oxford Clay, the Weald appears to have formed an embayment to the main depocentre, which lay to the southwest. The Weald Basin received the thinner, dominantly carbonate sediments of the Great Oolite, in contrast to the thicker, argillaceous Frome Clay facies deposited to the southwest. The lack of a clastic source in the Weald area throughout most of the Lower and Middle Jurassic means that there are few potential sand reservoirs in rocks of this age. However, the extensive carbonate ramp which was developed over much of the Weald during Middle Jurassic times gave rise to the important, dominantly oolitic, potential reservoirs of the Inferior Oolite and Great Oolite. This carbonate sequence is overlain by the thick argillaceous sediments of the Oxford Clay, which provide a good regional seal and are in turn followed by the Corallian sequence, which includes two major influxes of coarse clastics.

Active fault movements at the beginning of the Kimmeridgian led to the establishment of the Weald Basin as the major depocentre of southern England, with in excess of 2300 feet (700 metres) of Upper Corallian to Portlandian sediments being laid down in the centre of the basin, in contrast to the much thinner beds of equivalent age further to the west. Seismic evidence indicates that contemporaneous fault activity occurred throughout the deposition of the Kimmeridge Clay, with widespread growth across faults apparent in this interval. Clastic deposition again became important and sandstones are developed within the Kimmeridge Clay itself and in the Portlandian, as well as in the Corallian. The quality of these sands is very variable: those of the Corallian were deposited as a series of sand waves in a mid-shelf position, whilst the Portland Sand appears to have been deposited in a number of stacked bars, passing both laterally and vertically into lagoonal limestones. The Portland Sands are overlain by sabkha-type evaporites of the lower Purbeck, reflecting the general lowering of sea level at this time (Haq *et al.* 1987). Both the Corallian and Portlandian sands are best developed in the northern part of the Weald Basin and it is presumed that the clastic source lay to the north, on the London–Brabant Massif. The sands within the Kimmeridge Clay are confined to the eastern part of the Weald and may have had a local source.

The development of the Weald as a discrete basin continued into Lower Cretaceous time and deposition of the Wealden facies appears to have been almost confined to the Weald Basin and the Central Channel Basin, although erosion during Lower Greensand (Aptian) times makes it difficult to determine the original extent of these beds. With the rejuvenation of the Cornubian and London–Brabant Massifs and widespread lowering of sea level in the Valanginian (Haq *et al.* 1987), major, basin-wide clastic influx occurred. Seismic evidence again indicates considerable growth across faults during the deposition of the Wealden sequence, which supports the suggestion of Allen (1975) that Wealden deposition was controlled by active fault movements. Although seismic and well data suggests that movement on the major faults ceased prior to Aptian times (Fig. 3), both Arkell (1939) and Middlemiss (1975) consider that there was major erosion immediately north of the Hog's Back during the deposition of the Bargate Beds. Owen (1975) indicates zones of thinning within the Gault and Upper Greensand of the Weald Basin which appear to be related more to underlying structures of Jurassic age than to the later, Tertiary uplifts. It is believed that a full Upper Cretaceous cover was deposited in a gentle downwarp which extended far beyond the confines of the Weald Basin. There is little seismic evidence of contemporaneous fault movement during Upper Cretaceous times, although over most of the Weald area the Chalk is either too close to the surface to be clearly defined on seismic data or has been removed altogether. Mortimore (1986) demonstrates thickness variations in the upper part of the Chalk; unfortunately, the spacing of his control points is inadequate to enable firm correlations to be made with seismic structures but his thins appear to be related more to underlying Jurassic structure than to the later Tertiary folds. Chadwick (1985b) was able to demonstrate that basement subsidence was greater under the London Platform than it was under the Weald Basin during the deposition of the Chalk, the greater total Chalk thickness in the Weald area being related to compaction of the underlying sediments rather than subsidence, indicating the onset of the development of the London Basin and the inversion of the Weald Basin.

The lowermost Tertiary rocks lie unconformably on the Chalk and include much reworked flint: it is thought that almost the entire area of southern England became emergent at the end of the Cretaceous, associated with a worldwide graduall fall in sea level at this time, and that considerable thicknesses of Chalk may have been eroded in some areas prior to Palaeocene deposition. There is rarely any clear angularity to the unconformity at this level and it is possible that the earliest Tertiary sediments preserved represent the remnants of deposits which once extended across a large part of the Weald Basin (King 1981). The major depocentres lay, however, in the newly formed London and Hampshire Basins, both gentle downwarps overlying areas which had remained relatively high during Jurassic and Cretaceous deposition. The youngest Tertiary sediments preserved in southern England are of Oligocene age.

Tertiary Earth movements

In order to understand the changes in basin architecture which took place during the Tertiary uplift and inversion, it is first necessary to reconstruct the form of the Weald Basin prior to the uplift. The present day form of the basin at base Jurassic level is shown in a simplified manner in Fig. 4a. Seismic and outcrop information suggests that the fault movements which actively controlled Jurassic and Lower Cretaceous sedimentation had died out by Upper Cretaceous times and it has therefore been assumed that a map made on a level within the Chalk sequence will show little effect of earlier structures, whilst showing the full effect of the Tertiary movements. A study was carried out on the large number of water wells which penetrate the Chalk in southern England and of the published mapping carried out in the past for hydrogeological purposes and it was decided to make a regional map at base Upper Chalk level, or near base Senonian (Fig. 4b). This map is well controlled over the extensive areas of Chalk outcrop, but becomes conjectural where projected across the Weald uplift. Outside the present extent of Upper Chalk outcrop and subcrop, the map was constructed by building isopach layers of the various older formations and utilizing surface- or seismically-derived structure maps (making no adjustment for decompaction). The structure shown by Fig. 4b is considered to give a reasonable approximation of the morphology of the Tertiary uplift over

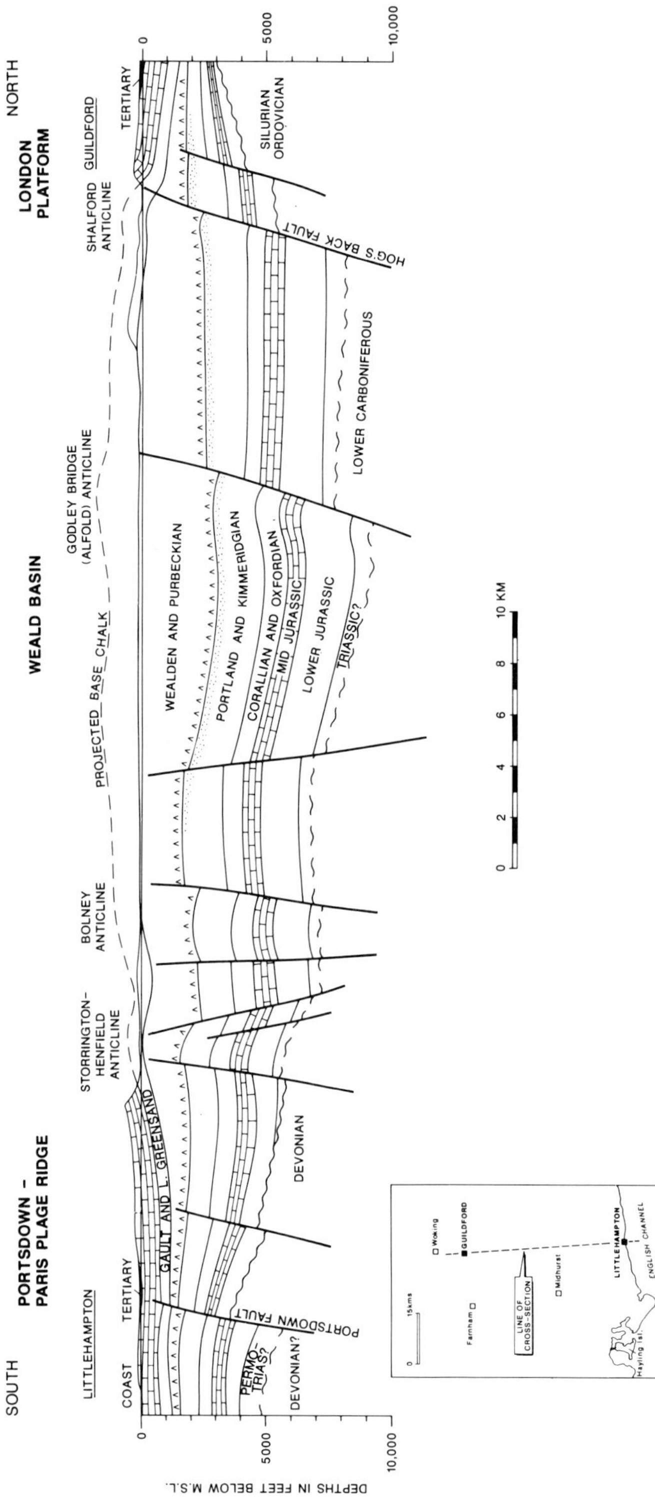

Fig. 3. Simplified north–south geological cross section through the central Weald Basin.

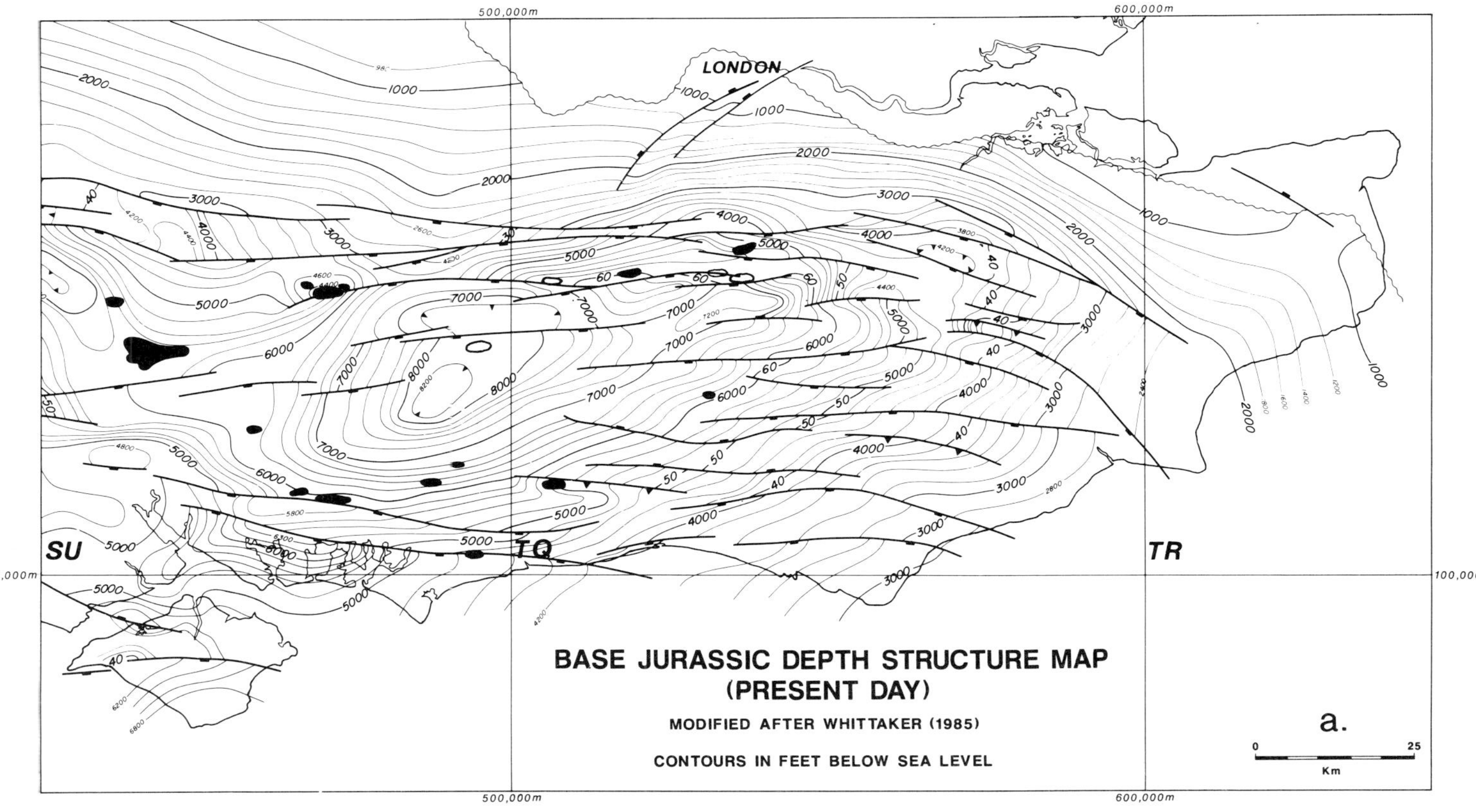

Fig. 4.(a) Present day base Jurassic depth structure map.

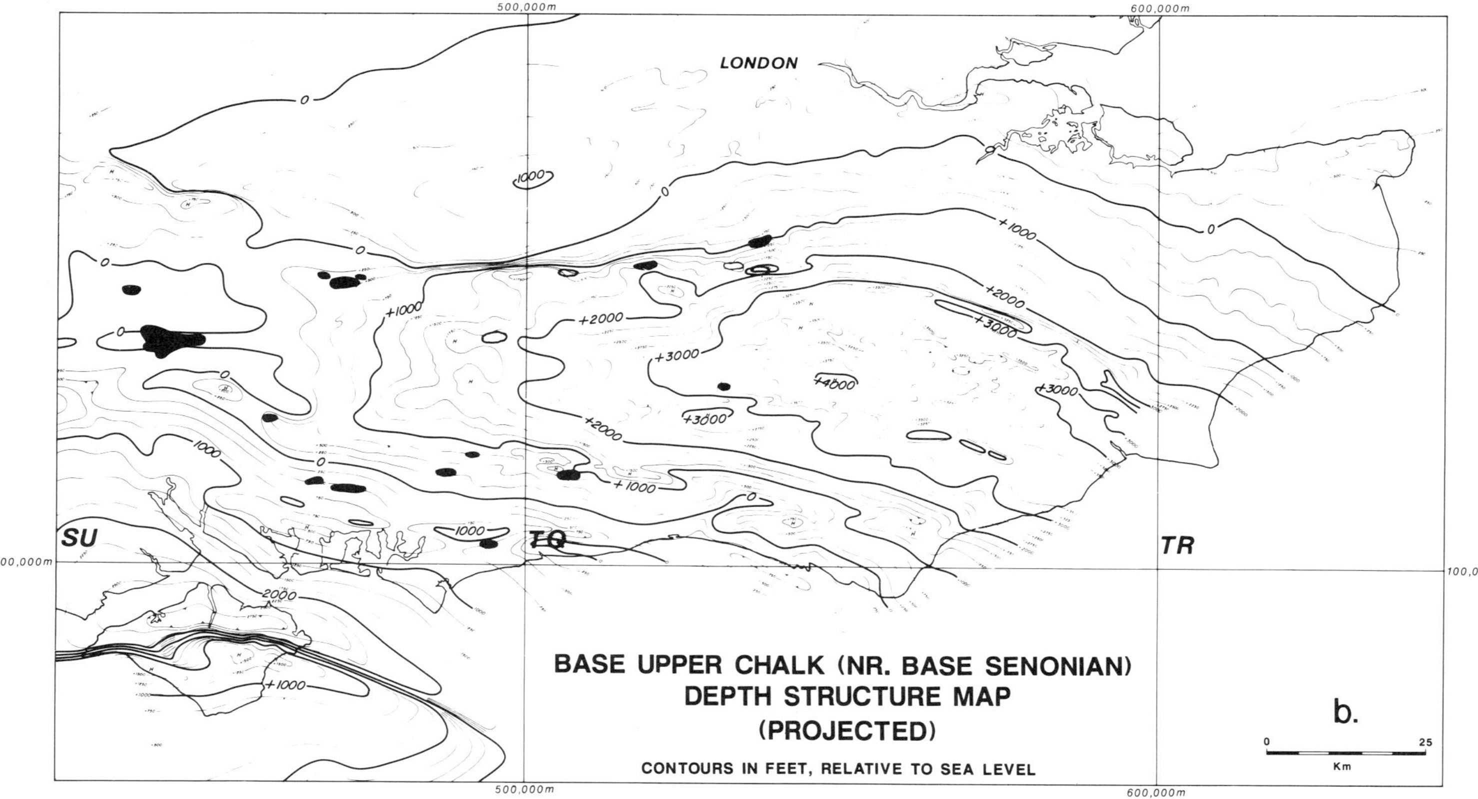

Fig. 4.(b) Present day base Upper Chalk (near base Senonian) depth structure map, projected across the Wealden uplift.

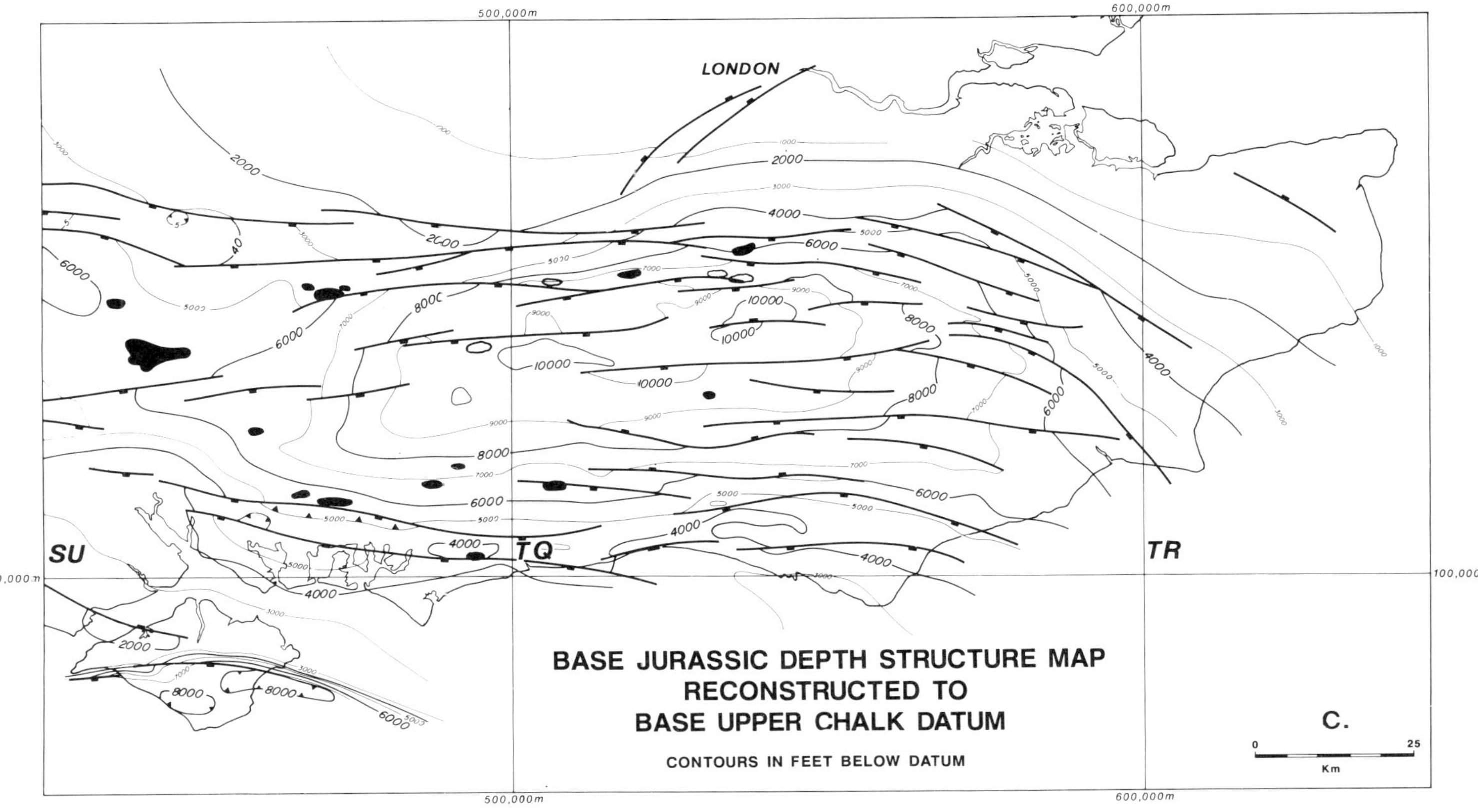

Fig. 4.(c) Base Jurassic depth structure map reconstructed to base Upper Chalk datum.

southern England and shows a maximum uplift in excess of 4000 feet (1220 metres) in the core of the Weald Anticlinorium. To arrive at an estimate of total uplift, however, the thickness of Upper Chalk and any Lower Tertiary sediments which were deposited over the Weald must be added: this is likely to provide an additional 700 to 1500 feet (215 to 455 metres) of uplift, giving a probable overall Tertiary uplift of the eastern part of the basin in excess of 5000 feet (1525 metres). An attempt was then made to remove the effects of this inversion from the present day base Jurassic map (Fig. 4a) by adding the base Upper Chalk topography (Fig. 4b) to it and effectively reconstructing the base Jurassic to a flattened base Upper Chalk datum. The result of this exercise (Fig. 4c) demonstrates the original extension of the Weald Basin to the east, towards the Boulonnais area of northern France, and indicates a depth of burial at base Upper Chalk times in excess of 10 000 feet (3050 metres) in the centre of the Weald. In order to compute a maximum burial depth, this map must be adjusted for later compaction and burial must again be increased by the thickness of the Upper Chalk and by any Lower Tertiary rocks which were deposited here. For comparison purposes and as an additional control on the reconstruction, studies were carried out on the distribution of Oxford Clay interval velocities in wells throughout the Weald Basin (Fig. 5a, b). The shape of the contours on Fig. 5a should accurately reflect the original depth of burial of the Oxford Clay and therefore the morphology of the basin prior to the Tertiary uplift. An attempt was then made to quantify uplift by comparing interval velocities with an 'undisturbed' burial depth against interval velocity curve, derived from wells around the margin of the Weald in areas which have not been strongly uplifted (Fig. 5b). Figure 5c sets out the estimated amount of uplift computed from study of the Oxford Clay interval velocities. A comparison of Figs 4 & 5 demonstrates that there is very good agreement between the morphology of the pre-Tertiary basin and of the subsequent uplift derived by both studies.

Prior to the Tertiary uplift the Weald took the form of a slightly asymmetric basin (Fig. 3), the northern margin of which is sharply defined by a series of down-to-the-south faults in the Hogs Back area, but which becomes less well defined further east. The basin was traversed by a number of major, east–west oriented fault trends, which had a strong effect on sedimentation. The southern margin was formed by a fairly gentle dip slope which was broken often by a large number of minor, antithetic faults. In response to the Alpine earth movements far to the south, the area of the Weald Basin appears to have progressively risen through the Lower Tertiary, giving rise to the major inversion feature of the 'Wealden Anticlinorium', which is about 140 km long and 60 km wide, covering an area of some 4500 km^2 and running from Petersfield (SU 750230) in the west out into the English Channel at Hastings (TQ 820090). This anticlinorium takes the form of a gentle regional uplift, which in the eastern part of the basin, as previously discussed, probably amounted to more than 5000 feet (1525 metres – equivalent to a 50% uplift of the base Jurassic), on to which are superimposed intense local uplifts. Although in general the Upper Jurassic and Lower Cretaceous basin was uplifted during the inversion, the zone of greatest uplift lies in the east of the Weald Basin and does not coincide with the zones of greatest depositional thickness of the Upper Jurassic and Lower Cretaceous (which conform closely to the reconstructed basin shown in Fig. 4c).

Superimposed on the anticlinorium are zones of intense uplift, the locations of which appear to have been controlled by the underlying structure; where large down-to-the-south faults were present in the Jurassic and Lower Cretaceous, intense inversion features often formed when the downthrown basins were uplifted and compressed against the higher blocks, with the formation of tight folds and with reversal of movement on pre-existing normal faults. Significantly, these zones of intense uplift are present throughout southern England and are not restricted to the Weald Basin, suggesting that their origin may be distinct from that which gave rise to the regional inversion. These zones of deformation typically manifest themselves at the surface by a series of east–west trending periclines, such as are seen along the southern margin of the London Platform. These folds generally face northwards, but vary in width and amplitude depending on their position in the basin and their orientation. Tight folds, with widths in the order of 2 km and amplitudes in the order of 500 feet (150 metres), are also found associated with northwest–southeast trending faults which appear to have undergone significant transpression, with reversal of movement. A typical example of this occurs at Detention (TQ 748402, Fig. 6), in Kent, where a pre-existing normal fault demonstrates up to 1100 feet (335 metres) of present-day reverse movement, resulting in the repetition of the Middle Jurassic section in the well. The resolution of the seismic line in Fig. 6 does not allow exact definition of the Detention structure,

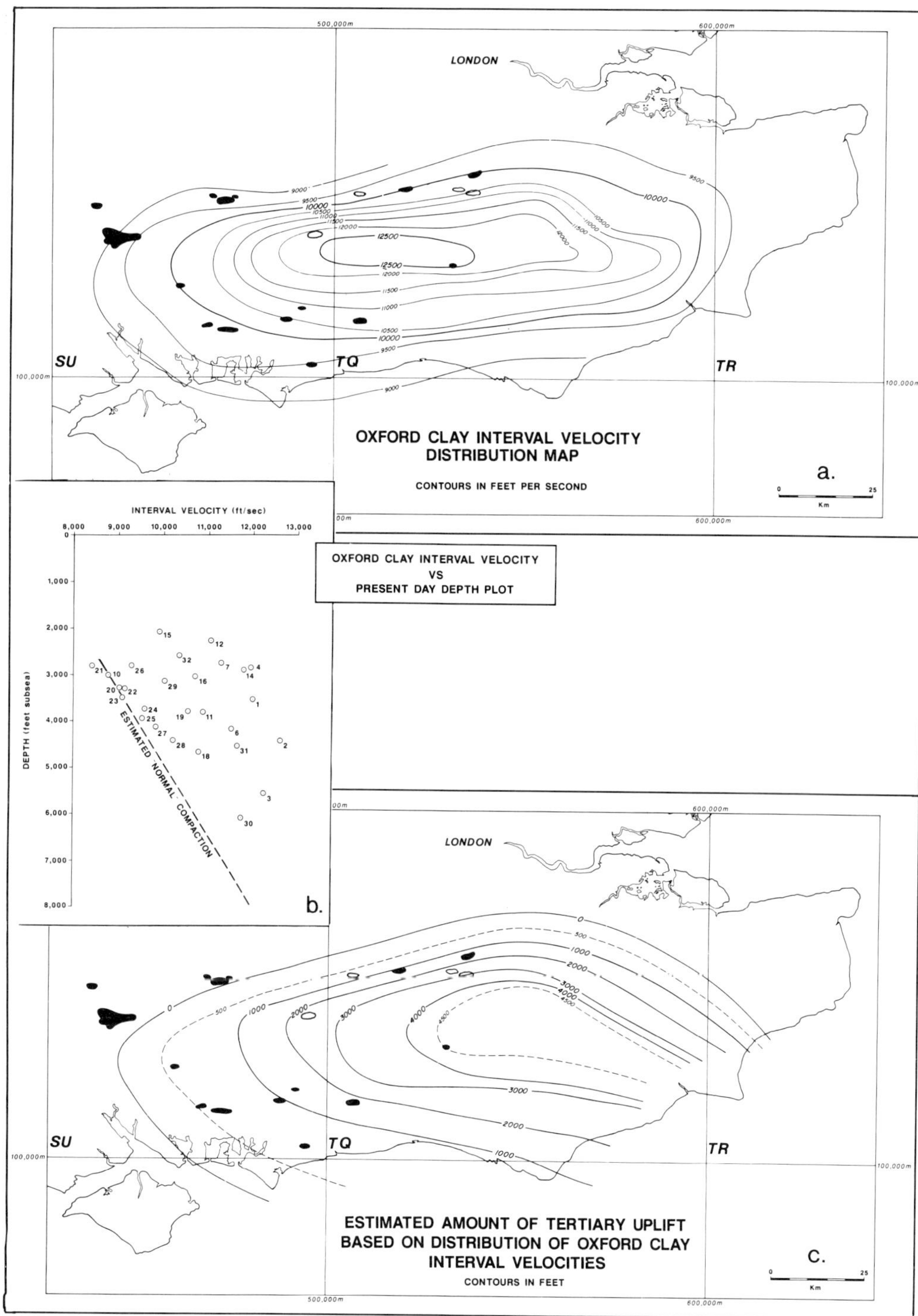

Fig. 5. (a) Oxford Clay interval velocity contour map. (b) Plot of Oxford Clay interval velocity vs present day depth, with line indicating estimated 'normal' compaction. (c) Map showing estimated Tertiary uplift of Oxford Clay, computed by reconstructing depths to the "normal" compaction line of Fig. 5b.

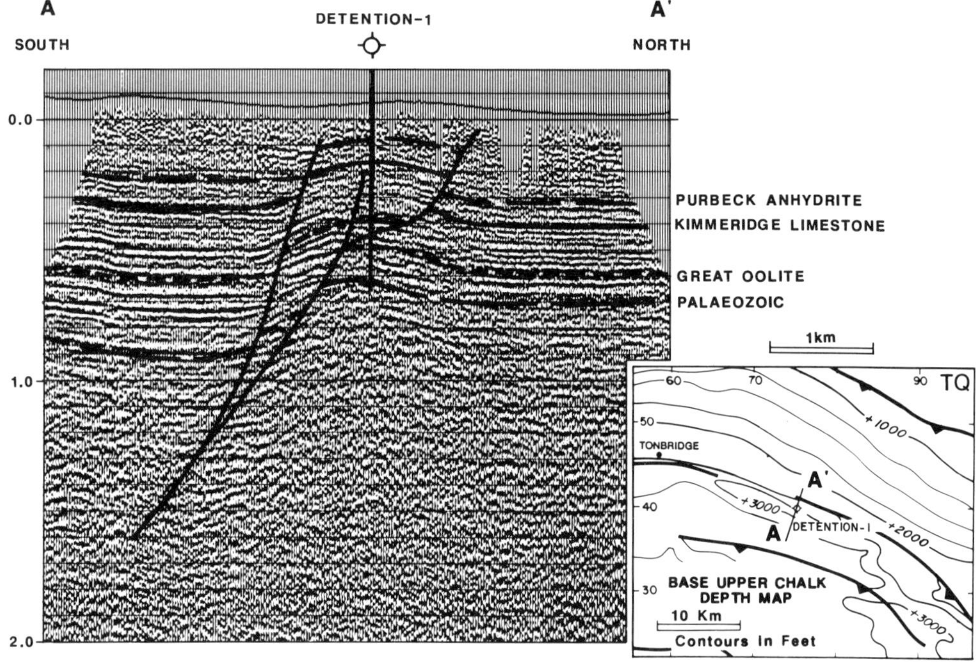

Fig. 6. Seismic cross section through Detention-1.

which analysis of the well indicates is complicated by a series of normal faults lying immediately above the reverse fault and cutting out some 500 feet (150 metres) of section.

More gentle folds are found in the deepest part of the basin, as for example at Godley Bridge (SU 952366, Fig. 7), where the fold has an amplitude of 250 feet (75 metres) and a width of 5 km. Between these uplifts there are extensive zones, averaging 20 km wide, which have remained relatively undeformed.

Normal faults within the Weald Basin are often low angle: those which were reactivated in the Tertiary appear to have deep roots and, although they may become listric at depth, appear to sole out in the basement rather than in the Mesozoic rocks. It is, however, interesting to note that many of the faults flatten as they pass through the Purbeck anhydrite, steepening again below this zone.

It is considered that the regional uplift probably began at the end of the Cretaceous, but it is postulated that the more intense local deformation may have been associated with a pulse of Alpine movements in the Miocene. It seems likely that uplift of the Weald continued into the Pliocene, at which time the margins of the basin, at least, were flooded during a general rise in sea level, and there is evidence for further uplift since that time. Relaxation during the Pleistocene in the eastern part of the Weald Basin led to movement along the Faille du Pas de Calais (Robasynski & Amedro 1986), which gave rise to the Straits of Dover and separated the English Weald Basin from its eastern margins in the Bois Boulonnais.

Reservoir potential

The Weald Basin contains a number of clastic and carbonate reservoirs and wells therefore often have multiple objectives. Reservoirs in the Triassic, Lower, Middle and Upper Jurassic and Lower Cretaceous have been proven hydrocarbon-bearing in various parts of the Basin (Figs 2 & 10). Within the Triassic of the Weald Basin, only the 'Rhaetic' has so far proved productive, but there remains potential for sands of the Sherwood Sandstone Group, particularly in the west, sealed by the Mercia Mudstone Group and the Lias. The Bridport Sands of the Upper Lias are poorly developed in the Weald and there does not, in any case, appear to be an effective seal between this level, the Inferior Oolite and the Great Oolite. For this reason, although there have been interesting shows in the Inferior Oolite, the Great Oolite forms the major objective in the Middle

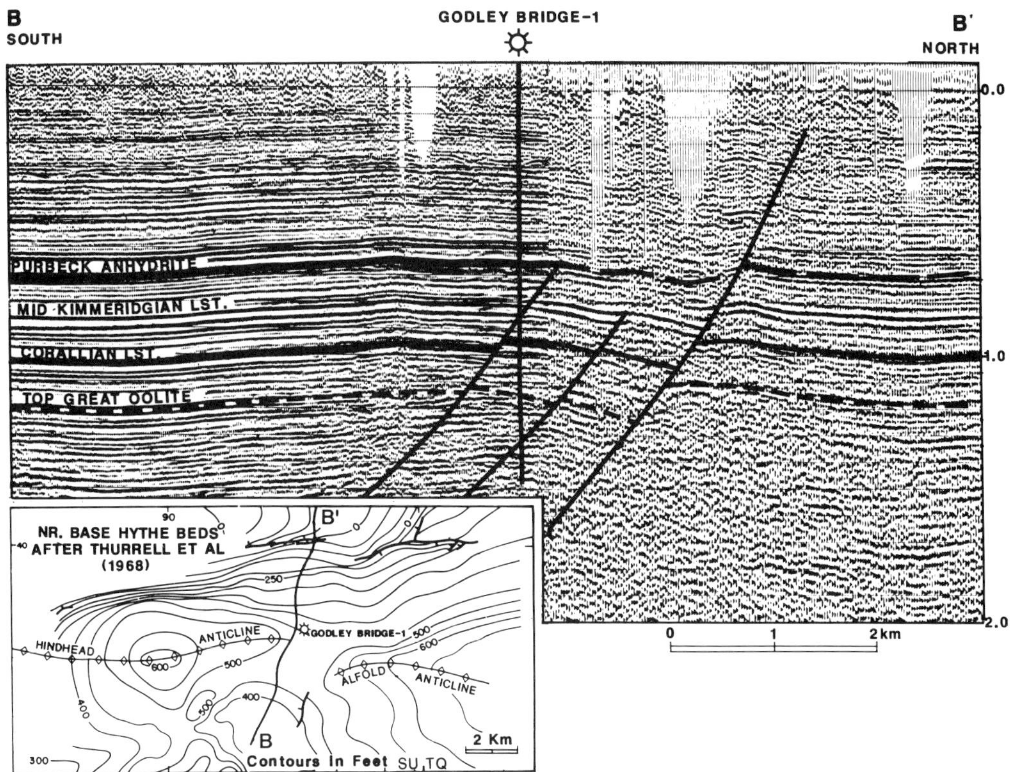

Fig. 7. Seismic cross section through Godley Bridge-1.

Jurassic and is the productive horizon at Horndean (SU 715126), Humbly Grove (SU 712448), Stockbridge (SU 450360) and Storrington (TQ 069115), in each case with the reservoir sealed by the overlying Oxford Clay. Corallian production was established in the 1960s from Bletchingley gasfield, where the reservoir is formed by a carbonate, but the more recent discovery at Palmers Wood (TQ 364520) has established the Corallian sand as a major reservoir; both reservoirs are ultimately sealed by the overlying thick Kimmeridgian shale sequence. The Portland Sand has been proved productive of gas at Godley Bridge, where the Purbeck anhydrite forms an excellent seal, and sands within the overlying Purbeck sequence form a gas reservoir at Albury (TQ 070475). There have been numerous shows of oil and gas within the overlying Wealden sequence, including Bolney-1 (TQ 280243) and Heathfield (TQ 580213), which produced gas to light the local railway station for some years. It is noteworthy that most of the Jurassic reservoirs are of better quality around the margins of the Weald Basin, with significant deterioration of quality towards the centre; these trends are well summarized by Penn *et al.* (1987).

The original depositional environment formed the major control on the quality of reservoirs within the Weald, as a function of sediment type, grain size, sorting and early diagenesis. The majority of Jurassic sediments laid down in the Weald Basin were deposited in low energy, wave-dominated environments and throughout most of the period no good clastic source was available. The occurrence of thick sands in both the Corallian and Portlandian, however, points to the existence of a clastic source on the London Platform towards the end of the Jurassic. Subsequent burial has resulted in considerable porosity deterioration and it is important to understand the distribution of the Tertiary inversions, since reservoirs with poor porosity and permeability characteristics now lie at quite shallow depths. Work on the Great Oolite shows a general relationship between porosity and depth of burial in the Weald Basin (Figs 8a, b): however, where early hydrocarbon emplacement occurred, retention of porosity has taken place.

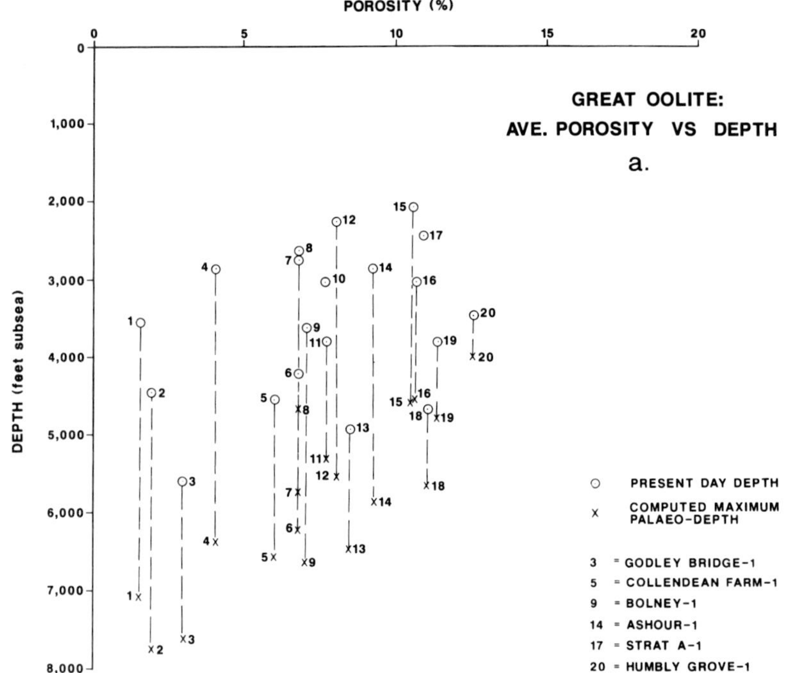

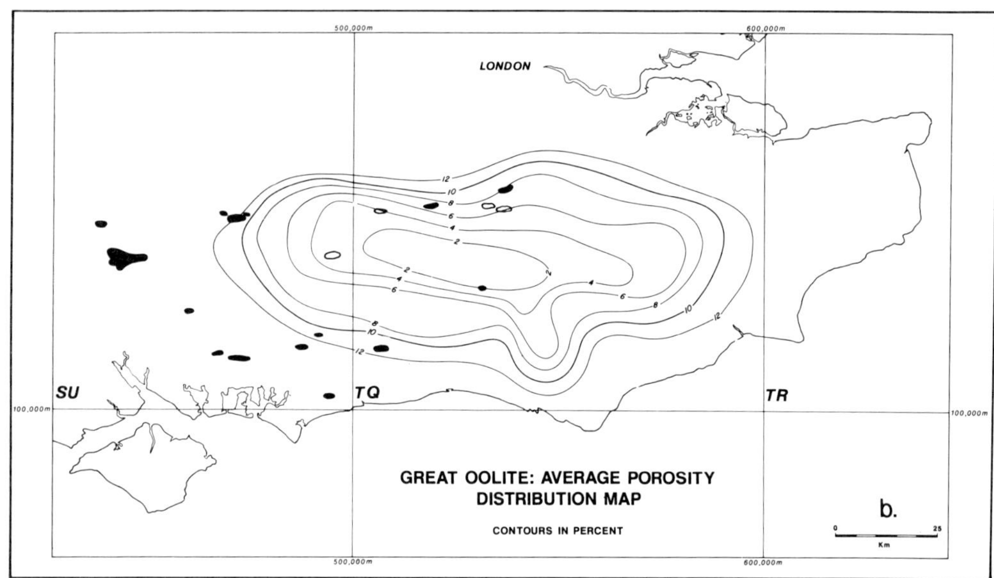

Fig. 8. (a) Plot of Great Oolite average porosity vs present day depth and reconstructed maximum palaeo-depth. (b) Great Oolite average porosity contour map.

Oil generation, migration and entrapment

There are three major Jurassic source rock intervals in the Weald Basin: Lias, Oxford Clay and Kimmeridge Clay. The Lias clays, particularly those of the Lower Lias, have Total Organic Carbon ('TOC') in the range of 0.5% to 2.1% and are considered to be a fair to good oil and gas source, although the interval shows considerable vertical and lateral variation in richness, deteriorating in quality in the eastern part of the basin. Basal parts of the Oxford Clay have TOCs up to 5%, being considered a moderately rich oil-prone source, and the shales of the Kimmeridge Clay form an extremely rich oil-prone source, with TOCs in excess of 10%. Coals of the Westphalian Coal Measures may also form a potential secondary source for gas in parts of the Weald Basin.

The application of the vitrinite reflectance method as a maturity indicator for the Weald Basin seems to give low estimates: studies indicate that this may be related to the type of organic material present in the sediments. However, the widespread occurrences of oil and gas, combined with other indicators of maturity, clearly show that oil and gas generation has occurred. In order to date the time of hydrocarbon generation, burial profiles have been constructed for a number of wells in the basin. Figure 9 shows those for Godley Bridge-1 and Bolney-1, both lying within what were originally the deepest parts of the Weald Basin (Fig. 4c). Modelling the subsidence history of the basin is complicated by the effects of the Tertiary tectonism and by erosion, however an estimation of the amount of inversion can be made from the reconstruction studies (Figs 4b

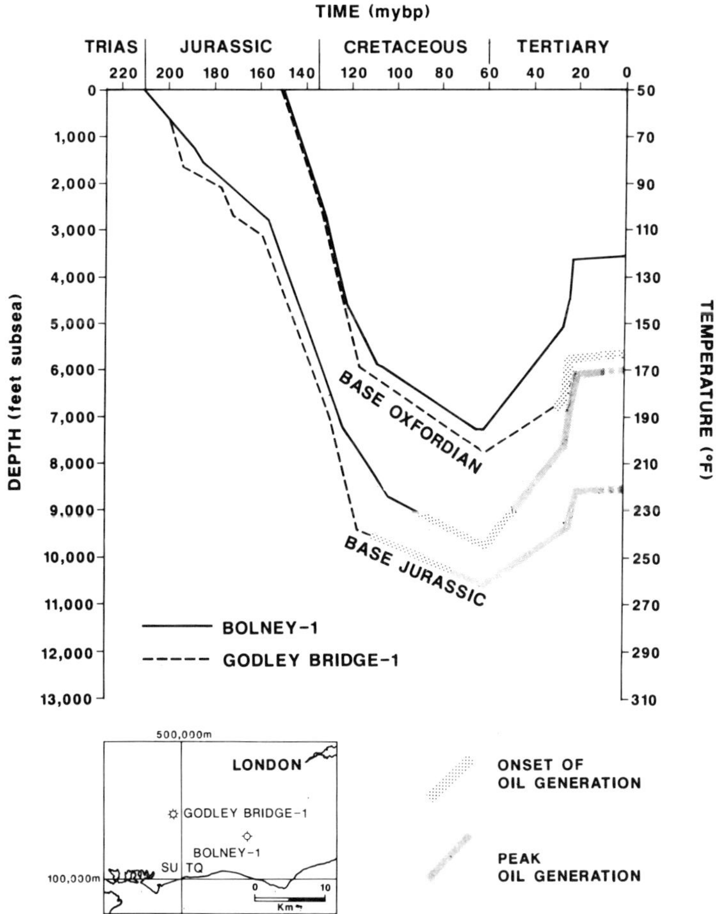

Fig. 9. Computed (non-decompacted) Weald Basin burial depth curves for Godley Bridge-1 and Bolney-1, with maturity estimates.

& 5c). The present average geothermal gradient in the basin, computed from bottom-hole temperatures and drill stem test results, is 1.8°F/100 feet, although Gale *et al.* (1984) demonstrate that there are considerable lateral variations in present day subsurface temperatures. This computed gradient is somewhat conservative, being considerably less than the 2.62°F/100 feet quoted by Ebukanson & Kinghorn (1986). It is thought that variations of heat flow in the basin have occurred over time, particularly during periods of active tectonism. Some control on palaeotemperatures within the Great Oolite has been derived from diagenetic studies by Sellwood *et al.* (1989), who suggest the possibility that there could have been a higher geothermal gradient in southern England during the late Jurassic and early Cretaceous.

From the application of Lopatin's method (Waples 1980) to the derived subsidence and temperature data, a conservative estimate for the timing of the onset of hydrocarbon generation can be derived. This work suggests that hydrocarbon generation from the Lias began in the deepest parts of the Weald Basin in early Cretaceous times (cf. Penn *et al.* 1987), with peak generation in the mid to Upper Cretaceous. The areal extent of mature Liassic shale in late Upper Cretaceous times is demonstrated by Fig. 10. Oxford Clay maturity was reached in the deepest parts of the basin in late Upper Cretaceous times and it is probable that the Kimmeridge Clay reached maturity in the very centre of the basin at this time. Burial depth studies indicate that the Lower Lias could have entered the gas window in the deepest part of the Weald Basin towards the end of the Cretaceous. The effect of the Tertiary uplift was to bring the source rocks progressively out of the optimum temperature and pressure window, such that generation had been effectively brought to a halt over much of the basin by the Miocene, although it is possible that it continued at a much reduced rate in the western part of the Weald Basin, which was not strongly uplifted. The uplift of the Weald gave rise to a second important phase of migration and remigration of both oil and gas as a result of tilting of the fluid conduits and the destruction of many earlier traps.

All the Jurassic oils studied from the Weald Basin are light crudes, with API gravities in the range of 35° to 42°, and are isotopically similar. Studies have shown no good correlation between oils and source rocks: the closest match is with the Lower Liassic shales but the results suggest that some degree of mixing is likely, with contributions from more than one source interval. The origin of the gas found to date is enigmatic: the gas is dry and from isotope work appears to have been generated coevally with the oil. The majority of gas discovered lies in Upper Jurassic reservoirs, whereas most of the oil lies in Middle Jurassic reservoirs. It is possible that this gas may have had its origin as methane exsolved from pore water at relatively shallow depths as a result of uplift; alternatively, it may have originated in deeper reservoirs and have preferentially migrated to higher levels than oil during the Tertiary uplift as a result of its greater mobility: in this context, it is interesting to note that gas in Upper Jurassic and Lower Cretaceous reservoirs has so far only been found in structures of Tertiary age.

Constraints may be placed on the timing of migration by petrographic and fluid inclusion studies carried out on the Great Oolite reservoir (for example by McLimans and Videtich 1987 and Sellwood *et al.* 1989). Three principal phases of cementation are seen by Sellwood *et al.*: early calcite rim cements, later saddle dolomite, with associated sphalerite, and still later mildly ferroan calcite. The ferroan calcite forms a pervasive cement throughout much of the Great Oolite and appears to have been the major cause of porosity reduction in the Weald Basin. Fluid inclusions occur within both the second two phases of cement, although the small size of inclusions in the saddle dolomite and sphalerite makes them difficult to study. Globules of oil were noted trapped between the early calcite cement and the saddle dolomite and numerous inclusions containing oil, gas or both were noted in the ferroan calcite cement. Interpretation of these results suggests that there were at least two phases of hydrocarbon migration: the first pre-dating or associated with the saddle dolomite, which was inferred to have precipitated from hot, saline brines at temperatures between 140° and 280°F, and the second associated with the ferroan calcite, which was precipitated from less saline brines at significantly lower temperatures. Sellwood *et al.* also confirm previous observations (e.g. Hancock & Mithen 1987) that ferroan calcite cement appears to have been inhibited from forming in certain older structures because of the presence of migrated hydrocarbons, again supporting the concept that there were at least two major phases of migration.

It is probable that the first phase of migration was associated with expulsion of fluids from Liassic shales during the latest Jurasic and early Cretaceous. The timing of the second phase of migration or remigration, with the associated precipitation of ferroan calcite is more difficult

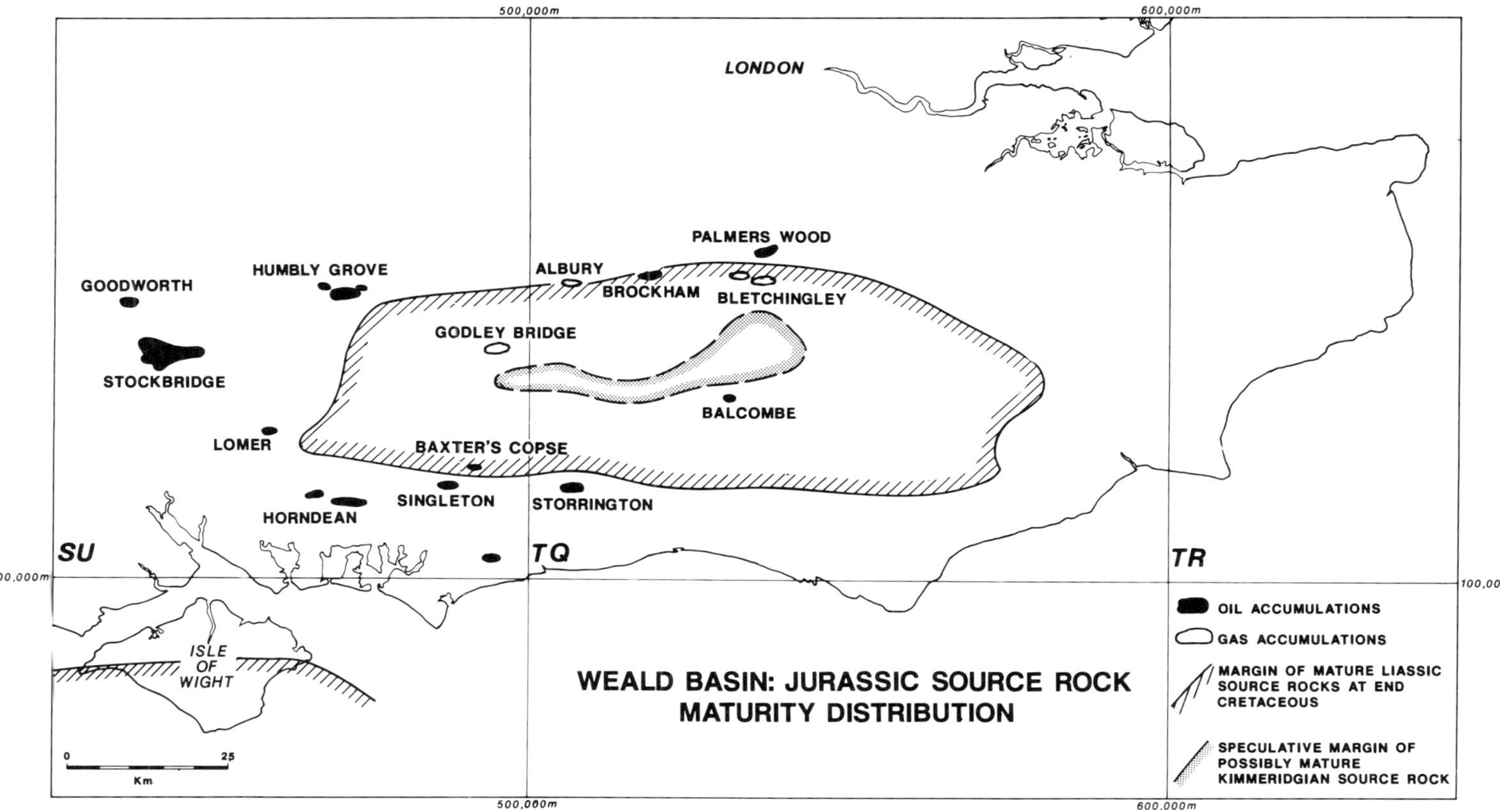

Fig. 10. Postulated areas of mature source rock at end Cretaceous times, showing known hydrocarbon occurrences.

to determine. The ferroan calcite cement has been inhibited from forming in the crests of pre-Tertiary structures, but seems to occur in the crests of structures of Tertiary age. It seems most likely that the migration and cementation episode corresponds to a major phase of change of physical equilibrium within the Weald Basin and it therefore may well have taken place as the Tertiary inversion uplifted the centre of the basin. It is interesting to note that, although Sellwood *et al.* discuss models which involve mixing of fluids expelled from deeply-buried shales with various ages of Cretaceous and early Tertiary seawaters in order to explain the radiogenic character of these ferroan calcite cements, the actual $^{87}Sr/^{86}Sr$ ratio of 0.7081–0.7085 corresponds to late Oligocene to early Miocene seawater, or that of the period which is assumed to have seen the most intense uplift in the Weald. Sellwood *et al.* note that this observation must, however, be tempered by the fact that measured salinities within the fluid inclusions are several times higher than that of seawater. On the basis of the evidence presently available, we believe the precipitation of the ferroan calcite cement (and the second phase of hydrocarbon migration) took place during the early to mid Tertiary.

The first phase of migration, in the late Jurassic to early Cretaceous, led to the accumulation of an oil phase in structures of pre-Upper Cretaceous age, as for example the Palmers Wood (Fig. 11) and Humbly Grove structures. The nature of the Humbly Grove trap has been discussed in some detail by Sellwood *et al.* (1985) and Hancock & Mithen (1987). Other traps were also formed by tilted fault blocks on which growth had occurred throughout the Jurassic and Lower Cretaceous. Those around the margins of the basin have remained relatively intact through the Tertiary uplift, although they were almost certainly affected by regional tilting, whereas those in the centre of the basin were most likely breached during the Tertiary movements. The second phase of hydrocarbon generation and migration, probably in the early Tertiary, led to existing traps being modified or breached and to new traps being formed. Godley Bridge gas accumulation (Fig. 7) has occurred in a trap which owes its origin mainly to Tertiary movements, whereas Storrington oilfield (Fig. 12) is an example of an existing trap which was strongly modified during the Tertiary. It is noteworthy that the traps which appear to have been formed during the Tertiary contain mostly gas. No accumulations have yet

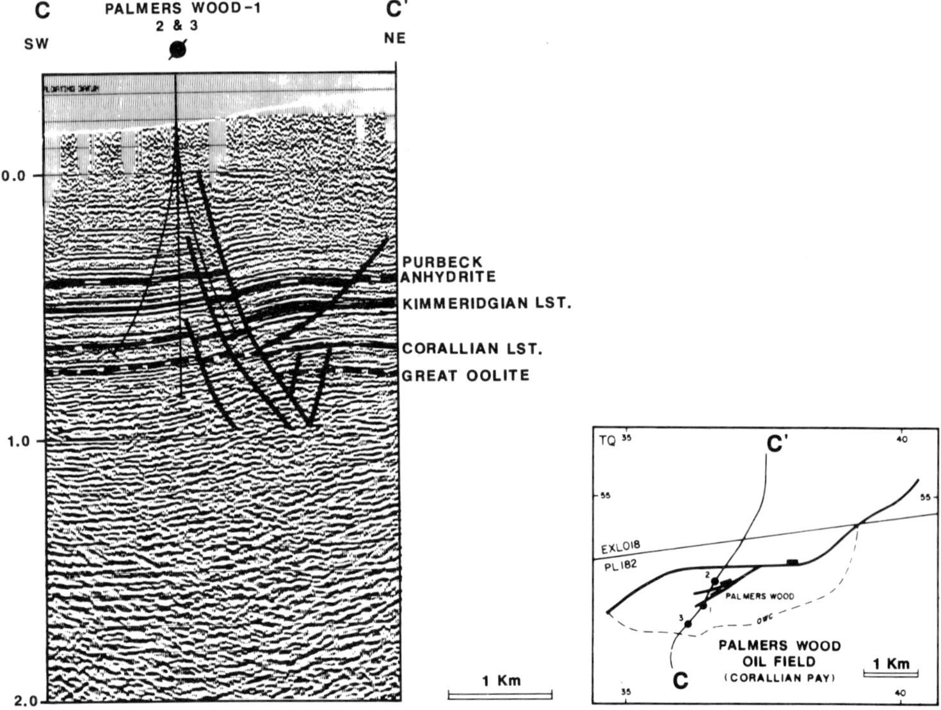

Fig. 11. Seismic cross section through Palmers Wood oilfield.

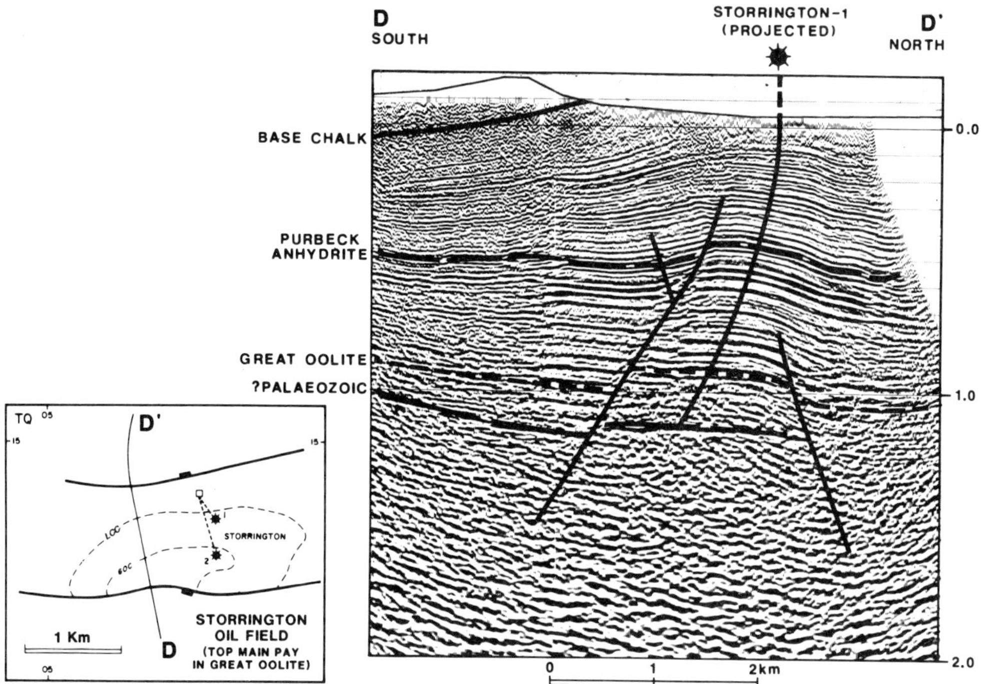

Fig. 12. Seismic cross section through Storrington oilfield.

been found in any of the tight folds associated with the intense local inversion movements: this may be due to a lack of closure at depth or to sealing problems along faults. Although Tertiary structures contain mainly gas, there is evidence that a minor, distinct oil phase also exists, suggesting that they may have received hydrocarbon charges in two phases: the first phase of oil and the second of dry gas. At Godley Bridge, evidence exists for two phases of migration into both Middle and Upper Jurassic reservoirs. There is a possibility that many of the Tertiary folds were in fact initiated as rollover anticlines associated with growth faults in the Upper Jurassic and Lower Cretaceous and that oil migrated into them prior to the Tertiary movements. Where these traps were not already present prior to the inversion, it is postulated that oil migrated into the incipient structures during the gentle early Tertiary uplift and the gas then entered the structures later in a second phase of migration caused by significant uplift, probably during the Miocene. In the case of existing traps, gas break-out may have occurred during Tertiary uplift and tilting, leading to the formation of gas caps (as at Storrington and Humbly Grove oilfields).

The 'plumbing system' of migration pathways which controlled the major vertical and horizontal movements of fluids in the basin were influenced by both sedimentary and tectonic factors. The sedimentary control is provided by the presence of the three thick, widespread shale sections of the Lias, the Oxford Clay and the Kimmeridge Clay, which would have created three, vertically separated, fluid regimes with the reservoir intervals of the Triassic, the Middle Jurassic and the Upper Jurassic providing the migration pathways. The best documented of these pathways is the Middle Jurassic Great Oolite, which appears to have allowed movement well away from the mature source area (for example, as far as Stockbridge and Goodworth (SU 370420), some 30 km west of the postulated edge of maturity in the Weald Basin, see Fig. 10) and demonstrates the effectiveness of the Oxford Clay seal. However, the effectiveness of the Great Oolite migration pathway would have been considerably reduced through time as progressive cementation occurred.

Tectonic control on migration was provided by faulting. It can be observed that hydrocarbons are distributed throughout the stratigraphic column in areas strongly affected by Tertiary inversion and there is clearly a relationship between major faults, which cut the entire section, and the occurrence of multiple pays with hydrocarbons at both shallow and deeper levels. It appears likely that fluid movement occurred directly up the fault planes, since throws often appear insufficient to have allowed vertical migration as a result of reservoir-to-reservoir juxtaposition across the faults. Additional evidence for this is provided by the fact that oil and gas shows are commonly encountered while drilling through fault planes. Movement up these fault planes may have permitted migration of hydrocarbons originally trapped in the Great Oolite into the better, sandstone reservoirs of the Corallian and Portland. It is significant that no overpressure exists within the basin and there is no relict evidence of palaeo-overpressure, suggesting that fault movement also provided a pressure release mechanism and that the faults are, in general, not sealing.

Conclusions

The Tertiary movements led to major changes in the petroleum geology of the Weald Basin. The structural style changed from that of an extensional, block-faulted basin to a compressional regime, with major uplift and inversion. This compression created a number of large structural traps and enabled the remigration of hydrocarbons into shallower, and in some cases better, reservoirs. The absence of a regional seal above the Jurassic led, however, to the loss of hydrocarbons to the surface during the late Tertiary. Unlike many inverted basins, the central inverted high of the Weald Anticlinorium has not as yet proved to contain major hydrocarbon accumulations: it is felt that this is due primarily to the lack of good reservoirs in this central area, although the magnitude of the uplift has almost certainly led to the breaching of many of the large structures. Additionally, much diagenetic damage had occurred prior to the later Tertiary uplifts, creating a further impediment to large-scale remigration into many of the Tertiary structures.

Although Tertiary structures were largely ignored once modern seismic demonstrated the presence of offset, older structures, recent successes have demonstrated that they remain a valid play: those Tertiary structures drilled prior to the advent of modern seismic can often be shown to have been tested off-structure. The most promising prospects are, however, likely to occur around the flanks of the Weald Basin, where uplift has not been great and traps of both Tertiary and earlier age should provide good targets. Finally, it should be noted that exploration of the Weald Basin is still at a very early stage: further drilling will assist immensely in our understanding of the tectonic and diagenetic problems and can also be expected to lead to the discovery of significant new reserves.

The authors are grateful to the Directors of Cairn Energy plc and their partners, Monument Oil and Gas plc and Arco British Ltd, for permission to publish much of the information contained in this paper. Fruitful discussions with many colleagues over several years are also gratefully acknowledged. The authors would like to make it clear that the interpretations and opinions expressed in this paper are their own and do not necessarily reflect those of any of the companies which have so kindly allowed the use of their confidential information.

References

Allen, P. 1975. Wealden of the Weald: a new model. *Proceedings Geologists' Association*, **86**, 389–437.

Arkell, W. J. 1933. *The Jurassic System in Great Britain*, Oxford.

—— 1939. Derived ammonites from the Lower Greensand of Surrey and their bearings on the tectonic history of the Hog's Back. *Proceedings Geologists' Association*, **50**, 22–25.

Chadwick, R. A. 1985a. Permian, Mesozoic and Cenozoic structural evolution of England and Wales in relation to the principles of extension and inversion tectonics. *In*: Whittaker, A. (ed.), 1985. *Atlas of Onshore Sedimentary Basins in England and Wales: Post-Carboniferous Tectonics and Stratigraphy*. Blackie, Glasgow.

—— 1985b. Cenozoic sedimentation, subsidence and tectonic inversion. *In*: Whittaker, A. (ed.) 1985. *Atlas of Onshore Sedimentary Basins in England and Wales: Post-Carboniferous Tectonics and Stratigraphy*. Blackie, Glasgow.

——, Kenolty, N. & Whittaker, A. 1983. Crustal structure beneath southern England from deep seismic reflection profiles. *Journal of the Geological Society, London,* **140**, 893–911.

Ebukanson, E. J. & Kinghorn, R. R. F. 1986. Maturity of organic matter in the Jurassic of southern England and its relation to the burial history of the sediments. *Journal of Petroleum Geology*, **9**, 259–280.

Gale, I. N., Rollin, K. E., Smith, N. J. P., Lee, M. K., Burley, A. J., Wheildon, J. & Brown, G. C. 1984. *Geothermal Map of the United Kingdom*. British Geological Survey.

Hancock, F. R. P. & Mithen, D. P. 1987. The geology of the Humbly Grove oil-field, Hampshire, U.K. *In*: Brooks, J. & Glennie, K. W. (eds) *Petroleum Geology of North West Europe*, Graham & Trotman, London, 161–170.

Haq, B. U., Hardenbol, J. & Vail, P. R. 1987. Chronology of fluctuating sea levels since the Triassic. *Science*, **235**, 1156–1166.

King, C. 1981. The stratigraphy of the London Clay. *Tertiary Research Special Paper*, **6**, 1–158.

McLimans, R. K. & Videtich, P. E. 1987. Reservoir diagenesis and oil migration: Middle Jurassic Great Oolite Limestone, Wealden Basin, southern England. *In*: Brooks, J. & Glennie, K. W. (eds) *Petroleum Geology of North West Europe*, Graham & Trotman, London, 119–128.

Middlemiss, F. A. 1975. Studies in the sedimentation of the Lower Greensand of the Weald, 1875–1975: a review and commentary. *Proceedings Geologists' Association*, **86**, 457–473.

Mortimore, R. N. 1986. Controls on Upper Cretaceous sedimentation in the South Downs, with particular reference to flint distribution. *In*: Sieveking, G. de C. & Hart, M. B. (eds), *The scientific study of flint and cherts – proceedings of the fourth international flint symposium, Brighton April 1983.* Cambridge University Press, Cambridge.

Owen, H. G. 1975. The stratigraphy of the Gault and Lower Greensand of the Weald. *Proceedings Geologists' Association*, **86**, 475–498.

Penn, I. E., Chadwick, R. A., Holloway, S., Roberts, G., Pharoah, T. C. & Allsop, J. M. 1987. Principal features of the hydrocarbon prospectivity of the Wessex Channel Basin, U.K. *In*: Brooks, J. & Glennie, K. W. (eds) *Petroleum Geology of North West Europe*, Graham & Trotman, London, 109–118.

Robasynski, F. & Amedro, F. 1986. The Cretaceous of the Boulonnais (France) and a comparison with the Cretaceous of Kent (United Kingdom). *Proceedings Geologists' Association*, **97**, 171–208.

Sellwood, B. W., Scott, J., Mikkelsen, P. & Ackroyd, P. 1985. Stratigraphy and sedimentology of the Great Oolite Group in the Humbly Grove Oilfield, Hampshire. *Marine and Petroleum Geology*, **2**, 44–55.

——, Shepherd, T., Evans, M. R. & James, B. 1989. Origin of late cements in oolitic reservoir facies: a fluid inclusion and isotopic study (Mid-Jurassic, S. England). *Sedimentary Geology*, **61**, 223–237.

——, ——, James, B., Evans, M. R. & Marshall, J. 1987. Regional significance of "dedolomitization" in Great Oolite reservoir facies of southern England. *In*: Brooks, J. & Glennie, K. W., (eds) *Petroleum Geology of North West Europe*, Graham & Trotman, London, 129–137.

Smalley, S. & Westbrook, G. K. 1982. Geophysical evidence concerning the southern boundary of the London Platform beneath the Hog's Back, Surrey. *Journal of the Geological Society, London*, **139**, 139–146.

Smith, I. F. 1986. Mesozoic Basins. *In*: Downing, R. A. & Gray, D. A. (eds). *Geothermal Energy-the potential in the United Kingdom*. H.M.S.O., London.

Thurrell, R. G., Worssam, B. C. & Edmonds, E. A. 1968. Geology of the country around Haslemere. *Memoir Geological Survey U.K.*

Waples, D. W. 1980. Time and temperature in petroleum formation: Application of Lopatin's method to petroleum exploration. *Bulletin of the American Association of Petroleum Geologists*, **64**, 916–926.

Whittaker, A. (ed.) 1985. *Atlas of Onshore Sedimentary Basins in England and Wales: Post-Carboniferous Tectonics and Stratigraphy.* Blackie, Glasgow.

Tectonic events responsible for Britain's oil and gas reserves: a summary

B. VAN HOORN

Shell Internationale Petroleum Mij. B.V., P.O. Box 162, 2501 AN The Hague, The Netherlands

In 1964, following the implementation of the 'Continental Shelf Act', the first offshore licenses in the United Kingdom were awarded. Subsequent exploration activities have resulted in the discovery of some 20 billion barrels of oil reserves and 70 trillion cubic feet of recoverable gas, most of which are located in the North Sea Basin. It is no surprise, therefore, that most of the papers presented in this volume concentrate on this area, where a particular and favourable tectonic evolution has lead to the coincidence of good source rocks, sufficiently deeply buried; good reservoir development; and timely trap formation in relation to hydrocarbon generation and migration. It is the purpose of this contribution to summarize the various tectonic phases which have contributed to the complex geological evolution of the British Isles and surrounding territorial waters and trace their impact upon the development of the hydrocarbon habitat.

Caledonian framework

The basic tectonic framework of NW Europe was created during Ordovician–early Devonian times, as a result of the closure of the Iapetus Ocean. Whereas in Norway this led to a NW–SE directed collision between the Laurentian and Fennoscandian Shields, in Britain a more complex accretion of Precambrian continental blocks and island arcs onto the N American craton occured (**Coward**). The resulting heterogeneity of this basement complex had a critical influence on subsequent tectonic episodes, during which existing basement lineaments were reactivated to varying degrees. In the Northern North Sea and the Atlantic Margin west of the Shetlands, there is an important SW–NE structural grain, reflecting reactivation of Caledonian thrusts. In the Central and Southern North Sea, the predominant NW–SE lineaments probably are related to a diffuse transform zone between the Fennoscandian Shield and accreted terrains.

Devono-Carboniferous

During the Devonian the post-orogenic collapse of the Caledonides was accompanied by a major sinistral translation between the Laurentian and Fennoscandian Shields. In the Northern North Sea these movements gave rise to the rapid subsidence of pull-apart basins (Midland Valley, Orcadian Basin) in which thousands of metres of Old Red Sandstone accumulated, locally providing lacustrine conditions with consequent deposition of organic-rich clays. The early Carboniferous evolution of the British Isles was governed by north–south extension along a series of NW–SE and NE–SW trending faults located along zones of Caledonian weakness (Northumberland, Stainmore, Craven, Gainsborough and Widmerpool Gulf Basins). From the Namurian onwards, crustal stretching ceased and broad regional subsidence was initiated, coincident with a major climatic change from semi-arid to monsoonal conditions. This resulted in the development of major south to southwest directed deltaic drainage systems with extensive coal deposition, forming both excellent gas-prone source rocks and potential reservoirs (**Leeder & Hardman**, **Fraser & Gawthorpe**).

Variscan orogeny

As a result of the closure of the Proto-Tethyan Ocean south of Britain and the subsequent collision of Laurasia and Gondwana, intra-plate stresses were transmitted into the Variscan foreland resulting in extensive deformation. In the East Midlands area, uplift and folding of the hanging walls of the main extensional fault trends created the principal traps (**Fraser & Gawthorpe**). Elsewhere, in the Silverpit area of the Southern North Sea, gas accumulations in Carboniferous reservoirs are also located in areas of Variscan inversion. Subsequently, during the latest Carboniferous–early Permian,

From Hardman, R. F. P. & Brooks, J. (eds), 1990, *Tectonic Events Responsible for Britain's Oil and Gas Reserves*, Geological Society Special Publication No 55, pp 393–395.

NW Europe was transected by a late Variscan post-orogenic system of conjugate shear faults which, in the North Sea area, triggered widespread magmatism and deformation of the sedimentary fill of the Variscan foreland basin.

Permian subsidence

By the end of Autunian times, wrench fault and volcanic activity had abated in NW Europe. The broad Northern and Southern Permian Basins, as well as the Moray Firth Basin, began to subside during the Saxonian, presumably in response to the decay of thermal anomalies that were induced during the Stephanian–Autunian phase of wrench faulting. Widespread arid conditions lead to the deposition of Rotliegend fluvial and aeolian sands, which now form major reservoirs in the Southern North Sea (**Glennie**). At the transition to the late Permian, the Zechstein sea advanced southward from the Norwegian–Greenland Sea rift and flooded the Rotliegend basins of NW Europe, resulting in extensive evaporite deposition. These evaporites not only controlled subsequent reservoir and trap distribution in the Central North Sea, but also form a highly effective seal to the Southern Permian Basin gas fields.

Triassic–early Jurassic rifting stage

At the transition from the Permian to the Triassic, rifting activity accelerated in the Norwegian-Greenland Sea area and in the Tethys domain. By early Triassic time NW and Central Europe were subjected to regional stresses causing the differential subsidence of a complex set of multidirectional grabens and troughs (**Steel & Ryseth**). Although the prime extensional direction was east–west, many of the fault-bounded grabens formed along reactivated Caledonian trends. In the Central North Sea, Triassic sedimentation and syndepositional faulting triggered diapiric deformation of Zechstein salts during the middle and late Triassic. This led to a complex distribution of Triassic reservoir sands in this area, in contrast to the extensive fluvial and aeolian sands of the Southern Permian Basin (Bunter Sand Fm.). Subsequent marine incursions into this basin resulted in the deposition of saliferous and anhydritic formations which now provide an effective seal to a number of gas accumulations.

Mid-Jurassic thermal doming

At the transition from the Lower to the Middle Jurassic, the Central North Sea became uplifted wards over the Ringkøbing–Fyn High. Uplift of this dome was accompanied by the development of a large volcanic complex at the triple junction between the Viking, Central and Moray Firth–Witch Ground Grabens (**Latin** *et al.*) Subsidiary volcanic centres occured in the southern Viking Graben, the Egersund Basin, in coastal Norway and in the Central Graben. Volcanics display the bimodal mafic-felsic alkaline chemistry that is typical for intracratonic rifts (**Latin** *et al.*). The bulk of these volcanics were extruded during the Bajocian. As a result of the thermal updoming, erosion products consisting mainly of recycled Triassic sandstone were deposited in adjacent grabens as major deltaic complexes, such as the main hydrocarbon-bearing Brent Group in the northern Viking Graben (**Richards**).

Late Jurassic–early Cretaceous rifting

Coeval with rapid sea-floor spreading in the Central Atlantic, the rate of crustal extension across the North Sea rift system accelerated during Kimmeridgian to Berriasian–Valangian times. This lead to major fault block rotation and salt diapirism and represents the principal trap-forming event in the Northern North Sea, Moray Firth and Central North Sea areas (**Roberts** *et al.*). Uplift of the Shetland Platform resulted in clastics being shed into the southern Viking Graben and Witch Ground Trough where they accumulated as submarine fan complexes, now of great economic importance (Kimmeridge, Brae and Magnus Sandstone Members). Elsewhere, in the Outer Moray Firth and Central North Sea, sheet sands of the Piper and Fulmar Formations constitute important economic targets (**O'Driscoll** *et al.*). These reservoirs were all charged by the most prolific source rock in NW Europe, the Kimmeridgian to Berriasian kerogenous shales; deposited during a sea-level high stand, these constitute the principal source of hydrocarbons in the Central and Northern North Sea. In the Southern North Sea crustal distension was accomodated by a dense series of reactivated NW–SE dextral oblique-slip faults along which differential subsidence occurred, resulting in the local onset of gas generation (Sole Pit Basin). Similar high rates of subsidence along reactivated Variscan faults have also been recognized in Southern England (**Butler & Pullan**).

Cretaceous–Tertiary subsidence and inversion

Throughout the Cretaceous and Tertiary the

by the combined effects of crustal separation in the North Atlantic and the onset and further development of Alpine collision between Europe and Africa. During the Aptian large parts of NW Europe were affected by differential fault block movements, coincident with major shearing and opening of the Bay of Biscay. As a result, submarine fan and turbidite sands were locally emplaced and form attractive reservoirs in the Outer Moray Firth (**Bisewski**) or in the West Shetland Basin. From Senonian times onwards, basins in the Southern North Sea and Southern England suffered large-scale inversion, possibly the result of intra-plate stress propagation from the Alpine deformation to the NW Europe foreland. Such inversion movements, also evident in the Dutch and Danish parts of the Central Graben, provided potential vertical migration paths for hydrocarbons. Moreover, they also may have triggered the huge mass flows of Chalk which now form porous reservoirs in the Norwegian Central Graben. In the Central and Northern North Sea areas, rifting ceased during the late Cretaceous and thermal subsidence prevailed, possibly resulting in the onset of oil generation in the graben axes. Igneous activity increased along the Faeroe–West Shetland and mid-Norway rift zones and culminated during the Paleocene in a major volcanic event (Thulean phase) that affected large areas of the Arctic–North Atlantic borderlands. Doming, centred over NW Scotland, resulted in large-scale erosion and shedding of clastics into the deepwater basins of the Central and Northern North Sea thermal sag (**Milton** *et al.*; **Harding** *et al.*). Subsequent compressional movements, during the Oligocene, partly destroyed old traps and created new traps in Southern England (**Butler & Pullan**), whilst simultaneously a second phase of inversion affected the Sole Pit and triggered salt diapirism in the Central Graben.

Subsequent evolution of the North Sea through recent times has been characterised by tectonic quiescence, with sedimentation and subsidence rates increasing from the Oligocene, as a consequence of which the area of mature Upper Jurassic source rock has been considerably extended.

Conclusions

The papers presented in this volume constitute a very useful overview of the tectonic evolution of Britain and its intimate relation to hydrocarbon distribution. Although the broad structural controls are now very well established, it is likely that the increased use of 3D seismic, and especially the introduction of new interpretation technologies, will provide a more profound understanding of hydrocarbon traps and reservoir distribution. Examples of such work are already displayed in this volume, e.g. **Boldy & Brealey**'s discussion on the timing, nature and sedimentary response to late Jurassic tectonics in the Outer Moray Firth; and the stratigraphic and structural framework of the Upper Jurassic and Lower Cretaceous reservoirs in the Witch Ground Graben (**O'Driscoll** ***et al.***). One may assume that such 3D based investigations will become more frequent in the future and will lead to a vastly improved appreciation of the tectonic controls on oil and gas distribution in Britain.

Index

Deformation Mechanisms, Rheology and Tectonics

Geological Society Special Publication No. 54

Edited by R.J. Knipe (Leeds University, UK) and E.H. Rutter (Manchester University, UK)

The papers in this book are gathered into groups that aim to reflect current research themes ranging from geologically-orientated rock mechanics, through structural and microstructural studies of naturally-deformed rock masses to large-scale tectonics. Some of the thematic groups contain a paper with a substantial review component to provide an introductory framework for those new to the subject, however, the book is dominated by original research papers.

- International field of contributors
- Published December 1990
- 528 pages
- 46 papers
- 326 illustrations
- List price* £85/US$142
- ISBN 0-903317-58-3

Principal Authors

N.L. Carter (Texas A&M University, USA)
R.H. Sibson (University of California, USA)
J.P. Evans (Utah State University, USA)
S.M. Agar (Leeds University, UK)
M. Casey (ETH-Z, Switzerland)
S.J.D. Cox (CSIRO Geomechanics, Australia)
S.J. Hippler (Leeds University, UK)
I.G. Main (Edinburgh University, UK)
G. Zulauf (Goethe Universitat, FRG)
I.S. Stewart (Bristol University, UK)
Teng-Fong Wong (State University of New York, USA)
R.K. Davies (Texas A&M University, USA)
H.W. Green II (University of California, USA)
B.E. Hobbs (CSIRO Geomechanics, Australia)
E.M. Klaper (Bern University, Switzerland)
D.L. Olgaard (ETH, Switzerland)
A. Ord (CSIRO Mechanics, Australia)
E. Carrio-Schaffhauser (IRIGM, France)
R.C.M.W. Franssen (Utrecht University, The Netherlands)
C.J. Spiers (Utrecht University, The Netherlands)
J.A. Gilotti (Uppsala University, Sweden)
M. Burkhard (Institut de Geologie, Switzerland)
A.N. Walker (Imperial College, UK)
J.H.P. De Bresser (Institute of Earth Sciences, The Netherlands)
M.S. Paterson (Australian National University, Australia)
D.J. Prior (Liverpool University, UK)
W. Skrotzki (Universitat Gottingen, FRG)
J.C. White (University of New Brunswick, Canada)
R.D. Law (Virginia Polytechnic, USA)
M.W. Jessell (Monash University, USA)
Jin-Han Ree (State University of New York, USA)
N.Ø. Olesen (Geologisk Institut, Denmark)
H. Schaeben (University of Bonn, FRG)
D.E. Karig (Cornell University, USA)
N.A. Yassir (University of Waterloo, Canada)
P.A.R. Nell (British Antarctic Survey, UK)
K.T. Pickering (Leicester University, UK)
Shumin Liu (Queen's University, Canada)
K.R. McClay (University of London, UK)
C.J.L. Wilson (Melbourne University, Australia)
J.P. Gratier (Universite Joseph Fourier, France)
M. Coli (Universita Firenze, Italy)
F. Sani (University of Florence, Italy)
R.E. Holdsworth (Reading University, UK)
J.E. Iliffe (University of South Carolina, USA)
J.L. Urai (Institute of Earth Sciences, The Netherlands)

Outline of Contents

Role of fluids in rock deformation: Conditions for fault-valve behaviour • Textures, deformation mechanisms and the role of fluids in the cataclastic deformation of granitic rocks • **Fracture and faulting:** Fracture evolution in the upper ocean crust: evidence from DSDP hole 504B • Calculation of bulk rheologies of structured materials and application to brittle failure in shear • Damage development during rupture of heterogeneous brittle materials: a numerical study • Velocity-dependent friction in a large direct shear experiment on gabbro • The evolution of cataclastic rocks from a pre-existing mylonite • Influence of fractal flaw distributions on rock deformation in the brittle field • Brittle deformation and graphitic cataclasites in the pilot research well KTB-VB • Brecciation and fracturing within neotectonic normal fault zones in the Aegean region • Mechanical compaction and the brittle-ductile transition in porous sandstones • **Instabilities and localization:** Shear bands in a plastic layer at yield subjected to combined shortening and shear: a model for the fault array in a duplex • The failure mechanism for deep-focus earthquakes • Instability, softening and localization of deformation • Reaction-enhanced formation of eclogite facies shear zones in granulite facies anorthosites • A case study of the role of second phase in the localization of deformation • Mechanical controls on dilatant shear zones • Propagation and localization of stylolites in limestones • **Flow mechanisms and flow laws** • Deformation of polycrystalline salt in compression and shear at 250-350°C • Experimental determination of constitutive parameters governing creep of rock salt by pressure solution • Phenomenological superplasticity in rocks • Ductile deformation mechanisms in micritic limestones naturally deformed at low temperatures, 150-350°C • Experimental study of grain-size sensitive flow of synthetic, hot-pressed calcite rocks • High-temperature deformation of calcite single crystals by r+ and f+ slip • Quartz rheology under geological conditions • Estimates of the rates of microstructural change in mylonites • Microstructure in hornblende of a mylonitic amphibolite • Albite deformation within a basal ophiolite shear zone • **Rocks fabrics:** Crystallographic fabrics; a selective review of their applications to research in structural geology • A simulation of the temperature dependence of quartz fabrics • High temperature deformation of octachloropropane; dynamic grain growth and lattice reorientation • The SEM/ECP technique applied on twinned quartz crystals • Practical application of entropy optimization in quantitative texture analysis • **Deformation of weak sediments:** Experimental and observational constraints on the mechanical behaviour in the toes of accretionary prisms • The undrained shear behaviour of fine-grained sediments • Deformation in an accretionary melange, Alexander Island, Antarctica • Vein structure and the role of pore fluids in early deformation of late Miocene volcanoclastic rocks, Miura group, SE Japan • **Experimental modelling using analogue materials:** Centrifuge modelling of thrust faulting; strain partitioning and sequence of thrusting in duplex structures • Deformation mechanisms in analogue models of extensional fault systems • Slickenside lineations due to ductile processes • **Deformation mechanisms and tectonics:** Transition between seismic and aseismic deformation in the upper crust • Vein distribution in a thrust sheet, a case history from N. Apennines, Italy • Extensional veining and shear joint systems in a trust-fold zone (N. Apennines, Italy) • Convergence-related dynamic spreading in a mid-crustal thrust zone: an orogenic wedge model • Structural implications of compactional strain caused by fault block rotation: evidence from two-dimensional numerical analogues • Alpine deformation on Naxos (Greece)

* FGS price available on request provided that the correct member number is quoted.

The Geometry of Normal Faults

Geological Society Special Publication No. 56

Edited by A.M. Roberts, G. Yielding and B. Freeman (Badley, Ashton & Associates, UK)

During the 1980s a resurgent interest in extensional tectonics resulted, to a large extent, from the ever-increasing non-proprietary availability of seismic-reflection data. In the early-to-mid 1980s much work on extensional fault-systems focused on the innovative application of thrust-belt-type models to extensional basins. In particular the concepts of section-balancing were introduced to those investigating normal faults.

By the late 1980s it was becoming apparent that the universal application of such models was fraught with difficulty. In particular the evidence of both earthquake seismology and detailed field studies began to indicate that faults involved in crustal extension may be, on all scales, essentially planar structures, not linked to a 'listric' thrust-type array.

It is now clear from the geological record that both planar and listric normal faults exist. This book aims to discuss the geological setting and interpretation of these structures.

Principal Authors

R.F.P. Hardman (Amerada Hess Ltd, UK)
D. Barr (BP, UK)
J. Cartwright (Imperial College, UK)
N.J. Kusznir (Liverpool University, UK)
A.M. Roberts (Badley, Ashton & Associates, UK)
G. Yielding (Badley, Ashton & Associates, UK)
M.P. Coward (Imperial College, UK)
A.G. Koestler (GEORECON, Norway)
S. Roberts (Bullard Laboratories, UK)
R. Westaway (Durham University, UK)
A. Beach (Alastair Beach Associates, UK)
T.J. Chapman (GECO, UK)
J.J. Walsh (Liverpool University, UK)
G. Dresen (Geologisches Institut des THD, Germany)
R.W. Krantz (Université de Rennes, France)
K.R. McClay (Royal Holloway & Bedford New College, UK)
B. Vendeville (Texas University at Austin, USA)
N. White (Bullard Laboratories, UK)

Outline of Contents

The significance of normal faults in the exploration and production of North Sea hydrocarbons • **Seismic and subsurface studies** • Subsidence and sedimentation in semi-starved half-graben: a model based on North Sea data • The kinematic evolution of the Coffee Soil Fault • A flexural-cantilever simple-shear/pure-shear model of continental lithosphere extension: applications to the Jeanne d'Arc Basin, Grand Banks and Viking Graben, North Sea • Deformation around basin-margin faults in the North Sea/mid-Norway rift • Seismic reflections from normal faults in the northern North Sea • **Field-based studies** • Extensional structures and their tectonic inversion in the Western Alps • Description of brittle extensional features in chalk on the crest of a salt ridge (NW Germany) • Active normal faulting in central Greece: an overview • Continental extension on sets of parallel faults: observational evidence and theoretical models • **Fault-displacement studies** • The geometry of normal faults in a sector of the offshore Nile Delta, Egypt • The displacement patterns associated with a reverse-reactivated, normal growth fault • Geometric and kinematic coherence and scale effects in normal fault systems • **Analogue-modelling and section-balancing** • Numerical and analogue modelling of normal fault geometry • Normal fault geometry and fault reactivation in tectonic inversion experiments • Physical and seismic modelling of listric normal fault geometrics • Mechanisms generating normal fault curvature: a review illustrated by physical models • Calculating normal fault geometries at depth: theory and examples

- Published January 1991
- 264 pages
- 17 papers
- 185 illustrations
- ISBN 0-903317-59-1
- List price* £58/US$115

* FGS price available on request provided that the correct member number is quoted.

Developments in Sedimentary Provenance Studies

Geological Society Special Publication No. 57

Edited by A.C. Morton (British Geological Survey, UK), S.P. Todd (BP, UK) and P.D.W. Haughton (Glasgow University, UK)

The study of sedimentary provenance interfaces several of the mainstream geological disciplines. Its remit includes the location and nature of sediment source areas, the pathways by which sediment is transferred from source to basin of deposition, and the factors that influence the composition of sedimentary rocks. Materials subject to study are as diverse as recent muds in the Mississippi River basin, Archaean shales, and soils on the Moon.

Provenance data can play a critical role in assessing palaeogeographic reconstructions, in constraining lateral displacements in orogens, in characterising crust which is no longer exposed, in testing tectonic models for uplift of fault block or orogen scale, in mapping depositional systems, in sub-surface correlation and in predicting reservoir quality. On a global scale, the provenance of fine-grained sediments have been used to monitor crustal evolution.

The aim of the book is to achieve a better understanding of how grain components which comprise a sedimentary rock were assembled, to reconstruct source areas with greater confidence, and to use provenance data more effectively to test tectonic models.

Principal Authors

P. Allen (Reading University, UK)
G.G. Zuffa (Universita di Bologna, Italy)
A.C. Morton (British Geological Survey, UK)
A. Tortosa (Universidad Complutense de Madrid, Spain)
A. Basu (Indiana University, USA)
A.J. Hurford (University of London, UK)
D.J. Batten (University of Wales, UK)
M.A. Velbel (Michigan State University, USA)
A.E. Milodowski (British Geological Survey, UK)
R. Valloni (Universita de Parma, Italy)
R.A. Cliff (Leeds University, UK)
P.A. Floyd (Keele University, UK)
C.M. Gerrard (Cotswold Archaeological Trust, UK)
J.R. Graham (Trinity College, Ireland)
D. Pirrie (Camborne School of Mines, UK)
B. Humphreys (Lemigas, Indonesia)
J. Arribas (Universidad Complutense de Madrid, Spain)
I.R. Garden (Badley, Ashton & Associates, UK)
G. Nichols (Royal Holloway & Bedford New College, UK)
P.A. Cawood (Memorial University of Newfoundland, Canada)
J.A. Evans (British Geological Survey, UK)
T. McCann (University College, Ireland)
M.J. Evans (BP, UK)
S.J. Cuthbert (Glasgow, UK)

Outline of Contents

Provenance research: Torridonian and Wealden, examples • Turbidite arenites in provenance studies • Geochemical studies of detrital heavy minerals: applications to provenance research • Quartz grain types in Holocene deposits from the Spanish Central System • Detrital opaque Fe-Ti oxide minerals in provenance determination • Fission track dating in discrimination of provenance • Reworking of plant microfossils and sedimentary provenance • Triassic rift-valley redbeds of eastern North America: a case study • Redistribution of rare earth elements during diagenesis of turbidite/hemipelagic mudrock sequences • Selective alteration of arkose framework on Oligo-Miocene turbidites • Sourcelands for Carboniferous Pennine river system: constraints from sedimentary evidence and U-Pb geochronology • Geochemistry and provenance of Rhenohercynian synorogenic sandstones • Sedimentary petrology for the archaeologist • A local source for the Ordovician Derryveeny Formation, western Ireland • Controls on the petrographic evolution of an active margin sedimentary sequence • An integrated approach to provenance studies: a case example from the Central Graben, North Sea • Petrographic evidence of different provenance in two alluvial fan systems (Palaeogene of the northern Tajo Basin, Spain) • Changes in the provenance of pebbly detritus in southern Britain and northern France associated with basin rifting • Petrological and geochemical determination of provenance in the southern Welsh Basin • Nature and record of igneous activity in the Tonga arc, SW Pacific • Isotopic characteristics of Ordovician greywacke provenance in the Southern Uplands of Scotland • Sandstones of arc and ophiolite provenance in a backarc basin, Halmahera, eastern Indonesia • The provenance of sediments in the Barrême thrust-top basin, Haute-Provence, France • Evolution of the Devonian Hornelen Basin, West Norway

- Published January 1991
- 312 pages
- 25 papers
- 202 illustrations
- Includes colour plates
- ISBN 0-903317-56-7
- List price* £66/US$130

* FGS price available on request provided that the correct member number is quoted.

Special Publications of The Geological Society

54 Deformation Mechanisms, Rheology and Tectonics
53 Glacimarine Environments: Processes and Sediments
52 Phosphorite Research and Development
51 The Cadomian Orogeny
50 Classic Petroleum Provinces
49 The Geology and Tectonics of the Oman Region
48 Geological Applications of Wireline Logs
47 Origins and Evolution of the Antarctic Biota
46 Phanerozoic Ironstones
45 Alpine Tectonics
44 Inversion Tectonics
43 Evolution of Metamorphic Belts
42 Magmatism in the Ocean Basins
41 Deltas: Sites and Traps for Fossil Fuels
40 Lacustrine Petroleum Source Rocks
39 Early Tertiary Volcanism and the Opening of the NE Atlantic
38 The Caledonian-Appalachian Orogen
37 Gondwana and Tethys*
36 Diagenesis of Sedimentary Sequences
35 Desert Sediments: Ancient and Modern
34 Fluid Flow in Sedimentary Basins and Aquifers
33 Geochemistry and Mineralization of Proterozoic Volcanic Suites
32 Coal and Coal-bearing Strata: Recent Advances
31 Geology and Geochemistry of Abyssal Plains
30 Alkaline Igneous Rocks
29 Deformation of Sediments and Sedimentary Rocks
28 Continental Extensional Tectonics
27 Evolution of the Lewisian and Comparable Precambrian High Grade Terrains
26 Marine Petroleum Source Rocks
25 Sedimentation in the African Rifts
24 The Nature of the Lower Continental Crust
23 Habitat of Palaeozoic Gas in North West Europe (Scottish Academic Press)
22 The English Zechstein and Related Topics
21 North Atlantic Palaeoceanography
20 Palaeoecology and Biostratigraphy of Graptolites
19 Collision Tectonics
18 Sedimentology: Recent Developments and Applied Aspects
14 Variscan Tectonics of the North Atlantic Region
10 Trench-Forearc Geology
8 The Caledonides of the British Isles — reviewed
6 Geological Background to Fossil Man
1 The Phanerozoic Time-Scale

Special Reports of The Geological Society

19 Magnetostratigraphy
18 Geophysical Logs in British Stratigraphy
17 Acritarchs in British Stratigraphy
16 Trilobites in British Stratigraphy
15 A Correlation of Jurassic Rocks in the British Isles, Pt 2, Middle and Upper Jurassic
14 A Correlation of Jurassic Rocks in the British Isles, Pt 1, Lower Jurassic
13 A Correlation of Triassic Rocks in the British Isles
12 A Correlation of Tertiary Rocks in the British Isles
10 A Correlation of Silesian Rocks in the British Isles
9 A Correlation of Cretaceous Rocks in the British Isles
8 A Correlation of Devonian Rocks in the British Isles
7 A Correlation of Dinantian Rocks in the British Isles
6 A Correlation of Precambrian Rocks in the British Isles
2 A Correlation of Cambrian Rocks in the British Isles
1 A Correlation of Silurian Rocks in the British Isles

Memoirs of The Geological Society

12 Palaeozoic Palaeogeography and Biogeography
11 The Ophiolite of Northern Oman
10 The Chronology of the Geological Record
9 The Nature and Timing of Orogenic Activity in the Caledonian Rocks of the British Isles
8 A Palaeogeological Map of the Lower Palaeozoic Floor below the cover of Upper Devonian, Carboniferous and Later Formations
6 Late Pre-Cambrian Glaciation in Scotland
5 Shallow-water Sedimentation, as illustrated in the Upper Devonian Baggy Beds
4 The Geology of Portuguese Timor

Geological Society Engineering Geology Special Publications

6 Field Testing in Engineering Geology
5 Engineering Geology of Underground Movements
4 Planning and Engineering Geology
3 Groundwater in Engineering Geology
2 Site Investigation Practice: Assessing BS 5930

Titles not listed are out of print
* *Available from Oxford University Press*